INTRODUCTION TO ELECTRONICS DESIGN

INTRODUCTION TO ELECTRONICS DESIGN

F. H. MITCHELL, Jr.
Associate Professor
Gonzaga University

F. H. MITCHELL, Sr.
Professor Emeritus
University of South Alabama

PRENTICE HALL, Englewood Cliffs, New Jersey 07632

Library of Congress Cataloging-in-Publication Data

MITCHELL, F. H. (Ferdinand Haverman) (date)
  Introduction to electronics design.

  Bibliography: p.
  Includes index.
  1. Electronic apparatus and appliances.   2. Electronic
circuit design.   I. Mitchell, F. H. (Ferdinand Haverman)
(date).   II. Title.
TK7870.M54   1988   621.381   87–17449
ISBN 0–13–481276–X

Editorial/production supervision: Joan L. Stone
Interior and cover design: Christine Gehring-Wolf
Manufacturing buyer: Margaret Rizzi

© 1988 by Prentice-Hall, Inc.
A Division of Simon & Schuster
Englewood Cliffs, New Jersey 07632

Printed in the United States of America
10 9 8 7 6 5 4 3 2 1

ISBN 0-13-481276-X   025

PRENTICE-HALL INTERNATIONAL (UK) LIMITED, *London*
PRENTICE-HALL OF AUSTRALIA PTY. LIMITED, *Sydney*
PRENTICE-HALL CANADA INC., *Toronto*
PRENTICE-HALL HISPANOAMERICANA, S. A., *Mexico*
PRENTICE-HALL OF INDIA PRIVATE LIMITED, *New Delhi*
PRENTICE-HALL OF JAPAN, INC., *Tokyo*
SIMON & SCHUSTER ASIA PTE. LTD., *Singapore*
EDITORA PRENTICE-HALL DO BRASIL, LTDA., *Rio de Janeiro*

CONTENTS

CHAPTER 5
DIODE MODELS
AND CIRCUIT APPLICATIONS 146

PART 3 TRANSISTORS AS LINEAR DEVICES 201

CHAPTER 6
BIPOLAR JUNCTION
TRANSISTOR OPERATION 202

CHAPTER 7
BJT BIASING TRADEOFFS 244

CHAPTER 8
FIELD-EFFECT TRANSISTOR OPERATION AND BIASING TRADEOFFS 280

PART 4   EQUIVALENT CIRCUITS, FREQUENCY RESPONSE, AND DISTORTION 329

CHAPTER 9
MIDRANGE AC AMPLIFIER DESIGN 331

CHAPTER 10
FREQUENCY RESPONSE OF AC AMPLIFIERS   377

CHAPTER 11
DISTORTION AND AMPLIFIER PERFORMANCE   434

PART 5   LINEAR APPLICATIONS   471

CHAPTER 12
OPERATIONAL AMPLIFIERS   472

CHAPTER 13
DESIGN WITH IC OP AMPS   515

CHAPTER 17
LOGIC
AND INTERFACE IMPLEMENTATION   682

PART 7   DESIGN AND FABRICATION TRADEOFFS   713

CHAPTER 18
TECHNOLOGY AND PERFORMANCE
IN MICROELECTRONICS   714

CHAPTER 19
COMPUTER-INTEGRATED DESIGN AND MANUFACTURING IN ELECTRONICS   759

APPENDIX 1
AC CIRCUIT ANALYSIS   771

APPENDIX 2
RC FILTERS   777

APPENDIX 3
RESONANT CIRCUITS   780

APPENDIX 4
LOGARITHMIC SCALES   783

APPENDIX 5
SIGNIFICANT FIGURES   785

APPENDIX 6
EXPERIMENTAL PROCEDURE   787

PREFACE

In recent years there has been a strong move among engineering schools to produce design-oriented curricula. Continual efforts are being made to strengthen the links between course content and the practice of engineering.

This text provides an introduction to the design process in the field of electronics. The intent is to provide ways of seeing and understanding problems and searching for solutions that are appropriate to the development of electronic circuits and systems in complex settings.

SYSTEMS APPROACH

Modern electronics uses a wide range of strategies to satisfy diverse and demanding applications. Many different types of building block elements are brought together, using complex interconnection and packaging techniques, to satisfy combined technical and nontechnical performance objectives. *Computer-aided design (CAD)* has become essential to the effective and efficient design of both the building block elements and the final circuits and systems, and *computer-aided manufacturing (CAM)* is often required to meet fabrication needs.

Further, the trend is toward integration of the design and manufacturing functions through computer network technology. Such *computer-integrated manufacturing (CIM)* systems function with design and fabrication as closely related activities. A flexible manufacturing system provides fabrication alternatives, and a "design-to-production" strategy makes the best use of resources.

In such a setting, the engineer must manage and direct the use of materials, technology, manufacturing, and resources. A systems approach to design emphasizes that the engineer must be knowledgeable concerning all areas of decision making that affect the meeting of performance objectives and must be able to creatively guide the tradeoffs between opportunities and constraints in all areas. CAD and CIM are major tools for this activity.

SYSTEM ELEMENTS

Electronics is dominated today by integrated circuits (ICs) as fundamental devices. For the most part, these ICs are fabricated from single-crystal silicon. Many different proprietary geometries and processes produce ICs with the most competitive performance parameters. As an alternative material, gallium arsenide is growing rapidly in use, and many other materials are being intensely pursued in research-and-development settings. Given the history of the field, it is reasonable to assume that in the near future the variety of IC types will continue to increase and the designer will be faced with additional selection processes.

Discrete diodes, transistors, and passive circuit elements (resistors, capacitors, and inductors) are also widely used as building block components. A complex circuit can mix ICs and discrete elements to achieve design objectives. Final circuit performance depends on the parameters associated with the various devices used and on the interconnection and packaging techniques that combine the devices into a functional unit. Innovative hybrid microcircuits and surface mount technologies represent evolutionary movements away from the typical printed-circuit-board integration strategies that have been in common use. In high-performance circuits, the method of system integration is as important as the device characteristics in terms of system performance, so further rapid change may be expected in this area.

Optoelectronics is also growing rapidly, giving the designer new options and decision opportunities for approaches to amplification, communication, and signal processing. Circuits are being configured with mixed optical and electronic devices to take advantage of innovative design.

Given the competitive pressures and the levels of research and development around the world, the rate of change in electronics can only increase. New building block devices will continue to become available, along with new system integration strategies, leading to more complex and difficult choices. Computer-aided design and computer-integrated manufacturing will be required if the designer is to make best use of the available resources.

STRATEGY

It is essential that the electronics student be well prepared to design circuits and systems in an environment characterized by the continual introduction of new building block alternatives and sophisticated computer networks. The skills that are taught must be oriented toward future settings. The result is an adaptive systems approach to meet performance objectives.

One of the major issues in preparing an electronics curriculum is how much internal device detail should be included versus the degree of emphasis on system design using devices and components described by external parameters. There is no single answer to the questions raised in such discussion, but there are many differing opinions.

Given the rapid introduction of new materials and devices into the field, the evolution toward CIM, and the role of the designer in managing how performance objectives are met, students must develop a hierarchical understanding of the entire design process. External device parameters will always be limited in the information they convey, particularly when efforts are made to apply available devices in innovative ways. The designer must "look inside" the device and understand its operation at a sufficiently fundamental level to correctly interpret the external parameters and to resolve performance questions that extend beyond the information that is typically available. The internal device detail thus supplements the use of external parameters in system design.

As new materials and technologies become available, the designer must develop a coordinated understanding of the strengths and weaknesses of the alternatives. If each device is treated as a special case, the designer loses the ability to relate the devices at a more fundamental level and therefore loses the ability to understand the common features among them. Since this understanding must come from an introductory exposure to semiconductor physics, this text starts by introducing the concepts and parameters that describe the properties of all semiconductor devices.

This educational approach is consistent with the present trends in the computer-aided design of electronic circuits and systems. As software systems become available for many different device types, based on alternative operational principles, the designer will naturally seek to perform tradeoffs among the device options. Lacking a fundamental understanding of device physics and the ways in which both discrete and integrated devices are created, the designer will be less effective in interacting with these sophisticated design systems.

This strategy is further reinforced by the growing importance of the integration (interconnection and packaging) technique used. The characteristics of the final system are dependent on the method of integration, so complete system simulations require that the designer extend beyond the examination of idealized external device parameters. The same foundation applies to understanding both internal device characteristics and the properties of complete circuits.

This text uses a step-by-step developmental strategy, leading from device physics to discrete components to ICs, with a continuing concern regarding system considerations. This strategy enables the student to develop an orderly understanding of the hierarchical relationships that typify electronics circuits and systems. In addition, the many aspects of design can be illustrated using the hierarchical setting, giving the student a broadened appreciation for the design processes that will be effective and efficient.

There are also continuing debates as to whether analog or digital circuits should be addressed first in an introductory text in electronics and as to the relative priorities that should be given to bipolar junction transistors (BJTs) and field-effect transistors (FETs). In preparation of this text, the guide has been the broad systems orientation discussed above. It was decided to treat analog circuits first in order to allow a sequential approach to many of the important modeling concepts that must be introduced to the student. The treatment of digital circuits flows smoothly out of the foundation established. Similarly, BJTs are treated before FETs to provide for a sequential transition from basic semiconductor physics to pn junctions and then to three-terminal devices. These decisions have been made with an emphasis on the progression that will best serve the needs of students for coherence in the material being covered and to provide a structure that links all the constituent elements of the models being developed.

SCOPE

In order to become proficient in electronics design, the designer must be able to draw on:

1. A fundamental understanding of the design processes that will be associated with a broad range of decision-making choices regarding all aspects of circuit production
2. A sufficient background in physical electronics to understand the operation of different building block devices and to appreciate the inherent design opportunities and constraints associated with a variety of elements and technologies
3. An understanding of the links that develop between device physics, the operation of single (discrete) devices, and complex ICs that make use of many devices in integrated form; as new technologies emerge, this understanding of hierarchical relationships enables the designer to maintain a coordinated framework for the available design elements and the ways in which they relate
4. An insight into the simplifying models that are used in the design process and the regions over which these models are valid and invalid
5. The mastery of the graphical and analytic methods of circuit design that provide a foundation for the selection of design strategies, evaluation of alternatives, and effective interaction with CAD systems no matter what type of implementation strategy is adopted
6. An appreciation for the realistic experimental nature of electronics
7. An orientation toward the use of computer support for the design and manufacture of electronic systems
8. A systems management approach to the design task

This book provides a coordinated learning process that relates to these areas.

As computer interaction becomes more integral to the design and manufacture of electronic circuits and systems, the engineer must develop a higher level of awareness regarding the design process itself and the multidimensional tradeoffs that result. For maximum use of computer support, the individual must balance competing performance objectives and constraints and the methods used to achieve the preferred design strategy, while interacting with a computer system on a continual basis. This text is intended to help students develop the skills and problem-solving insights that are necessary for electronics design in a computer-integrated format.

In subsequent chapters, the concepts required for understanding semiconductor devices and circuits are presented with an orientation toward building-block strategies, tradeoffs, and achieving performance objectives through an organized design process. The techniques of graphical and analytic problem solving are developed as resources for achieving performance objectives in circuit design. Students are led toward a mastery of the skills that are required to develop problem statements, consider alternative solution strategies, interact productively with CAD systems, and produce an optimum system design.

The development that follows emphasizes the setting of performance objectives for a circuit, then using problem-solving methods to find the circuit components that are required to obtain the objectives. The design orientation involves setting the objectives, then solving the problem to satisfy the objectives subject to device constraints. This may be contrasted with an alternative approach that involves assigning values to components, then finding what performance a given circuit will have. Both text discussion and examples include the emphasis on achieving objectives.

DESIGN CONCEPTS

There are many ways to interpret the meaning of engineering design. The following definition is by the Accreditation Board for Engineering and Technology (ABET):

> Engineering design is the process of devising a system, component, or process to meet desired needs. It is a decision-making process (often iterative), in which the basic sciences, mathematics, and engineering sciences are applied to convert resources optimally to meet a stated objective. Among the fundamental elements of the design process are the establishment of objectives and criteria, synthesis, analysis, construction, testing, and evaluation. The engineering design component of a curriculum must include at least some of the following features: development of student creativity, use of open-ended problems, development and use of design methodology, formulation of design problem statements and specifications, consideration of alternative solutions, feasibility considerations, and detailed system descriptions. Further, it is desirable to include a variety of realistic constraints such as economic factors, safety, reliability, aesthetics, ethics, and social impact.

—1985 Annual Report (Accreditation Board for Engineering and Technology, 1985)

The synthesis of the basis sciences, mathematics, and the engineering sciences has been emphasized in this text. The basic sciences provide an understanding of the materials and devices used in electronics, whereas mathematics provides the means for analysis, and the engineering sciences encompass the devising of circuits to achieve performance objectives.

As noted above, the design process includes the establishment of objectives and criteria, synthesis, analysis, construction, testing, and evaluation. Considerable effort has been spent in describing the nature of design objectives and how they arise in realistic engineering settings. (Introductory concepts are presented in Chapter 1 and issues associated with the setting of objectives are discussed throughout. Chapter 19 examines the larger context in which design objectives are determined.)

Exercise of the synthesis process has been sought by emphasizing the interrelationships and tradeoffs that arise among competing objectives. Various features of circuit performance are studied individually and then merged to provide a sense of the links that develop. The development in the text illustrates how circuit characteristics (for example, operating point stability and gain, or dc and ac circuit performance) can be studied separately, then merged to provide an understanding of composite circuit performance.

Methods of electronic circuit design and analysis are introduced throughout and illustrated in a variety of ways. Examples in each chapter are used to clarify and extend the text discussion. At the end of each chapter, questions with answers emphasize some of the key concepts presented. Exercises with solutions demonstrate design and analysis techniques introduced in the chapter and introduce optional concepts that extend chapter coverage. Additional problems provide the student with opportunities to expand understanding and build needed skills. Solutions to all problems are included in the *Solution Manual*.

To emphasize the importance of computer-aided analysis and design, computer applications have also been included at the end of most chapters. Some of these applications provide opportunities for solving specific problems, with computer programming required, whereas others provide the results of computer efforts that are complete and available for analysis and interpretation. Both types of problems serve to familiarize students with the use of computers in everyday engineering practice. (If computer access is not available, the programming problems may be omitted and emphasis placed on those applications whose results are provided without interfering with the flow of text material.)

Included in Chapter 5 and the computer applications is an introduction to the SPICE circuit simulation computer program. The utility of SPICE is demonstrated through worked-out simulations that the student may analyze without requiring actual programming effort and through other assignments requiring programming where appropriate computer capability is available.

A PC-based version of SPICE, called PSPICE (Tuinenga, 1988), is available in demonstration form from the publisher without charge, as part of the course materials provided with adoption of this text. The demonstration form of PSPICE has limited memory size (accommodating circuits of up to 10 transistors or equivalent memory requirements), but it includes all program capabilities and can adequately handle the simulation problems encountered in this text.

The construction, testing, and evaluation of circuits takes place in a laboratory setting. In order to enhance the coherence of the design process covered here, at least one laboratory experiment is included at the end of most chapters. These experiments emphasize the setting of objectives, circuit selection, and prediction of circuit performance, followed by circuit construction, test, and evaluation. The experiments apply the discussions of the text, using equipment and supplies that are generally available. Appendix 6 provides procedures for conducting the experiments.

The ABET definition of engineering design points out the importance of the development of student creativity, use of open-ended problems, development and use of design methodology, formulation of design problem statements and specifications, consideration of alternative solutions, and detailed system descriptions. Students usually experience the full nature of this cycle within the context of a required senior design project. The display of creativity and open-ended problems are best handled in such a setting. However, this text provides an introductory exposure to some of these concepts in preparation for a more complete design experience. Particularly, the development and use of design methodology and the formulation of design statements and specifications are considered from a range of perspectives. Background skills in feasibility study and system descriptions are also developed.

Finally, the ABET definition emphasizes the optimum use of resources and the consideration of realistic constraints, such as economic factors. This text brings to the student's attention a concern with resources and a preliminary look at nontechnical factors as they influence the design process. There are exercises throughout that illustrate how technical and nontechnical factors interact in electronics design.

The design of electronic circuits is approached on a sequential basis that begins with introductory concepts and proceeds to the operation of specific circuits to achieve performance objectives. The content of the book is divided into seven parts and nineteen chapters.

Chapters 1 through 8 provide an introduction to electronics that emphasizes design strategies, the properties of semiconductor devices, dc biasing, fundamental building block circuits, and performance tradeoffs. Alternative strategies for modeling components and circuits are considered throughout, and a foundation is provided for the more complex topics that follow.

Chapters 9 through 11 introduce ac-coupled amplifiers, the use of equivalent circuit models, frequency-dependent amplifier performance, and distortion. A broader range of building block circuit types, extended performance characteristics, and additional system considerations are also discussed.

Chapter 12 develops an understanding of the internal performance of operational amplifiers (op amps) by starting with simplified circuit versions and proceeding to modern IC devices. This progression serves as a bridge between the circuit concepts and design methods introduced in the earlier chapters and the increased level of difficulty associated with op amps. Chapter 13 illustrates how op amps may be applied as fundamental linear amplifier elements to achieve a wide range of circuit performance characteristics. Chapter 14 expands the discussion to include oscillators and considerations that enter into communication systems performance.

Chapters 15 through 17 present a detailed consideration of digital circuits and systems. Again using an evolutionary approach, Chapter 15 leads from simple gate circuit concepts to the families of circuits commonly applied today. Chapter 16 provides an overview of computer systems and an introduction to digital networking concepts. Chapter 17 describes some of the commercially-available families of devices (or chip-sets) that are available to implement system requirements.

Chapter 18 draws on the previous discussions of semiconductor properties to illustrate the issues and tradeoffs that arise in the design of semiconductor devices to achieve circuit and system performance objectives. This chapter emphasizes the relationships between fundamental material characteristics, device fabrication, and circuit parameters. It also contrasts the various types of device fabrication and integration strategies that are widely used in electronics. Chapter 19 applies many of the concepts of earlier chapters to an electronics design and manufacturing setting.

The text may be used to meet a variety of different curriculum needs. All chapters may be covered in a two-semester or three-quarter course sequence at the junior level. Typically, Chapters 1 through 9 might be included in the first semester and Chapters 10 through 19 in the second semester.

Selected use may be made of the exercises with solutions at the end of each chapter. The scope of the topics being covered in each chapter may be expanded or limited as desired by choosing among the available exercises. Problems may also be assigned on a selective basis. By reducing or expanding the use of the exercises and problems, the text may be adapted to specific student needs and schedule requirements.

If course emphasis is to be placed on circuits and systems and an introduction to physical electronics is provided elsewhere, coverage of the initial chapters can be reduced, providing time for a detailed consideration of later topics. On the other hand, extra time may be spent on these chapters if the material is new to students. Digital electronics may be covered in detail or reduced, depending on the preferred scope.

The text may also be adapted to one-semester or two-quarter courses by choosing the desired course emphasis, leaving out those chapters or sections that are covered elsewhere in the curriculum, and modifying the use of exercises and problems. The text can thus meet course requirements in a variety of settings.

It is assumed that students have completed an introductory circuits course and understand the use of phasors in ac circuit analysis. A basic understanding of calculus is also presumed.

For students who may need additional review of ac circuits, semilog graphs, and the uses of significant figures, Appendices 3, 4, and 5, respectively, are available for reference. These and other appendices can help students link between other courses and electronics applications, and may be drawn upon as appropriate.

ACKNOWLEDGEMENTS

The authors would like to express their appreciation for the assistance received during the preparation of this text. The comments and suggestions from reviewers have provided essential guidance during evolution of the manuscript. Particularly helpful were the efforts of Professors Wendell Cornetet, Jr. of Ohio State University, Patricia D. Daniels of the University of Washington, Frank H. Hielscher of Lehigh University, Alan H. Marshak of Louisiana State University, R. Fabian Pease of Stanford University, and William Sayle of the Georgia Institute of Technology.

Nick Gray and John Tero of Signetics Corporation provided additional support for the development of the materials on IC operational amplifiers. Professors Ray Birgenheier, Gail Allwine, Masao Shimoji, and Jerry McCoy of Gonzaga University provided much-appreciated assistance in evaluating draft materials. Assistance in preparation of the computer applications was received from Bill Reed and Paul Graham, while support for preparation of device characteristic curves was received from Kevin Williams, Bill Gaines, and T. J. Mills.

As identified throughout the text, many different companies have kindly given permission for use of their materials. The School of Engineering at Gonzaga University has been supportive of the production needs during development of the manuscript, and feedback from students in the School of Engineering has provided necessary input at every stage of development. Essential support for the preparation of this text has also been provided by Tim Bozik, Engineering and Computer Science Editor for Prentice Hall.

The authors express their appreciation to the above individuals and organizations. As is always the case, this final product represents the decisions made by the authors, who accept full responsibility for the materials as presented.

F. H. Mitchell, Jr.
F. H. Mitchell, Sr.

INTRODUCTION TO ELECTRONICS DESIGN

PART 1
INTRODUCTORY CONCEPTS

Electronic circuit design is a complex process that involves tradeoffs among many different performance objectives. The designer must thoroughly understand the technical choices involved and their implications. In addition, such factors as cost, reliability, and ease of production are important. The designer must manage and direct the achieving of performance objectives in the most effective and efficient way.

Part 1 provides the foundation to understand the electronics design process in a setting characterized by many possible circuit configurations, multiple materials and technologies for use in circuit fabrication, computer-aided design (CAD) support for use in considering the tradeoffs among the available choices, computer-integrated manufacturing (CIM) to link the design and manufacturing processes, and a systems management approach to engineering design.

Chapter 1 presents the issues in electronic circuit design and outlines the problem-solving methods that are widely used. This chapter delineates the tasks faced by the designer.

Chapter 2 provides a detailed introduction to the electrical properties of microelectronic materials. These properties determine the performance characteristics of the devices and components used to create complex electronic circuits and are thus key factors when tradeoffs are considered among different materials and technologies.

Chapter 3 explores the complexity of the resistance, capacitance, and inductance functions that are found in real electronic circuits. This chapter provides the perspective required to understand model limitations and the results obtained from circuit simulations.

The concepts and methods of analysis used in Part 1 will be drawn upon throughout the text. They provide the starting point for the chapters that follow.

Appendices 1 to 3 provide a brief review of ac circuit analysis, RC filters, and resonant circuits. These are intended as a convenient reference and to outline the key aspects of circuit analysis that are drawn upon in the text.

CHAPTER 1
THE DESIGN PROCESS IN ELECTRONICS

Engineering design is a creative process that applies knowledge and resources to solve problems and meet performance objectives. Effective design activity thus requires an understanding of the problems to be addressed and the settings in which they arise, combined with knowledge of the field, available resources, and insight into problem-solving strategies.

This text considers the issues associated with the design of electronic circuits. It is concerned with the types of problems that lead to the use of electronic devices and circuits and with the various solution strategies that may be followed. Given the complexity of the task and the continual rapid growth of technology, circuit designs are constantly changing. Therefore, the emphasis here is on an understanding of *general* design principles and the linking of these principles to design applications. The general principles are presented in the body of each chapter whereas the examples, exercises, and problems provide the applications.

1-1 THE SYSTEMS APPROACH

Many different considerations enter into realistic electronic circuit design. Multiple technical objectives usually exist and the satisfying of these objectives can require extreme ingenuity on the part of the designer. At the same time, the cost of producing the circuit is usually a major concern as are reliability, serviceability, and a range of other factors.

When the designer attempts to merge these objectives, a *systems* approach is being considered. Figure 1-1 illustrates the problem-solving approach that can result.

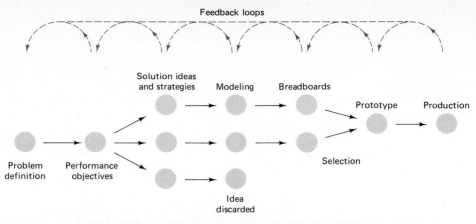

Fig. 1-1 The systems approach to electronics design, showing the importance of feedback loops.

The design process begins with a definition of the problem to be addressed and the selection of performance objectives. There must be a clear understanding of why a circuit or system is to be developed, combined with criteria by which progress toward the objectives may be determined. The problem definition and performance objectives arise from the setting in which design is taking place. *Problem definitions* arise when there are needs to be met whereas *performance objectives* provide qualitative and quantitative measures by which the satisfying of these needs may be assessed.

Solution ideas and strategies as based on education and experience, discussions with colleagues, and studies of the current literature. This is a creative step that makes use of a synthesis of concepts, bringing together the many ideas in a problem-solving framework. Potential solution strategies are contrasted and evaluated against one another and modifications are sought to create a solution approach that will best meet performance objectives. It is essential to consider the broadest range of possible strategies, along with the advantages and disadvantages of each.

Once the alternative ideas are in hand, various types of *modeling* are used to study how well each strategy can meet the design objectives. The modeling will usually involve the use of sets of equations and will often make use of computer simulations. Analysis is used to explore, predict, and evaluate how well alternative solution strategies will work.

All or some of the circuit choices may then be constructed in *breadboard* form, useful for laboratory test and evaluation. The breadboard is built by the design engineer using readily available components and equipment to test whether or not the design is fundamentally sound.

Based on the modeling and the breadboard testing effort, a preferred selection (or combination of models) can be made and a *prototype* constructed. The prototype is fabricated to be an exact replica of the production version to follow. Finally, when the designer is sure that an appropriate solution to the original problem has been found, *production* and final testing can begin.

Design takes place through a complex feedback process, as shown in Fig. 1-1. If difficulties are encountered at any phase, the designer must go back to earlier phases and reevaluate decisions. Problems with prototype development can lead to new breadboarding efforts, additional simulation activities, and even a reexamination of potential solution ideas and concepts. Similar loops may be established between all stages. Feedback may even require a return to the problem definition and performance objectives to determine if changes must be made at these fundamental levels. In summary, electronics design is an *iterative* process that may require repeated efforts through the design cycle.

Figure 1-1 makes it clear that the design of a production-oriented electronics circuit is a complex task. In order to make a satisfactory selection, it is necessary to deal with a broad mix of objectives and with relationships among different types of constraints.

Computer-aided design can help improve the reliability of the modeling stage and reduce the costs of breadboards and prototypes. As more sophisticated computer support becomes available, with the ability to realistically simulate all aspects of device and circuit performance, designers can produce such effective predictive models that the breadboard stage can be eliminated entirely in some cases.

This text provides an overview of the types of problems and solution approaches that are widely recognized and used in the field of electronics. The information can provide the basis for the development of solution ideas and strategies when new applications are found.

The information can also provide guidance in the development of models that may be used to compare design alternatives. Experience with model-building for electronics circuits and devices will often be transferable to new problems and strategies.

One result of a systems approach to problem-solving is that multiple performance objectives often lead to competing design constraints. Performance objectives can be in conflict with one another in such a way that compromise must be accepted. *Tradeoffs* are performed among these competing objectives to arrive at the preferred solution for a given application. As the application changes, so will the outcome of the tradeoff process. Proficiency in electronics design requires the ability to understand the principal tradeoffs that must be considered to produce a desired circuit, and adoption of problem-solving strategies that are appropriate to the circumstances.

Some of the tradeoffs experienced in electronics are illustrated in Examples 1-1 to 1-4. Each example is linked to a portion of the text that further explores the issues that are raised.

Example 1-1 **Tradeoffs in DC Circuit Design**

Variations on the voltage divider circuit in Fig. 1-2 are widely applied in circuit design. Assume that two competing performance objectives are set up for this circuit:

1. V_{OUT} should hold approximately constant as R_x is varied over a defined range;
2. The power dissipated in the circuit should be minimized.

What design strategy can be applied to this problem?

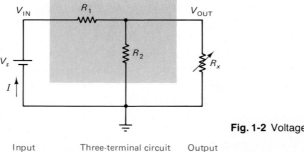

Fig. 1-2 Voltage divider circuit for Example 1-1.

Input   Three-terminal circuit   Output

Discussion   Objectives 1 and 2 are in direct conflict with each other. In order to have V_{OUT} held approximately constant, the current through R_2 must be much greater than the current through R_x. For this case,[1]

$$V_{OUT} = V_s\left(\frac{R_2\|R_x}{R_1 + R_2\|R_x}\right) \cong V_s\left(\frac{R_2}{R_1 + R_2}\right) \qquad (1\text{-}1)$$

so that R_2 must be much smaller than the minimum allowed value of R_x to achieve objective 1. If V_s and the ratio R_2/R_1 are held constant (to produce the desired nominal output voltage) while R_1 and R_2 are reduced in value, V_{OUT} will become progressively more independent of R_x.

However, since the power dissipated in the circuit is given by

$$P = V_s^2/(R_1 + R_2\|R_x) \qquad (1\text{-}2)$$

the dissipation will increase rapidly as R_1 and R_2 are decreased. To satisfy objective 2, R_1 and R_2 should be made as large as possible.

A tradeoff exists between small values of R_1 and R_2 to hold V_{OUT} approximately constant and large values of R_1 and R_2 to minimize P. The preferred compromise solution depends on the application.

One possible design strategy is based on defining the maximum variation in V_{OUT} that is allowed as R_x changes. Maximum values for R_1 and R_2 can then be chosen consistent with this requirement. The power dissipation is then calculated and the procedure is iterated as necessary to produce a balanced result. Example 1-2 illustrates the application of this strategy. ■

Figure 1-2 also introduces the concept of a three-terminal circuit with an input and an output. As the voltage V_{IN} is varied, the output V_{OUT} also varies. The relationship between V_{IN} and V_{OUT} depends on the detailed circuit structure. Three-terminal circuits are important in electronics and are discussed in the following chapters. Circuits of this type are often found in settings that require tradeoffs among various performance characteristics.

Example 1-2   Design of a Voltage Divider Circuit

Assume for the circuit of Fig. 1-2 that $V_s = 12.0$ V and V_{OUT} is to be 3.00 V when R_x is very large. The minimum value of R_x is 1.00 kΩ, and V_{OUT} must vary by no more than 10 percent as R_x varies over its maximum allowed range. The power dissipated in the circuit is to be minimized within the above constraints. Find the values for R_1 and R_2.

Solution   From Eq. (1-1), for large values of R_x,

$$3.00 \text{ V} = 12.0 \text{ V}\left(\frac{R_2}{R_1 + R_2}\right)$$

$$R_1 = 3.00R_2$$

---

[1]The representation $R_2\|R_x$ means that R_2 is in parallel with R_x, so that

$$R_2\|R_x = \frac{R_2 R_x}{R_2 + R_x}$$

and the variation in V_{OUT} is given by

$$12.0 \text{ V}\left(\frac{R_2}{R_1 + R_2} - \frac{R_2\|1.00 \text{ k}\Omega}{R_1 + R_2\|1.00 \text{ k}\Omega}\right) \leq (0.10)(3.00 \text{ V})$$

Eliminating R_1 between the two equations,

$$3.00 \text{ V} - 12.0 \text{ V}\left(\frac{R_2\|1.00 \text{ k}\Omega}{3.00R_2 + R_2\|1.00 \text{ k}\Omega}\right) \leq 0.300 \text{ V}$$

$$\frac{3.00R_2}{R_2\|1.00 \text{ k}\Omega} \leq \frac{12.0 \text{ V}}{2.70 \text{ V}} - 1 = 3.44$$

$$R_2 + 1.00 \text{ k}\Omega \leq 1.00 \text{ k}\Omega(3.44/3.00) = 1.15 \text{ k}\Omega$$

$$R_2 \leq 0.15 \text{ k}\Omega$$

To minimize the power consumption, choose the maximum possible value $R_2 \cong$ 0.15 kΩ. Then,

$$R_1 = 3R_2 \cong 0.45 \text{ k}\Omega$$

This example illustrates one way in which competing performance objectives can be resolved by compromise. A small variation in V_{OUT} has been allowed to manage the circuit power dissipation. Note that as the restriction on V_{OUT} tightens, the values of R_1 and R_2 must drop, raising the power dissipation. ∎

The above examples are concerned with a tradeoff between the accuracy with which a voltage V_{OUT} is determined and the power dissipated in the circuit. In order to obtain additional insight into the design process in electronics, it is helpful to explore other areas in which compromise solutions are required.

One of the important design problems in electronics is associated with efforts to reduce circuit size. Over the past 30 years, there has been a transition from the use of discrete components to printed circuit boards and then to hybrid microelectronics and integrated circuits. This evolution has resulted because more dense circuits result in smaller, lower-cost, more reliable products. (This trend is discussed further later in this chapter and in Chapter 18.)

One of the design considerations is that the equivalent resistances in electronic circuits dissipate energy and produce heat. In low-density systems, the heat can be removed by normal air flow. For high-density systems, heat removal is an important consideration.

Example 1-3 **Circuit Density and Heat Dissipation**

Complex electronic systems are often considered as combinations of large numbers of building block subunits. These subunits can be defined at many different levels of aggregation. The selection of an optimum technology in terms of subunit density depends on a variety of factors, and consideration of heat dissipation often helps determine the preferred solution.

A simplified version of the problem to be faced is illustrated in Fig. 1-3. Two types of data are shown here on a semilog plot. The basic subunit variable chosen here is the average component density D in the system of study.[2]

---

[2]The reasonableness of the range of values for D is discussed in Exercise 1-3.

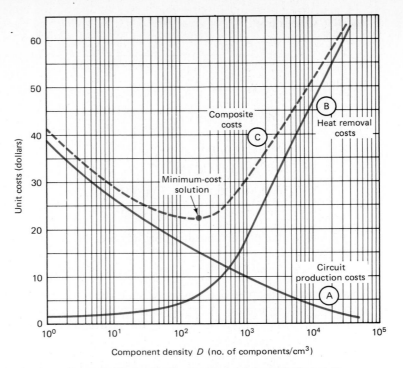

Fig. 1-3 Design tradeoff between assumed circuit production costs (curve A) and heat removal costs (curve B). Curve C (composite costs) is the sum of curves A and B, obtained graphically.

The logarithmic scale of the horizontal axis enables covering a wide range of values for D while still spreading out the details of the function for examination. (See Appendix 4 for a discussion of logarithmic scales.)

Curve A shows how circuit production costs decrease as a system becomes more dense. Based on this curve alone, the best design strategy is to maximize the component density D. On the other hand, curve B indicates that the costs of removing the heat generated by the system increase with component density. Based on curve B alone, the preferred strategy would be to use low-density circuits with inexpensive heat removal techniques.

The result of these two curves is another design tradeoff. In such situations, it is necessary to assess design objectives and make a decision regarding the best technology for a given application. Therefore, the problem is as follows: based on curves A and B in Fig. 1-2, what value of D should be selected for the system of interest to minimize the combined costs?

Solution   By graphically combining these two costs (curves A and B), a composite curve (curve C) is plotted with a cost minimum as shown. If only these two costs are considered, an optimum system density is defined and circuit components and configurations can be chosen to meet this density. The solution for D is 2.0×10^2 components/cm^3 at a unit cost of \$22.

In a real setting, other factors are important. The technical performance of the circuit can vary with circuit size and method of fabrication, so different predictive models are required over various density ranges. ∎

Important design constraints are also encountered in ac circuit analysis. The simple three-terminal circuit of Fig. 1-4 has many important applications. This is a *high-pass filter* that produces a frequency-dependent output. For a given effective input voltage $\mathbf{v}_i$, the output voltage $\mathbf{v}_o$ varies with the frequency of the source.[3] High- and low-pass filters and their characteristics are discussed in Chapter 10 and Appendix 2.

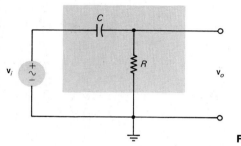

Fig. 1-4 High-pass filter for Example 1-4.

The simple RC filter shown in Fig. 1-4 can be described by a cutoff frequency

$$f_1 = \frac{1}{2\pi RC} \tag{1-3}$$

Frequencies above f_1 pass through the filter relatively unaffected whereas frequencies below f_1 are attenuated. The following example illustrates how the technical performance objectives of ac circuits can relate to nontechnical objectives.

Example 1-4   High-Pass Filter

In order to maximize the range of frequencies passed through the circuit of Fig. 1-4, f_1 should be made as small as possible. Thus, R and C should be large.

Assume that for a given application, R is fixed by other circuit considerations, so C must be increased to decrease f_1. Figure 1-5 describes the problem that can be faced under such circumstances.

As shown in Fig. 1-5(A), cutoff frequency f_1 of the high-pass filter decreases as C increases, providing a wider frequency band. However, Fig. 1-5(B) indicates that circuit size and cost also increase with C. This problem leads to a three-way tradeoff among frequency response, size, and cost. There is no clear indication of a preferred solution. In this situation, what value of C should be selected to produce a circuit with the preferred values of f_1, size, and cost?

---

[3] Boldface symbols ($\mathbf{v}_i$, $\mathbf{v}_o$) represent effective-value phasors in this text. Instantaneous time-varying parameters are represented by lowercase italic symbols (v_i, v_o), whereas dc voltages and currents are represented by capital italic symbols (V, I).

Some ambiguity is experienced for arbitrary time-varying signals that can take on dc values. In general, capital letters are used to represent both dc values and signals that are considered to vary among different dc levels.

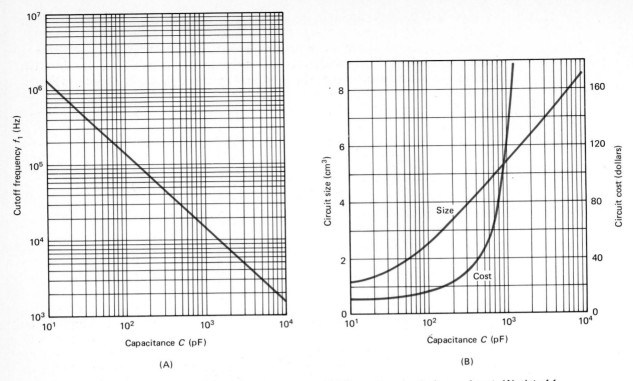

Fig. 1-5 Design tradeoff involving cutoff frequency, circuit size, and cost. (A) plot of $f_1 = 1/2\pi RC$ with $R = 10$ kΩ. (B) assumed data for the example.

Discussion To choose a value of C for this situation, the designer must assign *relative values* to the frequency response, cost, and size of the high-pass filter. If a low value of f_1 is most important, then size and cost must be accommodated. If size is critical, then a higher value of f_1 may be necessary.

A selection is made based on the relative importance of these three factors in a given design situation. The design objectives determine which solution is most appropriate. ■

What can be concluded from these four examples and the general systems approach to design? It is reasonable to conclude that the electronics engineer needs to be thoroughly familiar with the technical aspects of the field and must understand how to model the technical performance of alternative circuit design options. These skills are essential to successful design. However, these core skills must be matched with a diverse mix of additional insights and analytic skills necessary to formulate objectives, perform tradeoff studies, and select proposed designs.

This text is primarily concerned with meeting the technically oriented needs of electronics designers. However, the broader design issues will be considered in some of the problems as a reminder of the larger context in which real-world design takes place.

1-2 TECHNICAL DESIGN STRATEGIES

In order to apply the systems approach to electronic circuit design, it is necessary to understand how each component or device in the circuit works and how the different parts of the circuit interact to produce the total operation. This can be a complicated task when many different components and devices are involved. To simplify the task, a variety of modeling and approximation methods can be used as practical design tools.

Four specific strategies will be used to develop skills at working with electronic circuits:

1. Graphical methods for solving equations
2. Development of simplified, approximate models
3. Schematic drawings that can be used to visually compare different circuits and identify common circuit designs
4. Computer-aided design techniques

1-3 GRAPHICAL METHODS

Graphical methods of analysis are particularly useful in working with mixtures of linear and nonlinear functions. By graphing two functions on the same axes and looking for intercept points, simultaneous solutions to the two equations may be found. The graphical approach may also be used to extrapolate from one device to another. Two graphs may be compared to estimate how a change in device characteristics will affect circuit operation.

The following example demonstrates the utility of the graphical method for converting a fairly complex problem to one that can be solved without difficulty.

Example 1-5   Graphical Solution

Figure 1-6 shows a three-terminal circuit that includes a nonlinear device D, resistor R, and battery V_s. Assume that D is a semiconductor diode and that the performance of the circuit is described through the two equations:

$$I = I_0(e^{V/V_0} - 1) \qquad (1\text{-}4)$$

$$V_s = V + IR \qquad (1\text{-}5)$$

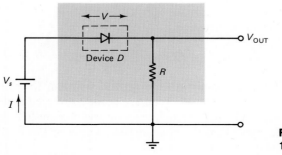

Fig. 1-6   Nonlinear circuit for Example 1-5.

(These equations are derived in detail in Chapters 4 and 5 and are used here to provide a relevant example of the importance of graphical analysis.)

Given values for I_0, V_0, V_s, and R, the values for V and I can be determined from the above two equations or by a graphical method. The advantages of a graphical solution are that the operation of the circuit can be seen in context and tradeoffs can be more clearly identified.

Assume that the constants associated with these two equations are defined through the plots of Fig. 1-7 (with V defined as positive). Your design objective is to achieve $V_{OUT} = 0.20$ V. Will these constants produce the desired result for the circuit shown in Fig. 1-6?

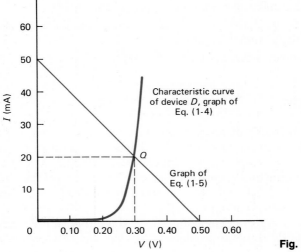

Fig. 1-7 Data for Example 1-5.

Solution The intersection of the two plots provides the solution point $Q(V = 0.30$ V, $I = 20$ mA). Further, the slope of the straight line given by Eq. (1-5) is

$$\text{slope} = \frac{dI}{dV} = \frac{d}{dV}\left(\frac{V_s - V}{R}\right) = -\frac{1}{R} \tag{1-6}$$

so

$$R = \frac{0.50 \text{ V}}{50 \text{ mA}} = 10 \ \Omega$$

Applying the above information,

$$V_{OUT} = IR = (20 \text{ mA})(10 \ \Omega) = 0.20 \text{ V}$$

Graphical analysis has confirmed the desired solution. In addition, observe that the general features of this type of circuit are readily revealed by the curves of Fig. 1-7. By adjusting the parameters associated with the straight-line graph of Eq. (1-5), the point Q can be moved on the characteristic curve of the device to obtain different operating characteristics for the circuit. ∎

Simplified models can change a problem that seems hopelessly complex into one that is workable. However, they always introduce approximations into the design effort, and ways must be found to evaluate how the approximations are reflected in the solutions.

Circuit analysis uses simplified models of components. It is common to assume that ideal or pure resistors, capacitors, and inductors are available. However, such ideal components do not exist. As will be discussed in Chapter 3, all real components are mixtures containing these three properties. The adequacy of the component model depends on the technical application (particularly on the frequencies involved) and on the design objectives.

Example 1-6   Resistor Model

Suppose that a given resistor is modeled by combining the components shown in Fig. 1-8, where $R = 10$ kΩ and $C = 0.10$ pF (for simplicity, assume $L = 0$).

Assume that the effect of C is negligible in a given application if the current through the capacitor is less than 1 percent of the current through the resistor. Your design objective is to select a resistor that will have a negligible capacitance at 1.0 MHz. Will the given resistor be a satisfactory choice?

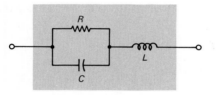

Fig. 1-8   Resistor model.

Solution

From the data given, the maximum frequency at which the capacitance is important is defined by the relationship $X_C = 100R$, so that

$$f = \frac{1}{2\pi (1.0 \times 10^2)(1.0 \times 10^4)(1.0 \times 10^{-13})} \text{ MHz} \cong 1.6 \text{ MHz}$$

For frequencies below 1.6 MHz, the equivalent capacitance can be neglected. The given resistor meets the performance specifications.

A similar calculation can be performed to find the frequency for which the inductive reactance is important (see Problem 1-21). Interactions between the L and C values are considered in Problem 1-20. The pure-resistor model is useful at lower frequencies, but its limits must be remembered by the designer. ■

There are other complexities that are also neglected in ideal component models. Applied voltages can create nonlinear responses, and temperature and other environmental effects can cause changes in the values of components. The discussions of semiconductor devices in Chapters 6, 8, and 18 introduce many different models that are useful for electronic circuit design. However, the applicable ranges of these models are often quite limited, so the models must be used with caution.

Schematics are used as a way to quickly understand a circuit of interest. A *schematic* is a symbolic representation of a circuit that can make design a much more straightforward process. Schematics are relatively easy to remember and can describe the essential information of circuit elements and interconnections.

With practice, the operation of a circuit can often be understood just by looking at its schematic. Certain arrangements of components and devices produce typical operational patterns that can be associated with the schematic.

Electronics engineers depend on schematics as shorthand to develop circuit understanding. More complex circuits can often be viewed as building blocks made up of simpler schematic units that are well-understood. Once these units are recognized, the designer can consider the entire circuit as a linked set of functions. Schematics are also useful in estimating how a circuit will respond to a design change before detailed equations are considered.

However, it must always be remembered that schematics are incomplete renderings of a circuit and imply a range of modeling assumptions. Schematics often do not adequately indicate circuit behavior at high frequencies, for which wires and other interconnections begin to function as circuit components. Electrical properties of bonds and contacts are also often not described by a schematic. Device geometry, nonlinear resistance effects, and environmental effects are all likely to be important factors not included in many schematics. Chapters 2 and 3 provide some background to evaluate schematics and assess their limits. Further discussions of these effects are included in Chapters 14 and 18.

1-6 COMPUTER-AIDED DESIGN TECHNIQUES

The design and fabrication of electronic circuits is today increasingly dominated by computer technology. At the materials science level, computers help design research and development experiments, operate laboratory equipment, collect and analyze data, and interpret results. At the functional element level, computers help plan how the elements will be fabricated to produce desired performance characteristics, then are used to control production processes and automatically test the resultant products.

The electronic circuit designer uses computer-aided design (CAD) for a wide range of support, from plotting equations so they can be studied and interpreted to making use of interactive software to provide an iterative design process. Interaction with a computer can draw on both the expert knowledge available in the system and system support in the geometric layout of circuits. Finally, production and testing are often dependent on computer-integrated manufacturing (CIM) with automated work units (robots), networks of communicating computer systems, and automatic test equipment.

A computer terminal can be a great help to both the student and the electronics designer. Uses range from graphing functions and exploring circuit performance on a parametric basis to using CAD packages to study circuit topography and function.

The computer application problems presented in this text will help the student become familiar with the advantages of interacting with a computer during the design process. These problems provide simple, relevant examples of the ways in which understanding can be improved through computer support. Many include the results of simulations performed on a computer and are available for study without computer access. In other cases, the problems require access to a computer. Many of the graphs contained in this text have been produced with computer assistance.

CAD has been a key resource in the rapid growth of the microelectronics industry. Since a single integrated circuit (IC) chip (measuring a few millimeters on each side) can contain thousands of diodes and transistors as well as complex interconnections, manual design of such circuits would be highly laborious. A resulting product might have numerous errors and would be prohibitively expensive.

ICs are now routinely fabricated through computer-interfaced equipment. The designer describes the circuit functions that are desired, and the computer system implements the functions in terms of detailed fabrication processes. The necessary circuit design rules are automatically performed by the software and can be applied to a range of technologies. This high-order functional approach to IC design has expanded the scope of circuit options available, reduced design time, and made possible the development of the custom-IC industry.

The computer-aided design of circuits is discussed throughout this text. Chapter 18 introduces the dominant microelectronics technologies in use today and illustrates how these technologies relate to design parameters for semiconductor devices. Chapter 19 explores how computer-integrated design and manufacturing can be useful in satisfying design objectives.

1-7 DATA COLLECTION

In order to apply the above design methods, the electronics engineer must make use of many different kinds of data. Information is needed to describe passive components (resistors, capacitors, and inductors), active devices (such as transistors and ICs), and instrumentation used as part of the circuit and to test the circuit.

Finding the particular information needed in a given situation can be quite a challenge. Some components come labeled with values and other information, and the designer must learn how to read, interpret, and apply such information. Devices and instruments often are marked with a code number so that design information can be found from a *data sheet* provided by the manufacturer.

Several different types of data sheets are included in the following chapters. They are typical of those found in electronics, and they will be used in solving problems at the end of the chapters. Unfortunately, there is no one format in general use for such sheets. The information they contain is usually limited. The manufacturer includes only the data that apply to common applications of the product.

The designer often finds that the information desired is not readily available. There are several strategies that can be followed at this point. The designer can choose to

1. Decide how to calculate the needed information from the information contained on the data sheet
2. Contact a manufacturer's representative for further information
3. Make a series of laboratory measurements to obtain first-hand data

Each of these strategies can require significant time and energy. The electronics engineer will often find the compiling of needed design information to be a major (and possibly frustrating) task.

Practice in this area is important. The satisfactory design of electronics circuits depends on a familiarity with the sources of data and their correct application.

The appropriate collection and use of data also requires a concern with significant figures. The difficulties of data collection and the accuracy of circuit prediction relate to the number of significant figures involved at each stage of analysis. In electronics design, many approximations are introduced in model development, and two to three significant figures are often appropriate for prediction of circuit performance, with additional significant figures used for special applications. The use of graphical analysis is also consistent with two to three significant figures.

Most of the examples and problems given conform to this level of accuracy. Data are usually assumed to be given in two to three significant figures, and most analyses use the same level of accuracy. Exceptions are noted and explained. Further details on significant figures are included in Appendix 5.

Example 1-7 Device Data

For the circuit described by Eqs. (1-4) and (1-5), assume that a data sheet shows that the value of I_0 for a given family of devices can vary over the range 0.1 to 0.5 nA. Assume $V_0 = 0.026$ V, $V = 0.500$ V, and $R_s = 50\ \Omega$. For a given application, V_s must be predicted within ± 2.0 V. Is the available database sufficient?

Solution

$$I_{min} = (0.1 \times 10^{-9}\ \text{A})(e^{0.500/0.026} - 1) \cong 22\ \text{mA}$$

$$I_{max} = (0.5 \times 10^{-9}\ \text{A})(e^{0.500/0.026} - 1) \cong 110\ \text{mA}$$

and

$$(V_s)_{min} = (0.500\ \text{V}) + (22\ \text{mA})(50\Omega) = 1.6\ \text{V}$$

$$(V_s)_{max} = (0.500\ \text{V}) + (110\ \text{mA})(50\Omega) = 6.0\ \text{V}$$

The value of V_s is not determined within the desired range.

The design requirements of this circuit can thus vary widely, depending on the value of I_0 that describes a particular device. More information is needed regarding the device of interest and how it relates to the family of devices being described to produce a reasonable design process. ■

1-8   DESIGNING ELECTRONIC CIRCUITS

The strategies described above can help the designer see the patterns that exist in electronic circuit design. Certain types of circuits occur over and over again in many different uses. Once the basic operation of a circuit is understood and can be recognized in several variations, a composite pattern becomes apparent. Then, when such a circuit is encountered as part of a larger system, it will be possible to recognize its function and perform any analyses that are necessary. Circuits can also be designed by starting with these common building block modules and modifying them to fit specific design objectives.

There is a clear need to develop a set of different circuits that are recognizable, along with the skills to modify and rearrange these circuits to meet new design objectives. This task will be approached on a step-by-step basis, working from simple circuit concepts to more complex ones.

Over the past decades, the electronics industry has been dominated by a trend toward smaller and smaller electronic components and circuits. Driving this process has been the search for ever-improving performance characteristics for electronic circuits and systems.

As illustrated in Fig. 1-9, electronics technology has evolved from hand-wired circuits to microelectronics, where most emphasis is placed today. At the extreme in terms of small size are the monolithic integrated circuits (ICs), where complex circuits are fabricated onto a single semiconductor crystal or chip.

The design and development of ICs is a capital-intensive task that is best suited to a high-volume market. Figure 1-10 shows a linear IC chip (an operational amplifier) that is used as a building block for more complex circuits. Figure 1-11(A) shows a very-large-scale integration (VLSI) digital IC chip that contains almost 70,000 tran-

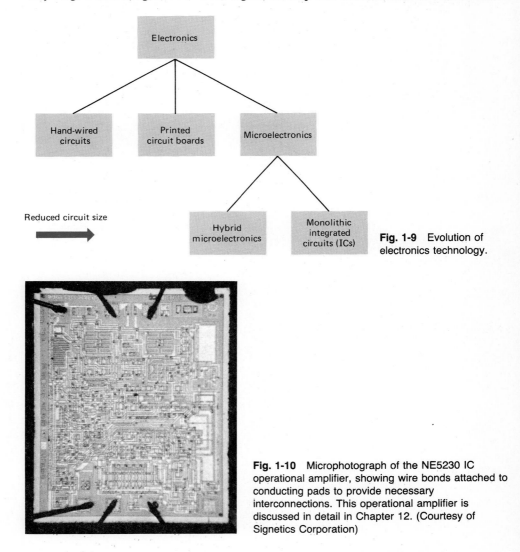

Fig. 1-9 Evolution of electronics technology.

Fig. 1-10 Microphotograph of the NE5230 IC operational amplifier, showing wire bonds attached to conducting pads to provide necessary interconnections. This operational amplifier is discussed in detail in Chapter 12. (Courtesy of Signetics Corporation)

Fig. 1-11(A) This very-large-scale integration (VLSI) digital IC is used in the MC68000 microprocessor and contains almost 70,000 transisters. This chip measures 246 × 281 mils (1 mil = 1/1000 in.). Further discussion on microprocessors is presented in Chapter 17. (Courtesy of Motorola, Inc.)

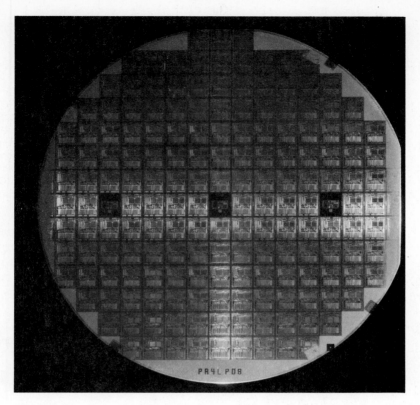

Fig. 1-11 (B) An array of MC68000 microprocessors on a thin wafer. (Courtesy of Motorola, Inc.)

sistors. Figure 1-11(B) shows how an array of such ICs is manufactured on a thin wafer of semiconductor material.

In order to create a circuit using ICs, a method must be used for interconnecting the ICs into the larger system. Hybrid microelectronics is used for this task. The word *hybrid* means that many different technologies bring passive and active elements together in an operating microcircuit. Hybrid microelectronics technology can package individual ICs for remounting into a circuit using the dual in-line package (DIP) shown in Fig. 1-12. Many different DIPs can be mounted on a printed circuit board to produce a complete circuit, as shown in Fig. 1-13.

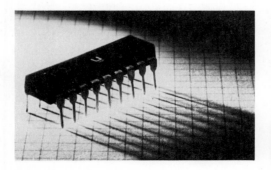

Fig. 1-12 This 18-pin dual in-line package (DIP) is used to mount ICs on printed circuit boards. (Courtesy of Unitrode Corporation)

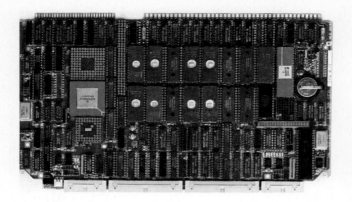

Fig. 1-13 Printed circuit board used in a microprocessor. The individual DIP carriers contain ICs. (Courtesy of Matrox Electronic Systems.)

Hybrid microelectronics can also be used to fabricate a circuit that combines unpackaged ICs and many other building block elements, as shown in Fig.1-14. Note the many elements brought together in a very small area. The growing interdependence between integrated circuits, printed circuits, and hybrid circuit manufacture is leading toward a more integrated viewpoint of microelectronics.

Fig. 1-14 Hybrid microcircuit fabricated with thick-film technology, indicating the many types of building block components used in such circuits. (Courtesy of Integrated Circuits Inc.)

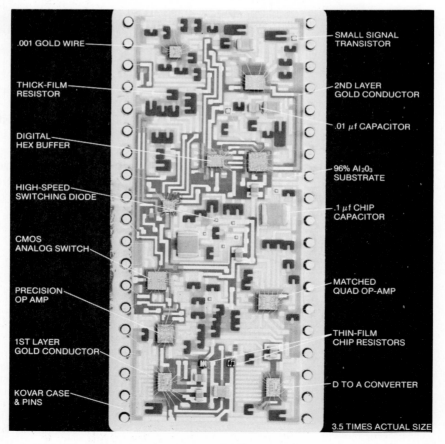

.001 GOLD WIRE

THICK-FILM RESISTOR

DIGITAL HEX BUFFER

HIGH-SPEED SWITCHING DIODE

CMOS ANALOG SWITCH

PRECISION OP AMP

1ST LAYER GOLD CONDUCTOR

KOVAR CASE & PINS

SMALL SIGNAL TRANSISTOR

2ND LAYER GOLD CONDUCTOR

.01 μf CAPACITOR

96% Al_2O_3 SUBSTRATE

.1 μf CHIP CAPACITOR

MATCHED QUAD OP-AMP

THIN-FILM CHIP RESISTORS

D TO A CONVERTER

3.5 TIMES ACTUAL SIZE

This chapter has introduced a systems approach to electronics design. Discussion has focused on the nature of tradeoffs that arise, various useful strategies, problems associated with data collection, and important technologies. The next step is to take a more detailed look at the scope of the process as the emphasis for this text.

A systems approach requires that the designer develop a working understanding of materials science and physical electronics, the characteristics of the building block elements to be used in design, and the combining of these elements into complete circuits and systems (see Fig. 1-15). The designer is required to coordinate decision-making that affects all areas. The tradeoffs involved in the production of electronic circuits are so complex that an understanding of all areas is required if the best decisions are to be made.

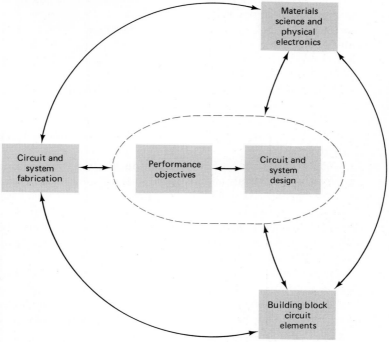

Fig. 1-15 Systems approach to circuit design (emphasizing the management and integration functions associated with the role of the designer; the connecting lines are supported by CAD and CIM applications of computer networks).

As discussed above, modeling is used extensively in electronics. A grounding in materials science and physical electronics is required if the designer is to understand the limitations associated with the various models that are being used. There are always operational regions in which these models break down and without an appreciation of the more fundamental properties and processes, the designer is limited by the models provided by someone else. Computer simulations of building block circuit elements often use complex models that require familiarity with more fundamental concepts. Additionally, as new building block elements become available, adequate modeling and circuit design will depend on the ability to accommodate new applications of the basic sciences.

Complete circuits are produced by combining many different types of elements. The performance of a circuit depends not only on the elements used, but also on the ways in which these elements are assembled. As noted above, one important method for circuit assembly is hybrid microelectronics. Without grounding in the relevant sciences and technologies, designers will not be able to understand the requirements and opportunities of such assembly procedures.

This text follows the systems viewpoint of electronics design that is illustrated in Fig. 1-15. Emphasis is on obtaining a broad appreciation for all the important factors that enter into electronics and for the methods that can perform necessary tradeoffs.

The text uses a hierarchical approach to explain how circuits and systems are developed, starting with the basic building block elements, then integrating upward to higher and higher functional levels. Passive components (resistors, capacitors, and inductors) and active devices (semiconductor diodes and transistors) are the most basic building block elements. In turn, integrated circuits (ICs) are combinations of diodes, transistors,, and other semiconductor elements. To understand IC operation, both individual (discrete) building block elements and the combinations of such elements into ICs must be considered.

Hybrid microcircuits are combinations of passive components, diodes and transistors, ICs, and other elements assembled into a complete circuit. Again, the emphasis is on learning about building block elements and then about the process of combination. A hybrid microcircuit may itself be used as building block in a larger electronics system.

Chapter 2 provides a fundamental introduction to the electrical properties of microelectronic materials and gives the starting point for an evolutionary study of electronics. Chapter 3 describes the properties of passive circuit elements and how physical considerations affect the models that are used and their limitations. Chapter 4 considers how semiconductor diodes of various types may be created as building block elements for circuit design. Chapter 5 then shows how these diodes may be applied to achieve circuit performance objectives.

Chapters 6 to 8 introduce transistors as additional building block elements and demonstrate some of the key considerations and tradeoffs that are encountered in circuit design. Chapters 9 to 11 expand on the level of complexity, explore how equivalent circuits may be used to aid in circuit design, and describe the types of constraints that may be encountered.

Chapters 12 to 14 provide a basic understanding of linear circuit applications. Chapter 12 introduces the operational amplifier, which is the IC building block that is most widely used in linear electronic circuits. Chapter 13 discusses practical linear system design with ICs and Chapter 14 describes sine-wave oscillator circuits and related systems applications.

Chapters 15 to 17 proceed to a study of digital circuit applications. Chapter 15 discusses digital circuits, and Chapter 16 indicates how these circuits can be combined into subsystems and systems. Chapter 17 describes practical digital system design with ICs.

Chapters 18 to 19 address microcircuit design and fabrication. Chapter 18 shows how materials, technologies, processes, and fabrication methods determine the circuit design parameters associated with semiconductor devices. Chapter 19 provides an overview of computer integrated design and manufacturing concepts.

QUESTIONS AND ANSWERS

Questions

1-1. What is the importance of a systems approach to design?

1-2. Why is it important to view electronics design as an iterative process?

1-3. In what ways can computer-aided design (CAD) have an important impact on breadboarding as part of the design process?

1-4. How do tradeoffs arise in electronics design?

1-5. How do relative values associated with performance objectives enter into the design process?

1-6. What are some of the advantages of graphical design methods?

1-7. Why are simplified models important?

1-8. Why do schematics provide limited representations of a circuit?

1-9. Why is CAD important in electronics design?

1-10. Why is a consideration of data collection an important part of the design process?

1-11. Why does microelectronics dominate the electronics industry today?

1-12. Why is the electronics designer required to develop generalist skills that include a fundamental understanding of devices, components, and manufacturing limitations?

Answers

1-1. The systems approach requires that the designer consider all types of performance objectives that affect the design process. Technical requirements, cost requirements, and manufacturing capabilities must all be considered to produce an optimum solution to a design problem. The systems approach requires a broad perspective and an emphasis on the best possible solution to a problem.

1-2. By viewing electronics design as an iterative process, the designer makes conscious use of feedback and learning. Iteration requires the formation of a trial design, the evaluation of this design from many points of view, and continual modification of the design until a best compromise solution is obtained. The design process is cyclical and repeated until the preferred solution is found.

1-3. If sophisticated CAD systems can adequately predict the performance of a breadboard, then this expensive and time-consuming stage of development is not required. Such CAD systems also facilitate the ability to consider manufacturing constraints as part of the design process.

1-4. Tradeoffs result from problem-solving in which different performance objectives are in conflict.

1-5. Relative values are required when there is no clear indication of a preferred solution and the relative importance of the various performance objectives must be weighed.

1-6. Graphical methods can be used to quickly gain oversight into the functional relationships involved in a given design activity. Plots of the various functions visually demonstrate the implications of all decisions. Simultaneous solutions to equations along with extrapolations among special cases may also be found by use of graphical methods.

1-7. Simplified models can transform a complex problem based on exact relationships into a simple problem based on approximate relationships. The result allows the designer to gain insight into the nature of important system relationships. However, since simplified models always include assumptions, their range of validity is limited and the designer must understand these limits.

1-8. Schematics use idealized device and component representations and neglect many types of distributed coupling. They are thus special-purpose models that must constantly be evaluated in terms of limitations.

1-9. The efficiency and effectiveness of all aspects of electronics design and manufacturing may be enhanced by CAD. The end result is to improve the design cycle and obtain products that can satisfy performance objectives in the best way.

1-10. All design methods depend on an adequate database for input to the models used. A thorough understanding of the limitations associated with data collection and application is essential to evaluate the validity of all model outputs.

1-11. The electronics industry is dominated by a trend toward microelectronics since high-density circuits are associated with improved performance objectives.

1-12. A broad, general understanding of all aspects of the design process is required to best satisfy the combined performance objectives experienced in realistic settings.

EXERCISES AND SOLUTIONS

Exercises

1-1. Given $R_1 = 2.0$ kΩ, $R_2 = 5.6$ kΩ, and $R_3 = 8.8$ kΩ, find $R_1 \| R_2 \| R_3$.

(1-1) **1-2.** Assume for the circuit of Fig. 1-2 that $V_s = 15.0$ V and V_{OUT} is to be 8.00 V for large values of R_x. The minimum value of R_x is 5.00 kΩ, and V_{OUT} must not drop below 7.50 V as R_x varies over its maximum allowed range. Find values for R_1 and R_2 to minimize the power dissipation in the circuit.

(1-1) **1-3.** A typical IC DIP configuration requires a surface area of 2.0 by 0.5 cm on a printed circuit board. When mounted on the board, the required thickness for the IC and board (with associated clearances) is 1.0 cm for a multiboard system. What ranges of values for the system component density D are associated with small-scale integration (SSI), midscale integration (MSI), large-scale integration (LSI), and very-large-scale integration (VLSI) levels of IC integration (as defined in Fig. 1-16) if each digital gate circuit consists of 10 components?

(1-1) **1-4.** Based on the information shown in Fig. 1-3:
(a) What is the unit heat removal cost for a component density $D = 60$ components/cm^3.
(b) For D of 2.0×10^3 components/cm^3?
(c) What is the average component volume for a density D of 4.0×10^2 components/cm^3?

(1-1) **1-5.** In Fig. 1-3, find the rate of change of the unit costs for both heat removal (curve B) and circuit production (curve A) with respect to component density D at the minimum-cost value of $D = 2.0 \times 10^2$ components/cm^3. What does this result tell you about the sensitivity of the solution to changes in costs?

(1-1) **1-6.** For a simple high-pass filter, you desire to obtain a value of f_1 of 1.0×10^2 Hz. Given that $C = 2.0$ μF, what value of R is required?

(1-1) **1-7.** For an RC high-pass filter, $R = 6.6$ kΩ and C = 0.01 μF. What value of f_1 results?

(1-1) **1-8.** Given the R and C values and costs of Fig. 1-17, what values of f_1 are available, and what costs are associated with each value?

FIGURE 1-17
Data for Exercise 1-8

R (kΩ)	Costs ($)	C (μF)	Costs ($)
1.0	0.25	0.01	0.45
2.2	0.35	0.001	0.65

1-9. As described in Appendix 2, a simple RC low-pass filter has a cutoff frequency $f_2 = 1/2\pi RC$.
(a) What is the significance of f_2?

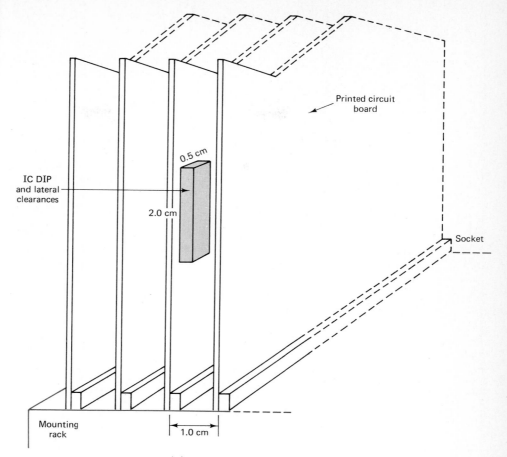

IC DIP and lateral clearances

Printed circuit board

0.5 cm

2.0 cm

Socket

1.0 cm

Mounting rack

(A)

Level of Integration	Gates/IC Chip
SSI	$10^0 - 10^1$
MSI	$10^1 - 10^2$
LSI	$10^2 - 10^3$
VLSI	$> 10^3$

(B)

Fig. 1-16 Printed circuit board configuration and data for Exercise 1-3.

(b) Given $R = 24$ kΩ and $C = 20$ pF, find f_2.

(c) Describe the graph of this function on the axes of Fig. 1-5(A) for $R = 24$ kΩ and the range of values for C that are shown.

(1-3) **1-10.** Given the characteristic curve of a tunnel diode, as shown in Fig. 1-18, find the solution(s) that exist for this curve and the equation

$$0.6V = (30 \ \Omega)I + V$$

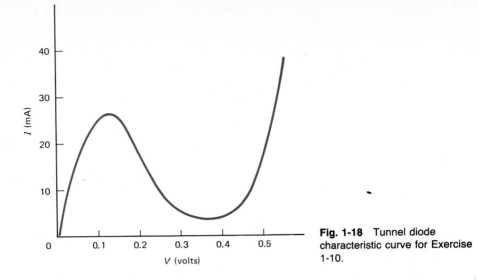

Fig. 1-18 Tunnel diode characteristic curve for Exercise 1-10.

(1-3) **1-11.** The two functions in Fig. 1-19 represent plots of Eqs. (1-4) and (1-5). Determine I_0, V_0, V_s, and R graphically.

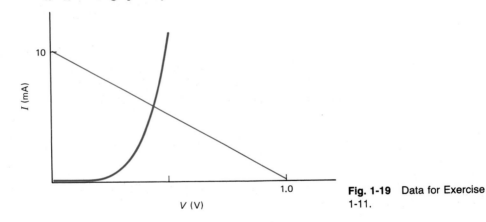

Fig. 1-19 Data for Exercise 1-11.

(1-4) **1-12.** For the resistor model in Fig. 1-8, assume that $R = 1.1$ kΩ, $C = 0$, and $L = 0.10$ μH. Find the frequency f for which the reactance of the inductance equals 1 percent of the resistor value.

(1-7) **1-13.** For a voltage divider, $V_{OUT} = V_{IN}[R_2/(R_1 + R_2)]$. If $R_2 = 5.60$ kΩ, $R_1 = 6.20$ kΩ, and $V_{IN} = 4$ V, how should V_{OUT} be written to demonstrate the proper use of significant figures?

(1-7) **1-14.** For Example 1-7, determine the values of $(V_s)_{min}$ and $(V_s)_{max}$ if V is known within 10 percent (neglecting any effects due to the uncertainty of $I_0 = 0.1$ nA and $V_0 = 0.026$ V).

1-15. A company guarantees to replace a given circuit if it fails within four years. One of two different testing methods (A or B) can be used. The cumulative number of failures over time is shown in Fig. 1-20. Every time a circuit fails, the replacement cost is $120. Test Method A costs $12.00/unit and test Method B costs $2.00/unit.
(a) From these data, which test method is the lowest cost strategy for the company?
(b) What other factors might enter into a choice between Methods A and B?

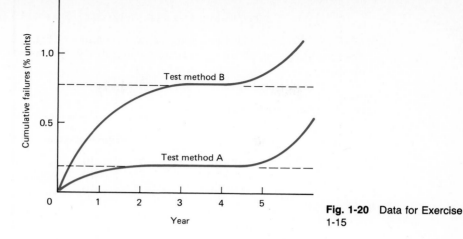

Cumulative failures (% units)

Test method B

Test method A

Year

Fig. 1-20 Data for Exercise 1-15

Solutions **1-1.** By definition, $R_1 \| R_2 = R_1 R_2 / (R_1 + R_2)$, so that

$$R_1 \| R_2 \| R_3 = 2.0 \text{ k}\Omega \| 5.6 \text{ k}\Omega \| 8.8 \text{ k}\Omega$$

$$= \frac{(2.0)(5.6)}{7.6} \text{ k}\Omega \| 8.8 \text{ k}\Omega$$

$$= 1.47 \text{ k}\Omega \| 8.8 \text{ k}\Omega = 1.3 \text{ k}\Omega$$

The calculation of parallel resistances can be performed in any order. An alternative solution strategy can use the equivalency $1/R_\| = 1/R_1 + 1/R_2 + 1/R_3$.

1-2. From the exercise, the given information includes $V_s = 15.0$ V, and

(a) $$V_{\text{OUT}} = 8.00 \text{ V for } R_x \longrightarrow \infty$$

(b) $$V_{\text{OUT}} = 7.50 \text{ V for } R_x = 5.00 \text{ k}\Omega$$

From Fig. 1-2 and Eq. (1-1),

(a) $$V_{\text{OUT}} = V_s \left(\frac{R_2}{R_1 + R_2} \right)$$

$$8.00 \text{ V} = (15.0 \text{ V}) \left(\frac{R_2}{R_1 + R_2} \right)$$

$$R_1 = 0.876 R_2$$

(b) $$V_{\text{OUT}} = V_s \left(\frac{R_2 \| R_x}{R_1 + R_2 \| R_x} \right)$$

$$7.50 \text{ V} = (15.0 \text{ V}) \left(\frac{R_2 \| 5.00 \text{ k}\Omega}{R_1 + R_2 \| 5.00 \text{ k}\Omega} \right)$$

$$\frac{R_2 (5.00 \text{ k}\Omega)}{R_2 + 5.00 \text{ k}\Omega} = R_1 = 0.876 R_2$$

Solving for R_2 and R_1,

$$R_2 = 0.71 \text{ k}\Omega \qquad \text{and} \qquad R_1 = 0.62 \text{ k}\Omega$$

1-3. From the information given,

$$\text{Volume/IC chip} = 0.5 \text{ cm} \times 1.0 \text{ cm} \times 2.0 \text{ cm} = 1.0 \text{ cm}^3$$

From Fig. 1-16, the definitions and component densities of Fig. 1-21 may be found. Note the importance of packaging and integration on the system-wide average component density.

FIGURE 1-21
Component Densities for Exercise 1-3

Level of Integration	Gates/IC Chip	Components/Gate	Components/cm^3
SSI	10^0–10^1	10	10^1–10^2
MSI	10^1–10^2	10	10^2–10^3
LSI	10^2–10^3	10	10^3–10^4
VLSI	$>10^3$	10	$>10^4$

1-4. (a) From Fig. 1-3, the heat removal cost for $D = 60$ components/cm^3 is approximately $3.50/unit.
(b) For $D = 2.0 \times 10^3$ components/cm^3, the heat removal cost is approximately $26.00/unit.
(c) The average component volume for $D = 4.0 \times 10^2$ components/cm^3 is

$$\text{Volume/component} = 1/D \cong 2.5 \times 10^{-3} \text{ cm}^3$$

1-5. By graphical determination,

(a)
$$\text{Slope for curve A} \cong \frac{-\$4}{(3 \times 10^2 - 1 \times 10^2)/\text{cm}^3} \cong -\$0.02 \text{ cm}^3$$

(b)
$$\text{Slope for curve B} \cong \frac{\$4}{(3 \times 10^2 - 1 \times 10^2)/\text{cm}^3} \cong \$0.02 \text{ cm}^3$$

At the minimum-cost value, the slopes of the component cost curves are the same.

1-6. From Eq. (1-3),

$$R = \frac{1}{2\pi f_1 C}$$

$$= \frac{1}{(6.28)(1.0 \times 10^2 \text{ Hz})(2.0 \times 10^{-6} \text{ F})} \cong 8.0 \times 10^2 \text{ } \Omega$$

1-7. From Eq. (1-3),

$$f_1 = \frac{1}{2\pi RC}$$

$$= \frac{1}{(6.28)(6.6 \times 10^3 \text{ } \Omega)(0.01 \times 10^{-6} \text{ F})} \cong 2.4 \text{ kHz}$$

1-8. By applying Eq. (1-3), the results of Fig. 1-22 are obtained. Note that the relationship between f_1 and cost is not so straightforward for this exercise.

FIGURE 1-22
Solution for Exercise 1-8

R (kΩ)	C (μF)	f_1 (kHz)	Cost ($)
1.0	0.01	16	0.70
1.0	0.001	160	0.90
2.2	0.01	7.3	0.80
2.2	0.001	73	1.00

1-9. (a) For a low-pass filter, frequencies below f_2 will pass through the filter relatively unaffected and frequencies above f_2 will be attenuated.

(b) By substitution,

$$f_2 = \frac{1}{2\pi RC}$$

$$= \frac{1}{(6.28)(24 \times 10^3\,\Omega)(20 \times 10^{-12}\,\text{F})} \cong 0.33\ \text{MHz}$$

(c) The result is a straight line parallel to the line shown, passing through the points

$$C = 1.0 \times 10^2\ \text{pF}, \quad f_2 = 6.6 \times 10^4\ \text{Hz}$$
$$C = 1.0 \times 10^3\ \text{pF}, \quad f_2 = 6.6 \times 10^3\ \text{Hz}$$

1-10. By plotting the equation as given on Fig. 1-18, three possible solutions are found, as shown in Fig. 1-23.

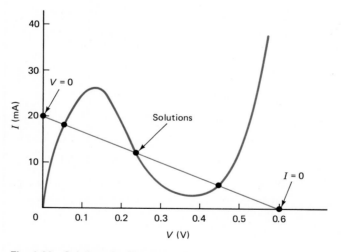

Fig. 1-23 Solutions for Exercise 1-10.

1-11. Using the data points A, B, C, and D, as shown in Fig. 1-24,

(a) For $I = 0$, $V = V_s$ (point A), so $V_s = 1.0$ V.

(b) The slope of the line AB is -10 mA/1.0 V, so $R = 1.0$ V/10 mA $= 0.10$ kΩ

(c) At points C and D,

$$1.0\ \text{mA} = I_0(e^{0.30\,V/V_0} - 1) \qquad \text{and} \qquad 10\ \text{mA} = I_0(e^{0.50\,V/V_0} - 1)$$

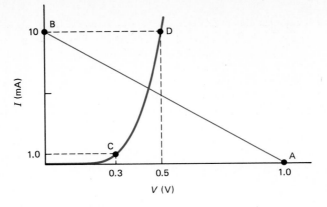

Fig. 1-24 Solution data points for Exercise 1-11.

By simultaneous solution,

$$I_0 = 32 \ \mu A$$

$$V_0 = 0.087 \ V$$

1-12. For the given specification,

$$X_L = 2\pi fL = 0.01R$$

$$f = \frac{(0.01)(1.1 \times 10^3 \ \Omega)}{2\pi (0.10 \times 10^{-6} \ H)} \cong 18 \ MHz$$

1-13. Following the guidelines in Appendix 5,

$$V_{OUT} = (4 \ V)\left(\frac{5.60 \ k\Omega}{6.20 \ k\Omega + 5.60 \ k\Omega}\right)$$

$$= (4 \ V)\left(\frac{5.60 \ k\Omega}{11.80 \ k\Omega}\right) \cong 2 \ V$$

1-14. By direct substitution,

$$I_{min} = (0.1 \times 10^{-9})(e^{0.450/0.026} - 1) \cong 3.3 \ mA$$

$$I_{max} = (0.1 \times 10^{-9})(e^{0.550/0.026} - 1) \cong 150 \ mA$$

$$(V_s)_{min} = 0.450 \ V + (3.3 \ mA)(50 \ \Omega) = 0.62 \ V$$

$$(V_s)_{max} = 0.550 \ V + (150 \ mA)(50 \ \Omega) = 8.05 \ V$$

1-15. (a) Within four years, there will be 8 percent failures using Method B and 2 percent failures using Method A. Let N be the number of circuits sold. Then, as shown in Fig. 1-25, B is the best strategy.
(b) Customer dissatisfaction with failures might be more important than the cost differential.

FIGURE 1-25
Solution Data for Exercise 1-5

Method	Replacement Cost	Test Cost	Total Cost
A	$(0.02)(\$120)N = \$2.40N$	$12.00N	$14.40N
B	$(0.08)(\$120)N = \$9.60N$	$2.00N	$11.60N

PROBLEMS

1-1. This chapter describes some of the general considerations that enter into electronics design. Make a list of the comments, ideas, observations, and questions that occur to you as you read this material. Base your list on your personal observations of the electronic devices that you encounter in everyday life and on any special experiences. Come to class prepared to discuss your list.

1-2. Consider the different ways in which computers can improve the systems approach to electronics design that is illustrated in Fig. 1-1. List the advantages and disadvantages of computer application at each stage of design. Come to class prepared to discuss your list.

(1-1)   **1-3.** Given $R_1 = 5.0$ kΩ and $R_2 = 15$ kΩ, find $(R_1 \| R_2)R_1/R_2 + R_2$.

(1-1)   **1-4.** Given $R_a = 10$ kΩ and $R_b = 2.2$ kΩ, find R_c if the relationship among the resistors is determined by the design equation

$$\frac{R_a}{R_a \| R_b + R_c} = 4.0$$

(1-1)   **1-5.** Assume for the circuit of Fig. 1-2 that the minimum value of R_x is 3.3 kΩ and the maximum value is 6.8 kΩ. If $V_s = 12.0$ V and V_{OUT} is to vary no more than ± 5 percent from 6.0 V, find the values for R_1 and R_2 that keep the power dissipated in the circuit as small as possible.

(1-1)   **1-6.** Assume for the circuit of Example 1-2 that V_{OUT} must vary by no more than 5 percent. Find the values for R_1 and R_2 and the power dissipated in the circuit.

(1-1)   **1-7.** The unit cost of heat removal for a complex electronic system is given by

$$C = \$2.00(1 + e^{D/D_0})$$

where D is the density (components/cm^3) and $D_0 = 1.0 \times 10^3$ components/cm^3. Using the cicuit production costs (curve A) of Fig. 1-3, find the component density that will result in the lowest-cost design.

(1-1)   **1-8.** Assume that other tradeoff considerations result in a selected component density of $D = 8 \times 10^3$ components/cm^3 for the systems described in Example 1-3. What circuit production costs and heat removal costs result?

(1-1)   **1-9.** For the simple high-pass filter of Fig. 1-4, a cutoff frequency of $f_1 = 0.16$ MHz is required. Which components in Fig. 1-26 should be selected?

FIGURE 1-26
Data for Problem 1-9

R (kΩ)	Costs ($)	C (pF)	Costs ($)
10	0.10	100	0.20
20	0.15	50	0.25

(1-1)   **1-10.** For an RC high-pass filter, $f_1 < 1 \times 10^3$ Hz is required. If $R = 100$ kΩ, what is the smallest capacitor that can be used?

(1-1)   **1-11.** For an RC low-pass filter, $R = 10$ kΩ and $C = 10$ pF. What value of f_2 results?

(1-1)   **1-12.** A low-pass filter is to be designed with $f_2 > 1.0$ MHz. If the distributed capacitance is 100 pF, what is the maximum series R that can be used?

(1-1)   **1-13.** Based on the information shown in Fig. 1-5:
(a) What cutoff frequency is associated with $C = 4.0 \times 10^2$ pF?
(b) Plot the average cost/cm^3 for the capacitors described in Fig. 1-5(B) as a function of C.

(1-3)   **1-14.** For Example 1-5, $V_{OUT} = 0.15$ V is desired, with the same resistor and diode characteristic curves shown in Fig. 1-7. What value of V_s is required?

(1-3) **1-15.** For Eq. (1-4), let $I_0 = 100$ pA, $V_0 = 0.026$ V, and $V = 0.450$ V. Find the current I that flows in the device (a semiconductor diode). What percentage uncertainty in V doubles I?

(1-3) **1-16.** You are given the characteristic curve of Fig. 1-27 (encountered in design with transistors) and the equation below. Find the solution(s) that exist for V and I.

$$12 \text{ V} = (120 \text{ } \Omega)I + V$$

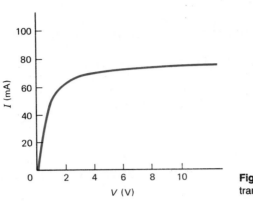

Fig. 1-27 Characteristic curve associated with transistors, used for Problem 1-16.

(1-3) **1-17.** Find the solution(s) that exists for the curve of Fig. 1-18 and the equation

$$0.8 \text{ V} = (40 \text{ } \Omega)I + V$$

(1-3) **1-18.** Find the solution(s) that exists for the curve of Fig. 1-27 and the equation

$$10 \text{ V} = (200 \text{ } \Omega)I + V$$

(1-4) **1-19.** The resistor of Fig. 1-8 is modeled with $R = 50$ kΩ, $C = 1.0$ pF, $L = 0$. For what frequencies will the current through the capacitor be less than 5 percent of the current through the resistor?

(1-4) **1-20.** For the resistor model in Fig. 1-8, assume that $R = 1.1$ kΩ, $C = 10$ pF, and $L = 0.10$ μH. What will be the resonant frequency for the resistor?

(1-4) **1-21.** For the resistor model in Fig. 1-8, $R = 6.8$ kΩ, $C = 0$, and $L = 10$ μH. Find the frequency f for which the reactance of the inductance is equal to 5 percent of the resistor value.

(1-7) **1-22.** For Example 1-7, I_0 is narrowed to the range 0.1 to 0.2 nA. What range for V_s results?

(1-7) **1-23.** How many significant figures are present in the following numbers?

$$4.6653$$

$$2.5 \times 10^3$$

$$0.0066$$

(1-7) **1-24.** What is the correct form in which to write the result of the operation

$$V = (1.6 \text{ k}\Omega)(5.566 \text{ mA}) = ?$$

COMPUTER APPLICATIONS

Throughout the text, computer applications that require programming and computer access are preceded by asterisks.

1-1. Example 1-5 shows how graphical methods can obtain a simultaneous solution to the equations

$$I = I_0(e^{V/V_0} - 1)$$

$$V_s = V + IR$$

that are often encountered in electronics design. A solution may also be obtained by writing a computer program that will find the values of I and V that satisfy both equations simultaneously (using an iterative procedure). Figure 1-28 shows the results of a FORTRAN computer simulation for the case

$$I_0 = 1.0 \times 10^{-11} \text{ A} \qquad V_s = 1.5 \text{ V}$$

$$V_0 = 0.03 \text{ V} \qquad R = 50 \text{ }\Omega$$

```
    SUPPLY VOLTAGE — BIAS
  1.5
    VALUE OF RESISTOR USED IN K–Ohms
  0.05
    TYPE OF DIODE — SI=1,GE=2,GAAS=3 ?
  1
  SI

        TYPE          LEAKAGE CURRENT
         SI           1.0000000E–11


    INITIAL VALUE FOR Q–POINT VOLTAGE?
  0.9


  OPERATING POINT OF THE DIODE AND LOAD LINE
```

SUPPLY VOLTAGE (volts)	Q-POINT VOLTAGE (volts)	Q-POINT CURRENT1 (mA)	Q-POINT CURRENT2 (mA)	LOAD RESISTOR (Ohms)
1.50	0.6380	17.2393	17.2393	50.

Y-INTERCEPT CURRENT (mA)	X-INTERCEPT VOLTAGE (volts)	Q-POINT VOLTAGE (volts)	Q-POINT CURRENT1 (mA)
30.000	1.50	0.6380	17.24

```
  FORTRAN STOP
  $
```

Fig. 1-28

Show by graphical means that this computer program has produced the correct results. From the data given, to what percentage accuracy has the simultaneous solution been found? Compare this with typical graphical results.

 1-2. Assume that the tunnel diode characteristic curve of Fig. 1-18 can be represented by the equation

$$I = I_0(e^{V/V_0} - 1) + I_p(V/V_p)e^{(1-V/V_p)}$$

where (I_p, V_p) is the approximate location of the peak.

Let

$$I_0 = 1.00 \times 10^{-12} \text{ A} \qquad I_p = 1.5 \text{ mA}$$

$$V_0 = 0.026 \text{ V} \qquad V_p = 0.5 \text{ V}$$

Write a computer program that produces a table of values for I as a function of V. Plot your results (by hand or using available software) and compare with Fig. 1-18. You are to produce a program listing, a table of values, and a graph.

* **1-3.** Write a computer program that calculates the percentage variation in V_{OUT} for the voltage divider of Fig. 1-2 as a function of R_1 and R_2 and the range of values for R_x. Use the program to solve Example 1-2, and compare your findings. Develop a set of parametric output curves that show how the percentage variations in V_{OUT} are affected by the design parameters of the circuit. Discuss your findings.

* **1-4.** Figures A2-1 and A2-2 in Appendix 2 illustrate the frequency response characteristics of simple high-pass and low-pass filters. Write a computer program that will generate these curves. Plot your results and discuss.

* **1-5.** Figures A3-1 and A3-2 in Appendix 3 illustrate how the (complex) currents through simple series and parallel resonant circuits vary with frequency. Write a computer program that will generate these curves. Plot your results and discuss.

EXPERIMENTAL APPLICATION

Most chapters in this book include at least one laboratory experiment that relates to the chapter materials. Appendix 6 provides suggestions as to how to plan and conduct these experiments, and can be used for reference.

1-1. Figure 1-5(A) for Example 1-4 shows how the cutoff frequency f_1 for a simple high-pass RC filter circuit is related to the R and C values selected. This type of circuit is further discussed in Appendix 2.

The objective of this experiment is to confirm the plot of Fig. 1-5(A). Design the needed high-pass filter circuit using a sine-wave source to produce the input and an oscilloscope to observe the input and output waveforms. Prepare a procedure that can satisfy the desired experimental objective. Establish error ranges for your measurements, and plot data points with these ranges indicated. Compare your theoretical predictions and experimental findings, and interpret the results of your experiment.

Discussion
The schematic to be used is shown in Fig. 1-29. The output of this circuit is given by

$$\frac{|\mathbf{v}_o|}{|\mathbf{v}_i|} = \frac{\omega RC}{\sqrt{1 + (\omega RC)^2}}$$

$$\theta = \tan^{-1} 1/\omega RC$$

At frequency f_1, $\omega RC = 1$ (or $f_1 = 1/2\pi RC$), so that

$$\frac{|\mathbf{v}_o|}{|\mathbf{v}_i|} = \frac{1}{\sqrt{2}}$$

$$\theta = \tan^{-1} 1 = 45°$$

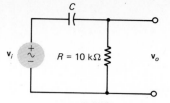

Fig. 1-29 Schematic
for Experimental Application 1-1.

A resistor $R = 10$ kΩ and different values of C are to be used. The general functional relationships for the attenuation and phase shift of the circuit as a function of f should be experimentally determined for a selected value of C to provide familiarity with the properties of high-pass filters. Then different values of C can be used and f varied for each to find the frequency for which the attenuation is $1/\sqrt{2}$ and the phase shift is 45° (defining f_1). Frequency f_1 can then be plotted as a function of C to confirm Fig. 1-5(A). Care should be taken to prepare to indicate on the graph the error ranges for each measurement. For small values of C, distributed capacitances in the circuit and measuring instruments will limit the accuracy of the measurements.

The collected data can be used to create a table similar to that of Fig. 1-30, showing the value of C, the measured value of f_1, and estimates of the measurement uncertainty with each value of f_1. Data may then be plotted for comparison with the theoretical curve of Fig. 1-5(A).

FIGURE 1-30
Typical Data Collected for Experimental Application 1-1

C (pF)	Experimental Value of f_1 (kHz)	Experimental Range Error (kHz)	% Error Range
1.0×10	Not measurable due to distributed capacitance effects		
4.7×10	1.2×10^3	$\pm 0.8 \times 10^3$	± 70
1.15×10^2	1.7×10^2	0.4×10^2	± 20
4.7×10^2	3.9×10	0.4×10	± 10
1.0×10^3	1.4×10	0.1×10	± 7
4.7×10^3	3.7	0.2	± 5
1.0×10^4	1.6	0.5	± 3

CHAPTER 2
ELECTRICAL PROPERTIES OF MICROELECTRONIC MATERIALS

Semiconductor devices are the principal building block elements of electronic circuits. These devices can be fabricated in discrete form or as integrated circuits (ICs). Many complex systems today use ICs as the starting point for circuit design and add discrete devices to achieve the final configuration.

There is constant competitive pressure to improve all types of building block components so as to enhance circuit performance. New materials and technologies are continually being introduced. As can be expected, many different tradeoffs exist among the implementation options that are available for a circuit.

Germanium (Ge), silicon (Si), and gallium arsenide (GaAs) are the most common materials used for the fabrication of semiconductor devices. Many other materials are being explored in research laboratories. The selection of a given material type helps determine the circuit performance options available to the designer. Device performance (including electrical and thermal properties) depends on the material characteristics in fundamental ways. Understanding circuit and model limits requires an awareness of these fundamental properties.

In addition, as noted in Chapter 1, electronic systems generally require the assembly of multiple ICs and devices into complex hybrid microcircuits and printed circuits. (Fiber optics systems further require optical interfaces between circuits and fiber cables.) The evolution of new hybrid and surface-mount assembly techniques has opened additional system design pathways. An understanding of the alternatives available here is necessary if the designer is to follow the system design strategy of Fig. 1-1 and evaluate all system alternatives and tradeoffs.

Recent research-and-development (R&D) efforts in electrical engineering have identified the importance of system interconnections and packaging to achieving performance objectives. It is no longer acceptable to treat circuit design as an idealized process, neglecting material properties, and still be able to optimize circuit design for each application. As sophisticated computer-aided design (CAD) software begins to provide alternative decision choices for the designer, a fundamental understanding of the tradeoffs that are being presented is essential to arrive at the preferred selection.

CAD is rapidly developing to the point where the designer will be faced with multiple tradeoffs regarding device selection, circuit design, and the technologies and materials to be used. These factors will interrelate, so that design will depend on the method of implementation. Iterative procedures will be required to explore all choices and maximize the performance of the final system. This is a more extended and complex approach to design than is often used today. It is reasonable to expect the choices to grow rapidly in the near future, so that additional computer support will be necessary to consider the choices and tradeoffs and to guide the computer-integrated manufacturing (CIM) process.

This chapter describes the atomic-level properties of materials commonly used in electronics. It introduces the resistivity ρ as one of the macroscopic material properties that drives device characteristics. Basic models of semiconductor physics are then used to show how ρ depends on fundamental atomic variables and how tradeoffs among materials may be understood in terms of these variables.

There are several examples that illustrate how atomic properties can directly affect circuit performance. Then, the final section describes some of the important ways in which these atomic-level variables are affecting the design process today.

The analysis starts with an idealized model of conduction in a *nonmetallic single-crystal material* and then extends to metals and other types of materials. The objective is to develop an adequate background of the dominant processes that determine the electrical characteristics of crystals and to introduce a range of quantitative methods that can be used in electronics design.

2-1 RESISTIVITY AND RESISTANCE

One of the most important properties of a material is its resistivity ρ. The resistance R of a sample of length ℓ and cross-sectional area A is given by

$$R = \rho \frac{\ell}{A} \tag{2-1}$$

The resistivity of the materials used to produce electronic building block elements and circuits helps determine the performance parameters associated with the elements and the operation of the final circuit configuration.

Figure 2-1 illustrates experimental values of ρ at 300°K[1] for a range of materials. As can be seen, most metals have low resistivity values and are therefore conductors. Most ceramics and polymers have high resistivity values and are therefore insulators. Semiconductors and mixtures of conductors and insulating materials can produce intermediate resistivity values.

---

[1] Throughout this text, room temperature is taken as 300°K.

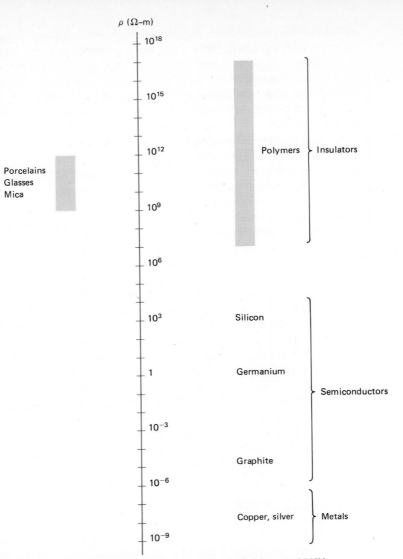

Fig. 2-1 Typical experimental values of the resistivity ρ at 300°K.

In order to understand the electrical properties of materials and the operation of semiconductor devices, it is necessary to gain a theoretical appreciation for the physical processes that produce these resistivity values. The following examples describe *why* materials have such a wide range of resistivity values and how to anticipate the resistivity based on material properties.

Example 2-1   Range of Variations for Resistivity and Resistance

Since resistance R of a sample is a function of both resistivity ρ and the geometric ratio ℓ/A, it is useful to gain insight into the relative importance of these two factors and how they interrelate. Assume that a graphite sample is 1.00 mm in diameter. What should be the length of the sample to produce $R = 3.8 \times 10^{-2}\ \Omega$?

Solution From Fig. 2-1, $\rho \cong 1.0 \times 10^{-5}\ \Omega \cdot m$, so that

$$\ell = \frac{RA}{\rho} = \frac{(3.8 \times 10^{-2}\ \Omega)[\pi (0.50)^2\ mm^2 \times (1.0 \times 10^{-3}\ m/mm)]}{1.0 \times 10^{-5}\ \Omega \cdot m} \cong 3.0\ mm$$

Since R is proportional to ρ, and ρ can vary by a factor of 10^{27} (from Fig. 2-1), the resistance of material samples can also be expected to vary by many orders of magnitude. The dimensional dependency of the resistance on the length and cross-sectional area enables the designer to obtain desired R values for a given ρ. ∎

Example 2-2 **Accuracy of Device Resistance**

Two graphite samples with the dimensions given in Example 2-1 are used in the voltage-divider circuit of Fig. 2-2. The output V_{out} must be known within 10 percent for R_1 and R_2 both having nominal values R_{nom}. If the length and diameter of each sample are known to ± 1.0 percent, will V_{out} be known with sufficient accuracy?

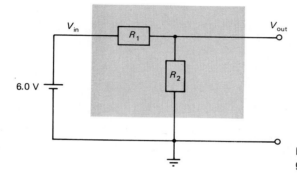

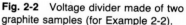

Fig. 2-2 Voltage divider made of two graphite samples (for Example 2-2).

Solution

$$R_{min} = R_{nom}\frac{0.990}{1.01^2} = 0.970 R_{nom}$$

$$R_{max} = R_{nom}\frac{1.01}{0.990^2} = 1.03 R_{nom}$$

$$(V_{out})_{min} = \frac{R_{min}}{R_{min} + R_{max}}6.0\ V = \frac{0.970}{2.00}6.0\ V = 2.9\ V$$

$$(V_{out})_{max} = \frac{R_{max}}{R_{min} + R_{max}}6.0\ V = \frac{1.03}{2.00}6.0\ V = 3.1\ V$$

The performance objective has been satisfied. This example describes the relationships between dimensional uncertainties and the resultant modification in circuit performance. ∎

2-2 CRYSTALLINE MATERIALS

The resistivity is determined by the properties of materials at the atomic level. A starting point for discussion is to consider the nature of crystalline structures.

A single crystal is formed by atoms bound together in an orderly structure. The individual atoms are interconnected by *bonds* that lock the atoms into a pattern that is repeated throughout the crystal. The *lattice structure* of a crystal consists of a set of points located in space where the atoms are placed, and it serves to define the pattern.

The bonds that hold the atoms together are electrostatic in nature and can be described as metallic, covalent, or ionic. At the moment, the discussion relates only to covalent and ionic bonds.

In covalent bonding, electrons circulate between atoms, producing a negative cloud between the positive ions, as shown in Fig. 2-3(A). Ionic bonds are created through an electron transfer between different types of atoms, one with a loosely attached electron and the other with a strong affinity to add an extra electron. The electron transfers from one atom to the other, producing oppositely charged ions that are attracted to one another, as shown in Fig. 2-3(B).

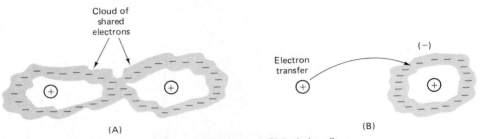

Fig. 2-3 Types of bonding: (A) covalent bonding and (B) ionic bonding.

The atoms inside a crystal are constantly vibrating due to their thermal energy. Under equilibrium conditions, these vibrations define the temperature of the crystal. A small fraction of the electrons collects sufficient energy to leave the valence shells (on the outside of the atom) and becomes free to move around inside the crystal. The interior of the crystal thus consists of vibrating atoms that are bound in position, mobile electrons in constant motion, and positively charged ions.

The resistivity of a material is determined by those electrons that can move in response to an externally applied electric field E. Such a field results if a voltage V is applied across the material.

2-3 ENERGY BANDS

A quantitative understanding of resistivity requires drawing on the energy-level concept. Electrons in an isolated atom exist in orbits around the nucleus, with only certain energies (radii) allowed. When these atoms join to form a crystal, the situation changes. Electrons are subject to interactions with the array of atoms located in the lattice structure. As a result, instead of existing at specific energy levels, electrons are found in energy *bands*.

In general, the forces exerted by the lattice on an electron restrict the range of energies that the electron can have as it moves about. The crystal-electron interactions define *allowed* energy bands for an electron. Other values are *forbidden*. No electron can take on an energy value in the forbidden range.

The interaction between the periodic structure of the crystal and the electrons in the crystal thus results in the concept of allowed and forbidden energy bands. Electrons can take on certain energy values (within allowed bands) and are forbidden from taking on energy values in forbidden energy bands. This result is illustrated using an *energy-level* diagram as shown in Fig. 2-4. The diagram is a one-dimensional illustration where only the vertical axis has significance. Plotted on the vertical axis are possible energy values for an electron in a crystal. The diagram graphically indicates which energies are allowed and which are forbidden and can thus identify all possible energy states for electrons in the crystal.

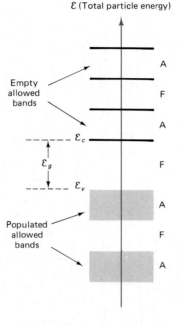

A = Allowed energy region or band
F = Forbidden energy region or band

Fig. 2-4 Energy band diagram (shown at a very low temperature, $T \to 0$).

Some of the allowed bands correspond to bound electron states, where the electron is attached to the atom. Other allowed bands correspond to electrons that can move in the crystal. The allowed bands can thus be divided into those corresponding to bound states and those corresponding to mobile states.

At very low temperatures, as T approaches $0°K$, lattice vibrations become small since there is little thermal energy in the crystal. The mobile electrons lose their energy and fall back into bound states. In this limit, the energy bands corresponding to bound states are full and the mobile bands are empty. The electrons in the crystal exactly fill a number of allowed bands. These bands are then fully populated. In the fully populated bands, the number of electrons matches the number of available energy states.

The uppermost (highest-energy) filled band is called the *valence band* because it corresponds to electrons that have been captured by ions into valence shells. The electrons are bound to host atoms and are no longer free to move in the crystal. Since the number of available valence-band energy states is equal to the number of available electrons, all states are filled and all free electrons are captured. As shown in Fig. 2-4, the upper energy marking the top of the valence band is called $\mathcal{E}_v$.

All allowed bands at higher energies are empty in this very-low-temperature limit. The first empty allowed band is called the *conduction band* because electrons that gain enough energy to jump from the valence band to this next higher band are then free to move in the crystal and contribute to electrical conduction. The lower energy marking the bottom of the conduction band is called $\mathcal{E}_c$. The valence and conduction bands are separated by a forbidden band, or gap, of width $\mathcal{E}_g = \mathcal{E}_c - \mathcal{E}_v$.

At very low temperatures, then, all electrons in the crystal are bound into valence-band energy states and are unable to move through the crystal. If a voltage is applied to such a material, no current flows.

2-4   THERMAL EFFECTS

Now consider what happens as the crystal is heated. The thermal vibration of atoms increases and the electrons begin to gain energy. Some of the electrons gain enough energy to jump across the energy gap $\mathcal{E}_g$ between the valence and conduction bands and to move around in the crystal. As the temperature increases, more and more electrons make this transition.

In the absence of an applied voltage, the electron motions are random and cancel one another. If a voltage is applied to the material, the electrons experience an electric field E that produces an average drift velocity. A current flow is observed. As the temperature increases, producing more carriers, the observed current also increases.

The excitation process that lifts electrons from the valence band to the conduction band can be symbolized as shown in Fig. 2-5. When an electron leaves its valence state, the host atom is left short an electron, producing an ion. The ion tends to reach out and capture an electron from a nearby atom, leaving this neighbor now short an electron. The process repeats itself over and over, and the missing electron location moves around in the crystal.

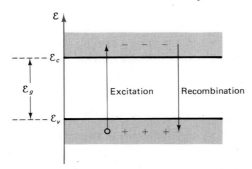

Fig. 2-5   Electron excitation and recombination.

Rather than keep track of the exchange process, it is more useful to consider a missing electron to be a positively charged *hole* and to track the progress of this hole through the crystal. The crystal thus contains two types of carriers, negative electrons and positive holes. Since a hole is a missing electron, the numbers of electrons and holes are equal and the total electrostatic charge must be zero.

Electrons have a limited lifetime in the conduction band and periodically fall back to the valence band in a recombination process. The crystal is thus filled with electrons and holes that are in constant random motion, driven by thermal energy, and participating in constant excitation and recombination processes as their energy states change.

As will be discussed in Chapter 4, excitation can also occur when electromagnetic radiation (for example, a beam of light) is incident on an energy-gap material. The energy in an electromagnetic field is carried by photons, where each photon has an energy $\mathcal{E}_{ph}$ proportional to its frequency. If there are photons available with $\mathcal{E}_{ph} > \mathcal{E}_g$, then these photons can be absorbed. Each photon transfers its energy to an electron, which is excited across the energy gap. Light shining on an energy-gap material can thus increase the number of electrons in the conduction band.

When electrons recombine with holes, the energy of excitation can go into heat (thermal vibrations in the crystal) or into photons (electromagnetic radiation). The relative probability of each type of decay process depends on the detailed structure of the material. Germanium (Ge) and silicon (Si) are *indirect* semiconductors, in that the electron momentum must change during recombination (achieved through coupling to crystal vibrations). This makes the probability of photon emission low for these materials. On the other hand, gallium arsenide (GaAs) is a *direct* semiconductor, with transitions that do not require momentum changes. Therefore, GaAs is often used in semiconductor devices that are intended to produce radiated light. Photon-semiconductor interactions and applications are discussed in Chapter 4.

As they move around in the crystal, both types of *carriers* (electrons and holes) are constantly interacting with the crystal lattice. Because of these interactions, both electrons and holes are found to have positive effective masses. The effective mass of the negative carriers (electrons) m_n^* and the effective mass of the positive carriers (holes) m_p^* have an important effect on current flow in the crystal.

This section is concerned with *intrinsic* semiconductors that have properties associated with pure crystals. Chapter 3 considers *extrinsic* properties that arise due to the planned addition of impurities to the crystal. In intrinsic semiconductors, the numbers of electrons (n) and holes (p) per unit volume are equal ($n = p = n_i$, where n_i represents the number of *carriers* per unit volume of either type).

There are two important relationships that are used as a basis for a quantitative discussion of materials and semiconductor devices. The first relationship provides a means for calculating the number of carriers (mobile electrons or holes) that exist in an intrinsic energy-gap material. From Appendix 7,

$$n_i = \frac{2}{h^3}(2\pi m_0 kT)^{3/2}\left(\frac{m_n^*}{m_0}\frac{m_p^*}{m_0}\right)^{3/4} e^{-(\mathcal{E}_c - \mathcal{E}_v)/2kT} \qquad (2\text{-}2)$$

where the following constants are being used:

Planck's constant: $\qquad\qquad h = 6.62 \times 10^{-34} \text{J} \cdot \text{s} = 4.14 \times 10^{-15} \text{ eV} \cdot \text{s}$

Boltzmann's constant: $\qquad k = 1.38 \times 10^{-23} \text{J}/°\text{K} = 8.62 \times 10^{-5} \text{ eV}/°\text{K}$

Rest mass of the electron: $\quad m_0 = 9.1 \times 10^{-31} \text{ kg}$

Note that h and k are given in two sets of units. The first set is in mks units and produces correct results when combined with other mks units. For the second set of values, the eV (electron-volt) has been introduced as a unit of energy. The eV represents the energy an electron gains by falling through a potential of 1 V. To convert J to eV, divide by the charge of the electron $(1.6 \times 10^{-19}$ C).

Since $\mathcal{E}_c - \mathcal{E}_v = \mathcal{E}_g$, n_i is explicitly dependent on the width of the energy gap $\mathcal{E}_g$ measured with respect to the thermal energy kT. For small values of T $(kT \ll \mathcal{E}_g)$, n_i tends toward zero and there are no free carriers, as expected. For large values of T $(kT \gg \mathcal{E}_g)$, n_i becomes large and approaches an asymptotic value. For this case, a large fraction of the total electrons in the crystal reaches the conduction band; the material can change from its solid phase and pass to a liquid or gas phase.

The second relationship to be used (also from Appendix 7) can be written in the form

$$\mathcal{E}_F = \tfrac{1}{2}(\mathcal{E}_c + \mathcal{E}_v) - \tfrac{3}{4}kT \ln \left(\frac{m_n^*}{m_p^*}\right) \tag{2-3}$$

which introduces a new parameter, the *Fermi energy* $\mathcal{E}_F$, which is the chemical energy of the material and is a constant throughout a material that is in equilibrium. As seen below, $\mathcal{E}_F$ is an important parameter.

Since m_n^* and m_p^* are of the same order of magnitude and typically $\mathcal{E}_F \gg kT$, Eq. (2-3) can be approximated (in an intrinsic material) by

$$\mathcal{E}_F \cong \tfrac{1}{2}(\mathcal{E}_c + \mathcal{E}_v) \tag{2-4}$$

This equation provides the useful information that the Fermi energy occurs near the center of the energy gap in an intrinsic semiconductor. $\mathcal{E}_F$ can thus be associated with the average energy of the mobile carriers in the crystal.

Example 2-3 **Intrinsic Density of Carriers**

Evaluation of a semiconductor device requires a prediction of the intrinsic density of carriers n_i in a pure crystal of Si at 300°K. Show how the database of Fig. 2-6 can be applied, and evaluate the usefulness of the model that is used.

FIGURE 2-6
Semiconductor Properties (300°K)

	$\mathcal{E}_g$	m_n^*/m_0	m_p^*/m_0	$\mu_n(\text{m}^2/\text{V} \cdot \text{s})$	$\mu_p(\text{m}^2/\text{V} \cdot \text{s})$	$D_n(\text{cm}^2/\text{s})$	$D_p(\text{cm}^2/\text{s})$
Ge	0.67	0.55	0.37	0.39	0.19	100	50
Si	1.11	1.1	0.59	0.14	0.05	35	12.5
GaAs	1.43	0.07	0.54	0.85	0.04	220	10

Solution The characteristics of Si are given in Fig. 2-6. By substitution

$$\frac{\mathcal{E}_g}{2kT} = \frac{1.11 \text{ eV}}{2 \times 8.62 \times 10^{-5} \text{ eV/°K} \times 300°K} \cong 21.46$$

and

$$n_i = \frac{2}{(6.62)^3 \times 10^{-102} \text{ (J} \cdot \text{s)}^3}$$

$$\times [2\pi (9.1 \times 10^{-31} \text{ kg})(1.38 \times 10^{-23} \text{ J/°K})(300°K)]^{3/2}(1.1 \times 0.59)^{3/4}e^{-21.46}$$

$$\cong 8.7 \times 10^{15} \text{ carriers/m}^3$$

This value of n_i is based on the theoretical model that led to Eq. (2-2). From the literature, it can be found that several considerations that relate to the nature of the possible transitions across the energy gap and the second-order temperature effects have been neglected in this model. The experimental value of n_i is about 1.6×10^{16} carriers/m³, higher than the model value by approximately a factor of 2 (Thurmond, 1975). Given the many assumptions built into the model (and discussed in Appendix 7), this is a reasonably good agreement between theory and experiment. In subsequent examples and problems in this book, the experimental value will be used.

In a pure crystal with a typical separation distance of 5 Å between atoms, a total of

$$\frac{1}{(5 \times 10^{-10} \text{ m})^3} \cong 8 \times 10^{27} \text{ atoms/m}^3$$

is found in the material. Therefore, at room temperature (300°K), only a very small percentage of the atoms contributes to the conduction band. ∎

Example 2-4   **Temperature Dependence of n_i**

By inspection, you conclude that the temperature dependence of n_i is dominated by the exponential term. You are to develop an estimate for the fractional change in n_i associated with a 1.0 percent increase in T at 300°K, using a linear approximation.

Solution   (a) First calculate the fractional change in n_i with T:

$$\frac{\partial n_i}{\partial T} \cong (\text{constant})\frac{\partial}{\partial T} (e^{-\mathcal{E}_g/2kT})$$

$$= (\text{constant})(e^{-\mathcal{E}_g/2kT})(\mathcal{E}_g/2kT^2)$$

$$\frac{1}{n_i}\frac{\partial n_i}{\partial T} = \frac{\mathcal{E}_g}{2kT^2} = \frac{1}{T}\left(\frac{\mathcal{E}_g}{2kT}\right)$$

$$= \frac{21.46}{300°\text{K}} \cong 0.072/°\text{K}$$

(b) Then for small changes in T, using a linear approximation,

$$\frac{\Delta n_i}{n_i} \cong \left(\frac{1}{n_i}\frac{\partial n_i}{\partial T}\right)\Delta T$$

$$\cong (0.072/°\text{K})(3.0°\text{K})$$

$$\cong 0.22$$

This is a significant result. A small percentage increase in T (1 percent) produces a much larger percentage increase (22 percent) in n_i due to the exponential relationship. Care must be taken not to exceed the narrow range of values for T for which this approximation is valid ($\Delta n_i/n_i \ll 1$). ∎

2-5   CONDUCTIVITY AND RESISTIVITY

The number of carriers in a crystal has now been found as a function of the energy gap $\mathcal{E}_g$ and temperature T. Deriving expressions for the conductivity σ and resistivity $\rho = 1/\sigma$ of the material starts with the definition (in one dimension):

$$J = \sigma E \qquad (2\text{-}5)$$

where J is the current density (current I divided by the cross-sectional area A), and E is the electric field intensity (in V/m). By considering that J consists of a flow of charged electrons and holes, the macroscopic variable J can be related to atomic-level properties

$$J = nqV_{dn} + pqV_{dp} \qquad (2\text{-}6)$$

where n and p are the number of negative and positive carriers per unit volume, respectively, with charge q, and V_d is the average *drift velocity* of the carrier type.

Finally, the *mobility* μ of a carrier can be introduced. By definition, the mobility is the ratio of the magnitude of the drift velocity to the applied electric field intensity:

$$\mu = \frac{|V_d|}{|E|} \qquad (2\text{-}7)$$

For low-field strengths, $\mu \cong$ constant, and for high-field strengths μ varies as $1/E$. Velocity V_d *saturates* and becomes approximately independent of E. This effect is important in the discussion of Chapter 8.

The ability of a carrier to move around in a crystal is limited by two effects. Thermal energy in the crystal causes the crystal atoms to vibrate from their equilibrium positions, and these vibrations interfere with the movement of carriers through the crystal. And crystals are never perfect but have many types of impurities and flaws that affect carrier motion. The carriers in the crystal experience scattering processes from the vibrating atoms and crystal impurities that limit the drift velocity V_d for a given field E. The mobility μ thus measures the relative strengths of these scattering processes. A large μ means that mobility is high because scattering is weak, and a small μ means that mobility is low because scattering is strong.

Combining the above equations,

$$\sigma E = q(n\mu_n + p\mu_p)E$$

or

$$\sigma = q(n\mu_n + p\mu_p) \qquad (2\text{-}8)$$

where μ_n is the mobility associated with the negative carriers (electrons), and μ_p is the mobility associated with the positive carriers (holes).

It is also useful to define the resistivity ρ of a material:

$$\rho = \frac{1}{\sigma} = \frac{1}{q(n\mu_n + p\mu_p)} \qquad (2\text{-}9)$$

Example 2-5   **Calculating ρ**

You are to determine the resistivity ρ for intrinsic Si at 300°K using the experimental value of n_i from Example 2-3 in order to validate the above model development.

Solution   From Example 2-3, $n_i \cong 1.6 \times 10^{16}$ carriers/m^3. From Eq. (2-9)

$$\rho = \frac{1}{qn_i(\mu_n + \mu_p)}$$

and from Fig. 2-6, $\mu_n = 0.14$ m^2/V·s and $\mu_p = 0.05$ m^2/V·s. Thus,

$$\rho = \frac{1}{(1.6 \times 10^{-19}\text{C})(1.6 \times 10^{16}/\text{m}^3)(0.19 \text{ m}^2/\text{V·s})} \cong 2.1 \times 10^3 \ \Omega\cdot\text{m}$$

This result is in approximate agreement with the typical experimental value as given in Fig. 2-1, helping demonstrate the effectiveness and validity of the energy-gap model of conduction in a crystal.

Combining the results of Examples 2-3 to 2-5 gives insight into the conduction properties of energy-gap materials. The large range in values for ρ is primarily associated with the exponential dependence of n_i on the temperature T. The number of carriers is extremely sensitive to the energy gap $\mathcal{E}_g$ divided by the reference thermal energy kT. ∎

Example 2-6   Temperature Sensitivity

Reconsider the voltage divider circuit of Fig. 2-2. Assume that the two resistors have the same dimensions and are made from intrinsic Si and that the temperature of the two resistors can differ by 3°K at an ambient temperature of 300°K. The output V_{out} must remain below 3.2 V for a given application. Will this divider perform in an acceptable way? (Assume that the linear approximation of Example 2-4 can still be used.)

Solution   From Example 2-4, n_i varies by about 22 percent. Therefore,

$$R_{\text{max}} = 1.22R_{\text{nom}}$$

$$R_{\text{min}} = 0.78R_{\text{nom}}$$

$$V_{\text{out}} = 6.0 \text{ V} \frac{R_{\text{max}}}{R_{\text{min}} + R_{\text{max}}}$$

$$= 6.0 \text{ V} \frac{1.22}{2} \cong 3.7 \text{ V}$$

The circuit will not meet the performance objectives. Extreme sensitivity to T is noted if Si samples are used as resistors. A 1 percent difference in temperature has changed V_{out} from 3.0 V (for equivalent resistor values) to 3.6 V. ∎

At low temperatures, impurity scattering dominates and μ is proportional to $T^{3/2}$. At high temperatures, thermal scattering dominates and μ is proportional to $T^{-3/2}$.

Near room temperature, where thermal scattering dominates, Eq. (2-9) can be written as

$$\rho = \left(\frac{1}{q}\right)\left(\frac{1}{n_i}\right)\left(\frac{1}{\mu_n + \mu_p}\right) \qquad (2\text{-}10)$$

$$= \frac{1}{q} \frac{h^3}{2(2\pi m_0 kT)^{3/2}} \left(\frac{m_0}{m_n^*} \frac{m_0}{m_p^*}\right)^{3/4} e^{\mathcal{E}_g/2kT}\left(\frac{1}{\text{constant} \times T^{-3/2}}\right)$$

If $\rho = \rho_0$ at a reference temperature $T = T_0$ (where ρ_0 is the resistivity at room temperature, $T_0 = 300°K$), then the ratio

$$\frac{\rho}{\rho_0} = \exp\left[\mathcal{E}_g\left(\frac{1}{2kT} - \frac{1}{2kT_0}\right)\right] \qquad (2\text{-}11)$$

can be formed.

This is an important and useful result. Equation (2-11) can be used to find the resistivity ρ of an energy-gap material as a function of T if data are available for the resistivity ρ_0 at any temperature T_0. Figure 2-1 indicates values of ρ_0 at 300°K for many different materials, enabling application of Eq. (2-11). Note that since n_i and ρ are reciprocally related, increases in T produce increases in n_i and decreases in ρ in about the same proportion (neglecting the effect of the dependence of μ on T).

Example 2-7   **Temperature Effect on ρ**

In a given circuit, a Ge sample is to be cooled to 200°K. In order to predict circuit operation, ρ must be found at this temperature. (From Fig. 2-1, $\rho \cong 2.0 \times 10^{-2}$ $\Omega \cdot$ m for Ge at 300°K.)

Solution   Over this wide temperature range, a linear approximation cannot be used. From Eq. (2-11),

$$\frac{\rho}{\rho_0} = \exp\left[\mathcal{E}_g\left(\frac{1}{2kT} - \frac{1}{2kT_0}\right)\right]$$

where

$$\rho_0 = 2.0 \times 10^{-2} \ \Omega \cdot \text{m}$$

$$T_0 = 300°\text{K}$$

$$T = 200°\text{K}$$

$$\mathcal{E}_g = 0.70 \text{ eV}$$

so that

$$2kT = 2(8.62 \times 10^{-5} \text{ eV/°K})(200°\text{K}) = 0.0345 \text{ eV}$$

$$2kT_0 = 2(8.62 \times 10^{-5} \text{ eV/°K})(300°\text{K}) = 0.0517 \text{ eV}$$

$$\rho = (2.0 \times 10^{-2} \ \Omega \cdot \text{m}) \exp\left[(0.70 \text{ eV})\left(\frac{1}{0.0345 \text{ eV}} - \frac{1}{0.0517 \text{ eV}}\right)\right]$$

$$\cong 17 \ \Omega \cdot \text{m}$$

Given the reference values of ρ at 300°K in Fig. 2-1, Eq. (2-11) can be used to find ρ at any other T. This simple result allows many different problems to be worked from a limited database.   ∎

Note from Fig. 2-1 that materials are divided into insulators, semiconductors, and metals, depending on their value of ρ at 300°K. Insulators are materials with a large energy gap (4 to 6 eV). Because of the gap width, relatively few electrons are excited into the conduction band at room temperature. Semiconductors are characterized by a typical energy gap of 0.5 to 1.5 eV, with many more electrons excited into the conduction band. Metals, which are discussed in the next section, do not have an energy gap and thus have a large number of electrons free to contribute to conduction processes.

The above characteristics are familiar ones around 300°K. Since ρ is a strong function of T, the characteristics of these materials change as T is varied over a wide range.

2-6 METALS

Metals are widely used to interconnect building block semiconductor elements. In addition, an understanding of metal properties is necessary to appreciate the nature of Schottky diodes and ohmic contacts, which are discussed in Chapter 4. The prior analysis for covalent and ionically bonded materials described a conduction process that is a function of electron excitation across an energy gap. The situation for metals is quite different.

In metals, valence electrons are easily removed from their host atoms and are free to move throughout the crystal structure. A cloud of negatively charged particles is distributed among the positively charged host ions and produces the necessary binding to create a stable structure, as shown in Fig. 2-7(A).

The allowed energy bands overlap so that no energy gap exists, as shown in Fig. 2-7(B). The electrons in a metal thus have energy states available that will allow them to move in response to an externally applied electric field.

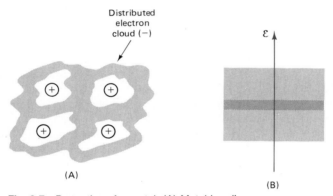

Fig. 2-7 Properties of a metal. (A) Metal bonding. (B) Overlapping energy bands.

In considering conduction processes, only electrons are carriers and the carrier density n is approximately independent of the temperature T.

Since

$$\rho = 1/qn\mu$$

and q and n are independent of T, the total temperature dependence of metals is a function of μ.

The electrons in a metal experience scattering from both thermal vibrations and impurities. At room temperature, where thermal vibrations dominate, μ is proportional to T^{-1}. At low temperatures, where impurity scattering dominates, μ is approximately independent of T. Thus,

$$\rho = A + BT \tag{2-12}$$

where A and B are constants.[2] If the resistivity ρ_0 at temperature T_0 is used a reference, then for temperature regions where thermal effects dominate

$$\rho/\rho_0 \cong T/T_0 \tag{2-13}$$

Equation (2-13) for metals is comparable to Eq. (2-11) for energy-gap materials. Figure 2-1 provides values for ρ at 300°K.

Example 2-8 **Resistivity for Metals**

From Fig. 2-1, note that $\rho \cong 1.0 \times 10^{-8}\ \Omega \cdot m$ for Cu at 300°K. For a circuit operating at 400°K, what will be the value of ρ for the Cu portions of the circuit?

Solution From Eq. (2-13)

$$\rho = (1.0 \times 10^{-8}\ \Omega \cdot m)\frac{400°K}{300°K} \cong 1.3 \times 10^{-8}\ \Omega \cdot m$$

A 100°K variation in T has led to only a small change in ρ. In comparison with semiconductors, the resistivity for metals is relatively insensitive to changes in T. ■

2-7 MORE COMPLEX MATERIALS

At this point it is useful to extend the application of the above findings to materials as they are usually found in nature. The topics covered here will be helpful in understanding some of the design constraints experienced in fabricating transistors, ICs, and hybrid microcircuits. The broad range of materials used to fabricate microelectronic circuits requires a brief introduction to the range of material properties that can be encountered.

The discussion begins with the periodic chart of the elements in Fig. 2-8. In an

Fig. 2-8 Periodic chart of the elements.

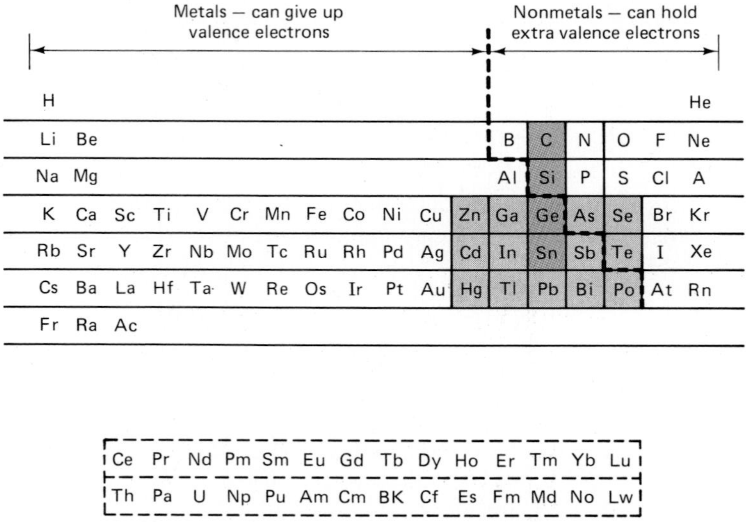

[2] This relationship is known as Matthiessen's rule.

approximate way, the periodic chart can be divided into metals, nonmetals, and semiconductors, as shown. Metals are generally to the left of columns III to IV and nonmetals are generally to the right. The overlap area around column IV defines the semimetals, or semiconductors.

Metals are characterized by valence electrons that are easily removed. Therefore, in a metal sample at room temperature, many electrons are removed from their atoms and are free to move around in the crystal. As a result, metals are good electrical conductors.

Nonmetals are characterized by strongly bound valence electrons and have a tendency to attract and hold additional electrons when they come near. Therefore, a nonmetal has very few electrons free to move around and the element is a poor conductor.

All materials can thus be described as metals, nonmetals, or semiconductors, or as combinations of these elements. These various materials can be organized as shown in Fig. 2-9.

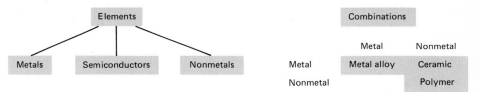

Fig. 2-9 Types of materials. All of these materials are used in microelectronics.

Combinations of metals are generally called metal *alloys,* and combinations of metals and nonmetals are called *ceramics*. Combinations of nonmetals are called *polymers*. All natural and manmade organic materials fall into the polymer category, with the latter including *plastics*. The fabrication of electronic circuits makes use of both pure materials and combinations of materials of many different types.

Most metals and ceramics are *polycrystalline* in form. When a liquid metal or ceramic is cooled and hardened, it usually forms a polycrystalline (or multicrystalline) structure, as shown in Fig. 2-10. When the liquid begins to harden, small growth nuclei form throughout the material. Crystal growth then occurs around each nucleus. Eventually, the growth patterns from different nuclei run together and pro-

Fig. 2-10 Polycrystalline material.

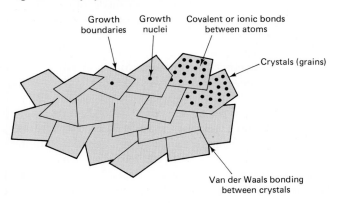

duce boundaries between the different single crystals. The material consists of many different crystals or grains, bound within each grain by metal, ionic, or covalent bonding and between grains by *Van der Waals,* or polarization, bonds. As discussed in Chapter 18, such bonds are important in hybrid microelectronics when many different types of materials must be bonded together.

The creation of large single crystals of a material, such as those used to fabricate semiconductor devices, starts with one growth nucleus and controlled growth outward from this point. Figure 2-11 illustrates one common way used to develop single crystals of semiconductor material. A seed crystal is touched to a pool of liquid material (the *melt*). The crystal is then rotated and pulled up slowly, causing the atoms to grow onto the solid interface in a way that maintains a single-crystal structure. Single crystals of other types of materials can also be formed using this method and other techniques.

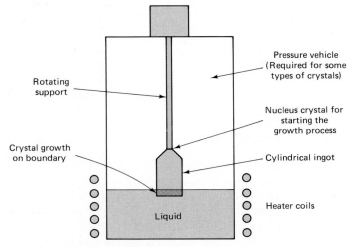

Fig. 2-11 Growth of a single-crystal material.

Where most metals and ceramics are polycrystalline and semiconductors consist of single crystals, some materials are amorphous, or disordered—there is no long-term pattern imposed on the atoms in the material. Glass can be considered a special type of ceramic that is amorphous in form. Localized order exists among molecules, but the long-term arrangement is random.

Polymers include all natural and man-made organic materials. They consist of chains of identical groupings of atoms called *mers*. Atoms within each mer are bound in a fixed relationship, and then the mers are linked. Bonding within each mer and between mers is covalent. A polymer consists of chains of mers, often very long, that are attracted by Van der Waals forces. A range of different chain lengths is observed in any material. Polymers can be crystalline, amorphous, or in a mixed form.

The electrical properties of a material depend on its atomic form. Earlier discussions of the conduction processes in crystals used as a model a pure single-crystal structure. If these results are expanded to a range of different material forms, a number of significant changes are observed in carrier behavior.

In polycrystalline forms, electrons experience some scattering along the grain boundaries that are formed. Therefore, it is reasonable to expect to find the resistivity ρ greater in a polycrystalline material at a given T. A similar effect is experienced if impurities or dislocations among different types of atoms can increase the resistivity of the material.

While most of the semiconductor devices discussed in this text are based on conduction processes in single crystals as described above, semiconductor devices are also produced from amorphous materials. Amorphous silicon is used to create solar cells, thin-film transistors, and other types of devices.

In amorphous silicon, localized order combines with long-term disorder to produce conduction properties that are a function of energy. Where single-crystal silicon develops an energy gap because it has carrier states available only in valence and conduction bands, amorphous silicon develops an energy gap because of variations in electron mobility with energy (Madan, 1986). Since amorphous silicon is much less expensive to produce and can be produced to cover large areas, it is commercially effective in applications associated with solar cells and large liquid-crystal displays. In the latter case, thin-film transistors can be formed to drive the crystal display.

2-8   MATERIALS FOR ELECTRONICS

This chapter is concerned with the electrical properties of materials as they affect the construction of semiconductor building block elements for electronic circuits. The energy-band concept has been developed, and equations have been derived for the resistivity of pure single-crystal nonmetals and metals. As has been indicated, conduction properties are influenced by impurities and dislocations and the degree to which the material is crystalline or amorphous.

The above development has introduced several important atomic-level parameters, including the energy gap ($\mathcal{E}_g$), density of carriers (n_i), and mobility (μ). These and other characteristics of semiconductor materials determine the most suitable application strategies for a range of materials. At present, most semiconductor devices are made from silicon. Many billions of dollars have been spent in developing a very detailed understanding of the properties of this common element in both single-crystal and polycrystalline form. Silicon is largely preferred over germanium because the larger $\mathcal{E}_g$ for silicon produces improved device properties at higher temperature, it is a better thermal conductor (to remove unwanted circuit heat), and its oxide (SiO_2) is a natural insulator that is useful in device fabrication. Germanium is used for specialized applications. For example, its narrower $\mathcal{E}_g$ makes it suitable for an infrared detector in fiber-optic communication systems (discussed in Chapter 4).

There is at present a major shift taking place toward the use of gallenium arsenide, particularly in high-speed digital applications. Despite significant fabrication problems due to its more complex crystal structure, the toxic nature of the constituent elements, and the brittle character of its crystals, GaAs has many attractive features, including higher mobility (μ) for charge carriers, leading to faster switching times; and a larger $\mathcal{E}_g$, leading to further operating improvement at high temperatures.

The nonsemiconductor materials discussed in this chapter provide an introduction to some of the considerations that are important in interconnections and packaging. As models of alternative approaches to system assembly become available in CAD systems, an awareness of these factors is required to understand why some design strategies are preferred over others.

Questions

2-1. Why is it important for the electronics designer to become familiar with the electrical properties of microelectronic materials?

2-2. How does a current flow develop in a crystal?

2-3. What determines the energy gap $\mathcal{E}_g$?

2-4. What crystal properties dominate the density of intrinsic carriers found in the material?

2-5. What is the significance of the Fermi energy?

2-6. What crystal parameters determine the resistivity of a material?

2-7. How are insulators, semiconductors, and conductors usually defined?

2-8. Why is the conductivity of a metal relatively insensitive to temperature?

2-9. What are some of the important electrical properties of materials that affect component and device properties, and thereby the design tradeoffs performed?

Answers

2-1. The electrical properties of microelectronic materials determine the operational characteristics of the components and devices used as building block elements in circuits and systems. An adequate understanding of the models used in design requires a fundamental understanding of the material characteristics that influence operating parameters.

2-2. A current flow develops in a crystal because thermal energy creates mobile carriers (electrons and holes) that are free to move in response to an externally applied voltage across the structure.

2-3. Interactions between the atoms and free electrons in a crystal prevent the electrons from moving through the crystal with energies that fall in the forbidden band. The energy gap $\mathcal{E}_g$ develops because the wave nature of the electron couples to the lattice spacing of the crystal and prevents electrons from taking on certain energy values. The crystal prevents these energy values from occurring by exerting electrostatic forces on the electrons as they move about the crystal.

2-4. The energy gap $\mathcal{E}_g$, the temperature T, and the effective masses of the positive and negative carriers in the crystal determine the intrinsic carrier density, which is a sensitive function of the ratio $\mathcal{E}_g/2kT$.

2-5. The Fermi energy $\mathcal{E}_F$ represents the chemical energy of a material and is constant throughout a material that is in equilibrium. $\mathcal{E}_F$ can be associated with the average energy of the mobile carriers in a crystal. Thus, when the energy distribution of carriers is modified, the value of $\mathcal{E}_F$ shifts. The relationship between $\mathcal{E}_F$ and the energy gap $\mathcal{E}_g$ helps determine the density of carriers and conduction properties of a crystal.

2-6. The resistivity of a material is determined by the density of intrinsic carriers, the unit charge on these carriers, and the carrier mobility, which describes how easily the carriers can move about the crystal.

2-7. The terms *insulator, semiconductor,* and *conductor* are usually associated with material properties at room temperature (300°K). At other temperatures, the electrical characteristics of materials can change substantially.

2-8. Since metals do not have energy gaps, the density of carriers is a weak function of the temperature T, so that the conductivity is not a strong function of T.

2-9. The energy gap $\mathcal{E}_g$, density of carriers n_i at room temperature, and mobility μ are important descriptors of all semiconductor devices. In addition, the type of crystal structure and nature of the bonds formed between materials are also important determinants of component and device performance.

Exercises

(2-1)

(2-1)

(2-1)

(2-4)

(2-4)

(2-4)

(2-5)

(2-5)

(2-5)

(2-6)

(2-7)

2-1. A material sample is 1.0 cm long and 2.0 mm² in cross-sectional area. Given the range of values for ρ in Fig. 2-1, what range of values for R is possible for the sample?

2-2. For a cylindrical sample of Ge, the length of the sample and the diameter are known with an accuracy of $\pm$ 5 percent. If $\ell = 2.5$ cm and $d = 0.10$ cm, what range of values for R is possible, neglecting any uncertainties in the value of $\rho = 1.0\ \Omega \cdot m$?

2-3. A sample of Si is 2.5 mm long and 1.0 mm² in cross-sectional area. What is the approximate resistance of the sample at 300°K, using $\rho = 1.0 \times 10^3\ \Omega \cdot m$? What can you conclude about the resistance of small pieces of intrinsic Si?

2-4. Calculate the magnitude of the term $kT \ln (m_n^*/m_p^*)$ for Ge, Si, and GaAs at 300°K. What can you conclude about the magnitude of this term with respect to $\mathcal{E}_g$?

2-5. Find the intrinsic density of carriers n_i in pure Ge and GaAs crystals at 300°K. Compare with the carrier density of Si at 300°K. What variable dominates the difference? (Use the model equations developed in this chapter.)

2-6. Find the ratio between the carrier densities in a pure Si crystal at 0 and 50°C. Is n_i a sensitive function of temperature?

2-7. Find the resistivity ρ of intrinsic Ge at 300°K. Compare your results with the data in Fig. 2-1.

2-8. A material has an energy gap of 0.50 eV. If the temperature changes from 300 to 400°C, what will be the percentage change in ρ?

2-9. A pure semiconductor material is to be selected so that ρ/ρ_0 changes by less than 50 percent for a temperature change of from 300 to 400°K. What restriction exists for the energy gap that must be associated with this material?

2-10. For a given application, ρ for a pure metal must change by less than 10 percent if a circuit is to perform as expected. Assuming an initial operating temperature of 300°K, find the lowest temperature for which this performance objective can be met.

2-11. Hybrid microelectronic circuits are often fabricated on a base or substrate made of alumina (Al_2O_3). What type of material is alumina and what electrical properties can it be expected to have?

2-12. The metal ruthenium (Ru) is used in the manufacture of thick-film resistors, which are discussed in Chapter 18. Consider a circuit for which a Ru-based thick-film resistor measures 50 mils $\times$ 20 mils $\times$ 25 μm (1 mil = 1/1000 in, 1 μm = 10^{-6} m, and 1 in = 2.54 cm). If each resistor is 20 percent Ru and Ru costs \$40/g, what is the cost of Ru for 10,000 resistors? (Assume a wastage of 20 percent and that the density of Ru is 12.4 g/cm³.)

2-13. Silver (Ag) and gold (Au) are also used to fabricate microelectronic circuits. Assume that a given company uses 1000 oz of silver and 200 oz of gold each month with a wastage rate of 10 percent. Half of this waste can be reclaimed through recycling. If silver costs \$30/oz and gold costs \$300/oz, what savings can be realized per month through recycling?

Solutions

2-1. By applying Eq. (2-1),

$$R = \rho \frac{\ell}{A} = \rho \frac{1.0 \times 10^{-2}\ m}{2.0\ mm^2 \times 10^{-6}\ m^2/mm^2}$$

$$= \rho(10^4/2\ m) = \rho(5 \times 10^3)\ m^{-1}$$

$$R_{min} \cong 5 \times 10^3 \times 10^{-9}\ \Omega \cong 5 \times 10^{-6}\ \Omega$$

$$R_{max} \cong 5 \times 10^3 \times 10^{18}\ \Omega \cong 5 \times 10^{21}\ \Omega$$

2-2. From Eq. (2-1),

$$R_{nom} = \rho\frac{\ell}{A} = (1.0\ \Omega \cdot m)\frac{2.5\ cm}{\pi\,(0.05\ cm)^2 \times 10^{-2}\ m/cm} \cong 32\ k\Omega$$

$$R_{min} = R_{nom}\frac{0.95}{1.05^2} = 0.86R_{nom} \cong 28\ k\Omega$$

$$R_{max} = R_{nom}\frac{1.05}{0.95^2} = 1.16R_{nom} \cong 37\ k\Omega$$

2-3. From Eq. (2-1),

$$R = \rho\frac{\ell}{A} = (1.0 \times 10^3\ \Omega \cdot m)\frac{2.5\ mm}{1.0\ mm^2 \times 10^{-3}\ m/mm}$$

$$\cong 2.5\ M\Omega$$

For the dimensions given here, the sample of intrinsic Si has a very large resistance.

2-4. By making use of the data in Fig. 2-6,

$$kT = (8.62 \times 10^{-5}\ eV/°K)(300°K) \cong 0.026\ eV$$

Ge: $kT \ln(0.55/0.37) \cong 0.010\ eV \ll 0.67\ eV$

Si: $kT \ln(1.1/0.59) \cong 0.016\ eV \ll 1.11\ eV$

GaAs: $kT \ln(0.07/0.54) \cong 0.053\ eV \ll 1.43\ eV$

2-5. Based on the data in Fig. 2-6, (a) for Ge,

$$\frac{\mathcal{E}_g}{2kT} = \frac{0.67\ eV}{2(8.62 \times 10^{-5}\ eV/°K)(300°K)} \cong 12.95$$

$$n_i = \frac{2}{(6.62)^3 \times 10^{-102}(J \cdot s)^3}$$
$$\times (2\pi \times 9.1 \times 10^{-31}\ kg \times 1.38 \times 10^{-23}\ J/°K \times 300°K)^{3/2}(0.55 \times 0.37)^{3/4}e^{-12.95}$$
$$\cong 1.8 \times 10^{19}\ carriers/m^3$$

(b) For GaAs,

$$\frac{\mathcal{E}_g}{2kT} = \frac{1.43\ eV}{2(8.62 \times 10^{-5}\ eV/°K)(300°K)} \cong 27.65$$

$$n_i = \frac{2}{(6.62)^3 \times 10^{-102}(J \cdot s)^3}$$
$$\times (2\pi \times 9.1 \times 10^{-31}\ kg \times 1.38 \times 10^{-23}\ J/°K \times 300°K)^{3/2}(0.07 \times 0.54)^{3/4}e^{-27.65}$$
$$\cong 2.1 \times 10^{12}\ carriers/m^3$$

2-6. From Eq. (2-2),

$$\frac{n_i(0°C)}{n_i(50°C)} = \left(\frac{273°K}{323°K}\right)^{3/2}\frac{e^{-23.58}}{e^{-19.93}} \cong 0.02$$

This is a large change, with a decrease of 98 percent at 0°C over 50°C.

2-7. From Eq. (2-9),

$$\rho = \frac{1}{qn_i(\mu_n + \mu_p)}$$

From Exercise 2-5, $n_i \cong 1.8 \times 10^{19}/\text{m}^3$, so that

$$\rho = \frac{1}{(1.6 \times 10^{-19} \text{ C})(1.8 \times 10^{19}/\text{m}^3)(0.39 + 0.19)\text{m}^2/\text{V} \cdot \text{s}}$$
$$\cong 0.6 \ \Omega \cdot \text{m}$$

which is in approximate agreement with Fig. 2-1.

2-8. From Eq. (2-11),

$$\frac{\rho}{\rho_0} = \exp\left[\mathcal{E}_g\left(\frac{1}{2kT} - \frac{1}{2kT_0}\right)\right]$$

where ρ is the resistivity at $T = 673°\text{K}$ and ρ_0 is the resistivity at $T_0 = 573°\text{K}$

$$\frac{\mathcal{E}_g}{2kT_0} = \frac{0.50 \text{ eV}}{2(8.62 \times 10^{-5} \text{ eV}/°\text{K})(573°\text{K})} \cong 5.06$$

$$\frac{\mathcal{E}_g}{2kT} = 5.06\frac{573°\text{K}}{673°\text{K}} \cong 4.31$$

$$\rho/\rho_0 = \exp(4.31 - 5.06) \cong 0.47$$

$$\frac{\rho - \rho_0}{\rho_0} \times 100\% \cong -53\%$$

2-9. From Eq. (2-11),

$$\frac{\rho}{\rho_0} < \exp\left[\mathcal{E}_g\left(\frac{1}{2kT} - \frac{1}{2kT_0}\right)\right]$$

so that

$$\mathcal{E}_g > \frac{\ln \rho/\rho_0}{1/2kT - 1/2kT_0} = \frac{\ln 0.5}{1/2k(400°\text{K}) - 1/2k(300°\text{K})} \cong 0.14 \text{ eV}$$

2-10. From Eq. (2-13),

$$\rho/\rho_0 = T/T_0 = T_{\min}/T_0$$

$$0.9 = T_{\min}/300°\text{K}$$

$$T_{\min} = (300°\text{K})(0.9) = 270°\text{K}$$

2-11. Since alumina is formed by a metal combined with a nonmetal, it is a ceramic. As expected, it is an insulating material.

2-12. The volume of each thick-film resistor (TFR) is

$$\mathcal{V} = (0.05 \text{ in})(0.02 \text{ in})(2.54 \text{ cm/in})^2(25 \times 10^{-6} \text{ m})(10^2 \text{ cm/m})$$
$$\cong 1.6 \times 10^{-5} \text{ cm}^3$$

so the Ru content per resistor is

$$0.20 \ \mathcal{V} = 0.32 \times 10^{-5} \text{ cm}^3$$

The mass of Ru per resistor is thus

$$m = \rho \mathcal{V} = (12.4 \text{ g/cm}^3)(0.32 \times 10^{-5} \text{ cm}^3) \cong 4.0 \times 10^{-5} \text{ g/TFR}$$

so the cost for 10,000 resistors without waste is

$$C = (\$40/\text{g})(4.0 \times 10^{-5} \text{ g})(10^4) \cong \$16.00$$

and with waste

$$C(\text{with waste}) = 1.2 \times \$16.00 \cong \$19.20$$

2-13. From the exercise,

$$\text{Gold waste/mo} = 200 \text{ oz/mo} \times 0.1 = 20 \text{ oz/mo}$$

$$\text{Silver waste/mo} = 100 \text{ oz/mo} \times 0.1 = 100 \text{ oz/mo}$$

so that the savings may be found

20 oz/mo × \$300/oz = \$6,000	savings in gold/mo
100 oz/mo × \$30/oz = 3,000	savings in silver/mo
\$9,000	total savings/mo

PROBLEMS

(2-1) **2-1.** Given $\ell = 1.0$ cm and $A = 1.0 \times 10^{-3}$ cm^2, find the resistance of a sample of graphite at 300°K. ($\rho \cong 1.0 \times 10^{-5} \Omega \cdot$ m from Fig. 2-1.)

(2-1) **2-2.** A sample of copper is 0.1 mm in diameter and has resistance 1.0 Ω. What is the length of the sample?

(2-4) **2-3.** Intrinsic Si at 300°K changes temperature by 0.5 percent. What must be the percentage change in the carrier density n_i?

(2-4) **2-4.** For Si, the average separation distance between atoms is 5.43 Å. How many atoms are found in 1.0 m^3 of Si?

(2-4) **2-5.** For Ge, the average distance between atoms is 5.66 Å. How many atoms are found in a wafer that is 1 mm thick and 3 in. in diameter?

(2-4) **2-6.** Find the intrinsic density of carriers n_i in a GaAs pure crystal at 400°K.

(2-4) **2-7.** A sample of Si is stored at 300°K. On removal from storage, the temperature drops by 2°C. Estimate the percentage change in carrier density.

(2-4) **2-8.** Intrinsic GaAs at 300°K changes temperature by −0.5 percent. What is the percentage change in the carrier density n_i?

(2-4) **2-9.** Intrinsic Ge at 300°K changes temperature by −1.5 percent. What is the percentage change in carrier density n_i?

(2-4) **2-10.** The carrier density of Si at 300°K changes by 20 percent. What is the percentage change in temperature?

(2-4) **2-11.** A sample of Si undergoes a 10 percent change in carrier density n_i. What is the corresponding change in temperature from a reference of 250°K?

(2-5) **2-12.** Find the ratio between the density of carriers n_i in a pure Ge crystal at 25 and 50°C.

(2-5) **2-13.** An electric field of 50 V/m exists in a sample of Si. Using the data of Fig. 2-1, find the current density J.

(2-5) **2-14.** An electric field of 1.0×10^4 V/m is applied internally to a Si crystal at 300°K. What are the electron and hole drift velocities?

(2-5) **2-15.** A current of 100 mA flows through a material with a cross-sectional area of 1.0 mm². Given $n = p \cong 1.0 \times 10^{19}/m^3$, find the average drift velocity of the carriers $(V_{dn} + V_{dp})/2$.

(2-5) **2-16.** Find the resistivity ρ for GaAs at 300°K using the value of n_i from Exercise 2-5. Using Fig. 2-1, compare with similar results for Si and Ge.

(2-5) **2-17.** Assume for Example 2-6 that the two resistors are fabricated from Ge instead of Si. Will the circuit meet the given performance objective?

(2-5) **2-18.** For Example 2-6, what is the maximum temperature range that may be allowed to meet the design objective?

(2-5) **2-19.** A 50 percent change in ρ is observed when a semiconductor material changes from 350 to 320°K. What is the energy gap for the material?

(2-5) **2-20.** Given $\rho = 5.0 \times 10^2$ $\Omega \cdot m$ at 300°K for an energy-gap material with $\mathcal{E}_g = 0.20$ eV, find ρ at 225°K.

(2-5) **2-21.** Given $\rho_0 = 5.0 \times 10^2$ $\Omega \cdot m$ at 300°K for a material with $\mathcal{E}_g = 0.55$ eV, find the temperature for which $\rho = 8.0 \times 10^2$ $\Omega \cdot m$.

(2-5) **2-22.** A material has an energy gap of 2.0 eV. If the temperature changes from 300 to 150°K, what is the fractional change in ρ?

(2-5) **2-23.** Find the resistivity ρ for GaAs at −150°C.

(2-5) **2-24.** Find the conductivity σ for Ge and Si at 300°K.

(2-5) **2-25.** A sample of material has conductivity σ_0 at 300°K and $\sigma = 10\sigma_0$ at 350°K. Find the energy gap of the material.

(2-5) **2-26.** The energy gap for a pure semiconductor material is defined by the requirement that ρ/ρ_0 be reduced by >20 percent for a temperature increase of 5 percent. If $T_0 = 300$°K, what range of values for $\mathcal{E}_g$ can be used?

(2-6) **2-27.** Find ρ for Ag at 75°C.

(2-6) **2-28.** For Eq. (2-12), assume that $A = 50$ $\Omega \cdot m$ and $B = 1.5$ $\Omega \cdot m/$°K. For what temperatures will the effect of A contribute 10 percent or more to the value of ρ?

COMPUTER APPLICATIONS

2-1. Equations A7-6 and A7-9 in Appendix 7 describe models for the density of states function $g_c(\mathcal{E})$ and probability function $P_c(\mathcal{E})$ associated with energy-gap materials. Computer support may be used to develop more detailed information regarding the actual shapes of these curves and provide additional insight into such analysis.

Absolute values for $\mathcal{E}_c$ and $\mathcal{E}_v$ may be estimated by setting the energy difference $(0 - \mathcal{E}_c)$ equal to the photoelectric work function of the material. The results of such a computer study for Si at 300°K are shown in Figs. 2-12 and 2-13. In Fig. 2-12 (p. 60), the differing values for m_n^* and m_p^* produce the asymmetry in the density of states. In Fig. 2-13 (p. 61), the probability function transition from 1.0 to 0.0 takes place as a step function within the resolution of the graph. The effect of a higher temperature (1000°K) on the probability function is shown in Fig. 2-14 (p. 62).

In all figures, energies are in eV. The values for the density of states per unit volume have been obtained by measuring $\mathcal{E}_c$ and $\mathcal{E}_v$ in eV, h in eV $\cdot$ s, and m_o in kg.

(a) Calculate the conversion factor that should be used to express the density of states results in J^{-1}. (b) Show by substitution into equations A7-6 and A7-9 that the computer results are correct.

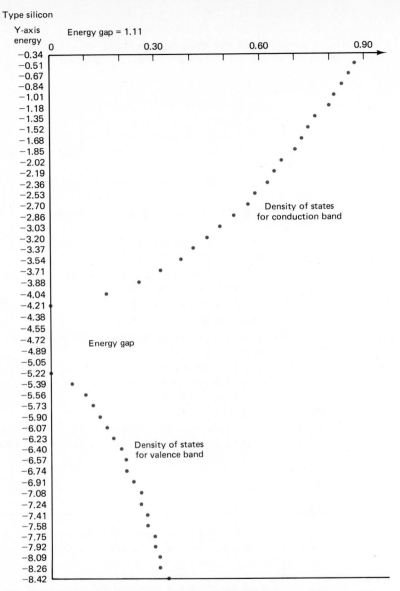

Fig. 2-12 Computer graph of the density of states functions $g_c(\mathcal{E})$ and $g_v(\mathcal{E})$.

Write a computer program that can confirm the value of the standard integral of Eq. (A7-13) in Appendix 7.

Write a computer program that produces a table of values for the resistivity of Si as a function of $x = 1/2kT$, for T ranging from 77°K (the temperature of liquid nitrogen) to

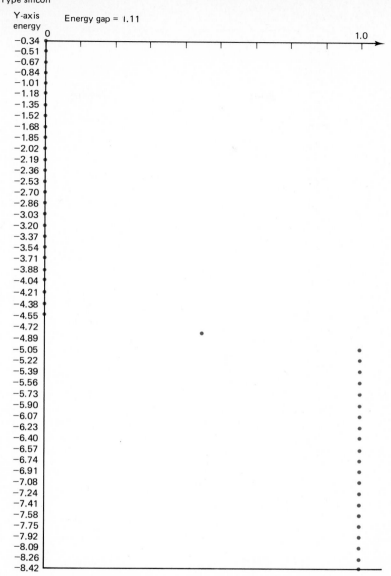

Type silicon

Y-axis
energy

Energy gap = 1.11

Fig. 2-13 Computer graph of the probability distribution function $P_c(\mathcal{E})$ at 300°K.

373°K (the boiling temperature of water). Take $T_0 = 300°$K as your reference temperature (with $x_0 = 1/2kT_0$) and use the ρ value from Example 2-3. Show that if ρ is plotted as a function of x or $x - x_0$ on semilog scales, a straight line results. Plot your results on semilog scales (by hand or by using software). Interpret your findings. Your assignment is to produce a program listing, a table of values, a graph, and interpretive comments.

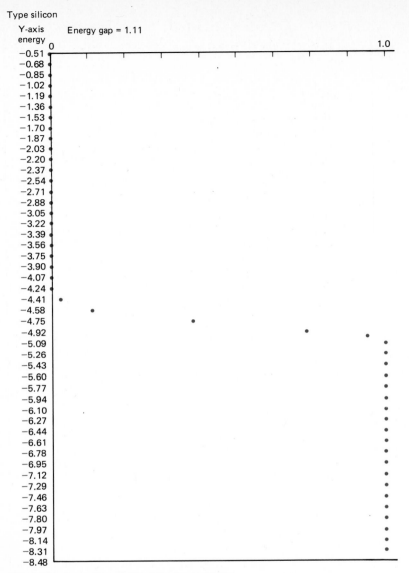

Fig. 2-14 Computer graph of $P_c(\mathcal{E})$ at 1000°K.

EXPERIMENTAL APPLICATION

2-1. The objective of this experiment is to measure the energy gap $\mathcal{E}_g$ of a semiconductor material. As discussed in this chapter, the resistivity ρ of an energy-gap material varies strongly with temperature. Equation (2-11) can be rewritten in the form

$$R = R_0 \exp\left[\mathcal{E}_g(x - x_0)\right]$$

where R_0 is the resistance of the material at temperature T_0 (usually room temperature), R is the resistance at any other temperature T, and

$$x = 1/2kT \qquad \text{and} \qquad x_0 = 1/2kT_0$$

If R/R_0 is plotted as a function of x or $(x - x_0)$ on semilog scales, a straight line is expected to result.

As discussed in Section 3-3, samples of semiconductor material are readily available in the form of *thermistors*. Set up a series circuit consisting of a digital voltmeter and a thermistor. The temperature of the thermistor can be varied using a hot-air blower and recorded using a thermometer or thermocouple.

Develop an experimental procedure and collect data giving R as a function of T. Plot R/R_0 as a function of $x - x_0$, show how well your data fit a straight-line approximation, and determine $\mathcal{E}_g$ at several points using this straight-line fit to reduce scatter measurement error. (Compare your experimental results and the results of Computer Application 2-3, if available.) Discuss and interpret your findings.

Discussion

Initial readings should be made of the reference resistance R_0 and temperature T_0. Then the temperature of the thermistor can be increased, while measurements are made of R and T at several points, staying within the device ratings. Typical data are shown in Fig. 2-15. Values for T (°K), x (eV^{-1}), $(x - x_0)$(eV^{-1}), and R/R_0 can be calculated as shown in Fig. 2-16.

FIGURE 2-15
Typical Data Collected for Experimental Application 2-1

Data Point	T (°C)	R (kΩ)	
1	21.6	11.8	(T_0, R_0)
2	26.2	9.74	
3	32.3	7.48	
4	38.6	5.91	
5	45.4	4.19	
6	49.6	3.36	
7	55.8	2.39	
8	63.8	1.92	
9	65.7	1.63	
10	69.5	1.19	
11	76.5	1.11	

FIGURE 2-16
Data Analysis for Experimental Application 2-1

Data Point	T(°K)	x (eV^{-1})	x − x$_0$(eV^{-1})	R/R$_0$
1	294.6	19.68	0	1.0
2	299.2	19.39	−0.89	0.83
3	305.3	19.00	−0.68	0.63
4	311.6	18.61	−1.07	0.50
5	318.4	18.22	−1.46	0.36
6	322.6	17.98	−1.70	0.28
7	328.8	17.64	−2.04	0.20
8	336.8	17.22	−2.46	0.16
9	338.7	17.12	−2.56	0.14
10	342.5	16.93	−2.75	0.10
11	349.5	16.60	−3.08	0.09

The results can then be plotted as shown in Fig. 2-17. Since R/R_0 as a function of $x - x_0$ should produce a straight line on semilog scales, a best-fit line can be drawn to average out measurement error. Values of $\mathcal{E}_g$ can then be determined at several locations (and should give approximately the same result).

From Fig. 2-17,

At point 1:
$$\mathcal{E}_g = \frac{\ln (R/R_0)}{x - x_0} \cong \frac{-2.3}{-3.0} \cong 0.8 \text{ eV}$$

At point 2:
$$\mathcal{E}_g \cong \frac{\ln 0.20}{-2.0} \cong \frac{-1.6}{-2.0} \cong 0.8 \text{ eV}$$

At point 3:
$$\mathcal{E}_g \cong \frac{\ln 0.50}{-1.0} \cong \frac{-0.70}{-1.0} \cong 0.7 \text{ eV}$$

Typical data for $\mathcal{E}_g$ are shown in Fig. 2-18.

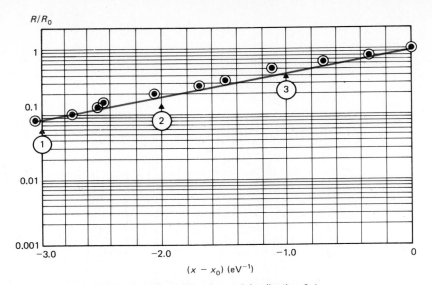

Fig. 2-17 Graph of the results of Experimental Application 2-1.

FIGURE 2-18
**Energy Gap $\mathcal{E}_g$ for
Selected Semiconductor
Materials (after Streetman,
1980)**

Material	$\mathcal{E}_g$ (eV)
Si	1.11
Ge	0.67
SiC(α)	2.86
AlP	2.45
AlAs	2.16
AlSb	1.6
GaP	2.26
GaAs	1.43
GaSb	0.7
InP	1.28

CHAPTER 3
MODELING RESISTANCE, CAPACITANCE, AND INDUCTANCE IN ELECTRONIC CIRCUITS

As sophisticated CAD systems become available to support circuit design and tradeoff analysis, the actual properties of all circuit elements must be realistically modeled. If choices are to be adequately evaluated by simulation and if the breadboard stage is to correlate highly with the simulation (or even be omitted from the design procedure), the elements cannot be treated in an idealized form.

All electronic circuits are affected by resistance, capacitance, and inductance. Device models use these functional concepts, as do interconnection and packaging models. Discrete components are much more complex than indicated by simple models, and distributed R, C, and L functions are important to an understanding of circuit performance. For these reasons, the designer must develop a sophisticated appreciation for the ways in which to appropriately represent these basic circuit functions.

This chapter explores the R, C, and L functions by considering the difficulties associated with the realistic modeling of discrete devices. This strategy develops an awareness that these functions are always found together and that limitations are inherent in models of circuit performance. The foundation established in this chapter enables the designer to look at schematics and equivalent circuits from a new perspective, aware that only selected device features are included. The objective is to build an understanding of the types of model limits that exist, thereby enabling the designer to interpret the results of complex simulations and to limit the use of the models to a range of validity.

As discussed in Chapter 1, real resistors always include small capacitive and inductive effects. The circuit application (particularly the frequency of operation) determines the allowable limits for the C and L involved.

The dependence of R on temperature and other environmental effects is important in assessing circuit performance over a range of conditions. Nonlinearities due to power dissipation are also important for some applications. These and other considerations enter into deciding how to model a resistor in order to satisfactorily predict circuit performance.

As noted in Chapter 2, the basic resistor-design equation can be written

$$R = \rho \frac{\ell}{A} \tag{3-1}$$

The value R of a resistor depends on the resistivity ρ of the material used and the geometry of the design.

An understanding of the relationship between ρ and the temperature T is essential to assess whether or not the resistor will function adequately over the required range of temperatures and to model resistor performance.

Insulating materials are poorly suited for resistor design since ρ is too large for the values of R of usual interest. Pure metals are also inadequate to the task since ρ is so small that a resistor would require a long length of material.

At first thought, a semiconducting material might seem a logical candidate as a resistor since intermediate values of ρ are available. However, a problem arises with respect to the temperature dependence of ρ for semiconductors. The variation of ρ with T is often discussed in terms of the *temperature coefficient of resistance (TCR)*, which is defined as

$$\text{TCR} = \frac{1}{\rho} \frac{\partial \rho}{\partial T} \tag{3-2}$$

A common way to measure TCR is in parts per million per °C (ppm/°C) or parts per million per °K (ppm/°K). Note that per °C and per °K values are the same since the width of a degree is the same for each. The ratio in Eq. (3-2) is multiplied by 10^6 to produce the appropriate value in ppm.

From Eqs. (2-11) and (2-13), the TCR for insulators and semiconductors (energy-gap materials) is

$$\text{TCR} = -\mathcal{E}_g / 2kT^2 \tag{3-3}$$

and for metals

$$\text{TCR} = 1/T \tag{3-4}$$

The TCR for a semiconductor material is the negative of the n_i temperature dependence observed in Chapter 2. This is to be expected since ρ and n_i are reciprocally related (neglecting the temperature dependence of μ). In the following discussion, as more complex materials are considered, the TCR is no longer simply related to an identifiable carrier density.

Example 3-1   **TCR**

As a design aid, it is desired to find the TCR for pure Si and a metal at 300°K.

Solution   At 300°K, for pure Si,

$$\text{TCR} = \frac{-1.11 \text{ eV}}{2(0.026 \text{ eV})(300°\text{K})} \cong -7.2 \times 10^4 \text{ ppm/°K}$$

and for a metal

$$\text{TCR} = 1/300°\text{K} \cong 3.3 \times 10^3 \text{ ppm/°K}$$

The TCR for an energy-gap material is typically a large negative number that is proportional to $\mathcal{E}_g$ and inversely proportional to T^2. The TCR for a metal is typically a smaller positive value that is inversely proportional to T. ■

The semiconductor material can be immediately ruled out as a candidate resistor material because of its high TCR. The pure metal is ruled out because ρ is too small and the TCR is still too high.

However, a little thought indicates that the semiconductor material has an important use for the design of temperature-sensitive devices, which are called *thermistors*. These devices enable small changes in T to be measured through the corresponding change in ρ. The semiconductor TCR is well suited for this purpose and Eq. (2-11) becomes the design equation for thermistors.

In order to select a suitable resistor material, several different strategies can be tried:

1. Use a material with a small energy gap $\mathcal{E}_g$ (introduced in Chapter 2) to reduce the TCR value given by Eq. (3-3)
2. Use a metal alloy that has a higher, nontemperature-sensitive value of ρ due to internal scattering processes
3. Use a mixture of metals and nonmetals that average out negative and positive TCRs.

All three approaches are used in resistor design to achieve TCRs in the range of 50 to 100 ppm/°C. Carbon has long been used as a resistor material because of its low $\mathcal{E}_g$. Nichrome (Ni + Cr) and Manganin (Cu + Ni + Mn) are common metal alloys that have a low TCR due to internal scattering processes. Values of ρ and the TCR for these materials are given in Fig. 3-1.

For a resistor with resistance R at temperature T and resistance R_0 at reference temperature T_0, the ratio R/R_0 can be written as a function of ρ/ρ_0 in the form

FIGURE 3-1
Resistivity and TCR for Resistor Materials at 300°K

Material	ρ ($\times 10^{-5}$ $\Omega \cdot$ m)	TCR (ppm/°C)
Carbon (graphite)	1	−300
Manganin (Cu + Ni + Mn)	5	±20
Nichrome (Ni + Cr)	10	70

$$\frac{R}{R_0} = \frac{\rho}{\rho_0} = \frac{\rho_0 + (\rho - \rho_0)}{\rho_0} = 1 + \frac{\rho - \rho_0}{\rho_0}$$

$$= 1 + \frac{1}{\rho_0}\left(\frac{\rho - \rho_0}{T - T_0}\right)(T - T_0) \qquad (3\text{-}5)$$

For small temperature variations $T - T_0$, a linear approximation produces

$$R/R_0 \cong 1 + (\text{TCR})(T - T_0) \qquad (3\text{-}6)$$

This equation can be useful in estimating how the value of a resistor changes as it is subjected to temperature change (for small temperature changes).

Example 3-2   **Temperature Dependence**

A 10-kΩ resistor has a TCR of 200 ppm/°C at 300°K. Estimate the change in resistor value at 350°K using a linear approximation.

Solution   From Eq. (3-6),

$$\Delta R = (\text{TCR})(T - T_0)R_0$$
$$= (200 \times 10^{-6}/°C)(350\ °K - 300\ °K)(1.0 \times 10^4\ \Omega)$$
$$= 100\ \Omega$$

Fig. 3-2 Significant figures for a resistor with a ± 5 percent tolerance

This is a small variation, of the order of 1 percent. If a circuit design requires more accuracy in the known value of R, it will be strongly affected by restrictions on the component characteristics.   ∎

| 10 |
| 11 |
| 12 |
| 13 |
| 15 |
| 16 |
| 18 |
| 20 |
| 22 |
| 24 |
| 27 |
| 30 |
| 33 |
| 36 |
| 39 |
| 43 |
| 47 |
| 51 |
| 56 |
| 62 |
| 68 |
| 75 |
| 82 |
| 91 |

There are many other reasons for variations in R that may need to be modeled. One of these relates to a decision-making process based on fabrication tolerances. Due to statistical variations in fabrication processes, discrete resistor values are distributed about a nominal value. The variations in resistance about the nominal value can be described in terms of the tolerance δ, which is chosen so that all resistors with a nominal value of R have actual values in the range $R(1 \pm \delta)$. Nominal values of resistors are chosen so that when tolerances are considered, the R values do not overlap. Thus, for a ±5% tolerance level, R values exist with a gap of 10 percent between them.

Figure 3-2 gives the significant figures[1] available with the ±5 percent tolerance level. Nominal resistor values are found by multiplying the two significant figures in Figure 3-2 by a power of 10. Resistor multipliers of 10^0 to 10^8 are commonly used in electronics. In order to model resistor performance, the effective tolerance δ associated with the component must be known.

---

[1] See Appendix 5 for more on significant figures.

A resistor is *linear* when the current flowing through it is proportional to the voltage across it. For a linear resistor, a plot of I as a function of V produces a straight line, as shown in Fig. 3-3(A). The slope of the line is a constant equal to $1/R$.

If the ratio V/I is not constant, as shown in Fig. 3-3(B), the resistor is *nonlinear*. The value of R changes due to heating effects and other factors.

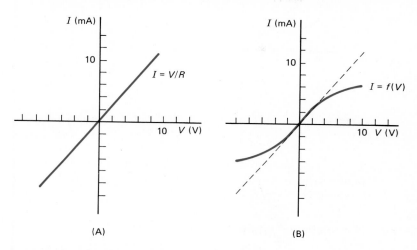

Fig. 3-3 Characteristic curves for (A) linear and (B) nonlinear resistors.

A plot of the current I through a circuit component as a function of the voltage V across the component is called the *characteristic curve* of the component. Characteristic curves are used extensively in the design process.

For future design purposes, it is often helpful to define a quantity r that is related to the slope of the characteristic curve at any point Q. For an applied voltage V, two resistances can be defined

$$R = \left.\frac{V}{I}\right|_Q \qquad \text{and} \qquad r = \left.\frac{\Delta V}{\Delta I}\right|_Q \tag{3-7}$$

In Fig. 3-3(B), both R and r vary with the selection of the operating point Q. The quantity R is the *dc resistance* and r is the *ac resistance* of a device.[2] Observe that R can be found graphically by reading values for V and I on any point on the curve, whereas r at any point on the curve is given by the reciprocal of the derivative at that point.

For a nonlinear device or component, the values of R and r, in general, are quite different. Given the characteristic curve of Fig. 3-3(B), R and r vary as shown in Fig. 3-4. Near the origin, the line is approximately straight so that $R \cong r$. However,

---

[2] The parameter r can be referred to as the ac or differential resistance of a device. The former reference is used here to emphasize the relationships between this parameter and the ac equivalent circuit parameters introduced in Chapter 9.

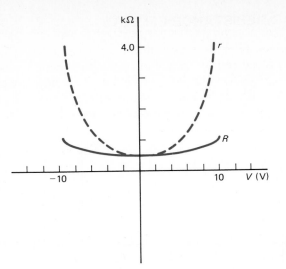

for large and small values of V, the characteristic curve begins to flatten out. The value of R rises slowly while the value of r (given by the reciprocal of the derivative) becomes very large.

Example 3-3   Nonlinear Resistance

A device is described by the characteristic curve in Fig. 3-3(B). For correct circuit operation, R and r must be greater at $V = 5.0$ V than at $V = 10$ V. Will this device function in a satisfactory way?

Solution   Refer to the graph. At $V = 5.0$ V,

$$R \cong 5.0 \text{ V}/4.5 \text{ mA} = 1.1 \text{ k}\Omega \qquad \text{and} \qquad r \cong 10 \text{ V}/5.5 \text{ mA} = 1.8 \text{ k}\Omega$$

and at $V = 10$ V,

$$R \cong 10 \text{ V}/6.0 \text{ mA} = 1.7 \text{ k}\Omega \qquad \text{and} \qquad r \cong 10 \text{ V}/1.0 \text{ mA} = 10 \text{ k}\Omega$$

Therefore, both R and r are greater at 10 V for this particular curve, and the device does not meet design needs.

In describing the operation of circuits with nonlinear resistances, values of both R and r are needed to describe the properties of the resistor near a particular operating point (determined by values for V and I). With the definitions above, both R and r can be found for any arbitrary device, given the characteristic curve.   ■

Example 3-4   Operating Points for Nonlinear Circuits

Assume that R_1 and R_2 in the three-terminal circuit of Fig. 3-5 are given by Fig. 3-3(A) and 3-3(B), respectively. (a) Find V_{out} for $V_{in} = 10$ V. (b) Find the dc and ac resistance for R_2.

Solution   (a) From Fig. 3-5 and Fig. 3-3(A),

$$I = \frac{V_{in} - V_{out}}{R_1} \qquad \text{and} \qquad R_1 = \frac{10 \text{ V}}{12 \text{ mA}} = 0.83 \text{ k}\Omega$$

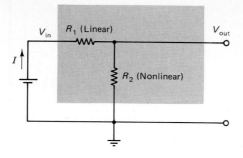

Fig. 3-5 Circuit for Example 3-4.

so that

$$I = \frac{10 \text{ V} - V_{\text{out}}}{0.83 \text{ k}\Omega}$$

If this equation for a straight line is plotted on the graph of Figure 3-3(B), the results of Fig. 3-6 are obtained. Therefore,

$$V_{\text{out}} \cong 6.0 \text{ V}$$

(b) From Fig. 3-6,

$$R \cong 6.0 \text{ V}/5.0 \text{ mA} \cong 1.2 \text{ k}\Omega \quad \text{and} \quad r \cong 10 \text{ V}/5.0 \text{ mA} \cong 2.0 \text{ k}\Omega$$

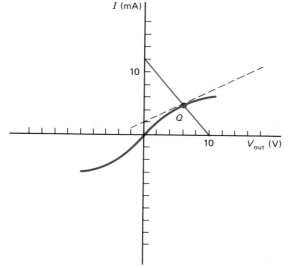

Fig. 3-6 Solution for Example 3-4.

This technique for the graphical solution of nonlinear problems was introduced in Chapter 1 and is further applied in the following chapters. ■

3-3 CAPACITOR MODELS

As will be seen in subsequent chapters, capacitors are widely applied in electronic circuits to develop frequency-dependent performance. For example, a series capacitor used to link two portions of a circuit blocks the dc voltages and passes the ac voltages above a certain frequency. Such high-pass filters serve a useful function in meeting circuit objectives.

On the other hand, inherent device capacitances can be a major problem in achieving design objectives. Equivalent capacitances between a changing voltage and ground can limit the high-frequency performance of a circuit.

A capacitor is a device (or arrangement of materials) that stores energy in an electric field. For the simple (ideal) parallel-plate capacitor, as shown in Fig. 3-7, envision two metal plates each with an area A separated by a dielectric with permittivity ϵ.

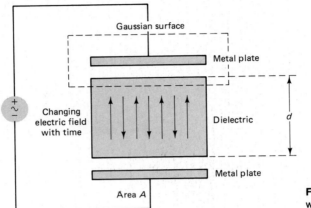

Fig. 3-7 Parallel-plate capacitor with a time-varying electric field.

The capacitance is defined as

$$C = dQ/dV$$

where Q is the charge on either capacitor plate, and V is the voltage placed across the plates. The charge on each plate can be found by applying Gauss' law, which states that

$$\int \bar{E} \cdot \hat{n} \, da = Q/\epsilon$$

where $\bar{E}$ is the electric field at every point over a closed surface, $\hat{n}$ is the unit vector perpendicular to each surface area da, Q is the total charge inside the surface, and ϵ is the permittivity of the medium.

For the parallel-plate capacitor of Fig. 3-7, it is assumed that d is much smaller than the plate dimensions. Therefore, the electric field is approximately constant everywhere inside the capacitor and zero outside. Using the Gaussian surface shown produces

$$EA \cong Q/\epsilon$$

The potential difference between the two plates is given by

$$V = \int E \, dx$$

where x runs perpendicular to the plates. Therefore,

$$V = Ed = \frac{Qd}{\epsilon A} \qquad \text{and} \qquad Q = \frac{\epsilon A}{d} V$$

and

$$C = \frac{dQ}{dV} = \epsilon \frac{A}{d} = \epsilon_0 \epsilon_r \frac{A}{d} \tag{3-8}$$

where $\epsilon_0 = 8.85 \times 10^{-12}$ F/m and ϵ_r is the relative permittivity.[3] The energy is stored in the electric field in the dielectric layer.

When a capacitor is modeled for circuit use, attention focuses on the dielectric that is used. It must withstand the highest circuit voltages applied to the plates without *breakdown,* allowing a current path to develop between the conducting plates. In order to create large values of C in a small volume, ϵ is made as large as possible. Large values of ϵ can be achieved through careful materials design.

However, every real dielectric displays an effective resistance when an alternating field is applied. As the field shifts back and forth, the atoms in the material are polarized in first one direction, then another. The atoms tend to move back and forth and experience friction due to their interactions. This friction produces an equivalent resistance across the plates of the capacitor and internal power consumption.

It is often difficult to obtain a large value for ϵ_r without producing undesirable resistive side effects in the dielectric. As is demonstrated below, the resistive effect can be represented by treating the relative permittivity of the dielectric as a complex number $\epsilon_r = k' - j\alpha$, where k' contributes to the large value of C (usually a desirable result), and α contributes to internal loss processes (usually not desirable). In general, both k' and α are functions of frequency.

Consider the properties of a capacitor formed with a dielectric that is represented by a complex number. By definition, the admittance of this functional element is

$$Y = j\omega C$$

From Eq. (3-8), letting $C_0 = \epsilon_0(A/d)$,

$$Y = j\omega C_0(k' - j\alpha) \tag{3-9a}$$

$$Y = j\omega k' C_0 + \omega\alpha C_0 \tag{3-9b}$$

For most reasonably designed capacitors, $|\alpha| \ll |k'|$. (Otherwise, the component is more of a resistor than a capacitor.) The admittance Y of the element consists of a capacitive term and a resistive term. To model the real capacitor, an equivalent circuit can be developed with an R and C in series or parallel. By convention, the parallel representation is used.

The admittance of a parallel RC combination can be written

$$Y = \frac{1}{R} + j\omega C \tag{3-10}$$

---

[3] This method of analysis is also helpful in Chapter 4 for the discussion of the capacitances associated with the junction diode.

From Eqs. (3-10) and (3-9a), the parallel RC model can be seen to work satisfactorily if

$$C = k'C_0$$

$$\frac{1}{R} = \omega\alpha C_0 \qquad (3\text{-}11)$$

The ratio α/k' is often set equal to a term $\tan\delta$, which is called the *loss tangent*. By definition,

$$\frac{1}{R} = \omega\alpha C_0 = \omega\left(\frac{\alpha}{k'}\right)(k'C_0) = \omega C \tan\delta \qquad (3\text{-}12)$$

For very small losses, α/k' becomes small, δ then becomes small, and R becomes large. Since R is in parallel with C, a large value of R has only a small circuit effect. As the losses become more important, R is reduced in value.

If an effective voltage $\mathbf{v}$ applied to the capacitor, the only power loss that takes place is in R (a pure capacitor produces no loss). Therefore,

$$P = \frac{|\mathbf{v}|^2}{R} = |\mathbf{v}|^2 \omega C_0 k' \tan\delta \qquad (3\text{-}13)$$

The term $k' \tan\delta$ is referred to as the *loss factor* for the capacitor. When using a capacitor as a functional element in a circuit, a decision must be made as to the magnitude of the effective resistance that is acceptable.

Typical values for k' and $\tan\delta$ are given in Fig. 3-8. The values of k' can be used to design capacitors having the desired capacitance C, and the $\tan\delta$ term can be used to find how much power is being dissipated in an insulating material with an applied ac voltage.

FIGURE 3-8
Typical Values for the Relative Dielectric Constant k' and the Loss Tangent ($\tan\delta$) for $f = 1\text{MHz}$

Material	k'	$\tan\delta$
Polyethylene	2	2×10^{-4}
Ceramic	9	3×10^{-3}
Glass	7	5×10^{-3}
Nylon	4	3×10^{-2}
Special-purpose dielectrics[1]	1–2	1×10^{-3}
	(1×10^1)–(5×10^1)	1×10^{-2}
	(5×10^1)–(1×10^3)	1×10^{-1}

[1] Dielectric thick-film materials used in hybrid microelectronics.

Example 3-5 Capacitor Design

The geometrical relationships for a capacitor can be illustrated by example. Assume that you wish to manufacture a capacitor with $C = 0.1$ μF using aluminum foil plates of 0.10 m^2 separated by a polyethylene material. You plan to form the parallel-plate capacitor, then roll it up in a small cylindrical shape to save space and provide mechanical integrity. What material thickness is needed?

Solution

$$C = \epsilon_0 k' \frac{A}{d}$$

$$d = \frac{\epsilon_0 k' A}{C} = \frac{(8.85 \times 10^{-12} \text{ F/m})(2)(0.10 \text{ m}^2)}{1.0 \times 10^{-7} \text{ F}}$$

$$\cong 0.02 \text{ mm} \qquad \blacksquare$$

Example 3-6 **Power Loss in a Capacitor**

A voltage $v = (12 \text{ V}) \sin (1.0 \times 10^5 t)$ is applied to the capacitor of Example 3-5. What power is dissipated in the capacitor?

Solution

$$P = \left(\epsilon_0 \frac{A}{d}\right) |\mathbf{v}|^2 \omega (k' \tan \delta)$$

$$= \left(\epsilon_0 k' \frac{A}{d}\right) |\mathbf{v}|^2 \omega \tan \delta$$

The effective value of the applied voltage is $12/\sqrt{2}$ V, so that

$$P = (1.0 \times 10^{-7} \text{ F})(12/\sqrt{2} \text{ V})^2 (1.0 \times 10^5 s^{-1})(2 \times 10^{-4})$$

$$\cong 0.1 \text{ mW}$$

An ideal capacitor model would not be able to predict this actual power loss. $\blacksquare$

Capacitors are manufactured using many different dielectric materials. Figure 3-9 compares several different capacitor types.

Fig. 3-9 Comparison of available values of C for several capacitor types.

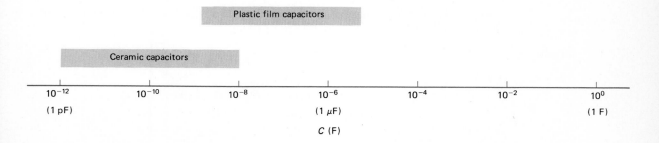

Plastic film capacitors are widely used for intermediate values of C, and ceramic capacitors are available with and without leads for intermediate and small values of C. Electrolytic capacitors are often used to block voltages and for low-frequency bypassing and filtering (applications of this type are considered in Chapter 5). They have a relatively high dissipation factor due to the electrolyte, but they also have a high volumetric efficiency (capacitance/unit volume). Leakage current can increase rapidly with T and the *temperature coefficient of capacitance (TCC)* is generally positive and can change by 1 percent/°C.

In selecting a capacitor for a circuit application and in predicting circuit performance, the designer must consider many different factors. The success of the circuit in meeting design objectives depends on the care used during the design process.

Capacitors also exist in circuits where they are unwanted. Any collection of conductors on an insulating layer has *distributed capacitances* between the conductors. Such capacitances can significantly affect the high-frequency response of circuits. Amplifier bandwidth and digital switching times can both be adversely affected.

Example 3-7   Distributed Capacitance

Assume that a distributed capacitance of 10 pF exists between two points in a circuit. The designer has determined that the capacitance will affect circuit performance if $X_C < 100$ kΩ. Above what frequency is the distributed capacitance significant?

Solution   The problem means finding the frequency f for which

$$\frac{1}{2\pi f C} < 1.00 \times 10^5 \ \Omega$$

and

$$f > 0.16 \text{ MHz}$$

A capacitance of >10 pF can be associated with typical wiring distances in a breadboard circuit. Therefore, at frequencies greater than 0.16 MHz, the designer can expect to find circuit function dependent on the physical arrangement of conductors that is used. ■

Example 3-8   High-Pass Filter with a Nonideal Capacitor

The performance of the high-pass filter introduced in Chapter 1 depends on the properties of the capacitor that is used. For the circuit of Fig. 3-10:

(a) Show that at frequency $f_1 = 1/2\pi RC$, the phase angle of $\mathbf{v}_0$ with respect to $\mathbf{v}_i$ is 45° if C is assumed ideal.

(b) Find the actual phase angle that is observed at this frequency if an $R_C = 5R$ is considered.

Solution

(a)
$$\frac{\mathbf{v}_0}{\mathbf{v}_i} = \frac{R}{-jX_C + R} = \frac{(R + jX_C)R}{R^2 + X_C^2}$$

$$\tan \theta = X_C/R$$

At $f_1 = 1/2\pi RC$, $R = X_C$ and $\theta = 45°$.

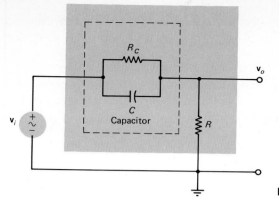

Fig. 3-10 Circuit for Example 3-8.

(b)

$$\frac{\mathbf{v}_0}{\mathbf{v}_i} = \frac{R}{(-jX_C)\,\|\,R_C + R} = \frac{R(R_C - jX_C)}{R(R_C - jX_C) - jR_C X_C}$$

$$= \frac{R[RR_C^2 + X_C^2(R + R_C) + jX_C R_C^2]}{(RR_C)^2 + X_C^2(R + R_C)^2}$$

$$\tan\theta = \frac{X_C R_C^2}{RR_C^2 + X_C^2(R + R_C)}$$

At $R = X_C$ and $R_C = 5R$,

$$\tan\theta = 0.806 \quad\text{and}\quad \theta \cong 39°$$

In circuits for which R and R_C are the same order of magnitude, the effective loss resistance associated with a capacitor can significantly change circuit performance. In most circuits discussed in the following chapters, the effect of R_C is usually neglected. However, a complete design analysis should include a quick check to see if R_C is important to the circuit of interest.

The resistive properties of the capacitors to be used in a circuit are often specified. A typical case is encountered when using the type of IC discussed in Section 17-6 (a dual-slope analog-to-digital converter circuit). For this device, the manufacturer specifies the type of capacitor to be used to minimize the effects of R_C. ■

3-4 INDUCTOR MODELS

Inductors are used in a variety of applications in electronic circuits. They are applied to the design of filters that provide frequency-dependent linkages from one part of a circuit to another and to the creation of resonant circuits that respond to a narrow band of frequencies. Further, the magnetic field effects associated with inductors are used to create memory storage devices.

An inductor is a circuit component that stores energy in a magnetic field. Inductors are fabricated by winding multiple loops of wire around a *core*. Each loop contributes to the total magnetic field that is created in the core. The magnetic field can store energy and interact with the external circuit.

Every inductor has associated with it a winding resistance and a distributed capacitance between the windings. A *coil* is an inductor for which the effects of the resistance and capacitance are considered.

To understand the behavior of a coil, start with the statement of Faraday's law

$$v = \frac{d}{dt}(N\Phi) \tag{3-14}$$

which relates the instantaneous voltage v across an inductor to the magnetic flux Φ (discussed below) and to the number of turns N in the coil.

The inductance L is defined by

$$L = \frac{d}{di}(N\Phi) \tag{3-15}$$

so that

$$v = \frac{d}{dt}(N\Phi) = \frac{d}{di}(N\Phi)\frac{di}{dt} = L\frac{di}{dt} \tag{3-16}$$

This equation describes the properties of an ideal inductor.

In order to calculate L for an inductor, several new variables must be introduced. The magnetic field H, measured in amperes/meter (A/m), and the magnetic flux density B, measured in webers/meter^2 (Wb/m^2) or tesla (1 Wb/m^2 = 1 T), describe the operation of inductors. The magnetic field H is created by the current flow in the windings of the inductor and does not depend on the core material. As a result of the magnetic field H that is set up, a magnetic flux density B develops in the inductor. The magnitude of B is dependent on the core material. The permeability μ of the core can be used to relate H and B:

$$B \equiv \mu H \tag{3-17}$$

For free space (or air), $\mu_0 = 4\pi \times 10^{-7}$ H/m or Wb/A·m, which is a constant and B is proportional to H. As will be seen below, this relationship becomes more complex with magnetic cores.

The total magnetic flux Φ inside the inductor is found by $\Phi = BA$, where A is the cross-sectional area of the inductor. Where B denotes the magnetic flux per unit area, Φ gives the total flux through the inductor.

The inductance L depends on the geometry of the coil, the number of turns N, and the core material. For the toroid of Fig. 3-11 (Winch, 1963),

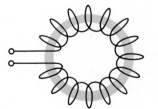

A = Cross-sectional area
ℓ = Circumference
N = Number of turns

Fig. 3-11 Toroidal inductor. The radius of the toroid is assumed large with respect to the radial distance across its core.

$$\Phi = AB \qquad H = \frac{N}{\ell}i$$
$$\tag{3-18}$$
$$d\Phi = A\,dB \qquad dH = \frac{N}{\ell}\,di$$

so that from Eq. (3-15)

$$L = \frac{N^2 A}{\ell} \left(\frac{dB}{dH} \right) \qquad (3\text{-}19)$$

For an air-core coil,

$$B = \mu_0 H \qquad \text{and} \qquad dB = \mu_0 \ dH \qquad (3\text{-}20)$$

For this case, Eq. (3-19) becomes

$$L = \mu_0 \frac{N^2 A}{\ell} = \mu_0 \mathcal{V} \left(\frac{N}{\ell} \right)^2 \qquad (3\text{-}21)$$

where $\mathcal{V}$ is the volume of the core, and N/ℓ is the number of turns per unit length. A high value of L requires maximizing $\mathcal{V}$, which leads to a large component size, and/or maximizing N/ℓ, which leads to a fine wire used for the windings.

The situation becomes more complicated when the nonideal inductor (or coil) is considered. Equation (3-16) describes an ideal inductor with no losses. However, every coil has resistance associated with the loops of wire that produce the inductance. For air-core coils and for other types of cores that are nonmagnetic, the resistance associated with the inductor winding produces the dominant loss effect.

A coil model for this case uses an inductor L and a resistor R in series. For a coil, then, Eq. (3-16) must be modified to read

$$v = L \frac{di}{dt} + iR \qquad (3\text{-}22)$$

For a sinusoidal applied voltage and an air-core coil, the current that flows in the coil is obtained from Eq. (3-22),

$$\mathbf{v} = (j\omega L)\mathbf{i} + \mathbf{i}R$$

$$\mathbf{i} = \frac{\mathbf{v}}{R + j\omega L} \qquad (3\text{-}23)$$

Example 3-9   Air-Core Coil

Consider an air-core coil that is 5.0 mm in diameter and has $N/\ell = 10 \times 10^3$ turns/m. What coil length should be used to produce an $L = 150 \ \mu\text{H}$?

Solution   From Eq. (3-21),

$$\ell = \frac{L}{\mu_0 A (N/\ell)^2}$$

$$= \frac{150 \times 10^{-6} \ \text{H}}{(4\pi \times 10^{-7} \ \text{H/m})[\pi (2.5 \times 10^{-3})^2 \ \text{m}^2](10 \times 10^3 \ \text{m}^{-1})^2} \cong 6.1 \ \text{cm} \qquad \blacksquare$$

Cores made from *ferromagnetic* and *ferrimagnetic* materials (the latter are often called *ferrites*) are used to obtain larger inductance in a small volume.

Ferromagnetic or ferrimagnetic inductor coils produce an inductance that varies with the applied magnetic field. No longer is B proportional to H, creating a straight-line relationship. Instead, the relationship between B and H takes on the form shown in Fig. 3-12, which is called a *hysteresis loop*.

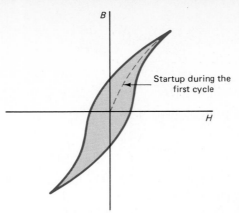

Startup during the
first cycle

Fig. 3-12 Hysteresis loop.

Note that as the H field cycles from positive to negative values, the B field also cycles. However, the core has a memory. When H decreases from its maximum positive value to zero, B does not go to zero. Instead, a residual magnetization remains. The domains in the core remain partically aligned in a preferential direction. The same type of process occurs when H changes from its most negative value back to zero. The memory property illustrated by the hysteresis loop is used to create magnetic data storage systems for both analog and digital systems (discussed in Chapter 15).

Inductor ratings involve nominal values of L, information regarding the composite losses R, frequency response, temperature response, size, and weight.

An inductance exists whenever a current flow sets up a magnetic field. Thus, any conductor has a self-inductance and any combination of conductors develops a mutual inductance. Distributed inductances can be important in circuits at high frequencies. Unfortunately, it is difficult to anticipate the value of a distributed inductance since the magnetic fields for the whole circuit interact.

Example 3-10 Distributed Inductance

Assume that an estimated distributed inductance of 1.0 μH exists between two points in a circuit. The designer has determined that the inductance will affect circuit performance if $X_L < 0.10$ kΩ and desires for the circuit to operate at 2 MHz. Will the distributed inductance be a problem?

Solution The solution requires finding the values of f for which

$$2\pi f L < 0.10 \text{ k}\Omega$$

so that for

$$f < 16 \text{ MHz}$$

the inductance is significant. A design problem exists with respect to this objective. ∎

Example 3-11 Inductor Filter

An LR combination can also be used to create a high-pass filter, as shown in Fig. 3-13. (a) Find the phase angle associated with this circuit if $X_L = R$ and $r = 0$. (b) If $R = X_L$ and $r = 0.10R$, find the phase shift associated with the circuit.

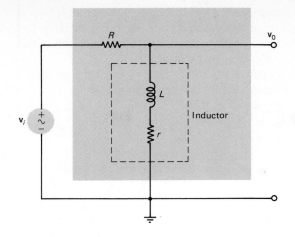

Fig. 3-13 Circuit for Example 3-11.

Solution

$$\frac{\mathbf{v}_0}{\mathbf{v}_i} = \frac{r + jX_L}{R + r + jX_L}$$

$$\tan \theta = \frac{X_L R}{(R + r)r + X_L^2}$$

(a) If $r = 0$,

$$\tan \theta = 1 \quad \text{and} \quad \theta = 45°$$

(b) If $r = 0.10R$

$$\tan \theta = \frac{1}{(1 + 0.10)(0.10) + 1} = 0.90$$

$$\theta \cong 42°$$

Again, the nonideal nature of a component can change circuit performance. ∎

3-5 TRANSFORMER MODELS

If two coils are arranged around a common core, so that most of the magnetic flux links both coils, a *transformer* is obtained. Transformers are often designed with a magnetic core to assure maximum flux linkage between the two coils. The magnetic flux tends to stay inside the core, and almost complete coupling can be obtained. Figure 3-14 shows the geometry for a toroidal transformer with a magnetic core.

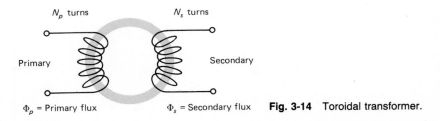

Φ_p = Primary flux Φ_s = Secondary flux   **Fig. 3-14** Toroidal transformer.

Transformers can increase or decrease the magnitudes of ac voltages and can be used for impedance matching and coupling between circuit stages. Equivalent circuits for transformers are often complex because of the combined inductive, resistive, and capacitive effects that exist.

For an ideal transformer, all of the flux from the primary (exciting) coil passes through the secondary (load) coil and there are no power losses.

From Eq. (3-14) and Fig. 3-14

$$v_p = \frac{d}{dt}(N_p\Phi_p) \qquad \text{and} \qquad v_s = \frac{d}{dt}(N_s\Phi_s) \tag{3-24}$$

and since Φ is the same for both the primary and secondary coils,

$$\frac{v_p}{v_s} = \frac{\mathbf{v}_p}{\mathbf{v}_s} = \frac{N_p}{N_s} \tag{3-25}$$

Since the power losses are assumed negligible,

$$\mathbf{P}_p = \mathbf{P}_s$$

$$\mathbf{v}_p\mathbf{i}_p = \mathbf{v}_s\mathbf{i}_s$$

$$\frac{\mathbf{i}_s}{\mathbf{i}_p} = \frac{\mathbf{v}_p}{\mathbf{v}_s} = \frac{N_p}{N_s} \tag{3-26}$$

The input and output impedances have the relationship

$$\frac{Z_s}{Z_p} = \frac{\mathbf{v}_s/\mathbf{i}_s}{\mathbf{v}_p/\mathbf{i}_p} = \frac{\mathbf{v}_s}{\mathbf{v}_p}\frac{\mathbf{i}_p}{\mathbf{i}_s} = \left(\frac{N_s}{N_p}\right)^2 \tag{3-27}$$

The performance of real transformers is complicated by imperfect flux linkages and losses due to winding resistance, eddy currents, and hysteresis. These factors modify all of the above relationships.

For transformers showing these more complex properties, the primary current is affected by the load in the secondary. An increasing secondary current $\mathbf{i}_s$ reduces the total flux in the core, reducing the effective inductance L of the primary and increasing the primary current $\mathbf{i}_p$. Frequency effects are important in determining the nature of the flux linkages that are formed and their effect on performance.

Example 3-12   Nonideal Transformer

For the transformer of Fig: 3-15, 90 percent of the primary flux links to the secondary. Find v_0.

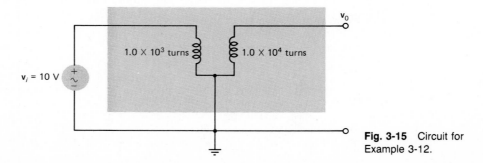

Fig. 3-15   Circuit for Example 3-12.

Solution

$$\Phi_s = 0.90\Phi_p$$

$$v_p = \frac{d}{dt}(N_p\Phi_p) \quad \text{and} \quad v_s = \frac{d}{dt}(N_s\Phi_s)$$

Therefore,

$$\frac{\mathbf{v}_p}{\mathbf{v}_s} = \frac{N_p}{0.90N_s}$$

$$\mathbf{v}_0 = \mathbf{v}_s = \mathbf{v}_p \frac{0.90N_s}{N_p}$$

$$= 10 \text{ V} \frac{0.90(1.0 \times 10^4)}{1.0 \times 10^3} = 90 \text{ V}$$

For an ideal transformer, the result would be $\mathbf{v}_0 = 100$ V. ∎

3-6   MODELS FOR PASSIVE CIRCUIT COMPONENTS

This chapter develops a set of models for resistors, capacitors, inductors, and transformers that show how the R, C, and L functions are interrelated and dependent on the materials and fabrication processes used to create a circuit of interest. Figure 3-16 summarizes some of the characteristics of realistic devices.

The intent of this chapter is to alert the student to the many approximations used in the component descriptions that follow. In general, these descriptions assume idealized elements and neglect distributed R, C, and L functions. A number of exceptions are noted and explained where particularly important.

Some of the problems also provide reminders of the approximate nature of the models being used. The designer should bear in mind the more complex models developed here and use them as a frame of reference for evaluating predicted circuit performance. A design process is not complete until the circuit has been evaluated in terms of the impact of the nonideal elements that are included. This type of evaluation is becoming available in the more sophisticated CAD systems and serves as a reminder during the study of alternative circuit configurations and tradeoffs.

FIGURE 3-16
Properties of Passive Circuit Elements

Property	Resistor	Capacitor	Inductor	Transformer
Idealized property	R	C	L	$\frac{V_p}{V_s} = \frac{N_p}{N_s}$
Distributed properties	L, C	R (or δ), L	R and C due to windings. Core losses	
Other important properties	Linearity (R and r)	Linearity	Linearity (hysteresis effects)	
	Maximum power dissipation	Voltage breakdown	Saturation current levels	
	TCR	TCC	Coupling efficiency	
	Tolerances	Tolerances	Tolerances	

Questions   **3-1.** Why is it important to consider how resistance, capacitance, and inductance are modeled in electronic circuits?

3-2. What assumption is usually made for resistor, capacitor, and inductor models during the initial design iteration?

3-3. Why does the TCR for a semiconductor material change with temperature?

3-4. What are some of the considerations that enter into the selection of a material to form a resistor?

3-5. What are some of the considerations that enter into the selection of a material to form a capacitor?

3-6. When used in a high-pass filter, how can the real properties of a capacitor material affect the filter output?

3-7. How does a core material affect the operation of an inductor?

3-8. How can the circuit effects of capacitors and inductors be contrasted?

Answers   **3-1.** When different types of models are used to represent resistance, capacitance, and inductance in electronic circuits, the ranges of validity of these models affect the design conclusions that are drawn. Similarly, the ways in which distributed resistance, capacitance, and inductance are represented help determine the limits of model validity. The designer must understand the nature of the models being used and their limits in order to assess the utility of initial design concepts.

3-2. During the initial design cycle, resistors, capacitors, and inductors are usually modeled in terms of discrete, ideal components and distributed effects are neglected. Further iterations are then used to test the appropriateness of these simple initial models.

3-3. Since the resistance R of a semiconductor material is a strong nonlinear function of T, the TCR (which is the fractional rate of change of ρ with T) also changes as a function of T.

3-4. A material to be used for a resistor in a circuit can be described in terms of its resistivity ρ over the temperatures of interest, manufacturing and fabrication properties, power dissipating capability, response to humidity and vibration, and other factors.

3-5. A material to be used to form a capacitor in a circuit can be described in terms of its complex permittivity as a function of temperature and frequency, its ability not to break down under applied voltage, response to humidity and vibration, and other factors.

3-6. The output (both magnitude and phase) of a high-pass filter depends on the (frequency-dependent) losses present in the capacitor.

3-7. The core material changes the relationships between the applied magnetic field H and the resultant magnetic flux density B and can introduce nonlinearities into this relationship.

3-8. Capacitor and inductor effects on a circuit can be contrasted in many different ways. In addition to introducing different phase relationships between the currents through the components and voltages across them ($-j/\omega C$ versus $j\omega L$), they can be compared in terms of energy storage (electric field versus magnetic field), material effects (ϵ versus μ) and coupling to the external circuit (a confined E field versus an unconfined H field).

EXERCISES AND SOLUTIONS

Exercises **3-1.** Find the TCR for pure Ge at 300°K.

(3-1)   **3-2.** Find the TCR for a typical pure metal at -100°C.

(3-1)   **3-3.** From the data in Fig. 3-1, estimate the energy gap for graphite at 300°K.

(3-1)   **3-4.** A 5.6-kΩ resistor has a TCR of -50 ppm/°C at 50°C. Estimate the change in the value of the resistor at 0°C.

(3-2)   **3-5.** When 5.0 V is applied across a resistive device, a current of 10 μA results. When a current of 15 V is applied, a current of 20 μA results. Is this a linear or nonlinear resistor?

(3-2)   **3-6.** For the resistor described by Fig. 3-3(B), find R and r at (a) $V = 2$ V and (b) $V = 8$ V.

(3-2)   **3-7.** In a given circuit, you need a device with a characteristic curve for which r increases with V. Which curve in Fig. 3-17 applies?

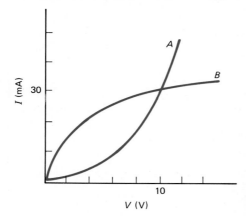

Fig. 3-17 Data for Exercises 3-7 and 3-8, and for Problem 3-11.

(3-2)   **3-8.** For Fig. 3-17, find R and r for curves A and B at (a) $V = 4$ V and (b) $V = 10$ V.

(3-3)   **3-9.** A 0.001-μF capacitor is designed with $\epsilon_r = 3.0$ and $d = 50$ μm. Find the effective plate area A that is required.

(3-3)   **3-10.** The voltage across a capacitor doubles and the frequency of the signal through the capacitor increases by a factor of 10. Find the change in the power dissipated in the dielectric of the capacitor (expressed as a ratio).

(3-3)   **3-11.** A distributed capacitance of 50 pF exists between two points in a circuit. For what frequencies is the effective impedance between these two points greater than 0.50 MΩ?

(3-3)   **3-12.** A lossy capacitor is used to make a high-pass RC filter. Given $R = 100$ kΩ, $R_C = 0.10$ MΩ, and $C = 0.10$ μF, at what frequency is the impedance of the capacitor equal to 10 percent of R_C? How will this circuit perform for frequencies much above and below this value?

(3-4)   **3-13.** Consider a toroid with a silicon-iron core ($\mu = 1.0 \times 10^3$ μ_0) wound with 1000 turns/cm and having a current flow $I = 0.3$ A. What value of B exists in the core?
   Note: The anticipated value of B from Eq. (3-16) may not be attainable since a core will saturate at a maximum value of B_s (for example, $B_s = 15$ T for silicon-iron alloys, 7 to 16 T for nickel-iron alloys, 20 T for cobalt-iron alloys, and 5 T for ferrites).

(3-4)   **3-14.** A coil and resistor are used to create a high-pass filter, as shown in Fig. 3-13. If $R = 2X_L$ and $r = 0.10R$ at the frequency of operation, find the phase shift associated with the circuit.

(3-5) **3-15.** For the transformer of Fig. 3-15, 85 percent of the primary flux links the secondary. Find v_0.

3-16. The production cost of a circuit depends on the number of circuits manufactured according to the curve in Fig. 3-18(A) (which is called a *learning curve* because costs drop as learning increases). The marketing costs vary as shown in Fig. 3-18(B). Marketing costs per circuit decline as volume increases until the market nears saturation, after which costs begin to rise. Assume that all other costs are independent of the volume of circuits produced. By graphically adding both curves in the figure, find the optimum volume of circuits that should be produced to minimize costs.

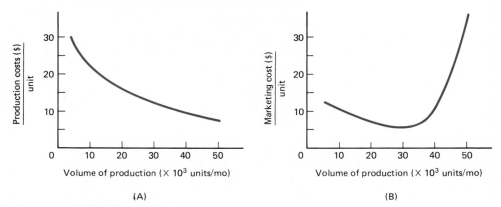

(A)   (B)

Fig. 3-18. Data for Exercise 3-16.

Solutions **3-1.** From Eq. (3-3),

$$\text{TCR} = -\frac{\mathcal{E}_g}{2kT^2} \cong \frac{-0.67 \text{ eV}}{2(8.62 \times 10^{-5} \text{ eV/°K})(300°\text{K})^2}$$

$$\cong -4.3 \times 10^4 \text{ ppm/°K}$$

3-2. From Eq. (3-4),

$$\text{TCR} \cong 1/T \cong 1/173°\text{K} \cong 5.8 \times 10^3 \text{ ppm/°K}$$

3-3. From Eq. (3-3),

$$\text{TCR} = \frac{-\mathcal{E}_g}{2kT^2}$$

$$\mathcal{E}_g \cong (300 \times 10^{-6}/°\text{K})(2)(8.62 \times 10^{-5} \text{ eV/°K})(300°\text{K})^2$$

$$\cong 4.7 \times 10^{-3} \text{ eV}$$

3-4. Using a linear approximation (because of the low TCR value and limited ΔT),

$$\Delta R \cong (-50 \times 10^{-6}/°\text{K})(0 - 50°\text{C})(5.6 \times 10^3 \ \Omega)$$

$$\cong 14 \ \Omega$$

3-5. Calculating the resistance for each Q point,

$$5.0 \text{ V}/10 \ \mu\text{A} = 0.50 \text{ M}\Omega \quad \text{and} \quad 15 \text{ V}/20 \ \mu\text{A} = 0.75 \text{ M}\Omega$$

This is not a linear resistor.

3-6. As shown in Fig. 3-19,

(a) At $V = 2.0$ V,

$$R \cong \frac{2.0\ \text{V}}{2.0\ \text{mA}} = 1.0\ \text{k}\Omega \qquad \text{and} \qquad r \cong \frac{2.0\ \text{V} - 0\ \text{V}}{2.0\ \text{mA} - 0\ \text{mA}} \cong 1.0\ \text{k}\Omega$$

(b) At $V = 8.0$ V,

$$R \cong \frac{8.0\ \text{V}}{5.0\ \text{mA}} = 1.6\ \text{k}\Omega \qquad \text{and} \qquad r \cong \frac{8.0\ \text{V} - 0\ \text{V}}{5.0\ \text{mA} - 4.0\ \text{mA}} \cong 8.0\ \text{k}\Omega$$

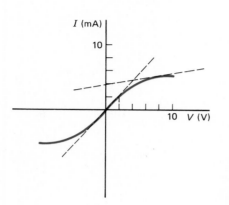

Fig. 3-19 Solution for Exercise 3-6.

3-7. Given $r = 1/\text{slope}$, curve B applies.

3-8. As shown in Fig. 3-20, for curve A,

(a) At $V = 4.0$ V,

$$R \cong \frac{4.0\ \text{V}}{4.0\ \text{mA}} \cong 1.0\ \text{k}\Omega \qquad \text{and} \qquad r \cong \frac{4.0\ \text{V} - 2.0\ \text{V}}{4.0\ \text{mA} - 0\ \text{mA}} \cong 0.50\ \text{k}\Omega$$

(b) At $V = 10$ V,

$$R \cong \frac{10\ \text{V}}{25\ \text{mA}} \cong 0.40\ \text{k}\Omega \qquad \text{and} \qquad r \cong \frac{10\ \text{V} - 6\ \text{V}}{25\ \text{mA} - 0\ \text{mA}} \cong 0.16\ \text{k}\Omega$$

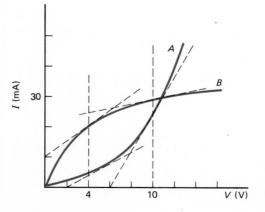

Fig. 3-20 Solution for Exercise 3-8.

For curve B,

(a) At $V = 4.0$ V,

$$R \cong \frac{4.0 \text{ V}}{20 \text{ mA}} \cong 0.20 \text{ k}\Omega \quad \text{and} \quad r \cong \frac{4.0 \text{ V} - 0 \text{ V}}{20 \text{ mA} - 10 \text{ mA}} \cong 0.40 \text{ k}\Omega$$

(b) At $V = 10$ V,

$$R \cong \frac{10 \text{ V}}{30 \text{ mA}} \cong 0.33 \text{ k}\Omega \quad \text{and} \quad r \cong \frac{10 \text{ V} - 4.0 \text{ V}}{30 \text{ mA} - 25 \text{ mA}} \cong 1.2 \text{ k}\Omega$$

3-9. From Eq.(3-8),

$$C = \epsilon \frac{A}{d} = \epsilon_0 \epsilon_r \frac{A}{d}$$

$$A = \frac{Cd}{\epsilon_0 \epsilon_r} = \frac{(1.0 \times 10^{-9} \text{ F})(50 \times 10^{-6} \text{ m})}{(8.85 \times 10^{-12} \text{ F/m})(3.0)}$$

$$\cong 1.9 \times 10^{-3} \text{ m}^2 \cong 19 \text{ cm}^2$$

3-10. From Eq. (3-13),

$$\frac{P_2}{P_1} = \frac{|\mathbf{v}_2|^2 \omega_2 C_0 k' \tan \delta}{|\mathbf{v}_1|^2 \omega_1 C_0 k' \tan \delta}$$

and since

$$|\mathbf{v}_2| = 2|\mathbf{v}_1| \quad \text{and} \quad \omega_2 = 10\omega_1$$

$$\frac{P_2}{P_1} = 40$$

3-11. For the specification given,

$$X_C = 1/\omega C > 0.50 \text{ M}\Omega$$

$$\omega < \frac{1}{(0.50 \times 10^6 \text{ }\Omega)(50 \times 10^{-12} \text{ F})}$$

$$\omega < 40 \text{ krad/s} \quad \text{and} \quad f < 6.4 \text{ kHz}$$

3-12. For the specification given,

$$X_C = 0.10 R_C$$

$$\omega = \frac{1}{(0.10 R_C)(C)}$$

$$= \frac{1}{(0.10)(0.10 \times 10^6 \text{ }\Omega)(0.10 \times 10^{-6} \text{ F})} \cong 1.0 \text{ krad/s}$$

For $\omega \ll 1.0$ krad/s, $X_C \to \infty$ and

$$\mathbf{v}_0 \to \frac{R}{(R + R_C)} \mathbf{v}_i \cong \frac{0.10 \times 10^6 \text{ }\Omega}{2 \times 0.10 \times 10^6 \text{ }\Omega} \mathbf{v}_i = \frac{\mathbf{v}_i}{2}$$

For $\omega \gg 1.0$ krad/s, $X_C \to 0$ and

$$\mathbf{v}_0 \cong \mathbf{v}_i$$

3-13. From Eq. (3-18),

$$H = N\frac{i}{\ell} = (10^3/\text{cm})(10^2 \text{ cm/m})(0.3 \text{ A}) = 0.3 \times 10^5 \text{ A/m}$$

Without saturation,

$$B = \mu H = (1.0 \times 10^3)(4\pi \times 10^{-7} \text{ H/m})(0.3 \times 10^5 \text{ A/m})$$

$$\cong 38 \text{ T}$$

With saturation,

$$B \leq B_s = 15 \text{ T}$$

$$B \cong 15 \text{ T}$$

3-14. Using the results of Example 3-11,

$$R = 2X_L \qquad \text{and} \qquad r = 0.10R$$

$$\tan\theta = \frac{X_L R}{(R + r)r + X_L^2} = \frac{2X_L^2}{(1.10)(2X_L)(0.10)(2X_L) + X_L^2}$$

$$= \frac{2X_L^2}{0.44X_L^2 + X_L^2} \cong 1.4$$

$$\theta \cong 54°$$

3-15. Following the model of Example 3-12,

$$\Phi_s = 0.85\Phi_p$$

$$\frac{\mathbf{v}_p}{\mathbf{v}_s} = \frac{N_p}{0.85N_s}$$

$$\mathbf{v}_0 = \mathbf{v}_p\frac{0.85N_s}{N_p} = (10 \text{ V})\frac{(0.85)(1.0 \times 10^4)}{(1.0 \times 10^3)} = 85 \text{ V}$$

3-16. By graphical addition, the solution of Fig. 3-21 results.

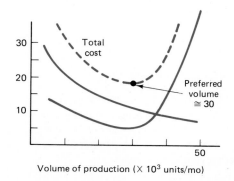

Volume of production ($\times 10^3$ units/mo)   **Fig. 3-21** Solution for Exercise 3-16.

PROBLEMS

(3-1)   **3-1.** Find the TCR for GaAs at 0°C.

(3-1)   **3-2.** Find the TCR for Ge at 100°C.

(3-1)   **3-3.** Find the TCR for a pure metal at 100°C.

(3-1)   **3-4.** A 1.0-kΩ resistor has a TCR of -400 ppm/°C at 0°C. Estimate the change in value of the resistor at 10°C.

(3-1)   **3-5.** A 5.0-kΩ resistor increases its value by 1 percent as the temperature increases by 20°K. What is the TCR for the resistor?

(3-1)   **3-6.** A 5.6-kΩ resistor has a TCR of -250 ppm/°C at 300°K. Using a linear estimate, find the change in value of the resistor at 280°K.

(3-2)   **3-7.** A change in voltage across a device from 0.80 to 0.90 V results in an increase in current through the device from 10 to 80 mA. Estimate the ac resistance of the device in this region.

(3-2)   **3-8.** A 10-V dc power supply is placed across a linear 12-kΩ resistor. What current flows?

(3-2)   **3-9.** The voltage across a 9.6 kΩ linear resistor is changed from 8.0 to 12.0 V. The initial current is 5.0 mA. What is the final current?

(3-2)   **3-10.** The voltage across a resistor is changed from 8.0 to 12.0 V and the current changes from 5.0 to 7.0 mA. Find the dc resistance at each point and the ac resistance between these points.

(3-2)   **3-11.** For Fig. 3-17 the choice of V biases the device at a particular point (called the Q, or quiescent, point) on the characteristic curve. For each curve, find the Q points described by (a) $I = 15$ mA, (b) $R = 0.25$ kΩ, and (c) $r = 0.4$ kΩ.

(3-2)   **3-12.** Assume that resistances R_1 and R_2 for the three-terminal circuit of Fig. 3-5 are given by Figs. 3-22(A) and (B), respectively. (a) Find V_{out} for $V_{in} = 4.0$ V. (b) Find the dc and ac resistances for R_2 at the operating point.

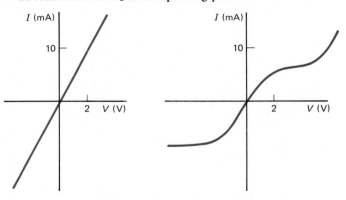

Fig. 3-22   Characteristic curves for Problem 3-12.

(3-3)   **3-13.** Find the typical complex permittivity ϵ_r of ceramic and glass at 1.0 MHz.

(3-3)   **3-14.** You are to produce a 1.0×10^3-pF capacitor using a polyethylene dielectric. Given $d = 1.0 \times 10^{-2}$ mm, find the effective capacitor area required.

(3-3)   **3-15.** For a parallel-plate capacitor, $C = 0.001$ μF and d = 0.01 mm. If a nylon insulating layer is used, what effective plate area is required?

(3-3)   **3-16.** The effective resistance of a 0.001 μF capacitor is 1.0 MΩ at 1.0 MHz. Estimate the material used to fabricate the capacitor.

(3-3) **3-17.** Find the effective resistance R associated with a ceramic $0.01\text{-}\mu F$ capacitor at 1.0 MHz.

(3-3) **3-18.** A 4-V effective ac voltage at 1.0 MHz is applied to a nylon film capacitor with $C = 1.0 \ \mu F$. Find the power dissipated in the capacitor.

(3-3) **3-19.** A $0.01\text{-}\mu F$ capacitor with a nylon dielectric is operated at $f = 1.0$ MHz. If $v = (20 \text{ V}) \cos \omega t$, estimate the power dissipated in the capacitor.

(3-3) **3-20.** The operating temperature of a $0.01\text{-}\mu F$ electrolytic capacitor changes from 0 to $100°C$. What change in the capacitor value can be expected?

(3-3) **3-21.** A distributed capacitance of 25 pF exists between two connector leads in a circuit. Above what frequency does the condition $X_C < 50$ kΩ hold?

(3-3) **3-22.** Distributed capacitance affects the performance of a specific circuit operating at 1.0 MHz if $X_C < 50$ kΩ. What is the largest value of C that can be allowed?

(3-3) **3-23.** For Example 3-8, find the phase angle at frequency $2f_1$ with R_C very large.

(3-3) **3-24.** Find the change in $|v_{out}|/|v_{in}|$ that is associated with the high-pass filter of Example 3-8 when comparing filter performance with (a) $R_C = X_C = R$, and (b) $R_C = R/2$, $X_C = 2R$.

(3-3) **3-25.** Consider the circuit of Fig. 3-10 with $C = 5.0 \times 10^{-7}$ F and $R = 12$ kΩ. Find the phase of the output voltage v_0 with respect to v_i at $f = 60$ Hz if R_C is very large.

(3-4) **3-26.** The current through an (ideal) $10\text{-}\mu H$ inductor changes by 10 mA in 0.01 ms. What voltage is observed across the inductor?

(3-4) **3-27.** A magnetic field of 10 A/m is observed in an air-core coil. Find the magnetic flux density in the coil.

(3-4) **3-28.** An air-core toroidal coil has a volume of 2.5 cm^3 and an inductance of 1.5 mH. Find the number of turns per unit length required to fabricate the coil.

(3-4) **3-29.** Consider an air-core coil that is 3.5 mm in diameter and has $N/\ell = 200$ turns/cm. What coil length should be used to produce an inductance $L = 0.2$ mH?

(3-4) **3-30.** A 4-V (effective) ac source is applied across a $100\text{-}\mu H$ coil at $f = 10$ kHz. If the coil resistance is 100 Ω, find the phasor that can be used to represent the current in the coil.

(3-4) **3-31.** A distributed inductance of 2.5 mH exists between two points in a circuit. At what frequency does the effective impedance between these two points become 0.50 MΩ?

(3-4) **3-32.** Distributed inductance affects the performance of a given circuit operating at 1.0 MHz if $X_L > 10$ kΩ. What is the largest value of L that can be allowed?

(3-4) **3-33.** A distributed inductance of 100 μH is estimated to exist between two connectors in a circuit. Will the design goal $X_L < 10$ kΩ be observed at $f = 10$ MHz?

(3-4) **3-34.** For the LR high-pass filter in Fig. 3-13, find the phase angle associated with the circuit if $R = X_L$ and $r = 0.20R$.

(3-4) **3-35.** For the LR high-pass filter in Fig. 3-13, find the ratio $|v_{out}|/|v_{in}|$ for $R = X_L$ and (a) $r = 0$ and (b) $r = 0.20$ R.

(3-5) **3-36.** For an ideal transformer, $N_s = 0.10 \ N_p$ and $Z_s = 1.0$ kΩ. If $v_p = 120$ V, (a) find v_s and (b) find Z_p.

(3-5) **3-37.** For a given transformer, $v_p = 25$ V and $v_s = 45$ V. If the secondary has twice as many turns as the primary, what percentage of the flux from the primary is linking to the secondary?

(3-5) **3-38.** Two circuits are to be matched in impedance using a transformer. The first circuit has an output impedance of 50 Ω, and the second circuit has an input impedance of 100 Ω. (a) What should be the turns ratio for the transformer? (b) What (ideal) changes in voltage levels occur between the two circuits?

3-39. In a production setting, a designer finds that the use of 1-percent resistors produces a circuit that can be sold for $25.00 and the use of 5-percent resistors produces a circuit that can be sold for $20.00. There are 20 resistors in the circuit and the costs are as follows:

1-percent resistors: $100/1000

5-percent resistors: $50/1000

Assuming all other costs are the same, which resistors should be selected?

3-40. In a given application, you are to decide whether to use an expensive capacitor costing $220/1000 or an inexpensive one costing $80/1000. Each circuit contains 50 capacitors. You predict that the major impact of the choice will be on circuit failures. The expensive capacitors will fail every 500 h (hours) of operation and the inexpensive capacitors will fail every 200 h of operation. If repair costs are $5.00 for each case and the circuit lifetime is about 2000 h, which type of capacitor should be selected?

COMPUTER APPLICATIONS

3-1. A computer program may be used to find the magnitude of the impedance $|Z|$ and phase angle $\theta = \tan^{-1} Z_{im}/Z_{real}$ as a function of frequency f for the resistor model of Fig. 3-23. The results of such a computer simulation for a (wirewound) resistor are shown in Figs. 3-24 and 3-25, using a logarithmic scale for f and measuring $|Z|/R$ in decibels (as discussed in Appendix 10) with $C = 1.0$ pF and $L = 1.0$ μH. (a) Explain the asymptotic behavior of the graph for large and small values of f. (b) Discuss the properties of the series resonance that results. (c) Theoretically predict the value of f that should correspond to the minimum value of $|Z|$. Compare with the prediction of the computer model. (d) Find the value of the resistor R. (e) What percentage change in R and phase shift will be observed at $f = 1.0$ MHz, $f = 5.0$ MHz, and $f = 10$ MHz? (f) Show that the results of Fig. 3-25 may be *scaled* if the products $\omega L/R$ and ωRC are held constant. Given this finding, how can the data of Fig. 3-25 be used to illustrate the frequency-response characteristics of a resistor with $C = 10$ pF and $L = 0.10$ μH?

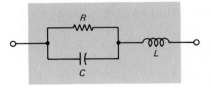

Fig. 3-23 Resistor model for Computer Application 3-1.

* **3-2.** Write a computer program to calculate V_{out} for the voltage divider of Fig. 2-2 given V_{in}, the values of R_1 and R_2 at reference temperature T_0, the TCR values at T_0, and the temperature T of operation. (Assume that $T - T_0$ is sufficiently small that the TCR can be treated as approximately constant over the range of interest.) Given the TCR for R_1 of 200 ppm/°C and the TCR for R_2 of -200 ppm/°C, use your program to plot V_{out}/V_{in} as a function of $T - T_0$ for $T_0 = 300$°K and T varying from 250 to 350°K.

* **3-3.** For Example 3-2, assume that the linear approximation used is not sufficiently accurate. Write a computer program that can solve the problem, using the relationship of Eq. (2-11). Compare the results of your program with those obtained using the linear approximation.

PHASE	MAG-dB	MAGNITUDE ($\times 10^{-1}$)	FREQUENCY ($\times 10^3$)	OMEGA ($\times 10^3$)
−0.36	0.0	10.0E+00	10.00E+01	6.3E+02
−0.72	0.0	10.0E+00	20.00E+01	1.3E+03
−1.08	0.0	10.0E+00	30.00E+01	1.9E+03
−1.44	0.0	10.0E+00	40.00E+01	2.5E+03
−1.80	0.0	10.0E+00	50.00E+01	3.1E+03
−2.16	0.0	99.9E−01	60.00E+01	3.8E+03
−2.52	0.0	99.9E−01	70.00E+01	4.4E+03
−2.88	0.0	99.9E−01	80.00E+01	5.0E+03
−3.24	0.0	99.8E−01	90.00E+01	5.7E+03
−3.60	0.0	99.8E−01	10.00E+02	6.3E+03
−7.16	−0.1	99.2E−01	20.00E+02	1.3E+04
−10.67	−0.2	98.3E−01	30.00E+02	1.9E+04
−14.10	−0.3	97.0E−01	40.00E+02	2.5E+04
−17.42	−0.4	95.4E−01	50.00E+02	3.1E+04
−20.63	−0.6	93.6E−01	60.00E+02	3.8E+04
−23.70	−0.8	91.5E−01	70.00E+02	4.4E+04
−26.63	−1.0	89.3E−01	80.00E+02	5.0E+04
−29.41	−1.2	87.0E−01	90.00E+02	5.7E+04
−32.04	−1.5	84.6E−01	10.00E+03	6.3E+04
−51.04	−4.2	61.7E−01	20.00E+03	1.3E+05
−61.19	−6.8	45.6E−01	30.00E+03	1.9E+05
−66.99	−9.1	35.0E−01	40.00E+03	2.5E+05
−70.55	−11.2	27.6E−01	50.00E+03	3.1E+05
−72.82	−13.1	22.3E−01	60.00E+03	3.8E+05
−74.26	−14.8	18.1E−01	70.00E+03	4.4E+05
−75.09	−16.6	14.8E−01	80.00E+03	5.0E+05
−75.43	−18.4	12.1E−01	90.00E+03	5.7E+05
−75.27	−20.3	97.1E−02	10.00E+04	6.3E+05
82.18	−26.7	46.2E−02	20.00E+04	1.3E+06
88.81	−17.4	13.5E−01	30.00E+04	1.9E+06
89.57	−13.5	21.1E−01	40.00E+04	2.5E+06
89.79	−11.0	28.2E−01	50.00E+04	3.1E+06
89.88	−9.1	35.0E−01	60.00E+04	3.8E+06
89.93	−7.6	41.7E−01	70.00E+04	4.4E+06
89.95	−6.3	48.3E−01	80.00E+04	5.0E+06
89.97	−5.2	54.8E−01	90.00E+04	5.7E+06
89.98	−4.3	61.2E−01	10.00E+05	6.3E+06
90.00	1.9	12.5E+00	20.00E+05	1.3E+07
90.00	5.5	18.8E+00	30.00E+05	1.9E+07
90.00	8.0	25.1E+00	40.00E+05	2.5E+07
90.00	9.9	31.4E+00	50.00E+05	3.1E+07
90.00	11.5	37.7E+00	60.00E+05	3.8E+07
90.00	12.9	44.0E+00	70.00E+05	4.4E+07
90.00	14.0	50.2E+00	80.00E+05	5.0E+07
90.00	15.0	56.5E+00	90.00E+05	5.7E+07
90.00	16.0	62.8E+00	10.00E+06	6.3E+07
90.00	22.0	12.6E+01	20.00E+06	1.3E+08
90.00	25.5	18.8E+01	30.00E+06	1.9E+08
90.00	28.0	25.1E+01	40.00E+06	2.5E+08
90.00	29.9	31.4E+01	50.00E+06	3.1E+08
90.00	31.5	37.7E+01	60.00E+06	3.8E+08
90.00	32.9	44.0E+01	70.00E+06	4.4E+08
90.00	34.0	50.3E+01	80.00E+06	5.0E+08
90.00	35.0	56.5E+01	90.00E+06	5.7E+08
90.00	36.0	62.8E+01	10.00E+07	6.3E+08
90.00	42.0	12.6E+02	20.00E+07	1.3E+09
90.00	45.5	18.8E+02	30.00E+07	1.9E+09
90.00	48.0	25.1E+02	40.00E+07	2.5E+09
90.00	49.9	31.4E+02	50.00E+07	3.1E+09

Fig. 3-24 Data results of computer simulation for Computer Application 3-1.

* **3-4.** Write a computer program that can calculate the ac resistance for a tunnel diode that is described by the characteristic equation of Computer Application 1-2 in Chapter 1. Use your program to produce a plot of r as a function of V for several values of the constants I_0, I_p, and V_p.

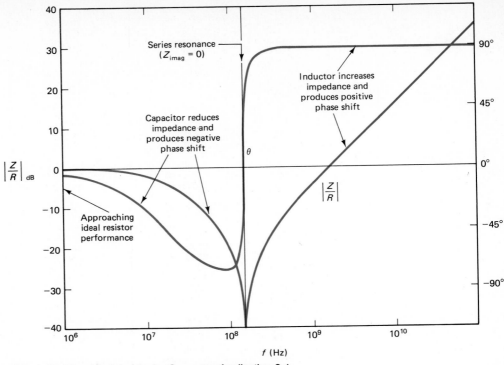

Fig. 3-25 Graphical results for Computer Application 3-1.

EXPERIMENTAL APPLICATION

3-1. The objective of this experiment is to measure the capacitance C and equivalent loss resistance R_C for an electrolytic capacitor as a function of f over a range of frequencies. The circuit of Fig. 3-26 can be used, with a dc source to maintain the correct polarity across the capacitor. Show that

$$\left|\frac{\mathbf{v}_{out}}{\mathbf{v}_{in}}\right| = \frac{|\mathbf{v}_{out}|}{|\mathbf{v}_{in}|} = \frac{1}{\sqrt{(\omega C R_{ext})^2 + (1 + R_{ext}/R_C)^2}}$$

$$\tan\theta = \frac{-\omega C R_{ext}}{(1 + R_{ext}/R_C)}$$

Fig. 3-26 Schematic for Experimental Application 3-1.

with $\mathbf{v}_{in}$ as the (real) reference voltage. For each value of f selected, show that C and R_C can be determined in terms of the measurable quantities $|\mathbf{v}_{out}|/|\mathbf{v}_{in}|$ and θ. Develop a procedure describing how you plan to collect and analyze the data and display your results to meet the experimental objectives. You may use a sine-wave generator and oscilloscope to determine C and R_C over the widest frequency range for which useful data can be collected. Plot and interpret your findings.

Discussion

Aluminum and tantalum electrolytic capacitors are represented in Fig. 3-27. Two conducting layers are separated by an electrolyte that can be in liquid, paste, or solid form. During fabrication, a large dc voltage is applied across the capacitor in the direction shown, building up a thin film of oxide on the positive end (anode). This oxide functions as the dielectric for the capacitor, with its thickness controlled by the magnitude and duration of the applied voltage and the temperature during formation.

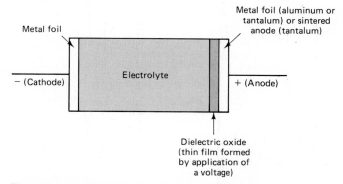

Fig. 3-27 Construction of an electrolytic capacitor.

This is a polarized capacitor and must be used with the anode connected to a positive voltage. Some oxide builds up on the negative conductor (cathode) whenever the applied voltage has a negative slope, reducing the effectiveness of the capacitor. Large reverse voltages can also cause excessive heating of the electrolyte and can destroy the capacitor.

The electrical properties of the capacitor are affected by the nature of the oxide and electrolyte. The value C can be a strong function of the temperature and frequency of operation.

The required analysis can be performed as follows:

$$\frac{\mathbf{v}_{out}}{\mathbf{v}_{in}} = \frac{(-jX_C) \parallel R_C}{R_{ext} + (-jX_C) \parallel R_C}$$

$$= \frac{X_C R_C [-jR_C R_{ext} + X_C(R_{ext} + R_C)]}{(R_C R_{ext})^2 + X_C^2(R_{ext} + R_C)^2}$$

$$\left| \frac{\mathbf{v}_{out}}{\mathbf{v}_{in}} \right| = \frac{|\mathbf{v}_{out}|}{|\mathbf{v}_{in}|} = \frac{R_C}{\sqrt{(\omega C R_C R_{ext})^2 + (R_{ext} + R_C)^2}}$$

$$\tan \theta = \frac{-R_C R_{ext}}{X_C(R_{ext} + R_C)} = \frac{-\omega C R_C R_{ext}}{R_{ext} + R_C}$$

so that

$$-\omega C = -\tan \theta (1/R_C + 1/R_{ext})$$

$$\frac{|\mathbf{v}_{out}|^2}{|\mathbf{v}_{in}|^2} = \frac{1}{(\tan^2 \theta)(1/R_C + 1/R_{ext})^2 R_{ext}^2 + (1 + R_{ext}/R_C)^2}$$

$$= \frac{\cos^2 \theta}{(1 + R_{ext}/R_C)^2}$$

$$R_C = \frac{R_{ext}}{[\cos \theta/(|\mathbf{v}_{out}|/|\mathbf{v}_{in}|) - 1]}$$

Collected data can be recorded in a table with columns labeled f (Hz), $|\mathbf{v}_{out}|$, $|\mathbf{v}_{in}|$, and θ, as shown in Fig. 3-28. The data analysis table can then include the headings $\cos \theta$, $|\mathbf{v}_{out}|/|\mathbf{v}_{in}|$, R_C, and C, as shown in Fig. 3-29. The component parameters C and R_C can then be plotted as a function of frequency, as shown in Fig. 3-30.

FIGURE 3-28
Typical Data Collected for Experimental Application 3-1

Data Point	f (kHz)	$|\mathbf{v}_{out}|$ (V)	$|\mathbf{v}_{in}|$ (V)	θ (degrees)
1	0.10	0.58	0.80	36
2	0.20	0.38	0.76	54
3	0.30	0.28	0.72	60
4	0.40	0.21	0.71	65
5	0.50	0.18	0.70	67
6	0.60	0.15	0.69	69
7	0.70	0.13	0.69	69
8	0.80	0.12	0.69	72
9	0.90	0.10	0.68	70
10	1.00	0.09	0.68	72

$R_{ext} = 100 \ \Omega$

FIGURE 3-29
Analysis for Experimental Application 3-1

Data Point	$\cos \theta$	$\dfrac{|\mathbf{v}_{out}|}{|\mathbf{v}_{in}|}$	R_C $(10^2 \Omega)$	C (μF)
1	0.809	0.725	8.6	12.9
2	0.593	0.500	5.4	13.1
3	0.508	0.382	3.0	11.9
4	0.426	0.298	2.3	12.1
5	0.397	0.250	1.7	11.7
6	0.358	0.220	1.6	11.3
7	0.355	0.193	1.2	11.0
8	0.309	0.169	1.2	11.2
9	0.335	0.150	0.8	11.1
10	0.309	0.138	0.8	11.0

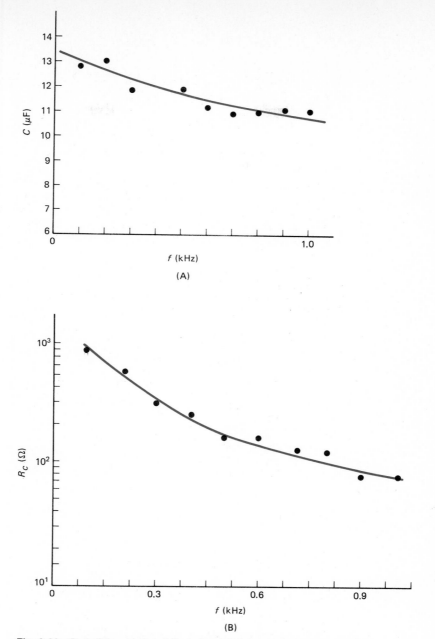

Fig. 3-30 Plot of the results of Experimental Application 3-1: (A) *C* as a function of frequency and (B) R_C as a function of frequency.

Half Wave Rectifier
Output

PART 2
SEMICONDUCTOR DIODES AND APPLICATIONS

Part 2 introduces some fundamental (two-terminal) semiconductor devices, shows how different models can be used to represent these devices, and demonstrates how the prediction of circuit performance depends on the models and applications. Chapters 4 and 5 are primarily concerned with devices and circuits that are used for waveform shaping. Such shaping operations are integral to both analog and digital electronic circuits, and are encountered in many different settings.

Chapter 4 describes the operation of several types of semiconductor diodes and develops models for use in circuit design. Junction, zener, and Schottky diodes are introduced as important nonlinear devices that can be used to shape waveforms to achieve performance objectives. Tunnel diodes are introduced to explore the significance of negative resistance. And photodiodes, light-emitting diodes, and laser diodes are discussed as building block elements for fiber-optic communication systems.

In Chapter 5, diodes are used in circuit applications. Emphasis is on developing a family of application models and on understanding the significance of waveform shaping. Rectifiers and filters are covered to provide insight into the properties of dc power supplies.

Fourier analysis is introduced in Chapter 5 as a way to understand how nonsinusoidal signals can be represented and to explicitly demonstrate how nonlinear devices and circuits change the frequency components present in a waveform. This ability to view circuit operation in the frequency domain is further applied in Chapter 11.

Chapter 5 also provides an introduction to SPICE, a commonly used computer program that can be a significant aid in evaluating the performance of electronic circuits. Throughout the text, problem solutions using SPICE (Simulation Program with Integrated Circuit Emphasis) are available to assist in understanding both circuit performance and the utility of CAD.

Chapters 4 and 5 provide preparation for the later discussion of transistors. Junction semiconductor properties are important in understanding both bipolar and junction field-effect devices.

CHAPTER 4
DIODE OPERATION

Chapter 2 provided an overview of the relationships between the atomic-level properties of pure materials and the resultant resistivity. As noted, the parameters that were introduced are important in determining tradeoffs between materials and technologies.

In order to fabricate semiconductor devices, the pure materials introduced in Chapter 2 must be modified by the controlled introduction of small quantities of selected impurities. These impurities change the macroscopic properties of the crystals under study and produce useful device characteristics. Where Chapter 2 describes the *intrinsic* properties of pure crystals, this chapter is concerned with *extrinsic* properties determined by the use of impurities added to the crystal.

Chapter 4 describes how various types of diodes can be produced by modifying the atomic-level properties of materials. Chapter 5 explores different types of models that can represent these diodes, and some of the model limitations.

4-1 SEMICONDUCTOR MATERIALS

An important class of materials is formed with atoms that have four valence electrons that link to form covalent bonds. Such atoms can be identified by again studying the periodic chart of the elements, as shown in Fig. 2-8. Observe that germanium (Ge) and silicon (Si) fall into this category. (Although other elements in this column also have four valence electrons, they do not have properties that are as useful for present purposes.) Such crystals can also be manufactured by combining equal parts of elements with three and five outer electrons, providing gallium arsenide (GaAs) and indium antimonide (InSb).

Ge, Si, and GaAs are used extensively in semiconductor devices today. They have different electrical properties because of the differences in their bonds and provide a wide range of performance characteristics for the designer.

Chapter 2 includes a description of the conduction processes in such materials and a calculation of intrinsic properties. The characteristics of pure crystals are now modified by considering what happens when various types of impurities are added.

4-2 DOPING

A useful range of semiconductor materials can be obtained by adding impurities to a crystal to obtain different numbers of mobile electrons and holes at a given T. With a process called *doping*, crystals can be designed with the desired electrical properties.

If a *donor* impurity with an extra (fifth) valence electron is added, the extra electron is weakly bound to the host atom and is easily excited to the conduction band. The situation is illustrated by the energy-level diagram in Figure 4-1(A). The energy level of the extra (doped) valence electrons is close to the conduction band, so that at room temperature almost all of the doped electrons are excited into the conduction band. If a donor doping level of N_d electrons per unit volume is used, the total electron density in the conduction band is increased above the intrinsic value n_i to $n = n_i + N_d$. Because of the increased number of electron carriers, this is an *n-type* (negative-carrier) semiconductor.

Since the Fermi level represents the average energy of the mobile carriers, it shifts upward toward the edge of the conduction band. Let $\mathcal{E}_F$ now be the doped

Fig. 4-1 Doping semiconductor materials: (A) n-type semiconductor and (B) p-type semiconductor.

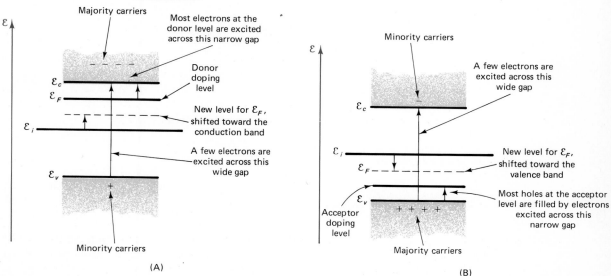

Fermi level and $\mathcal{E}_i$ be the original (intrinsic) Fermi level. The upward shift in the Fermi level, $\mathcal{E}_F - \mathcal{E}_i$, is a function of how much the doped electron density has exceeded the intrinsic value. From Appendix 7,

$$n = n_i e^{(\mathcal{E}_F - \mathcal{E}_i)/kT} \qquad \text{(n-type)} \qquad (4\text{-}1)$$

where n is the total electron carrier density.

If an *acceptor* impurity with a missing valence electron (only three present) is added to the crystal, a different situation arises. Each added atom is missing a bond and adds a hole to the crystal. As shown in Figure 4-1(B), the energy level of this impurity-caused hole is just above the valence band, and at room temperature almost all of the holes are filled with electrons from the valence band. If an acceptor doping level of N_a holes per unit volume is used, the total hole density in the valence band is increased above the intrinsic hole density n_i to $p = n_i + N_a$. Because of the increased number of hole carriers, this is a *p-type* (positive-carrier) semiconductor.

For this case, the downward shift in the Fermi level is related to the change in hole density by the relation

$$p = n_i e^{(\mathcal{E}_i - \mathcal{E}_F)/kT} \qquad \text{(p-type)} \qquad (4\text{-}2)$$

where p is the total hole-carrier density.

The product of the electron and hole densities in a semiconductor remains constant as the material is doped. From Eqs. (4-1) and (4-2),

$$np = n_i^2 \qquad (4\text{-}3)$$

From this equation, observe that doping of one type greatly reduces the density of the other type of carrier. In an n-type semiconductor, n is large so that p is small. The electrons are called the *majority* carriers, and the holes are called the *minority* carriers. In a p-type semiconductor, p is large and n is small. The holes are called the *majority* carriers, and the electrons are the *minority* carriers for this case. Figure 4-2 illustrates the charged particles that are present in both types of semiconductor.

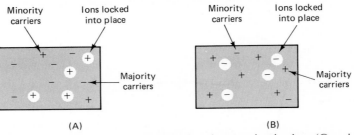

(A)   (B)

Fig. 4-2 (A) n-type and (B) p-type semiconductors, showing ions ($\oplus$ and $\ominus$) and carriers ($+$ and $-$).

Example 4-1   Doping

A sample of Si is doped with $N_d = 1.0 \times 10^{22}$ donor atoms per m³. What are the majority carriers in the resultant material, and what type of semiconductor has been created? What are the minority carriers, and what is their density at 300°K?

Solution   Donor atoms result in an increased number of electron carriers, which are the majority carriers, and an n-type semiconductor is obtained. The minority carriers are holes and the total carrier densities are related by

$$p = n_i^2/n$$

From Example 2-3 in Chapter 2, n_i at 300°K is $\cong 1.6 \times 10^{16}$ carriers/m³. Since the doping density is much greater than n_i, n is approximately equal to the doping level and

$$p \cong \frac{(1.6 \times 10^{16})^2}{1.0 \times 10^{22}} \cong 2.56 \times 10^{10} \text{ holes/m}^3$$

The electron density $n = 1.0 \times 10^{22}/\text{m}^3$ exceeds the hole density $p = 2.56 \times 10^{10}/\text{m}^3$ by many orders of magnitude. By reference to Example 2-3 in Chapter 2, it can be seen that both n and p are still many orders of magnitude below the density of atoms in the crystal. ■

4-3 DIFFUSION

Now consider what happens when an inhomogeneous (spatially varying) carrier density is introduced into a crystal through doping. When the concentration or density of one type of particle is much higher at one place in the crystal than in another, the particle distribution tends to spread out until a uniform concentration is obtained. The *diffusion* process takes place through thermal collisions and is encountered in everyday life. For example, when odors are introduced into a room, they quickly spread throughout the room. The spreading of molecules is due to the concentration gradient and thermal collisions.

The process is illustrated in Fig. 4-3. In both Figs. 4-3(A) and 4-3(B), carriers diffuse from higher to lower density locations. A diffusion current is generated in each case. In Fig. 4-3(A), the diffusion current density J_p is in the $+x$ direction, since this is the direction of hole flow. In Fig. 4-3(B), the diffusion current density J_n is in the $-x$ direction, opposite the direction of electron flow.

The diffusion current densities are given by

$$J_n = qD_n \frac{dn}{dx}$$

$$\tag{4-4}$$

$$J_p = -qD_p \frac{dp}{dx}$$

Fig. 4-3 Inhomogeneous carrier distributions: (A) holes and (B) electrons. (Continued on p. 104.)

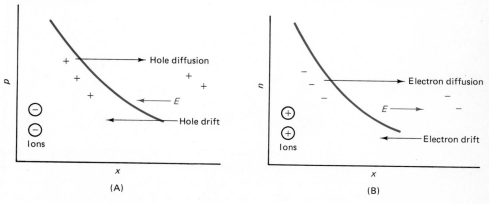

(A)

(B)

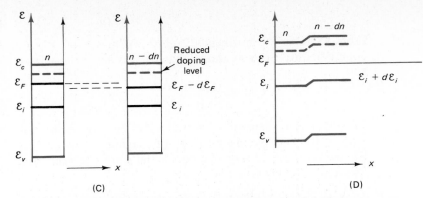

(C)   (D)

Fig. 4-3 (cont.)   Inhomogeneous carrier distributions: (C) thin slices of two homogeneous materials (with different doping levels) separated from one another. (D) Thin slices of homogeneous materials (with different doping levels) in contact with one another. As the material becomes very thin, these slices can be viewed as portions of a single inhomogeneous material. The Fermi level $\mathcal{E}_F$ must be constant throughout, so $\mathcal{E}_i$ becomes a function of location. An initial carrier flow between regions serves to equalize $\mathcal{E}_F$ values and changes the effective intrinsic carrier density level.

In each case, the current density is proportional to the rate of change of the carrier density in space (which is negative for the cases shown in Fig. 4-3). The diffusion constants D_n and D_p describe how rapidly the carriers diffuse as a result of the carrier variation. Typical values are given in Fig. 2-6.

As may be seen in Fig. 4-3, the carrier diffusion process produces a net electrostatic charge that gives rise to an electric field E. This electric field acts to produce a drift flow of carriers. The drift current densities are given by

$$J_n = \sigma_n E = qn\mu_n E$$
$$J_p = \sigma_p E = qp\mu_p E$$
(4-5)

Since no external electric field is present, the total current density must be equal to zero. This requires that

$$J_n \text{ (diffusion)} + J_n \text{ (drift)} = 0$$
$$qn\mu_n E = -qD_n \frac{dn}{dx}$$
(4-6)

and

$$J_p \text{ (diffusion)} + J_p \text{ (drift)} = 0$$
$$qp\mu_p E = qD_p \frac{dp}{dx}$$
(4-7)

An equilibrium doping distribution is maintained (with variations around an average value due to thermal collisions).

Figures 4-3(C) and (D) indicate how an inhomogeneous carrier density changes the energy level diagram of a material. Figure 4-3(C) shows the energy level diagrams for two (separate) materials with different doping. Figure 4-3(D) illustrates what happens when the two materials are joined. The Fermi level must be constant throughout the resultant material, so the edges of the valence and conduction bands must shift as shown.

From Eq. (4-1),

$$dn = n_i e^{(\mathcal{E}_F - \mathcal{E}_i)/kT}\left(-\frac{d\mathcal{E}_i}{kT}\right) = \frac{n}{kT}(-d\mathcal{E}_i) \tag{4-8}$$

Substituting into Eq. (4-6),

$$E = -\frac{D_n}{\mu_n}\frac{1}{n}\frac{dn}{dx} = \frac{D_n}{\mu_n kT}\frac{d\mathcal{E}_i}{dx} \tag{4-9}$$

The work done in moving a charge from one region to another must be equal to the average energy difference between the regions. Since

$$F\, dx = qE\, dx = d\mathcal{E}_i \tag{4-10}$$

the relationship

$$E = \frac{D_n}{\mu_n kT}qE \tag{4-11}$$

is obtained, so that

$$D_n/\mu_n = kT/q \tag{4-12}$$

This important result is called the *Einstein relation*. The same equation holds for both hole and electron carriers. Given a value of μ, D can be found through use of this equation. At room temperature,

$$D_n/\mu_n = D_p/\mu_p = kT/q = 0.026 \text{ V} \tag{4-13}$$

Example 4-2 Diffusion Constant

In a sample of Si at 300°K, the electron mobility is given by $\mu_n = 0.14 \text{ m}^2/\text{V}\cdot\text{s}$. Find the diffusion constant D_n.

Solution

$$D_n = \mu_n(kT/q) = (0.14 \text{ m}^2/\text{V}\cdot\text{s})(0.026 \text{ V}) = 3.6 \times 10^{-3} \text{ m}^2/\text{s}$$

which may be compared with the value in Fig. 2-6. ∎

Example 4-3 Effect of Temperature on the Diffusion Constant

In the region where thermal scattering dominates, $\mu \propto T^{-3/2}$. Given this relationship, how does D vary with T? If the temperature increases from 300 to 400°K, how does D change?

Solution From Eq. (4-12),

$$D = \mu\left(\frac{kT}{q}\right) \propto T^{-3/2}\left(\frac{kT}{q}\right) \propto T^{-1/2}$$

so that

$$\frac{D\ (400°\text{K})}{D\ (300°\text{K})} = \sqrt{\frac{300}{400}} = 0.87$$

Thus, the diffusion constant decreases at higher temperatures. ■

4-4 PN JUNCTION

One of the most important electronic devices is created by forming a junction between n- and p-type materials from a single crystal. One end of the crystal is doped with donor material, and the other end is doped with acceptor material. The n and p types must be joined in a single crystal to allow the carriers to move across the boundary region. The resulting device is a *junction semiconductor diode*.

An energy level diagram for the junction is shown in Fig. 4-4. Since the Fermi energy $\mathcal{E}_F$ must be constant throughout the material, the valence and conduction levels shift with respect to one another. A transition region is created between the n-type and p-type regions, where the carrier energy varies with location. This energy level diagram thus has two dimensions of significance (along both coordinate axes).

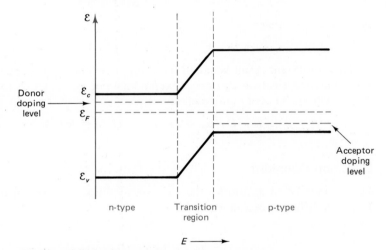

Fig. 4-4 Energy levels across a pn junction.

Consider what happens at the pn junction, as shown in Fig. 4-5. There are two basic processes that take place across the transition region. Since the n-type material has a higher density of electrons than does the p-type, electrons diffuse from n-type to p-type. Similarly, since the p-type material has a higher density of holes than does the n-type, holes diffuse from p-type to n-type.

This migration does take place. When electrons and holes meet in the transition region, they combine. The electron fills the hole and both carriers vanish. The result is a *depletion* region where the equilibrium negative- and positive-carrier densities are many magnitudes smaller than those in the n- and p-type regions.

As the charges combine, the material is left with ions locked into place at both ends of the depletion region. These ions create an electric field E across the deple-

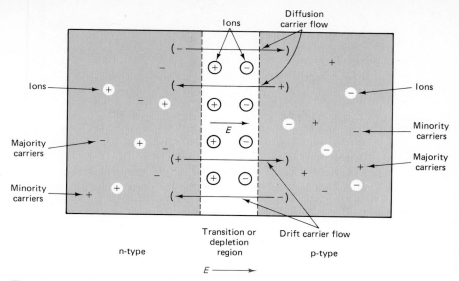

Fig. 4-5 Carrier flow across an unbiased pn junction.

tion region. Now consider the second process taking place across the junction, the drift of charge carriers. There are a few minority carriers in each semiconductor type. Whenever these minority carriers wander into the depletion region (through random thermal motion), the electric field catches them and pushes them across the junction, producing a drift current.

Since the crystal must be at equilibrium when there is no external source of energy, and the doping levels are arbitrary, the following relationships must hold:

Majority carrier diffusion current due to *holes* (p → n)

= Minority carrier drift current due to *holes* (n → p)

Majority carrier diffusion current due to *electrons* (n → p)

= Minority carrier drift current due to *electrons* (p → n)

The result is no net current flow.

In mathematical terms,

$$qD_p \frac{dp}{dx} = q\mu_p p E(x)$$

$$qD_n \frac{dn}{dx} = -q\mu_n n E(x) \tag{4-14}$$

By definition, the electric field intensity and the potential at any point are related by $E = -dV/dx$. Making this change of variable and using the Einstein relation [Eq. (4-12)], Eq. (4-14) becomes

$$\frac{-dV}{dx} = \frac{kT}{q} \frac{1}{p} \frac{dp}{dx} \tag{4-15}$$

Integrating from the n-type to the p-type, with equilibrium hole densities p_n and p_p,

$$-\int_{V_n}^{V_p} dV = \frac{kT}{q} \int_{p_n}^{p_p} \frac{dp}{p} \tag{4-16}$$

and

$$-(V_p - V_n) = \frac{kT}{q} \ln \frac{p_p}{p_n} \tag{4-17}$$

The integration has proceeded from the n-type to the p-type, in the direction of the electric field. Since the electric field is in the $+x$ direction, $dV/dx = -E$ must be negative. Therefore, $(V_p - V_n)$ is negative.

The difference in potential across the depletion region is defined as the *contact potential* $V_t = V_n - V_p$. The contact potential arises because the electric field in the depletion region produces different energy levels on the two sides of the region. The contact potential can be related to the energy levels in the diode through the relationship

$$V_t = \frac{(\mathcal{E}_F)_n - (\mathcal{E}_F)_p}{q} \tag{4-18}$$

where $(\mathcal{E}_F)_n$ and $(\mathcal{E}_F)_p$ are the Fermi levels for the n-type and p-type regions, respectively, as they would exist if the two regions were not in contact. For sufficient doping to cause $(\mathcal{E}_F)_n$ and $(\mathcal{E}_F)_p$ to approach the edges of the conduction and valence bands, respectively, V_t approaches $\mathcal{E}_g/q$ as an upper limit.

Using the definition of V_t,

$$p_n = p_p e^{-qV_t/kT} \tag{4-19}$$

This equation gives the relationship between the hole density in the n-type, the hole density in the p-type, and the contact potential between these two regions.

The contact potential can be found by combining Eqs. (4-3) and (4-17):

$$V_t = \frac{-kT}{q} \ln \frac{p_p}{p_n} = \frac{kT}{q} \ln \frac{p_p n_n}{n_i^2} \tag{4-20}$$

If the doping densities N_a and N_d and the intrinsic carrier density n_i are known, V_t can be calculated.

Example 4-4   Diffusion

Given a density of $1.0 \times 10^{22}/m^3$ for the majority carriers in the p-type and n-type regions of a Si junction diode (due to doping), estimate the hole density in the n-type region at 300°K (with no external voltage applied) by making use of the contact potential.

Solution   The objective is to find

$$p_n = p_p e^{-qV_t/kT}$$

The contact potential V_t is given by Eq. (4-20),

$$V_t = (0.026 \text{ V}) \ln \frac{(1.0 \times 10^{22})^2}{(1.6 \times 10^{16})^2} \cong 0.694 \text{ V}$$

So for a result,

$$p_n \cong 1.0 \times 10^{22}(e^{-0.694/0.026}) \cong 2.56 \times 10^{10}/m^3$$

which is in agreement with Example 4-1.

The doping densities N_d and N_a are determined during diode fabrication and are controllable parameters. The intrinsic carrier density n_i is a property of the crystal type and temperature. The desired contact potential V_t can be produced by varying n_n and p_p to meet design objectives. ■

With no applied potential, the equilibrium described above results from equalities in diffusion and drift currents. What happens to this balance if an external potential V is applied to the junction? If the positive polarity is applied to the p-type, the electric field across the depletion region is reduced. Diffusion then increases. The drift current is dominated by the availability of minority carriers in the depletion region. Since the electric field is essentially confined to the depletion region, the applied potential V does not bring additional minority carriers into the depletion region. The drift current thus remains approximately constant.

The effective potential across the junction becomes $(V_t - V)$, and the carrier density in the n-type next to the depletion region becomes

$$p_n' = p_p e^{-q(V_t - V)/kT} \qquad (4\text{-}21)$$

The result of the applied bias V is illustrated in Fig. 4-6. The applied voltage reduces the barrier height of the depletion region so that the diffusion current increases.

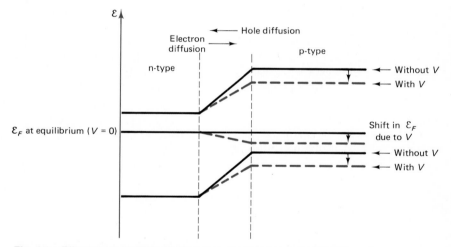

Fig. 4-6 Effect of an applied potential V on the junction energy levels.

The greater diffusion current moves more electrons from the n-type to the p-type and more holes from the p-type to the n-type. As carriers are *injected* across the depletion region, *excess* minority carrier densities develop close to the edges of the region. The injected carriers have lifetimes τ_p and τ_n before they recombine with the majority carriers in each region bulk region. These recombination processes take place as the carriers diffuse into the bulk regions, producing exponential decays in the injected carrier densities and limitations on penetration by these carriers into the bulk regions. (For a further discussion, see Exercise 4-6.)

As the carriers recombine, the n-type is left with a net positive charge on its ions and the p-type is left with a net negative charge on its ions. The electrons present in the wires to the applied voltage provide a flow of electrons to cancel out this building electrostatic charge, as shown in Fig. 4-7.

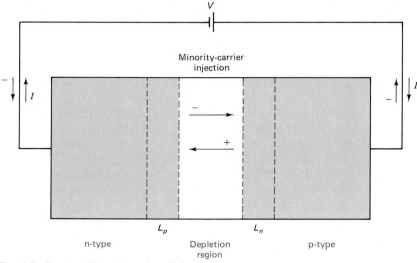

Fig. 4-7 Forward-biased junction diode.

The net result of all this activity is a flow of electrons between the voltage source and diode, producing a net current I. The current I that flows when V is applied depends on the excess numbers of carriers that find their way across the junction. With $V = 0$ (no applied voltage), Eq. (4-21) takes the form

$$p_n = p_p e^{-qV_i/kT} \qquad (4\text{-}22)$$

When V is applied, a net injection of holes into the n-type region takes place, so that next to the depletion region

$$p_n' = p_p e^{-qV_i/kT} e^{qV/kT} \qquad (4\text{-}23)$$

The excess density of minority carriers ($\Delta p = p_n' - p_n$) due to injection is thus given by

$$\Delta p_n = p_p e^{-qV_i/kT}(e^{qV/kT} - 1) \qquad (4\text{-}24)$$

In a pn junction, the diffusion currents due to electrons and holes are in the same direction (since the gradients are in the opposite directions). So from Eq. (4-4)

$$I = JA = qA\left(D_p\frac{\Delta p}{L_p} + D_n\frac{\Delta n}{L_n}\right) \qquad (4\text{-}25)$$

where L_p and L_n are characteristic diffusion lengths for the carriers. The diffusion lengths are a measure of how far the (excess) injected carriers diffuse into the p-type and n-type regions.

By the substitution of Eq. (4-24) for Δp_n and an equivalent expression for Δn_p into Eq. (4-25),

$$I = qA\left(D_p\frac{p_p}{L_p} + D_n\frac{n_n}{L_n}\right)e^{-qV_t/kT}(e^{qV/kT} - 1) \tag{4-26}$$

The parameter I_0 is defined as

$$I_0 = qA\left(D_p\frac{p_p}{L_p} + D_n\frac{n_n}{L_n}\right)e^{-qV_t/kT} \tag{4-27a}$$

which may also be written in the form

$$I_0 = qA\left(\frac{D_p}{L_pn_n} + \frac{D_n}{L_np_p}\right)n_i^2 \tag{4-27b}$$

since $\exp(-qV_t/kT) = n_i^2/p_pn_n$ from Eq. (4-20).

If V_0 is defined so that

$$V_0 = kT/q$$

then

$$I = I_0(e^{V/V_0} - 1) \tag{4-28}$$

which is the theoretical curve that describes the electrical behavior of the ideal pn *junction*. As will be seen, an accurate description of the complete pn junction *diode* behavior requires several modifications to this equation.

In the above derivation, Eqs. (4-19) and (4-21) were used to describe the increase in minority carrier (hole) density in the n-type region due to an increased flow of holes across the junction. The associated decrease in holes that takes place in the p-type region was neglected. For low levels of carrier flow, this is an acceptable approximation. However, for high levels of flow, the change in majority carrier density in the p-type is important.

Appendix 8 outlines a more general analysis that results in the following modification to Eq. (4-28):

$$I \cong I_0\frac{e^{qV/kT} - 1}{1 - e^{-2q(V_t-V)/kT}} \tag{4-29}$$

For $V \ll V_t$, the exponential term in the denominator can be neglected. However, as V approaches V_t, the denominator tends toward zero and I increases without bound. Thus, for any junction, the voltage V applied to the junction *can never exceed* the contact potential V_t. This is an important limiting case to remember.

By inspection of Eq. (4-29) it can be observed that the denominator becomes important when

$$(V_t - V) \lesssim kT/q$$

Therefore, this modification is small except when V is within $\lesssim 0.026$ V of V_t (at 300°K). This term is assumed negligible under normal conditions and usage.

4-5 DIODE CHARACTERISTIC CURVE

A plot of Eq. (4-28) is shown in Fig. 4-8. For $V/V_0 \gg 1$, the equation becomes

$$I \cong I_0e^{V/V_0} \qquad V/V_0 \gg 1 \tag{4-30}$$

so the current across a forward-biased junction varies exponentially with the applied

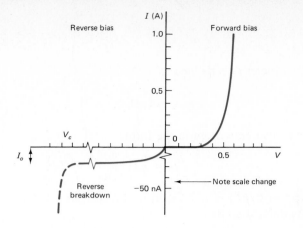

Fig. 4-8 Theoretical characteristic for an ideal pn junction.

voltage V. When the bias is reversed ($V/V_0 \ll 1$), the exponential term becomes very small so that

$$I \cong -I_0 \qquad V/V_0 \ll 1 \qquad (4\text{-}31)$$

The current across a reverse-biased junction is a small, constant value based on this equation.

Experimental values for I_0 can be found from diode data sheets that specify reverse (leakage) currents. Typical values of I_0 for small reverse-bias voltages (and at 27°C) are

$$\text{Ge: } I_0 \cong 10^{-6} \text{ to } 10^{-8} \text{ A}$$

$$\text{Si: } I_0 \cong 10^{-9} \text{ to } 10^{-16} \text{ A}$$

The theoretical pn junction equation provides only an approximate introduction to the characteristic curve of the junction diode. Two adjustments to Eq. (4-28) are important.

First note that during diffusion, carrier recombination can take place in both the bulk material of the diode and in the depletion region. Based on theoretical analysis, allowance for this effect can be made by introducing an *ideality factor n* that varies between 1 and 2, depending on the current level in the diode. The parameter V_0 can be replaced by the term nV_0 to account for recombination in the depletion region.

Carrier flow effects can also be noted outside of the depletion region. In the above discussion, there has been no mention of the *bulk resistance* of the n-type and p-type regions and the resistance caused by metal (wire)-semiconductor contacts. The IR_b drop across the bulk regions and the contacts reduces the current through the junction for a given applied voltage V below that predicted for Eq. (4-28). The voltage across the junction is $V - IR_b$.

By combining the impact of the ideality factor n and the bulk resistance R_b, Eq. (4-28) can be modified to

$$I = I_0(e^{(V-IR_b)/nV_0} - 1) \qquad (4\text{-}32)$$

which becomes the theoretical equation for a *junction diode*[1] (remembering also the modification to Eq. (4-28) that applies as V approaches V_t).

---

[1] This is a nonlinear equation since I appears on the left side of the equation and in the exponent.

To understand the significance of n and R_b, it is helpful to find out how these two parameters affect the slope of the characteristic curve shown in Fig. 4-8. To do this, first solve for V:

$$V = IR_b + nV_0 \ln (I/I_0 + 1) \qquad (4\text{-}33)$$

For $I \gg I_0$ (a situation that usually holds in circuit applications),

$$dV/dI \cong R_b + nV_0/I \qquad (4\text{-}34)$$

so that the slope of the characteristic curve is given by

$$\frac{1}{r} = \frac{dI}{dV} = \frac{1}{dV/dI} \cong \frac{1}{R_b + nV_0/I} \qquad (4\text{-}35)$$

Values of $R_b > 0$ and $n > 1$ reduce the slope of the characteristic curve for all current levels. The cumulative effect is to shift the characteristic curve to higher voltage levels and to decrease the slope of the characteristic curve at higher current levels, as shown in Fig. 4-9. The total voltage across the diode appears partially across the junction itself and partially across the bulk resistance R_b.

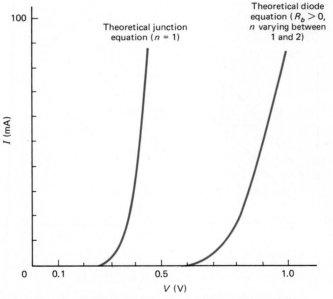

Fig. 4-9 The effects of R_b and n on the diode characteristic.

As noted in Eq. (4-13), $V_0 = kT/q = 0.026$ V at 300°K. This result was derived from an analysis of semiconductor device properties and is thus based on the models as developed. When devices are applied to circuit design and an attempt is made to predict circuit performance, a concern must be raised as to the number of significant figures to be used. Due to the various uncertainties that exist in the shape of the diode curve, it can often be reasonable to limit values of I_0 and V_0 to one significant figure. However, by convention, the value $V_0 = 0.026$ V is used in this text.

Example 4-5   Estimating AC Resistance

For a given circuit, it is desired to keep I sufficiently low to have the bulk resistance R_b of a device negligible [less than 10 percent of the second term in Eq. (4-34)]. If $R_b \cong 5.0\ \Omega$ and $n \cong 1.2$, what is the maximum current that can be used?

Solution   From Eq. (4-34),

$$R_b < 0.10(nV_0/I)$$

$$I < \frac{0.10V_0}{R_b} = \frac{(0.10)(0.026\text{ V})}{5.0\ \Omega} \cong 0.52\text{ mA}$$

The relative importance of the two terms in Eq. (4-34) depends on the current level. At lower current levels, the contribution due to the second term dominates, and R_b is negligible. At higher current levels, R_b dominates. The importance of R_b in circuit design thus depends on the current levels used. ∎

Since diode currents can cover a wide range of values, from microamperes to many amperes, the semilog plot of Fig. 4-10 is sometimes an effective way to study the characteristic curve.

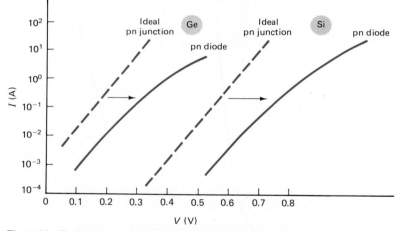

Fig. 4-10   Typical characteristics of pn junctions and pn diodes. The junction curves cannot extend beyond the values of V_t in each case.

Equation (4-28) can be written in the form:

$$\ln (I/I_0) = V/V_0 \qquad I \gg I_0 \tag{4-36}$$

Therefore, if a semilog plot is used, I is linearly related to V, as shown. The effects of n and R_b are also illustrated in this form with typical experimental characteristic curves.

Circuit design often requires an estimate for the approximate forward voltage drop across a diode. It is clear from the previous analysis that this drop depends on the current level through the diode. However, over a limited current range, the diode forward voltage can be approximated as a constant value. Based on the theoretical junction equation (with $n = 1$ and $R_b = 0$), the voltage across a Si junction is approximately 0.4 to 0.5 V for current levels in the 1 to 100 mA range. With the addition of the n and R_b factors as shown in Fig. 4-9, the typical voltage drop might be 0.6 to 0.8 V for a diode over this current range.

For Ge junctions, theoretical drops of 0.1 to 0.2 V over this current range (with $n = 1$ and $R_b = 0$) become 0.3 to 0.4 V with the n and R_b factors considered.

In electronics design, the voltage across a forward-biased Si diode is often taken to be $\cong 0.7$ V and the voltage across a Ge diode to be $\cong 0.3$ V. However, it is important to recognize that these values are valid over a limited current range and must be adjusted as necessary, depending on the circuit application.

4-6   DIODE CAPACITANCE

In the depletion region of a diode, the positive and negative mobile carriers have combined to produce a very small (equilibrium) carrier density. The dominant charge density is produced by the uncompensated ions in the lattice. The depletion region thus consists of two planes of charge, as shown in Fig. 4-11(A). From Chapter 3 this charge distribution functions as a capacitor with a value of

$$C = dQ/dV \qquad (4\text{-}37)$$

The electric field inside the depletion region takes the form shown in Fig. 4-11(B), with its maximum value on the interface between the n and p regions. The field reduces linearly to zero on the edges of the depletion region. (These electric field properties are discussed in Exercise 4-12.)

The field can be related to the voltage across the junction:

$$(V_t - V) = \int E \, dx = \text{Area under the } E\text{-versus-}x \text{ curve} \qquad (4\text{-}38)$$

Fig. 4-11   Calculating the width of the depletion region: (A) charge distribution in the depletion region ($N_a > N_d$) and (B) resultant electric field.

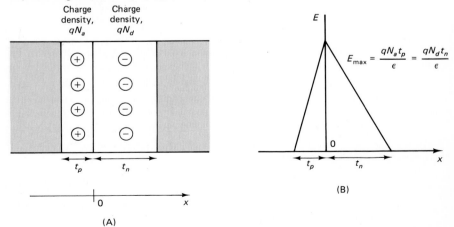

Therefore, using the definitions of Fig. 4-11,

$$(V_t - V) = \frac{qN_at_p}{\epsilon}\left(\frac{t_p}{2}\right) + \left(\frac{qN_dt_n}{\epsilon}\right)\left(\frac{t_n}{2}\right) \tag{4-39}$$

To maintain charge neutrality

$$N_at_p = N_dt_n, \tag{4-40a}$$

so that

$$(V_t - V) = \frac{qN_at_p^2}{2\epsilon}\left(1 + \frac{N_a}{N_d}\right)$$

$$t_p^2 = \frac{2\epsilon(V_t - V)}{qN_a(1 + N_a/N_d)} \tag{4-40b}$$

and similarly

$$t_n^2 = \frac{2\epsilon(V_t - V)}{qN_d(1 + N_d/N_a)} \tag{4-40c}$$

Equations (4-40 a to c) have many important implications. It may be observed that the total width of a depletion region ($t_p + t_n$) narrows under forward bias (approaching zero as V approaches V_t) and widens under reverse bias ($V < 0$). The component widths are inversely proportional to doping levels, so the depletion region will extend furthest into the least doped side of the junction. Transistor design and performance, as discussed in Chapters 6, 8, and 18, will be affected by these results. The total charge in either region is given by

$$Q = q(N_at_pA)$$

$$= qN_aA\left[\frac{2\epsilon(V_t - V)}{qN_a}\left(\frac{N_d}{N_a + N_d}\right)\right]^{1/2}$$

$$= A\left[2q\epsilon(V_t - V)\left(\frac{N_aN_d}{N_a + N_d}\right)\right]^{1/2}$$

so that

$$C = \frac{dQ}{dV} = A\left[\frac{q\epsilon}{2}\left(\frac{N_aN_d}{N_a + N_d}\right)\frac{1}{(V_t - V)}\right]^{1/2} \tag{4-41}$$

The result of plotting C as a function of V is shown in Fig. 4-12. In the region $V > 0$, the capacitance C of the diode is accompanied by high current levels. For $V < 0$, the diode can provide a tunable capacitance C that requires only a very small current flow. This device, called a *varactor*, is used in microelectronic circuits to create a variable capacitance C. A changing V across the device can produce a range of values for C and can thereby adjust circuit performance.

Example 4-6   Diode Capacitance

The effective capacitance of a (reverse-biased) diode junction must change by a factor of 1.5 in a given circuit application. Given $V_t = 1.0$ V and an initial voltage of 2.0 V across the diode, what must be the second (larger) voltage to produce the desired ratio?

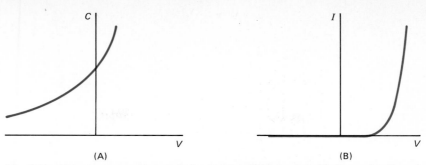

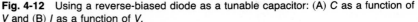

Fig. 4-12 Using a reverse-biased diode as a tunable capacitor: (A) C as a function of V and (B) I as a function of V.

Solution   From Eq. (4-41),

$$\left(\frac{C_1}{C_2}\right)^2 = \left(\frac{V_t - V_2}{V_t - V_1}\right) = 1.5^2$$

$$V_1 \cong -2.0 \text{ V}$$

$$[1.0 \text{ V} + (-V_2)] = 1.5^2(1.0 \text{ V} + 2.0 \text{ V})$$

$$V_2 \cong -5.8 \text{ V}$$

The above capacitance is associated with the ion charge distribution in the depletion region of the diode. For a forward-biased diode, an additional capacitive effect is observed due to the buildup of excess minority carriers near the depletion region. (This effect is described in Exercise 4-13.) In many forward-biased cases, the effective capacitance due to the excess charge buildup is much greater than the effective capacitance due to the ion charge layer.

The total effective capacitance of the diode limits the speed with which the diode changes from one bias state to another. For analog applications, the result can be a high-frequency cutoff associated with the device. For digital applications, when the voltage across a junction diode shifts from forward to reverse bias, the diode experiences a delay in moving from the *on* state to the *off* state. This delay is due to the time it takes to change the charge distribution in the device (see Exercises 4-6 and 4-13). This type of recombination delay is important in digital circuit applications and in other transient circuit applications. Switching delay times are further discussed in Chapter 15. ∎

4-7   TEMPERATURE EFFECTS

Both V_0 and I_0 are functions of temperature T. The effect of temperature on the junction diode can be described by finding how V changes with T for a constant current I, as shown in Fig. 4-13. If the temperature variation due to the term in parentheses in Eq. (4-27b) and the T^3 dependence of n_i^2 are neglected (a good approximation), then

$$I \cong I_c(\text{constant}) \times e^{-\mathcal{E}_g/kT} e^{qV/kT} \tag{4-42}$$

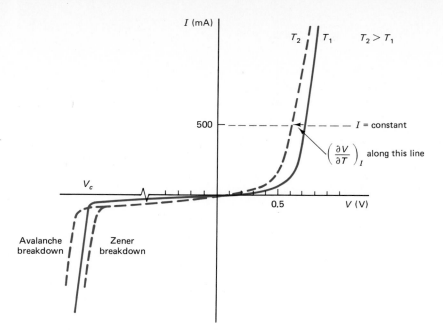

Fig. 4-13 Temperature dependence of the junction diode characteristic curve.

and

$$\left(\frac{\partial V}{\partial T}\right)_I \cong \frac{\partial}{\partial T}\left(\frac{\mathcal{E}_g}{q} + \frac{kT}{q}\ln\frac{I}{I_c}\right)_I \tag{4-43}$$

$$\cong \frac{K}{q}\ln\frac{I}{I_c} = \frac{V - \mathcal{E}_g/q}{T} \tag{4-44}$$

The value of $(\partial V/\partial T)_I$ can range from a maximum at low junction voltages to a minimum for larger junction voltages. The subsequent shift is as shown in Fig. 4-13.

Example 4-7 **Shift in Q Point Due to a Change in T**

Find $(\partial V/\partial T)_I$ for a Si diode at 300°K. Assume that $V = 0.5$ V.

Solution At room temperature,

$$\left(\frac{\partial V}{\partial T}\right)_I \cong \frac{0.5 \text{ V} - 1.1 \text{ V}}{300°\text{K}} \cong -2 \text{ mV/°K} \tag{4-45}$$

4-8 DIODE DATA SHEET

Figure 4-14 shows a data sheet for a typical diode. Diodes are often given a code number that begins with the characters $1N$, followed by several identifying digits. As illustrated, data sheets usually provide many different types of information. Absolute maximum ratings are used to set boundaries in circuit design. The mechanical specifications describe the mounting package and can include some indications of the ability of the diode to withstand external stress. The electrical specifications give

Features

· Metallurgical bond
· Planar passivated chip
· DO-7 package

Description

General-purpose low-current diode with high reliability characteristics

Absolute Maximum Ratings, at 25°C   **1N456**

Reverse working voltage . . . . 25 V . . . .
Peak reverse voltage . . . . 30 V . . . .
Average output current . . . . 90 mA . . . .
Surge current, 8.3 mS . . . . 700 mA . . . .
Operating temperature range . . . . −65 to +150°C . . . .
Storage temperature range . . . . −65 to +200°C . . . .

Mechanical Specifications

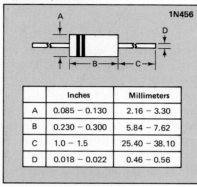

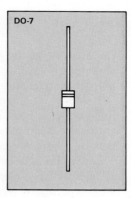

DO-7

	Inches	Millimeters
A	0.085 − 0.130	2.16 − 3.30
B	0.230 − 0.300	5.84 − 7.62
C	1.0 − 1.5	25.40 − 38.10
D	0.018 − 0.022	0.46 − 0.56

Electrical Specifications (at 25°C unless noted)

Type	Forward Voltage	Reverse Current	Reverse Current @ T_A = 150°C	Peak Reverse Voltage @ 100 μA
1N456	1.0 V @ 40 mA	25 nA @ 25 V	5 μA @ 25 V	30 V

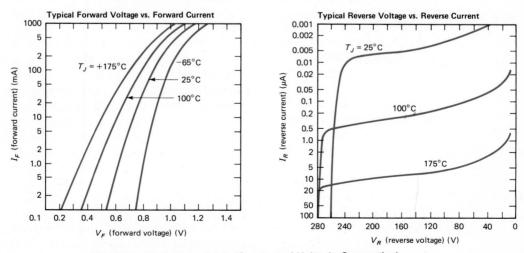

Fig. 4-14   Diode data sheet. (Courtesy of Unitrode Corporation).

typical operating-point conditions for use by the designer in determining the appropriateness of the diode for the planned design.

The characteristic curves illustrate the nonideal (real) characteristics of the diode in both the forward- and reverse-bias directions. The forward-bias curves would be straight lines based on Eq. (4-28). The changes in slope are due to the bulk resistance R_b and changes in the ideality factor n.

Based on the theoretical diode equation, the reverse current is a constant independent of V. As shown on the data sheet, this is an oversimplification. The current I_0 increases somewhat in the reverse-bias region until breakdown occurs and a large reverse current flows. This breakdown property is further described next.

4-9   ZENER DIODES

The analysis of the junction diode that was completed above predicts that for a reverse bias a small, constant current flow should be expected. This current is due to minority-carrier injection across the junction.

However, the experimental diode curves illustrated in Fig. 4-14 show an additional effect at large reverse voltages. When a critical bias voltage $-V_c$ is reached, a large increase in reverse current flow is suddenly observed. This flow arises from effects that were not considered in the earlier conduction model of the diode.

Zener breakdown occurs when the externally applied potential creates a sufficiently large electric field across the depletion region to pull bound electrons in the p-type region out of their host atoms and across the junction into the n-type region. The large reverse bias raises the potential energy of the electrons in the p-type region until they are able to tunnel across the depletion region to available states in the n-type region. A sudden increase in current is observed when the applied voltage is sufficient to produce the required ionization energy across the gap.

Another type of breakdown occurs due to the avalanche effect. If the electric field is large enough, electrons moving in the depletion region can gain enough energy between collisions to ionize neutral atoms when collisions do occur. The original mobile electron has now produced a second mobile electron and both carriers are capable of repeating the ionization process with new atoms. The result is a rapid increase in current at the critical-voltage level.

As illustrated on the diode data sheet shown in Fig. 4-14, a diode in the breakdown region behaves almost as a constant-voltage source. The voltage across the diode is approximately independent of the current through it.

In electronics, diodes that are intended for use in the breakdown region are often called zener diodes, with the name used in a general sense to indicate the region of the characteristic curve that is of interest. Zeners can be purchased with V_c values ranging from a few volts to several hundred volts. As shown in Chapter 5, zener diodes provide a useful way to develop a known reference voltage for a circuit.

4-10   SCHOTTKY DIODES AND OHMIC CONTACTS

Another type of diode can be created from metal-semiconductor junctions. Consider the device shown in Fig. 4-15(A), consisting of a metal bonded to an n-type crystal.

As noted earlier, all materials can be classified in terms of the average carrier en-

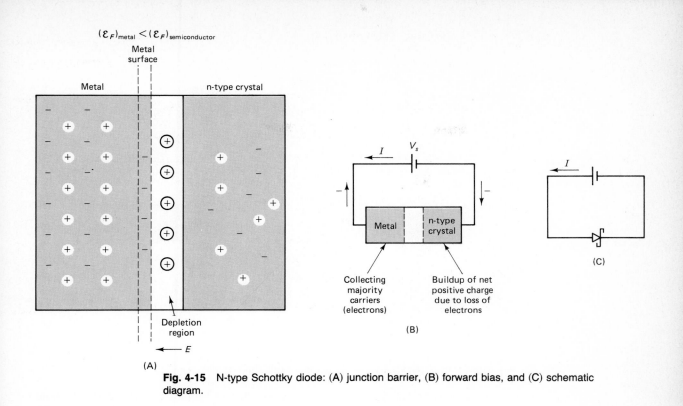

Fig. 4-15 N-type Schottky diode: (A) junction barrier, (B) forward bias, and (C) schematic diagram.

ergy in the material, called the Fermi energy $\mathcal{E}_F$. Assume that the materials in Fig. 4-15(A) are selected so that before they are connected, $\mathcal{E}_F$ in the semiconductor is greater than $\mathcal{E}_F$ in the metal.

When the bond is formed, the electron majority carriers in the n-type crystal start to flow from the higher-energy semiconductor states to the lower-energy metal states. A layer of electrons builds up on the metal surface. The portion of the semiconductor near the metal surface is *depleted* of carriers, leaving only a layer of positive ions locked into place. A depletion region is created with no carriers available.

An electric field develops from the positive ions to the electron layer. This field resists the electron flow and produces an equilibrium state with a high-resistivity region along the edge of the semiconductor. The Fermi levels in the two materials are the same when the equilibrium state is reached.

A forward bias causes an exponentially increasing current to flow, as shown in Fig. 4-15(B). A battery reduces the electric field in the transition region, allowing an increased flow of electron majority carriers from the semiconductor to the metal.

The reverse-bias current I_0 for the Schottky diode is due to minority carriers in the semiconductor and is usually larger than the reverse current for the pn junction diode. The symbol in Fig. 4-15(C) is used to indicate a Schottky diode.

A similar Schottky diode can be constructed from a metal and a p-type crystal with the Fermi level $\mathcal{E}_F$ initially higher in the metal, as shown in Fig. 4-16(A). Electrons flow from the higher-energy states in the metal to the lower-energy states in the semiconductor.

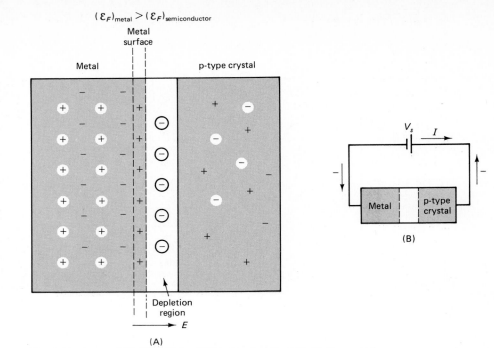

Fig. 4-16 P-type Schottky diode: (A) junction barrier and (B) forward bias.

As the electrons enter the p-type region, they combine with the majority hole carriers and produce a depletion region. A net positive charge builds up on the metal surface and a net negative ion layer builds up in the semiconductor. The resulting E field resists the carrier flow and produces an equilibrium state where the Fermi levels are the same in both materials.

A forward bias again causes an exponentially increasing current to flow. A battery reduces the electric field, allowing an increased flow of electrons from the metal to the p-type region. Majority carriers again produce the forward current in the Schottky diode. And, as before, the reverse current I_0 is larger than that for the pn junction diode. As described in Exercise 4-16, Schottky diodes have a much faster turn-off time than do pn junction diodes. Schottky diodes are commonly applied in digital circuits, as discussed in Chapter 15.

In order to make connections to semiconductor materials, it is important to be able to form ohmic metal-semiconductor contacts that do not behave like diodes. Figure 4-17 shows metal-semiconductor junctions that allow a symmetric flow of current in both directions. In each case, the Fermi levels are adjusted through a flow of electrons. In Fig. 4-17(A), electrons from the metal increase the number of majority carriers in the n-type region and no depletion region is formed. In Fig. 4-17(B), electrons flow from the p-type region to the metal surface and increase the number of holes in the p-type region. Again, no depletion region is formed.

No barrier develops to the current flow and the junctions behave as resistors. Small changes in the applied voltage V in either direction produce immediate carrier readjustments as the average energy levels shift. An applied voltage produces a flow of majority carriers in either direction without the nonlinear effects noted for the Schottky junctions.

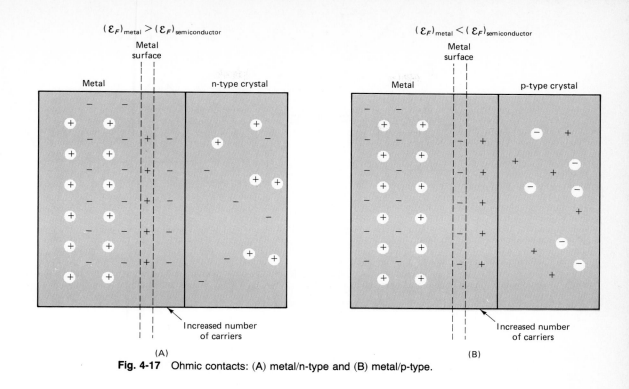

$(\mathcal{E}_F)_{\text{metal}} > (\mathcal{E}_F)_{\text{semiconductor}}$

$(\mathcal{E}_F)_{\text{metal}} < (\mathcal{E}_F)_{\text{semiconductor}}$

Fig. 4-17 Ohmic contacts: (A) metal/n-type and (B) metal/p-type.

4-11 TUNNEL DIODES

For typical semiconductor materials, as illustrated in Fig. 4-1, the Fermi level $\mathcal{E}_F$ lies just below $\mathcal{E}_c$ in the n-type and just above $\mathcal{E}_v$ in the p-type. However, when n-type and p-type semiconductors are very heavily doped, the situation can change. If the impurity atoms come sufficiently close to interact, they form a band structure that can overlap with the edges of the conduction and valence band. In such cases, $\mathcal{E}_F$ can be above $\mathcal{E}_c$ in the n-type and $\mathcal{E}_F$ can be below $\mathcal{E}_v$ in the p-type. When the two regions are in contact and the Fermi levels equalize, the results of Fig.4-18(A) (p. 124) are obtained.

In the n-type semiconductor, the energy levels between $\mathcal{E}_F$ and $\mathcal{E}_c$ are occupied by electrons since there are available states below $\mathcal{E}_F$. Similarly, the energy levels between $\mathcal{E}_v$ and $\mathcal{E}_F$ in the p-type semiconductor are occupied by holes. The result is called a *degenerate* semiconductor.

When a forward voltage is applied, the barrier height is reduced. There will be conduction band electrons in the n-type semiconductor with energies greater than the energies of available states (holes) in the p-type. If the depletion region is very narrow, these electrons can *tunnel* through the barriers, providing a net current flow.

As the forward bias increases, this current grows. Then a maximum tunneling current is obtained. If the bias is further increased, the available conduction states begin to decrease, reducing the current flow.

The result is the characteristic curve shown in Fig. 4-18(B). The tunneling cur-

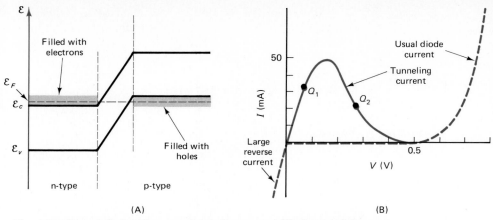

(A)

(B)

Fig. 4-18 Tunnel diode: (A) energy level diagram and (B) characteristic curve.

rent is added to the usual diode current (produced by diffusion and drift) to provide the combined forward-bias curve for the tunnel diode. The large reverse current flow develops because under reverse bias the number of states available for tunneling grows rapidly.

4-12 PHOTODIODES

A photodiode is a pn junction diode that is illuminated with light falling on the depletion region, as shown in Fig. 4-19(A). If the light is of frequency f, it consists of photons with energy $\mathcal{E}_{ph} = hf$, where h is Planck's constant $= 6.62 \times 10^{-34}$ J·s or 4.14×10^{-15} eV·s.

Fig. 4-19 Photodiode: (A) light incident on the pn junction, (B) characteristic curve, and (C) controlling current flow with a light source.

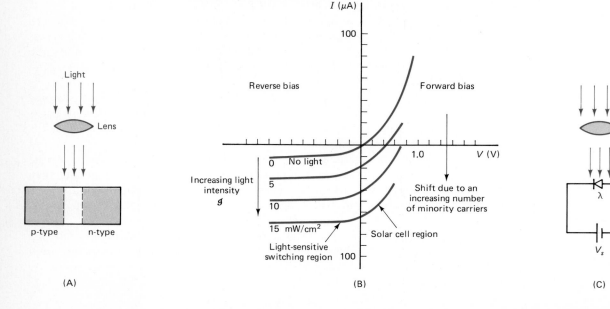

(A)

(B)

(C)

To understand what happens when the photons enter the crystal, first recall that thermal energy is constantly producing electron-hole pairs (this process is illustrated in Fig. 2-5). Thermal vibrations provide enough energy to excite electrons from the valence band to the conduction band.

If the photon energy $\mathcal{E}_{ph}$ is greater than the energy gap $\mathcal{E}_g$ between the valence and conduction bands, the photons also succeed in creating electron-hole pairs. As photons collide with bound electrons, they give up their energy (and vanish), propelling the electrons into the conduction band. In each case, a hole is left behind in the valence band.

Figure 4-5 shows the types of carriers that exist in a pn junction. When light shines on the depletion region, the number of carriers of both types increases. The majority-carrier (forward) current flow in the diode is only slightly changed, since there are already so many majority carriers present due to doping. However, the thermally generated number of minority carriers is very small, and the optically generated minority carriers can greatly exceed the thermally generated minority carriers.

The characteristic curve of the device takes on the form shown in Fig. 4-19(B). As the number of minority carriers increases, the characteristic curve is offset by an increasing reverse current flow.

A major use of the photodiode is based on the fact that a higher light intensity produces a greater reverse current flow. This proportional relationship enables the development of circuits that are sensitive to the amount of light incident on the device, as shown in Fig. 4-19(C). Light meters and light-sensitive switching circuits make use of this effect.

The power per unit area (P/A) incident on the depletion region can be related to the flow of photons through the equation

$$\mathcal{I} = P/A = \begin{array}{c} \text{number of photons} \\ \text{per second incident} \\ \text{on a unit area} \end{array} \times \begin{array}{c} \text{energy of} \\ \text{each photon} \end{array} \qquad (4\text{-}46)$$

To calculate P/A in mW/cm^2 as shown in Fig. 4-19, the photon energy must be given in Joules.

The photodiode can also be used as a power-generating device or solar cell if the battery is removed from the circuit of Fig. 4-19(C). The electron-hole pairs that are produced by the light create a difference in Fermi levels between the p-type and the n-type semiconductor, causing a net current to flow out of the p side of the diode. The voltage drop across the diode is in the usual direction, but the current flow is reversed. As a result, power is supplied from the diode to the circuit. The voltage associated with such a source is limited by the usual forward voltage for a diode (0.5 to 1.0 V for Si). Many diodes or solar cells must be connected to produce a significant power-generating capability.

4-13 LIGHT-EMITTING DIODES AND LASER DIODES

Light-emitting diodes (LEDs) and laser diodes are widely used as sources of electromagnetic radiation. When used in conjunction with photodiode detectors and fiber-optic transmission cables, these sources provide the means to create optical communication systems.

When a current flows through a diode, some of the electrons and holes combine in the depletion region. This recombination process was introduced in Section 2-4. The recombination releases energy that can be in the form of thermal vibrations in the crystal or as radiated photons. The probability of producing photons is much higher in semiconductors with direct transitions, such as GaAs. In these materials, light radiates outward from the depletion region when the diode conducts.

One of the problems with this type of source is that the photons tend to be radiated in many different directions, so that much of the generated light cannot be linked into a fiber-optic cable. This problem can be partially addressed by using various types of doping to limit the region over which recombination takes place and to produce changes in the index of refraction that can partially guide the radiated light. An efficient LED thus becomes quite complicated in geometry in order to maximize the fraction of light that is captured by the fiber-optic cable.

The optical power P radiated by an LED is proportional to the driving current I. If η is the fraction of the electrons that recombine in the depletion region, then

$$P = \begin{array}{c} \text{number of carriers} \\ \text{recombining per} \\ \text{second} \end{array} \times \begin{array}{c} \text{energy per} \\ \text{recombination} \end{array} \qquad (4\text{-}47)$$

$$= \eta(I/q)\mathcal{E}_g$$

The light from the LED can be changed by providing changes in I. For small changes in current around an average value, a linear analog source can be produced. If I is turned *off* and *on,* a digital source results.

In order to maximize coupling between the LED and a fiber cable, various types of lens arrangements can be used. The LED can be placed in a metal package with a glass lens on top, as shown in Fig. 4-20(A). However, much of the radiated power is still lost in such a configuration. An alternative is to attach a very small microlens to the LED and package the entire arrangement with a clear window.

The light from an LED is *incoherent,* in that the photons are randomly emitted due to uncorrelated recombination processes. A different type of source, with *coherent* radiation, results from the laser diode.

In a laser diode, the current flow through the diode is increased until the number of electrons in the conduction band becomes larger than the number left in the valence band. This produces a *population inversion,* with more electrons at the higher-energy levels and fewer at the lower levels.

If an optical cavity is created by cleaving the ends of the semiconductor crystal, so that most of the radiation is contained, the conditions for lasing can be met. The excited electrons couple and transition back to the lower-energy state in a coordinated process that results in photons being emitted *in phase.* The laser diode thus functions as a coherent-light source.

Because of the directional nature of the cavity and the coherent nature of the radiation, the laser diode output diverges much less than the light from an LED. Thus, coupling can be made quite efficient.

A typical laser radiates at several different frequencies at the same time, with most output power combined to a narrow frequency band. A lens can be used to couple to the cable, or the cable can be inserted directly into the package, as shown in Fig. 4-21.

The output power of a laser diode takes the form shown in Fig. 4-21(B). Below

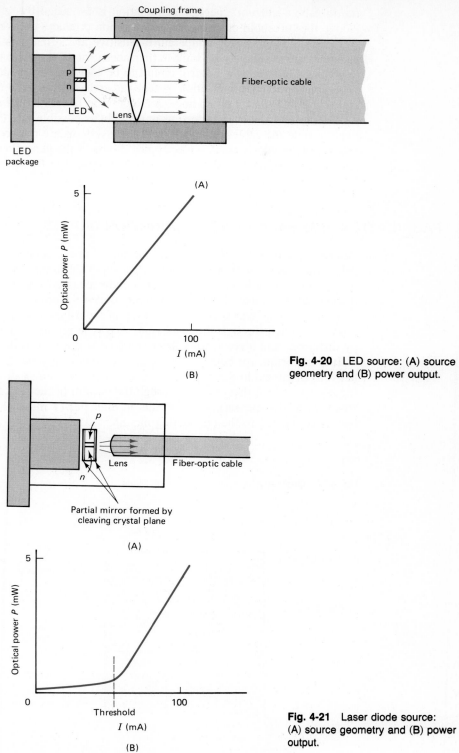

Fig. 4-20 LED source: (A) source geometry and (B) power output.

Fig. 4-21 Laser diode source: (A) source geometry and (B) power output.

the threshold current, incoherent radiation is produced by random recombination. Above the threshold current, the lasing action produces coherent radiation.

Laser diodes are very temperature-sensitive and often must be cooled. Complex packages are available with built-in thermoelectric coolers and thermocouple sensors to enable a temperature-control circuit to be established outside of the device.

A fiber-optic communication system can consist of an LED or laser diode source, the fiber cable, and a photodiode detector. The fabrication of such systems is complex because of the need to feed as much light as possible into the cable and to effectively couple the cable to the detector. Such systems are now commercially available and are being widely used in electronics. Some of the general considerations that enter into communication system design are discussed in Chapters 14 and 16.

4-14 FABRICATION AND PACKAGING OF JUNCTION DIODES

Discrete junction diodes are usually fabricated in the geometry shown in Fig. 4-22. Fabrication starts with a highly doped (n-type) substrate. By exposing the substrate to a heated gas, a layer of n-type semiconductor is grown expitaxially (maintaining the crystal structure) on top of the highly doped material.

An insulating layer is developed on top of the n-type region through oxidation and a hole is etched through the insulation. A p-type region is created through the hole by diffusion, and a conducting layer is added to connect to the p-type region.

It is important for both of the metal-semiconductor junctions to form ohmic contacts (as discussed in Section 4-10). A wire bond is used to attach the upper conducting layer to a lead that extends through the bottom of the package. This wire is very small, typically measuring a few mils in diameter (1 mil = 1/1000 in). The other lead is connected to the conducting mount.

Fig. 4-22 Geometry for a discrete junction diode.

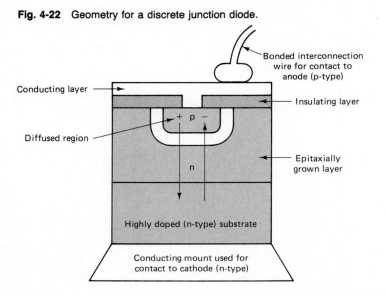

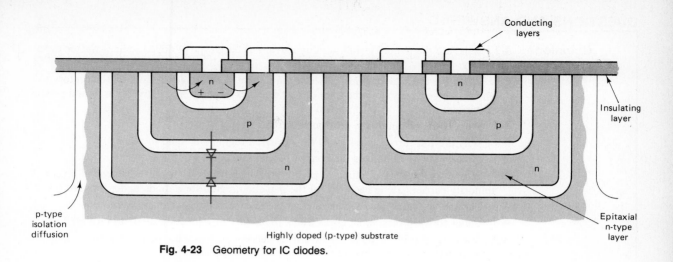

Fig. 4-23 Geometry for IC diodes.

The entire unit is hermetically sealed to protect it from environmental effects. The result can be regarded as a very simple hybrid microcircuit, with the junction diode integrated into a package that can then be inserted into a printed circuit board or used for other applications. In actual fabrication, hundreds of diodes are created simultaneously on the same substrate. The devices are then cut apart and individually packaged.

When diodes are used as building block elements in ICs, the arrangement of Fig. 4-22 is modified. If numerous devices of the type shown in this figure are formed on the same crystal, interference takes place among the devices. In order to provide the necessary isolation, back-to-back pn junctions are created around each diode, as shown in Fig. 4-23. Different p- and n-type diffusions are then used to create the diode itself.

After the epitaxial layer is grown, a p-type diffusion is used to separate the device regions. Conducting layers above the insulating layer are used to interconnect devices that are on the same IC chip. To connect the chip itself to external leads, wire bonds are again used.

The methods used for fabricating these diode geometries and the techniques of hybrid microelectronics are further discussed in Chapter 18. Figure 4-24 shows several different types of discrete diodes in their final packages.

Fig. 4-24 Various types of packages for discrete semiconductor diodes. (Courtesy of Motorola, Inc.)

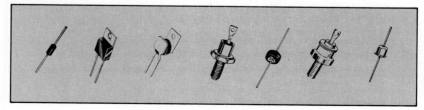

Questions
4-1. What are the majority and minority carriers in each type of doped semiconductor material?

4-2. What are the relationships between the doping densities for donors and acceptors (N_d and N_a, respectively) and the total carrier densities (n and p, respectively) in n- and p-type materials?

4-3. What is the origin of the diffusion current?

4-4. Why are the carrier densities in the depletion region not equal to zero?

4-5. Under what conditions is the contact potential V_t equal to $\mathcal{E}_g/q$?

4-6. Why does the finite lifetime of each carrier type result in the development of layers of minority carriers along each side of the depletion region for a diode under forward bias?

4-7. Can the voltage across a real junction diode exceed the contact potential?

4-8. Under forward bias, what effect limits the ability of a junction diode to turn *off*?

4-9. Does the shift in the diode characteristic curve with T produce a uniform effect along all I (= constant) curves? (See Fig. 4-13.)

4-10. What breakdown processes can take place in a zener diode?

4-11. What is the principal advantage of the Schottky diode?

4-12. What is a degenerate semiconductor?

4-13. What are the basic elements of a fiber-optic communication system?

Answers
4-1. In an n-type semiconductor, the electrons are majority carriers and the holes are minority carriers. In a p-type semiconductor, the holes are majority carriers and the electrons are minority carriers.

4-2. The total carrier densities are related to the doping densities through the relationships

$$n = N_d + n_i$$
$$p = N_a + n_i$$

For typical doping levels much greater than n_i, $p \cong N_a$ and $n \cong N_d$.

4-3. The diffusion current arises from the thermal energy of the carriers that tends to distribute them homogeneously in space.

4-4. Carrier densities exist in the depletion region due to the carriers in transit between the n- and p-type regions and the continuing thermal creation of carriers.

4-5. In general, the contact potential V_t is equal to the difference in the Fermi levels in the n- and p-type regions divided by the carrier charge. For heavy doping, the Fermi levels approximately coalign with the edges of the conduction and valence bands, resulting in $V_t \cong \mathcal{E}_g/q$.

4-6. Carriers that diffuse across the depletion region continue to diffuse into the n- and p-type regions until they recombine. The finite lifetimes determine how deeply the carriers penetrate and develop the layers shown in Fig. 4-25.

4-7. The voltage across a real junction diode can exceed the contact potential only to the degree that an IR_b voltage drop develops across the bulk resistance and contacts associated with the diode.

4-8. Under forward bias, the buildup of excess minority carriers in each region limits the diode turn-off switching time. (See Exercises 4-6 and 4-14 for further discussion.)

4-9. The shift in the diode characteristic curve with T is less for high current levels [as indicated by Eq. (4-44)].

4-10. A zener diode can make use of zener breakdown or avalanche breakdown.

4-11. The switching time of the Schottky diode is not limited by carrier recombination times (as is discussed in Exercise 4-4), so more rapid turn-off transitions can be accomplished.

4-12. In a degenerate semiconductor produced by heavy doping, the Fermi levels occur inside the conduction and valence bands.

4-13. The basic elements of a fiber-optic communication system are a light source (such as an LED or laser diode), a transmission medium (such as a fiber-optic cable), and a light-sensitive receiver (such as a photodiode).

EXERCISES AND SOLUTIONS

Exercises
(4-2)

4-1. When an intrinsic Si crystal is doped, the Fermi level $\mathcal{E}_F$ shifts toward the edge of the conduction band by 0.1 eV. What type of semiconductor results, and what is the doping density (at 300°K)?

(4-2)

4-2. Intrinsic Si at 300°K is doped with 10^{22} donors/m³. What types of majority and minority carriers result? Find the density of the minority carriers.

(4-3)

4-3. In a Si sample, the hole density doping varies with location according to the equation

$$p = p_0 e^{-x/L_p}$$

where L_p is a constant. Find the magnitude of the diffusion current density J_p associated with this gradient at $x = 0$ given

$$L_p = 1.0 \times 10^{-3} \text{ cm}$$

$$D_p = 12 \text{ cm}^2/\text{s}$$

$$p_0 = 1.0 \times 10^{22}/\text{m}^3$$

(4-3)

4-4. A Si diode is designed with 1.0×10^{22} acceptors/m³ and 1.0×10^{23} donors/m³. Find the contact potential at 300°K.

(4-4)

4-5. In a Si junction diode, the actual contact potential is measured to be 0.60 V. If the hole density of $1.0 \times 10^{20}/\text{m}^3$ is observed on the p-type side of the depletion region, what will be the hole density on the n-type side? Assume that no external voltage is applied and T = 300°K.

(4-4)

4-6. Graphically describe the various carrier density relationships that exist in a junction diode.

(4-4)

4-7. In order to calculate an approximate value for I_0, choose the values

$$D_p \cong D_n \cong 10 \text{ cm}^2/\text{s}$$

$$V_t \cong 0.60 \text{ V}$$

$$A \cong 0.10 \text{ mm}^2$$

$$p_p = n_n \cong 1.0 \times 10^{22}/\text{m}^3$$

$$L_p = L_n \cong 1.0 \times 10^{-4} \text{ cm}$$

Find I_0 at 300°K. Repeat the calculation for $V_t \cong 0.50$ V and note the sensitivity of the results to this parameter.

(4-4)

4-8. By direct substitution, find the value of $V_0 = kT/q$ at 400°K.

(4-4)

4-9. From the theoretical junction equation, find I at 300°K if $I_0 = 1.0 \times 10^{-8}$ A, $V = 0.40$ V, $R_b \cong 0$, and $n = 1.0$. Explain the results obtained.

(4-4) **4-10.** Assume that $I_0 = 1.0 \times 10^{-8}$ A, $V = 0.490$ V, and $V_t = 0.500$ V at 300°K. Find I, taking allowance of the effects caused by the general analysis of Appendix 8.

(4-5) **4-11.** For a junction diode at 300°K, $I_0 = 5.0 \times 10^{-7}$ A, $I = 10$ mA, $n = 1.2$, $R_b = 10$ Ω, and $V = 0.400$ V. (a) Find the voltage across the diode. (b) Find the ac resistance of the diode at this operating point.

(4-6) **4-12.** Derive the necessary equations to describe the electric field E in the depletion region of a diode.

(4-6) **4-13.** Apply the results of Exercise 4-6 to describe why a forward-biased diode can have an effective capacitance much greater than the capacitance produced by the depletion region.

(4-7) **4-14.** For a Ge diode, the junction contribution to V is usually $\cong 0.20$ to 0.30 V for currents of interest. Taking $V = 0.20$ V, estimate $(\partial V / \partial T)_I$ at 300°K for a diode with these properties.

(4-8) **4-15.** Estimate n and R_b for the 1N456 diode at 175°C (assuming n is approximately constant over the range shown in Fig. 4-14).

(4-10) **4-16.** Describe the charge distribution for an n-type Schottky diode and explain why this device has a much shorter *on-off* switching time than does the pn junction.

(4-11) **4-17.** Find the ac resistance r for the tunnel diode of Fig. 4-18 at points Q_1 and Q_2.

(4-12) **4-18.** For the photodiode of Fig. 4-19, with reverse bias and $\mathcal{I} = 10$ mW/cm², how many photons/s are incident on the photodiode if the frequency of the incident beam is $f = 1.0 \times 10^{15}$ Hz and $A = 1$ mm²?

(4-13) **4-19.** An LED is to produce the optical power of 1.0 mW. If $\eta = 0.10$ and $\mathcal{E}_g = 0.80$ eV, what current must flow through the diode?

(4-13) **4-20.** The laser diode of Fig. 4-21 is operated with a 70-mA current. If 5 percent of the radiated light passes through a fiber-optic cable and is incident on the photodiode of Fig. 4-19 (with an active area of 1 mm²), what current will flow on the (reverse-biased) photodiode?

Solutions **4-1.** The shift toward $\mathcal{E}_c$ requires a donor, producing an n-type semiconductor.

$$n \cong n_i e^{(\mathcal{E}_F - \mathcal{E}_i)/kT}$$

$$= (1.6 \times 10^{16}/\text{m}^3)(e^{0.10/0.026}) \cong 7.5 \times 10^{17}/\text{m}^3$$

4-2. Doping with donors produces an n-type semiconductor. The majority carriers are electrons, and the minority carriers are holes.

$$p = n_i^2/n = (1.6 \times 10^{16})^2/10^{22} \cong 2.56 \times 10^{10}/\text{m}^3$$

4-3. From Eq. (4-4),

$$J_p = -qD_p \frac{dp}{dx} = -qD_p \frac{d}{dx}(p_0 e^{-x/L_p}) = \frac{qD_p p_0}{L_p} e^{-x/L_p}$$

$$J_p\big|_{x=0} = \frac{(1.6 \times 10^{-19}\ \text{C})(12\ \text{cm}^2/\text{s})(1.0 \times 10^{-2}\ \text{m/cm})(1.0 \times 10^{22}/\text{m}^3)}{1.0 \times 10^{-3}\ \text{cm}}$$

$$\cong 1.9 \times 10^5\ \text{A/m}^2$$

4-4. From Eq. (4-20),

$$V_t = \frac{kT}{q} \ln \frac{p_p n_n}{n_i^2}$$

$$= (0.026\ \text{V}) \ln \frac{(1.0 \times 10^{22})(1.0 \times 10^{23})}{(1.6 \times 10^{16})^2}$$

$$\cong 0.75\ \text{V}$$

4-5. From Eq. (4-19),

$$p_n = p_p e^{-qV_t/kT}$$

$$qV_t/kT = 0.60 \text{ V}/0.026 \text{ V} = 23.08$$

$$p_n = (1.0 \times 10^{20}/\text{m}^3)e^{-23.08} \cong 9.5 \times 10^9/\text{m}^3$$

4-6. Refer to Fig. 4-25. With doping, n_n increases above n_i and $p_n = n_i^2/n_n$ decreases. Similarly, p_p increases above n_i and $n_p = n_i^2/p_p$ decreases. As shown in the figure, the p-type material is more heavily doped (indicated by the notation p^*).

For zero bias, the diffusion and drift currents exactly cancel, preventing excess carrier buildup on either side of the depletion region. For forward bias, the diffusion current grows and the drift current stays about the same. Layers of excess carrier density develop as shown. The $1/e$ thickness of the excess carrier layers (L_p, L_n) is determined by the recombination times (τ_p, τ_n). Since the layers decrease exponentially (for long diodes with-

Fig. 4-25 Graphical illustration of the carrier densities in a forward-biased junction diode. Under forward bias, the drift current components remain approximately the same, but the diffusion current components grow exponentially. The diffusion currents are not shown to scale because hole diffusion is much larger than electron diffusion for a p*n junction. The forward bias produces layers of excess charge carriers on both sides of the depletion region, as shown. For long p* and n regions (with lengths much longer than L_p and L_n), the density of excess carriers decreases exponentially away from the depletion region. For short p* and n regions (with lengths smaller than L_p and L_n), a substantial excess carrier density builds up throughout the regions. (Further detail on this latter case is presented in Chapter 6.) Under reverse bias, the diffusion currents rapidly decrease to zero, but a residual drift current remains due to minority-carrier flow. The result is a net decrease in the minority-carrier densities within a distance (L_pL_n) of the depletion region. This effect can be visualized in the figure by letting Δp and Δn take on negative values.

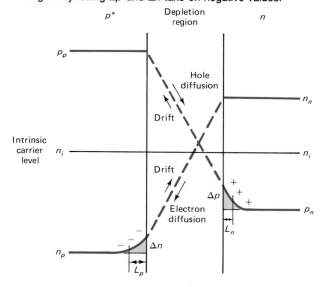

out end effects), then $\Delta p = (\Delta p)_0 e^{-x/L_p}$. From the continuity requirements, it can be shown that $L_p = (D_p \tau_p)^{1/2}$ and $L_n = (D_n \tau_n)^{1/2}$.

For reverse bias, the diffusion and drift currents caused by the doping decrease together due to the buildup of an electric field from the n to the p^* region. The diffusion current approaches zero and a residual drift current remains due to the available minority carriers in each region. Those minority carriers within a distance (L_p, L_n) of the depletion region may be caught by the reverse-bias electric field and pulled across the depletion region. The result is a small reverse-bias current (I_0).

4-7. From Eq. (4-27a),

(a)
$$I_0 = qA\left(D_p\frac{p_p}{L_p} + D_n\frac{n_n}{L_n}\right)e^{-qV_t/kT}$$

$$qV_t/kT = 0.60\text{ V}/0.026\text{ V} = 23.08$$

$$e^{-23.08} = 9.47 \times 10^{-11}$$

Then
$$I_0 = (1.6 \times 10^{-19}\text{ C})(0.10\text{ mm}^2)(1.0 \times 10^{-3}\text{ m/mm})^2$$
$$\times \frac{2 \times [10\text{ cm}^2/\text{s} \times 1.0 \times 10^{22}\text{ m}^3](9.47 \times 10^{-11})}{1.0 \times 10^{-4}\text{ cm}}$$

$$\cong 3.0 \times 10^{-9}\text{ A}$$

(b)
$$\frac{(I_0 \text{ at } 0.60\text{ V})}{(I_0 \text{ at } 0.50\text{ V})} = \frac{e^{-0.60/0.026}}{e^{-0.50/0.026}} = \frac{e^{-23.08}}{e^{-19.23}} = e^{-3.86} \cong \frac{1}{47}$$

$$I_0 \text{ at } 0.50\text{ V} = 47(I_0 \text{ at } 0.60\text{ V})$$

4.8. By direct substitution,
$$V_0 = \frac{kT}{q} = \frac{1.38 \times 10^{-23}\text{ J/}^\circ\text{K} \times 400^\circ\text{K}}{1.6 \times 10^{-19}\text{ C}} \cong 0.035\text{ V}$$

4-9. From Eq. (4-28),

(a)
$$I = I_0(e^{V/V_0} - 1)$$
$$= (1.0 \times 10^{-8})(e^{0.40/0.026} - 1)\text{ A}$$
$$\cong 0.48\text{ mA}$$

(b)
$$\frac{kT}{q} = (0.026)\frac{373^\circ}{300^\circ} \cong 0.032\text{ V}$$
$$I = (1.0 \times 10^{-8})(e^{0.40/0.032} - 1)$$
$$\cong 2.7\text{ mA}$$

Since I_0 is the same for both calculations, despite the change in T, the results of parts (a) and (b) must refer to two different devices. If T increases for a given device, I will always increase for a given V.

4-10. From Eq. (4-29),

$$I \cong \frac{I_0(e^{V/V_0} - 1)}{1 - e^{-2(V_t - V)/V_0}}$$

$$\cong \frac{(1.0 \times 10^{-8})(e^{0.490/0.026} - 1)}{1 - e^{-2(0.500 - 0.490)/0.026}} \cong \frac{1.5}{1 - 0.46}$$

$$\cong 2.8 \text{ A}$$

4-11. (a) From Eq. (4-32),

$$I = I_0(e^{(V-IR_b)/V_0} - 1)$$

$$\ln\left(\frac{I}{I_0} + 1\right) = \frac{V - IR_b}{V_0}$$

$$V = IR_b + V_0 \ln\left(\frac{I}{I_0} + 1\right)$$

$$\cong (10 \text{ mA})(10 \ \Omega) + (0.026 \text{ V}) \ln\left(\frac{10 \times 10^{-3}}{5.0 \times 10^{-7}}\right)$$

$$\cong 0.36 \text{ V}$$

(b) From Eq. 4-35

$$r \cong R_b + \frac{nV_0}{I}$$

$$= 10 \ \Omega + \frac{(1.2)(0.026 \text{ V})}{10 \times 10^{-3} \text{ A}} \cong 13 \ \Omega$$

4-12. Figure 4-26 illustrates the electric fields that arise due to planar charge distributions of thickness t_p and t_n. The charge densities are taken to be equal to the donor and acceptor doping densities ($N_d \cong n_n$, $N_a \cong p_p$). From Gauss' law, the vector electric field $\overline{E}$ outside of a region can be related to the charge contained in the region.

Fig. 4-26 Electric fields due to planar charge distributions.

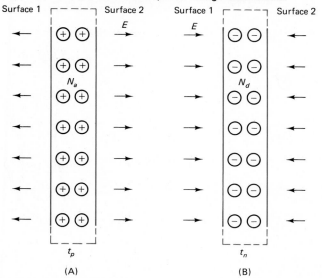

(A)   (B)

$$\int_{\text{Surface}} \overline{E} \cdot \hat{n} \, dA = Q/\epsilon$$

where $\hat{n}$ is a unit vector normal to the surface at each small surface area dA. Consider the surface shown in Fig. 4-26(A). Assume that $t = t_p$ or t_n is very small with respect to the plate dimensions. Then, the resultant fields are approximately parallel to the x axis and constant in magnitude as shown. Performing the above integration,

$$\int_{\text{Surface}} \overline{E} \cdot \hat{n} \, dA = \int_{\text{Surface 1}} E \, dA + \int_{\text{Surface 2}} E \, dA = 2EA$$

Given the enclosed charge is $Q = qN\mathcal{V} = (qN)(At)$,

$$2EA = \frac{q}{\epsilon} NAt$$

$$E = \frac{qN}{2\epsilon} t$$

so that

$$E_p = \frac{qN_a t_p}{2\epsilon} \qquad \text{and} \qquad E_n = \frac{qN_d t_n}{2\epsilon}$$

In order to maintain charge neutrality in the depletion region, $N_a t_p = N_d t_n$ and the maximum electric field is given by

$$E = E_p + E_n = \frac{qN_a t_p}{\epsilon} = \frac{qN_d t_n}{\epsilon}$$

Figure 4-27 shows the component fields that will result in the depletion region.

Fig. 4-27 Component fields in the depletion region (solution for Exercise 4-12).

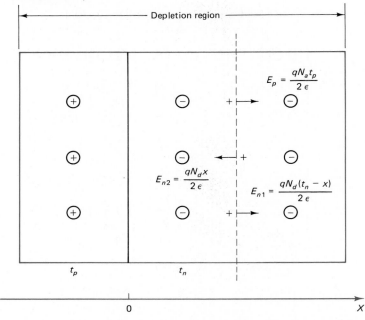

136   Chapter 4   Diode Operation

At an arbitrary location $x > 0$, the positive charge and that portion of the negative charge on the positive side of position x are added together, and the portion of the negative charge on the negative side of position x is in the opposite direction. As a result,

$$E(x) = \frac{qN_a t_p}{2\epsilon} + \frac{qN_d}{2\epsilon}(t_n - x) - \frac{qN_d x}{2\epsilon}$$

$$= \frac{qN_a t_p}{2\epsilon} + \frac{qN_d}{2\epsilon}(t_n - 2x) \qquad x > 0$$

The field takes the linear form shown in Fig. 4-11(B).

4-13. As may be noted in Fig. 4-25, thin layers of minority carriers build up on each side of the depletion region. Under forward bias, these layers of charge can be sufficiently dense to produce the dominant capacitive effect for the diode. The delay time associated with diode turn-off is often due to the presence of these carriers. The distributions decay with characteristic recombination times τ_p and τ_n until the diode returns to its *off* condition.

4-14. From Eq. (4-44),

$$\left(\frac{\partial V}{\partial T}\right)_I \cong \frac{V - \mathcal{E}_g/q}{T} \cong \frac{0.20 \text{ V} - 0.67 \text{ V}}{300°\text{K}} \cong -1.6 \text{ mV/°K}$$

4-15. From Eq. (4-35),

$$r \cong R_b + \frac{nV_0}{I}$$

Refer to Fig. 4-14 to obtain the following data points:

$$I_F = I_1 = 500 \text{ mA} \qquad \text{and} \qquad r_1 \cong 0.10 \text{ V}/400 \text{ mA} \cong 0.25 \text{ }\Omega$$

$$I_F = I_2 = 1 \text{ mA} \qquad \text{and} \qquad r_2 \cong 0.10 \text{ V}/1.5 \text{ mA} \cong 67 \text{ }\Omega$$

Therefore,

$$nV_0 = (r_1 - R_b)I_1 = (r_2 - R_b)I_2$$

$$R_b = \frac{I_1 r_1 - I_2 r_2}{I_1 - I_2}$$

$$R_b = \frac{(0.25)(0.50) - (1.0 \times 10^{-3})(67)}{0.50} \cong 0.12 \text{ }\Omega$$

$$nV_0 \cong (r_1 - R_b)I_1 \cong (r_2 - R_b)I_2$$

$$nV_0 \cong 0.065 \text{ V} \cong 0.067 \text{ V}$$

confirming a constant n as a reasonable approximation.

$$n \cong 0.065/0.026 \cong 2.5$$

4-16. As shown in Fig. 4-15, the electrons in the n-type semiconductor have a higher $\mathcal{E}_F$ than do those in the metal. Equilibrium is obtained when electron flow from the semiconductor to the metal builds up a charge layer on the metal surface. Since the metal has no energy gap and no holes, there is no buildup of excess carriers on the semiconductor side of the depletion region. Under forward bias, the electron density on the metal surface increases. When this bias is removed, the recombination time for the (majority carrier) electrons in the metal is very fast. Therefore, the *on-off* switching time for the diode is much reduced.

4-17. From the graphical measurements shown in Fig. 4-28:

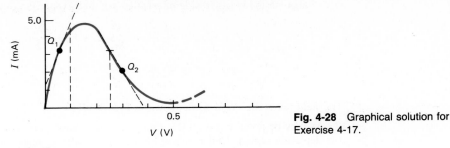

Fig. 4-28 Graphical solution for Exercise 4-17.

At Q_1,

$$r \cong \frac{0.10 \text{ V}}{3.5 \text{ mA}} \cong 29 \ \Omega$$

At Q_2,

$$r \cong \frac{-0.15 \text{ V}}{3.5 \text{ mA}} \cong -43 \ \Omega$$

4-18. From Eq. (4-46),

$$\mathcal{I} = P/A = \frac{\text{photons}}{(\text{sec})(\text{area})} \times \text{energy}$$

$$\frac{\text{photons}}{\text{sec}} = \frac{\mathcal{I}(\text{area})}{\text{energy}} = \frac{10 \times 10^{-3} \text{ W/cm}^2/(0.1 \text{ cm})^2}{(6.62 \times 10^{-34} \text{ J} \cdot \text{s})(1.0 \times 10^{15}/\text{s})}$$

$$\cong 1.5 \times 10^{14}/\text{s}$$

4-19. From Eq. (4-47),

$$P = \eta \left(\frac{I}{q}\right) \mathcal{E}_g$$

$$I = \frac{P}{\eta (\mathcal{E}_g/q)} = \frac{1.0 \text{ mW}}{(0.10)(0.80 \text{ V})} \cong 12 \text{ mA}$$

4-20. From Fig. 4-21,

$$\text{Optical power} \cong 1.5 \text{ mW}$$

$$\mathcal{I} \cong \frac{(0.05)(1.5 \text{ mW})}{0.01 \text{ cm}^2} \cong 7.5 \text{ mW/cm}^2$$

$$I \cong 40 \ \mu\text{A}$$

PROBLEMS

(4-2)   **4-1.** Doped n-type Si at 300°K has a density of carriers $n = 1.0 \times 10^{22}/\text{m}^3$. Find the shift in the Fermi level in eV due to the doping.

(4-2)   **4-2.** Doped p-type Si at 300°K has a density of carriers $p = 1.0 \times 10^{22}/\text{m}^3$. Find the shift in the Fermi level in eV due to the doping.

(4-2) **4-3.** Find the density of minority carriers for the materials in problems 4-1 and 4-2.

(4-2) **4-4.** You observe in doped Si that the density of minority carriers is 1.0×10^{12} electrons/m³. What is the density of majority carriers, and what type of semiconductor material has been fabricated?

(4-3) **4-5.** An electric field $E = 50$ V/m is set up in acceptor-doped Si with $p = 5.0 \times 10^{21}/m^3$ at 300°K. Find the drift current density.

(4-3) **4-6.** Find the diffusion constant D_p for a sample of Ge at 300°K, using the Einstein relation and $\mu_p = 0.19$ m²/V·s. Compare your results with the data in Fig. 2-6.

(4-3) **4-7.** Calculate the ratio D_p/μ_p for each semiconductor material in Fig. 2-6. How would these ratios change at $T = 100°C$?

(4-4) **4-8.** For a pn junction, $(\mathcal{E}_F)_n - (\mathcal{E}_F)_p = 0.10$ eV. Find the contact potential.

(4-4) **4-9.** A junction is created by doping the n- and p-type regions so that $p_p = n_n = 1.0 \times 10^3 n_i$. Find the contact potential at 300°K.

(4-4) **4-10.** A Si diode is doped with 10^{22} acceptors/m³ and 10^{23} donors/m³. Find the contact potential at −50°C.

(4-4) **4-11.** The following data are obtained for the operation of a junction diode. Estimate I_0 for the diode.

$$D_p = 10 \text{ cm}^2/\text{s} \qquad D_n = 5 \text{ cm}^2/\text{s} \qquad V_t \cong 0.80 \text{ V}$$

$$p_p = 10^{23}/\text{m}^3 \qquad n_n = 10^{22}/\text{m}^3 \qquad A = 0.20 \text{ mm}^2$$

$$L_p = 2.0 \times 10^{-4} \text{ cm} \qquad L_n = 5.0 \times 10^{-4} \text{ cm}$$

(4-4) **4-12.** By direct substitution, find the value of kT/q at 100°C.

(4-5) **4-13.** From the theoretical junction equation, find the current I at 80°C if $I_0 = 1.0$ nA, $V = 0.55$ V, $R_b \cong 0$, and $n = 1.2$.

(4-5) **4-14.** Given

$$I_0 = 5 \times 10^{-11} \text{ A} \qquad R_b = 5.0 \text{ }\Omega$$

$$I = 1.0 \text{ mA} \qquad n = 1.2$$

find V for a junction diode at 0°C.

(4-5) **4-15.** Graphically find R and r for the theoretical curves in Fig. 4-9 at $I = 40$ mA. If n is approximately 1.0 in both cases, what value for R_b results?

(4-6) **4-16.** For a junction diode,

$$N_a = 1.0 \times 10^{21}/\text{m}^3 \qquad t_p = 1.0 \times 10^{-5} \text{ cm} \qquad \epsilon_r = 12$$

Find the maximum value of the electric field in the depletion region.

(4-6) **4-17.** For a Si varactor at 300°K, you desire to produce $C = 100$ pF, given

$$N_d = N_a = 1.0 \times 10^{22}/\text{m}^3 \qquad \epsilon = 12\epsilon_0$$

$$A \cong 1.0 \text{ mm}^2$$

What voltage should be applied across the varactor?

(4-6) **4-18.** The voltage across a varactor of Problem 4-17 changes to −10 V. Find the percentage change in capacitance C with respect to the initial value.

(4-7) **4-19.** Show that Eq. (4-44) follows from Eq. (4-42).

(4-7) **4-20.** For a Si diode, $V_t \cong 0.8$ V. Estimate $(\partial V/\partial T)_I$ at 100°C.

(4-7) **4-21.** For a given diode, $(\partial V/\partial T)_I \cong -1.5$ mV/°C for $V = 0.60$ V and $T = 300°K$. Find the energy gap $\mathcal{E}_g$.

(4-8) **4-22.** Find the leakage current for the 1N456 diode at $T = 100°C$ and a reverse voltage of -150 V.

(4-8) **4-23.** Find the ac resistance r for the 1N456 diode for $I = 100$ mA and $T = 100°C$.

(4-9) **4-24.** Estimate the ac resistance r of the 1N456 diode at $T = 100°C$ in the reverse-biased breakdown region.

(4-9) **4-25.** A zener diode has a rating $V = 6.0$ V. Draw the (ideal) characteristic curve that would be associated with such a device, and show the correct biasing for the device.

(4-12) **4-26.** What is the minimum frequency for light that can be used to create electron-hole pairs in Si?

(4-12) **4-27.** Assume that 1.5×10^{18} photons/s are incident on a photodiode. If each photon has the minimum frequency necessary to create electron-hole pairs in Ge, what power is incident on the area?

(4-13) **4-28.** An LED with $\eta = 0.15$ and $I = 10$ mA has $\mathcal{E}_g = 1.11$ eV. What optical power is produced?

(4-13) **4-29.** The laser diode of Fig. 4-21 produces a coherent output power of 4.0 mW. (a) What current must flow through the diode? (b) If the current drops to 40 mA, how much coherent power is produced?

4-30. A standard design package is available to produce a hybrid microcircuit (discussed in Chapter 18) if I_0 is known within 20 percent. For a greater variation in I_0, custom design is necessary. Circuit cost is dominated by the engineering time required to produce the circuit. Using the standard package requires 40 hours of staff time at $20/h. The custom design requires 60 hours at the same rate. The circuit production run totals 1000 circuits. How much design cost must be added to the circuit in each case to recover this cost?

4-31. The estimated lifetime for a new electronics product line depends on its design as shown in Fig. 4-29. Version A peaks sooner but loses market share more rapidly. Version B rises more slowly but lasts longer. (a) Which version will sell the largest number of circuits? (b) Assume that each product sells for $80.00 in 1987. If the inflation rate i averages 0.06/yr (6 percent), and the selling cost increases with inflation, what will be the selling price in 1997? (The selling price grows over n years according to the factor $(1 + i)^n$.)

Fig. 4-29 Data for Problem 4-31.

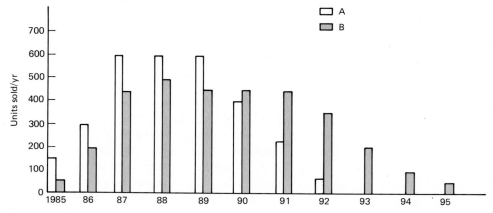

4-1. The nonlinear diode characteristic curve of Eq. (4-32)

$$I = I_0(e^{(V-IR_b)/nV_0} - 1)$$

may be solved by using (iterative) numerical means. The results of such a computer simulation are shown in Figure 4-30 with $I_0 = 1.0 \times 10^{-11}$ A, $n = 1.2$, $R_b = 10$ Ω, and $V_0 = 0.03$ V. (a) Confirm by direct substitution that the computer result is correct. (b) Approximately what current and voltage levels define the boundary between regions for which R_b is and is not important? (c) Why can such a large voltage (up to 10 V) be dropped across the diode?

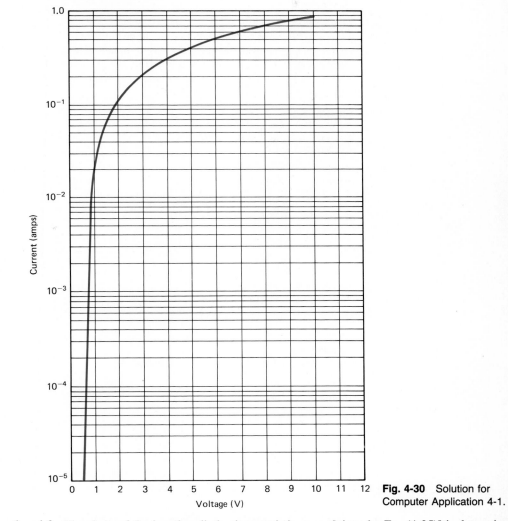

Fig. 4-30 Solution for Computer Application 4-1.

* **4-2.** The slope of the junction diode characteristic curve [given by Eq. (4-35)] is determined by the parameters I_0, V_0, and n. Write a computer program that can perform a parametric study of the effects of each parameter on the properties of the curve produced. Graph and interpret your results.

4-1. The objective of this experiment is to explore the characteristic curves of semiconductor diodes and to gain an understanding of the ranges of validity and limitations associated with various diode models. Plan to collect data that will enable you to plot the characteristic curves for selected Si and Ge diodes. Several values of R_b and n for each device can be estimated by applying Eq. (4-35) at two nearby locations on the characteristic curve. (Computer programs of the type described above may also be helpful in analysis.)

Estimates for I_0 for each diode can be obtained once values for R_b and n have been established. Develop a procedure describing how you will conduct this experiment and analyze your findings. Discuss and explain all results.

Discussion

Figure 4-31 shows typical data that might be collected to define the characteristic curve for a Ge power diode (with a large, measurable value of R_b). Figure 4-32 indicates how r can be found as a function of I. The data of Fig. 4-31 should be plotted on both linear and semilog scales in order to support analysis and consideration of the conclusions to be drawn.

FIGURE 4-31
Typical Data Collection for a Ge Power Diode for Experimental Application 4-1

Data Point	I (A)	V (V)
1	5.0×10^{-7}	1.5×10^{-2}
2	1.0×10^{-6}	2.0×10^{-2}
3	5.0×10^{-6}	6.3×10^{-2}
4	1.0×10^{-5}	8.5×10^{-2}
5	5.0×10^{-5}	1.4×10^{-1}
6	1.0×10^{-4}	1.8×10^{-1}
7	1.0×10^{-3}	3.4×10^{-1}
8	1.0×10^{-2}	8.3×10^{-1}
9	1.0×10^{-1}	3.0

FIGURE 4-32
Data Analysis for Experimental Application 4-1

Data Points	ΔI (A)	ΔV (V)	$r = \Delta V/\Delta I$	I_{av} (A)
1–2	5.0×10^{-7}	0.5×10^{-2}	10 K	0.75×10^{-6}
2–3	4.0×10^{-6}	4.3×10^{-2}	11 K	3.0×10^{-6}
3–4	5.0×10^{-6}	2.2×10^{-2}	4.5 K	0.75×10^{-5}
4–5	4.0×10^{-5}	5.5×10^{-2}	1.4 K	3.0×10^{-5}
5–6	5.0×10^{-5}	0.4×10^{-1}	0.80 K	0.75×10^{-4}
6–7	9.0×10^{-4}	1.6×10^{-1}	0.18 K	5.0×10^{-4}
7–8	9.0×10^{-3}	4.9×10^{-1}	54	5.0×10^{-3}
8–9	9.0×10^{-2}	2.2	24	5.0×10^{-2}

There are several ways in which R_b and n can be identified. One approach is to prepare a log-log plot of r as a function of I, as shown in Figure 4-33, and to use the two asymptotes for large and small current values to identify R_b and n.

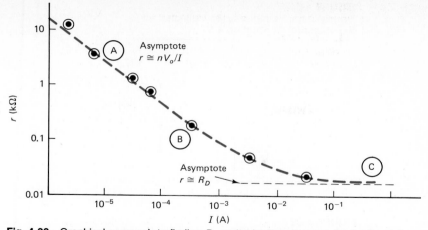

Fig. 4-33 Graphical approach to finding R_b and n for Experimental Application 4-1.

For the given data, the following results are obtained.
(a) At point A,

$$n \cong rI/V_0 \cong 1.3$$

(b) At point B,

$$n \cong \frac{rI}{V_0} \cong 3.5$$

(c) At point C,

$$R_b \cong 20 \ \Omega$$

(d) Near data point 3,

$$I_0 \cong \frac{I}{e^{V/nV_0} - 1} \cong 0.9 \times 10^{-6} \ \text{A}$$

4-2. The objective of this experiment is to determine the temperature dependence of the characteristic curves for Ge and Si diodes and find $\mathcal{E}_g$ for each diode. Design a circuit (using available equipment) that allows you to record the characteristic curves for the diodes at room temperature and at an elevated temperature. Use a hot air blower and thermometer or thermocouple to measure T. Build your circuit and collect the necessary data. Graphically calculate $(\partial V/\partial T)_I$ for several locations on each diode characteristic curve. Find $\mathcal{E}_g$ and interpret your findings.

Discussion
From Eq. (4-44),

$$\left(\frac{\partial V}{\partial T}\right)_I = \frac{V - \mathcal{E}_g/q}{T}$$

Therefore, a study of the temperature dependence of the junction diode characteristic curve can be used to find $\mathcal{E}_g$. Typical data elements for data collection and analysis are shown in Figs. 4-34 and 4-35. Figure 4-36 illustrates how a plot of $(\partial V/\partial T)_I$ as a function of V can be used (by extrapolation) to determine $\mathcal{E}_g$.

FIGURE 4-34
Typical Data Elements to Be Collected for
Experimental Application 4-2

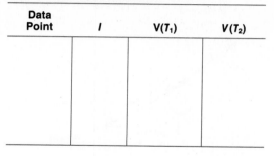

Data Point	I	$V(T_1)$	$V(T_2)$

FIGURE 4-35
Data Analysis Table for Experimental
Application 4-2

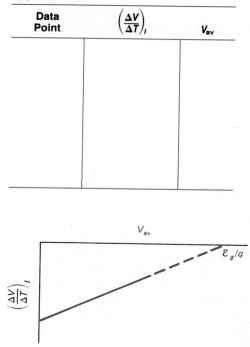

Data Point	$\left(\dfrac{\Delta V}{\Delta T}\right)_I$	V_{av}

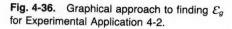

Fig. 4-36. Graphical approach to finding $\mathcal{E}_g$ for Experimental Application 4-2.

4-3. The objective of this experiment is to experimentally determine the characteristic curve for a photodiode. Design a circuit that is appropriate for this purpose, using a photographic light meter to measure the intensity of a light source at the photodiode. Develop a procedure that can be used to plot the photodiode characteristic curve, as shown in Fig. 4-19(B). Build your circuit, collect the necessary data, and interpret your findings.

Discussion
The characteristic curve to be obtained for the photodiode is affected by the spectral properties of both the source and diode. In turn, the intensity of the light

as measured by the light meter depends on the spectral properties of the sensor used in the meter. Thus, the results of this experiment are valid specifically for the equipment and materials being used. The experiment is a useful way to study the linearity of the photodiode with the relative intensity of the light source and to measure the angular resolution (beam width) of the lens assembly that is part of the diode.

CHAPTER 5
DIODE MODELS AND CIRCUIT APPLICATIONS

Chapters 2 and 4 outline the operational principles for several types of semiconductor diodes. A number of fundamental parameters are introduced to provide a means for comparing device characteristics, and these parameters are linked to models that can be used to explore circuit tradeoffs. These two chapters provide a sufficient introduction to the electrical properties of materials and the characteristics of semiconductor junctions to allow discussion of diode properties from a more fundamental basis and to be able to engage in productive design tradeoffs.

Chapter 3 approaches the design problem from a different direction, with a concern for the R, C, and L functions that appear in electronic circuits. The concepts of Chapter 3 serve to place a boundary on the accuracy that can be associated with circuit schematics and provide a reference framework that must be considered in any design effort.

Chapter 5 begins the application process, building on the concepts that have been developed previously. The diodes of Chapter 4 are now used to create waveform-shaping circuits that are important in many electronics settings. The utility of various devices for the applications that are developed can be interpreted in terms of the background provided in the earlier discussions. As will be seen, the nonlinear properties of these devices are essential to electronics design. It will be observed that the nonideal aspects of these devices must be constantly considered in circuit models if the resultant performance is to be adequately predicted.

Chapter 5 also introduces SPICE (Simulation Program with Integrated Circuit Emphasis), a computer simulation program, as a method for the computer-aided design of circuits. Several of the applications in this chapter are treated by using SPICE, both to illustrate the value of the software support and to continue the building of an array of recognized circuit types.

As derived in Chapter 4, the theoretical diode equation can be written

$$I = I_0(e^{(V-IR_b)/nV_0} - 1) \tag{5-1}$$

The properties of this equation have been explored using graphical methods. Because of its complexity, this model is of limited practical use in circuit design. However, the equation for dV/dI that was derived earlier [Eq. (4-34)] can often be useful.

Consider a situation for which R_b is negligible and $n \cong 1$ (the conditions associated with lower-current levels). Then the ac resistance r of the diode is given by

$$r = dV/dI \cong V_0/I \tag{5-2}$$

Under these circumstances, if V_0 and the dc current I through a diode are known, an estimate can be made of the ac resistance of the diode at the operating point. This property is used in later chapters.

Example 5-1   DC and AC Resistance for a Diode

For the Si diode curve in Fig. 5-1, find R and r for (a) $V = 0.60$ V and (b) $V = 0.70$ V.

Fig. 5-1 Typical diode curves.

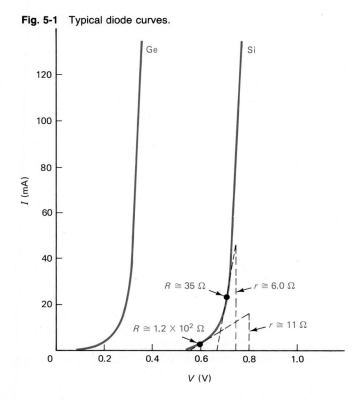

Solution   Applying graphical methods to Fig. 5-1:

(a) At 0.60 V,

$$R = \frac{V}{I} \cong \frac{0.60 \text{ V}}{5.0 \text{ mA}} \cong 1.2 \times 10^2 \text{ } \Omega$$

$$r = \frac{\Delta V}{\Delta I} \cong \frac{0.22 \text{ V}}{20 \text{ mA}} \cong 11 \text{ } \Omega$$

(b) At 0.70 V,

$$R \cong \frac{0.70 \text{ V}}{20 \text{ mA}} \cong 35 \text{ } \Omega$$

$$r \cong \frac{0.30 \text{ V}}{50 \text{ mA}} \cong 6.0 \text{ } \Omega$$

Or using Eq. (5-2),

(a) At 0.60 V,

$$r \cong \frac{0.026 \text{ V}}{5.0 \text{ mA}} \cong 5.2 \text{ } \Omega$$

(b) At 0.70 V,

$$r \cong \frac{0.026 \text{ V}}{20 \text{ mA}} \cong 1.3 \text{ } \Omega$$

The accuracy limitations on estimating r are evident in this example. If r must be known with accuracy for a specific device, a direct graphical determination can be made. However, since r can vary among devices that have a common code number, it is important to design circuits that can accommodate a range of r values when they are assembled. In some ways, then, it is best to estimate r, remember the tolerances involved, and create circuits that can perform with such variations. ∎

Because of the importance of the junction diode, other "working models" need to be developed for circuit design. By starting with experimental observations, four models for the junction diode can be developed, as shown in Fig. 5-2. Figure 5-2(A) is the equivalent circuit for an "ideal" diode; if the diode is forward-biased, it has no resistance, and if the diode is reverse-biased, it has infinite resistance. The ideal diode is thus a switch that turns *off* or *on,* depending on the polarity of the applied voltage.

Figure 5-2(B) gives an equivalent circuit that is somewhat closer to the actual diode characteristic. The switch does not come *on* until the applied voltage exceeds a certain threshold (typically 0.2 to 0.3 V for Ge diodes and 0.6 to 0.7 V for Si diodes). Figure 5-2(C) adds further realism by providing the diode with a finite resistance when it is *on.* The slope of the *on* portion of the curve is equal to the reciprocal of the average ac resistance r of the forward-biased diode. Finally, Figure 5-2(D) illustrates a model with several *piecewise linear* steps. Over a small range of forward voltages, the diode is assumed to have an average value of ac resistance r_{av}. The value of r_{av} changes when different sections of the model are used.

The models shown in Fig. 5-2 progress from more simple to more complex (and

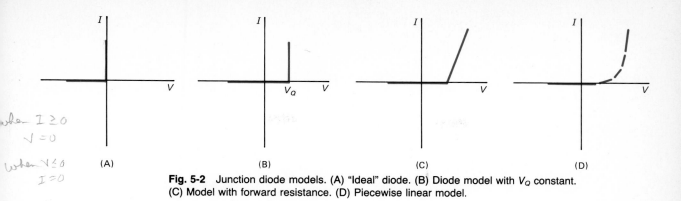

Fig. 5-2 Junction diode models. (A) "Ideal" diode. (B) Diode model with V_Q constant. (C) Model with forward resistance. (D) Piecewise linear model.

exact) models. At first look, it might seem that Fig. 5-2(D) would always be the best model selection. However, the more accurate models are also the most complex to use in circuit design efforts, so a tradeoff exists between model accuracy and utility. For this reason, the models of Fig. 5-2(B) and (C) are in broad use as compromise selections. The model in Fig. 5-2(A) finds use in very approximate calculations, and the model in Fig. 5-2(D) is applicable to detailed (often computer-based) analyses.

5-2 GRAPHICAL DESIGN OF OPERATING POINTS

One of the recurrent problems in electronics design involves the selection of a linear resistor R and a source voltage V_s that will produce a desired current level I_Q in a nonlinear resistive device, as shown in Fig. 5-3.[1] The voltage V_Q and current I_Q define the operating point, or Q point, of the nonlinear device in a specific circuit. As the circuit parameters R and V_s change, the Q point shifts.

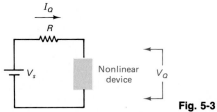

Fig. 5-3 Nonlinear device design problem.

Assume that the nonlinear device is a junction diode, as shown in Fig. 5-4, and the design objective is to select R and V_s so that a current I_Q flows in the circuit. Summing voltages around the loop in Fig. 5-3,

$$V_s = IR + V \tag{5-3}$$

where V is the voltage across the diode.

The curve $I = f(V)$ is given in the graph of Fig. 5-4. To obtain the desired current I_Q, these two equations must be solved simultaneously. This can be done graphically by plotting both equations on the same axes; they will intersect at point Q as

---

[1] The subscript Q indicates the *quiescent* or *resting* operating point for a circuit.

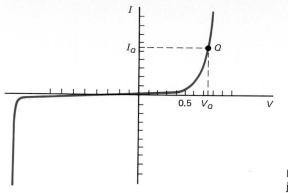

Fig. 5-4 Characteristic curve for a junction diode.

shown in Fig. 5-5. The plot of Eq. (5-3) is called the *load line,* and, as noted before, the plot of $I = f(V)$ is called the device *characteristic.*

From Fig. 5-5, the slope of the load line is equal to $-(1/R)$. As discussed in Chapter 3, resistors are often manufactured in standard values for which only limited choices of R are available.

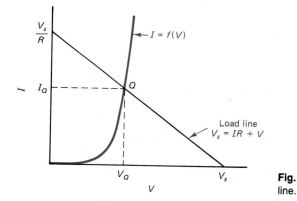

Fig. 5-5 Graphical solution using a load line.

Observe that for the load line, when $I = 0$, $V = V_s$, so the x-intercept gives the value of V_s. The load line must be selected so that the resultant value of V_s is within available power supply specifications and so that V_Q is less than V_{max}, the maximum allowable voltage across the nonlinear device. Additionally, consider that the y-intercept of the load line occurs at $I = V_s/R$. V_s and R must be chosen so that I_Q is less than I_{max}, the maximum current value allowed for the device.

In addition to the above constraints, the power dissipated in the device must always be less than its maximum power rating P_{max}. If the equation

$$P_{max} = IV \tag{5-4}$$

is plotted on the same axes as the other curves, an upper boundary to the possible Q points can be established, as shown in Fig. 5-6.

Given a Q point and V_s, the Q point variables V_Q and I_Q are related to the intercepts V_s and V_s/R as follows:

$$\frac{V_Q}{(V_s/R) - I_Q} = R \quad \text{and} \quad \frac{V_s - V_Q}{I_Q} = R \tag{5-5}$$

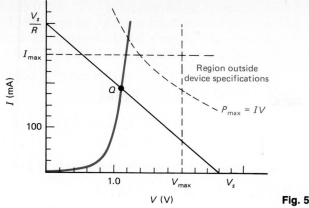

Fig. 5-6 Setting limits on the Q point.

These equations will be useful in connecting the various design parameters. If the Q point variables (I_Q, V_Q) and the value of R are specified, the vertical and horizontal load line intercepts can be found.

$$V_s/R = I_Q + V_Q/R \quad \text{and} \quad V_s = V_Q + I_Q R \qquad (5\text{-}6)$$

Example 5-2 **Selecting a Q Point**

A junction diode has the characteristic curve and maximum device ratings shown in Fig. 5-7. Assume that you have available a power supply with $V_s \le 4.0$ V and wish to produce a current $I_Q = 2.0$ mA as the circuit objective. Using discrete resistors (tolerance ± 5 percent), find several appropriate load lines.

Fig. 5-7 Selecting a standard resistor to produce a desired Q point (possible values for R are determined from Fig. 3-2).

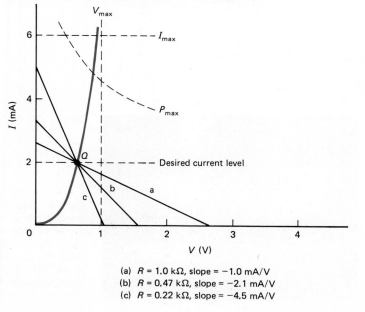

(a) $R = 1.0$ kΩ, slope = -1.0 mA/V
(b) $R = 0.47$ kΩ, slope = -2.1 mA/V
(c) $R = 0.22$ kΩ, slope = -4.5 mA/V

Solution   First, draw the horizontal line corresponding to $I_Q = 2.0$ mA, as shown in Fig. 5-7. The load line must pass through the point Q, intercept the V axis at 4.0 V or less, and have a slope corresponding to a standard resistor value. Several possible selections are shown. One of the following combinations of V_s and R can be chosen:

(a)
$$V_s = V + IR = 0.70 \text{ V} + (2.0 \text{ mA})(1.0 \text{ k}\Omega) = 2.7 \text{ V}$$

$$I_{max} = 2.7 \text{ V}/1.0 \text{ k}\Omega = 2.7 \text{ mA}$$

(b)
$$V_s = V + IR = 0.70 \text{ V} + (2.0 \text{ mA})(0.47 \text{ k}\Omega) = 1.6 \text{ V}$$

$$I_{max} = 1.6 \text{ V}/0.47 \text{ k}\Omega = 3.4 \text{ mA}$$

(c)
$$V_s = V + IR = 0.70 \text{ V} + (2.0 \text{ mA})(0.22 \text{ k}\Omega) = 1.1 \text{ V}$$

$$I_{max} = 1.1 \text{ V}/0.22 \text{ k}\Omega = 5.0 \text{ mA}$$ ■

In the above example, the design process begins with the characteristic curve of the device, $I = f(V)$. This curve can be obtained from the circuit of Fig. 5-8(A). The voltages V_R and V_Q across the resistor and diode can be varied by changing V_s and R. If V_s and R are varied and the resultant values of V_R and V_Q measured, a number of data points can be obtained for $I = V_R/R$ and V_Q as shown in Fig. 5-8(B).

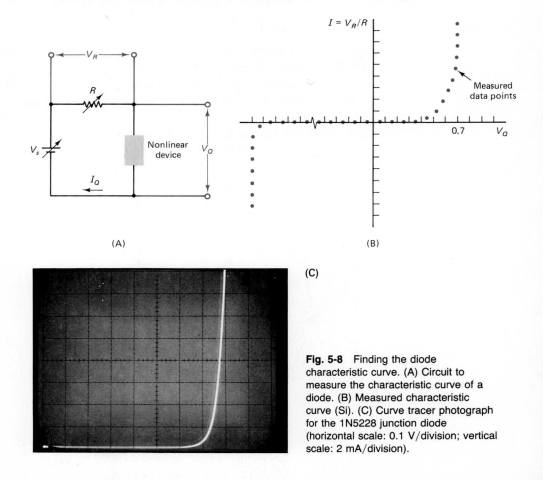

(A)

(B)

(C)

Fig. 5-8 Finding the diode characteristic curve. (A) Circuit to measure the characteristic curve of a diode. (B) Measured characteristic curve (Si). (C) Curve tracer photograph for the 1N5228 junction diode (horizontal scale: 0.1 V/division; vertical scale: 2 mA/division).

In many electronics design settings, a *curve tracer* is available. This instrument automatically scans a range of voltages for a given choice of R and displays the characteristic curve directly on an oscilloscope screen as shown in Fig. 5-8(C).

Based solely on the boundaries of Fig. 5-6, there is no limit on the value of R that can be used to determine Q. However, other factors come into consideration in an actual design case. The slope of the load line is equal to $-1/R$. For small values of R, the slope becomes very large and I_Q shifts substantially with a change in the characteristic curve of the diode. A useful lower limit is placed on R by requiring that

$$R > \Delta V_Q / \Delta I_Q$$

where ΔV_Q is the maximum uncertainty or shift in the diode characteristic (due to variation among devices and temperature variations), and ΔI_Q is the maximum allowable change in I_Q that is acceptable to achieve the design objective.

An upper limit is placed on R by the available power supply (and by power dissipation limits):

$$R < \frac{(V_s)_{\max} - V_Q}{I_Q}$$

so that

$$\frac{\Delta V_Q}{\Delta I_Q} < R < \frac{(V_s)_{\max} - V_Q}{I_Q} \tag{5-7}$$

Example 5-3   **Limits on the Load Resistor R**

You have available a power supply for which $V_s < 20$ V and a Si diode. The objective is to choose V_s and R to achieve a load line that produces $I = 10$ mA ($\pm 5\%$) with a temperature range of operation of $\Delta T = 100°C$. What values can be chosen for R?

Solution   From Example 4-7 of Chapter 4 and the above development,

$$\Delta V_Q \cong (-2 \text{ mV/°C}) \times 100°C = -0.2 \text{ V}$$

and

$$\Delta I_Q = (10 \text{ mA})(0.05) = 0.5 \text{ mA}$$

Thus, R must fall in the range

$$\frac{0.2 \text{ V}}{0.5 \text{ mA}} < R < \frac{20 \text{ V} - 0.7 \text{ V}}{10 \text{ mA}}$$

$$0.4 \text{ k}\Omega < R < 1.9 \text{ k}\Omega$$

   ■

The original problem faced in this section was to find the values of V_s and R required to produce a desired current flow I in the circuit of Fig. 5-4(A). The solution given here uses graphical methods for solving simultaneous equations [Eqs. (5-1) and (5-3)]. Computer methods can also be used, although the nonlinear nature of Eq. (5-1) requires an interative solution.

The graphical strategy for problem-solving to find operating points is useful in electronics because the actual device characteristics can be used instead of approximate equations and solutions are quickly and easily found. Since many electronics design problems are limited to two to three significant figures in accuracy, the graphical method is also appropriate in terms of realistic predictions of circuit performance.

5-3 TEMPERATURE DEPENDENCE OF A DIODE Q POINT

As noted in the discussion of Chapter 4, the characteristic curves of semiconductor devices are generally a function of temperature. The relationship between I and V changes as the temperature increases or decreases. In order to predict how a circuit will operate, the designer must be concerned with the specific changes that occur as a result of temperature effects.

For junction diodes, a typical change is shown in Fig. 5-9. As T increases, the voltage across the device decreases along the load line. The increase in temperature shifts the quiescent operating point from Q_1 to Q_2.

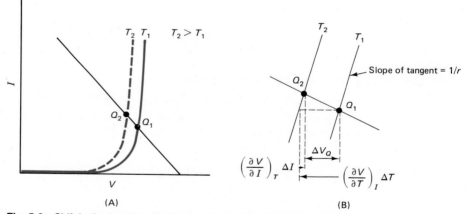

Fig. 5-9 Shift in the junction diode curve due to increased temperature.

In order to quantify this change, a first-order approximation[2] is to let

$$\Delta V \cong \left(\frac{\partial V}{\partial T}\right)_I \Delta T + \left(\frac{\partial V}{\partial I}\right)_T \Delta I$$

$$\cong \left(\frac{\partial V}{\partial T}\right)_I \Delta T + r\Delta I \tag{5-8}$$

Since

$$\Delta V/\Delta I = -R$$

---

[2] The approximation involves taking ΔV and ΔT to be finite intervals.

then

$$\Delta V \cong \left(\frac{\partial V}{\partial T}\right)_I \Delta T - r\frac{\Delta V}{R}$$

$$\Delta V = \left(\frac{1}{1 + r/R}\right)\left(\frac{\partial V}{\partial T}\right)_I \Delta T \tag{5-9}$$

The geometric interpretation of these terms is shown in Fig. 5-9(B). If $r \ll R$ for the forward-biased diode (the diode is well past the knee of the curve and R is not too small), then

$$\Delta V \cong (\partial V/\partial T)_I \Delta T \tag{5-10}$$

where from Chapter 4,

$$(\partial V/\partial T)_I \cong -2 \times 10^{-3} \text{ V/°C}$$

for a typical Si diode.

The change in ΔT can result from two factors: from a shift in the ambient temperature of the air around the diode and/or from the power being dissipated internal to the diode. Thus,

$$\Delta T = (\Delta T)_{\text{amb}} + P_Q \theta_{JA} \tag{5-11}$$

where $P_Q = I_Q V_Q$ is the power dissipated in the diode, and θ_{JA} is the thermal resistance from the junction of the device to the ambient air. $(\Delta T)_{\text{amb}}$ is usually measured with respect to room temperature (27°C).

The thermal resistance might typically be 50 to 300°C/W for a diode without a heat sink and 5 to 30°C/W with a large heat sink. Combining Eqs. (5-10) and (5-11),

$$\Delta V = \left(\frac{\partial V}{\partial T}\right)_I [(\Delta T)_{\text{amb}} + P_Q \theta_{JA}] \tag{5-12}$$

Example 5-4   **Shift in Q Point Due to Internal Heating**

Consider a forward-biased Si diode for which $V_Q = 0.7$ V, $I_Q = 100$ mA, and the surrounding air is at 27°C. What shift in Q point should be expected due to internal heating if $\theta_{JA} = 50$°C/W and $r \ll R$?

Solution   For the device values given above,

$$\Delta V \cong (-2 \times 10^{-3} \text{ V/°C})(0.07 \text{ W})(50°\text{C/W}) \cong -7 \text{ mV}$$

For this application, the percentage shift in the voltage across the device is small, of the order of 1 percent. The study of three-terminal devices in the next section reveals some configurations for which much larger effects can be observed. ■

For a diode in a small enclosed area, the power dissipated can also raise the ambient temperature. If the diode and its enclosure are almost a closed system (with limited heat flow out of the system),

$$\Delta Q \cong mC\Delta T = P_Q t \tag{5-13}$$

for the closed system, where ΔQ = heat being developed in the system, m = system mass, and C = average specific heat of the system.

Therefore,

$$\Delta V \cong \left(\frac{\partial V}{\partial T}\right)_I \left(\frac{P_Q t}{mC} + P_Q \theta_{JA}\right) \tag{5-14}$$

The temperature of the closed system increases with time, producing *thermal drift* in the Q point until the device current is limited by the load line. If the temperature continues to rise, the device will be destroyed.

5-4 RECTIFIER CIRCUITS

A useful circuit can be created by placing an ac voltage source, a diode, and a resistor in series, as shown in Fig. 5-10(A). Assume that the input ac voltage is given by $v = V_m \sin \omega t$.

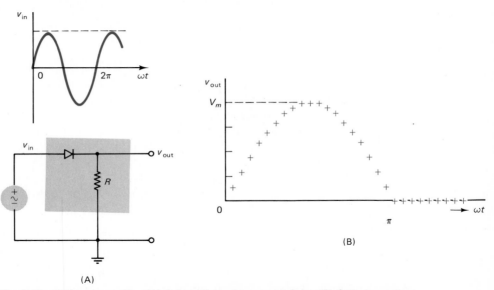

(A)

(B)

Fig. 5-10 Half-wave rectifier. (A) Schematic and input waveform. (B) Output waveform generated with the ideal diode model, using a computer simulation (see Computer Application 5-3).

The output waveform v_{out} to be expected can be initially predicted by making use of the ideal diode model. When the diode is forward-biased, $v_D = 0$, and the input v_{in} appears across R without change. When the diode is reverse-biased, $I_D = 0$ and $v_{\text{out}} = 0$. The result is an output that consists of halves of sine waves, as shown in Fig. 5-10(B).

The *Fourier expansion*[3] of such a waveform is given by

$$v = V_m\left(\frac{1}{\pi} + \frac{1}{2}\sin \omega t - \frac{2}{3\pi}\cos 2\omega t - \frac{2}{15\pi}\cos 4\omega t + \cdots\right) \tag{5-15}$$

---

[3] The Fourier series representation is discussed in Appendix 9.

The total voltage is written as a sum of many different sinusoidal terms. The frequency of each ac term is higher than the ones before (the angular frequencies are ω, 2ω, 4ω, . . .). The coefficients in front of each sine or cosine term become smaller with higher-order terms, so the series converges rapidly. A dc voltmeter across R measures the average voltage V_m/π given by the dc term.

This circuit is a *half-wave rectifier*. It uses as its input an ac signal with an average value of zero and produces as its output a series of voltage pulses with an average value different from zero. The nonlinear characteristic curve of the diode has resulted in the production of new frequency components in the rectifier output.

To determine how the use of an actual diode characteristic model would affect the findings of this simple model, refer to Figs. 5-11 and 5-12. The voltage v_{out} can be found as a function of v_{in} by using the method shown.

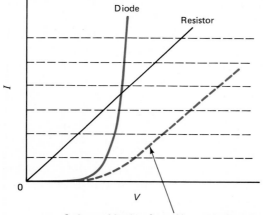

Series combination, formed by a point-by-point addition of the diode and resistor characteristic curves along I = constant lines

Fig. 5-11 Finding a composite characteristic curve.

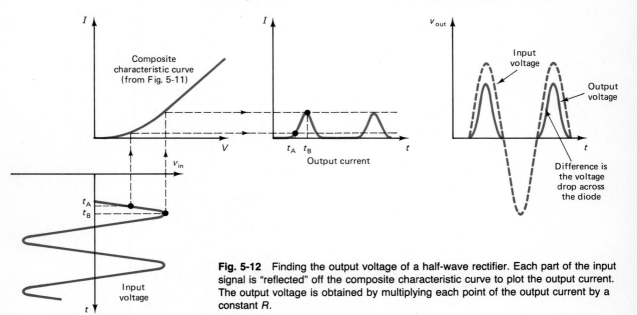

Fig. 5-12 Finding the output voltage of a half-wave rectifier. Each part of the input signal is "reflected" off the composite characteristic curve to plot the output current. The output voltage is obtained by multiplying each point of the output current by a constant R.

A composite curve is developed to represent the series combination of the diode and resistor. The characteristic curve of each component is plotted. Since the same current flows through each device, a series of horizontal lines can be drawn and the two curves added on a point-by-point basis. The result is a combined characteristic curve for the two devices. The voltage V now represents the combined voltage across the diode and resistor, and is equal to the input voltage v_{in}.

Given the composite curve, I is related to v_{in} as shown in Fig. 5-12. The input voltage is plotted vertically below the characteristic curve with time increasing in the downward direction. The current is plotted horizontally beside the characteristic curve. For selected times (t_A, t_B, . . .), the magnitude of the current is as shown. The output voltage is equal to the current times R. Since the combined characteristic is not linear, the output is distorted portions of a sine wave. The exact form of the output can be found from a step-by-step graphical analysis.

The difference between the input and output waveforms is due to the varying voltage drop across the diode. If V_m is much larger than the maximum voltage drop across the diode, the composite characteristic curve is approximately a straight line and the output wave approaches halves of sine waves. If V_m is of the magnitude of the maximum voltage drop across the diode, the composite characteristic is strongly nonlinear and substantial distortion in each sine-wave half will be observed (refer to Problem 5-39).

A *full-wave rectifier* is produced by using two diodes in the configuration shown in Fig. 5-13. The transformer enables the amplitude of the input voltage to be stepped up or down to produce the desired output (refer to the discussion of Section 3-5). For an ideal transformer, the voltage across each part of the transformer secondary is given by the turns ratio, as shown in Fig. 5-13(A). The two diodes take turns conducting, based on the polarity of the input waveform. For an ideal diode, the Fourier expansion of the output is now given by

Fig. 5-13 Full-wave rectifier. (A) Schematic and input waveform.

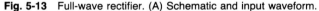
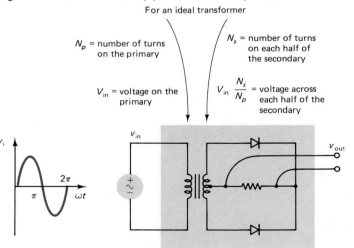

For an ideal transformer

N_p = number of turns on the primary

N_s = number of turns on each half of the secondary

V_{in} = voltage on the primary

$V_{in} \dfrac{N_s}{N_p}$ = voltage across each half of the secondary

(A)

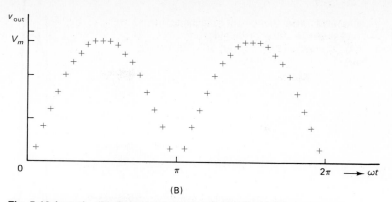

Fig. 5-13 (cont.) (B) Output waveform generated with the ideal diode model, using a computer simulation (see Computer Application 5-3).

$$v = V_m\left(\frac{2}{\pi} - \frac{4}{3\pi}\cos 2\omega t - \frac{4}{15\pi}\cos 4\omega t + \cdots\right) \qquad (5\text{-}16)$$

The average value is given by $2V_m/\pi$. The exact form of each output pulse can be found by using the graphical methods of Fig. 5-12.

Example 5-5 **Rectifier Output**

The dc component of a full-wave rectifier is designed to be 12 V. What amplitude input wave should be selected (using the ideal diode model)?

Solution From Eq. (5-16),

$$V_{dc} = V_m(2/\pi)$$

$$V_m = 12 \text{ V}(\pi/2) \cong 19 \text{ V}$$

Observe that the dc component of a full-wave rectifier output is just twice that of a half-wave rectifier output. The use of a more complex circuit results in improved circuit performance. ■

5-5 BRIDGE RECTIFIERS

The bridge rectifier is another type of full-wave rectifier, as shown in Fig. 5-14(A). The input is applied across points a and b, and the output is taken across resistor R. When v_{in} is positive, diodes D_1 and D_4 are *on* and the current through R flows from c to d. When v_{in} is negative, diodes D_2 and D_3 are *on* and the current again flows from c to d.

For the circuit of Fig. 5-14(A) the voltage across R is given by

$$v_R(t) = |v_{in}(t)| - 2v_D(t) \qquad (5\text{-}17)$$

With the ideal diode model, $v_D(t) = 0$ for forward-biased diodes. Then $v_R(t) = v_{in}(t)$ for positive v_{in} and $v_R(t) = -v_{in}(t)$ for negative v_{in}, as shown in Figs. 5-14(C) and 5-15(A).

If the model in use considers the forward-voltage drop across the diodes to be important, then the forward-bias condition results in $v_R(t) = |v_{in}(t)| - 2v_D$, where

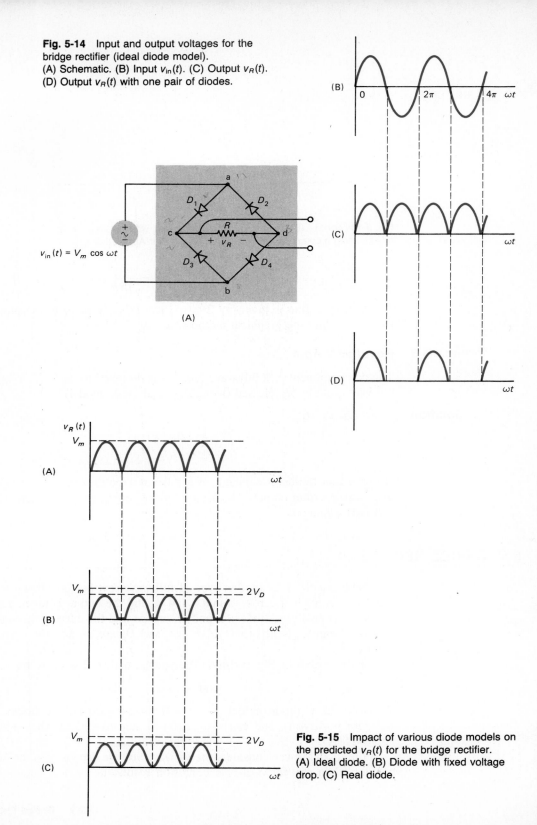

Fig. 5-14 Input and output voltages for the bridge rectifier (ideal diode model).
(A) Schematic. (B) Input $v_{in}(t)$. (C) Output $v_R(t)$.
(D) Output $v_R(t)$ with one pair of diodes.

$v_{in}(t) = V_m \cos \omega t$

(A)

(B)

(C)

(D)

Fig. 5-15 Impact of various diode models on the predicted $v_R(t)$ for the bridge rectifier. (A) Ideal diode. (B) Diode with fixed voltage drop. (C) Real diode.

$v_D \cong 0.3$ or 0.7 V for $v_{in} > 2v_D$ and $v_D = 0$ for $v_{in} < 2v_D$, as shown in Fig. 5-15(B). If the actual characteristic curve for the diodes is used, the wave shape changes, as shown in Fig. 5-15(C).

Example 5-6   **Bridge Rectifier Using an Actual Diode Characteristic**

Consider a full-wave bridge rectifier with $v_{in} = (5 \text{ V}) \sin \omega t$ and $R = 100 \ \Omega$. Using the actual diode characteristic curve, find v_{out}.

Solution   The effects of the two diodes and the resistor R can be combined by noting that the current I through the three components is always the same. The graphical technique follows the approach used above. If $I = f(V)$ and $I = V/R$ are plotted on the same scales and a series of horizontal lines drawn corresponding to values of I, then a composite characteristic curve $2f(V) + V/R$ results, as shown in Fig. 5-16. A series of times t_A, t_B, . . . , is selected and values of the input waveform are fixed for each time. If these values are "reflected" off the composite curve, then the output current I can be plotted for each selected value. To find v_{out}, multiply each value of I by a constant R, creating an additional plot (not shown). Note from the plot of the output current that the output is now distorted for small values of V and no longer consists of parts of a sine wave. ■

Fig. 5-16   Output current for a bridge rectifier using the actual diode curve.

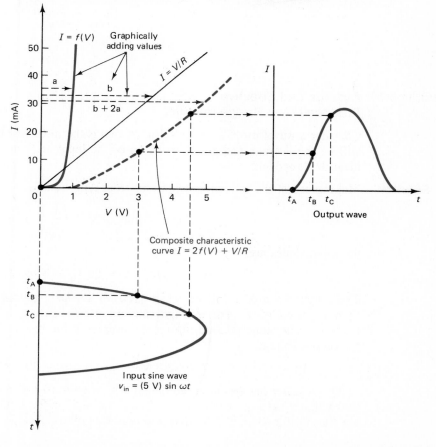

When a rectifier is analyzed, appropriate diode model selection depends on the relative magnitude of the voltages involved, the application, and the desired accuracy of prediction. The selection of R depends on two factors. It is usually desirable to make sure that the diodes spend most of their operating time well past the knee of the diode curve in order to treat them as approximately linear devices. As can be seen in Fig. 5-17, lower values of R move the operating point Q further above the knee. However, care must be taken not to exceed the maximum allowed values of I, V, and P. The appropriate choice for R lies in the range defined by the above two considerations.

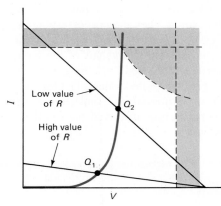

Fig. 5-17 Impact of high and low values of R on the Q point.

Example 5-7 Average and Effective Values Produced by a Bridge Rectifier

Consider a waveform $v_{in} = (20\text{ V})\sin \omega t$ that is the input to a bridge rectifer. What will be the average and effective values of the input and output voltages using the ideal diode model?

Solution Making use of the following definitions from Appendix 1,

$$V_{av} = \frac{1}{T}\int_0^T v\,dt \quad \text{and} \quad \mathbf{v} = \left(\frac{1}{T}\int_0^T v^2\,dt\right)^{1/2}$$

the input values are

$$V_{av} = 0 \quad \text{and} \quad \mathbf{v} = 20\text{ V}/\sqrt{2} \cong 14\text{ V}$$

The output waveform will be $v_{out} = |(20\text{ V})\sin \omega t|$, giving a positive waveform with a period of π. The average value of a sine wave from 0 to π is equal to $(2/\pi)V_m$. The same relationship for the average value results from Eq. (5-16). For the output voltage,

$$V_{av} \cong (20\text{ V})0.637 = 12.7\text{ V} \quad \text{and} \quad \mathbf{v} = 20\text{ V}/\sqrt{2} \cong 14\text{ V}$$

The rectifier has produced an output with a nonzero average voltage. The effective voltage has not been changed (using the ideal diode model). Rectifier circuits are thus useful when it is desired to produce an output that has a dc component. ∎

Since there are few electronic circuits in which the pulsating output of a rectifier can be used directly, smoothing filters are often placed between the rectifier and the load resistor R_L. Their purpose is to reduce the ac components of v_{out} in comparison with the dc (average) value of v_{out}.

Consider the half-wave rectifier circuit of Fig. 5-18(A), using an ideal diode model. (This is not a practical circuit, but is used as a starting point for discussion.)

Start with the case for which there is no charge on the capacitor ($v_{out} = 0$) and no input voltage ($v_{in} = 0$). As v_{in} increases in value, diode D is forward-biased. The capacitor is connected directly across v_{in} and builds up a charge so that v_{out} tracks v_{in}.

Fig. 5-18 Half-wave rectifier with smoothing filter. (A) Ideal circuit. (B) Addition of a load resistor R_L. (C) Addition of a charging resistance R_f.

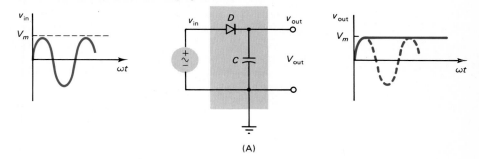

(A)

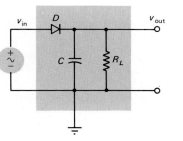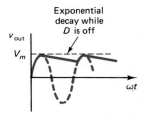

(B)

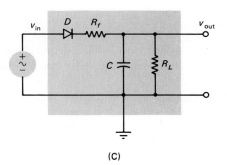

(C)

As v_{in} peaks, a maximum voltage of V_m is developed across the capacitor. As v_{in} then begins to drop, diode D is reverse-biased and the output remains at V_m, as shown in Fig. 5-18(A).

In practical applications, the output of the filter is applied as a voltage source for another circuit. The circuit to be connected across v_{out} can be represented by placing a load resistor R_L in parallel with C, as shown in Fig. 5-18(B).

The waveform v_{out} is now changed. When v_{in} begins to decline from its peak value V_m, the diode is reverse-biased. The charged capacitor is in series with load R_L and isolated from the rectifier circuit. The voltage across C does not stay constant, but decays exponentially. Note that load R_L is an integral part of the circuit and can strongly affect v_{out}. When several components are all connected to the common (ground) terminal, the schematic is often drawn as shown in Fig. 5-18(B). This is still a three-terminal circuit since the bottom input and output leads are connected.

As a final consideration, the finite resistance of the voltage source and diode must be recognized. These resistors limit the flow of current to the capacitor and load resistor. Additional resistance can also be added in series with the diode. All of these possible sources of resistance can be represented with a single resistor R_f in series with D, as shown in Fig. 5-18(C). The circuit in Fig. 5-18(C) is a practical one that will be analyzed in detail.

The rectifier shown in Fig. 5-18(C) is a *nonlinear* circuit due to the operation of the diode. The diode behaves like a switch, connecting the source v_{in} to R_f when D is *on* and isolating v_{in} from R_f when D is *off*. In order to apply linear analysis to this nonlinear circuit, two linear circuits must be defined, one that exists when D is *on* and one when D is *off*.

Figure 5-19 illustrates the periodic solution for Fig. 5-18(C) and the two solution regions that are formed. Thus, t_1, t_2, t_3, . . . define the boundaries between regions. On each boundary, $v_{out} = v_{in}$ and $i_D = 0$, marking the switching condition for the diode.

Between positive input peaks (region II, when D is *off*), the output is an exponential discharge through R_L. When D is *on* (region I), a charging current flows and v_{out} rises toward v_{in}. The relationship between v_{out} and v_{in} is determined by the charging time constant of the capacitor. In Fig. 5-19, v_{out} never reaches V_m because of the current limiting produced by R_f.

In order to calculate the v_{out} waveform, differential equations must be developed for each region and solutions obtained. Then the boundary conditions between regions must be applied to evaluate the solution constants. This is a complex path to follow for a circuit solution, even when using an ideal diode model. For a realistic diode model, numerical integration methods must be used.

Nonlinear problems of this type are often encountered in electronics and for many years solutions could be obtained only by very tedious means. However, with the development of standard software packages to simulate circuit performance, a powerful method of study became available to explore the behavior of complex circuits.

One of the most broadly available simulation programs, called SPICE, is discussed in Section 5-12 and in a number of computer applications in this and following chapters. SPICE can be used directly to determine v_{out} for half-wave and full-wave rectifiers with a capacitor smoothing filter. Figure 5-20 (see pp. 166–67) shows typical results for SPICE simulations of half-wave and full-wave rectifiers

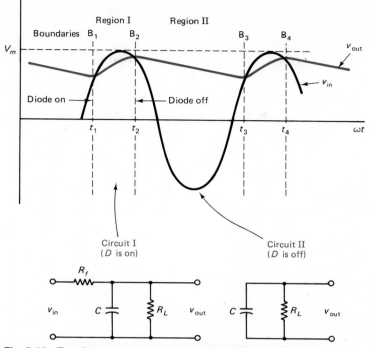

Fig. 5-19 Two linear circuits representing total circuit performance (steady-state condition).

with filters. (Refer to Computer Application 5-2 at the end of this chapter for further details.)

As R_L is reduced, the output becomes less smooth. As R_f is increased, the output becomes unable to reach the maximum input voltage value. It is also important to note that as R_f is decreased, the peak diode current increases. This can present a design problem, since D must be able to withstand this pulse of current. A compromise solution is to make R_f just small enough to produce a maximum output of about V_m, but not to reduce it any further than necessary.

The effectiveness of a filter is determined by examining the ripple factor (RF) of a rectifier with and without the filter. The ripple factor is defined as

$$\text{RF} = \frac{\text{effective value of the ac components of the output voltage}}{\text{dc or average value of the output voltage}} \tag{5-18}$$

Large values of RF mean that a large variation exists in the voltage waveform. A small value of RF means that the output approximates a battery with a small voltage variation around an average value.

For the *unfiltered* rectifier output, the ripple factor can be found by making use of the Fourier expansions in Eqs. (5-15) and (5-16). The leading term in each equation

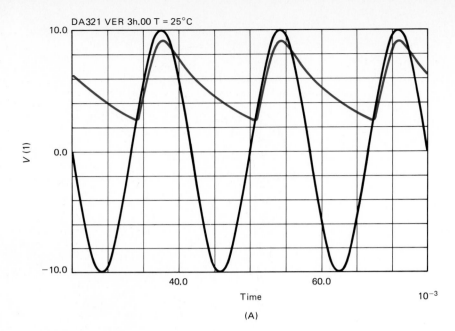

(A)

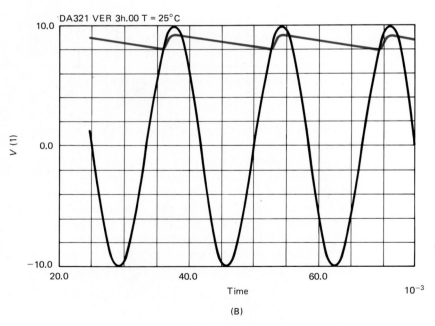

(B)

Fig. 5-20 SPICE simulation of half-wave and full-wave rectifiers with capacitor filters, showing input and output waveforms. A typical Si diode characteristic curve is used in each case. Half-wave rectifier with capacitor filter: (A) $C = 10 \ \mu F$, $R_L = 1.0 \ k\Omega$. (B) $C = 10 \ \mu F$, $R_L = 10 \ k\Omega$.

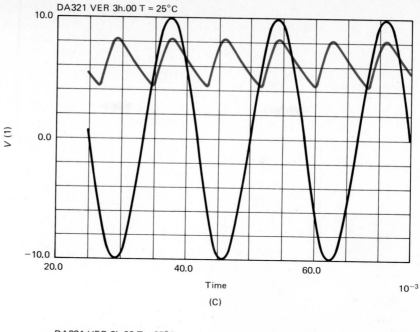

(C)

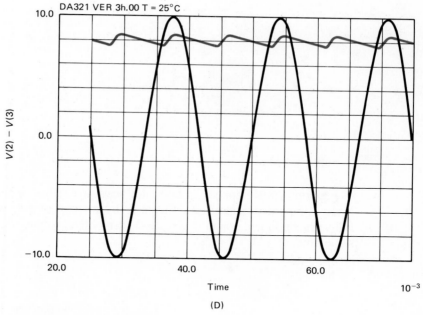

(D)

Fig. 5-20 (cont.) Full-wave rectifier with capacitor filter: (C) $C = 10\ \mu F$, $R_L = 0.8\ k\Omega$. (D) $C = 10\ \mu F$, $R_L = 5.0\ k\Omega$.

is the dc or average value V_{av} of the output, and the remaining terms constitute the ac components v_{ac}. Thus

$$\text{RF} = \frac{\left(\dfrac{1}{T}\displaystyle\int_0^T v_{ac}^2\, dt\right)^{1/2}}{V_{av}} \tag{5-19}$$

The ripple factors for the filtered outputs of Fig. 5-20 can be found from the approximation in Fig. 5-21. For a full-wave rectifier,

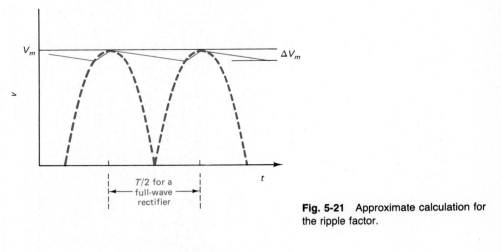

Fig. 5-21 Approximate calculation for the ripple factor.

$$\text{RF} \cong \frac{\left\{\dfrac{1}{T/2}\displaystyle\int_0^{T/2}\left[\left(V_m - \Delta V_m \dfrac{t}{T/2}\right) - V_m\right]^2 dt\right\}^{1/2}}{V_{av}} \tag{5-20}$$

$$= \frac{1}{\sqrt{3}}\frac{\Delta V_m}{V_{av}} \tag{5-21}$$

Equation (5-21) applies to both half-wave and full-wave rectifiers. However, for given component values, ΔV_m increases by a factor of two when changing from a full-wave smoothed output to a half-wave smoothed output, increasing the RF by this same amount.

Example 5-8 **Smoothing Filter**

Determine the ripple factor for the rectifier used for Fig. 5-20 with an ideal diode model and no smoothing filter. Then, using the output waveform of Fig. 5-20(D), find the RF with the filter.

Solution With no filter and an ideal diode model, the rectifier output is given by Eq. (5-16)

$$v = V_m\left(\frac{2}{\pi} - \frac{4}{3\pi}\cos 2\omega t - \frac{4}{15\pi}\cos 4\omega t + \cdots\right)$$

The RF can be found in the form

$$\text{RF} = \frac{\left[\dfrac{1}{T}\displaystyle\int_0^T \left(-\dfrac{4}{3\pi}\cos 2\omega t - \dfrac{4}{15\pi}\cos 4\omega t + \cdots\right)^2 dt\right]^{1/2}}{2/\pi}$$

$$= \frac{\dfrac{1}{\sqrt{2}}\left[\left(\dfrac{4}{3\pi}\right)^2 + \left(\dfrac{4}{15\pi}\right)^2\right]^{1/2}}{2/\pi} \cong 0.48$$

(5-22)

where the higher-order terms are neglected. The cross product terms have vanished when they are integrated over a period.

The ripple factor of the smoothed output is about $(1/\sqrt{3})(1.2\text{V}/8\text{V}) \cong 0.09$. The filter has reduced the ripple factor of the rectifier output from 0.48 to 0.09. ■

The above discussion has emphasized a single *RC* smoothing filter. There are many other possible filter designs, including those shown in Fig. 5-22. The basic solution approach used above can be extended to these more complex cases. In addition, other design approaches are covered in texts on filters.

Fig. 5-22 Various filter circuits.

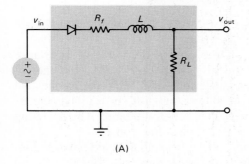

(A)

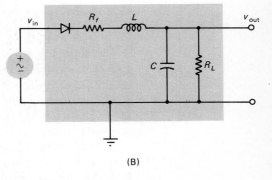

(B)

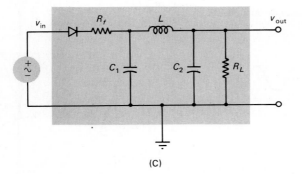

(C)

Junction diodes can be designed for use in a reverse-bias mode, as shown in Fig. 5-23. Such *zener diodes* are intended to make use of the sudden change in the ac resistance of the device that occurs at the critical (positive) bias voltage $V_C = V_z$.

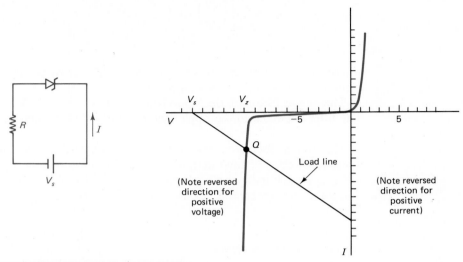

Fig. 5-23   Zener diode characteristic.

For $V < V_z$, very little current flows in the diode. Both the dc resistance $R = V/I$ and the ac resistance $r = \Delta V / \Delta I$ are very large. For $V > V_z$, the characteristic curve is almost vertical so that ac resistance shifts to a very small value. All possible operating points for the zener now lie along an almost vertical characteristic, so any load line produces about the same voltage across the zener. The operating point of the device can be readily selected using the load-line concept.

Zener diodes are used in the design of reference voltage circuits, as shown in Figure 5-24. First consider what v_{out} would be with the zener removed. Resistors R_1 and R_2 form a voltage divider so that

Fig. 5-24   Reference voltage circuit.

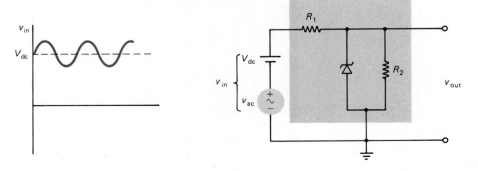

$$v_{out}\big|_{\text{no diode}} = \left(\frac{R_2}{R_1 + R_2}\right)v_{in} \qquad (5\text{-}23)$$

Now replace the zener. If the zener firing voltage V_z is greater than $v_{out}|_{\text{no diode}}$, the zener does not fire and v_{out} is still equal to $v_{out}|_{\text{no diode}}$. If the zener firing voltage is less than $v_{out}|_{\text{no diode}}$, the zener does fire and forces $v_{out} = V_z$. The current through the zener "adjusts itself" so that the voltage across R_1 is $v_{in} - V_z$. This is possible because given the small value of r when the zener is *on*, a wide range of currents I_z produces about the same value of V_z, as shown in Fig. 5-25. R_1 must be large enough to limit I_z to safe values.

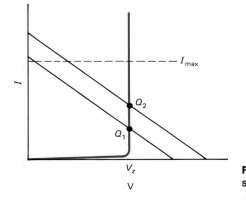
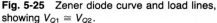

Fig. 5-25 Zener diode curve and load lines, showing $V_{Q1} \cong V_{Q2}$.

In typical design applications, R_1 and R_2 are selected to achieve the desired output waveform with a given input. If $v_{out}|_{\text{no diode}} > V_z$ for all values of v_{in}, then the zener is always *on* and v_{out} is locked at V_z, as shown in Fig. 5-26.

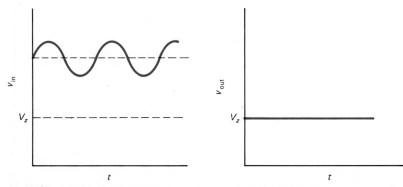

Fig. 5-26 Input and output signals for the circuit of Fig. 5-24 with $v_{out|\text{no diode}} > V_z$ for all values of v_{in}.

This application could be used in conjunction with the rectifier and filter circuit discussed in Section 5-6. By placing the reference voltage circuit of Fig. 5-24 across the filtered output, the small remaining ripple can be removed, as shown in Fig. 5-27.

A tradeoff exists between using a large capacitor filter to reduce the RF of a power supply versus use of a smaller capacitor and a zener clamp to produce a lower

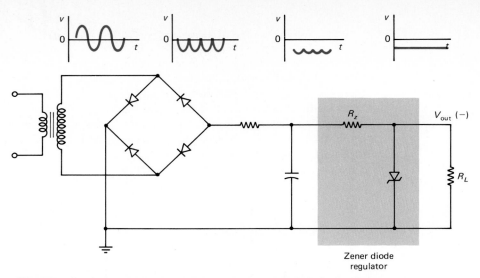

Fig. 5-27 Negative-voltage power supply using a zener diode for regulation.

RF. Large capacitors require substantial circuit volume and can be expensive. However, the capacitor filter causes energy to be dissipated only when it is charging or discharging.

The zener diode can be quite small in size, but it must be able to absorb current levels ranging from zero to the maximum current delivered to the load (assuming that the load can change from very large to a minimum value). Power dissipation is high if the zener must absorb large currents over much of the operating cycle, and the capacitor filter may be preferred. If the current load is fairly stable, so that power dissipation in the zener can be limited, it may represent the preferred choice.

Example 5-9 Zener Diode Reference-Voltage Circuit

For the circuit of Fig. 5-26, assume that $V_z = 6.0$ V, $R_1 = R_2$, and $v_{in} = 8.0$ V$(1 + \sin \omega t)$. Find v_{out}.

Solution From Eq. 5-23,

$$v_{out}\big|_{no\ diode} = \frac{1}{2} v_{in}$$

so when $v_{in} > 12$ V, the zener *fires*. For $v_{in} < 12$ V, the zener is *off*. The output follows the input voltage when the zener is off. The resulting output is shown in Fig. 5-28.

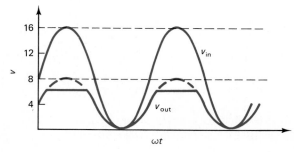

Fig. 5-28 Input and output of Zener circuit for Example 5-9.

If the dc component of the input voltage is increased so that $v_{in} = 20$ V + (8.0 V) sin ωt, the output is fixed at 6.0 V for all portions of the cycle. ∎

5-8 REGULATED POWER SUPPLIES

The rectifier and filter circuits described above provide outputs that are approximations to the desired "ideal" dc source. However, typical electronic circuits require sources that are almost independent of the line voltage, load resistance and temperature, and almost free of ripple without the use of excessively large filter capacitors. For many applications, the output voltage must be adjustable. Practical power supplies require more complex circuits to achieve the required performance objectives.

As described in Section 5-7, a zener diode can be used as a voltage regulator for a fixed output voltage. A higher performance strategy is to use an integrated circuit (IC) voltage regulator circuit to convert the rectifier output to a near-ideal power supply source, as shown in Fig. 5-29. A zener diode is used as an absolute voltage reference, while the control circuit senses the regulated output, compares it with the reference voltage, and provides a control voltage to modify the rectifier output as required and produce the stable dc output level.

In many applications, it is important to protect the power supply output against inadvertant short circuits that might destroy either the circuit under test or operation of the supply itself. Thus, current-limiting circuits are often incorporated into the regulator design.

Fig. 5-29 Positive-voltage power supply using an IC voltage regulator. (A) Applying an IC voltage regulator circuit. (Continued on p. 174.).

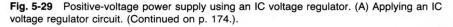

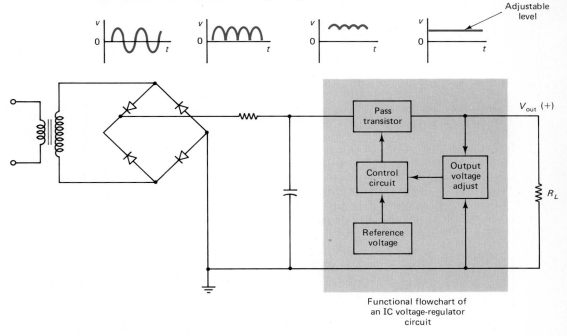

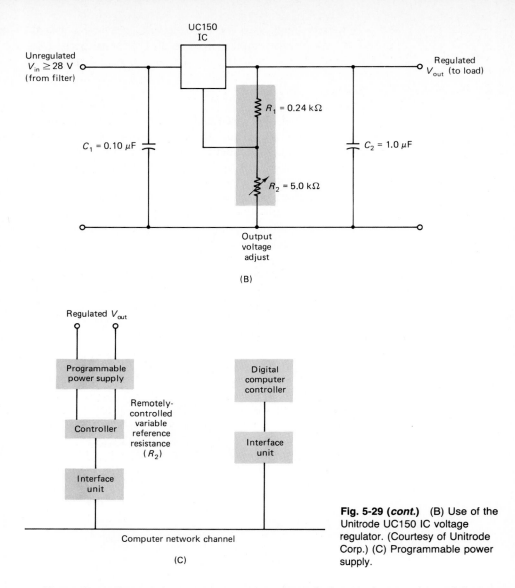

(B)

Regulated V_{out}

Programmable power supply

Remotely-controlled variable reference resistance (R_2)

Controller

Interface unit

Digital computer controller

Interface unit

Computer network channel

(C)

Fig. 5-29 (*cont.*) (B) Use of the Unitrode UC150 IC voltage regulator. (Courtesy of Unitrode Corp.) (C) Programmable power supply.

Since the entire supply current must pass through the regulator, substantial power dissipation can take place within this circuit. The package must be designed to produce the minimum possible thermal resistance between the *pass transistor* used for this purpose and the environment. For high current levels, an external pass transistor may be required, with *heat sinks* to reduce the effective thermal resistance.

Figure 5-29(B) shows how the Unitrode UC150 IC voltage regulator can be applied. Capacitor C_1 is included if the regulator is at some distance from the power supply filter capacitor, while C_2 is optional and improves transient response. For this circuit, the IC is internally designed so that

$$V_{out} = V_{ref}(1 + R_2/R_1) \qquad (5\text{-}24)$$

where $V_{ref} = 1.25$ V.

A variable supply voltage can be obtained by changing the value of R_2. This circuit can provide 3 A over a 1.2 to 25 V output range. Line regulation is typically 0.005 percent and load regulation is typically 0.1 percent. The IC is thermally compensated and has full overload protection. The internal operation of IC voltage regulators is discussed in Exercise 13-15.

Power supplies are becoming steadily more sophisticated in terms of performance objectives and application strategies. A commercial power supply is typically a complex system that makes maximum use of ICs to reduce ripple, improve regulation, and broaden control options.

Programmable power supplies are available to allow remote operation that is useful in many settings. Figure 5-29(C) illustrates how a digital computer operating over a computer network could be used to remotely control v_{out}. Using the circuit in Figure 5-29(B), a digital code might be used to produce a change in the effective R_2 seen by the regulator circuit. A field effect transistor (of the type introduced in Chapter 8) can provide one approach to a voltage-controlled variable resistor.

This system requires the application of analog amplifiers, analog-to-digital converters, digital-to-analog converters, and computer networks. These types of circuits are discussed in future chapters. As will be observed in these chapters, implementation requirements for such a system can often be largely met by using available IC combinations.

5-9 CLIPPING AND CLAMPING CIRCUITS

Diodes are often used to design *clipping circuits* that "clip off" waveforms above a selected input and *clamping circuits* that shift the dc value of a waveform. Many circuits produce both clipping and clamping effects. Such circuits can be useful in producing wave shapes that meet circuit design needs and will be applied in Chapter 15. Consider the circuit of Fig. 5-30 with an ideal diode, assuming that the magnitude of the battery voltage V_s is less than the amplitude of the input voltage V_m. Without the diode in place, the voltage at point A would always be $v_{\text{in}}(t)$. The voltage at point B is $-V_s$.

When $v_{\text{in}}(t)$ is more positive than $-V_s$, diode D is reverse-biased and *off*. The diode behaves as an open circuit and the battery does not affect the circuit.

When $v_{\text{in}}(t)$ is more negative than $-V_S$, the diode is forward-biased and *on*. The diode behaves as a short circuit, connecting the negative side of the battery to the output. The resulting output waveform is clipped off as shown.

When $v_{\text{in}}(t) = 0$ (point a), the applied voltage across D is $-V_s$. With the ideal diode model, no current flows. For all values of $v_{\text{in}}(t)$ up to point b, the diode remains reverse-biased and the current remains at zero. With D *off* during this segment, this portion of the circuit can be neglected and the input and output are directly connected through R. For this situation, $v_{\text{out}} = v_{\text{in}}$.

When v_{in} reduces to $-V_s$, the applied voltage across D is zero. For $v_{\text{in}} < -V_s$, the diode comes *on* and v_{out} becomes locked at $-V_s$. At point d, v_{in} increases above $-V_s$ and the diode turns *off* again. The circuit has clipped off the input voltage below the voltage $v_{\text{in}} = -V_s$.

The result changes if a load resistor R_L is placed as shown in Figure 5-30(B). The voltage across the diode is zero when $v_{\text{in}}[R_L/(R + R_L)] = -V_s$. The clipping level remains at $-V_s$, but the output amplitude is reduced by the voltage divider. The ef-

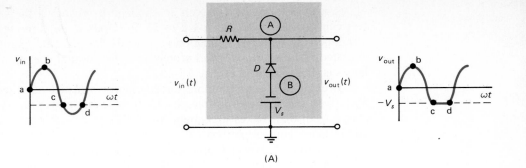

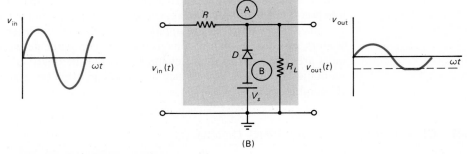

Fig. 5-30 Sample clipping circuit.

fect on the output depends on the resistor ratio. If the output amplitude is greater than $-|V_s|$, the diode has no effect since the diode is always *off*. If the output amplitude is less than $|V_s|$, the output is clipped as shown.

There are many different combinations of batteries, resistors, and diodes that can be used to shape a waveform. A general strategy for working such problems is as follows (again assuming the ideal diode model for simplicity):

1. Identify the voltage that would exist on both sides of the diode terminals without the diode being in the circuit.
2. Find the voltage levels and times for which $v_D = 0$. These levels and times define the boundaries between regions in which D is *on* and regions in which D is *off*.
3. Explore the voltage levels in the regions on both sides of $v_D = 0$ to identify where D is *on* and where it is *off*.
4. When D is forward-biased, it can be replaced by a short circuit. When it is reverse-biased, it can be replaced by an open circuit.
5. In each region, the output can be determined for the given input waveform. The solutions match across the boundary defined by $v_D = 0$.

Example 5-10 Clipping Circuits

Find the output waveform for the circuit and input shown in Fig. 5-31 assuming an ideal diode and $R_L \gg R$.

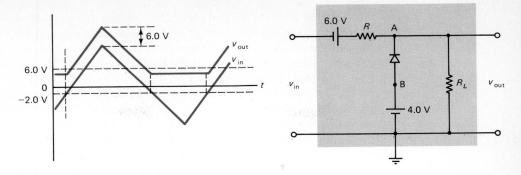

Fig. 5-31 Clipping circuit for Example 5-10.

Solution When $v_{in} = 0$, $v_A = 6.0$ V, and $v_B = 4.0$ V, the diode is reverse-biased so that $v_{out} \cong 6.0$ V. For $v_{in} > 0$, v_A becomes more positive and the diode remains reverse-biased. The output is $v_{out} \cong v_{in} + 6.0$ V. When v_{in} drops to -2.0 V, the voltage across the diode is zero, and when $v_{in} < -2.0$ V, the diode is forward-biased and v_{out} clamps to the 4.0-V battery. The waveform has been clipped and clamped to a different dc level. ■

The method described above can be applied to study the outputs of many different clipping circuits and to use circuit design strategies to achieve a desired output waveform for a given input.

5-10 PHOTODIODE CIRCUITS

The photodiode characteristic curve relates the intensity of the light incident on a diode "window" (in mW/cm^2) to the current through the device and the voltage across the device, as shown in Fig. 5-32. The characteristic curve depends on the specific light frequencies present in the source and the frequency response of the detector.

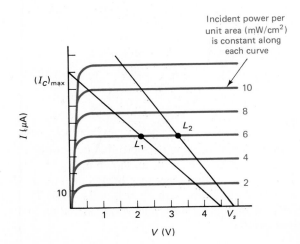

Fig. 5-32 Photodiode characteristic.

Note that above a certain threshold voltage, the photodiode current is approximately constant and independent of V. The ac resistance r is very high, giving rise to the flat curve. For a given intensity of the incident light, the current through the device is almost independent of the load line selected. The photodiode thus behaves like a constant current generator.

The characteristic curve of the photodiode can be determined by shining a light on the device window, varying V, and measuring I. Each time the frequency spectrum of the source is changed, a new characteristic curve must be measured. After the family of curves has been found, a simple circuit can be designed to develop an output voltage v_{out} that is proportional to the energy density of the incident light, as shown in Figure 5-33. Note that the photodiode is operated in the reverse-bias direction.

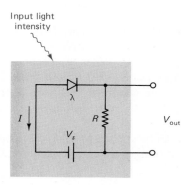

Fig. 5-33 Photodiode circuit.

Example 5-11   Photodiode Circuit

Assume that for a given photodiode, $I = 20$ μA with an incident light intensity of 10 mW/cm^2. (a) Using the circuit of Fig. 5-33 with $I = 20$ μA, $V_s = 20$ V, and $R = 100$ kΩ, find V_{out} and V_{diode}. (b) If the incident light intensity keeps increasing, what is the maximum possible value for V_{out}?

Solution   (a)
$$V_{out} = (20\ \mu A)(100\ k\Omega) = 2.0\ V$$
$$V_{diode} = 10\ V - 2.0\ V = 8.0\ V$$

(b) The maximum voltage across R occurs when
$$V_{out}\big|_{max} = (V_s/R)R = V_s = 20\ V \qquad \blacksquare$$

5-11   TUNNEL DIODE CIRCUITS

The tunnel diode is one of the few two-terminal devices that has a negative value of r over part of the characteristic curve, as shown in Fig. 5-34. Design with the tunnel diode has several unique features because of this negative resistance region.

Consider an initial condition with $V_s = 0$ and $I = 0$. Then, let V_s increase to V_{s1} so that Q_1 becomes the operating point. As V_s increases to V_{s2}, the operating point goes to Q_2. If V_s continues to increase to V_{s3}, then the only operating point shifts to

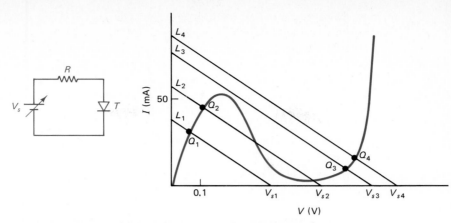

Fig. 5-34 Tunnel diode circuit and characteristic: increasing V_s.

Q_3. At this transition point, there is a large change in V produced by a relatively small change in V_s. Because of the energy requirements of the device, this shift takes a small but finite time. This time delay is important only at high frequencies or for circuits that switch very rapidly. An additional increase in V_s shifts the operating point to Q_4.

As V_s is now reduced to V_{s5}, the operating point moves to Q_5, as shown in Fig. 5-35. Then, when $V_s = V_{s6}$, the operating point must shift to Q_6. Another large shift in V ocurs for a small change in V_s, again requiring energy and taking a finite time for the change to take place. Further decreases in V_s produce the operating point Q_7.

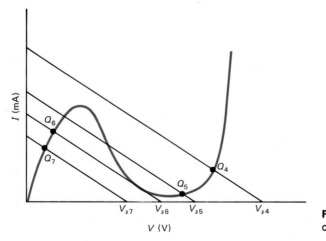

Fig. 5-35 Tunnel diode circuit and characteristic: decreasing V_s.

If the load line intersects the characteristic curve of the tunnel diode at three points (see line L_b in Fig. 5-36), the tunnel diode operates at Q_{b1} or Q_{b3}; it does not operate at Q_{b2}. Because of the negative resistance of r, the circuit slides from Q_{b2} to Q_{b3} on its own. If the load line intersects the characteristic curve only in the negative resistance region, point Q_a is obtained. This point is difficult to observe directly because of the instability produced by the negative resistance.

The operating points of a tunnel diode thus depend on the load line that is se-

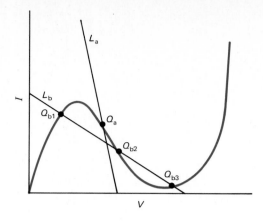

Fig. 5-36 Alternative load line intercepts.

lected. By selecting a certain family of load lines, the tunnel diode can be used as a two-state device. The tunnel diode can also be biased with a negative value of r and used to create an oscillator circuit (to be discussed in Chapter 14).

5-12 DIODE DESIGN

A range of design methods has now been developed for two-terminal devices. The load line concept is generally applicable to any new two-terminal device that is encountered. In addition, practice has been gained with the process of using idealized models to estimate circuit behavior. Experience with these models helps develop insight into patterns of analysis that can be used for these and similar circuits.

5–13 SPICE COMPUTER PROGRAM

Throughout this text, the usefulness of computers in electronics design is demonstrated through applications at the end of each chapter. It is important to recognize that standard software packages are available for computer-aided design. These can supplement and strengthen the detailed analysis required for a complex circuit that has to meet performance objectives. One of the most common packages generally available is SPICE (*S*imulation *P*rogram with *I*ntegrated *C*ircuit *E*mphasis). Several different versions of SPICE are described in the references for this chapter.

SPICE is a computer program that enables the designer to predict how a given circuit will function. It is not a design program, since it does not create a circuit to produce desired performance parameters; it is an analysis tool that predicts the behavior of a circuit that is fed into it.

SPICE is often best used in combination with the types of analysis described in this text. As noted in Chapter 1, the design of an electronic circuit begins with the setting of objectives. Approximate modeling and calculations are useful to determine whether or not the general strategy being followed is likely to be successful. SPICE can then be applied to obtain more precise data regarding detailed aspects of circuit performance. Through an iterative process, SPICE can greatly aid the design activity.

In order to use SPICE, the designer first creates a schematic using combinations of resistors, capacitors, inductors, diodes, and transistors of various types. Each node of the circuit is then assigned a number (with the ground node = 0). Each component in the circuit is assigned an alphanumeric name and a value. The input and output nodes are identified along with all nodes that require the application of dc voltage levels. The desired waveform is placed on the input, the dc voltages are applied, and the format of the output is specified. The complete circuit is used to develop a file that is input to the SPICE program, with instructions included regarding the type of analysis desired.

Figures 5-49 and 5-52 illustrate simple circuits that have been used as input for SPICE simulations of a half-wave rectifier (with and without a capacitor filter). The figures also provide listings of the data input to SPICE.

A standard format is used to describe each circuit component. Resistors are represented by an alphanumeric name beginning with the letter "R". The two numbers following the name give the node connections, and the entry is completed by giving the value for the resistor. Capacitors follow a similar format, using names beginning with the letter "C". Diodes are entered using an alphanumeric name beginning with the letter "D", followed by the connection nodes and the name of the diode model that is to be used. This model is then defined in terms of standard input variables, using another input line.

A similar approach is used to define input voltages (VIN) with appropriate node connections and specification of the waveform to be used. The remaining entries provide information regarding the desired output and its format. As can be seen, SPICE lends itself well to an iterative mode of operation due to the simple data format. User manuals (as listed in the references) specify the format to use for each type of component and how to achieve the type of simulation desired.

SPICE can be used to simulate dc circuit performance, transients, and linear ac analysis. Circuits can contain complex component and device models of many different types.

The built-in diode and transistor models that are available for use can be modified through variations in basic physical parameters defined by the user. Some of the available models are introduced in the following chapters.

A dc analysis is automatically performed at the beginning of a SPICE simulation to determine initial conditions and find the values for ac parameters (such as r) that change with the bias point. A linearized circuit model can then be used to study the performance of the circuit for small-amplitude input signals. (This type of application is discussed in Chapters 7 to 9.)

Transient analysis describes the output of a circuit using initial conditions determined by the dc analysis and a time frame set by the user. For large-signal waveforms, a Fourier analysis of the output waveform can be obtained directly from SPICE, yielding the Fourier coefficients discussed in this chapter.

SPICE is designed to simulate circuit operation at a nominal temperature of 300°K. The equations used in SPICE contain a variety of different temperature-dependent terms, so the effect of temperature change on circuit performance can also be found.

Some versions of SPICE operate in a batch processing mode, while others allow for an interactive mode using displays on a monitor. The user can feed in a circuit design, observe the performance graphed on a monitor, then modify the circuit and observe the changes in circuit operation. SPICE can thus be a flexible design aid.

As illustrated in Computer Application 5-1 at the end of this chapter, SPICE can be used to predict the output waveforms that result from half- and full-wave rectifiers, using realistic diode characteristic curves. A number of additional SPICE applications are included in the Computer Application sections in future chapters.

QUESTIONS AND ANSWERS

Questions
5-1. Why is it difficult to make direct use of the theoretical diode equation [see Eq. 5-1]?

5-2. What tradeoffs affect the selection of a diode model for use in circuit design?

5-3. What factors affect the selection of a load line?

5-4. What will happen to a diode that operates in an enclosure for which heat exchange with the environment is severely restricted?

5-5. What insight does Fourier analysis provide into the operation of rectifiers?

5-6. What is the significance of a waveform with a reduced ripple factor?

5-7. How do the semiconductor properties of zener diodes affect device performance in a circuit?

5-8. How do the semiconductor properties of photodiodes affect device performance in a circuit?

5-9. Why are computer programs like SPICE useful in circuit design?

Answers
5-1. Since the theoretical diode equation is nonlinear in the current I, it is difficult to apply in closed form.

5-2. When selecting a diode model for circuit design, model accuracy must be traded off against ease of application. Complex, more complete, and accurate models provide the potential for predicting circuit operation with confidence, but are often difficult to apply. On the other hand, approximate, simple models are limited in their predictive ability, but are easier to apply.

5-3. A load line is determined by the desired operating point Q, the desired slope $-1/R$, and the preferred values for the intercepts on the x-axis (V_s) and y-axis (V_s/R). The resultant Q point must be within acceptable operating ranges for the diode, and the slope of the load line must produce the desired circuit stability with temperature T and insensitivity to variations among devices. As will be seen below, all of these load line properties affect decisions regarding circuit design to achieve performance objectives.

5-4. If a diode operates without the ability to remove the heat that develops through thermal exchange with the environment, the temperature of the diode continues to increase until the device temperature exceeds the device rating and the diode is destroyed.

5-5. Fourier analysis can be used to describe the frequency components of the outputs from half-wave and full-wave rectifiers. This type of analysis explicitly demonstrates how the nonlinear diode characteristic curve introduces new frequency components into the output, thereby changing the nature of the wave shape. The ability to think of nonsinusoidal waveforms in terms of the (amplitude and phase) relationships among frequency components provides a powerful approach to the study of circuit performance, and will be used in several subsequent chapters.

5-6. The ripple factor (RF) is used to describe the smoothness of a waveform. For a flat (time-independent) wave, as achieved by using a battery, the ripple factor approaches zero. The voltage variations around the average voltage value are negligible. For dc power supplies consisting of rectifier-filter combinations, a small ripple is observed "riding on" the average dc output. As the ripple factor is reduced, the output approaches the idealized case of a perfect (unvarying) voltage source. The RF measures how close to this ideal a given circuit performs.

5-7. Zener diodes make use of zener or avalanche breakdown when large reverse-bias voltages are applied across the devices. An understanding of the types of carriers in the device and their properties, and the energy relationships among carriers, is required in order to anticipate where breakdown will occur and to predict the ac resistance of the characteristic curve in the breakdown region. The very low ac resistance produces the circuit behavior associated with a constant-voltage source.

5-8. An understanding of the properties of photodiodes requires insight into energy-band concepts, the formation of electron-hole pairs by photons, and the ways in which these pairs affect current flow in the device. Photodiode performance depends on the spectrum of the source and the band structure of the device, and the prediction of detector sensitivity in a fiber-optic communication system is based on the source properties, absorption mechanisms in the photodiode, and carrier motion in the device after excitation.

5-9. SPICE and similar computer programs enable the designer to more accurately predict how a given circuit will function. Given a more detailed predictive ability, the designer can better evaluate the tradeoffs that are involved and improve on decision-making. In a computer-integrated manufacturing (CIM) environment, computer simulations can reflect the properties of the specific manufacturing processes that are in use, providing the designer with a high level of correlation between design and product performance.

These programs do not design the circuit of interest, in that they do not evaluate the tradeoffs and select the final configuration. Rather, they provide improved information regarding the performance of potential circuit designs, enhancing the design process.

As expert systems (based on artificial intelligence techniques) become more widely available, circuit simulations are beginning to extend further into the decision-making process. It is reasonable to expect that circuit design will eventually be a truly interactive process between the individual and a computer system.

EXERCISES AND SOLUTIONS

Exercises

(5-1)   **5-1.** A diode at 300°K is biased so that a dc current $I = 20$ mA flows through it. Estimate the ac resistance r of the diode.

(5-1)   **5-2.** A diode has parameter values $R_b = 5.0 \ \Omega$ and $n = 1.5$ at 10 mA. (a) Find r based on these values. (b) Assume that an approximate calculation of r is made without knowing R_b or n, so the assumptions $R_b = 0$ and $n = 1.0$ are used. Find the value of r that results from this calculation. (c) What percentage error is observed by using the model in part (b) if the actual values are those given in part (a)?

(5-2)   **5-3.** You are given a circuit with a battery $V_s = 3.0$ V and wish to produce a current $I = 50$ mA with a series combination of a resistor and the forward-biased Si diode of Fig. 5-1. Find the required resistor value.

(5-2)   **5-4.** A load line intercepts the vertical axis at 100 mA. If $R = 25 \ \Omega$ for the load line, (a) find the Q point using the Ge diode of Fig. 5-1. (b) Find the necessary battery voltage to create a series circuit at the specified Q point.

(5-3)   **5-5.** For a forward-biased diode, $V_D \cong 0.30$ V and $I_D = 1.0$ A. The ambient temperature is 75°C above the reference level. If $\theta_{JA} = 30$°C/W, find the approximate voltage shift that occurs in the characteristic curve.

(5-4)   **5-6.** Draw a schematic for a half-wave rectifier, indicating the input and output. Sketch the output you expect using the ideal diode model with a sine wave input. Given $V_m = 100$ V for the input amplitude, find the average and effective output voltages.

(5-4) **5-7.** Draw a schematic for a full-wave rectifier using a transformer input, indicating the input and output. Sketch the output you expect using the ideal diode model with a sine-wave input. Given $V_m = 10$ V for the input amplitude and $V_s = 10N_p$, find the average and effective output voltages.

(5-4) **5-8.** Using a point-by-point graphical construction, add the individual terms in Eq. (5-15) to determine how close the result is to an expected half-wave rectified waveform.

(5-5) **5-9.** Given the input sine wave $v = (2.0 \text{ V}) \sin \omega t$, a 20-$\Omega$ load, and the Si diode of Fig. 5-1, (a) graphically determine the actual voltage waveform that is observed at the load for a bridge rectifier. (b) Graphically determine V_{av} for the output. (c) Compare the answer of part (b) with the value of V_{av} that is obtained using the ideal diode model. (d) Compare with the V_{av} obtained using a diode model with V_D constant at 0.7 V.

(5-6) **5-10.** By direct integration, find the ripple factor for a half-wave rectified output. Use the ideal diode model and the Fourier expansion of Eq. (5-15).

(5-7) **5-11.** You have an 8-V zener and a 20-V battery, as shown in Fig. 5-37. For what values of R_1/R_2 is the zener *on*?

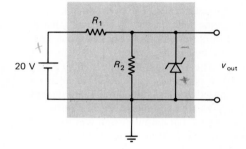

Fig. 5-37 Circuit for Exercise 5-11.

(5-7) **5-12.** A fixed-voltage power supply is to use the schematic shown in Fig. 5-27. Assume that $V_{out} = -6.0$ V and the filter produces a ripple factor of 0.10. Given Si diodes and an ac source $\mathbf{v} = 110$ V, find the turns ratio that should be used for the (ideal) transformer. Assume $R_z = (R_L)_{min} = 0.10$ kΩ.

(5-8) **5-13.** A variable-voltage power supply is to make use of the schematic shown in Fig. 5-29 with the UC150 IC voltage regulator. Assume that $V_{out} = 2$ to 20 V and the filter produces a ripple factor of 0.10. A minimum voltge drop of 4.0 V must be maintained across the pass transistor. Given $R_2 = 0$ to 5.0 kΩ and $R_1 = 0.50$ kΩ, what values should be selected for the zener diode voltage and the (ideal) transformer turns ratio, given the line voltage $\mathbf{v} = 110$ V?

(5-9) **5-14.** Design clipper circuits that transform the input signals in Fig. 5-38 to the desired outputs shown using the ideal diode model.

(5-10) **5-15.** The photodiode in Fig. 5-30, a 70-kΩ resistor, and a 5.0-V source are connected in series. (a) If the incident light is 6.0 mW/cm², what is the voltage across the resistor? (b) If the light changes to 5.0 mW/cm², what is the voltage change across the resistor?

(5-11) **5-16.** A tunnel diode has the characteristic curve shown in Fig. 5-34. It is placed in series with a resistor $R = 75$ Ω and a source $V_s = 3.0$ V so that it is forward-biased. (a) Find the possible operating point(s). (b) Find r at each operating point.

5-17. A company produces dc power supplies (consisting of a rectifier/filter combination). A new BA-4 model is introduced to increase market share. However, despite adequate breadboard testing, the BA-4 product begins to experience large numbers of failures. Based on the present return rates, the company expects that 20 percent of all power supplies will eventually be returned for repair during the warranty period, requiring the re-

Input | Output

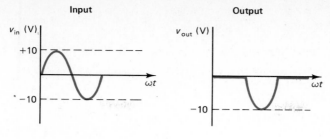

(A)

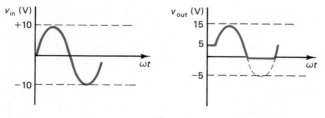

(B)

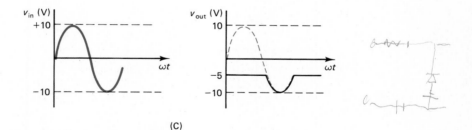

(C)

Fig. 5-38 Waveforms for Exercise 5-14.

placement of all diodes that are burning out unexpectedly. At the end of the first year, the company discontinues the BA-4 in favor of a new BA-5 model configuration. During the period the BA-4 was offered, 1000 power supplies were sold at $50.00 each. Repair costs are $60.00/unit and the total design and production costs were $40,000. (a) What net profit or loss was experienced by the company for the BA-4? (b) What would the profit have been without the rework and repair requirement?

Solutions   **5-1.** From Eq. (5-2), $r \cong 0.026$ V/20 mA $\cong 1.3$ Ω.

5-2. From Eq. (4-35),

(a)
$$r = R_b + \frac{nV_0}{I}$$

$$= 5.0 \ \Omega + \frac{(1.5)(0.026 \ \text{V})}{10 \ \text{mA}} \cong 8.9 \ \Omega$$

(b)
$$r \cong V_0/I = 0.026 \ \text{V}/10 \ \text{mA} \cong 2.6 \ \Omega$$

$$\frac{(8.9\ \Omega - 2.6\ \Omega)}{8.9\ \Omega} \times 100\% \cong 71 \text{ percent error is observed}$$

3.0 V

50 mA

Si diode of Fig. 5-1

Fig. 5-39 Schematic for Exercise 5-3.

5-3. The schematic of interest is shown in Fig. 5-39. From Fig. 5-1,

$$I_Q = 50 \text{ mA results in } V_Q \cong 0.7 \text{ V}$$

From Eq. (5-5),

$$R = \frac{V_s - V_Q}{I_Q} = \frac{3.0 \text{ V} - 0.7 \text{ V}}{50 \text{ mA}} \cong 46\ \Omega$$

Observe that it is not necessary to redraw Fig. 5-1 with an extended x-axis to accommodate the desired load line.

5-4. From Fig. 5-40,

$$V_Q \cong 0.32 \text{ V} \qquad \text{and} \qquad I_Q \cong 87 \text{ mA}$$

From Eq. (5-6),

$$V_s = V_Q + I_Q R \cong 0.32 \text{ V} + (87 \text{ mA})(25\ \Omega)$$

$$\cong 0.32 \text{ V} + 2.18 \text{ V} \cong 2.5 \text{ V}$$

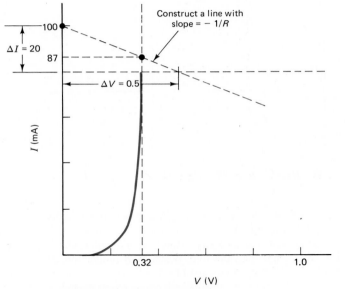

Fig. 5-40 Solution approach for Exercise 5-4.

5-5. From Eq. (5-12),

$$\Delta V \cong \left(\frac{\partial V}{\partial T}\right)_I [(\Delta T)_{\text{amb}} + P_Q\, \theta_{JA}]$$

$$\cong (-2 \text{ mV/}^\circ\text{K})[75^\circ\text{C} + (0.3 \text{ V})(1.0 \text{ A})(30^\circ\text{C/W})]$$

$$\cong (-2 \text{ mV/}^\circ\text{K})(84^\circ\text{C}) \cong -0.2 \text{ V}$$

Since the °K and °C units are of equal width, the units cancel.

5-6. The schematic and output take the form of Fig. 5-41. From Eq. (5-15),

$$V_{av} = V_m/\pi = 100 \text{ V}/\pi \cong 31.8 \text{ V}$$

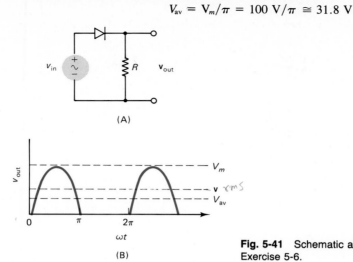

(A)

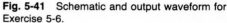

(B)

Fig. 5-41 Schematic and output waveform for Exercise 5-6.

From Appendix 1,

$$\mathbf{v} = \left[\frac{1}{2\pi}\int_0^{2\pi} v^2(\theta)\, d\theta\right]^{1/2}, \qquad \theta = \omega t$$

$$= \left[\frac{\int_0^\pi V_m^2 \sin^2\theta\, d\theta}{2\pi}\right]^{1/2} = V_m\left[\frac{\int_0^\pi \frac{1}{2}(1 - \cos 2\theta)\, d\theta}{2\pi}\right]^{1/2}$$

$$= V_m\left(\frac{\pi/2}{2\pi}\right)^{1/2} = V_m/2 = 100 \text{ V}/2 = 50 \text{ V}$$

5-7. The schematic and output take the form of Fig. 5-42. From Eq. (5-16),

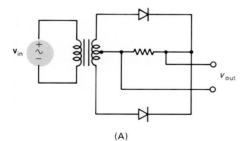

(A)

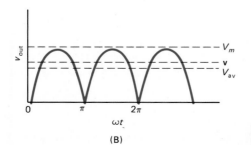

(B)

Fig. 5-42 Schematic and output waveform for Exercise 5-7.

$$V_{av} = 2V_m/\pi = (2 \times 100 \text{ V})/\pi \cong 63 \text{ V}$$

From Appendix 1,

$$\mathbf{v} = \left[\frac{1}{2\pi} \int_0^{2\pi} v^2(\theta) \, d\theta \right]^{1/2}, \qquad \theta = \omega t$$

$$= V_m \left[\frac{2 \int_0^{\pi} V_m^2 \sin^2 \theta \, d\theta}{2\pi} \right]^{1/2}$$

$$= V_m/\sqrt{2} \cong 71 \text{ V}$$

5-8. By graphical construction, the results of Fig. 5-43 are obtained. Note the rapid convergence of the series to the expected form.

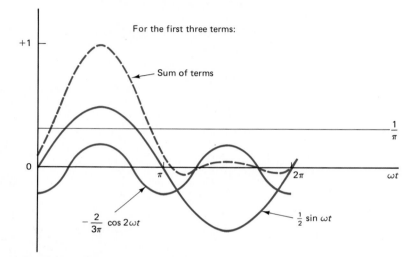

Fig. 5-43 Graphical solution for Exercise 5-8.

5-9. (a) The graphical analysis of this problem is shown in Figure 5-44. (b) The average value for the output can be determined by calculating

$$V_{av} = \frac{\text{number of squares under the curve}}{\text{number of squares from 0 to } \pi \text{ on the x-axis}}$$
$$\times \text{ voltage represented by each square along the y-axis}$$

$$\cong 58/8 \times (0.1/4) \text{ V} \cong 0.18 \text{ V}$$

(c) With the ideal diode model,

$$V_{av} = 2V_m/\pi = (2 \times 2.0 \text{ V})/\pi \cong 1.27 \text{ V}$$

(d) With a diode model using $V_D = 0.7$ V

$$V_{av} = \frac{2}{\pi}(V_m - 2V_D) = 2 \times \frac{0.6 \text{ V}}{\pi} \cong 0.38 \text{ V}$$

5-10. By definition,

$$\text{ac ripple} = \left[\frac{1}{2\pi} \int_0^{2\pi} \left(\frac{1}{2} \sin \omega t - \frac{2}{3\pi} \cos 2\omega t - \cdots \right) \right.$$

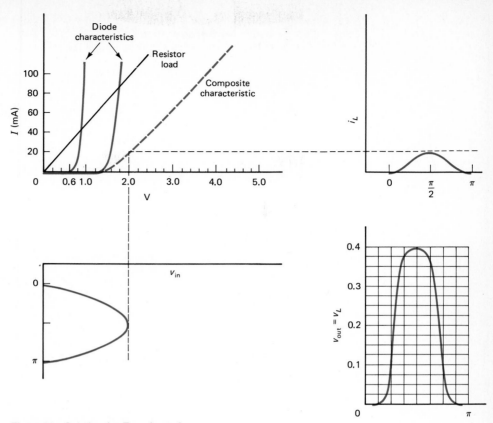

Fig. 5-44 Solution for Exercise 5-9.

$$\times \left(\frac{1}{2} \sin \omega t - \frac{2}{3\pi} \cos 2\omega t - \cdots \right) \times d(\omega t) \Bigg]^{1/2}$$

$$= \left[\frac{1}{2\pi} \int_0^{2\pi} \left(\frac{1}{2}\right)^2 \sin^2 \omega t \, d(\omega t) + \int_0^{2\pi} \left(\frac{2}{3\pi}\right)^2 \cos^2 2\omega t \, d(\omega t) + \cdots \right]^{1/2}$$

$$= \frac{1}{\sqrt{2}} \left[\left(\frac{1}{2}\right)^2 + \left(\frac{2}{3\pi}\right)^2 + \cdots \right]^{1/2}$$

$$\text{RF} = \frac{\dfrac{1}{\sqrt{2}} \left[\left(\dfrac{1}{2}\right)^2 + \left(\dfrac{2}{3\pi}\right)^2 + \cdots \right]^{1/2}}{1/\pi}$$

$$\cong 1.2$$

5-11. The zener will be *on* if

$$\left(\frac{R_2}{R_1 + R_2}\right) 20 \text{ V} > 8 \text{ V}$$

or

$$R_1/R_2 < 1.5$$

5-12. Choose $V_z = 6.0$ V. Given $R_z = R_L$, the most positive voltage out of the RC filter must be -12 V to assure that the zener stays *on*.

For RF $= 0.10$, Eq. (5-21) can be used to find the ratio

$$\frac{\Delta V_m}{V_m} = \sqrt{3}\,(0.10) \cong 0.17$$

For a small RF (with V_{min} as the most negative voltage out of the filter circuit),

$$V_{av} \cong \frac{1}{2}(V_{min} + V_{max})$$

Therefore, to achieve $V_{out} = -6.0$ V and to assure that the zener diode is always *on*,

$$\Delta V_m = -(V_{max} - V_{min}) = \frac{1}{2}(V_{min} + V_{max})(0.17)$$

and with $V_{max} = -12$ V,

$$V_{min} = -14.2 \text{ V}$$

With maximum voltage drops of 0.7 V across each diode, the peak amplitude of the transformer output must be

$$V_s = -14.2 \text{ V} - 1.4 \text{ V} = -15.6 \text{ V}$$

with an effective value

$$\mathbf{v}_s = 15.6 \text{ V}/\sqrt{2} \cong 11 \text{ V}$$

Thus, the turns ratio on the transformer should be

$$N_s/N_p = 11 \text{ V}/110 \text{ V} = 0.10$$

5-13. From Eq. (5-24), the minimum output voltage is obtained with $R_z = 0$ and the maximum with $R_z = 5.0$ kΩ. Therefore,

$$V_{max} = V_{ref}\!\left(1 + \frac{5.0 \text{ k}\Omega}{0.50 \text{ k}\Omega}\right) = 11V_{ref} = 20 \text{ V}$$

and

$$V_{ref} = 20 \text{ V}/11 \cong 1.8 \text{ V}$$

is the required value for the zener.

Given a minimum of 4.0 V across the pass transistor, the minimum input to the regulator circuit must be $V_{in} > 20 \text{ V} + 4.0 \text{ V} = 24$ V. Given RF $= 0.10$, from Eq. (5-21),

$$\Delta V_m = V_{max} - V_{min} = \sqrt{3}(0.10)\frac{1}{2}(V_{min} + V_{max})$$

and with $V_{min} = 24$ V,

$$V_{max} = 26.04 \text{ V}/0.915 = 28.5 \text{ V}$$

If maximum voltage drops of 0.7 V develop across each diode, the peak transformer output must be

$$V_s = 28.5 \text{ V} + 1.4 \text{ V} = 29.9 \text{ V}$$

with an effective value

$$v_s = 29.9 \text{ V}/\sqrt{2} \cong 21 \text{ V}$$

The turns ratio on the transformer should be

$$N_s/N_p = 21 \text{ V}/110 \text{ V} \cong 0.19$$

5-14. By applying the methods of Section 5-7, the results of Fig. 5-45 are obtained.

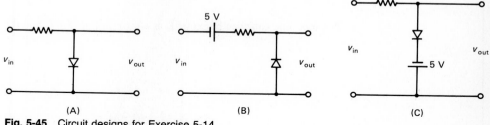

(A)   (B)   (C)

Fig. 5-45 Circuit designs for Exercise 5-14.

5-15. The schematic is shown in Fig. 5-46(A), which results in the load line shown in Fig. 5-46(B). (a) At the operating point, $V_D \cong 2.0$ V, so that $V_R \cong 3.0$ V. (b) $\Delta V_D \cong -\Delta V_R \cong 0.7$ V.

Fig. 5-46 Circuit and load line for Exercise 5-15.

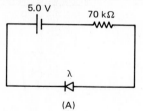

(A)

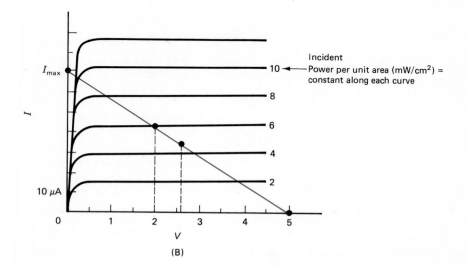

(B)

5-16. (a) For the given load line, the I-axis intercept is

$$I_{max} = \frac{3.0 \text{ V}}{75 \text{ } \Omega} \cong 40 \text{ mA}$$

At $V = 0.7$ V, the value of I is

$$I = \frac{3.0 \text{ V} - 0.7 \text{ V}}{75 \text{ } \Omega} \cong 31 \text{ mA}$$

The load line can then be plotted, as shown in Fig. 5-47, and the operating points found.

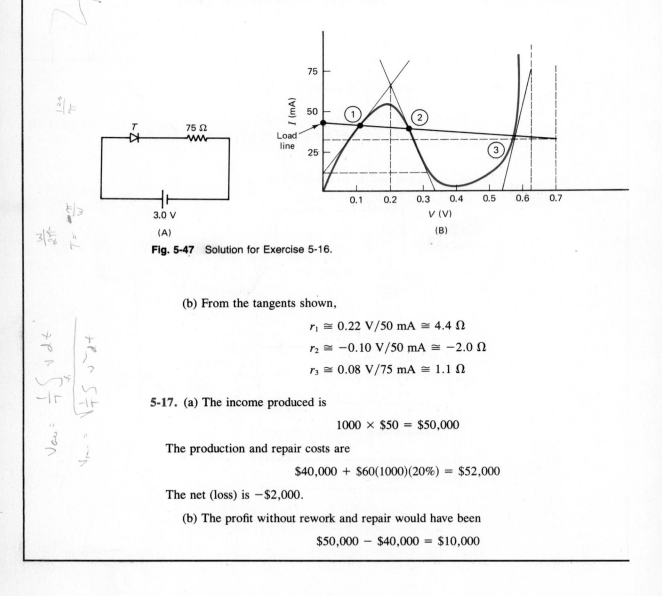

Fig. 5-47 Solution for Exercise 5-16.

(b) From the tangents shown,

$$r_1 \cong 0.22 \text{ V}/50 \text{ mA} \cong 4.4 \text{ } \Omega$$
$$r_2 \cong -0.10 \text{ V}/50 \text{ mA} \cong -2.0 \text{ } \Omega$$
$$r_3 \cong 0.08 \text{ V}/75 \text{ mA} \cong 1.1 \text{ } \Omega$$

5-17. (a) The income produced is

$$1000 \times \$50 = \$50,000$$

The production and repair costs are

$$\$40,000 + \$60(1000)(20\%) = \$52,000$$

The net (loss) is $-\$2,000$.

(b) The profit without rework and repair would have been

$$\$50,000 - \$40,000 = \$10,000$$

PROBLEMS

(5-1) **5-1.** A current of 25 μA flows in a diode at 300°K. Estimate the ac resistance of the diode.

(5-1) **5-2.** If the current through a diode increases by a factor of 10, how does the ac resistance change?

(5-1) **5-3.** Find the ac resistance for the diode in Fig. 5-8(C) at $I = 2.0$ mA by graphical means and by estimation (equation). Compare the results.

(5-2) **5-4.** A diode, resistor, and 6.0-V battery in series produce the operating point $V_Q = 0.60$ V, $I_Q = 35$ mA. Find the value of the resistor.

(5-2) **5-5.** A diode, a 2.0 kΩ resistor, and a dc power supply in series provide the operating point $V_Q = 0.40$ V, $I_Q = 100$ μA. Find the correct setting for the power supply.

(5-2) **5-6.** Consider the load line in Fig. 5-6. (a) How much can V_s be increased while still remaining in the allowed operating region? (b) What are V_Q and I_Q at the maximum value of V_s?

(5-2) **5-7.** A load line produces the operating point $V = 2.0$ V, $I = 6.0$ mA with $R = 50$ Ω. Find the value of V_s required.

(5-2) **5-8.** Find R and r for the diode characteristic shown in Fig. 5-8(C) at $V = 0.75$ V. If $V_s = 2.5$ V, what load resistor is required to produce a load line that passes through the operating point defined by $V = 0.75$ V?

(5-2) **5-9.** You wish to produce a current $I = 15$ mA(± 10 percent) in a series circuit consisting of a Si diode, a resistor, and a power supply. If the diode temperature can vary by 50°C and if $V_s < 15$ V for the power supply, what range of values for R is possible?

(5-3) **5-10.** A Si diode is mounted in a case with $\theta_{JA} = 100$°C/W. If $V_D = 0.60$ V and $I_D = 25$ mA, what change in temperature is observed due to internal heating?

(5-3) **5-11.** A Si diode is in series with a 30-Ω resistor and a power supply. If $I = 3.0$ mA flows in the diode, what change in the voltage across the diode is associated with a temperature change of 100°C?

(5-3) **5-12.** The ambient temperature for a Si diode increases from 25 to 100°C. Find the approximate voltage shift that occurs across a forward-biased diode given $r \ll R$.

(5-3) **5-13.** A Si diode is biased with a 100-Ω ac resistance and is in series with a 50-Ω resistor. When forward-biased, what change in the voltage across the diode is expected if T changes from 300 to 250°K?

(5-3) **5-14.** A Ge diode in an enclosure dissipates 0.5 W and θ_{JA} for the diode case is 50°C/W. (a) If the operating temperature of the entire system changes from 300 to 350°K, what is the temperature of the diode? (b) If $r \ll R$ for the diode bias point, what approximate voltage change is observed across the diode? (Assume $(\partial V / \partial T)_I \cong -1.5$ mV/°C).

(5-3) **5-15.** A Si diode dissipates 0.5 W in a container that allows very little heat to escape. If the (mc) product for the container is 2.5 J/°C and $\theta_{JA} = 100$°C/W for the diode, how long will it take for the diode temperature to increase by 100°C?

(5-4) **5-16.** What are the amplitudes for the first three terms in the Fourier expansion for a half-wave rectifier output?

(5-4) **5-17.** What are the amplitudes for the first three terms in the Fourier expansion for a full-wave rectifier output?

(5-4) **5-18.** A sine wave with a 20-V amplitude is input to a half-wave rectifier. Assuming an ideal diode, find the average and effective values for the output voltage.

(5-4) **5-19.** A sine wave with a 20-V amplitude is input to a full-wave rectifier. Assuming an ideal diode, find the average and effective values for the output voltage.

(5-4) **5-20.** Consider a half-wave rectifier with $v_{in} = (10$ V) cos ωt. Find the average and effective values of the output using the diode model with $V_D = 0.7$ V.

(5-4)   **5-21.** Using a point-by-point graphical construction, add the individual terms in Eq. (5-16) to determine how close the result is to an expected full-wave rectified form.

(5-5)   **5-22.** Consider a bridge rectifier with $v_{in} = (20 \text{ V}) \cos \omega t$ and $R = 1.1 \text{ k}\Omega$. Find the average value of the output current using the ideal diode model.

(5-5)   **5-23.** For a bridge rectifier, the amplitude of the input is 10 V. Using a diode model with $V_D = 0.7$ V, draw the expected output waveform.

(5-5)   **5-24.** An ac voltage with 50 V peak-to-peak is input to a bridge rectifier. Using the ideal diode model, find the average and effective values for the input and output voltages.

(5-5)   **5-25.** Draw a schematic for a bridge rectifier, indicating input and output. Assume an input square-wave signal of amplitude ±50 V. Using the ideal diode model, find the average output voltage.

(5-6)   **5-26.** Find the frequency and period for the waveforms in Figs. 5-20(A) to (D).

(5-6)   **5-27.** Graphically estimate the time constants associated with the exponential decays in Figs. 5-20(A) and (C). Are these values consistent with the defined circuit?

(5-6)   **5-28.** Calculate the ripple factor associated with the waveforms in Figs. 5-20(A) to (D).

(5-6)   **5-29.** An output waveform takes on the following approximate form:

$$v = [0.50 + 0.20 \cos 2\omega t] \text{ V}$$

Sketch the waveform and calculate the ripple factor.

(5-7)   **5-30.** Find the v_{out} waveforms for each of the circuits shown in Fig. 5-48 using the ideal diode model.

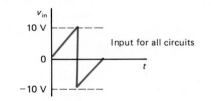

Fig. 5-48 Input waveform and clipping circuits for Problem 5-30.

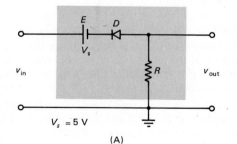

(A)

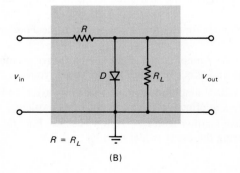

(B)

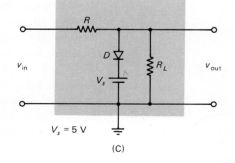

(C)

194   Chapter 5   Diode Models and Circuit Applications

(5-8) **5-31.** Find the peak power dissipated in the zener diode used in the regulator circuit of Exercise 5-12.

(5-8) **5-32.** For the circuit of Fig. 5-24 let $V_Z = 3.0$ V. The input is given by a 12-V battery in series with a 6-V peak-to-peak sine wave. Find V_{out} if $R_1 = 2R_2$.

(5-8) **5-33.** Show how IC voltage regulators can be used to create adjustable positive and negative voltages from a single bridge rectifier.

(5-8) **5-34.** You are given the power supply of Fig. 5-29, using the UC150 voltage regulator. Assume that the nominal output voltage is 12.0 V. What would you expect to observe for the maximum change in the regulated output voltage?

(5-8) **5-35.** An IC voltage regulator of the type illustrated in Fig. 5-29(B) is used with $V_{ref} = 2.5$ V. Find the maximum and minimum possible values for V_{out} if $R_2 = 5.0$ kΩ and $R_1 = 0.5$ kΩ.

(5-8) **5-36.** If $R_L = 0.10$ kΩ for the variable-voltage power supply of Exercise 5-13, what peak power is dissipated in the pass transistor for $V_{out} = 12$ V?

(5-9) **5-37.** The photodiode of Fig. 5-32 is in series with a 0.10 MΩ resistor and a 12-V power supply. If light of intensity 8.0 mW/cm^2 is incident on the photodiode, what are the voltages across the photodiode and the resistor?

(5-10) **5-38.** The tunnel diode described by Fig. 5-34 is in series with a 50-Ω resistor and a 1.5-V power supply. Find the possible operating point or points and r at each point.

COMPUTER APPLICATIONS

5-1. The SPICE computer program can be used to predict the output of half-wave and full-wave rectifiers with realistic diode curves. Figures 5-49 through 5-51 show the input data and output waveforms for SPICE simulations of half-wave and bridge rectifiers. Figure 5-49 includes the schematic and node definitions and the input data file used for the half-wave rectifier. A Si diode was assumed, with typical (default) parameters. The output is

Fig. 5-49 (A) Schematic and node definitions, and (B) input file for a SPICE simulation of a half-wave rectifier.

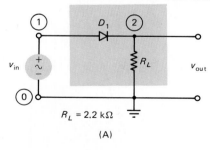

$R_L = 2.2$ kΩ

(A)

```
T HWRECT
/L ALL
    1        HALF WAVE RECTIFIER
    2        *
    3        * RESISTOR
    4        *
    5        RL 2 0 2.2K
    6        D1 1 2 DOD
    7        *
    8        * DIODE
    9        *
   10        .MODEL DOD D(PB=0.7 RS=10 CJO=2PF BV=40.0)
   11        *
   12        * INPUT SINE WAVE 60HZ
   13        *
   14        VIN 1 0 SIN(0 2 60)
   15        .TRAN 0.2778MS 25MS
   16        *
   17        * OUTPUT
   18        *
   19        .PRINT TRAN V(2) V(1)
   20        .PLOT TRAN V(2) V(1)
   21        .END
```

(B)

shown in Fig. 5-50 for an input amplitude of 2.0 V. The output of a SPICE simulation of a full-wave rectifier with $R_L = 2.2$ kΩ is shown in Fig. 5-51. Draw graphs of the voltage across each diode as a function of time for each circuit.

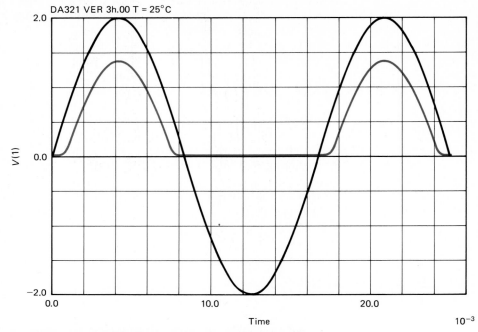

Fig. 5-50 Output of SPICE simulation for a half-wave rectifier.

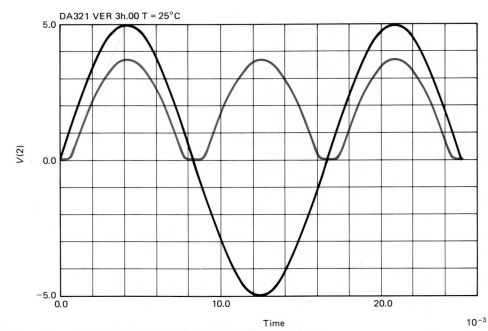

Fig. 5-51 Output of SPICE simulation for a full-wave rectifier.

5-2. The SPICE computer program can be used to predict the effect of a capacitor filter on the output of a half-wave or full-wave rectifier. Figure 5-52 shows the program input that was used to produce the graph for the half-wave rectifier/capacitor filter output of Fig. 5-20(A). Figure 5-53 (p. 198) shows comparable input data for the full-wave rectifier/capacitor filter of Fig. 5-20(C). (a) Compute the ripple factors obtained in each case. (b) Measure the time constant associated with the circuit of Fig. 5-20(A) while the diode is *off*, and confirm that the expected value is obtained. (c) Plot the diode current as a function of time for each circuit.

* **5-3.** The graph of Fig. 5-10 for an ideal half-wave rectifier output was produced by using a FORTRAN computer program to combine the terms given in Eq. (5-15). Develop a computer program that will enable you to recreate these results. Then modify your program to produce the full-wave rectifier results of Fig. 5-15 using the terms in Eq. (5-16).

* **5-4.** Assume that for the input to a computer-aided design program for the bridge rectifier, you are given the characteristic curve of the diodes being used and an arbitrary input waveform. Write the program required to calculate the output voltage waveform. Use the program to find the output waveform if $v_{in} = (4 \text{ V}) \sin \omega t$ and the diode of Fig. 5-8(C) is used. Turn in your program listing, graph, and interpretation.

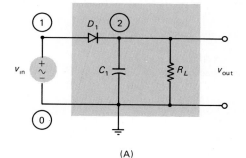

(A)

```
/L ALL
      1    HALF WAVE RECTIFIER WITH CAPACITOR FILTER
      2    *
      3    * RESISTOR
      4    *
      5    RL 2 0 1K
      6    *
      7    * CAPACITOR
      8    *
      9    C1 2 0 10UF
     10    *
     11    * DIODE
     12    *
     13    D1 1 2 DOD
     14    .MODEL DOD D(PB=0.7 RS=10 CJO=2PF BV=40.0)
     15    *
     16    * INPUT SINE WAVE AT 60HZ
     17    *
     18    VIN 1 0 SIN(0 10 60)
     19    .TRAN 0.25MS 75MS 25MS
     20    *
     21    * OUTPUT
     22    *
     23    .PRINT TRAN V(2) V(1)
     24    .PLOT TRAN V(2) V(1)
     25    .WIDTH IN=72 OUT=80
     26    .END
```

(B)

Fig. 5-52 (A) Schematic and node definitions, and (B) input file for a SPICE simulation of a half-wave rectifier with a capacitor filter.

```
/T FWCF
/L ALL
    1        FULL WAVE RECTIFIER WITH CAPACITOR FILTER
    2        *
    3        * RESISTORS
    4        *
    5
    6        RL 2 3 800
    7        *
    8        * CAPACITOR
    9        C1 2 3 10UF
   10        *
   11        * DIODES
   12
   13        D1 1 2 DOD
   14        D2 3 1 DOD
   15        D3 0 2 DOD
   16        D4 3 0 DOD
   17        .MODEL DOD D(PB=0.7 RS=10 CJO=2PF BV=40)
   18        *
   19        * INPUT SIGNAL
   20        VIN 1 0 SIN(0 10 60)
   21        .TRAN 0.25M 75MS 25MS
   22        *
   23        * OUTPUT
   24        *
   25        .PRINT TRAN V(2,3) V(1)
   26        .PLOT TRAN V(2,3) V(1)
   27        .END
```

Fig. 5-53 Input file for a SPICE simulation of a full-wave rectifier with capacitor filter.

* **5-5.** Develop SPICE programs to study the frequency response characteristics of the filter circuits of Fig. 5-22. Choose values and show the types of responses that are available. Based on your SPICE insights, compare the different types of filters discussed in this chapter.

* **5-6.** Apply SPICE to calculate the frequency response characteristics of the high-pass and low-pass filters discussed in Appendix 2.

EXPERIMENTAL APPLICATIONS

5-1. The objective of this experiment is to design a circuit with a diode D and resistor R in series with a dc source V_s so that a certain prescribed current I flows in the circuit. Develop a procedure that can be used to direct your activities. Choose as your objective $I = 10$ mA with less than a 5 percent variation for T changing from ambient to 75°C. Then select the appropriate parameter values for your circuit. Use a hot-air blower to change the temperature of the diode, and observe the change in the Q point. Compare your predicted and experimental Q points at several temperatures, and find $(\partial V / \partial T)_I$.

Discussion
The load line must be selected so that the current through the circuit changes by less than 5 percent over the designated temperature range. A theoretical analysis can be used to predict appropriate values for R and V_s. Once the circuit is constructed, I_Q and V_Q can be measured as a function of device temperature (see Fig. 5-54). A graph of ΔI_Q as a function of ΔT can be used to compare the relation-

Fig. 5-54 Typical data for
Experimental Application 5-1

T(°C)	I_Q	V_Q

ship between the data obtained and the maximum allowable range given as a design constraint, as shown in Fig. 5-55. A graph of ΔV_Q as a function of ΔT can be used to find $(\partial V/\partial T)_I$, as shown in Fig. 5-56. Analyze and interpret your results.

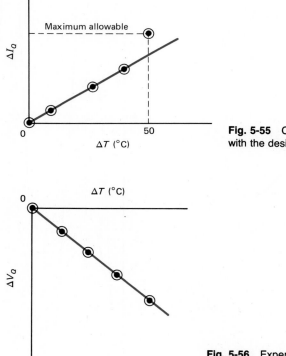

Fig. 5-55 Comparison of the actual changes in I_Q with the design maximum.

Fig. 5-56 Experimental determination of $(\partial v/\partial T)_I$.

5-2. The objective of this experiment is to design, build, and evaluate a bridge rectifier with parameter values that will illustrate the distortion produced by the nonideal diode characteristic curve. Design the rectifier with the input amplitude of $V_m = 4.0$ V and select the load resistor R_L to limit the current flow through the diodes as necessary. Develop a procedure to describe how you will set up and conduct the experiment. Record your output waveforms, then graphically determine V_{av} and **v** for the output. Refer to Exercise 5-9. Explain and discuss your results.

Discussion

The average output voltage V_{av} can be determined by using the graphical method shown in Exercise 5-9. By squaring each point on the output waveform, performing a similar averaging technique, and taking the square root of the result, **v** can also be obtained graphically. A typical output waveform is shown in Fig. 5-57. The method used for the graphical analysis is shown in Fig. 5-58. The experimental results should be compared with theoretical predictions using the ideal diode model, and the model based on V_D having a constant value of 0.7 V.

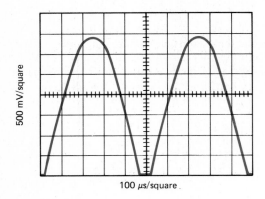

Fig. 5-57 Typical output waveform for a full-wave rectifier.

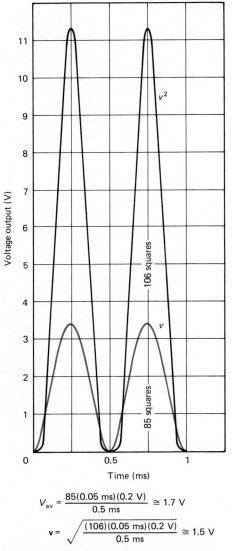

$$V_{av} = \frac{85(0.05 \text{ ms})(0.2 \text{ V})}{0.5 \text{ ms}} \cong 1.7 \text{ V}$$

$$\mathbf{v} = \sqrt{\frac{(106)(0.05 \text{ ms})(0.2 \text{ V})}{0.5 \text{ ms}}} \cong 1.5 \text{ V}$$

Fig. 5-58 Graphical analysis to find V_{av} and **v** for the rectified output.

MOSFET

PART 3
TRANSISTORS AS LINEAR DEVICES

Part 3 provides an introduction to transistors as building block elements for electronics. Emphasis is on understanding the operational principles for several types of transistors in common use, drawing on the material and junction properties previously developed and applying these devices for linear operation.

Chapters 6 to 8 introduce several types of bipolar and field-effect transistors. Emphasis is on learning how to bias transistors to achieve linear operation and on the performance tradeoffs that arise when such devices are used in amplifier circuits.

The dc-coupled amplifier is used as a simple circuit that illustrates a variety of important design considerations. Such single-stage amplifier circuits are commonly encountered as part of larger systems and provide a useful way to explore the operational characteristics of transistors. The detailed study of these stages provides familiarity with baseline configurations that are widely encountered.

A study of this type of circuit provides an introduction to the concepts of biasing, linear and nonlinear operation, thermal drift, and negative feedback, and provides an appropriate setting for a discussion of some of the tradeoffs that arise in electronics design.

CHAPTER 6
BIPOLAR JUNCTION TRANSISTOR OPERATION

In many design applications, a need exists for circuits that can *amplify* input voltages or currents so that changes in the input produce larger changes in the output. The devices and components that have been discussed so far do not have this capability. This chapter describes how bipolar junction transistors (BJTs) can provide the needed capability to create many different types of circuits that have an amplifying capability.

6-1 BIPOLAR JUNCTION TRANSISTORS AND THE COMMON-BASE CONFIGURATION

A bipolar junction transistor is created by forming two back-to-back junction diodes, using a single crystal for the entire combination, as shown in Fig. 6-1. This figure shows two resultant structures, one in the npn arrangement, as shown in Fig. 6-1(A), and the other in the pnp arrangement, as shown in Fig. 6-1(B). The npn arrangement is more widely used today, particularly in ICs, because of preferred fabrication processes and improved performance. The models of Fig. 6-1 are used to help understand the function of the BJT. Actual geometries are discussed in Section 6-9.

Refer to the npn device in Fig. 6-1(A). What happens as the junction between regions 1 and 2 is forward-biased, as shown?

In general, a majority flow of electrons results from the n-type to the p-type and a majority flow of holes results from the p-type to the n-type. However, suppose

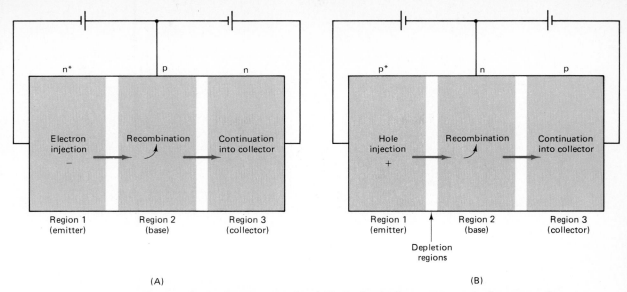

Region 1
(emitter)

Region 2
(base)

Region 3
(collector)

(A)

Region 1
(emitter)

Depletion
regions

Region 2
(base)

Region 3
(collector)

(B)

Fig. 6-1 Carrier flow due to emitter injection (neglecting reverse carrier flow across the emitter-base junction and minority-carrier flow across the collector-base junction).

that the n-type region is doped much more strongly than the p-type region (with the strong doping symbolized by n*). For this case, the electron flow greatly exceeds the hole flow.

Electrons are then being continuously *injected* from the n-type (where they are majority carriers) to the p-type (where they are minority carriers). If region 2 is made very narrow and the junction between regions 2 and 3 is reverse-biased, as shown, most of the electrons that are injected into region 2 diffuse to the depletion region between regions 2 and 3, are caught in the applied electric field, and are swept into region 3.

The result is that most of the electrons injected from region 1 to region 2 end up in region 3. A small fraction of the electrons recombines with the majority-carrier holes in region 2 and does not make it into region 3.

Region 1 is called the *emitter* (which emits electrons into region 2). Region 3 is the *collector* (which collects the electrons after they have diffused across region 2). Region 2 is called the *base* because of historical methods of fabrication.

A small flow of hole minority carriers results from region 3 to region 2, or from collector to base, due to the reverse-biased junction. Most of these holes diffuse across the base and end up in the emitter. The combined carrier flows shown in Fig. 6-2(A) are produced.

The dominant electron flow passes from emitter to base to collector whereas the small reverse hole flow passes from collector to base to emitter. These flows are supported by an external motion of electrons from the batteries applying the bias voltages. The total current out of the emitter and into the collector is large, whereas a small current flows in the base. The resulting device is called an npn BJT.

A similar analysis can be performed for the pnp BJT. The major flow is then due

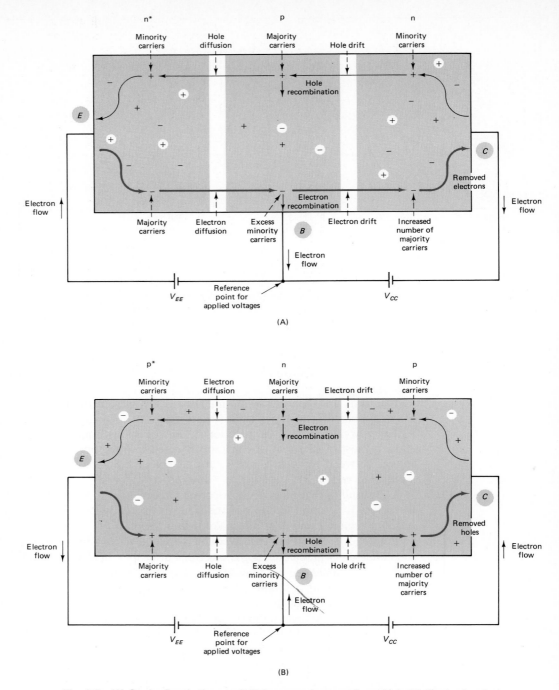

Fig. 6-2 (A) Carrier flow in the npn BJT (common-base configuration). (B) Carrier flow in the pnp BJT (common-base configuration). Further detail on carrier distributions is presented in Exercise 6-1.

to holes ejected from the emitter to the base, diffusing across the base, and swept into the collector. The reverse flow is due to minority-carrier electrons in the collector. The resultant carrier flow is shown in Fig. 6-2(B).

Exercise 6-1 provides a further description of the carrier-density distribution in a p*np BJT.

The transistors of Figs. 6-1 and 6-2 are in the *common-base* configuration. The voltages of the emitter and collector are measured with respect to the base. Note that in each figure there are two batteries, joined at the base. Other transistor configurations will be introduced as BJT applications are further explored.

The three connections to the BJT lead naturally to the concepts of an *input* terminal, an *output* terminal, and a *common* terminal, as shown in Fig. 6-3. For the common-base configuration, the input terminal is the emitter, the output terminal is the collector, and the common terminal is the base.

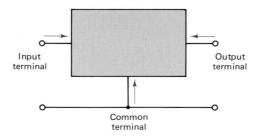

Fig. 6-3 Input/output model of the BJT. (This same model applies to any three-terminal device.) By convention, the arrows show positive current directions.

For the configuration and bias conditions shown, the magnitudes of the input and output currents are about the same and the common (base) current is small compared with the dominant flow between the emitter and collector. These relative current magnitudes are important to the properties of the common-base BJT amplifier.

6-2 NOTATION

The standard notation for BJT currents and voltages can be developed from the input/output model of Fig. 6-3. It is assumed that all currents flow *into* the device (defining the positive direction of flow). To apply this convention to the BJTs of Figs. 6-1(A) and (B), let I_E = the emitter current, I_C = the collector current, and I_B = the base current. These parameters then have the following signs:

$$\text{npn} \quad \begin{cases} I_E \text{ is negative} \\ I_C \text{ is positive} \\ I_B \text{ is positive} \end{cases}$$

$$\text{pnp} \quad \begin{cases} I_E \text{ is positive} \\ I_C \text{ is negative} \\ I_B \text{ is negative} \end{cases}$$

Any voltage written in the form V_{xy} refers to the voltage of terminal x relative to terminal y. If terminal x is negative with respect to terminal y, then $V_{xy} < 0$. If terminal x is positive with respect to terminal y, then $V_{xy} > 0$.

For the common-base configuration, voltages are always measured with respect to the base, so the subscript B appears in second place ($y = B$). For the common-emitter configuration discussed in the next section, voltages are measured with respect to the emitter, so the subscript E always appears in the second place ($y = E$).

The notation V_{ii} is used for the bias batteries, where $i = E$, B, or C. A battery V_{ii} determines the applied potential with respect to the common reference point.

The schematic symbols for the BJT are shown in Fig. 6-4. The emitter is always identified by an arrow pointing into or out of the base. The direction of the arrow is the direction of the current flow in the emitter and serves to distinguish between npn and pnp devices.

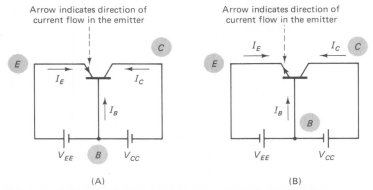

Fig. 6-4 Schematic symbols for the BJT (common-base configuration, linear bias): (A) pnp (I_B and I_C have negative values) and (B) npn (I_E has a negative value).

Based on the above notation, three regions of BJT operation can be defined. If the applied voltages have the polarities used in the above discussion, as shown in Fig. 6-4, the BJT is said to be in the *active* operating region, where approximately *linear* behavior is observed. Operation in the active region implies that *small* changes in the BJT currents (I_E, I_C, and I_B) and *small* changes in the voltages across the device (V_{EB} and V_{CB} or V_{BE} and V_{CE}) are linked in approximately linear relationships. This region is thus important when it is desired to have changes in one variable produce proportional changes in another.

If the emitter-base battery is reversed so that the emitter-base junction is reverse-biased, emitter injection ceases and the BJT is in the *cutoff* region. If both junctions are forward-biased, the BJT is in the *saturation* region. The significance of these operating regions becomes clear below.

The bias relationships in Fig. 6-5 can thus be obtained for the common-base configuration. Note that the battery providing the bias must be reversed for the shift from pnp to npn devices.

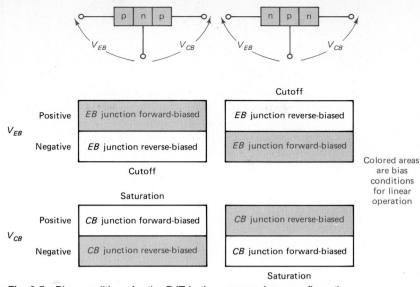

			Cutoff

	Positive	EB junction forward-biased	EB junction reverse-biased
V_{EB}	Negative	EB junction reverse-biased	EB junction forward-biased

Cutoff

Colored areas are bias conditions for linear operation

Saturation

	Positive	CB junction forward-biased	CB junction reverse-biased
V_{CB}	Negative	CB junction reverse-biased	CB junction forward-biased

Saturation

Fig. 6-5 Bias conditions for the BJT in the common-base configuration.

6-3 COMMON-EMITTER CONFIGURATION

Another BJT configuration is shown in Figs. 6-6 and 6-7. This is the *common-emitter* configuration with both base and collector voltages measured with respect to the emitter. For operation in the active region, V_{BB} and V_{CC} are chosen so that the emitter-base diode is forward-biased and the collector-base diode is reverse-biased. This

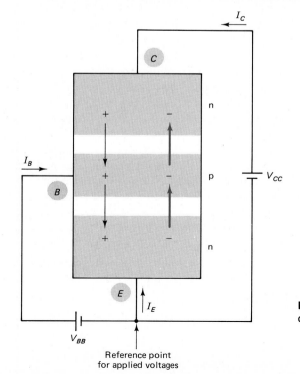

Fig. 6-6 Common-emitter configuration of the npn.

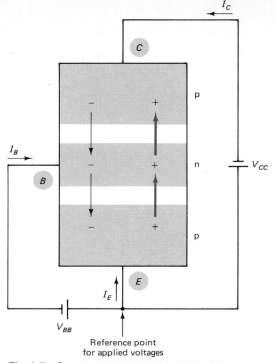

Fig. 6-7 Common-emitter configuration of the pnp.

linear operation is obtained under the conditions shown in Fig. 6-8. Note that the voltage across the collector-base junction is given by $V_{CB} = V_{CE} - V_{BE}$.

In the common-emitter configuration, the input current is I_B and the output current is I_C. The output current is much greater than the input current. The common-base and common-emitter configurations are thus quite different in the relationships that exist between the input and output currents. (The common-collector configuration has not yet been discussed. For this arrangement, all voltages are measured with respect to the collector. The common-collector configuration is introduced in Chapter 9.)

In Chapter 7, the voltage changes across the input terminals of a BJT are related to the voltage changes across the output terminals. As will be seen, a BJT can amplify input signals.[1] This amplification results from two important properties: (1) the low resistance of the forward-biased emitter-base junction and the high resistance of the base-collector junction, and (2) the ability to control the injection of carriers from the emitter to the base by externally determining how many electrons are allowed to flow into or from the base (from external sources).

---

[1] The term *signal* here refers to a general time-varying voltage or current that varies around a quiescent point Q.

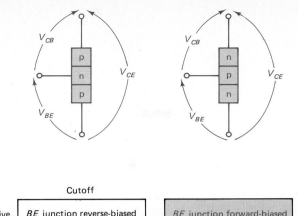

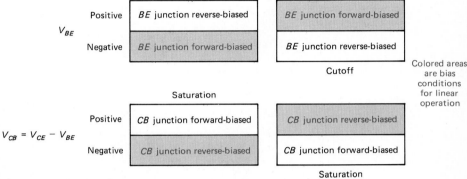

Fig. 6-8 Bias conditions for the BJT in the common-emitter configuration.

6-4 CHARACTERISTIC CURVES

The basics of BJT operation have been explored in terms of electron- and hole-carrier flow. Now consider how the characteristics of the BJT can be graphically illustrated, making use of the input/output concept.

Graphical descriptions are derived for the relationships between the input voltages and currents and the output voltages and currents for the BJT. These can be found by using the input and output characteristic curve formats of Figs. 6-9 through 6-12.

For the common-base configuration in Figs. 6-9 and 6-10, the input characteristic is found by plotting $I_{in} = I_E$ versus $V_{in} = V_{EB}$ for fixed values of $V_{out} = V_{CB}$. The output characteristic shows $I_{out} = I_C$ versus $V_{out} = V_{CB}$ for fixed values of $I_{in} = I_E$.

Observe the general properties of the common-base characteristics. The input characteristic shows the expected form of the emitter-base diode and indicates only a slight dependence on output voltage V_{CB}. For values of V_{CB} that reverse-bias the collector-base junction (achieving linear bias), the output characteristic consists of flat, equally spaced lines corresponding to $I_C \cong I_E$. For a forward-biased collector-base junction, the values of I_C drop rapidly toward zero. The reverse current flow through the collector-base diode reduces the total current I_C, which thus begins to drop. The

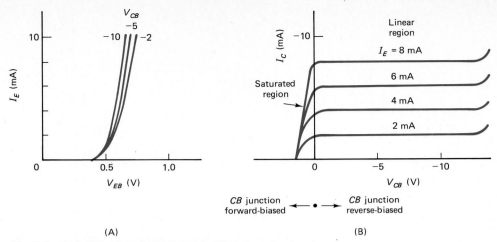

Fig. 6-9 Typical common-base characteristic curves (pnp).

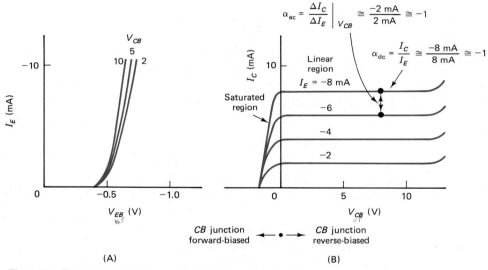

$$\alpha_{ac} = \frac{\Delta I_C}{\Delta I_E}\bigg|_{V_{CB}} \cong \frac{-2\text{ mA}}{2\text{ mA}} \cong -1$$

$$\alpha_{dc} = \frac{I_C}{I_E} \cong \frac{-8\text{ mA}}{8\text{ mA}} \cong -1$$

Fig. 6-10 Typical common-base characteristic curves (npn).

BJT enters the *saturation* region. For sufficiently large forward-bias values of V_{CB}, $I_C = 0$.

For the common-emitter configuration in Figs. 6-11 and 6-12, the input characteristic is found by plotting $I_{in} = I_B$ versus $V_{in} = V_{BE}$ for fixed values of $V_{out} = V_{CE}$. The output characteristic shows $I_{out} = I_C$ versus $V_{out} = V_{CE}$ for fixed values of $I_{in} = I_B$.

For the common-emitter input characteristic, the base-emitter diode again shows only a slight dependence on output voltage V_{CE}. For values of V_{CE} that reverse-bias the collector-base junction (achieving the linear region), the output characteristic

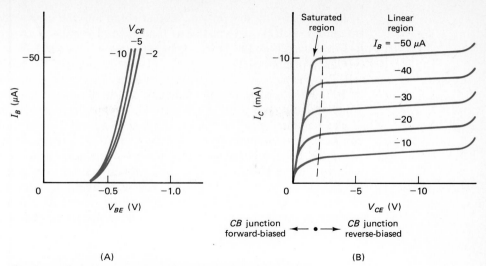

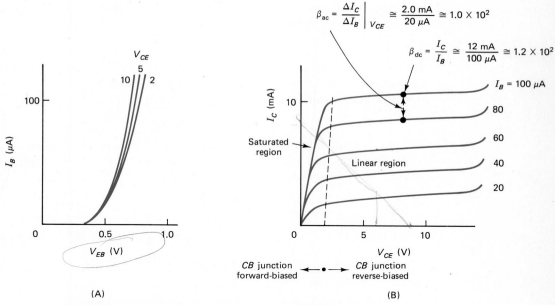

Fig. 6-11 Typical common-emitter characteristic curves (pnp).

$$\beta_{ac} = \frac{\Delta I_C}{\Delta I_B}\bigg|_{V_{CE}} \cong \frac{2.0\ \text{mA}}{20\ \mu\text{A}} \cong 1.0 \times 10^2$$

$$\beta_{dc} = \frac{I_C}{I_B} \cong \frac{12\ \text{mA}}{100\ \mu\text{A}} \cong 1.2 \times 10^2$$

Fig. 6-12 Typical common-emitter characteristic curves (npn).

consists of lines that are only *approximately* flat and equally spaced. The I = constant lines have a measurable value for ac resistance r.

In many cases, the collector current I_C can be taken as approximately constant for a given I_B over a defined range of interest, and the I_B = constant lines are approximately equally spaced. Under these conditions, nearly linear operation of the BJT results. In other cases, the effect of a finite r must be taken into account. The selection of an appropriate model depends on the accuracy requirements for the problem being studied.

Exercises 6-7 to 6-9 provide further detail into the ways in which the characteristic curves of BJTs are related to fundamental semiconductor parameters. Exercise 6-7 describes how the device geometry determines the BJT current gain. Building on this information, Exercise 6-8 explains the origin of the finite output resistance of a BJT in the common-emitter configuration. Exercise 6-9 summarizes the properties of BJT output characteristics, drawing on the developed materials. Chapter 18 provides a further extension of the introductory material in this chapter.

The saturated portions of Figs. 6-9 to 6-12 correspond to one another. For the common-base configuration (Figs. 6-9 and 6-10), the saturation region is entered when the collector-base junction is forward-biased (opposite the voltage polarity used for linear operation). For the common-emitter configuration (Figs. 6-11 and 6-12), the saturation region is entered when the collector-emitter voltage drop is smaller than the drop across the base-emitter diode. As a result, the collector-base junction is forward-biased. This region is an important one and is discussed further in Chapter 15.

6-5 BJT PARAMETERS

It is found useful to introduce several current ratios to describe BJT characteristics. The *common-base current gain* or *dc alpha* (α_{dc}) of a BJT is defined as

$$\alpha_{dc} = I_C/I_E \tag{6-1}$$

Given values of I_C and I_E, α_{dc} can be calculated. From the prior discussion in this chapter, α_{dc} is negative and $|\alpha_{dc}|$ is slightly less than unity.

The *common-emitter current gain* or *dc beta* (β_{dc}) is defined as

$$\beta_{dc} = I_C/I_B \tag{6-2}$$

Because I_C is usually much greater than I_B, β_{dc} has a large value (typically 50 to 250) in the linear operating region. By setting the total currents flowing into a BJT equal to zero, the relationship

$$I_E + I_C + I_B = 0 \tag{6-3}$$

is found for all transistor arrangements.

By substituting Eqs. (6-1) and (6-2) into (6-3), α_{dc} and β_{dc} can be related

$$I_C/\alpha_{dc} + I_C + I_C/\beta_{dc} = 0$$

or

$$\alpha_{dc} = -\frac{\beta_{dc}}{\beta_{dc} + 1} \tag{6-4}$$

As expected, $\alpha_{dc} \cong -1$ when $\beta_{dc} \gg 1$. The BJT can be further described in terms of an ac alpha (α_{ac}) and an ac beta (β_{ac}),

$$\alpha_{ac} = \left.\frac{\Delta I_C}{\Delta I_E}\right|_{V_{CB}=\text{constant}} \tag{6-5}$$

$$\beta_{ac} = \left.\frac{\Delta I_C}{\Delta I_B}\right|_{V_{CE}=\text{constant}} \tag{6-6}$$

These parameters describe how changes in one current level relate to changes in another level, and are used for the study of signal amplification.

An approximate relationship between α_{ac} and β_{ac} can be established by noting that in the common-base configuration

$$I_C = \text{function } (I_E, V_{CB})$$

so that to first order

$$\Delta I_C = \frac{\Delta I_C}{\Delta I_E}\bigg|_{V_{CB}} \Delta I_E + \frac{\Delta I_C}{\Delta V_{CB}}\bigg|_{I_E} \Delta V_{CB} \tag{6-7}$$

Similarly, in the common-emitter configuration,

$$I_C = \text{function } (I_B, V_{CE})$$

so that to first order

$$\Delta I_C = \frac{\Delta I_C}{\Delta I_B}\bigg|_{V_{CE}} \Delta I_B + \frac{\Delta I_C}{\Delta V_{CE}}\bigg|_{I_B} \Delta V_{CE} \tag{6-8}$$

In each case, the second term is proportional to the reciprocal of the slope of the device characteristic curve. If these curves are taken as approximately flat, neglecting slope effects, then

$$\frac{\Delta I_C}{\Delta I_E} \cong \frac{\Delta I_C}{\Delta I_E}\bigg|_{V_{CB}} = \alpha_{ac}$$

$$\frac{\Delta I_C}{\Delta I_B} \cong \frac{\Delta I_C}{\Delta I_B}\bigg|_{V_{CE}} = \beta_{ac} \tag{6-9}$$

Since

$$I_E + I_C + I_B = 0,$$

$$\Delta I_E + \Delta I_C + \Delta I_B = 0 \tag{6-10}$$

the result

$$\alpha_{ac} \cong -\frac{\beta_{ac}}{\beta_{ac} + 1} \tag{6-11}$$

is obtained.

The device parameters α_{dc}, α_{ac}, β_{dc}, and β_{ac} can be determined graphically from the characteristic curves shown in Fig. 6-9 through 6-12. See, for example, Figs. 6-10 and 6-12. The dc alpha (α_{dc}) is determined by picking a possible operating point on the output characteristic, measuring I_E and I_C, and calculating $\alpha_{dc} = I_C/I_E$. The dc beta (β_{dc}) is determined by picking a possible operating point on the output, measuring I_B and I_C, and calculating $\beta_{dc} = I_C/I_B$.

The parameter α_{ac} is found by selecting two possible operating points along a vertical line $V_{CB} = $ constant and calculating $\alpha_{ac} = \Delta I_C/\Delta I_E$. The parameter β_{ac} is found by selecting two possible operating points along a vertical line $V_{CE} = $ constant and calculating $\beta_{ac} = \Delta I_C/\Delta I_B$.

Example 6-1   BJT Parameters

From the BJT characteristic curves in Figs. 6-10 and 6-12, find α_{dc}, α_{ac}, β_{dc}, and β_{ac}.

Solution   Referring to Fig. 6-10, gather data along the vertical line $V_{CB} = 8.0$ V. Then $I_E \cong -8.0$ mA when $I_C \cong 8.0$ mA, so that

$$\alpha_{dc} \cong -1$$

Similarly, evaluate α_{ac} from the values

$$\alpha_{ac} = \left.\frac{\Delta I_C}{\Delta I_E}\right|_{V_{CB}} \cong \frac{2.0 \text{ mA}}{-2.0 \text{ mA}} \cong -1$$

As can be noted, the determination of α_{dc} or α_{ac} from the common-base characteristic curve is limited in accuracy.

From Fig. 6-12, select $V_{CE} = 8.0$ V and the point

$$I_B \cong 0.100 \text{ mA}$$

$$I_C \cong 12.0 \text{ mA}$$

so that

$$\beta_{dc} = I_C/I_B \cong \frac{12.0 \text{ mA}}{0.100 \text{ mA}} \cong 120$$

A more accurate value for α_{dc} is obtained by calculating

$$\alpha_{dc} = -\frac{\beta_{dc}}{1 + \beta_{dc}} = -\frac{120}{121} \cong -0.992$$

Finally, β_{ac} can be evaluated from the values

$$\beta_{ac} = \left.\frac{\Delta I_C}{\Delta I_B}\right|_{V_{CE}} \cong \frac{12.0 \text{ mA} - 8.0 \text{ mA}}{100 \text{ }\mu\text{A} - 80 \text{ }\mu\text{A}} = \frac{4.0 \text{ mA}}{20 \text{ }\mu\text{A}} \cong 2.0 \times 10^2 \qquad \blacksquare$$

6-6   BJT DATA SOURCES

There are several different ways in which data can be gathered to describe the operation of a particular BJT. One of the most common ways is to use a manufacturer's data sheet.

Figure 6-13 shows a typical pnp BJT data sheet. BJTs are often identified by a code starting with *2N* followed by several digits. The upper portion of the data sheet gives the absolute maximum ratings for important voltage, current, and power levels at an ambient temperature of 25°C, and the lower portion gives typical operating points at 25°C.

Several of the parameters on the data sheet use a third subscript "o." This means that the third device terminal (the one not appearing in either other subscript) is open when the measurement is made.

Observe that several different notations are being used. For example, the variable h_{FE} is used to represent the forward-current transfer ratio instead of β_{dc}. The *h* parameters provide a different way for describing transistor operation and are dis-

FIGURE 6-13
Data Sheet for the 2N5208 (Courtesy of General Electric Semiconductor)

The GE 2N5208 is a silicon pnp planar, passivated, epitaxial transistor designed for general-purpose amplifier applications.

C
B
E

Absolute Maximum Ratings (T_A = 25°C, unless otherwise specified):

		2N5208	UNITS
Voltages			
Collector-to-Emitter	V_{CEO}	25	V
Emitter-to-Base	V_{EBO}	3.0	V
Collector-to-Base	V_{CBO}	30	V
Current			
Collector	I_C	50	mA
Dissipation			
Total Power ($T_C \leq 25$°C)	P_T	1.0	W
Total Power ($T_A \leq 25$°C)	P_T	350	mW
Derate Factor ($T_C \geq 25$°C)		8.0	mW/°C
Derate Factor ($T_A \geq 25$°C)		2.8	mW/°C
Temperature			
Operating	T_J	−55 to +150	°C
Storage	T_{STG}	−55 to +150	°C
Lead ($\frac{1}{16}'' \pm \frac{1}{32}''$ from case for 10 s)	T_L	+260	°C

Electrical Characteristics (T_A = 25°C, unless otherwise specified):

		2N5208		
	SYMBOL	MIN.	MAX.	UNITS
Collector-Emitter Breakdown Voltage	BV_{CEO}			
(I_c = 1.0 mA, I_B = 0)		25	—	V
Collector-Base Breakdown Voltage	BV_{CBO}			
(I_C = 0.1 mA, I_E = 0)		30	—	V
Emitter-Base Breakdown Voltage	BV_{EBO}			
(I_E = 10 μA, I_C = 0)		3.0	—	V
Collector Cutoff Current	I_{CBO}			
(V_{CB} = 10 V, I_E = 0)		—	10	nA
Emitter Cutoff Current	I_{EBO}			
(V_{BE} = 2.0 V, I_C = 0)		—	100	nA
Forward Current Gain Ratio	h_{FE}			
(I_C = 2.0 mA, V_{CE} = 10 V)		20	120	
Base-Emitter On Voltage	$V_{BE(on)}$			
(I_C = 2.0 mA, V_{CE} = 10 V)		—	0.85	Volts
Current Gain-Bandwidth Product	f_T			
(I_c = 2.0 mA, V_{CE} = 10 V, f = 100 MHz)		300	1200	MHz
Collector-Base Capacitance	C_{CB}			
(V_{CB} = 10 V, I_E = 0, f = 1.0 MHz)		—	1.0	pF

cussed in Chapter 9. As can be noted, β_{dc} has values of 20 to 120 for the 2N5208. This variability in current-gain values is one of the problems to be faced in design with BJTs.

The allowable power dissipation levels are specified in terms of the temperature of the transistor case (T_C) and the temperature of the ambient air (T_A). The 2N5208 can dissipate 1 W if the case temperature is held below or equal to 25°C. For $T_C > 25$°C, the allowable power dissipation level drops by 8 mW/°C. If the temperature of the ambient air is below or equal to 25°C, this BJT can dissipate only 350 mW, and if $T_A > 25$°C, the allowable power dissipation level drops by 2.8 mW/°C. The significance of the device capacitances and the gain-bandwidth product are discussed in Chapter 10.

Example 6-2   **BJT Power Dissipation**

What maximum power dissipation is allowed for the 2N5208 if the surrounding air is at 75°C?

Solution   Refer to Fig. 6-13. T_C is the case temperature and T_A is the temperature of the ambient air. The maximum allowed power dissipation for $T_A \leq 25$°C is 350 mW. The derating factor is 2.8 mW/°C.

Therefore, for 75°C,

$$P_{max}(75°C) = 350 \text{ mW} - (2.8 \text{ mW/°C})(75 - 25)°C$$

$$\cong 210 \text{ mW}$$

The power-dissipating ability of a BJT drops rapidly with higher temperatures. This aspect of performance can restrict the options available to the designer. Because of the importance of this parameter, transistor data sheets usually provide information regarding the derating factor for maximum power dissipation. ■

In realistic design settings, it is often necessary to supplement the information provided on the data sheet with additional first-hand data. Sophisticated equipment is available to support such data collecting activities.

One of the most common needs is to examine the characteristic curve of a given device. A transistor curve tracer, or an equivalent instrument, can be used for this purpose. A curve tracer displays the device characteristic directly on an oscilloscope screen, where it can be conveniently photographed. Figure 6-14 illustrates typical results that can be obtained by using this strategy. Figure 6-14(A) shows the

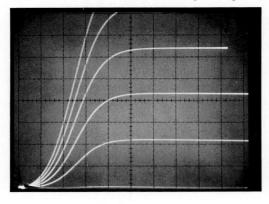

Fig. 6-14(A)   Common-emitter output characteristic for the 2N2907A. (Horizontal scale: 0.05 V/division; vertical scale: 1 mA/division; current steps: 20 μA/step.

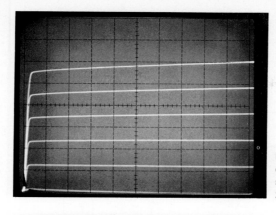

Fig. 6-14(B) Common-emitter output characteristic for the 2N2907A. (Horizontal scale: 1 V/division; vertical scale: 2 mA/division; current steps: 20 μA/division.

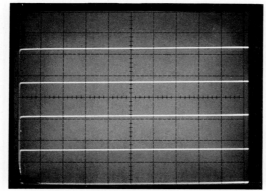

Fig. 6-14(C) Common-base output characteristic for the 2N2907A. (Horizontal scale: 2 V/division; vertical scale: 0.5 mA/division; current steps: 1 mA/division.

2N2907A common-emitter output characteristic for small values of I_B and V_{BE}. Figure 6-14(B) shows the same device and configuration at higher current and voltage levels, and Fig. 6-14(C) shows the device in the common-base configuration.

Alternatively, plotters are available to directly record the output of a curve tracer. Figure 6-15 shows common-emitter and common-base graphics plots for the 2N4810 BJT.

Fig. 6-15(A) Common-emitter output characteristic obtained from the Hewlett-Packard HP4145A semiconductor parameter analyzer. (Continued on p. 218.)

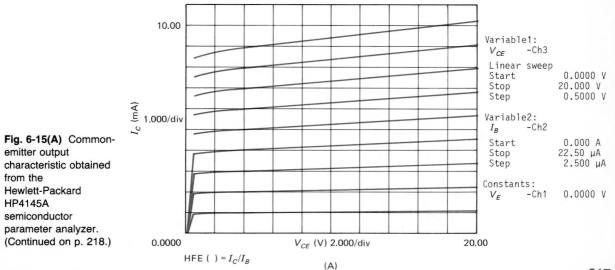

```
Variable1:
 V_CE    -Ch3
Linear sweep
Start        0.0000 V
Stop         20.000 V
Step         0.5000 V

Variable2:
 I_B     -Ch2

Start        0.000 A
Stop         22.50 µA
Step         2.500 µA

Constants:
 V_E     -Ch1   0.0000 V
```

HFE ( ) = I_C/I_B

(A)

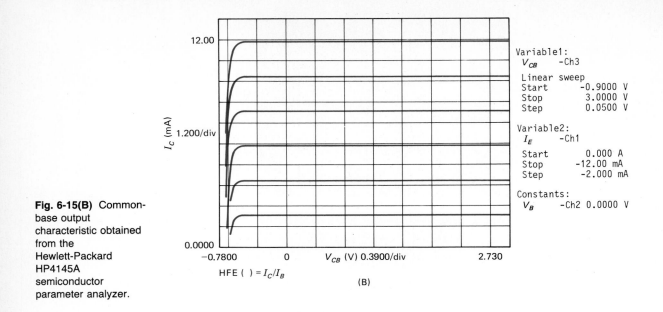

Variable1:
V_{CB} –Ch3
Linear sweep
Start   –0.9000 V
Stop   3.0000 V
Step   0.0500 V

Variable2:
I_E –Ch1

Start   0.000 A
Stop   –12.00 mA
Step   –2.000 mA

Constants:
V_B –Ch2 0.0000 V

I_C (mA)

1.200/div

12.00

0.0000

–0.7800   0 V_{CB} (V) 0.3900/div   2.730

HFE ( ) = I_C/I_B

(B)

Fig. 6-15(B) Common-base output characteristic obtained from the Hewlett-Packard HP4145A semiconductor parameter analyzer.

6-7 PARAMETER VARIATIONS

In the above discussion, α_{dc}, β_{dc}, α_{ac}, and β_{ac} are treated as approximately constant in the linear region for the BJT. The adequacy of this assumption depends on the application.

In general, these parameters vary with current level, applied voltage, and temperature. The variations do not have a major impact on the common-base characteristic curves (Figs. 6-9 and 6-10) because α_{dc} varies around unity by only a small percentage. Some circuit designs take advantage of this situation.

However, these effects are very noticeable for the common-emitter characteristic curves (Figs. 6-11 and 6-12). Over most of the linear operation range, the current-level dependency produces lines that are farther apart for higher values of I_C. The result is that β_{dc} and β_{ac} increase somewhat with I_C. The applied voltage dependency produces a positive slope in the I_B = constant lines, as illustrated.

The characteristic curve of the input (emitter-base) diode of the BJT varies with T according to the development of Chapter 4. In addition, the BJT current gain is sensitive to temperature change. As shown in Fig. 6-16, the I_B = constant lines "float up" with higher T. As a result, both β_{dc} and β_{ac} increase with T.

This temperature dependence of the current gain can be traced to the injection of carriers from the emitter to the base. At higher temperatures, the injection efficiency increases, producing higher values of α_{dc} and resulting in higher values of β_{dc}.

From Eq. (6-4),

$$\beta_{dc} = \frac{-\alpha_{dc}}{1 + \alpha_{dc}} \tag{6-12}$$

As the number of carriers crossing the base region increases, $|\alpha_{dc}|$ comes closer to unity and β_{dc} increases.

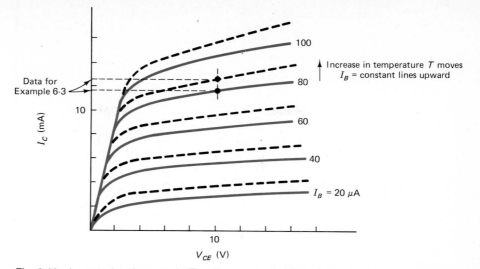

Fig. 6-16 Impact of an increase in T on the common-emitter characteristic curve (npn).

In design efforts, it is often useful to adopt models with constant α and β values. If the operating region is restricted, the temperature is held approximately constant, and if one to two significant figures are acceptable for desired accuracy, the model is adequate. If these assumptions are violated, there must be a further examination of the effect of parameter variations on the performance of the final circuit.

To make the situation more difficult, the α and β parameters can change substantially from device to device, even when the devices have the same manufacturer's code number. Design accuracy is limited not only by the uncertainty associated with the parameters of a given device, but also by variations among devices.

In some circuit applications, the uncertainties caused by both effects are quite acceptable. For example, a variable resistor can be added to a circuit and adjusted to obtain the operating point or gain that is desired. In other applications, performance must be predicted and controlled much more tightly. Possible strategies for this situation are discussed in the following chapters.

Example 6-3 **Temperature Dependence of β_{dc}**

Assume that the upward shift in the characteristic curve of Fig. 6-16 is associated with a temperature change of 5°C. Find the percentage change in β_{dc} per °C associated with the curve.

Solution The objective is to find

$$\frac{\Delta\beta}{\beta} \times 100/\Delta T \qquad (\%/°C)$$

where $\Delta\beta = \beta_2 - \beta_1$. If I_{C1} and I_{C2} are the collector currents associated with T_1 and T_2, then

$$\Delta\beta = \frac{I_{C2}}{I_B} - \frac{I_{C1}}{I_B} \qquad \text{and} \qquad \beta = \frac{I_{C1}}{I_B}$$

Since the same I_B curve is being used at I_{C1} and I_{C2},

$$\frac{\Delta\beta}{\beta} \times 100/\Delta T = \frac{I_{C2} - I_{C1}}{I_{C1}(T_2 - T_1)} \times 100 \qquad (\%/°C)$$

Substituting from Fig. 6-16,

$$\frac{\Delta\beta}{\beta} \times 100/\Delta T \cong \frac{11.2 - 10.7}{(10.7)(5)} \times 100 \cong 0.9 \ \%/°C$$

which is a typical value for a Si BJT near 300°K. ■

Example 6-4 Temperature Dependence of α_{dc}

For the device of Example 6-3, how will α_{dc} vary with T?

Solution Since

$$\alpha_{dc} = \frac{-\beta_{dc}}{\beta_{dc} + 1}$$

$$\frac{-d\alpha_{dc}}{d\beta_{dc}} = \frac{1}{(\beta_{dc} + 1)^2} \cong \frac{1}{\beta_{dc}^2}$$

$$\frac{\Delta\alpha_{dc}}{\alpha_{dc}} \times 100/\Delta T \cong \frac{\Delta\beta_{dc}}{\beta_{dc}^2} \times 100/\Delta T \cong \frac{1}{\beta_{dc}}\left(\frac{\Delta\beta_{dc}}{\beta_{dc}} \times 100/\Delta T\right)$$

Therefore, the fractional change $\Delta\alpha_{dc}/\alpha_{dc}$ is smaller than the fractional change $\Delta\beta_{dc}/\beta_{dc}$ by a factor of $1/\beta_{dc}$. For this example

$$\frac{\Delta\alpha_{dc}}{\alpha_{dc}} \cong \frac{0.9\%/°C}{200} \cong 5 \times 10^{-3}\%/°C$$

The variation in β_{dc} with T is much greater than the variation in α_{dc} with T. ■

6-8 EBERS-MOLL MODEL

The pn junction equation of Chapter 4 can be used to develop the Ebers-Moll model for the BJT (Ebers and Moll, 1954). This model can provide insight into the operation of the transistor and can be used to relate device parameters to the characteristic curves that will result. This model is widely used in computer simulation programs, where the junction equations are used in combination with series resistors that represent the bulk resistances of the device.

A useful approach is to first derive the equations for the model and then consider how the equations can be applied. Refer to the pnp in the common-base configuration shown in Fig. 6-17. The emitter-base and collector-base diodes can be described using the earlier theoretical diode curve, so that

$$I_F = I_{Fo}(e^{V_{EB}/V_o} - 1) \qquad \text{pnp, common-base}$$

$$I_R = I_{Ro}(e^{V_{CB}/V_o} - 1) \qquad \text{pnp, common-base} \qquad (6-13)$$

Positive currents are defined as those flowing into the base region.

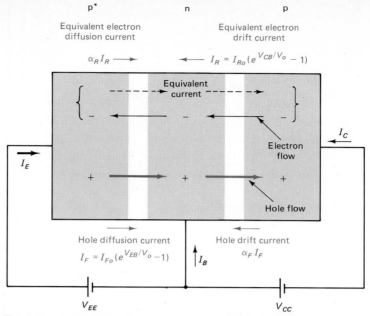

Fig. 6-17 The Ebers-Moll model of the (pnp) BJT (biased for linear operation). Positive current flows are taken into the base.

For linear operation, I_F is the current component that results across the emitter-base junction due to carrier injection from the emitter and I_R is the current component that results across the collector-base junction due to minority-carrier flow. As indicated in Fig. 6-17, most of the majority carriers that leave the emitter cross the base and end up in the collector. Similarly, most of the minority carriers that leave the collector cross the base and end up in the emitter. The continuations of these initial flows are described by $\alpha_F I_F$ and $\alpha_R I_R$, where $|\alpha_F|$ and $|\alpha_R|$ represent the fractions of the carriers that make it through the base region. From the above discussion, $\alpha_R \cong -1$ and $\alpha_R \cong -1$.

The total emitter, collector, and base currents can now be found by applying Kirchhoff's current law:

$$I_E = I_F + \alpha_R I_R$$
$$I_C = \alpha_F I_F + I_R$$
$$I_E + I_C + I_B = 0$$
(6-14)

For linear operation, V_{CB} is negative, so that $I_R \cong -I_{Ro}$. V_{EB} is positive, so that $I_F \cong I_{Fo} e^{V_{EB}/V_o}$. The component current flows in Fig. 6-17 add to produce the total currents I_E and I_C. For an npn BJT, this sign convention produces the same set of equations. The directions of all current flows reverse.

The Ebers-Moll model of the BJT can be represented by using two diodes that are described by the diode equation and coupled to dependent current generators, as shown in Fig. 6-18. The magnitude of each current generator is dependent on the current through the linked diode. The current generators represent the continuation of the majority- and minority-carrier flows across the base.

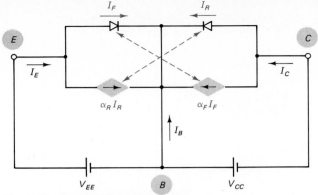

Fig. 6-18 Baseline Ebers-Moll model.

If $|V_{CB}|$ is sufficiently large so that $|V_{CB}/V_o| \gg 1$, then the above currents are (approximately) independent of the exact size of the battery V_{CC} (which sets V_{CB}). However, increases in V_{EE} increase V_{EB}, thus increasing I_B and I_C.

The above analysis can be extended to all BJT types and configurations. For an npn in the common-base configuration,

$$I_F = I_{Fo}(e^{-V_{EB}/V_o} - 1) \qquad \text{npn, common-base}$$
$$I_R = I_{Ro}(e^{-V_{CB}/V_o} - 1) \qquad \text{npn, common-base} \tag{6-15}$$

using the notation that has been introduced. The negative signs preceding the voltages are necessary to produce a forward-biased emitter-base junction and a reverse-biased collector-base junction for the npn, using the definitions given earlier.

For the common-emitter configuration, the diode equations take the form

$$I_F = I_{Fo}(e^{-V_{BE}/V_o} - 1) \qquad \text{pnp, common-emitter}$$
$$I_R = I_{Ro}(e^{(V_{CE} - V_{BE})/V_o} - 1) \qquad \text{pnp, common-emitter} \tag{6-16}$$

$$I_F = I_{Fo}(e^{V_{BE}/V_o} - 1) \qquad \text{npn, common-emitter}$$
$$I_R = I_{Ro}(e^{-(V_{CE} - V_{BE})/V_o} - 1) \qquad \text{npn, common-emitter} \tag{6-17}$$

in order to maintain the correct signs.

The symmetry of the Ebers-Moll model points out that a BJT can also be operated in an inverted mode, with the collector injecting carriers into the base. For most BJT applications, this is an inefficient and unused mode of operation. However, this inverse mode is essential in digital circuits using TTL technology, which is discussed in Chapter 15.

The Ebers-Moll model can be used in the SPICE simulation that was introduced in Section 5-12. Figure 6-19 shows how the baseline model can be modified in preparation for computer simulation. The current generators for I_E and I_C are combined. For the model shown in Fig. 6-19,

$$I_B = -[I_F(1 + \alpha_F) + I_R(1 + \alpha_R)] \tag{6-18}$$

The three resistors, r_E, r_C, and r_B, represent effective bulk resistance effects.

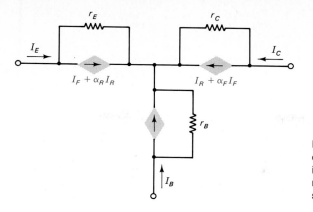

Fig. 6-19 Ebers-Moll model using combined constant-current generators in the linear operating region. The model has bulk resistances for SPICE simulations.

Example 6-5   Common-Base Currents

A pnp Si BJT is biased[2] so that $V_{EB} = 0.450$ V and $V_{CB} = -10.0$ V. Assuming that $I_{Fo} \cong I_{Ro} \cong 5.00$ nA and $\alpha_F \cong \alpha_R = -0.990$, find all of the current flows in the Ebers-Moll model of Fig. 6-18. Find α_{dc}.

Solution   First note that

1. The BJT is in the common-base configuration because B is the second subscript.
2. A positive value of V_{EB} for the pnp is of the correct polarity to forward-bias the emitter-base junction for linear operation.
3. A negative value of V_{CB} for the pnp is of the correct polarity to reverse-bias the collector-base junction for linear operation.

Then from Eq. 6-13,

$$I_F = (5.00 \times 10^{-9} \text{ A})(e^{0.450/0.026} - 1) \cong 164 \text{ mA}$$

$$I_R = (5.00 \times 10^{-9} \text{ A})(e^{-10.0/0.026} - 1) \cong -5.00 \text{ nA}$$

so that

$$\alpha_R I_R \cong 4.95 \text{ nA}$$

$$\alpha_F I_F \cong -0.990 \times 164 = -162 \text{ mA}$$

$$I_E = I_F + \alpha_R I_R \cong 164 \text{ mA}$$

$$I_C = \alpha_F I_F + I_R \cong -162 \text{ mA}$$

$$I_B = -(I_E + I_C) \cong -2 \text{ mA}$$

The resulting currents are shown in Fig. 6-20. For α_{dc},

$$\alpha_{dc} = I_C/I_E = -162 \text{ mA}/164 \text{ mA} = -0.988$$

The I_R terms are negligible in the linear operating region. In other regions of operation, these terms are important.   ■

---

[2] Three significant figures are required since differences are being taken between approximately equal numbers.

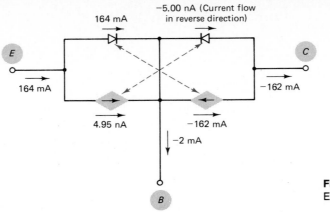

Fig. 6-20 Current flows for Example 6-5.

Example 6-6 Common-Emitter Currents

An npn Ge BJT is biased so that $V_{BE} = 0.250$ V and $V_{CE} = 10.0$ V. Assume that $I_{Fo} \cong I_{Ro} \cong -3.00$ μA and $\alpha_F \cong \alpha_R = -0.990$. Find β_{dc} for the BJT.

Solution Observe that

1. The BJT is in the common-emitter configuration because E is the second subscript.
2. A positive value of V_{BE} for an npn BJT means that the base-emitter diode is forward-biased.
3. A positive value of V_{CE} for an npn reverse-biases the collector-base junction by voltage $V_{CB} = V_{CE} - V_{BE} = -9.7$ V.

Therefore, from Eq. 6-17,

$$I_F = (-3.00 \times 10^{-6} \text{ A})(e^{0.250/0.026} - 1) \cong -45.0 \text{ mA}$$

$$I_R = (-3.00 \times 10^{-6} \text{ A})(e^{-9.70/0.026} - 1) \cong 3.00 \text{ } \mu\text{A}$$

$$\alpha_R I_R \cong -2.97 \text{ } \mu\text{A}$$

$$\alpha_F I_F \cong 44.6 \text{ mA}$$

$$I_E = I_F + \alpha_R I_R \cong -45.0 \text{ mA}$$

$$I_C = \alpha_F I_F + I_R = 44.6 \text{ mA}$$

$$I_B = -(I_C + I_E) = 0.4 \text{ mA}$$

and for β_{dc},

$$\beta_{dc} = I_C/I_B = 44.6 \text{ mA}/0.4 \text{ mA} \cong 1.0 \times 10^2$$

Again in this example, the I_R terms are negligible because linear operation is obtained. ∎

Example 6-7 Saturated BJT

For a pnp BJT,

$$\alpha_F \cong \alpha_R = -0.990$$

$$V_{EB} = 0.450 \text{ V}$$

$$V_{CB} = 0.400 \text{ V}$$

$$I_{Fo} \cong I_{Ro} = 5.00 \text{ nA}$$

The collector current I_C is to be found under these conditions, with forward-biased emitter-base and collector-base diodes.

Solution From the Ebers-Moll model,

$$I_C = -(0.990)(5.00 \times 10^{-9} \text{ A})(e^{0.450/0.026} - 1) + (5.00 \times 10^{-9} \text{ A})(e^{0.400/0.026} - 1)$$

$$= -163 \text{ mA} + 24 \text{ mA} = -139 \text{ mA}$$

The first term represents the current flow due to the carriers that are injected from the emitter to the base and reach the collector. The second term represents the current flow from collector to base that reaches the emitter because of the forward-biased collector-base junction. This is the saturation mode of operation. Note that both the I_F and I_R terms are important in the saturation region. ■

6-9 FABRICATION AND PACKAGING OF BJTs

BJTs are fabricated using an extension of the junction diode techniques discussed in Section 4-13. Figure 6-21 shows a discrete BJT. An additional diffusion has been used to produce the n-type region for the emitter, and the wire bonds are now at the

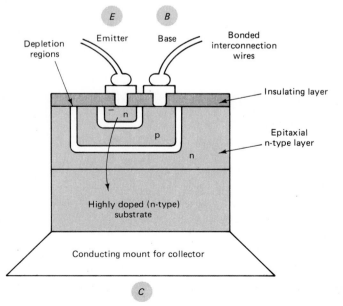

Fig. 6-21 Geometry for a discrete BJT.

top of the device for both base and emitter. The result is again a simple hybrid microcircuit.

A similar strategy is used to create IC transistors, as shown in Fig. 6-22. The epitaxial n-type layer is now used for the collector of the BJT. To maintain isolation, a reverse bias must be maintained on the boundary between the p-type substrate and the n-type epitaxial layer.

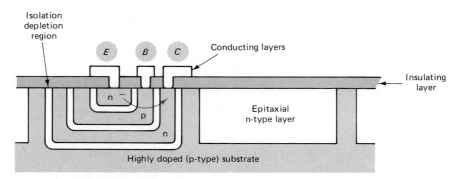

Fig. 6-22 Geometry for an IC BJT.

The tradeoffs involved in BJT design and fabrication are discussed in Chapter 18. The fundamental semiconductor properties introduced in previous chapters and in this chapter are used in Chapter 18 to describe the linkage between device operation and performance parameters. The characteristic curves of the BJT (as given by the Ebers-Moll model) and other device properties can then be directly related to fundamental device concepts. Figure 6-23 shows a variety of different BJT discrete packages.

Fig. 6-23 Various types of packages for discrete BJTs. (Courtesy of General Electric Semiconductor.)

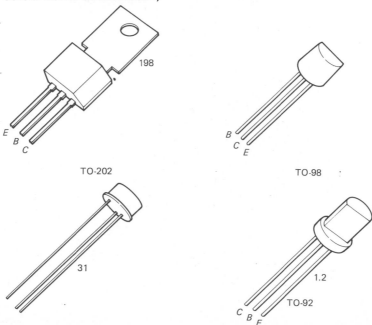

QUESTIONS AND ANSWERS

Questions

6-1. For the pnp BJT: (a) Which region is the most strongly doped? (b) What types of particles are injected from the emitter to the base to produce the dominant current flow across this junction? (c) What fundamental semiconductor property determines how many carriers recombine in the base?

6-2. What insights from the diode studies of Chapter 4 can help provide an understanding of the BJT?

6-3. What current flow convention is associated with the BJT?

6-4. What voltage convention is associated with the BJT?

6-5. When is a BJT in the active region?

6-6. What is the difference between the common-base and common-emitter configurations? What about a common-collector configuration?

6-7. What BJT properties result in the device's ability to provide voltage amplification?

6-8. How is the saturation region defined for a BJT?

6-9. Why is it difficult to design a BJT amplifier that will have exactly the desired performance characteristics?

6-10. What fundamental concepts are used to develop the Ebers-Moll model?

Answers

6-1. For the pnp BJT, the p-type emitter is the most strongly doped. The dominant current flow across the emitter-base junction is thus due to hole injection. Carrier recombination in the base is determined by the lifetime τ_p of the holes in the n-type region.

6-2. The emitter-base junction is a pn junction diode and thus can be understood based on the discussion of Chapter 4. The recombination process in the base region has been previously addressed by considering the exponential decay of carriers into a bulk region due to the finite carrier lifetime. Clearly, in order to maximize the number of carriers reaching the collector, the characteristic lengths L_n and L_p (for npn and pnp BJTs) must be long compared with the thickness of the base region.

The base-collector junction is a reverse-biased diode that is "force fed" minority carriers that diffuse across the base. Note that the BJT cannot be thought of as two diodes "back to back" because such a model does not allow for the linking carrier flow across the base region.

6-3. The BJT current flow convention defines all currents as positive when they flow into the device.

6-4. The BJT voltage convention V_{xy} refers to the voltage of terminal x with respect to terminal y. If $V_x > V_y$, then $V_{xy} > 0$, and if $V_x < V_y$, then $V_{xy} < 0$.

6-5. The active region is realized when the emitter-base junction is forward-biased and the base-collector junction is reverse-biased.

6-6. For the common-base configuration, the emitter and collector voltages are measured with respect to a common-base terminal. For the common-emitter configuration, the base and collector voltages are measured with respect to a common-emitter terminal. For the common-collector configuration (introduced in Chapter 9), the base and emitter voltages are measured with respect to the common-collector terminal.

6-7. The voltage amplification of the BJT results from the low resistance of the forward-biased emitter-base junction and the high resistance of the reverse-biased base-collector junction. Since approximately the same current flows across the two junctions (in the active region of operation), changes in voltage across the input base-emitter junction are much smaller than the changes across the common-base or common-emitter output voltage. The ability to control carrier flow injection from the emitter to base provides the external control mechanism for the device.

6-8. In the saturation region of the BJT, the collector-base junction becomes forward-biased, reducing the total current I_C. In the common-emitter configuration, the saturation effect is used in gate circuits to produce V_{CE} values that are approximately independent of base current (Chapter 15). In this way, protection can be obtained from variations in device parameters and noise in producing large systems of digital building block units.

6-9. The current gain of a given BJT type is not a constant, but depends on variations among devices introduced during fabrication, voltage and current levels of operation, and the temperature of operation. Circuit gain levels can vary widely as device properties change. As will be seen in Chapters 9 and 11, the use of negative feedback in circuit design provides a strategy for producing BJT amplifier circuits with predictable gain properties.

6-10. The Ebers-Moll model is developed using the concept of ideal diode junctions and dependent current generators linked to current flow in these diodes. It is the linkage between the diode currents and the current generators that prevents the BJT from being described as "two back-to-back diodes."

EXERCISES AND SOLUTIONS

Exercises
(6-1)

6-1. How can the carrier flow discussions for the pn junction diode as presented in Section 4-4 and Exercise 4-6 in Chapter 4 be extended to describe the observed properties of BJTs?

(6-1)

6-2. Assume that for a given BJT, $\alpha_{dc} = -0.990$ and $I_2 = (1.0 \times 10^{-3})I_1$. What fraction of the carriers injected into the base recombines before reaching the collector?

(6-1)

6-3. For a pnp transistor: (a) What are the majority carriers in the emitter? (b) When majority carriers are injected from the emitter into the base, what happens to the number of minority carriers in the base? (c) For linear operation, which BJT junction is forward-biased and which is reverse-biased?

(6-1)

6-4. For an npn transistor: (a) What are the minority carriers in the base? (b) What happens to the number of minority carriers in the base due to emitter injection? (c) For linear operation, which junction is forward-biased and which is reverse-biased?

(6-2)
(6-3)

6-5. Describe which junctions are forward-biased and which are reverse-biased, using the notation developed in Section 6-2.

(a)	$V_{CB} = -0.5$ V	and	$V_{EB} = -0.7$ V	npn
(b)	$V_{CB} = -8.0$ V	and	$V_{EB} = 0.7$ V	pnp
(c)	$V_{CE} = 6.0$ V	and	$V_{BE} = 0.7$ V	npn
(d)	$V_{CE} = -6.0$ V	and	$V_{BE} = 0.7$ V	pnp

(6-3)
(6-4)

6-6. Assume that the BJT in Fig. 6-24 is described by the characteristic curve in Fig. 6-11. In order to bias at $I_B = -30$ μA, what values of R, V_{CC} and I_C can be used?

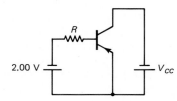

Fig. 6-24 Circuit for Exercise 6-6.

(6-4) **6-7.** Using the models developed in this and previous chapters, derive approximate relationships for I_C, I_B, β_{dc}, and β_{ac} as a function of basic BJT parameters.

(6-4) **6-8.** Using the concepts developed in this and previous chapters, discuss why the common-emitter output characteristic curves for the BJT have a finite output resistance in the active operating region.

(6-4) **6-9.** How can the previously developed properties of semiconductor junctions be applied to understand how the output characteristic curve of a BJT varies from the idealized device performance in the active region?

(6-4)
(6-5) **6-10.** (a) For the pnp common-base characteristic curves in Fig. 6-9, graphically determine the approximate value of r for the input at $V_{EB} = 0.60$ V. Use the $V_{CB} = -5.0$ V line. (b) From the output characteristic, find α_{dc} for $V_{CB} = -10$ V. Use the $I_E = 6.0$ mA line. (c) What conclusion can be drawn about r for the output characteristic in the linear region?

(6-4)
(6-5) **6-11.** (a) For pnp common-emitter characteristic curves in Fig. 6-11, graphically determine the approximate value of r for the input at $V_{BE} = -0.70$ V. Use the $V_{CE} = -5.0$ V line. (b) From the output characteristic, find β_{dc} for $V_{CE} = -8$ V. Use the $I_B = 40$ μA line. (c) What conclusion can be drawn about r for the output characteristic at the location given in part (b)?

(6-5) **6-12.** A BJT operates with $I_C = 5.000$ mA and $I_B = 30$ μA. What type of BJT is being discussed? Find α_{dc} and β_{dc}.

(6-5) **6-13.** Given $\beta_{dc} = 1.2 \times 10^2$ and $\beta_{ac} = 1.4 \times 10^2$ for a certain operating point, find α_{dc} and α_{ac}.

(6-5) **6-14.** Given r_{CE} as the ac resistance of the input characteristic of a BJT in the common-emitter configuration and r_{CB} as the ac resistance of the same BJT in the common-base configuration, show that

$$r_{CE} \cong \beta_{ac} r_{CB}$$

(6-6) **6-15.** What BJT properties give rise to effective capacitances that will restrict device operation?

(6-6) **6-16.** From Fig. 6-14 (A), find β_{dc}, β_{ac}, α_{dc}, and α_{ac} around $V = 0.15$ V and $I = 80$ μA.

(6-6)
(6-7) **6-17.** A Si BJT is rated at 300 mW at 25°C. If the derating factor is 4 mW/°C, at what temperature is the power dissipation level rated at 250 mW? Approximately what percentage change in β would you expect over this temperature range? Approximately what change in V_{BE} would you expect over this range?

(6-7) **6-18.** For a given BJT, $\beta_{dc} = 150$ at 30°C and $\beta_{dc} = 180$ at 50°C. Calculate the percentage change in β_{dc} and the percentage change in α_{dc} experienced over this temperature range with respect to the room temperature values (at 30°C).

(6-8) **6-19.** For an npn BJT,

$$\alpha_F = \alpha_R = -0.995 \qquad V_{BE} = 0.450 \text{ V}$$

$$I_{Fo} = I_{Ro} = -1.00 \text{ nA} \qquad V_{CE} = 0.050 \text{ V}$$

Find I_B, I_C, I_E, α_{dc}, and β_{dc}. What can you conclude about the region in which the BJT is operating?

6-20. For a given BJT, the production costs and market (selling) price vary as shown in Fig. 6-25. What tolerance level in β_{dc} produces the highest-profit product?

Solutions **6-1.** When majority carriers are injected from the emitter of a BJT into the base, an excess level of minority carriers builds up in the base region. (This process can be recalled by referring to Exercise 4-6 in Chapter 4, which considers the injection of carriers across a pn junction diode.) Once in the base, the excess carriers diffuse across the base until they

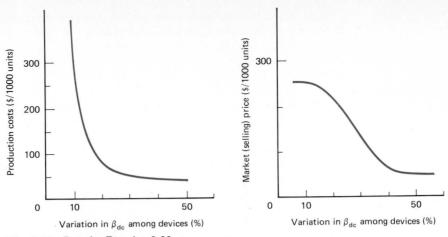

Fig. 6-25 Data for Exercise 6-20.

reach the base-collector junction, where the applied (reverse-bias) potential creates an electric field that pulls the carriers into the collector region. The result is illustrated in Fig. 6-26.

The number of carriers reaching the collector is reduced below the idealized number by the *injection efficiency ratio* γ of the emitter-base junction and the recombination of carriers as they cross the base region, experienced through the *base transport factor B*.

Fig. 6-26 Graphical illustration of the carrier densities in a BJT biased for active operation. The thin base region results in an approximately linear decrease in minority carriers across the region.

The emitter is doped more strongly than the base, and the base and collector are shown doped at about equal levels. As discussed in Exercise 8 and Chapter 18, it is sometimes desirable to reduce the collector doping in order to limit the effects of base narrowing.

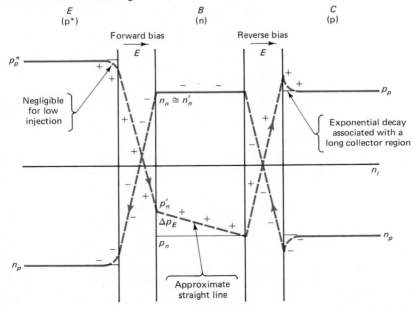

Let I_1 be the current that results by injecting carriers from the emitter to the base in a BJT and I_2 be the current flow that results from reverse-carrier flow from the base to the emitter. The injection efficiency ratio is then

$$\frac{I_1}{I_1 + I_2} = \frac{I_1}{I_E} = \gamma$$

If the emitter is doped much more heavily than the base, then γ approaches $+1$.

The fraction of the injected carriers that reaches the collector is called the *base transport factor* $B = I_C/I_1$. If the base is very thin, then carrier recombination is minimized and B tends toward -1. From the above, $\gamma B = \alpha_{dc}$ for the BJT. For a high-current-gain BJT, both $|\gamma|$ and $|B|$ must be within a fraction of a percentage of the ideal value of unity.

In order to obtain these high values, the emitter must be doped much more heavily than the base (to maximize γ) and the base region must be made very thin (to limit the opportunity for carrier recombination and maximize B). However, as doping levels increase, the doped atoms interact and cause the effective energy gap to decrease in value. This results in an increase in the number of thermally generated minority carriers, which is an undesirable result. Therefore, there is, in general, a preferred range of values for doping levels, balancing the desired γ with *band-gap narrowing*. Similarly, a very thin base increases B but subjects the base to *breakdown* (due to the electric field created across the region) and *punch-through* (due to the reduction in the effective width of the base region as the collector-base depletion region expands under reverse bias). These factors are discussed in further detail in Exercises 6-7 to 6-9 and in Chapter 18.

6-2. Using the sign conventions of this chapter, the fraction of carriers injected into the base that recombines before reaching the collector is given by

$$\frac{I_1 + I_C}{I_1} = 1 + \frac{I_C}{I_1} = 1 + B = 1 + \frac{\alpha_{dc}}{\gamma}$$

Therefore

$$\frac{I_1 + I_C}{I_1} = 1 + \frac{\alpha_{dc}}{[I_1/(I_1 + I_2)]} = 1 + \alpha_{dc}\left(1 + \frac{I_2}{I_1}\right)$$

and by substitution

$$\frac{I_1 + I_C}{I_1} = 1 - (0.990)(1 + 0.001) = 0.009 \text{ or } 0.9\%$$

6-3. (a) The majority carriers in the emitter of a pnp are holes (see Fig. 6-2(B). (b) When holes are injected into the base, the number of minority carriers increases substantially. (c) The emitter-base junction is forward-biased and the collector-base junction is reverse-biased for linear operation.

6-4. (a) The minority carriers in the base of an npn are electrons [see Fig. 6-2(A)]. (b) Emitter injection increases the number of minority carriers in the base. (c) The emitter-base junction is forward-biased and the collector-base junction is reverse-biased for linear operation.

6-5. From the notation developed: (a) the data describe a common-base npn with the collector-base forward-biased by 0.5 V and the emitter-base forward-biased by 0.7 V. This is a saturated condition. (b) The data describe a common-base pnp with the collector-base reverse-biased by 8 V and the emitter-base forward-biased by 0.7 V. This produces linear operation. (c) The data describe a common-emitter npn with the collector-base reverse-biased by $(6 - 0.7)$ V and the emitter-base forward-biased by 0.7 V. This produces linear operation. (d) The data describe a common-emitter pnp with the collector-base reverse-biased by $(6 + 0.7)$ V and the emitter-base reverse-biased by 0.7 V. This produces cutoff of the BJT.

6-6. A range of possible values for R and V_{CC} is possible. As shown in Fig. 6-27(A), the input load line can vary slightly, depending on the value of V_{CE}. Figure 6-27(B) indicates how the values for V_{CC} over the linear region produce small changes in I_C due to the measurable ac resistance r of the output characteristic.

(a) For $V_{CC} = 10$ V, $V_{CE} = -10$ V so that $I_C \cong 6.5$ mA and

$$R \cong 2.00 \text{ V}/45.0 \ \mu\text{A} \cong 44.4 \text{ k}\Omega$$

(b) For $V_{CC} = 5$ V, $V_{CE} = -5$ V so that $I_C \cong 6.0$ mA and

$$R \cong 2.00 \text{ V}/47.0 \ \mu\text{A} \cong 42.6 \text{ k}\Omega$$

(c) For $V_{CC} = 2$ V, $V_{CE} = -2$ V so that $I_C \cong 5.5$ mA and

$$R \cong 2.00 \text{ V}/49.0 \ \mu\text{A} \cong 40.8 \text{ k}\Omega$$

Fig. 6-27 Possible load lines for Exercise 6-6.

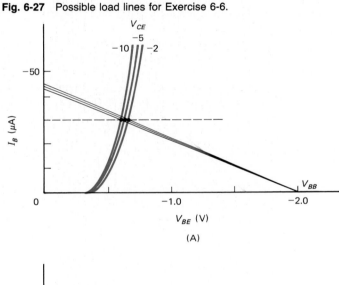

(A)

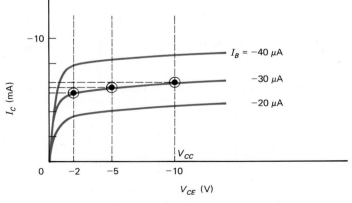

(B)

Given the defined linear range for the BJT, any of the above choices are acceptable to satisfy the $I_B = -30 \; \mu$A design objective. However, if it is important to be able to accommodate a range of values for V_{CE} around a given bias point, the extreme values of V_{CC} are less useful.

Observe that if a resistor is placed in series with V_{CC}, an output load line with finite slope replaces the $V_{CE} =$ constant lines used in Fig. 6-27(B). This more general problem, and the design constraints associated with it, are explored in detail in Chapter 7.

6-7. As can be seen from Fig. 6-26, the minority-carrier density in the base region is approximately given by a straight line with $p'_n \cong \Delta p_E$ on the emitter side and $p'_n \cong 0$ on the collector side. The total injected charge in the base at any given time is thus given by

$$Q_p \cong qAW \, (\Delta p_E/2)$$

where A is the cross-sectional area and W the width of the base. If the charge must be replaced in a time τ_p

$$I_B \cong Q_p/\tau_p \cong -\frac{qAW}{\tau_p}(\Delta p_E/2)$$

where the minus sign maintains the sign convention. This same result can be obtained from more detailed analysis (for example, see Streetman, 1980).

The collector current is approximately given by the emitter-base diffusion current due to forward bias. From Eq. (4-4b),

$$I_C \cong -qAD_p(\Delta p_E/W)$$

to first order. This result can also be obtained from more detailed analysis.

Combining the above equations,

$$\beta_{\text{dc}} = I_C/I_B \cong 2(L_p/W)^2$$

where $L_p^2 = D_p \tau_p$ as introduced in Exercise 4-6 in Chapter 4. Thus, a large value of β_{dc} requires a recombination length L_p much greater than the base width W, as expected. This result does not take into account the finite injection efficiency γ, which is assumed to be equal to unity. For this sample approximation, β_{dc} has a constant value everywhere in the active region and is a function only of the ratio L_p/W. Therefore, $\beta_{\text{dc}} = \beta_{\text{ac}}$ throughout the active region. The effects discussed in Exercises 6-8 and 6-9 cause β_{dc} and β_{ac} to vary with I_C and V_{CE}, giving a more accurate representation of actual devices.

6-8. The finite output resistance r of the common-emitter BJT is due to *base narrowing*.[3] For linear operation, as the voltage $|V_{CE}|$ increases, the reverse bias across the base-collector junction increases. As discussed in Section 4-6, the width of a depletion region increases as the junction becomes more reverse-biased.

From geometric considerations, a wider base-collector depletion region reduces the effective widths of the base and collector. From Eq. (4-40), the relative expansions of the depletion region into the base and collector will be determined by the doping levels. The expansions of the depletion region must take place in such a way that the total charge in the region is approximately zero. If $N_C \gg N_B$, then the expansion of the depletion region will most strongly affect the base. If $N_C \ll N_B$, then the expansion of the depletion region will most strongly affect the collector. The latter case is explored in Chapter 18.

For a narrower base, the base transport factor B and common-base current gain α_{dc} will increase, providing an increase in I_C with V_{CE} for a given I_B. The slope of the resultant curve is $1/r$. For more pronounced base narrowing, r will be further reduced to lower, more significant values.

---

[3] The discussion of Exercise 6-8 is based on an extension of concepts originally suggested by Early (1952).

From Exercise 6-7, for a pnp

$$I_C \cong -qAD_p(\Delta p_E/W)$$

The increase in carrier density due to injection is held constant if V_{BE} is constant. Given that the thickness of the base region is a function of V_{CE},

$$\left.\frac{\partial I_C}{\partial V_{CE}}\right|_{V_{BE}} \cong -qAD_p\Delta p_E\left(\frac{1}{W^2}\frac{\partial W}{\partial V_{CE}}\right) = \left.\frac{I_C}{W}\frac{\partial W}{\partial V_{CE}}\right|_{V_{BE}}$$

From Eq. (4-40),

$$W = W_o - K\sqrt{V_t - V_{CB}} \qquad (V_{CB} < 0 \text{ for pnp})$$

where

$$K = \sqrt{\frac{2\epsilon}{qN_d(1 + N_B/N_C)}}$$

and W_o is the (reference) base width that would exist with $V_t - V_{CB} = 0$.

Given

$$\frac{\partial W}{\partial V_{CE}} \cong \frac{\partial W}{\partial V_{CB}}$$

then

$$\left.\frac{\partial I_C}{\partial V_{CE}}\right|_{V_{BE}} \cong \frac{I_C(K/2)(V_t - V_{CB})^{-1/2}}{(W_o - K\sqrt{V_t - V_{CB}})}$$

As noted from the input characteristic curve, V_{CB} has only a weak effect on the relationship between V_{BE} and I_B. Therefore, $V_{BE} = $ constant can be associated with $I_B \cong$ constant, producing

$$\frac{1}{r} = \left.\frac{\partial I_C}{\partial V_{CE}}\right|_{I_B} \cong \frac{I_C}{2\left(\dfrac{W_o}{K}\right)^2\left[1 - \left(\dfrac{V_t - V_{CB}}{V_{PT}}\right)^{1/2}\right]\left(\dfrac{V_t - V_{CB}}{V_{PT}}\right)^{1/2}} \equiv \frac{|I_C|}{V_E}$$

where $V_{PT} = (W_o/K)^2$ is the *punch-through voltage*. The dependence of V_E on V_{CB} is illustrated in Fig. 6-28. As can be seen, V_E varies only slightly over the typical range of values for V_{CB}.

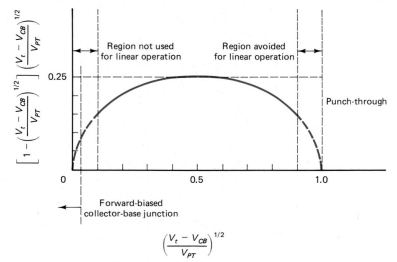

Fig. 6-28 Variation of V_E with V_{CB}, suggesting the reasonableness of an assumption $V_E \cong$ constant over typical ranges of linear operation for the BJT.

It is often convenient to treat the denominator of this result as being approximately constant over these typical values. Based on such a simplifying model, $V_E \cong$ constant and the slopes of the output characteristic curves are approximately independent of V_{CE}. For such a case,

$$r \cong V_E/|I_C|$$

for a pnp or npn, and all of the output characteristic curves cross the axis at the same point $V_{CE} = -V_E$. (This arrangement is shown in Fig. 9-3 and is further discussed in Section 9-3.) V_E is called the *Early voltage* (Early, 1952; and Gray and Meyer, 1984).

Combining the above results,

$$V_E = 2\left(\frac{W_o}{K}\right)^2\left[1 - \left(\frac{V_t - V_{CB}}{V_{PT}}\right)^{1/2}\right]\left(\frac{V_t - V_{CB}}{V_{PT}}\right)^{1/2}$$

where

$$K = \left(\frac{2\epsilon N_C}{qN_B^2}\right)^{1/2} \quad \text{and} \quad V_{PT} = \left(\frac{W_o}{K}\right)^2$$

for the case $N_B \gg N_C$.

In the center range of Fig. 6-28,

$$V_E \cong \frac{1}{2}\frac{W_o^2 qN_B^2}{2\epsilon N_C}$$

The approximation above for r is a useful one because it provides a way to estimate the value of r for a BJT if the dc collector current is known and if a reasonable value for V_E can be estimated. In typical cases, $V_E \cong 50$ to 100 V.

6-9. An idealized BJT, operated in the active region, has a constant value of $\beta_{dc} = \beta_{ac}$ and an infinite output resistance, giving rise to flat output curves. Realistic devices vary from this model in many different ways.

At low- and high-current levels, the values of β_{dc} and β_{ac} are reduced. The reduction at low-current levels occurs because the injection efficiency drops due to the relatively increasing importance of the recombination of injected carriers in the depletion region. The drop at high-current levels occurs when the number of injected carriers becomes sufficiently large to increase the effective conductivity of the base region, again reducing the injection efficiency.

The finite value of r has been discussed in Exercise 6-8. As the base-collector junction of the BJT becomes more reverse-biased, the thickness of the depletion region increases (as discussed in Section 4-6). Therefore, the effective base thickness decreases. As a result, the base transport factor B increases and the current gain of the BJT increases. For a given I_B, I_C increases with V_{CB}. This effect produces the finite output resistance r of the BJT (Figs. 6-11 and 6-12). Exercises 6-7 to 6-9 provide a link between semiconductor device properties and BJT performance. These relationships are further explored in Chapter 18.

6-10. (a) From Fig. 6-9(A), Fig. 6-29 can be constructed and

$$r \cong 0.2 \text{ V}/10 \text{ mA-} \cong 20 \ \Omega$$

(b) From Fig. 6-9(B), Fig. 6-30 can be constructed and

$$\alpha_{dc} = I_C/I_E \cong 6.0 \text{ mA}/6.0 \text{ mA} \cong 1.0$$

(c) The slope is so flat, r is too large to be graphically determined.

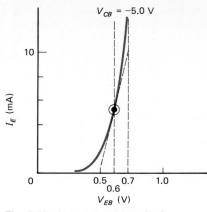

Fig. 6-29 Input operating point for Exercise 6-10.

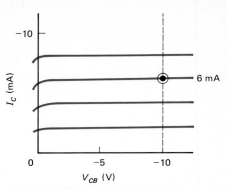

Fig. 6-30 Output operating point for Exercise 6-10.

6.11. (a) From Fig. 6-11(A), Fig. 6-31 can be constructed, and

$$r = 0.2 \text{ V}/50 \text{ } \mu\text{A} \cong 4 \text{ k}\Omega$$

(b) From Fig. 6-11(B), Fig. 6-32 can be constructed and

$$\beta_{dc} = I_C/I_B \cong -8.5 \text{ mA}/-40 \text{ } \mu\text{A} = 2.1 \times 10^2$$

(c) From Fig. 6-11(B),

$$r \cong 8 \text{ V}/2 \text{ mA} = 4 \text{ k}\Omega$$

From Exercise 6-8,

$$V_E \cong rI_C = (4 \text{ k}\Omega)(8.5 \text{ mA}) \cong 34 \text{ V}$$

which is a low value for this parameter.

Fig. 6-31 Input operating point for Exercise 6-11.

Fig. 6-32 Output operating point for Exercise 6-11.

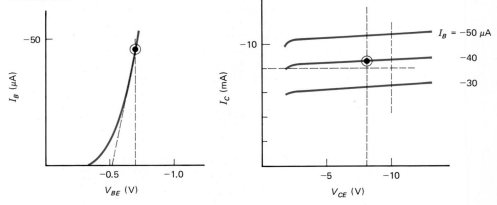

6-12. Since both I_C and I_B flow into the BJT, an npn is being described. By definition,

$$\alpha_{dc} = \frac{5.000 \text{ mA}}{-(5.000 + .030) \text{ mA}} \cong -0.994$$

From Eq. (6-4),

$$\beta_{dc} \cong \frac{-\alpha_{dc}}{1 + \alpha_{dc}} \cong \frac{0.994}{0.006} \cong 1.7 \times 10^2$$

6-13. From Eq. (6-4),

$$\alpha_{dc} = \frac{-\beta_{dc}}{\beta_{dc} + 1} \cong \frac{-1.2 \times 10^2}{(1.2 \times 10^2) + 1} \cong -1.0$$

$$\alpha_{ac} \cong \frac{-\beta_{ac}}{\beta_{ac} + 1} \cong \frac{-1.4 \times 10^2}{(1.4 \times 10^2) + 1} \cong -1.0$$

6-14. By definition,

$$r_{CB} \cong \frac{|\Delta V_{EB}|}{\Delta I_E} \quad \text{and} \quad r_{CE} \cong \frac{|\Delta V_{BE}|}{\Delta I_B}$$

Therefore, since $|\Delta V_{EB}| = |\Delta V_{BE}|$,

$$\frac{r_{CB}}{r_{CE}} \cong \frac{\Delta I_B}{\Delta I_E} \cong \frac{1}{\beta_{ac}}$$

$$r_{CE} \cong r_{CB} \beta_{ac}$$

6-15. The effective base-emitter capacitance is dominated by the buildup of excess carriers next to the depletion region (as discussed in Exercise 6-1 and illustrated in Fig. 6-26). This effective capacitance is proportional to the emitter current and increases rapidly for a thicker base region because a larger total charge is contained. The effective base-collector capacitance is due to the reverse-biased depletion region also discussed in Section 4-6). As is demonstrated in Chapter 9, a small base-collector capacitance can have a large effect on ac-amplifier performance.

6-16. From Fig. 6-14(A), Fig. 6-33 can be drawn, and

$$\beta_{dc} \cong 6.2 \text{ mA}/80 \text{ } \mu\text{A} \cong 78$$

$$\beta_{ac} \cong \frac{(6.8 - 5.2) \text{ mA}}{(90 - 70) \text{ } \mu\text{A}} = \frac{1.6 \text{ mA}}{20 \text{ } \mu\text{A}} = 80$$

$$\alpha_{dc} = \frac{-\beta_{dc}}{\beta_{dc} + 1} = \frac{-78}{79} \cong -0.987$$

$$\alpha_{ac} \cong \frac{-\beta_{ac}}{\beta_{ac} + 1} = \frac{-80}{81} \cong -0.988$$

6-17. (a) From Example 6-2,

$$P_D(T) = P_D(25°C) - (4 \text{ mW/°C})(T - 25°C)$$

$$250 \text{ mW} = 300 \text{ mW} - (4 \text{ mW/°C})(T - 25°C)$$

$$T = 25°C + 50 \text{ mW}/4 \text{ mW/°C} \cong 38°C$$

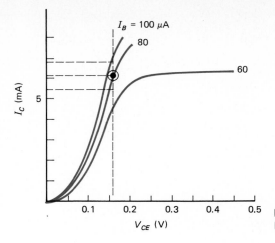

Fig. 6-33 Characteristic curves for Exercise 6-16.

(b) From Example 3, assuming

$$\Delta\beta/\beta \cong 0.9\%/°C$$

then

$$\frac{\Delta\beta}{\beta}(25°C \text{ to } 38°C) \cong 0.9\%/°C)(13°C) \cong 12\%$$

(c) From Chapter 4,

$$\left(\frac{\partial V_{BE}}{\partial T}\right)_I \cong -2 \text{ mV/°C}$$

so that

$$\left(\frac{\partial V_{BE}}{\partial T}\right)_I \Delta T \cong (-2 \text{ mV/°C})(13°C) \cong -26 \text{ mV}$$

6-18. From the data given,

$$\% \text{ change in } \beta_{dc} = \frac{\beta_{dc}(50°C) - \beta_{dc}(30°C)}{\beta_{dc}(30°C)} \times 100\%$$

$$= \frac{180 - 150}{150} \times 100\% = 20\%$$

$$\% \text{ change in } \alpha_{dc} = \frac{(-180/181) - (-150/151)}{-(150/151)} \times 100\%$$

$$= \frac{-0.9945 + 0.9934}{-0.9934} \times 100\% = 0.1\%$$

6-19. This is the common-emitter configuration. Therefore, from Eq. (6-17),

$$I_F = (-1.00 \text{ nA})(e^{0.450/0.026} - 1) \cong -32.9 \text{ mA}$$

$$I_R = (-1.00 \text{ nA})(e^{-(0.050-0.450) \text{ V}/0.026 \text{ V}} - 1) \cong -4.8 \text{ mA}$$

and from Eq. (6-14),

$$I_E = (-32.9 \text{ mA}) + (-0.995)(-4.8 \text{ mA}) \cong -28.1 \text{ mA}$$

$$I_C = (-0.995)(-32.9 \text{ mA}) + (-4.8 \text{ mA}) \cong 27.9 \text{ mA}$$

$$I_B = -(I_E + I_C) \cong 0.2 \text{ mA}$$

The resulting currents are shown in Fig. 6-34. The BJT is operating in the saturated region.

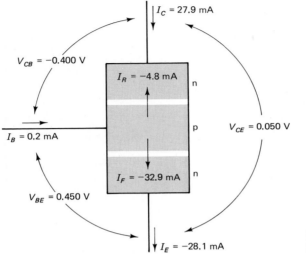

$I_C = 27.9$ mA

$V_{CB} = -0.400$ V

$I_R = -4.8$ mA   n

p $V_{CE} = 0.050$ V

$I_B = 0.2$ mA

$I_F = -32.9$ mA   n

$V_{BE} = 0.450$ V

$I_E = -28.1$ mA

Fig. 6-34 Current flows for Exercise 6-19.

6-20. The graph of Fig. 6-35 is obtained for the difference between market selling-price and production cost. A choice of 20 percent variation in β_{dc} among devices produces the highest-profit product.

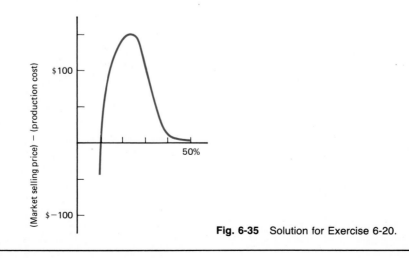

$100

50%

(Market selling price) − (production cost)

$−100

Fig. 6-35 Solution for Exercise 6-20.

PROBLEMS

(Exercise 6-1) **6-1.** For a BJT, the injected current from emitter to base is 9.50 mA and the reverse carrier flow from base to emitter is 0.05 mA. Find the injection efficiency.

(Exercise 6-1) **6-2.** For a BJT the injected current from emitter to base is 55.0 μA. If the base transport factor is 0.990, estimate the collector current assuming linear operation.

(6-1)
(6-2) **6-3.** Describe the configurations and types of BJT operation obtained for each of the circuits shown in Fig. 6-36.

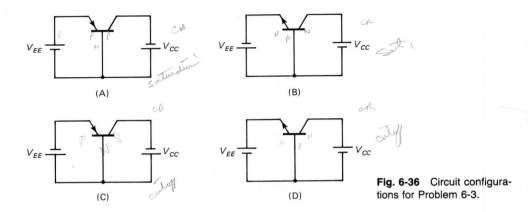

(A)   (B)

(C)   (D)

Fig. 6-36 Circuit configurations for Problem 6-3.

(6-2)
(6-3) **6-4.** Describe the configurations and types of BJT operation that can result for each of the circuits shown in Fig. 6-37.

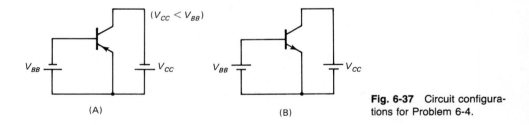

(A)   (B)

Fig. 6-37 Circuit configurations for Problem 6-4.

(6-3)
(6-4) **6-5.** Assume that the BJT in Fig. 6-38 is described by the characteristic curves of Fig. 6-11. The desired bias point is $I_B = -50 \mu$A. Find the values of R, V_{CC}, and I_C that can be used.

(6-3)
(6-4) **6-6.** Assume that the BJT in Fig. 6-39 is described by the characteristic curves in Fig. 6-12. The desired bias point is $I_B = 80 \mu$A. Find the values of R, V_{CC}, and I_C that can be used.

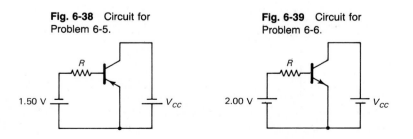

Fig. 6-38 Circuit for Problem 6-5.

Fig. 6-39 Circuit for Problem 6-6.

(6-5) **6-7.** Find I_B and the input and output resistance r for the characteristic curves in Fig. 6-11 for $V_{BE} = -0.7$ V and $V_{CE} = -10$ V.

(6-5) **6-8.** Find I_B and the input and output ac resistance r for the BJT characteristic curves in Fig. 6-12 for $V_{BE} = 0.7$ V and $V_{CE} = 10$ V.

(6-5) **6-9.** Given $I_C = 20$ mA for $I_B = 0.15$ mA and $V_{CE} = 6$ V, and $I_C = 40$ mA for $I_B = 0.28$ mA and $V_{CE} = 6$ V, calculate α_{dc}, α_{ac}, β_{dc}, and β_{ac}. If V_{CE} increases to 12 V, what types of change in these results can be expected?

(6-5) **6-10.** Given $\beta_{dc} = 200$ and $\beta_{ac} = 250$, find α_{dc} and estimate α_{ac}. Under what conditions can a large error be encountered with the value for α_{ac}?

(6-6) **6-11.** What power can the 2N5208 BJT dissipate if the case temperature is maintained at 50°C?

(6-6) **6-12.** What power can the 2N5208 BJT dissipate if the ambient air temperature is maintained at 50°C?

(6-6) **6-13.** What is the maximum value of V_{CC} that should be used with the 2N5208 BJT to assure that absolute maximum ratings cannot be exceeded?

(6-6) **6-14.** Find the output ac resistance for the 2N2907A BJT for $V_{CE} = 0.15$ V and $I_B = 40$ μA.

(6-6) **6-15.** Find the output ac resistance for the 2N2907A BJT for $V_{CE} = 5.0$ V and $I_B = 100$ μA.

(6-6) **6-16.** From Fig. 6-14(A), find β_{dc}, β_{ac}, α_{dc}, and α_{ac} around (a) $V_{CE} = 0.10$ V and $I_B = 60$ μA, (b) $V_{CE} = 0.20$ V and $I_B = 60$ μA, (c) $V_{CE} = 0.30$ V and $I_B = 60$ μA.

(6-6) **6-17.** From Fig. 6-14(B), find β_{dc}, β_{ac}, α_{dc}, and α_{ac} in the middle of the linear region shown.

(6-6) **6-18.** Find the range of values for β_{dc}, β_{ac}, and r over the linear BJT range shown in Fig. 6-15(A).
(6-7)

(6-7) **6-19.** A typical example of β_{dc} as a function of T for a BJT is shown in Fig. 6-40. Calculate the fractional change in β_{dc} with T at points A to F.

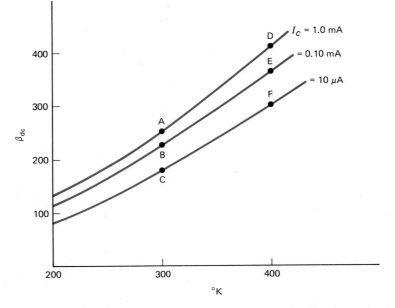

Fig. 6-40 Typical curves for β_{dc} vs T for the linear region. For devices with a lower value of β_{dc}, the fractional rate of change with temperature will increase. (Estimated data taken from Gray and Meyer, 1984.)

(6-8) **6-20.** An npn Si BJT is biased so that $V_{EB} = -0.4000$ V and $V_{CB} = 8.000$ V. Assume that $I_{Fo} = I_{Ro} = -1.000 \times 10^{-8}$ A and $\alpha_F = \alpha_R = -0.9950$. (a) Find I_C, I_E, I_B, and α_{dc}. (b) Find the same variables if the Ebers-Moll parameters are known to only two significant figures.

(6-8) **6-21.** A pnp Ge BJT is biased in the common-emitter configuration with $V_{BE} = -0.2500$ V and $V_{CE} = -10.00$ V. Assume that $I_{Fo} = I_{Ro} = 8.000 \times 10^{-7}$ A and $\alpha_F = \alpha_R = -0.9950$. (a) Find I_C, I_E, I_B, and β_{dc}. (b) Find the same variables if the Ebers-Moll parameters are known to only two significant figures.

COMPUTER APPLICATIONS

* **6-1.** Write a computer program to calculate α_{dc} given the input data for the common-base Ebers-Moll model (V_{EB}, V_{CB}, I_{Fo}, I_{Ro}, α_F, α_R). Use your program to check on the calculations of Example 6-1. Turn in your program listing and the value of α_{dc} that you find.

* **6-2.** Use SPICE to simulate the Ebers-Moll BJT model of Fig. 6-18. Use this simulation to develop idealized input and ouput characteristic curves for a BJT (with α_F and α_R constant). Compare these with Figs. 6-9 through 6-12.

* **6-3.** SPICE can be used to produce the characteristic curve for a BJT once the relevant device parameters are input as a data file. Using a SPICE program and a typical set of parameter values, produce the resultant curve for a typical npn in a common-emitter configuration. Determine β_{dc} and β_{ac} graphically at several locations in the linear region and on the saturation boundary. Compare these with the results of Problem 6-19.

EXPERIMENTAL APPLICATIONS

6-1. The objective of this experiment is to become familiar with the common-emitter characteristic curves of a selected BJT and to practice making measurements of important BJT parameters. At room temperature, gather data on the input characteristic (using a curve tracer or equivalent circuit) for a range of values for I_B that fall within device specifications. Then gather data on the output characteristic for values of V_{CE} and I_C within specification. Increase the temperature and obtain the input and output characteristics for the same device. Use a hot-air blower and a thermometer or thermocouple to measure temperatures.

Measure the ac resistance r of the input characteristic at several I_B levels, and compare with the approximation $r \cong V_o/I_B$. Is the temperature effect important? Then determine the fractional change in β_{dc} as a function of T by comparing the two output characteristic curves and calculating

$$\frac{1}{\beta_{dc}} \frac{\partial \beta_{dc}}{\partial T} \cong \frac{I_{C2} - I_{C1}}{\frac{1}{2}(I_{C2} + I_{C1})\Delta T}$$

where I_{C1} and I_{C2} are measured on the same I_B = constant line at temperatures separated by ΔT. It is helpful to mark and retain this BJT for future use.

Discussion
The BJT input characteristic curves follow the junction diode properties discussed in Chapter 4. A typical BJT output characteristic curve as a function of

temperature has been recorded in Fig. 6-41. If points 1 and 1A are used, the fractional change in β as a function of T is found to be

$$\frac{1}{\beta_{dc}} \frac{\partial \beta_{dc}}{\partial T} \cong \frac{I_{C2} - I_{C1}}{\frac{1}{2}(I_{C2} + I_{C1})\Delta T}$$

$$= \frac{8.2 \text{ mA} - 5.8 \text{ mA}}{\frac{1}{2}(8.2 \text{ mA} + 5.8 \text{ mA})(96°C - 34°C)}$$

$$\cong 0.6\%/°C$$

Fig. 6-41 Typical BJT output characteristic curve as a function of temperature for Experimental Application 6-1.

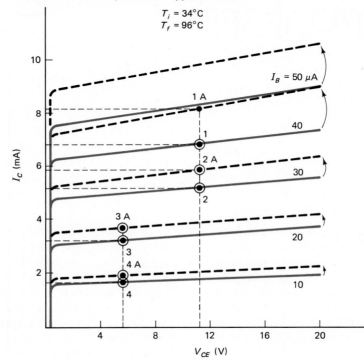

CHAPTER 7
BJT BIASING TRADEOFFS

Some of the basic issues involved in circuit design with BJTs are considered in this chapter, with a focus on BJT models, biasing, and the tradeoffs that are encountered in amplifier design. The BJT applications discussed here are often encountered as part of larger system designs.

Biasing tradeoffs are introduced by making use of simple dc-coupled amplifier circuits. When an input voltage to a dc-coupled amplifier is varied around a bias value, the output voltage waveform is a reduced or magnified replica of the input (for idealized amplifier performance). Such an amplifier can be used to amplify both very slowly varying signals and rapidly varying input waveforms.

The circuits introduced in this chapter provide entry to some essential design strategies. They also illustrate some of the difficulties inherent in designing a high-quality dc-coupled amplifier. The *operational amplifier,* covered in Chapter 12, is a widely used, high-quality dc-coupled amplifier that is available as an IC building block element. The operational amplifier uses complex circuits and many transistors to solve the problems associated with the building block circuits introduced here.

Much of this chapter is concerned with developing an organized approach to the design of circuits with specific performance objectives. Using the BJT dc-coupled amplifier as the circuit of interest and obtaining a given amplifier voltage gain as the primary performance objective, the following development outlines a step-by-step design process. It is important to understand the nature of this process, so that it can be explored and generalized as an approach for more complicated circuits to follow. Both the dc-coupled amplifier itself and the strategies developed here contribute to the understanding of electronics design.

The temperature dependence of the Q point is of explicit importance in dc-coupled amplifiers and is thereby treated with the detail it deserves. (As is shown in Chapter 9, this factor affects ac-coupled amplifiers less directly, so that an adequate foundation must be previously established.) In addition, the dc-coupled amplifiers that are used in this chapter provide the means for experimental studies of the performance parameters associated with BJTs.

7-1 BJT MODELS

Chapter 6 introduced characteristic curves for the BJT in the common-base and common-emitter configurations. It is helpful to develop approximate models for these curves for use in circuit design. Figures 7-1 and 7-2 illustrate models that are in common use in the linear region of operation.

For the input characteristic curves, the small effect due to the output voltage is often neglected. The family of input diode curves is replaced by a single diode curve,

Fig. 7-1 Common-base configuration (npn). (A) Actual and approximate input characteristics. (B) Actual and approximate output characteristics and output load line.

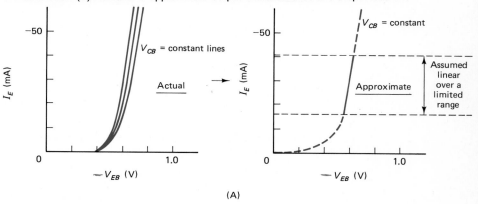

(A)

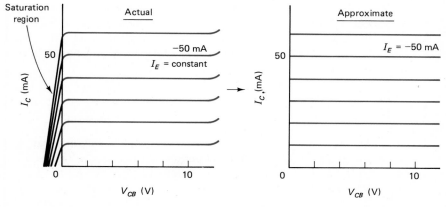

(B)

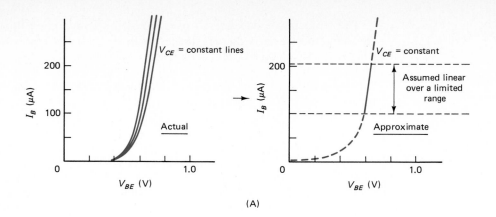

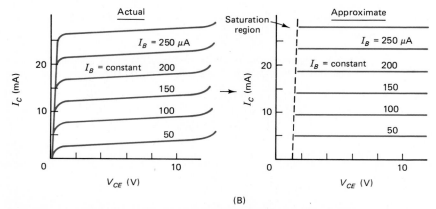

Fig. 7-2 Common-emitter configuration (npn). (A) Actual and approximate input characteristics. (B) Actual and approximate output characteristics.

as shown in Figs. 7-1(A) and 7-2(A). The curve is assumed to be approximately linear over the range of input current values that are to be used (that is, r is approximately constant over the range of possible operating points).

In the linear region, the output characteristic curves are approximated as a set of flat, equally spaced lines. This is a good approximation for the common-base curves, but a less-accurate approximation for the common-emitter curves. (A common-emitter model with a finite value of r is introduced in Chapter 9 as a more adequate representation.)

7-2 COMMON-BASE DC-COUPLED AMPLIFIER

The operation of a common-base dc-coupled amplifier can now be considered, as shown in Fig. 7-3. The input circuit consists of a voltage source V_{in} (negative to turn *on* the emitter-base diode) in series with a resistor R_E and the BJT input characteristic. The output circuit consists of a battery V_{CC} in series with a resistor R_C and the BJT output characteristic. V_{in} and R_E fix the input load line and V_{CC} and R_C fix the out-

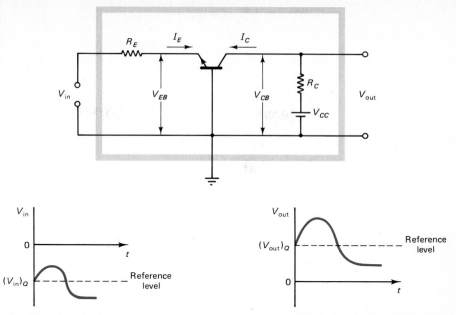

Fig. 7-3 Common-base dc-coupled amplifier (npn). (A) Circuit schematic. (B) Arbitrary input waveform. Voltage v_{in} must always be negative to correctly bias the BJT. (C) Corresponding output waveform. Voltage v_{out} must always be positive for a linearly biased BJT.

put load line. The operating points for these circuits can be found by combining the load lines with the input and output characteristics of the BJT.[1]

The approximate models for the input and output characteristics *and* the load lines are shown in Fig. 7-4. As V_{in} varies around $(V_{in})_Q$, the input load line shifts as shown. (The slope of the load line stays the same because R_E is fixed.) Changes in V_{in} cause changes in I_E. Since $I_C \cong -I_E$, $(\alpha_{dc} \cong -1)$, the two vertical axes can be aligned and the effect of the shifting input load line is to move the output operating point up and down the output load line as shown. As a result, $V_{CB} = V_{out}$ changes. All of these changes are proportional and linear operation is obtained (for the model being used).

Note that as V_{in} becomes more negative, I_E and I_C increase and V_{out} becomes less positive. As a result, the input and output waveforms track one another, increasing and decreasing together, as shown in Fig. 7-3. The input and output are in phase.

The input and output Q points are often chosen to accommodate a maximum change in the input voltage or current while remaining in the linear portion of the output curve. For equal variations around a reference level, this objective can be achieved by biasing the BJT with $(V_{out})_Q$ in the middle of the linear operating region when $V_{in} = (V_{in})_Q$. For unequal variations, the Q point can be set to one side to maximize the available signal range.

The circuit of Fig. 7-3 is called a common-base dc-coupled amplifier. The voltage gain of this circuit is defined as

---

[1] Operation of the dc-coupled amplifiers in Chapters 7 and 8 is considered to take place through a sequence of different dc operating points determined by the input voltage. Since emphasis is on variations among dc bias conditions, capital letters are used to represent circuit parameters.

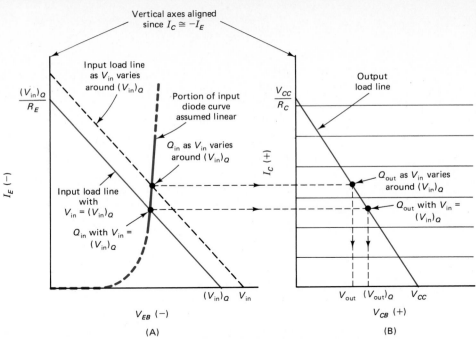

Fig. 7-4 Operating points for a common-base dc-coupled amplifier. The input and output circuits are linked by aligning the I_E and I_C axes (because $\alpha_{dc} \cong -1$). (A) Input circuit. (B) Output circuit.

$$A_V = \frac{\text{change in output voltage around } (V_{out})_Q}{\text{change in input voltage around } (V_{in})_Q} \qquad (7\text{-}1)$$

A simple equation for A_V can be found by starting with the schematic of Fig. 7-3. All voltages are measured with respect to *ground* to provide a common reference level.

From the input circuit,

$$V_{in} = I_E R_E + V_{EB} \qquad (7\text{-}2)$$

For arbitrary changes in V_{in},

$$\Delta V_{in} = (\Delta I_E)R_E + \Delta V_{EB} \qquad (7\text{-}3)$$

For the output circuit,

$$V_{out} = V_{CC} - I_C R_C \qquad (7\text{-}4)$$

and for related arbitrary changes,

$$\Delta V_{out} = -(\Delta I_C)R_C \qquad (7\text{-}5)$$

Therefore,

$$A_V = \frac{\Delta V_{out}}{\Delta V_{in}} = \frac{-(\Delta I_C)R_C}{(\Delta I_E)R_E + \Delta V_{EB}} \qquad (7\text{-}6)$$

$$= -\left(\frac{\Delta I_C}{\Delta I_E}\right)\frac{R_C}{R_E + \Delta V_{EB}/\Delta I_E} \qquad (7\text{-}7)$$

To proceed further, ΔI_C and ΔV_{EB} must be related to ΔI_E. In general, V_{EB} is a function of I_E and V_{CB}, so to first order

$$\Delta V_{EB} = \left.\frac{\Delta V_{EB}}{\Delta I_E}\right|_{V_{CB}} \Delta I_E + \left.\frac{\Delta V_{EB}}{\Delta V_{CB}}\right|_{I_E} \Delta V_{CB} \qquad (7\text{-}8)$$

Usually, the second term is small and is neglected. This is equivalent to replacing the family of input curves by a single curve, so that no V_{CB} dependence exists for the input. Therefore,

$$\Delta V_{EB} \cong r\Delta I_E \qquad (7\text{-}9)$$

where r is the ac resistance associated with the BJT input. Further, note that I_C is a function of I_B and V_{CB}, so that to first order

$$\Delta I_C = \left.\frac{\Delta I_C}{\Delta I_E}\right|_{V_{CB}} \Delta I_E + \left.\frac{\Delta I_C}{\Delta V_{CB}}\right|_{I_E} \Delta V_{CB} \qquad (7\text{-}10)$$

Usually, the second term is small. This is equivalent to assuming that the output characteristic curves are flat. Therefore,

$$\Delta I_C \cong \alpha_{ac} \Delta I_E \qquad (7\text{-}11)$$

The above assumptions are the same ones introduced in Fig. 7-1. Substituting from Eqs. (7-9) and (7-11) into Eq. (7-7),

$$A_V \cong -\alpha_{ac}\frac{R_C}{R_E + r} \qquad (7\text{-}12)$$

Since α_{ac} is negative, the input and output are in phase. The same result [Eq. (7-12)] is obtained for the pnp BJT.

Example 7-1 **Alternative Strategies for Input Bias Arrangements**

Assume that the input to a common-base dc-coupled amplifier is a signal V_{in} that varies around $(V_{in})_Q = 0$. How can the circuit of Fig. 7-3 be modified to obtain the correct bias for linear BJT operation?

Solution One possible approach is to place a battery (or power supply) in series with the input signal V_{in}, as shown in Fig. 7-5(A). Another approach is to introduce a parallel current flow, as shown in Fig. 7-5(B). The fixed current I_1 due to V_{EE} and R_1 produces

Fig. 7-5 Alternative bias circuits. (A) Series battery. (B) Parallel current source. (C) Alternative way to represent the schematic of part (B).

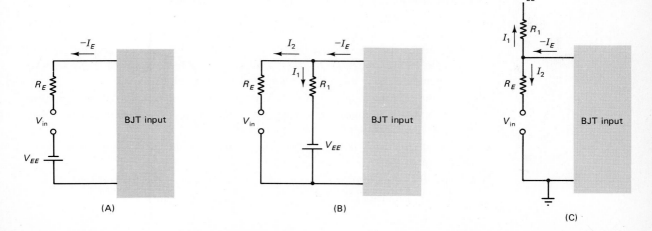

the desired reference bias condition. As V_{in} moves around $(V_{in})_Q$, I_E varies as $-I_E = I_1$ (fixed) $+ I_2$ (variable).

Figure 7-5(C) shows an alternative way for representing the circuit of Fig. 7-5(B). The positive side of the battery (or power supply) is assumed connected to the common terminal (which is the common ground for the circuit). This representation for voltage sources can be used with any portion of a circuit. ■

7-3 TRADEOFFS FOR COMMON-BASE AMPLIFIER DESIGN

DC-coupled amplifiers are often designed to accommodate a maximum variation in output signal V_{out} and to obtain a selected circuit voltage gain A_V. The significance of these objectives can be understood by remembering the gain equation [Eq. (7-12)] and by referring again to Fig. 7-4.

From Fig. 7-4(B), the maximum variation in output signal can be obtained by placing $(V_{out})_Q$ approximately at the middle of the available linear operating region (assuming equal signal variations around $(V_{out})_Q$. The reference output and input bias points can be fixed from this objective.

As can be seen from Eq. (7-12), the gain A_V depends on R_C, R_E, and r. Since $\alpha \cong -1$ and is only weakly dependent on temperature T, this circuit is stable as the temperature of the BJT is subject to change. The gain A_V thus depends on the slopes of the input and output load lines and on the slope of the input diode characteristic curve around the reference point $(V_{in})_Q$.

Inspection of Fig. 7-4 reveals that a more negative slope for the input load line (R_E smaller) increases the current swing in I_E associated with variations in V_{in} and thus increases the gain. Similarly, a more positive slope for the input diode characteristic curve (r smaller) increases the swing in I_E and increases the gain. A less negative slope for the output load line (R_C larger) produces a larger swing in V_{out} for a given variation in I_E and increases the gain. The reasonableness of Eq. (7-12) is thus confirmed and the equation can be interpreted in graphical terms.

One approach to the design of a common-base dc-coupled amplifier is to

1. Determine the input and output characteristics for the BJT to be used. (To simplify this step, the input characteristic can be taken as a typical diode and the output characteristic for the common-base configuration always takes the form shown in Fig. 7-1.)

2. Define a trial output load line. One possible strategy is to start by choosing $(I_C)_{max}$ and V_{CC} as large as possible within device and power supply ratings. This assures that the largest possible range of output currents and voltages is available to the amplifier. (These values may have to be adjusted downward to accommodate the availability of standard resistor values.)

3. Select the reference operating point for the output in the middle of the available linear operating range. (This strategy is modified if V_{in} varies around $(V_{in})_Q$ more in one direction than the other.)

4. Note that $|I_E| \cong |I_C|$, and determine r for the input at the input operating point (either graphically or by using $r \cong V_o/I_E$).

5. Choose the input load line (value of R_E) that produces the desired gain A_V.

6. Find $(V_{in})_Q$ required to produce the desired input Q point.

7. Modify the output load line if necessary and repeat the design cycle to achieve an acceptable combination of values. (If R_C is increased to produce a larger A_V, the available linear region is reduced.)

Iteration in the design process may be required to produce mutually consistent values for all resistors and voltage sources. Inspection of Fig. 7-4 reveals some of the biasing tradeoffs that exist. Larger values of R_C require larger values of V_{CC} and thus are limited by available supply voltages. Reductions in R_E below r produce only fractional increases in A_V and increase the sensitivity of the input operating point to temperature. Further, reductions in R_E limit the maximum swing in V_{in} that can be accommodated. The available values of A_V are bounded by device and supply considerations.

Example 7-2   Designing a Common-Base DC-Coupled Amplifier

Consider the design of a common-base dc-coupled amplifier with the performance objective $A_V = 10$, using the circuit and device characteristics shown in Fig. 7-6. In order to set the reference operating points, assume that the input voltage swings in the positive direction with three times the magnitude of the swings in the negative direction.

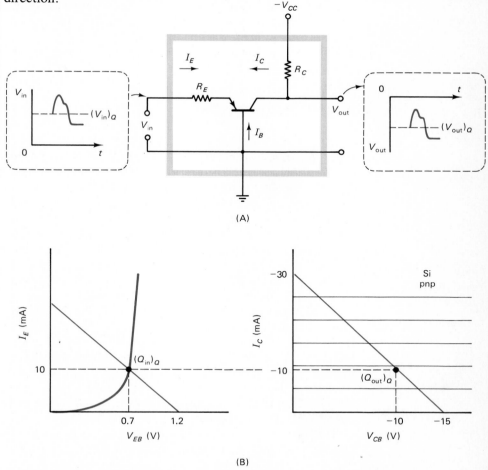

Fig. 7-6 Example of a common-base dc-coupled amplifier design. (A) Circuit schematic. (B) Device characteristics.

Solution A trial load line is selected with $V_{CC} = -15$ V and $(I_C)_{max} = -30$ mA. Therefore, $R_C = (15$ V$)/(30$ mA$) = 0.50$ kΩ.

The $(V_{out})_Q$ point is chosen to accommodate positive voltage variations that are about three times greater than the negative voltage variations. Thus,

$$(V_{CB})_Q = -10 \text{ V}$$

$$(I_C)_Q = -10 \text{ mA}$$

For the ac input resistance

$$r \cong 0.026 \text{ V}/10 \text{ mA} = 2.6 \ \Omega$$

From Eq. (7-12) and the design objective $A_V = 10$,

$$R_E \cong \frac{R_C}{A_V} - r = \frac{0.50 \text{ k}\Omega}{10} - 2.6 \cong 47 \ \Omega$$

To obtain the desired output Q point, select

$$(V_{in})_Q = (10 \text{ mA})(47 \ \Omega) + 0.7 \text{ V} \cong 1.2 \text{ V}$$

The above example illustrates many of the key features of dc-coupled amplifiers. Selection decisions must be made for several parameters to achieve the design objectives, and relationships among these parameters must be maintained. The given solution is an acceptable one, but represents only one of many possible solutions to the design problem. Iteration can be used to obtain standard resistor values as necessary.

∎

7-4 TRADEOFFS FOR COMMON-EMITTER AMPLIFIER DESIGN

The approximate characteristics of Fig. 7-2 can be used to design a common-emitter dc-coupled amplifier. Consider the device configuration shown in Fig. 7-7. Using an npn BJT, the bias batteries take on the polarities shown.

The vertical axes for the input and output characteristics cannot be aligned because the former is scaled to I_B and the latter is scaled to I_C. However, since these two parameters are related by $I_C = \beta_{dc} I_B$ and β_{dc} can be determined from the output characteristic, points can be indirectly transferred between the two characteristics. Figure 7-8 illustrates the design process. As before, the allowable input signal swing can be maximized by choosing an appropriate location for $(V_{out})_Q$. To map between $(V_{in})_Q$ and $(V_{out})_Q$, β_{dc} is used as a conversion factor from I_B to I_C.

As V_{in} varies around $(V_{in})_Q$, the input load line shifts as indicated. Changes in V_{in} cause changes in ΔI_B. Since $\Delta I_C = \beta_{ac}\Delta I_B$, the related changes ΔI_C can be calculated. The initial shift in V_{in} thus produces a known change in V_{out} as Q_{out} moves up and down the output load line. Using the BJT model shown, linear operation is obtained.

Observe that as V_{in} becomes more positive, V_{out} becomes less positive. As a result, the input and output waveforms are inverted, as shown in Fig. 7-7. The input and output are 180° out of phase.

The voltage gain A_V of this circuit can be found by again summing voltages around input and output circuit loops. From Fig. 7-7,

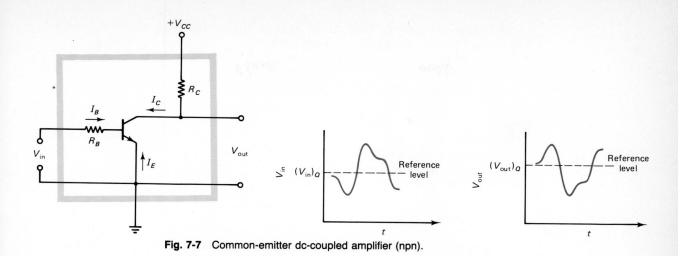

Fig. 7-7 Common-emitter dc-coupled amplifier (npn).

$$V_{\text{in}} = I_B R_B + V_{BE} \tag{7-13}$$

$$\Delta V_{\text{in}} = (\Delta I_B)R_B + \Delta V_{BE} \tag{7-14}$$

For the input circuit and for the output circuit,

$$V_{\text{out}} = V_{CC} - I_C R_C \tag{7-15}$$

$$\Delta V_{\text{out}} = -(\Delta I_C)R_C \tag{7-16}$$

so that

$$A_V = \frac{\Delta V_{\text{out}}}{\Delta V_{\text{in}}} = \frac{-(\Delta I_C)R_C}{(\Delta I_B)R_B + \Delta V_{BE}} \tag{7-17}$$

$$= -\left(\frac{\Delta I_C}{\Delta I_B}\right)\frac{R_C}{R_B + \Delta V_{BE}/\Delta I_B} \tag{7-18}$$

Fig. 7-8 Operating points for the common-emitter dc-coupled amplifier.

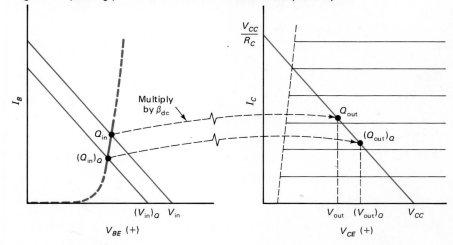

It is now necessary to relate ΔI_C to ΔI_B and ΔV_{BE} to ΔI_B. Using the same approach developed for the common-base configuration,

$$\Delta V_{BE} \cong r\Delta I_{EB} \qquad (7\text{-}19)$$

if the effect of V_{CE} on the input is neglected, and

$$\Delta I_C \cong \beta_{ac}\Delta I_B \qquad (7\text{-}20)$$

if the output characteristic curves are assumed to be approximately flat. The above assumptions are equivalent to those illustrated in Fig. 7-2. Substituting from Eqs. (7-19) and (7-20) into Eq. (7-18),

$$\beta_{ac} = \frac{A_v(R_b + r)}{R_C}$$

$$A_V \cong -\beta_{ac}\frac{R_C}{R_B + r} \qquad (7\text{-}21)$$

Be sure to note that the ac input resistance r of the circuit in the common-base configuration is not the same as the ac input resistance r in the common-emitter configuration. There are no subscripts on this parameter because its correct interpretation comes from an understanding of the problem being solved.

The voltage gain for the common-emitter configuration has the potential for being a large number because A_V is proportional to β_{ac}. The gain is also dependent as before on the slopes of the input and output load lines. The negative sign indicates that the input and output are out of phase.

Example 7-3   Designing a Common-Emitter DC-Coupled Amplifier

You wish to produce a common-emitter dc-coupled amplifier with $A_V = -10$, using a given npn Si BJT, as shown in Fig. 7-9.

Solution   A load line is selected with $V_{CC} = 15$ V and $(I_C)_{max} = 30$ mA. The chosen Q point is at $(V_{CE})_Q = 10$ V, $(I_C)_Q = 10$ mA, $(I_B)_Q = 50$ μA, and

$$R_C = 15\text{ V}/30\text{ mA} = 0.50\text{ k}\Omega$$

From the figure,

$$\beta_{dc} \cong \frac{10\text{ mA}}{50\ \mu\text{A}} \cong 2.0 \times 10^2 \qquad \text{and} \qquad \beta_{ac} = \frac{\Delta I_C}{\Delta I_B}\bigg|_{V_{CE}} \cong \frac{4.0\text{ mA}}{20\ \mu\text{A}} \cong 2.0 \times 10^2$$

and for r

$$r \cong 0.026\text{ V}/50\ \mu\text{A} \cong 0.52\text{ k}\Omega$$

From Eq. (7-21),

$$R_B = -\frac{\beta_{ac}R_C}{A_V} - r = -\left(\frac{2.0 \times 10^2}{-10}\right)(0.50\text{ k}\Omega) - 0.52\text{ k}\Omega \cong 9.5\text{ k}\Omega$$

For the reference input voltage,

$$(V_{in})_Q = (50\ \mu\text{A})(9.5\text{ k}\Omega) + 0.7\text{ V} \cong 1.2\text{ V}$$

and a satisfactory solution has been obtained. Many other solutions are also possible. ∎

Examples 7-2 and 7-3 can be compared to contrast the design constraints associated with common-base and common-emitter dc-coupled amplifiers. In each case,

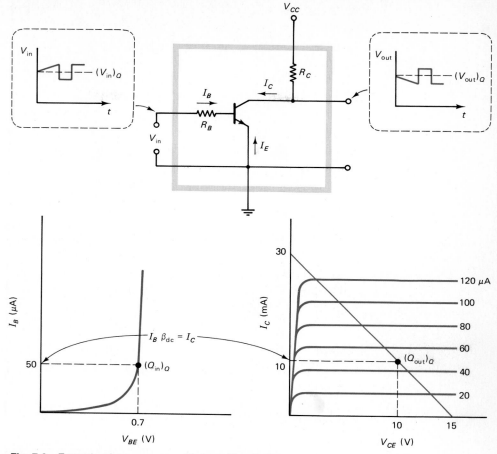

Fig. 7-9 Example of a common-emitter amplifier design.

$V_{CC} = 15$ V, $|(I_C)_{max}| = 30$ mA and $R_C = 0.50$ kΩ. The common-base amplifier has a gain $A_V = 10$ proportional to α_{ac} and the common-emitter amplifier has a gain $A_V = -10$ proportional to β_{ac}. The input resistor increases from 47 Ω in the former case to 9.5 kΩ in the latter, reducing current flow from the source. The significance of these differences is discussed further below.

7-5 Q-POINT TEMPERATURE DRIFT

The BJT/common-emitter configuration is much more temperature-sensitive than the circuits that have been discussed previously. Both V_{BE} and β shift with temperature, producing the effect shown in Fig. 7-10.

As T increases, the input diode voltage V_{BE} decreases, according to previous discussion (Sections 4-7 and 5-3). In addition β_{dc} and β_{ac} increase so that the output characteristic curves "float upward" and become more widely spaced. It turns out that the circuit changes caused by shifts in V_{BE} and β_{dc} are additive, so that the net result can be quite significant.

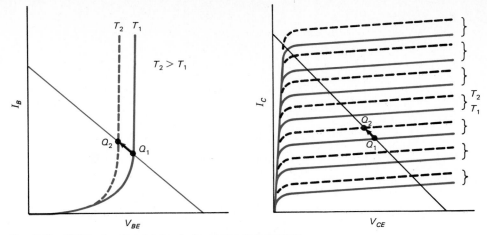

Fig. 7-10 Shift in the Q point due to increasing temperature.

A shift in the output voltage $(V_{out})_Q$ is observed for the common-emitter configuration due to a temperature change ΔT. This shift can be found by referring to the circuit of Fig. 7-7 and noting that

$$V_{CE} = V_{CC} - R_C I_C \qquad \text{and} \qquad V_{in} = I_B R_B + V_{BE} \tag{7-22}$$

Since

$$I_C = \beta_{dc} I_B \tag{7-23}$$

then

$$V_{CE} = V_{CC} - R_C(\beta_{dc} I_B)$$

Substituting for I_B,

$$V_{CE} = V_{CC} - R_C \beta_{dc}\left(\frac{V_{in} - V_{BE}}{R_B}\right) \tag{7-24}$$

Both β_{dc} and V_{BE} are functions of T. Since

$$(V_{out})_Q = (V_{CE})_Q$$

the first-order shift is found to be

$$\Delta(V_{out})_Q \cong \frac{\partial(V_{CE})_Q}{\partial T}\Delta T \tag{7-25}$$

$$\cong -\left(\frac{\beta_{dc} R_C}{R_B}\right)\left[(V_{in} - V_{BE})\left(\frac{1}{\beta_{dc}}\frac{\partial \beta_{dc}}{\partial T}\right) - \frac{\partial V_{BE}}{\partial T}\right]\Delta T \tag{7-26}$$

where all parameters are evaluated at the initial Q point of interest.

From previous study (Example 4-7, Chapter 4 and Example 6-3, Chapter 6), the estimates $\partial V_{BE}/\partial T \cong -2.0 \times 10^{-3}$ V/°C and $(1/\beta_{dc})(\partial \beta_{dc}/\partial T) \cong 1.0 \times 10^{-2}$/°C have been obtained for Si. Thus, the shifts in the input and output operating points combine to produce the total $\Delta(V_{out})_Q$.

The temperature shift ΔT may be due to changes in the ambient temperature $(\Delta T)_{amb}$, to internal heating P_D, or to a combination of these effects. Therefore, the relationship

$$\Delta T = (\Delta T)_{\text{amb}} + P_D\theta_{JA} \qquad (7\text{-}27)$$

can be used in conjunction with Eq. (7-26).

Example 7-4   *Q*-Point Drift Due to Internal Heating: Case I

Assume that you wish to find the Q-point shift due to internal power dissipation for the common-emitter circuit of Example 7-3, using the values

$$\left(\frac{\partial V_{BE}}{\partial T}\right) = -2.0 \times 10^{-3} \text{ V/°C}$$

$$\left(\frac{1}{\beta_{\text{dc}}}\frac{\partial \beta_{\text{dc}}}{\partial T}\right) = 1.0 \times 10^{-2}/\text{°C}$$

$$\theta_{JA} = 50\text{°C/W}$$

Solution   From Example 7-3,

$$P_D \cong (V_{CE})_Q(I_C)_Q$$
$$= (10 \text{ V})(10 \text{ mA}) = 0.10 \text{ W}$$
$$A_V = -10$$
$$(V_{\text{in}})_Q - (V_{BE})_Q = 0.50 \text{ V}$$

so that from Eq. (7-26),

$$\Delta(V_{\text{out}})_Q \cong \frac{-(2.0 \times 10^2)(0.50 \text{ k}\Omega)}{9.5 \text{ k}\Omega}(7.0 \times 10^{-3} \text{ V/°C})(0.10 \text{ W})(50\text{°C/W})$$
$$= -0.37 \text{ V}$$

There is a small but observable shift in the Q point of the circuit. If θ_{JA} is larger by a factor of two or three, the shift becomes more significant. ∎

Example 7-5   *Q*-Point Drift Due to Internal Heating: Case II

Reconsider the characteristic curves of Fig. 7-9 with all current values multiplied by a factor of ten. Find the effect on the problem of Example 7-4.

Solution   With all currents larger by a factor of ten,

$$(I_C)_Q = 0.10 \text{ A} \qquad r \cong 0.026 \text{ V}/0.50 \text{ mA} \cong 52 \ \Omega$$
$$(I_B)_Q = 0.50 \text{ mA} \qquad R_C = 15 \text{ V}/0.30 \text{ A} = 50 \ \Omega$$

To achieve the gain $A_V = -10$ requires that

$$R_B = \frac{2.0 \times 10^2}{-10}50 - 52 = 0.95 \text{ k}\Omega$$

$$(V_{\text{in}})_Q = (0.50 \text{ mA})(0.95 \text{ k}\Omega) + 0.7 \text{ V} \cong 1.2 \text{ V}$$

Therefore,

$$\Delta(V_{\text{out}})_Q \cong -10.5[(0.50 \text{ V})(1.0 \times 10^{-2}/\text{°C}) + 2.0 \times 10^{-3}\text{V/°C}](1.0 \text{ W})(50\text{°C/W})$$
$$\cong -3.7 \text{ V}$$

There is definitely a design problem for this case. The Q point migrates from an expected value of 10 V toward 6 V; it is possible under such circumstances to produce *thermal runaway,* when the Q point continues to drift until the BJT is driven into saturation. If θ_{JA} is larger by a factor of two or three, this drift to saturation occurs more rapidly. ∎

7-6 TEMPERATURE STABILIZATION FOR THE COMMON-EMITTER CIRCUIT

It is desirable to design common-emitter circuits that have reduced temperature dependence. If the Q point can be stabilized as the temperature changes, the behavior of the circuit can be better predicted and the circuit can be maintained in the linear region.

One typical approach to stabilization is to add an emitter resistor, as shown in Fig. 7-11. For this circuit, the operating equations take the form:

$$V_{in} = I_B R_B + V_{BE} - I_E R_E \tag{7-28}$$

$$V_{out} = V_{CC} - I_C R_C \tag{7-29}$$

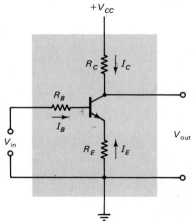

Fig. 7-11 Addition of an emitter resistor for temperature stabilization.

For changes in the parameters of Eqs. (7-28) and (7-29),

$$\Delta V_{in} = (\Delta I_B)R_B + \Delta V_{BE} - (\Delta I_E)R_E$$

$$\Delta V_{out} = -(\Delta I_C)R_C \tag{7-30}$$

Since $(\Delta I_B) \cong -\beta_{ac}(\Delta I_E)$ and is approximately independent of $(V_{in})_Q$,

$$\Delta V_{in} \cong (\Delta I_B)(R_B + \beta_{ac} R_E) + \Delta V_{BE} \tag{7-31}$$

and

$$A_V = \frac{\Delta V_{out}}{\Delta V_{in}} \cong \frac{-(\Delta I_C)R_C}{\Delta I_B(R_B + \beta_{ac} R_E) + \Delta V_{BE}}$$

$$\cong -\beta_{ac}\frac{R_C}{R_B + \beta_{ac} R_E + r} \tag{7-32}$$

The amplifier gain has been reduced by the addition of R_E. If $R_E = 0$, Eq. (7-32) reduces to Eq. (7-21) derived above.

Construction of the input and output load lines must also be modified by the addition of R_E. From Eq. (7-28),

$$V_{in} = I_B R_B + V_{BE} - I_E R_E$$

$$\cong I_B R_B + (\beta_{dc})I_B R_E + V_{BE}$$

On the input characteristic curve (with I_B versus V_{BE}), the load line intersects at

$$I_B = 0 \qquad V_{BE} = V_{in}$$

$$V_{BE} = 0 \qquad I_B = \frac{V_{in}}{R_B + (\beta_{dc})R_E} \tag{7-33}$$

From Fig. 7-11,

$$V_{out} = V_{CC} - I_C R_C = V_{CE} - I_E R_E$$

so that

$$V_{CC} \cong I_C R_C + I_C R_E + V_{CE} \tag{7-34}$$

On the output characteristic curve (with I_C versus V_{CE}), the load line intersects at

$$I_C = 0 \qquad V_{CE} = V_{CC}$$

$$V_{CE} = 0 \qquad I_C = \frac{V_{CC}}{R_C + R_E} \tag{7-35}$$

To find the shift $\Delta(V_{out})_Q$ due to a temperature change ΔT for this circuit, Eq. (7-26) must be extended. Returning to the circuit of interest,

$$V_{in} = I_B R_B + V_{BE} + I_E R_E \cong I_B(R_B + \beta_{dc} R_E) + V_{BE}$$

$$V_{out} = V_{CC} - I_C R_C \tag{7-36}$$

so that

$$V_{out} = V_{CC} - \beta_{dc} R_C I_B$$

$$\cong V_{CC} - \beta_{dc} R_C \left[\frac{V_{in} - V_{BE}}{R_B + \beta_{dc} R_E} \right] \tag{7-37}$$

The objective is to find

$$\Delta(V_{out})_Q \cong \frac{\partial(V_{out})_Q}{\partial T} \Delta T \tag{7-38}$$

and by differentiating,

$$\Delta(V_{out})_Q = \left(\frac{-\beta_{dc} R_C}{R_B + \beta_{dc} R_E} \right) \left[(V_{in} - V_{BE}) \left(\frac{R_B}{R_B + \beta_{dc} R_E} \right) \frac{1}{\beta_{dc}} \frac{\partial \beta_{dc}}{\partial T} - \frac{\partial V_{BE}}{\partial T} \right] \Delta T \tag{7-39}$$

Note that if $R_E = 0$, Eq. (7-39) reduces to Eq. (7-26) as it must. A comparison of these two equations reveals that R_E has the effect of reducing the thermal shift $\Delta(V_{out})_Q$. However, at the same time, Eq. (7-32) indicates that the voltage gain A_V is also reduced by R_E. A design tradeoff exists between these two parameters.

The addition of R_E has produced another important result. As discussed earlier, the values for β_{ac} can vary among devices with the same manufacturer's code number. Let β_{ac1} and β_{ac2} be the extreme values associated with a particular device type of interest. Then without R_E, the extreme circuit gains to be experienced with different devices are given by

$$A_{V1} = -\beta_{ac1} \frac{R_C}{R_B + r}$$

$$A_{V2} = -\beta_{ac2} \frac{R_C}{R_B + r} \tag{7-40}$$

so that

$$\frac{A_{V1}}{A_{V2}} = \frac{\beta_{ac1}}{\beta_{ac2}} \tag{7-41}$$

The circuit gain is proportional to the device gain. With the addition of an emitter resistor R_E, this ratio becomes

$$\frac{A'_{V1}}{A'_{V2}} = \frac{-\beta_{ac1}\left(\dfrac{R_C}{R_B + \beta_{ac1} R_E + r}\right)}{-\beta_{ac2}\left(\dfrac{R_C}{R_B + \beta_{ac2} R_E + r}\right)}$$

$$= \frac{1 + \dfrac{R_B + r}{\beta_{ac2} R_E}}{1 + \dfrac{R_B + r}{\beta_{ac1} R_E}} \qquad (R_E \neq 0) \tag{7-42}$$

The addition of R_E can reduce the dependence of the amplifier on β_{ac}, providing a circuit that is less affected by variations in β_{ac} due to different devices. The tradeoff is now more complex. By increasing R_E, high circuit gain can be exchanged for a combination of thermal stability [smaller $\Delta(V_{out})_Q$] and a lower gain A'_V that is less dependent on device characteristics. The incentive for circuits with a larger initial gain A_V is clear. By providing an emitter resistor R_E, some of the gain can be exchanged for other circuit performance characteristics, still leaving a gain A'_V that is adequate to meet design objectives.[2]

Now examine the limiting case for which $R_B \ll \beta_{dc} R_E$ and $(R_B + r) \ll \beta_{ac} R_E$ (consistent requirements). This result can be obtained by choosing to reduce R_B in value. However, at the same time, the current flow from the source must increase. Equations (7-32) and (7-39) reduce to

$$A_V \cong -R_C/R_E \tag{7-43}$$

$$\Delta(V_{out})_Q \cong \frac{R_C}{R_E}\left(\frac{\partial V_{BE}}{\partial T}\right)\Delta T \tag{7-44}$$

Equation (7-42) becomes

$$A'_{V1}/A'_{V2} \cong 1 \tag{7-45}$$

and the circuit properties are no longer dependent on β.

---

[2] The effect of R_E on the frequency response of a circuit is discussed in Chapter 10.

The gain of the amplifier is reduced in magnitude and independent of β_{ac} because such a large fraction of the output is being fed back in series with the input voltage (refer again to Fig. 7-11). Since the input and output voltages are 180° out of phase, the result is *negative feedback,* which is discussed in further detail in Chapters 9 and 11.

Equation (7-43) is an attractive result because it means that the Q point of the BJT is much more stable. The elimination of the β-dependent term occurs because with $R_B \ll \beta_{dc} R_E$ (or $I_B R_B \ll I_C R_C$), the voltage on the base is approximately equal to $V_{in} \cong V_{BE} + I_E R_E$. If any changes in V_{BE} are small with respect to the voltage across R_E, then the voltage across R_E is held constant with temperature as long as V_{in} is fixed. I_C is then fixed and cannot vary with T.

In some circumstances, the above tradeoff can be a desirable one, reducing the gain of the circuit in order to create a β-independent circuit. However, it can be difficult to obtain this limiting case in actual design practice. R_B can be constrained to a large value because of the current capability of the source, and R_E can be limited by the desired gain objective.

Figure 7-12 illustrates another way to hold the current through R_E constant with T without reducing R_B. However, the gain of the circuit must now be reduced to achieve this objective. The resistors R_{B1} and R_{B2} form a voltage divider across the

Fig. 7-12 Addition of an input voltage divider to achieve circuit stability. (A) Amplifier schematic. (B) Rearrangement of the input circuit. (C) Thevenin equivalent.

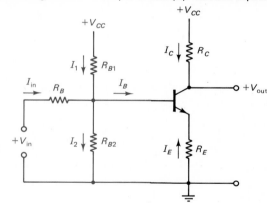

(A)

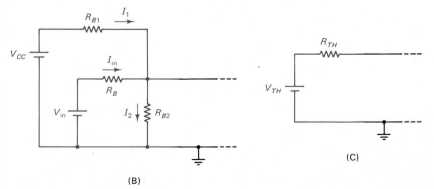

(B)

(C)

base of the BJT. The portion of the circuit indicated in color can be replaced by the Thevenin equivalent, as shown. From straightforward circuit analysis,

$$R_{TH} = R_B \| R_{B1} \| R_{B2} \qquad (7\text{-}46)$$

$$V_{TH} = V_{CC}\frac{R_{TH}}{R_{B1}} + V_{in}\frac{R_{TH}}{R_B} \qquad (7\text{-}47)$$

The first term for V_{TH} provides a fixed current to achieve the Q-point bias for the BJT, and the second term is the effective (reduced-magnitude) input signal. Previous results for A_V [Eq. (7-32)] and $\Delta(V_{out})_Q$ [Eq. (7-37)] can be applied to the circuit of Fig. 7-12 through the substitution:

$$V_{in} \longrightarrow V_{TH}$$
$$ \qquad\qquad\qquad (7\text{-}48) $$
$$R_B \longrightarrow R_{TH}$$

so that

$$A_V \cong -\beta_{ac}\left(\frac{R_C}{R_{TH} + \beta_{ac}R_E + r}\right)\left(\frac{R_{TH}}{R_{B1}}\right)$$

$$\Delta(V_{out})_Q \cong \left(\frac{-\beta_{dc}R_C}{R_{TH} + \beta_{dc}R_E}\right)\left[(V_{TH} - V_{BE})\left(\frac{R_{TH}}{R_{TH} + \beta_{dc}R_E}\right)\frac{1}{\beta_{dc}}\frac{\partial\beta_{dc}}{\partial T} - \frac{\partial V_{BE}}{\partial T}\right]\Delta T$$
$$ \qquad\qquad\qquad (7\text{-}49) $$

Thus, the condition $R_B \ll \beta_{dc}R_E$, to remove the β dependency from the expressions for A_V and $\Delta(V_{out})_Q$ becomes

$$R_{TH} \ll \beta_{dc}R_E \qquad (7\text{-}50)$$

Resistors R_{B1} and R_{B2} can be chosen to be sufficiently small for this relationship to hold. This is equivalent to requiring that currents I_1 and I_2 in Fig. 7-12 be much larger than I_{in} and I_B.

However, as the values of R_{B1} and R_{B2} are reduced, the gain is also reduced, as shown in Eq. (7-49). It is desirable to assure that the condition of Eq. (7-50) holds without making R_{B1} and R_{B2} any smaller than necessary.

The circuit of Fig. 7-12 produces the desired bias conditions by fixing the voltage across R_E. Since I_B is much smaller than I_1 or I_2, the voltage on the base is fixed by the divider. The voltage across R_E is thus determined, producing a constant I_E and I_C. The operating point of the BJT is determined by the combination of the load line and voltage divider. Once V_{CC}, R_C, and R_E are specified and the divider fixes I_E, an operating point in the linear region results. This circuit is a popular one to achieve circuit stabilization and is further utilized in Chapter 9.

Example 7-6   Temperature Stabilization with an Emitter Resistor

An emitter resistor with $R_E = 0.1\,R_C$ is added to the common-emitter amplifier of Example 7-5 (Fig. 7-11).[3] What will happen to the gain and stabilization of the circuit?

---

[3] See Exercise 7-15 for a discussion of this selection.

Solution For the circuit of Example 7-5,

$$A_V = -10 \qquad\qquad R_C = 50\ \Omega$$

$$(V_{in})_Q - (V_{BE})_Q = 0.5\ \text{V} \qquad R_B = 0.95\ \text{k}\Omega$$

$$P_D = 1.0\ \text{W} \qquad\qquad \beta_{dc} \cong \beta_{ac} \cong 2 \times 10^2$$

$$r \cong 52\ \Omega$$

and

$$\Delta(V_{out})_Q \cong -3.7\ \text{V}$$

If an emitter resistor is added to this circuit, with $R_E = 0.1R_C$, Eqs. (7-32) and (7-39) can be applied:

$$A_V = \frac{(-2 \times 10^2)(50)}{0.95\ \text{k}\Omega + 52\Omega + (2 \times 10^2)(5)\Omega} \cong -5.0$$

$$\Delta(V_{out})_Q \cong \frac{(-2 \times 10^2)(50)}{0.95\ \text{k}\Omega + (2 \times 10^2)(5)\Omega}$$

$$\times \left[(0.50\ \text{V})\left(\frac{0.95\ \text{k}\Omega}{1.95\ \text{k}\Omega}\right)(1.0 \times 10^{-2}/°\text{C}) + 2.0 \times 10^{-3}\text{V}/°\text{C} \right](1.0\ \text{W})(50°\text{C}/\text{W})$$

$$\cong -1.1\ \text{V}$$

The gain has been reduced by 50 percent and the temperature drift has been reduced by 70 percent. (The dc operating point has also shifted since R_E has been added to the circuit.)

Temperature drift can also be reduced by placing a heat sink on the BJT. If θ_{JA} is reduced from 50°C/W to 5°C/W, then $\Delta(V_{out})_Q$ drops by a factor of ten. ■

Example 7-7 **Effect of R_E on Sensitivity to Parameter Variations**

For Example 7-6, what effect does R_E have on the sensitivity of the circuit to a ± 20 percent change in β_{ac} among devices?

Solution With $R_E = 0$, the maximum variation in A_V due to device differences is

$$A_{V1}/A_{V2} = 240/160 = 1.5$$

With $R_E = 0.1R_C$

$$\frac{A'_{V1}}{A'_{V2}} = \frac{1 + \dfrac{1.0\ \text{k}\Omega}{(0.16\ \text{k}\Omega)(5)}}{1 + \dfrac{1.0\ \text{k}\Omega}{(0.24\ \text{k}\Omega)(5)}} = \frac{1 + 1.25}{1 + 0.83} \cong 1.2$$

The sensitivity has been reduced from a 50-percent variation to a 20-percent variation. As described above, the negative feedback introduced by R_E enables the designer to modify circuit operation to achieve desired performance objectives through several related tradeoffs. The adjustment of R_E is one useful method for modifying circuit operation for a given application. ■

The results of the above example are shown in Fig. 7-13. The comparison indicates how R_E shifts the tradeoff between gain, thermal stability, and sensitivity to device parameters.

	Amplifier Gain A_v	Thermal Drift $\Delta(V_{out})_Q$	Sensitivity to Device β_{ac} A_{v1}/A_{v2} (%)
$R_E = 0$	-10	-3.7 V$/-0.37$ V with heat sink	50
$R_E = 0.1\,R_C$	-5.0	-1.1 V$/0.11$ V with heat sink	20

7-7 BJT BIASING TRADEOFFS

In this chapter, dc-coupled amplifiers are used to illustrate many of the facets of BJT operation that are important in circuit design. The techniques developed here are useful in understanding more complex circuits.

Figure 7-14 summarizes the voltage gain equations for basic common-base and

Fig. 7-14 DC-coupled amplifier design equations using BJTs.

Common-base npn

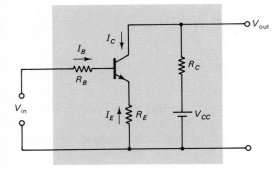
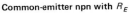

$$V_{in} = I_E R_E + V_{BE}$$
$$V_{out} = V_{CC} - I_C R_C = V_{CE}$$
$$A_V = \frac{\Delta(V_{out})}{\Delta(V_{in})} \cong -\alpha_{ac}\frac{R_C}{R_E + r}$$

Common-base pnp

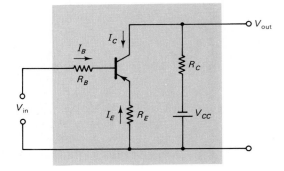

$$V_{in} = I_E R_E + V_{BE}$$
$$V_{out} = -V_{CC} - I_C R_C = V_{CE}$$
$$A_V = \frac{\Delta(V_{out})}{\Delta(V_{in})} \cong -\alpha_{ac}\frac{R_C}{R_E + r}$$

Common-emitter npn with R_E

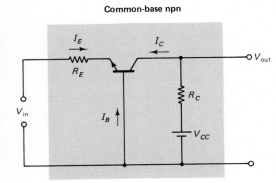

$$V_{in} = I_B R_B + V_{BE} - I_E R_E$$
$$V_{out} = V_{CC} - I_C R_C = V_{CE} - I_E R_E$$
$$A_V = \frac{\Delta(V_{out})}{\Delta(V_{in})} \cong -\beta_{ac}\frac{R_C}{R_B + r + \beta_{ac}R_E}$$

Common-emitter pnp with R_E

$$V_{in} = I_B R_B + V_{BE} - I_E R_E$$
$$V_{out} = -V_{CC} - I_C R_C = V_{CE} - I_E R_E$$
$$A_V = \frac{\Delta(V_{out})}{\Delta(V_{in})} \cong -\beta_{ac}\frac{R_C}{R_B + r + \beta_{ac}R_E}$$

common-emitter dc-coupled amplifiers using both npn and pnp BJTs. The addition of a voltage divider across the base of the common-emitter configuration can be accommodated by using the Thevenin equivalent circuit, as explained above.

Voltage gain has been emphasized in this chapter. However, the current gain ac of each amplifier can be quickly determined. By definition,

$$A_i = \frac{\Delta(I_{\text{out}})}{\Delta(I_{\text{in}})} \tag{7-51}$$

Since $(\Delta I_C) \cong -(\Delta I_E)$, $A_i \cong -1$ for the common-base configuration, and since $(\Delta I_C) \cong \beta_{\text{ac}}(\Delta I_B)$, $A_i \cong \beta_{\text{ac}}$ for the common-emitter configuration (without the presence of a voltage divider on the input). Voltage, current, and power gain are explored further in Chapter 9.

Amplifier design with BJTs in the common-base and common-emitter configurations can be contrasted, as shown in Fig. 7-15. In the common-base configuration, the current gain is about unity and the voltage gain is limited by the ratio $R_C/(R_E + r)$. The ac input resistance $r \cong V_o/I_E$ is small because I_E is large. The common-base amplifier is temperature-stable because α_{ac} is approximately independent of temperature effects and is approximately independent of variations among devices.

In the common-emitter configuration without stabilization, the current gain is much larger (proportional to β_{ac}) and the voltage gain is proportional to $R_C/(R_B + r)$. In comparison with the common-base configuration, the ac input resistance for this case $r \cong V_o/I_B$ is larger by the factor β_{ac}. The common-emitter amplifier is temperature-dependent because β_{dc} varies significantly with temperature.

The common-emitter configuration can be improved by introducing negative feedback through an emitter resistor R_E and using an input voltage divider. The resultant circuit trades off gain for temperature stabilization and a reduced dependence on the variations among devices.

The best configuration to select for a given application depends on the specific performance objectives that are set. By forming side-by-side evaluations, as shown in Fig. 7-15, the designer can perform a careful and complete review of the choices to be made and their implications.

FIGURE 7-15
Comparison of Common-base and Common-emitter Amplifier Circuits

Property	Common-Base Amplifier Circuit	Common-Emitter Amplifier Circuit	
		Without Stabilization	With Emitter Resistor Stabilization (see Fig. 7-11)
A_i	$\cong -1$	β_{ac}	Reduced[1]
A_V	$\cong \dfrac{R_C}{R_E + r}$	$\cong -\beta_{ac}\dfrac{R_C}{R_B + r}$	$-\beta_{ac}\dfrac{R_C}{R_{TH} + r + \beta_{ac}R_E}$
r	$\cong V_0/I_E$	$\cong V_0/I_B$	Increased[1]
Temperature dependence	Small	Large	Can be made small by design
Dependence on variations in device parameters	Small	Large	Can be made small by design

[1]The available base current into the BJT is reduced due to the current flow through R_{B1} and R_{B2}.

Questions   **7-1.** What topics covered in this chapter can be generalized for more complex circuit designs?

7-2. What approximations are required to produce a dc-coupled amplifier with a constant voltage gain A_V?

7-3. Why is the voltage gain of a common-base amplifier positive whereas the voltage gain of a common-emitter amplifier is negative?

7-4. Why is the Q point for a common-emitter amplifier a sensitive function of temperature?

7-5. Why is the Q point not a sensitive function of T for a common-base amplifier?

7-6. What is thermal runaway?

7-7. Why does an emitter resistor help stabilize the common-emitter dc-coupled amplifier?

7-8. Why does an emitter resistor reduce the voltage gain of the common-emitter dc-coupled amplifier?

7-9. What are the tradeoffs involved in reducing the temperature dependence of the Q point by reducing R_B versus placing a voltage divider across the base of the BJT?

Answers   **7-1.** This chapter describes how graphical and analytic design methods are interrelated, and then provides a needed way to conceptualize some of the more difficult methods of analysis to follow. The discussion of dc-coupled amplifiers illustrates how the voltage gain of an amplifier (an ac property) and its Q-point bias (a dc-bias property) are not independent, but must be addressed together as part of a comprehensive design process. The tradeoffs observed between stability with temperature, voltage gain, and device properties are commonly experienced with many types of circuits, and the use of negative feedback to achieve tradeoffs among these factors leads to more general applications of this powerful concept.

7-2. A dc-coupled amplifier produces a constant voltage gain A_V only if the parameters in the gain equation (α_{ac} and r, or β_{ac} and r) are approximately constant over the range of signal amplitudes in use.

7-3. For a common-base amplifier, the input and output waveforms are *in phase,* tracking one another, whereas for a common-emitter amplifier, the input and output waveforms move in opposite directions. If θ is designated as the phase shift associated with an arbitrary amplifier circuit, $\theta = 0°$ for the common-base dc-coupled amplifier and $\theta = 180°$ (producing inversion) for the common-emitter dc-coupled amplifier.

7-4. As the temperature of the BJT changes, both the input and output common-emitter characteristic curves are modified. These changes interact with the fixed voltages and resistors in the circuit, modifying the operating conditions for the BJT.

7-5. For the common-base configuration, the output characteristic of the BJT changes very little with T. Therefore, the operating conditions for the BJT in this configuration only change slightly with temperature.

7-6. Thermal runaway occurs when the Q point continues to drift until the BJT is driven into saturation. If the temperature continues to rise, the BJT may be destroyed.

7-7. As the temperature increases, the collector current I_C through the BJT also tends to increase (because β_{dc} is larger). A corresponding increase in the voltage drop across R_E *reduces* the net base-emitter voltage across the input diode, tending to reduce I_C in a stabilizing arrangement. The 180° phase shift associated with the common-emitter configuration is essential to achieving this negative feedback.

7-8. As the input voltage increases, the collector current I_C through the BJT tends to increase (because I_B is larger). A corresponding increase in the voltage drop across R_E *reduces* the net base-emitter voltage across the input diode, tending to reduce I_B and I_C.

7-9. For a reduced R_B, the source must provide more current, but the voltage gain is maximized. With a voltage divider, the source current does not increase and the voltage gain is reduced.

EXERCISES AND SOLUTIONS

Exercises *All exercises and problems assume $T = 300°K$ and a Si BJT unless otherwise stated.*

(7-2) **7-1.** For a dc-coupled amplifier, a change in the signal voltage from 0.030 to 0.045 V produces a change in output voltage from 8.0 to 5.5 V. What is the gain of the amplifier?

(7-2) **7-2.** For a common-base dc-coupled amplifier, $R_C = 1.0$ kΩ, $R_E = 50$ Ω, and $I_E = 5.0$ mA. Estimate the circuit gain A_V.

(7-2) **7-3.** The performance objective $A_V > 25$ is set for a common-base dc-coupled amplifier with $R_C = 1.5$ kΩ and $I_E = 1.0$ mA. For what range of R_E values does the amplifier perform as desired?

(7-2) **7-4.** Consider the graphical design of a dc-coupled amplifier, using the schematic of Fig. 7-3 and the common-base output characteristic curve in Fig. 7-1. Add load lines with $V_{CC} = 8.0$ V, $R_C = 0.16$ kΩ and $(V_{in})_Q = -1.2$ V, $R_E = 24$ Ω to give initial input and output Q points. (a) Prepare a graph of V_{out} versus V_{in}. Assume a new value for V_{in} and draw an input load line parallel to the original one. Find a new Q point, find I_E, and move to the output characteristic and find the corresponding V_{CB}. Repeat for a range of values for V_{in}, then graph V_{CB} as a function of V_{in}. (b) Determine A_V graphically (from the slope of the line in the linear region). (c) Calculate A_V using Eq. (7-12) for comparison. (Find r graphically or through estimation.) (d) From the graph, measure the maximum signal swing $\pm\Delta V_{out}$ that can be accommodated in the linear region.

(7-3) **7-5.** Design a common-base dc-coupled amplifier with a gain $A_V = 8.0$ using the pnp BJT characteristic in Fig. 6-9 (apply the design strategy outlined in Section 7-3). Obtain the maximum possible linear region for your design for equal signal variations. Assume that $V_{CC} = 12$ V and $(I_C)_{max} = -10$ mA and that all resistor values are available.

(7-3) **7-6.** Design a BJT common-base dc-coupled amplifier to achieve a gain of $A_V = 50$ if the Q point is at $V_{CE} = 20$ V, $I_C = 10$ mA; and $(I_C)_{max} = 30$ mA. Find R_E and V_{in}.

(7-4) **7-7.** The performance objective $|A_V| > 30$ is set for a common-emitter dc-coupled amplifier with $R_C = 1.5$ kΩ and $I_B = 10$ μA. For what range of R_B values does the amplifier perform as desired if $\beta_{ac} \cong 150$ for the BJT being used?

(7-4) **7-8.** The performance objective $|A_V| > 25$ is set for a common-emitter dc-coupled amplifier with $R_C = 1.5$ kΩ and $\beta_{ac} = 150$ for the BJT being used. What range of values for I_B can be used if $R_B = 0$?

(7-4) **7-9.** Design a common-emitter dc-coupled amplifier with a gain $A_V = -5.0$ using the pnp characteristics shown in Fig. 6-11 and the circuit of Fig. 7-7 modified for pnp operation. Choose your operating point [with equal variations around $(V_{out})_Q$] to obtain the largest possible linear operating region. Assume $V_{CC} = 12$ V, $(I_C)_{max} = -10$ mA and do not restrict the available resistor values.

(7-4) **7-10.** Consider the common-emitter dc-coupled amplifier of Fig. 7-7 with $R_C = 2.2$ kΩ, $R_B = 10$ kΩ, and $(I_E)_Q = -7.5$ mA. (a) Find the gain A_V with $\beta_{dc} \cong \beta_{ac} = 125$. (b) For a given transistor type, assume that β_{ac} can vary by ±50 percent among devices. Find the variation in circuit gain that can occur as devices are interchanged.

(7-5) **7-11.** For a Si BJT used in a common-emitter dc-coupled amplifier, $(V_{in})_Q = 2.0$ V, $(V_{BE})_Q = 0.7$ V, and $R_C = 2.5$ kΩ. The maximum allowance $\Delta(V_{out})_Q$ due to thermal drift is -1.0 V. If $\beta_{dc} = 150$, estimate the minimum value of R_B that can be used over a range of 10°C.

(7-5) **7-12.** For a common-emitter dc-coupled amplifier, $(V_{in})_Q = 2.0$ V, $\beta_{dc} = 2.0 \times 10^2$, $R_C = 1.0$ kΩ, and $R_B = 15$ kΩ. (a) Estimate the shift in Q point that accompanies a change of 50°C in the temperature of a Si device. (b) An emitter resistor $R_E = 0.10R_C$ is added. Find the new Q-point shift.

(7-6) **7-13.** A common-emitter dc-coupled amplifier has a gain of -20 with a BJT having $\beta_{ac} = 150$. What is the amplifier gain if a new BJT having $\beta_{ac} = 200$ is used to replace the original device?

(7-6) **7-14.** A common-emitter dc-coupled amplifier with a negative feedback resistor $R_E = 0.5$ kΩ has a gain of -20 with a BJT for which $\beta_{ac} = 150$. For this amplifier, $R_B = 25$ kΩ and $I_B = 5.0$ μA. (a) A new BJT with $\beta_{ac} = 200$ is introduced into the circuit. What is the new amplifier gain? (b) Assuming a Si BJT with $\beta_{ac} \cong \beta_{dc}$, approximately what bias voltage $(V_{in})_Q$ is required for the circuit (with the second device in place)? (c) If $R_E = 0.1R_C$, $\beta_{dc} \cong \beta_{ac}$, and $(V_{CE})_Q = V_{CC}/2$, what value for V_{CC} is required (with the second device in place)?

(7-6) **7-15.** The best ratio R_C/R_E for a common-emitter amplifier is based on the circuit application. If $R_C \gg R_E$, then very little negative feedback is taking place and little stabilization is obtained. If $R_C \ll R_E$, most of the output is being fed back in series with the input and the amplifier gain is sharply reduced. A balance is needed between stabilization and gain.

Assume that stabilization is to be obtained by assuring that when the base voltage V_B is held constant (by holding V_{in} = constant or by using an input voltage divider), the voltage across R_E and therefore I_E is approximately constant. Variations in device parameters and temperature will then have only a small effect. Show that $R_E \cong 0.1R_C$ is a reasonable choice to achieve this objective.

(7-6) **7-16.** For the circuit of Fig. 7-12(A), let $R_B = 25$ kΩ, $R_E = 1.0$ kΩ, $V_{CC} = 12$ V, $-(I_E)_Q \cong (I_C)_Q = 1.0$ mA, and $V_{BE} \cong 0.7$ V. Choose R_{B1} and R_{B2} so that $I_1 \cong I_2 \cong 20I_B$. Assume that $\beta_{dc} \cong 100$. What fractional reduction in amplifier gain is experienced due to the presence of R_{B1} and R_{B2} to provide temperature stabilization for the circuit?

7-17. The market for a dc-coupled amplifier circuit depends on the thermal stability of the circuit. The market is segmented into the three ranges, as shown in Fig. 7-16(A), while circuit cost is related to the thermal stability, as shown in Fig. 7-16(B). What value of $\Delta(V_{out})_Q$ should be selected to produce a circuit with the highest profit margin (PM)?

Fig. 7-16 Data for Exercise 7-17.

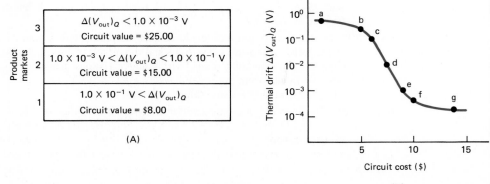

(A)

(B)

Solutions   **7-1.** From Eq. (7-1),

$$A_V = \frac{(5.5 - 8.0)\text{ V}}{(0.045 - 0.030)\text{ V}} = -1.7 \times 10^2$$

7-2. From Eq. (7-12),

$$A_V = -\alpha_{\text{ac}}\frac{R_C}{R_E + r}$$

using the approximation

$$r \cong \frac{0.026\text{ V}}{5.0\text{ mA}} \cong 5.2\ \Omega \quad \text{and} \quad \alpha_{\text{ac}} \cong -1$$

Then by substitution,

$$A_V \cong \frac{1.0\text{ k}\Omega}{(50\ \Omega + 5.2\ \Omega)} \cong 18$$

7-3. From Eq. (7-12),

$$R_E < \left(\frac{R_C}{A_V} - r\right)$$

With $r \cong 0.026\text{ V}/1.0\text{ mA} \cong 26\ \Omega$,

$$R_E < \left(\frac{1.5\text{ k}\Omega}{25} - 26\right) \cong 34\ \Omega$$

7-4. (a) The objective of this step is to prepare a graph of V_{out} as a function of V_{in} for given values of V_{CC}, R_C, $(V_{\text{in}})_Q$, and R_E. The specified load lines are shown in Figs. 7-17(A) and (B). As V_{in} varies around $(V_{\text{in}})_Q$, a family of input load lines is produced. The input operating point moves along the input characteristic curve, producing movement in the output operating point along the output load line. For each value of V_{in}, a value of $|I_E| \cong |I_C|$ and a value of $V_{CB} = V_{\text{out}}$ can be obtained. Thus, from the set of load lines illustrated, a graph of V_{out} as a function of V_{in} can be prepared, as shown in Fig. 7-18.

Fig. 7-17   Load lines for Exercise 7-4.

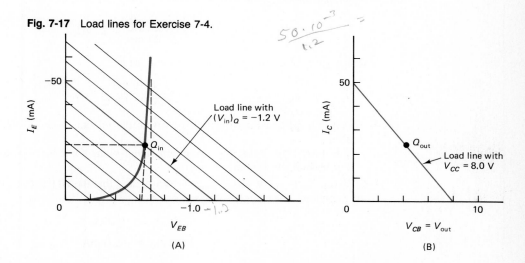

(A)

(B)

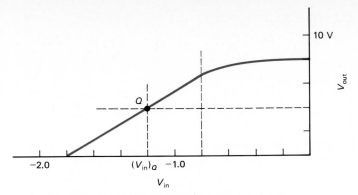

Fig. 7-18 Voltage v_{out} as a function of v_{in} for Exercise 7-4.

As V_{in} approaches zero, the BJT nears the cutoff region where the base-emitter diode turns *off* and the curve flattens. As V_{in} approaches -1.8 V, I_C approaches 50 mA and V_{out} is limited by the output load line. As I_E increases toward the maximum $(I_E)_{max} = V_{CC}/R_C$, $\alpha_{dc} = I_C/I_E$ drops and $\beta_{dc} = -\alpha_{dc}/(1 - \alpha_{dc})$ is also reduced. In this region, charge builds up in the collector to prevent additional base-collector carrier flow, reducing I_C and increasing I_B. As can be observed, the linear range extends from $V_{in} \cong -1.8$ V to $V_{in} \cong -0.8$ V. The chosen Q point is approximately in the middle of this region.

(b) At the Q point, the voltage gain A_V is approximately

$$A_V = \text{slope} = 7.0 \text{ V}/1.0 \text{ V} \cong 7.0$$

(c) From Eq. (7-12),

$$A_V \cong -\alpha_{ac}\frac{R_C}{R_E + r}$$

and

$$r \cong \frac{0.026 \text{ V}}{25 \text{ mA}} \cong 1.0 \ \Omega$$

or graphically,

$$r \cong \frac{0.1 \text{ V}}{50 \text{ mA}} \cong 2.0 \ \Omega$$

Substituting and using the graphical value for r,

$$A_V \cong \frac{0.16 \ k\Omega}{24 \ \Omega + 2 \ \Omega} \cong 6.2$$

which is in agreement with part (b) within the tolerances due to the approximations involved.

(d) From Fig. 7-18, the maximum signal swing for linear operation is about $+2$ to -4 V around the Q point.

7-5. Given $V_{CC} = 12$ V and $(I_C)_{max} = 10$ mA, and that the BJT is to be biased in the middle of the output linear region, choose

$$(V_{CB})_Q = -6.0 \text{ V}$$

$$(I_C)_Q = -5.0 \text{ mA}$$

Further,

$$R_C = 12 \text{ V}/10 \text{ mA} = 1.2 \text{ k}\Omega$$

Given $|I_C| \cong |I_E|$ at the Q point,

$$r \cong 0.026 \text{ V}/5.0 \text{ mA} \cong 5.2 \text{ }\Omega$$

and from Eq. (7-12),

$$R_E \cong \left(\frac{R_C}{A_V} - r\right) = \frac{1.2 \text{ k}\Omega}{8.0} - 5.2 \text{ }\Omega \cong 1.4 \times 10^2 \text{ }\Omega$$

From Fig. 6-9, the required $(V_{\text{in}})_Q$ is given by

$$(V_{\text{in}})_Q \cong 0.7 \text{ V} + R_E I_E = 0.7 \text{ V} + (1.4 \times 10^2)(5.0 \text{ mA}) \cong 1.4 \text{ V}$$

7-6. Given $I_C = 10$ mA,

$$r \cong 0.026 \text{ V}/10 \text{ mA} = 2.6 \text{ }\Omega$$

From the data given,

$$R_C = \frac{20 \text{ V} - 0 \text{ V}}{30 \text{ mA} - 10 \text{ mA}} = 1.0 \text{ k}\Omega$$

Therefore, from Eq. (7-12),

$$R_E \cong \left(\frac{R_C}{A_V} - r\right) = \frac{1.0 \text{ k}\Omega}{50} - 2.6 \cong 17 \text{ }\Omega$$

and

$$(V_{\text{in}})_Q \cong 0.7 \text{ V} + (10 \text{ mA})(17 \text{ }\Omega) \cong 0.9 \text{ V}$$

7-7. From Eq. (7-21),

$$A_V = -\beta_{\text{ac}}\frac{R_C}{R_B + r}$$

and with

$$r \cong \frac{0.026 \text{ V}}{10 \text{ }\mu\text{A}} \cong 2.6 \text{ k}\Omega$$

$$R_B < 150\left(\frac{1.5 \text{ k}\Omega}{30}\right) - 2.6 \text{ k}\Omega$$

$$R_B < 4.9 \text{ k}\Omega$$

7-8. From Eq. (7-21),

$$A_V \cong -\beta_{\text{ac}}\frac{R_C}{0.026 \text{ V}/I_B}$$

$$I_B > \frac{25(0.026 \text{ V})}{150(1.5 \text{ k}\Omega)} = 2.9 \text{ }\mu\text{A}$$

7-9. From the data given, $V_{CC} = 12$ V and $(I_C)_{\text{max}} = -10$ mA. To maximize the linear operation region, pick $(V_{CE})_Q \cong -7.0$ V, $(I_C)_Q \cong -4.8$ mA, and $(I_B)_Q \cong -22 \text{ }\mu\text{A}$ (see Fig. 7-19). At this point,

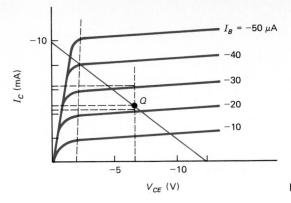

Fig. 7-19 Solution for Exercise 7-9.

$$\beta_{dc} \cong \frac{4.8 \text{ mA}}{22 \text{ }\mu\text{A}} \cong 2.2 \times 10^2$$

and around this point

$$\beta_{ac} \cong \frac{2.0 \text{ mA}}{10 \text{ }\mu\text{A}} \cong 2.0 \times 10^2$$

For the load line given,

$$R_C = 12 \text{ V}/10 \text{ mA} = 1.2 \text{ k}\Omega$$

From Eq.(7-21),

$$A_V = -\beta_{ac}\frac{R_C}{R_B + r}$$

with

$$r \cong 0.026 \text{ V}/22 \text{ }\mu\text{A} \cong 1.2 \text{ k}\Omega$$

The choice for R_B is

$$R_B = \frac{(2.0 \times 10^2)(1.2 \text{ k}\Omega)}{5.0} - 1.2 \text{ k}\Omega$$

$$\cong 48 \text{ k}\Omega - 1.2 \text{ k}\Omega \cong 47 \text{ k}\Omega$$

7-10. For the data given,

$$(I_B)_Q \cong -(I_E)_Q/\beta_{dc} = 7.5 \text{ mA}/125 \cong 60 \text{ }\mu\text{A}$$

$$r \cong 0.026 \text{ V}/60 \text{ }\mu\text{A} \cong 0.43 \text{ k}\Omega$$

Therefore, from Eq. (7-21),

$$A_V = (-125)\frac{2.2 \text{ k}\Omega}{10 \text{ k}\Omega + 0.43 \text{ k}\Omega} \cong -26$$

and

$$(A_V)_{min} = (-26)(0.5) = -13$$

$$(A_V)_{max} = (-26)(1.5) = -39$$

7-11. From Eq. (7-26),

$$\Delta(V_{out})_Q = \frac{-\beta_{dc} R_C}{R_B} \left[(V_{in} - V_{BE}) \left(\frac{1}{\beta_{dc}} \frac{\partial \beta_{dc}}{\partial T} \right) - \frac{\partial V_{BE}}{\partial T} \right] \Delta T$$

$$-1.0 \text{ V} > \frac{-(150)(2.5 \text{ k}\Omega)}{R_B} [(1.3 \text{ V})(1.0 \times 10^{-2}/°\text{C}) + 2.0 \times 10^{-3} \text{ V}/°\text{C}](10°\text{C})$$

so that

$$R_B > 56 \text{ k}\Omega$$

7-12. From Eq. (7-26) or Eq. (7-39) with $R_E = 0$,

(a) $\Delta(V_{out})_Q \cong \dfrac{(-2.0 \times 10^2)(1.0 \text{ k}\Omega)}{15 \text{ k}\Omega} [(1.3 \text{ V})(1.0 \times 10^{-2}/°\text{C}) + 2 \times 10^{-3} \text{ V}/°\text{C}](50°\text{C})$

$$\cong -10 \text{ V}$$

so that saturation occurs.

(b) $\qquad R_B + \beta_{dc} R_E = 15 \text{ k}\Omega + (2.0 \times 10^2)(0.1)(1.0 \text{ k}\Omega) = 35 \text{ k}\Omega$

So from Eq. (7-39),

$$\Delta(V_{out})_Q = \frac{(-2.0 \times 10^2)(1.0 \text{ k}\Omega)}{35 \text{ k}\Omega}$$

$$\times \left[(1.3 \text{ V}) \left(\frac{15 \text{ k}\Omega}{35 \text{ k}\Omega} \right) (1.0 \times 10^{-2}/°\text{C}) + 2 \times 10^{-3} \text{ V}/°\text{C} \right] (50°\text{C})$$

$$\cong -2.2 \text{ V}$$

Thermal drift is still a problem, so a heat sink should be considered to further reduce ΔT.

7-13. From Eq. (7-41),

$$-20/A_{V2} = 150/200$$

$$A_{V2} = (-20)(200/150) \cong -27$$

7-14. (a) From the data given,

$$r \cong 0.026 \text{ V}/5.0 \text{ }\mu\text{A} \cong 5.2 \text{ k}\Omega$$

$$R_B + r \cong 30 \text{ k}\Omega$$

So from Eq. (7-42),

$$\frac{-20}{A'_{V2}} = \frac{1 + \dfrac{30 \text{ k}\Omega}{(200)(0.5 \text{ k}\Omega)}}{1 + \dfrac{30 \text{ k}\Omega}{(150)(0.5 \text{ k}\Omega)}} \cong 0.93$$

$$A'_{V2} \cong -20/0.93 \cong -22$$

(b) $\qquad (V_{in})_Q \cong 0.7 \text{ V} + I_B R_B + I_E R_E$

$$\cong 0.7 \text{ V} + 0.1 \text{ V} + 0.5 \text{ V} \cong 1.3 \text{ V}$$

(c) $\qquad V_{R_E} = 0.5 \text{ V}$

$$V_{R_C} = (5.0 \text{ k}\Omega)(200)(5.0 \text{ }\mu\text{A}) = 5.0 \text{ V}$$

Therefore,

$$V_{CC} = 0.5 \text{ V} + 5 \text{ V} + (V_{CE})_Q$$

Given $(V_{CE})_Q = V_{CC}/2$,

$$V_{CC} \cong 11 \text{ V}$$

7-15. In general,

$$V_B = V_E + V_{BE}$$

Temperature and device variations affect V_{BE}. Assume that such variations produce $\Delta V_{BE} \lesssim 0.1$ V. Then stabilization requires that $V_E \gg \Delta V_{BE}$, and $V_E \cong 1$ V is reasonable. For operation in the middle of the linear operating region, $(V_{CE})_Q \cong V_{CC}/2$. Therefore, the voltage drop across $R_C + R_E$ should be about $V_{CC}/2$. For V_{CC} values ranging from 10 to 20 V, the voltage across R_C is 4 to 9 V at the Q point. Since $I_C \cong I_E$, this requires that

$$R_C/R_E = 4 \text{ to } 9$$

A solution $R_C = 10 \, R_E$ at the upper end of this range is appropriate for larger dc supply voltages.

7-16. From the data given,

$$(I_E)_Q = -1.0 \text{ mA} \quad \text{and} \quad (I_B)_Q \cong 1.0 \text{ mA}/100 \cong 10 \text{ } \mu\text{A}$$

Therefore,

$$I_1 \cong I_2 \cong 20 I_B = 0.2 \text{ mA}$$

and

$$V_B = V_{BE} + I_E R_E = 0.7 \text{ V} + (1.0 \text{ mA})(1.0 \text{ k}\Omega) \cong 1.7 \text{ V}$$

By substitution,

$$R_{B1} \cong \frac{12 \text{ V} - 1.7 \text{ V}}{0.2 \text{ mA}} \cong 51 \text{ k}\Omega$$

$$R_{B2} \cong \frac{1.7 \text{ V}}{0.2 \text{ mA}} \cong 8.5 \text{ k}\Omega$$

The reduction in gain is [from Eq. (7-47)],

$$\frac{R_{TH}}{R_B} = \frac{R_B \| R_{B1} \| R_{B2}}{R_B}$$

$$= \frac{25 \text{ k}\Omega \| 51 \text{ k}\Omega \| 8.5 \text{ k}\Omega}{25 \text{ k}\Omega} \cong 0.23$$

A fractional reduction in gain of 0.77 is noted.

7-17. Refer to Fig. 7-20. The following calculations result:
Region 1:

(a)
$$PM = \$8.00 - \$2.00 = \$6.00$$

(b)
$$PM = \$8.00 - \$5.00 = \$3.00$$

(c)
$$PM = \$8.00 - \$6.00 = \$2.00$$

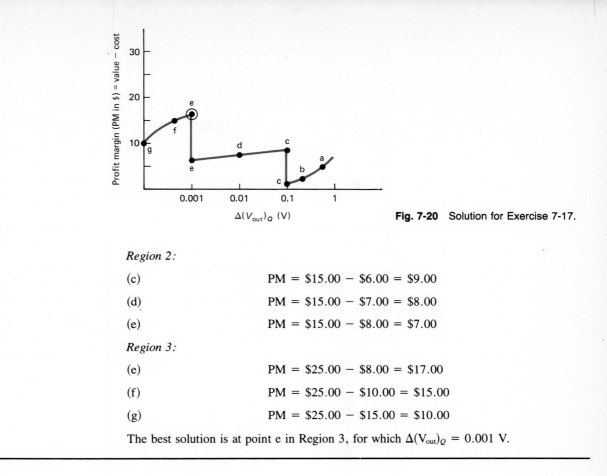

Fig. 7-20 Solution for Exercise 7-17.

Region 2:

(c)   PM = $15.00 − $6.00 = $9.00

(d)   PM = $15.00 − $7.00 = $8.00

(e)   PM = $15.00 − $8.00 = $7.00

Region 3:

(e)   PM = $25.00 − $8.00 = $17.00

(f)   PM = $25.00 − $10.00 = $15.00

(g)   PM = $25.00 − $15.00 = $10.00

The best solution is at point e in Region 3, for which $\Delta(V_{out})_Q$ = 0.001 V.

PROBLEMS

(7-2)   **7-1.** For a dc-coupled amplifier, a change in the input voltage from 0.20 to 0.30 V produces an output voltage change from 8.0 to 10.0 V. What is the amplifier gain?

(7-2)   **7-2.** For a common-base dc-coupled amplifier, R_C = 5.0 kΩ, R_E = 0.10 kΩ, and I_E = 0.60 mA. Estimate the circuit gain A_V.

(7-2)   **7-3.** The performance objective $A_V > 25$ is set for a common-base dc-coupled amplifier with R_E = 50 Ω and I_E = 5.0 mA. For what range of R_C values does the amplifier perform as desired?

(7-2)   **7-4.** Consider the graphical design of a dc-coupled amplifier using the schematic of Fig. 7-3 and the characteristic curves in Fig. 7-1. Choose V_{CC} = 10 V, $(I_C)_{max}$ = 50 mA, $(V_{in})_Q$ = −1.5 V, and R_E = 40 Ω. Prepare a graph of V_{out} as a function of V_{in} and determine A_V graphically at several points in the linear and cutoff regions.

(7-3)   **7-5.** Design a common-base dc-coupled amplifier with a gain A_V = 20 using the BJT characteristic in Fig. 6-10 with the Q point in the middle of the given linear operating region with $(I_C)_{max} \leq$ 10 mA. Find R_E and $(V_{in})_Q$.

(7-3) **7-6.** Apply the strategy of Section 7-3 to design a common-base dc-coupled amplifier with a gain $A_V = 10$ using the npn characteristic in Fig. 6-10. Assume that $V_{CC} = 10$ V and the Q point is located at $V_{CB} = 5.0$ V, $I_C = 4.0$ mA. Assume that all resistor values are available.

(7-4) **7-7.** Consider the common-emitter output characteristic curve in Fig. 7-2. Add an output load line with $V_{CC} = 8.0$ V and $R_C = 0.32$ kΩ. Let $R_B = 5.0$ kΩ. (a) Prepare a graph of V_{out} versus V_{in} following the procedure of Problem 7-2. (b) Determine A_V graphically. (c) Calculate A_V using Eq. (7-21) for comparison. (d) Measure the maximum signal swing $\pm\Delta(V_{out})$ that can be accommodated in the linear region (assuming equal positive and negative variations).

(7-4) **7-8.** Design a common-emitter dc-coupled amplifier using the npn characteristic in Fig. 6-12, $V_{CC} = 8.0$ V, and the Q point $(I_C)_Q = 8.0$ mA, $(V_{CE})_Q = 6.0$ V to achieve a gain $A_V = -8.0$.

(7-5) **7-9.** For a common-emitter dc-coupled amplifier, $(V_{in})_Q = 1.8$ V and $\beta_{dc} \cong 150$. If $R_C = 2.5$ kΩ and $R_B = 20$ kΩ, estimate the shift in $\Delta(V_{out})_Q$ that occurs for T increasing from 27 to 100°C.

(7-4)
(7-5)
(7-6) **7-10.** A common-emitter amplifier (see Fig. 7-7) has the values $R_C = 3.0$ kΩ, $R_B = 12$ kΩ, and $(I_C)_Q = 1.0$ mA. Values of $\beta_{dc} \cong \beta_{ac}$ can vary from 50 to 150. (a) What range of values is observed for the gain A_V? (b) If an emitter resistor $R_E = 0.10 \, R_C$ is introduced (see Fig. 7-11) and the original R_C is dropped to $R_C' = 0.90 R_C$, what range of gain values is observed due to the variations in β? (c) What is the percentage variation in A_V with respect to the average value before and after the introduction of R_E?

(7-6) **7-11.** Estimate A_V for the circuit of Fig. 7-21. Assume that r is small enough to be negligible and both β_{dc} and $\beta_{ac} > 100$.

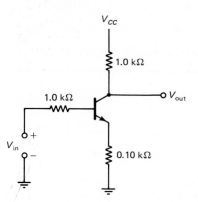

Fig. 7-21 Amplifier circuit for Problem 7-11.

(7-6) **7-12.** Design a BJT common-emitter dc-coupled amplifier with emitter resistor stabilization so that $A_V = -5.0$. Let $(I_C)_{max} = 20$ mA, $V_{CC} = 20$ V, and bias approximately in the middle of the linear operating region. If $R_C = 10 \, R_E$, find R_C, R_E, R_B, and $(V_{in})_Q$ (assume $\beta_{ac} \cong 150$).

(7-6) **7-13.** Use Eq. (7-25) to derive Eq. (7-26).

(7-6) **7-14.** Use Eq. (7-38) to derive Eq. (7-39).

(7-6) **7-15.** Show that Eqs. (7-46) and (7-47) give the correct Thevenin equivalent circuits.

(7-6) **7-16.** Given $R_{B1} = 30$ kΩ, $R_{B2} = 10$ kΩ, and $R_B = 5.6$ kΩ, find V_{TH} and R_{TH} for the circuit of Fig. 7-12 if $V_{in} = 6.0$ V and $V_{CC} = 12$ V.

(7-6) **7-17.** What problems can arise if you use heat sinks to limit the thermal drift in a circuit (as discussed in Example 7-6)?

(7-6) **7-18.** You wish to design a common-emitter dc-coupled amplifier with a gain of $A_V = -10$ independent of the transistor β. Assume that your source has a very low internal resistance. You estimate that for the given npn BJT, $(I_C)_Q = 10$ mA is appropriate. (a) Draw a schematic of the circuit you will use. (b) If you select $R_C = 1.0$ kΩ, what value should be assigned to R_E? (c) Show that r is probably negligible. (d) What source voltage is needed at the Q point?

(7-7) **7-19.** For the circuit of Fig. 7-12, assume that $R_B = 20$ kΩ, $R_{B1} = 30$ kΩ, and $R_{B2} = 5.0$ kΩ. Find how much the current gain of the amplifier is reduced by the voltage divider across the base of the BJT. (Assume $(r + \beta_{ac}R_E) = 50$ kΩ.)

COMPUTER APPLICATIONS

7-1. SPICE can be used to simulate the operation of a dc-coupled amplifier. Figures 7-22 to 7-25 show simulations for a common-emitter amplifier. Figures 7-23 and 7-25 indicate how a piecewise linear complex waveform can be input to the amplifier and the resulting output. What is the gain of each amplifier as determined from the simulation results? How does this finding compare with the gain prediction using the methods of analysis of this chapter?

* **7-2.** Write a computer program that can be used to solve Exercise 7-17. Use the program to calculate the profit margin for the dc-coupled amplifier circuit under discussion for a circuit cost from $0 to $25.00. Plot your results and interpret your findings. Turn in your program design, results, and conclusions.

* **7-3.** Use SPICE to find the output waveform for the amplifier circuit of Example 7-3 using the same input of Fig. 7-23.

Fig. 7-22 SPICE simulation of a BJT dc amplifier.

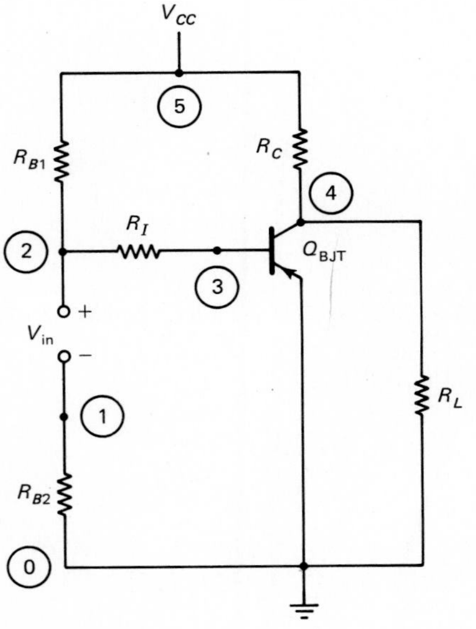

```
/T DCBJT
/L ALL
 1      DC BJT AMPLIFIER
 2      *
 3      * RESISTORS
 4      *
 5      RI 2 3 77K
 6      RB1 5 2 3.4K
 7      RB2 1 0 600
 8      RC 5 4 2.2K
 9      RL 4 0 2.2K
10      *
11      * POWER SUPPLY AND INPUT SIGNAL
12      *
13      VCC 0 5 DC 20
14      VIN 2 1 PWL(0 0 20US 1 40US 2 70US 2 100US -2 120US -2 125US
15      + -2 175US 0)
16      *
17      * TRANSISTORS
18      *
19      QBJT 4 3 0 MDA
20      .MODEL MDA PNP(BF=180 CJE=30PF CJC=8PF)
21      .TRAN 1.5US 200US
22      *
23      * OUTPUT
24      *
25      .PRINT TRAN V(2) V(4)
26      .PLOT TRAN V(2) V(4)
27      .END
```

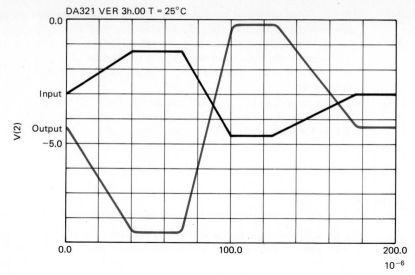

Fig. 7-23 Output waveform for SPICE simulation of a BJT dc-coupled amplifier. Observe that the effective capacitance of the BJT has resulted in a "rounding off" of the output waveform and a phase shift between the input and output waveforms. With an idealized BJT model, these effects would not be observed.

Fig. 7-24 SPICE simulation of a BJT dc-coupled amplifier with negative feedback.

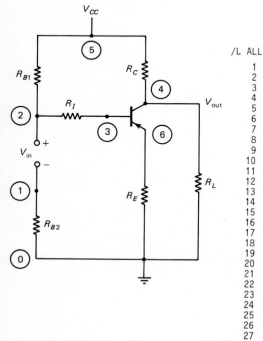

```
/L ALL
   1      DC BJT AMPLIFIER WITH NEGATIVE FEEDBACK
   2      *
   3      * RESISTORS
   4      *
   5      RI 2 3 77K
   6      RB1 5 2 3.4K
   7      RB2 1 0 600
   8      RC 5 4 2.2K
   9      RL 4 0 2.2K
  10      RE 6 0 220
  11      *
  12      * POWER SUPPLY AND INPUT
  13      *
  14      VCC 0 5 DC 20
  15      VIN 2 1 PWL(0 0 20US -1 40US -2 70US -2 100US 2 120US 2 125US 2
  16      + 175US 0)
  17      *
  18      * TRANSISTORS
  19      *
  20      QBJT 4 3 6 MDA
  21      .MODEL MDA PNP(BF=180 CJE=30PF CJC=8PF)
  22      *
  23      * OUTPUT
  24      *
  25      .TRAN 1US 200US
  26      .PRINT TRAN V(2) V(4)
  27      .PLOT TRAN V(2) V(4)
  28      .END
```

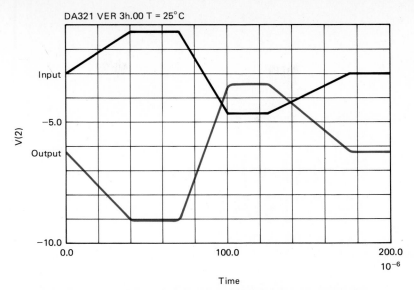

Fig. 7-25 Output waveform for SPICE simulation of a BJT dc-coupled amplifier with negative feedback.

EXPERIMENTAL APPLICATIONS

7-1. The objective of this experiment is to design, build, and test a BJT common-base dc-coupled amplifier with $A_V = 5(\pm 10\%)$. The design strategy to be used is discussed in detail in Section 7-3. Is your design dependent on the BJT that you select? Use available standard resistor values. When your circuit is constructed, check the dc bias voltages at each node. If these are satisfactory, then determine the small-signal amplifier gain $\Delta V_{\text{out}}/\Delta V_{\text{in}}$. Compare the predicted and actual bias voltages and percentage error between the predicted and observed values for A_V. Discuss and interpret your findings.

7-2. The objective of this experiment is to design, build, and test a BJT common-emitter dc-coupled amplifier with $A_V = -5(\pm 10\%)$. First check the dc bias voltages to make sure the desired operation point has been obtained. Then collect the data needed to plot V_{out} as a function of V_{in} following the method of Exercise 7-4. Find the gain in the linear region by measuring the slope of the curve. Heat the BJT and measure the shift in Q point $(\Delta V_{\text{out}})_Q$ in the linear region. Compare with the predicted value of $(\Delta V_{\text{out}})_Q$. Discuss and interpret your results.

7-3. The objective of this experiment is to redesign the common-emitter amplifier of Experimental Application 7-2 to use an emitter resistor R_E. The effect of R_E on the temperature stability of the circuit and gain of the circuit is to be evaluated. Select $R_E = 0.10R_C$ and keep $R_E + R_C$ fixed to maintain the same output load line with and without R_E. Predict the reduced gain and temperature drift expected for your circuit with R_E included. Measure A_V and $(\Delta V_{\text{out}})_Q$ and compare with the predictions for this experiment. Then compare the results of Experimental Applications 7-2 and 7-3.

CHAPTER 8

FIELD-EFFECT TRANSISTOR OPERATION AND BIASING TRADEOFFS

This chapter introduces semiconductor devices called field-effect transistors (FETs). The junction FET (JFET) is considered first in order to take advantage of earlier discussions of pn junctions. Both BJTs and JFETs make use of the fundamental pn junction properties, although these devices differ significantly in the ways in which amplification is obtained.

In practical applications, the JFET is less important than the metal oxide silicon FETs (MOSFETs) treated subsequently in this chapter. MOSFETs find particular application in high-density digital circuits, as discussed in Chapters 15 and 17.

The following sections describe how FETs operate, how to bias them for linear operation, and how biasing tradeoffs can be conducted. Since MOSFETs are typically used in large arrays, the individual devices described here can be viewed as building block elements.

Characteristic equations for FETs are derived by drawing on semiconductor properties as presented in earlier chapters. These equations provide insight into device operation and are useful in circuit design.

8-1 JUNCTION FIELD-EFFECT TRANSISTORS

The junction field-effect transistor (JFET, pronounced Jay-Fet) makes use of the properties of a reverse-biased pn junction. The properties of pn junctions are examined in Chapters 4 and 6. As established [in Eq. (4-40)], the effective width of the depletion region between the p-type and the n-type regions depends on the applied voltage. For positive bias, the depletion region narrows, and for reverse bias, the depletion region widens. The width of the depletion region can thus be varied by modifying the voltage that appears across it.

This effect is used in the JFET with the concepts shown in Fig. 8-1. In Fig. 8-1(A), an n-type semiconductor is shown connected to a source V_{DD}. Current flow in the semiconductor is dominated by the majority carriers (electrons). The electrons flow from the negative end of the n-type, which is called the *source,* to the positive end of the n-type, which is called the *drain.* The carriers enter by way of the source and leave by way of the drain. From previous discussion in Chapter 3, the magnitude of the current I is determined by the resistivity of the n-type material and the cross-sectional area A.

Now consider Fig. 8-1(B), with layers of p-type semiconductor on each side of the n-type region. The p-type regions are heavily doped, so that the voltage throughout these regions is about the same everywhere.

The entire voltage drop V_{DD} appears across the n-type region. At the source S, $V = 0$ for the n-type region. Further, $V = 0$ throughout the highly doped p-type regions because they are grounded. A net voltage drop $V = 0$ thus appears across the pn junction at location A.

At the drain D, $V = V_{DD}$ for the n-type region and $V = 0$ for the p-type regions. Therefore, a net reverse bias of V_{DD} appears across the pn junction at location B. The pn junctions experience a reverse bias that varies from 0 V at the source end of the device to V_{DD} at the drain end.

From Eq. (4-40), the depletion region associated with a pn junction becomes wider as the reverse bias increases in magnitude. The depletion regions therefore develop as shown in Fig. 8-1(B), approaching width $W = 0$ at the source and a maximum value of W at the drain.

What will be the effect of these depletion regions? As V_{DD} increases, the effective cross-sectional area of the n-type region is reduced and the resistance R of the region increases.

The width of these depletion regions can be varied by using an additional voltage V_{GG}, as shown in Fig. 8-1(C). The negative bias due to V_{GG} further reverse-biases the pn junctions, increasing the widths of the depletion regions. The cross-sectional area of the remaining n-type further decreases as shown.

Fig. 8-1 Basic operation of a JFET. (A) Carrier flow in an n-type semiconductor. (B) Effect of p-type regions. (C) Addition of a gate voltage V_{GG}.

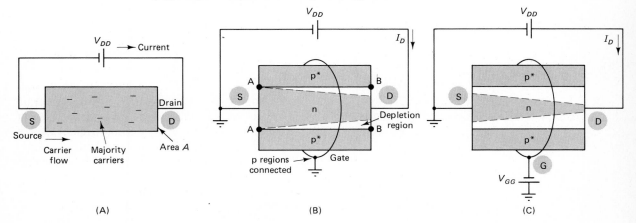

Figures 8-1(B) and (C) illustrate the basic features of a JFET. The n-type region is the *channel* and the combined p-type regions form the *gate*. Electrons flow through the channel, from the source to the drain, and the magnitude of the current flow is determined by V_{DD} and the gate voltage V_{GG}. The device shown in Fig. 8-1 is called an n-channel JFET.

A similar p-channel JFET can be fabricated by creating a p-type channel surrounded by heavily doped n-type regions and by reversing the polarities of V_{DD} and V_{GG}. Carrier flow in the p-channel JFET is due to holes, with the carriers again entering at the source S and leaving at the drain D. Figure 8-2 shows a p-channel JFET using a configuration that is closer in arrangement to actual devices.

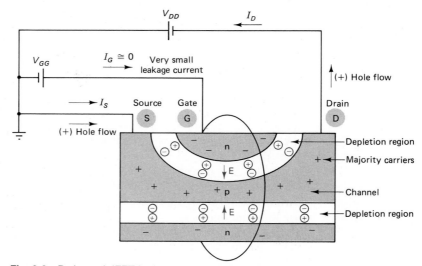

Fig. 8-2 P-channel JFET in the common-source configuration.

8-2 JFET CHARACTERISTIC CURVES

The voltage and current notations adopted for use with JFET design are similar to those introduced for the BJT. V_{DS} is the voltage of the drain with respect to the source, and V_{GS} is the voltage of the gate with respect to the source. Currents are positive into the device.

Now consider how the current through a JFET varies with V_{DS} for a fixed value of V_{GS}. Start with $V_{GS} = 0$ [Fig. 8-1(B)]. For small values of V_{DS}, the depletion region is thin and the channel has an initial resistance R_o. The JFET characteristic in this state can be represented by a straight line of slope $1/R_o$, as shown in Fig. 8-3. As V_{DS} increases, the pn junctions become more reverse-biased and the ac resistance of the channel increases. As a result, the slope of the characteristic curve decreases as shown.

As V_{DS} grows larger, the depletion regions finally expand to the point where they touch in the middle of the channel and *pinch-off* occurs. The resistance of the channel becomes very large due to the creation of a depletion region across the channel, and, as will be seen, I_D remains approximately constant for larger values of V_{DS}.

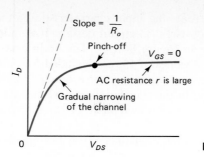

Fig. 8-3 Characteristic curve for a JFET with $V_{GS} = 0$.

The above discussion was for $V_{GS} = 0$. The more reverse-biased the junctions become due to an increase in the magnitude of V_{GS}, the more the current through the channel is reduced for any value of V_{DS}. The widened depletion regions achieve pinch-off for a reduced value of V_{DS}. The result is the family of curves shown in Fig. 8-4.

The dashed line in Fig. 8-4(B) marks the boundary between the *transition* region (where the channel has not yet closed) and the *saturation* region[1] (where pinch-off has occurred). When the gate becomes sufficiently negative, the channel is pinched off for all values of V_{DS} and no current flows. A small change in the voltage V_{GS} controls the flow of carriers through the channel by changing the channel width, so the JFET can be used for the design of amplifier circuits.

Where the BJT requires a significant input current, the JFET does not. The gate current through the reverse-biased pn junction is small, due only to minority carriers that produce a leakage current flow. The FETs are particularly suited to applications where it is desired to minimize the current flow into the amplifying device. This is an important design advantage in many cases.

Fig. 8-4 Characteristic curves for an n-channel JFET. (A) Transfer characteristic. (B) Output characteristic.

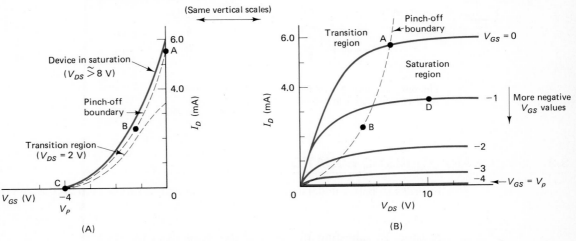

(A)

(B)

---

[1] The saturation region for the BJT [Figs. 6-9 to 6-12] and the saturation region for the JFET [Fig. 8-4(B)] have different meanings and apply to different regions of the characteristic curves.

Given the small input current, the input characteristic of the device gives only limited information and is not useful in circuit design. It is more helpful to use the *transfer characteristic* shown in Figure 8-4(A). The drain current I_D is shown as a function of V_{GS}. Only negative values of V_{GS} are included for an n-channel JFET because the gate-channel diode must remain reverse-biased. In the saturation region, there is only a slight variation in the transfer characteristic as V_{DS} changes. Observe that the transfer characteristic can be drawn based on the information provided in the output characteristic.

The above analysis applies to the n-channel JFET. For npn devices with a p-channel, similar effects are observed. Figure 8-5 shows the characteristic curves for a p-channel JFET. Figure 8-6 shows the schematic symbols for the two JFET types. The arrows define whether the gate is made of n-type or p-type material. The type of semiconductor used in the channel is of the opposite type in each case.

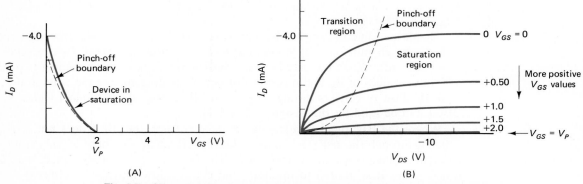

Fig. 8-5 Characteristic curves for a p-channel JFET. (A) Transfer characteristic. (B) Output characteristic.

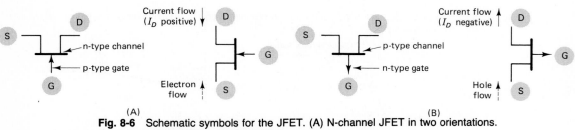

Fig. 8-6 Schematic symbols for the JFET. (A) N-channel JFET in two orientations. (B) P-channel JFET in two orientations.

8-3 JFET CHARACTERISTIC EQUATIONS

In order to better understand JFET operation and to support circuit design efforts, it is useful to develop equations that represent the characteristic curves described above. As illustrated in Fig. 8-1, the JFET is basically a voltage-controlled variable resistor. By adjusting V_{DS} and V_{GS}, the locations of the surfaces of the depletion regions are modified. In turn, these surfaces determine the effective channel geometry and the resultant device resistance.

Using the simple geometry of Fig. 8-1(C), the JFET characteristic curve can be defined through the equation

$$I_D = \frac{V_{DS}}{R} = \frac{V_{DS}}{\rho \int_0^L \dfrac{dx}{A(x)}} \tag{8-1}$$

using the definitions in Fig. 8-7. The channel is considered to consist of thin resistors of length dx and cross-sectional area $A(x)$; ρ is assumed constant in the channel; use has been made of Eq. (3-1) for each resistor element; and the total resistance is found by integration because the elements are in series. In general, $A(x)$ is a function of both V_{DS} and V_{GS}. V_{DS} produces the reduction in $A(x)$ from source to drain, with V_{GS} reducing $A(x)$ evenly along the length of the channel.

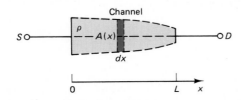

Fig. 8-7 Definition of terms for the JFET model.

Note that the JFET geometry enters into Eq. (8-1) only under the integral. Therefore, the resulting functional relationship is somewhat insensitive to the exact form of $A(x)$ that is used. This provides encouragement for the application of the simplified models.

The objective of this section is to gain insight into the general properties of the characteristic curves that describe JFET operation. Any models that are applied should be as simple as possible to avoid complexity that might mask an understanding of the physical properties that determine device performance.

To obtain a reasonable assumption for $A(x)$ over the channel length, first observe that voltage V_{DS} gives rise to the decrease in channel width from source to drain. The maximum reverse bias across the gate-channel junction occurs at the drain, maximizing the depletion region width and minimizing the channel width W.[2] Since the channel is narrower at the drain than at the source, the resistance per unit length is also higher at the drain. Therefore, most of the voltage V_{DS} drops across the channel in the vicinity of the drain. As a result, the narrowing of the channel width due to V_{DS} tends to be most important near the drain.

This observation suggests an exploratory model for use, with the closing of the channel due to V_{DS} restricted to a region $\Delta L \ll L$ near the drain as shown in Fig. 8-8. The effect of a more gradual transition can be considered once discussion of this introductory model is complete.

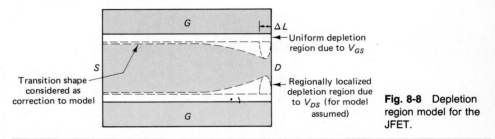

Fig. 8-8 Depletion region model for the JFET.

---

[2] The narrowing of the channel is analogous to base-narrowing as discussed in Chapter 6, Exercise 6-8.

If V_{DS} acts to close the channel over a localized length $\Delta L \ll L$, then the resistance of the channel is not strongly affected until the pinch-off condition is approached. Given

$$R = \left(\rho\frac{L_1}{A_1}\right)_{\text{(Over portion of the channel not affected by } V_{DS}\text{)}} + \left(\rho\frac{\Delta L}{A_2}\right)_{\text{(Over localized region affected by } V_{DS}\text{)}}$$

the second term is small until $A_2 \ll A_1$. In the limit for a small region $\Delta L \rightarrow 0$ over which V_{DS} acts, V_{DS} does not affect channel resistance until the voltage approaches the critical pinch-off value. For the sign convention being used, pinch-off occurs for $V_{GS} - V_{DS} = V_P$ in the region defined by ΔL. The effect due to V_{GS} throughout the channel will dominate until the critical value of V_{DS} is approached, and $A(x)$ can be taken as approximately constant for purposes of integration for V_{DS} below pinch-off.

For this simple model, Eq. (8-1) becomes

$$I_D \cong \frac{V_{DS}}{\rho\dfrac{L}{A(V_{GS})}} \tag{8-2}$$

The dependence of A on V_{GS} can be obtained by reference to Eq. (4-40), which describes how the width of the depletion region depends on the bias across it. Let W_o be the reference width[3] of the channel with $(V_t - V) = 0$. Then the width of the channel W becomes

$$W = W_o - 2K(V_t - V_G)^{1/2} \tag{8-3}$$

with

$$K = \left[\frac{2\epsilon}{qN_{CH}(1 + N_{CH}/N_G)}\right]^{1/2} \cong \left[\frac{2\epsilon}{qN_{CH}}\right]^{1/2} \tag{8-4}$$

If the channel is of thickness z, then

$$I_D \cong \frac{V_{DS}z}{\rho L}[W_o - 2K(V_t - V)^{1/2}] \tag{8-5}$$

If reverse bias is assumed and $V \gg V_t$, then

$$I_D \cong \left(\frac{V_{DS}zW_o}{\rho L}\right)_{\text{max}}\left[1 - \left(\frac{V_{GS}}{V_P}\right)^{1/2}\right] \tag{8-6}$$

where

$$|V_P| = \left(\frac{W_o}{2K}\right)^2 = \frac{W_o^2}{8}\frac{qN_{CH}}{\epsilon} \tag{8-7}$$

Equation (8-6) describes the operation of the JFET in the transition region. The current I_D is proportional to V_{DS}, with the slope of the linear relationship determined by V_{GS}. V_{GS} is the gate-to-source voltage; V_P is the pinch-off voltage, the value of V_{GS} that produces zero drain current; and I_D is the drain current for any bias value V_{GS}. For an n-channel JFET, I_D is positive and V_{GS} (and V_P) are negative. For a p-channel JFET, I_D is negative and V_{GS} (and V_P) are positive.

---

[3] For the notation used in Chapter 4, $V < 0$ for a reverse-biased pn junction and $V > 0$ for a forward-biased pn junction. The parameter W_o is defined for convenience, but $W = W_o$ is not physically obtainable.

The pinch-off condition is defined by $V_{GS} - V_{DS} = V_P$. Thus, along the pinch-off boundary,

$$I_D \cong -\left(\frac{V_P z W_o}{\rho L}\right)\left(1 - \frac{V_{GS}}{V_P}\right)\left[1 - \left(\frac{V_{GS}}{V_P}\right)^{1/2}\right] \qquad (8\text{-}8)$$

$$= I_{D0}\left(1 - \frac{V_{GS}}{V_P}\right)\left[1 - \left(\frac{V_{GS}}{V_P}\right)^{1/2}\right] \qquad (8\text{-}9)$$

As can be seen from Fig. 8-9, this equation defines the saturation transfer characteristic curve. Any increase in reverse bias $|V_{DS}| > |V_P|$ is dropped across the high-resistance depletion region at the pinch-off point, and I_D remains approximately constant. (The rationale for this assumption is further discussed in Exercise 8-5.)

Consideration of a more gradual profile in the channel and gradual changes in mobility μ as saturation occurs (Exercise 8-5) modifies the results by producing a "rounding off" between the solutions in the transition and saturation regions. The net result is to shift the point $V_{DS}/V_P = 1.0$ along the horizontal axis, thereby redefining the units along this axis. The new boundary between the transition and saturation regions can be constructed by requiring that $V_{GS} - V_{DS} = V_{P'}$ and originating the boundary at P'.

The resultant output and transfer characteristic curves bear a strong resemblance to the experimentally observed results of Figs. 8-4 and 8-5. The model has produced a characteristic equation with many of the features of an actual device.

Fig. 8-9 JFET characteristic curves for an n-channel device, as predicted by a simple model. The relationships between the output and transfer characteristics are shown. Because linear circuit operation uses the saturation region, the saturation transfer characteristic is usually important for amplifier design.

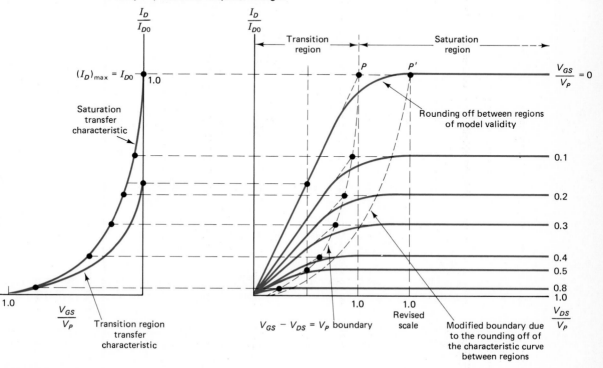

Although a highly approximate voltage profile has been used to describe the internal channel properties, the integration process has helped "average out" some of the model details. Equation (8-8) is useful because the terms can be understood by inspection. The ratio $\rho L / W_o z$ is the resistance R of the channel with zero reverse bias on the pn junctions. The term $I_{D0} = V_P / R$ is thus the current that flows in the JFET with voltage V_P from source to drain and no reduction in channel width. The term $1 - V_{GS}/V_P$ represents the reduction in current flow due to the closing of the channel by V_{DS} and V_{GS} and V_P is the voltage that results in $W = 0$ at the drain [Eqs. (8-3) and (8-7)]. The term $(1 - V_{GS}/V_P)^{1/2}$ represents the reduction in current due to V_{GS} applied uniformly through the channel. Despite the approximate nature of the calculations used, the above equations are useful to understand the essential features associated with JFET operation.

The above discussion of JFET operation illustrates how the properties of pn junctions and simple geometric models can be drawn upon to derive equations that describe device performance. Expressions for I_{D0} [Eq. (8-8)] and V_P [Eq. (8-7), modified as necessary for $V_P \rightarrow V_P'$] have been developed in terms of semiconductor parameters. Further application of these relationships is made in Chapter 18.

Clearly, many other types of geometric models can be used to produce JFET characteristic curves. Given the difficulties associated with actual geometries and voltage variations along the channel, computer simulation is often required where a high level of exactness is desired.

Middlebrook (1963) has proposed a model with V_{DS} varying linearly from the source to the drain. This assumption is associated with an average cross-sectional area and neglects the nonlinear "closing down" of the channel from the source to the drain. The model chooses to compensate for this assumption by letting the effective average width of the channel be a function of both V_{DS} and V_{GS}. This results in the characteristic equation

$$I_D = I_{D0}(1 - V_{GS}/V_P)^2 \tag{8-10}$$

which is in good agreement with experimental data despite the limitations in the assumptions made.

Streetman (1980) describes an alternative model, based on a linearly varying channel width, as illustrated in Fig. 8-1(C). This results in the characteristic equation

$$I_D = 3I_{D0}\left[-\frac{V_{GS}}{V_P} + \frac{2}{3}\left(\frac{V_{GS}}{V_P}\right)^{3/2} + \frac{1}{3}\right] \tag{8-11}$$

which again gives the expected shape for the saturated transfer characteristic curve.

The characteristic curves given by Eqs. (8-9) through (8-11) are compared in Fig. 8-10. Typical experimental data are often close to the simple prediction of Eq. (8-10), so this equation is widely used for the design of JFET amplifier circuits. In this text, Eq. (8-10) is used to describe the saturated transfer characteristic of the JFET.

The *transconductance* g_m of the JFET is defined as the slope of the transfer characteristic at any point. For the saturated transfer characteristic [Eq. (8-10)],

$$g_m = \frac{\partial I_D}{\partial V_{GS}} = \frac{-2I_{D0}}{V_P}\left(1 - \frac{V_{GS}}{V_P}\right) \tag{8-12}$$

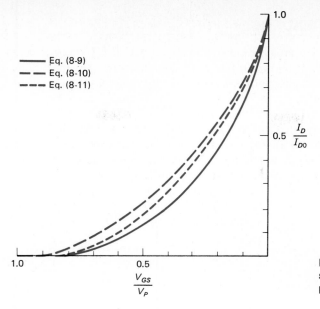

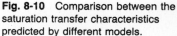

Fig. 8-10 Comparison between the saturation transfer characteristics predicted by different models.

Equations (8-10) and (8-12) apply to both types of JFET if the correct signs are used when substituting into the equations. For the convention used, g_m is always positive. The maximum value of g_m is obtained with $V_{GS} = 0$.

$$(g_m)_{max} = -2I_{D0}/V_P \tag{8-13}$$

and the minimum value is obtained with $V_{GS} = V_P$.

Example 8-1   JFET Operating Points

For an n-channel JFET with $V_P = -2.0$ V, a current of 5.0 mA is desired. If $I_{D0} = 10$ mA, what are the required value for V_{GS} and the resultant value for g_m?

Solution   From Eq. (8-10),

$$V_{GS}/V_P = 1 - (I_D/I_{D0})^{1/2}$$

so that

$$V_{GS} = (-2.0 \text{ V})[1 - (5.0 \text{ mA}/10 \text{ mA})^{1/2}] = -0.58 \text{ V}$$

From Eq. (8-12),

$$g_m = \frac{-2(10 \text{ mA})}{-2.0 \text{ V}}\left(1 - \frac{-0.58}{-2.0}\right) \cong 7.1 \text{ m}\mho$$

The transconductance is often expressed in units of *mhos* (1/ohms), symbolized $\mho$, or *Siemens* (S). The former unit is used in this text. Observe that Eqs. (8-10) and (8-12) can be used together to relate the parameters V_{GS}, V_P, I_D, I_{D0}, and g_m. Given all but two of these parameters, the others can be found. These relationships are used together throughout this and following chapters. ■

Changes in device temperature affect the JFET transfer characteristic due to two processes, as shown in Fig. 8-11. As the temperature increases, more collisions take place in the channel and carrier mobility decreases as

$$\mu = \mu_o(T/T_o)^{-n} \tag{8-14}$$

with n typically varying between 1.7 and 2.7 (Sze, 1981). The carriers are less free to respond to an external field, so there is a decrease in current I_D. This effect is most important for large values of I_D, where the depletion regions are narrow and collision processes dominate current flow.

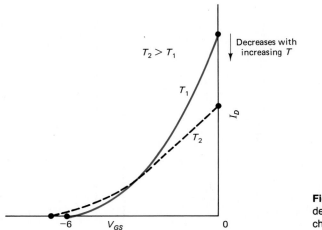

Fig. 8-11 Temperature dependence for the JFET transfer characteristic (n-channel).

In the same device, an increased temperature also produces a small decrease in the thickness of the depletion regions. Increased carrier energy leads to deeper penetration into the depletion region and a widened channel. (The result is increased contact potential.) In order to pinch off the channel, a more negative voltage V_{GS} must be applied. This process is most important for small values of I_D near pinch-off.

The transfer characteristics for two different temperatures are shown in Fig. 8-11. The curves cross at approximately $|V_P - V_{GS}| \cong 0.6$ to 0.7 V, indicating a point where I_D is not dependent on temperature.

8-5 DATA SOURCES FOR THE JFET

Typical data sheets for n-channel and p-channel JFETs are shown in Fig. 8-12(A) and 8-12(B). For the J210–212, the gate reverse current I_{GSS} (leakage current) is shown to be small, less than 100 μA. The pinch-off voltage V_P is called $V_{GS(off)}$ and varies over several volts, with the exact range depending on which device is selected. The maximum drain current I_{D0} is called I_{DSS} and varies by a significant amount for each device. (For example, I_{DSS} can be as small as 2 mA or as large as 15 mA for the J210.) Note that g_{fs} is used for g_m.

For the J270–271, the leakage current can increase to 200 μA. The pinch-off

FIGURE 8-12(A)
Data Sheet for N-Channel JFETs
(Courtesy of National Semiconductor Corporation)

J210-12 N-Channel JFETs
General Description

The J210 thru J212 series of n-channel JFETs is characterized for low- to medium-frequency amplifiers requiring high transconductance and low input capacitance.

Absolute Maximum Ratings (25°C)

Gate-drain or gate-source voltage	−25 V
Gate current	10 mA
Total device dissipation (25°C free-air temperature)	350 mW
Power derating (to +125°C)	3.5 mW/°C
Storage temperature range	−55 to +150°C
Operating temperature range	−55 to +150°C
Lead temperature (1/16 in. from case for 10 s)	300°C

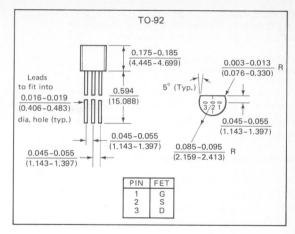

Electrical Characteristics (25°C unless otherwise noted)

	Parameter	Conditions	J210 Min	J210 Typ	J210 Max	J211 Min	J211 Typ	J211 Max	J212 Min	J212 Typ	J212 Max	Units
I_{GSS}	Gate Reverse Current	$V_{DS} = 0$, $V_{GS} = -15$ V (Note 1)			−100			−100			−100	pA
$V_{GS(off)}$	Gate-Source Cutoff Voltage	$V_{DS} = 15$ V, $I_D = 1$ nA	−1		−3	−2.5		−4.5	−4		−6	V
BV_{GSS}	Gate-Source Breakdown Voltage	$V_{DS} = 0$, $I_G = -1$ μA	−25			−25			−25			V
I_{DSS}	Saturation Drain Current	$V_{DS} = 15$ V, $V_{GS} = 0$ (Note 2)	2		15	7		20	15		40	mA
I_G	Gate Current	$V_{DG} = 10$ V, $I_D = 1$ mA		−10			−10			−10		pA
g_{fs}	Common-Source Forward Transconductance (Note 2)	$f = 1$ kHz	4000		12000	7000		12000	7000		12000	μmho
g_{os}	Common-Source Output Conductance	$V_{DS} = 15$ V, $V_{GS} = 0$		150			200			200		μmho
C_{iss}	Common-Source Input Capacitance	$f = 1$ MHz		5.0			5.0			5.0		pF
C_{rss}	Common-Source Reverse Transfer Capacitance			1.5			1.5			1.5		pF
e_n	Equivalent Short-Circuit Input Noise Voltage	$f = 1$ kHz		10			10			10		nV/Hz$^{1/2}$

Note 1: Approximately doubles for every 10° C increase in T_A.
Note 2: Pulse test duration = 2 ms.

FIGURE 8-12(B)
Data Sheet for P-Channel JFETs (Courtesy of National Semiconductor Corporation)

J270, J271 P-Channel JFETs
General Description

The J270 thru J271 series of p-channel JFETs is characterized for low- to medium-frequency small-signal amplifiers requiring high transconductance and low-input noise voltage.

Absolute Maximum Ratings   (25°C)

Gate-drain or gate-source voltage (Note 1)	30V
Gate current	−50 mA
Total device dissipation (25°C free-air temperature)	350 mW
Power derating (to +125°C)	3.5 mW/°C
Storage temperature range	−55 to +150°C
Operating temperature range	−55 to +150°C
Lead temperature (1/16 in. from case for 10 s)	300°C

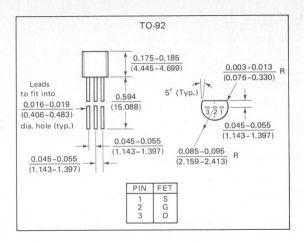

Electrical Characteristics (25°C unless otherwise noted)

Parameter		Conditions	J270 Min	J270 Typ	J270 Max	J271 Min	J271 Typ	J271 Max	Units
I_{GSS}	Gate reverse current	$V_{DS} = 0$, $V_{GS} = 20V$ (Note 2)			200			200	pA
$V_{GS(off)}$	Gate-source cutoff voltage	$V_{DS} = -15\,V$, $I_D = -1\,nA$	0.5		2.0	1.5		4.5	V
BV_{GSS}	Gate-source breakdown voltage	$V_{DS} = 0$, $I_G = 1\,\mu A$	30			30			
I_{DSS}	Saturation drain current	$V_{DS} = -15\,V$, $V_{GS} = 0$ (Note 3)	−2		−15	−6		−50	mA
I_G	Gate current	$V_{DG} = 15\,V$, $I_D = I_{DSS(min)}$		15			60		pA
g_{fs}	Common-source forward transconductance (Note 3)		6000		15000	8000		18000	μmho
g_{os}	Common-source output conductance	f = 1 kHz, $V_{DS} = -15\,V$, $V_{GS} = 0$			200			500	
C_{iss}	Common-source input capacitance			20			20		pF
C_{rss}	Common-source reverse transfer capacitance	f = 1 MHz		5			5		
e_n	Equivalent short-circuit input-noise voltage	$V_{DS} = -10\,V$, $I_D = I_{DSS(min)}$ f = 1 kHz		10			10		nV/Hz$^{1/2}$

Note 1: Geometry is symmetrical. Units may be operated with source and drain leads interchanged.
Note 2: Approximately doubles for every 10°C increase in T_A.
Note 3: Pulse test duration = 2 ms.

voltage V_P varies from 0.5 to 2.0 V for the J270 and from 1.5 to 4.5 V for the J271. The maximum drain current I_{D0} is negative, indicating flow out of the drain in this specific notation, and varies from 2 to 15 mA for the J270 and from 6 to 50 mA for the J271.

Example 8-2   JFET Power Dissipation

For the J270 JFET, what is the maximum power dissipation rating at 75°C?

Solution   From Fig. 8-12(B), the maximum device dissipation is 350 mW at 25°C and the de-rating factor is 3.5 mW/°C. Therefore, at 75°C

$$P_m \ (75°C) = 350 \text{ mW} - (3.5 \text{ mW/°C})(75 - 25)°C$$

$$= 175 \text{ mW}$$

As noted in Chapter 6, most transistor data sheets specify a derating factor. Because the temperature of operation is so important to device capabilities, thermal data are usually essential to circuit and system development.   ■

The photograph of Fig. 8-13(A) shows the output characteristic of a 2N5457 p-channel JFET. The graphics plots of Figs. 8-13(B) and 8-13(C) show other output and transfer characteristics for the 2N4416A JFET.

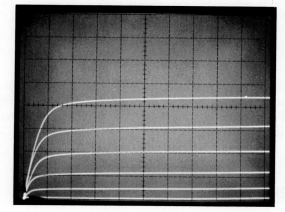

Fig. 8-13(A)   Output characteristic for the 2N5457 JFET. (Horizontal scale: 2 V/division; vertical scale: 1 mA/division; voltage steps: 0.5 V/step)

Fig. 8-13(B)   JFET output characteristic.

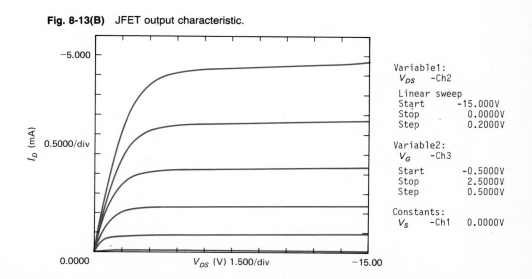

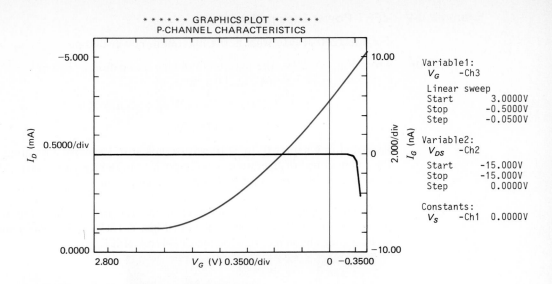

Variable1:
V_G −Ch3
Linear sweep
Start   3.0000V
Stop   −0.5000V
Step   −0.0500V

Variable2:
V_{DS} −Ch2
Start   −15.000V
Stop   −15.000V
Step   0.0000V

Constants:
V_S −Ch1   0.0000V

Fig. 8-13(C) JFET transfer characteristic.

Note that both graphics characteristics include a slight forward bias on the gate-channel diodes (maintained below the voltage for which the diodes will be strongly conducting). Figure 8-13(C) also illustrates how the gate current begins to rise with this forward-bias condition.

8-6 TRADEOFFS FOR JFET COMMON-SOURCE AMPLIFIER DESIGN

A typical dc-coupled amplifier circuit for the n-channel JFET and characteristic curves for the device are shown in Figs. 8-14 and 8-15. Since the gate current is small, the current drawn from the input usually does not enter into the design. Cir-

Fig. 8-14 JFET common-source dc-coupled amplifier with fixed bias (n-channel).

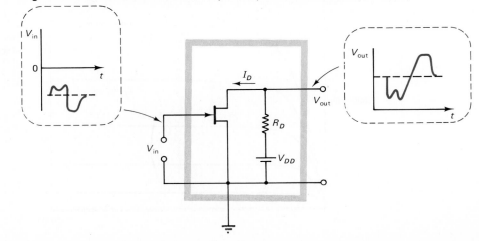

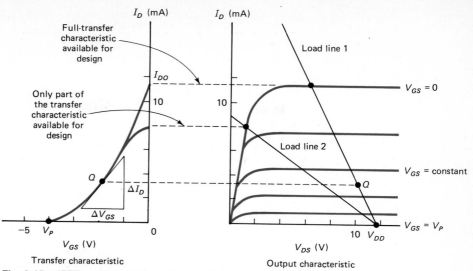

Full-transfer characteristic available for design

Only part of the transfer characteristic available for design

I_D (mA)

I_{DO}

10

Q

ΔI_D

ΔV_{GS}

-5 V_P

V_{GS} (V)

Transfer characteristic

I_D (mA)

Load line 1

10

Load line 2

$V_{GS} = 0$

V_{GS} = constant

Q

$V_{GS} = V_P$

10 V_{DD}

V_{DS} (V)

Output characteristic

Fig. 8-15 JFET common-source characteristic curves (n-channel).

cuit operation is thus determined by V_{in} and the output load line (given by R_D and V_{DD}).

The output and transfer characteristics have the same vertical axis (I_D). If drawn to the same scale, the two characteristics can be aligned as shown by the dashed lines.

If R_D is chosen so that the output characteristic load line intercepts the I_D axis above I_{D0} (load line 1 in Fig. 8-15), then the full-transfer characteristic is available for use during the design process. However, if R_D is large, so that the load line intercepts the I_D axis below I_{D0} (load line 2), only part of the transfer characteristic is available for the design. In the latter case, the transfer curve is *rounded off* as the JFET leaves the saturation region. When a JFET circuit is being designed, the transfer characteristic must be adjusted to reflect the influence of the output load line.

For the circuit of Fig. 8-14, the design of a given Q point can be achieved by satisfying the following two equations:

$$V_{in} = V_{GS} \tag{8-15}$$

$$V_{out} = V_D - I_D R_D \tag{8-16}$$

The effect of changes in input voltage on these equations is given by

$$\Delta V_{in} = \Delta V_{GS}$$
$$\Delta V_{out} = -(\Delta I_D)R_D \tag{8-17}$$

As V_{in} is varied, the voltage gain is given by

$$A_V = \frac{\Delta(V_{out})}{\Delta(V_{in})} = -R_D\left(\frac{\Delta I_D}{\Delta V_{GS}}\right) \tag{8-18}$$

To the first order,

$$\Delta I_D = \frac{\Delta I_D}{\Delta V_{GS}}\bigg|_{V_{DS}} \Delta V_{GS} + \frac{\Delta I_D}{\Delta V_{DS}}\bigg|_{V_{GS}} \Delta V_{DS}$$

In the current-saturation region, the second term is small so that

$$A_V \cong -R_D\left(\frac{\Delta I_D}{\Delta V_{GS}}\bigg|_{V_{DS}}\right) \tag{8-19}$$

The term

$$\frac{\Delta I_D}{\Delta V_{GS}}\bigg|_{V_{DS}}$$

is the slope of the slope of the transfer characteristic (with I_D plotted as a function of V_{GS}). This ratio has been defined as the forward transconductance g_m of the JFET, so that

$$A_V \cong -g_m R_D \tag{8-20}$$

The forward transconductance g_m can be determined by graphically measuring the slope of the transfer characteristic at the desired operating point. For the common-source configuration, A_V is negative. The output is 180 degrees out of phase with the input. The common-emitter BJT amplifier and the common-source JFET amplifier are similar in terms of this inversion that exists between input and output signals.

The design of a JFET amplifier can be simplified by means of the following approximate relationships that were introduced earlier:

$$I_D \cong I_{D0}(1 - V_{GS}/V_P)^2 \tag{8-21}$$

$$g_m \cong \frac{\partial I_D}{\partial V_{GS}} \cong \frac{-2I_{D0}}{V_P}\left(1 - \frac{V_{GS}}{V_P}\right) \tag{8-22}$$

These equations can be used to approximate the transfer characteristic and thus avoid the need to graphically find g_m.

The design of the JFET dc-coupled amplifier centers around use of the transfer characteristic. By combining Eqs. (8-21) and (8-22) for the transfer characteristic and its slope g_m and by making use of Eq. (8-20) for the gain A_V, the design process can be reduced to the simultaneous solution of three equations. In applying this strategy, care must be taken to seek solutions that are in the linear region of the device and that satisfy all design specifications.

Example 8-3 Common-Source DC-Coupled Amplifier

Consider the design of an n-channel JFET common-source dc-coupled amplifier to achieve $A_V = -10$ with the following device values:

$$I_{D0} = 10 \text{ mA}$$

$$V_P = -6.0 \text{ V}$$

and with

$$(I_D)_Q = 5.0 \text{ mA}$$

$$(V_{DS})_Q = 10 \text{ V}$$

where the values for $(I_D)_Q$ and $(V_{DS})_Q$ determine a Q point in the linear region for the JFET. Find the values for R_D and V_{DD}.

Solution Solving Eq. (8-21) for V_{GS},

$$V_{GS} = V_P[1 - (I_D/I_{D0})^{1/2}] \cong -1.7 \text{ V}$$

as the needed bias point.

From Eq. (8-22),

$$g_m \cong \frac{(-2)(10 \text{ mA})}{-6.0 \text{ V}}\left(1 - \frac{-1.7 \text{ V}}{-6.0 \text{ V}}\right) \cong 2.4 \text{ m}\mho$$

To attain a gain $A_V = -10$,

$$R_D = -A_V/g_m = 10/2.4 \text{ m}\mho \cong 4.2 \text{ k}\Omega$$

Once the Q point and R_D are specified, the intercepts V_D and V_D/R can be found

$$\frac{V_{DD} - (V_{DS})_Q}{(I_D)_Q} = R_D \quad \text{and} \quad \frac{(V_{DS})_Q}{V_{DD}/R_D - (I_D)_Q} = R_D$$

$$V_{DD} = (V_{DS})_Q + (I_D)_Q R_D$$

$$= 10 \text{ V} + 5.0 \text{ mA}(4.2 \text{ k}\Omega) \cong 31 \text{ V}$$

The load line selected rounds off the transfer characteristic, as shown in Fig. 8-16, limiting the available range of values for V_{GS}. Note that a large supply voltage $V_{DD} = 31$ V is required for this solution.

Such a high value for V_{DD} may not be a satisfactory selection in many circuit applications. Typical semiconductor circuits are operated at lower voltage levels to reduce

Fig. 8-16 Relationship between input and output signals for Example 8-3.

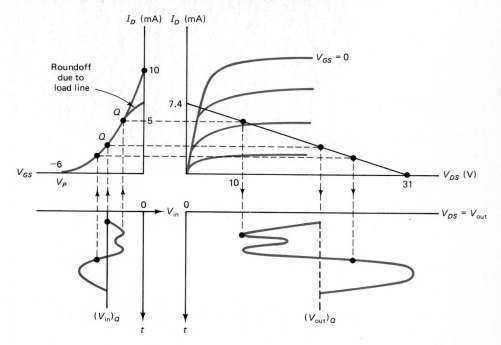

power consumption, and the maximum allowable voltage drop across the source-drain is limited.

If V_{DD} is unsatisfactory, then R_D must be reduced to result in an acceptable power-supply voltage. The gain of the amplifier will then decrease. A tradeoff exists between a low value of V_{DD} and a high amplifier gain A_V, with R_D adjusted to produce the desired compromise solution. ■

8-7 JFET AMPLIFIER WITH SELF-BIAS

The n-channel JFET used above requires that the input signal include a dc-voltage component that provides a constant negative bias to the gate. The signal consists of variations in the input voltage around this dc-bias level. A second battery may be required in series with the signal to produce the desired bias offset.

An alternative approach to achieving the needed bias is to use the self-bias approach shown in Fig. 8-17. A resistor in series with the *source* can produce the needed offset. The voltage drop across R_S is of the correct polarity to make the gate negative with respect to the source. The JFET transfer characteristic can still be described by Eq. (8-21). At the same time,

$$V_{GS} = -I_D R_S \qquad (V_{in} = 0) \qquad (8\text{-}23)$$

from Fig. 8-17. Both equations must hold true. The operating point Q is thus given by the intersection of the transfer characteristic curve [Eq. (8-21)] and the bias load line [Eq. (8-23)]. A graphical construction or Eq. (8-23) can be used to find the value of R_S that is needed to produce a desired Q point.

In order to find the gain A_V for the self-bias circuit, observe that

$$\begin{aligned} V_{out} &= V_{DD} - I_D R_D \\ V_{in} &= V_{GS} + I_D R_S \end{aligned} \qquad (8\text{-}24)$$

so that

$$\begin{aligned} \Delta V_{out} &= -(\Delta I_D)R_D \\ \Delta V_{in} &= \Delta V_{GS} + (\Delta I_D)R_S \end{aligned} \qquad (8\text{-}25)$$

and

$$\begin{aligned} A_V = \frac{\Delta(V_{out})}{\Delta(V_{in})} &\cong \frac{-(\Delta I_D)R_D}{\Delta V_{GS} + (\Delta I_D)R_S} \\ &= \frac{-R_D}{R_S + \dfrac{\Delta V_{GS}}{\Delta I_D}} \cong \frac{-R_D}{R_S + 1/g_m} \end{aligned} \qquad (8\text{-}26)$$

Note that if $R_S = 0$, the previous result [Eq. (8-20)] is obtained

$$A_V = -g_m R_D$$

In Section 7-6, an emitter resistor R_E is used to temperature stabilize the common-emitter BJT amplifier. Negative feedback caused by R_E reduces the temperature drift, reduces the total amplifier gain, and decreases the dependence of the circuit

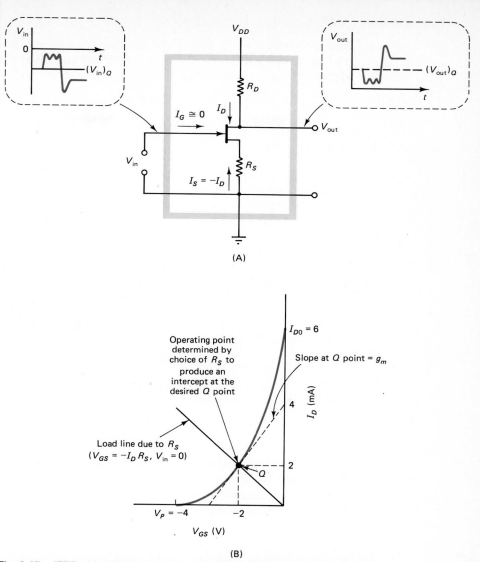

Fig. 8-17 JFET with self-bias. (A) Self-bias circuit. (B) Transfer characteristic.

gain on the BJT parameters. The negative feedback produced by R_E allows several important design tradeoffs to be made.

Equation (8-26) illustrates how negative feedback caused by R_S produces a different type of tradeoff for the JFET. The addition of R_S reduces any shift in I_D due to temperature variations in the transfer characteristic. As the characteristic varies, changes are along the load line due to R_S, not along a vertical line, as is the case with fixed bias. This advantage is discussed below.

The addition of R_S has also resulted in one less power supply, an important advantage. But a price has been paid through a reduced gain A_V. Further variations on this circuit are covered in Chapter 9.

Example 8-4   JFET Amplifier with Self-Bias

You are given a JFET having the transfer characteristic shown in Fig. 8-17. You wish to design a dc-coupled amplifier with self-bias using the Q point shown. Find appropriate values of R_S, R_D, and V_{DD} to achieve $A_V = -10$ V.

Solution   From Fig. 8-17(B), $(V_{GS})_Q = -2.0$ V and $(I_D)_Q = 2.0$ mA. Therefore,

$$R_S = -(V_{GS})_Q/(I_D)_Q \cong 2.0 \text{ V}/2.0 \text{ mA} = 1.0 \text{ k}\Omega$$

From the graph,

$$g_m \cong 4.0 \text{ mA}/3.0 \text{ V} \cong 1.3 \text{ m}\mho$$

Therefore, using Eq. (8-26),

$$-10 = \frac{-R_D}{1/1.3 \text{ m}\mho + 1.0 \text{ k}\Omega}$$

$$R_D \cong 18 \text{ k}\Omega$$

To find V_{DD}, add the voltages from V_{DD} to ground

$$V_{DD} = (I_D)_Q(R_D + R_S) + (V_{DS})_Q$$

For this problem, $(V_{DS})_Q$ is not specified. Assume that linear bias is obtained with $(V_{DS})_Q = 6$ V. Therefore,

$$V_{DD} \cong (2.0 \text{ mA})(18 \text{ k}\Omega + 1.0 \text{ k}\Omega) + 6 \text{ V} \cong 44 \text{ V}$$

which is often too large for many applications. Further, the transfer characteristic is rounded off at the intercept:

$$\frac{V_{DD}}{R_D + R_S} = \frac{44 \text{ V}}{19 \text{ k}\Omega} = 2.3 \text{ mA}$$

which significantly limits the available linear operating range [see Fig. 8-17(B)].

The result is a circuit design that is formally correct but not a reasonable design solution. This example illustrates how iterative solutions can be required when initial design calculations produce results that are not acceptable.   ■

8-8   OTHER COMMON-SOURCE DESIGN CONCEPTS

As noted above, the addition of R_S to achieve self-bias results in a decrease in the amplifier gain. An alternative design strategy is to replace R_S with a forward-biased junction diode (Figure 8-18). From Eq. (8-26), the gain now becomes

$$A_V = \frac{-R_D}{1/g_m + r} \tag{8-27}$$

where r is the ac resistance of the biasing diode near the Q point. If the diode is biased past the knee of the curve, $r \ll 1/g_m$ and $A_V \cong -g_m R_D$. Fixed bias and diode bias thus produce similar results.

A JFET amplifier can also be configured using a second JFET as an active load that approximates a constant current source, as shown in Fig. 8-18. The Q point is

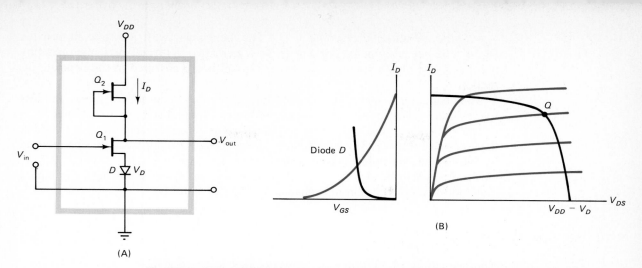

(A)

(B)

Fig. 8-18 JFET amplifier with diode bias and an active load. (A) JFET amplifier schematic. (B) Characteristic curves.

chosen so that the active load Q_2 is biased in its saturation region. The amplifier gain becomes

$$A_V \cong -g_m r_0 \tag{8-28}$$

The amplifier gain is thus very large. The use of constant-current sources for amplifier loads is further discussed in Chapter 12 with respect to operational amplifier design.

8-9 TEMPERATURE DEPENDENCE OF OPERATING POINTS

The temperature dependence of the JFET transfer characteristic is discussed above. Figure 8-19 shows how this dependence changes the output operating point for the common-source dc-coupled amplifier.

Fig. 8-19 Variation in Q point with changing temperature.

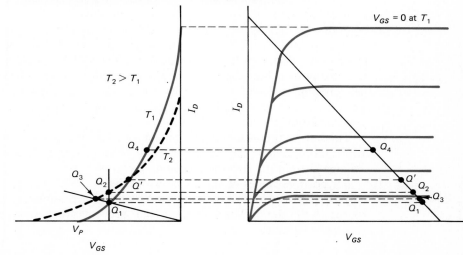

Assume an initial bias point at Q_1. For fixed bias [with a source to provide $(V_{in})_Q$], the operating point shifts to Q_2 as T increases. With self-bias, the operating point shifts to Q_3. If the initial bias is at Q_4, shifts are observed in the opposite directions.

The change in I_D due to temperature effects is a maximum for fixed bias and less for self-bias. Thus, by sacrificing gain, additional thermal stability is obtained.

It may be noted that little change in I_D occurs as T varies if the circuit is biased at the crossover point Q. The temperature dependence of the JFET can thus be reduced by biasing at this point, which is located at approximately $|V_{GS} - V_P| = 0.7$ V. However, g_m can be significantly below its maximum at this point, again forcing a reduction in amplifier gain in exchange for thermal stability.

8-10 MOSFET DEVICES

Another family of FET devices is well-suited to the design of integrated circuits. These devices take advantage of the field effect used in the JFET, but do not use a junction diode between the gate and channel. Instead, the gate and channel are separated by an insulating layer.

These devices can be generally called MISFETs (metal-insulator semiconductor field-effect transistors) or IGFETs (insulated gate FETs). However, since the most common insulating layer is made of silicon oxide (SiO_2), they are most often called MOSFETs (metal oxide silicon FETs). N-channel devices are referred to in terms of NMOS technology and p-channel devices in terms of PMOS technology.

8-11 DEPLETION OR D-MOSFETs

MOSFETs are fabricated in two basic forms. The first form, the depletion, or D-MOSFET, is illustrated in Fig. 8-20 for an n-channel device. The D-MOSFET is somewhat similar to the JFET except that the gate-channel diode for the JFET (consisting of a p-type and an n-type region separated by a depletion region due to reverse bias) has been replaced by a conducting layer and a semiconducting region separated by an insulating layer. As a result, V_{GS} can take on positive as well as negative values without causing a gate current to flow.

Figure 8-20 illustrates the operation of the n-channel depletion MOSFET. The channel width is controlled by the electric field created by the metal layer, acting through the insulating layer. For negative values of V_{GS}, as shown in Fig. 8-20(A), the field pushes away the majority carriers (electrons) from the top part of the channel, creating a depletion layer and reducing the current flow through the channel from the source to the drain. As E increases, the channel narrows and progresses toward pinch-off.

This effect produces output and transfer characteristics similar to those for the JFET for $V_{GS} < 0$. The dc gate current is zero except for the (usually negligible) leakage current through the insulating layer.

For positive values of V_{GS}, as shown in Fig. 8-20(B), the electric field attracts the majority carriers toward the top of the channel, increasing the current flow from the source to the drain. The insulating layer prevents the carriers from flowing to the gate and maintains a dc gate current near zero.

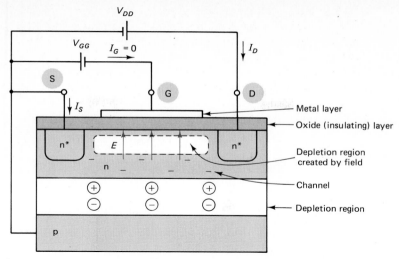

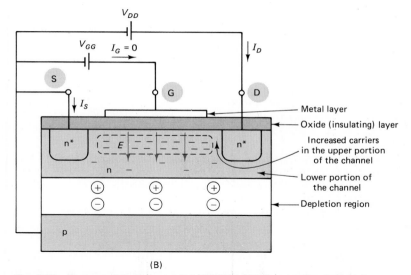

Fig. 8-20 N-channel depletion or D-MOSFET. (A) V_{GS} negative (reduced carrier density in the channel region reduces conduction until a depletion region results). (B) V_{GS} positive (increased carrier density in the channel region increases conduction).

The p-channel D-MOSFET operates in a similar way, except that the majority carriers in the channel are holes. Positive values of V_{GS} reduce the channel current, whereas negative values increase it. The schematic symbols and characteristic curves for the n-channel and p-channel depletion MOSFETs are shown in Fig. 8-21. The curves represent extensions of the JFET characteristics, allowing the device to be operated at zero and both positive and negative bias values.

This is a useful result because ac amplifiers (Chapter 9) and oscillators (Chapter 14) can be built using the MOSFET with no dc bias required for the gate. Bias at $V_{GS} = 0$ is acceptable.

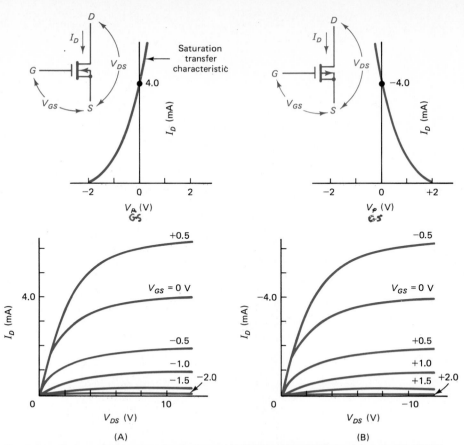

Fig. 8-21 Characteristic curves for depletion MOSFETs. (A) NMOS (n-channel). (B) PMOS (p-channel). As shown by the schematic symbols, the substrate is usually connected to the source, so that the pn junction between the substrate and channel is reverse-biased (see Figs. 8-27 and 8-28 for reference).

As will be seen below, the transfer characteristic for the D-MOSFET can be described using the JFET characteristic [Eq. (8-10)] extended to include both forward and reverse biases across the gate-channel junction. This extension of the JFET characteristic curve can be expected because the D-MOSFET also determines the current flow through itself by controlling the width of the depletion regions in the device. The electric field created by the gate-insulator combination is comparable to the field created by the ions in the depletion region, as shown in Fig. 8-22.

The potential difference V_{G-CH} between the gate and channel is due to a constant electric field in the insulator and a linearly decreasing electric field in the depletion region. (The latter relationship was derived in Section 4-6 and Exercise 4-12). Therefore,

$$V_{G-CH} = \int_{\text{Surface of conducting layer}}^{\text{Bottom of depletion region}} E \, dx = \left(\frac{qN_{CH}t_{CH}}{\epsilon}\right)\left(\frac{t_{CH}}{2} + t_{ox}\right) \quad (8\text{-}29)$$

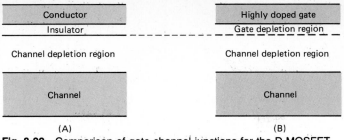

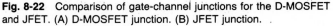

Fig. 8-22 Comparison of gate-channel junctions for the D-MOSFET and JFET. (A) D-MOSFET junction. (B) JFET junction.

where N_{CH} is the channel doping level, t_{CH} is the thickness of the depletion region in the channel, and t_{ox} is the thickness of the insulating oxide layer. From Fig. 4-11,

$$E_o = \frac{qN_{CH}t_{CH}}{\epsilon} \tag{8-30}$$

is the maximum value of the electric field along the boundary between the oxide and channel. Using the same model strategy as introduced above for the JFET, V_{G-CH} can be taken approximately equal to V_{GS} to establish asymptotic behavior in the transition region. Solving for t_{CH},

$$t_{CH} = \left(t_{ox}^2 + \frac{2\epsilon|V_{GS}|}{qN_{CH}}\right)^{1/2} - t_{ox} \tag{8-31}$$

If

$$t_{ox} \ll (2\epsilon|V_{GS}|/qN_{CH})^{1/2} \tag{8-32}$$

$$t_{CH} = (2\epsilon|V_{GS}|/qN_{CH})^{1/2} \tag{8-33}$$

which produces the same characteristic curve of Fig. 8-9 for the JFET [by comparison with Eqs. (8-3) and (8-4)]. This condition is associated with a thin oxide layer and most of the applied voltage V_{GS} appears across the depletion region.

If

$$t_{ox} \gg (2\epsilon|V_{GS}|/qN_{CH})^{1/2} \tag{8-34}$$

then

$$t_{CH} = t_{ox}\left(1 + \frac{2\epsilon|V_{GS}|}{qN_{CH}t_{ox}^2}\right)^{1/2} - t_{ox} \cong \frac{\epsilon|V_{GS}|}{qN_{CH}t_{ox}} \tag{8-35}$$

For this case, most of the applied voltage appears across the oxide, producing a field $|V_{GS}|/t_{ox}$ in the channel. For this latter case, Eqs. (8-3) and (8-4) become

$$W = W_o - 2K'|V_{GS}| \tag{8-36}$$

$$K' = \epsilon/qN_{CH}t_{ox} \tag{8-37}$$

and the equivalent to Eq. (8-6) takes the form

$$I_D \cong \frac{V_{DS}zW_o}{\rho L}\left(1 - \frac{V_{GS}}{V_P}\right) \tag{8-38}$$

where

$$|V_P| = (W_o/2K')^2 \tag{8-39}$$

Along the boundary $V_{DS} = -V_P(1 - V_{GS}/V_P)$,

$$I_D \cong I_{D0}(1 - V_{GS}/V_P)^2$$

$$I_{D0} = \frac{-V_P z W_o}{\rho L} \tag{8-40}$$

Since the insulating layer will not allow conduction between the gate and channel, this solution holds for both reverse- and forward-biased gate-channel junctions and is used here as an extension of Eq. (8-10) for D-MOSFET circuit design.

8-12 DEPLETION MOSFET AMPLIFIER

The depletion MOSFET can provide a dc-coupled amplifier that does not need either a source resistor R_S or a dc offset bias on the gate, as shown in Fig. 8-23. Since the depletion MOSFET can produce linear operation around $V_{GS} = 0$, this circuit has advantage of simplicity.

Fig. 8-23 (A) Circuit schematic, and (B) transfer characteristic for an n-channel D-MOSFET amplifier.

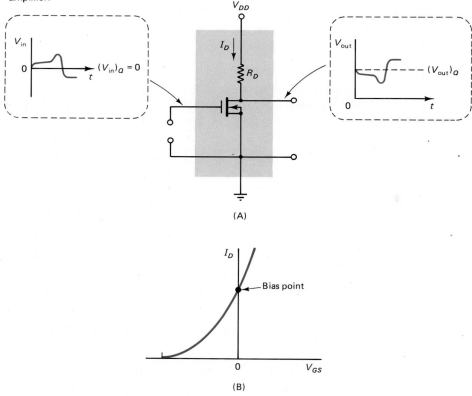

The operating point is shown in Fig. 8-18 at $I_D = I_{D0}$. From Eq. (8-22),

$$g_m \cong -2I_{D0}/V_P \tag{8-41}$$

at this point.

Example 8-5 **Depletion MOSFET Amplifier**

Design a dc-coupled amplifier with a gain of $A_V = -8.0$, using a depletion MOSFET with $I_{D0} = 8.0$ mA and $V_P = -4.0$ V.

Solution Values must be selected for R_D and V_D. From Eq. (8-41),

$$g_m \cong -2(8.0 \text{ mA})/-4.0 \text{ V} \cong 4.0 \text{ mʊ}$$

Then, from Eq. (8-20),

$$A_V = -R_D g_m$$

$$R_D \cong 8.0/4.0 \text{ mʊ} \cong 2.0 \text{ k}\Omega$$

V_{DD} must be selected to make sure the MOSFET stays in the linear region and to obtain the desired output voltage range. If $(V_{DS})_Q = 4$ V is chosen,

$$V_{DD} = I_D R_D + (V_{DS})_Q$$
$$= (8.0 \text{ mA})(2.0 \text{ k}\Omega) + 4 \text{ V} \cong 20 \text{ V}$$

The vertical intercept, limiting the transfer characteristic range, is

$$V_{DD}/R_D = 20 \text{ V}/2.0 \text{ k}\Omega = 10 \text{ mA}$$

Since $V_{DD}/R > I_{D0}$, no rounding off exists. ■

8-13 ENHANCEMENT OR E-MOSFETs

The second form of the MOSFET does not have a built-in channel, as shown in Fig. 8-24. The channel must be created by the applied field, which acts to pull carriers toward the insulating layer and thus create a significant carrier density. This *enhancement, or E-MOSFET*, has no drain current ($I_D = 0$) for negative values of V_{GS} and for positive values of V_{GS} up to a critical level V_T. This turns out to be a useful property for the design of digital circuits, as discussed in Chapter 15. V_T is the *threshold voltage*.

The operation of the n-channel enhancement MOSFET is shown in Fig. 8-24. For negative values of V_{GS}, holes in the p-type substrate are attracted toward the insulating layer. However, no current can flow since the drain-to-substrate pn diode is reverse-biased, as shown in Fig. 8-24(A).

For small positive values of V_{GS}, up to a threshold value V_T, the same condition holds. The p-region between the two n-regions prevents any current from flowing.

Then, as V_{GS} approaches V_T, a new situation develops, as shown in Fig. 8-24(B). The electric field becomes strong enough to push away the holes near the insulating layer, forming a depletion layer between the source and drain. Electrons from the n-type regions around the source and drain are drawn by the electric field into this extended depletion region, producing an n-channel as shown. For voltages

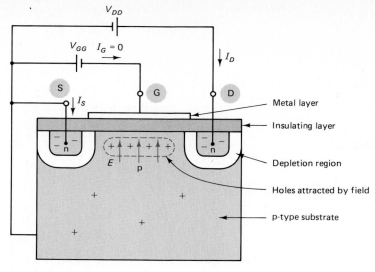

(A)

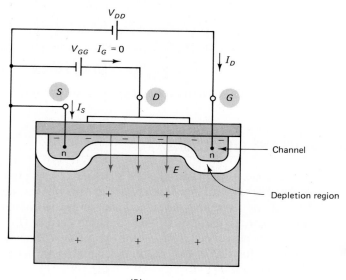

(B)

Fig. 8-24 N-channel enhancement or E-MOSFET. (A) V_{GS} negative (the depletion regions prevent conduction). (B) V_{GS} positive and greater than V_T (a channel forms, allowing conduction).

$|V_{GS}| > |V_T|$, the enhancement MOSFET has an increasing current flow as V_{GS} increases.

The output and transfer characteristic curves for n- and p-channel E-MOSFETs are shown in Fig. 8-25. Also shown are the schematic symbols that are used for this device.

A simple approximate relationship for the transfer characteristic of the E-

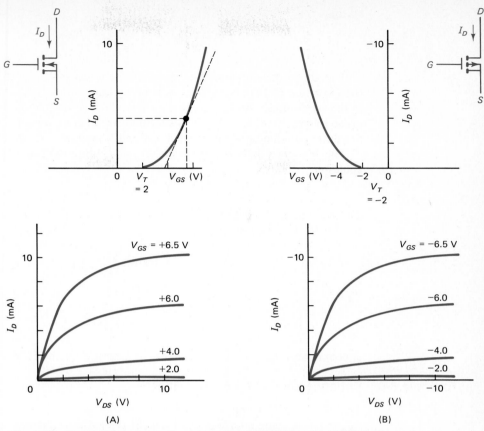

Fig. 8-25 Characteristic curves for enhancement MOSFETs. (A) NMOS (n-channel). (B) PMOS (p-channel).

MOSFET can be found by considering the channel to consist of a thin sheet of charge (of thickness Δt) below the insulating layer. For the moment, neglect the effect of V_{DS} on the channel and consider an n-channel device (V_{GS} and V_T both positive). Once a channel is formed ($V_{GS} > V_T$), the charge in this layer is given by

$$Q = \epsilon EA = \epsilon \frac{V_{GS} - V_T}{t_{ox}} A \qquad (8\text{-}42)$$

from Gauss' law (previously applied in Exercise 4-12, Chapter 4). The resistance of the channel is

$$R = \rho \frac{L}{z(\Delta t)} = \frac{L}{(N_{CH}q\mu)z\,\Delta t} \qquad (8\text{-}43)$$

Since

$$(N_{CH}q)zL\,\Delta t = Q \qquad (8\text{-}44)$$

$$R = \frac{L^2}{\mu Q} = \frac{Lt_{ox}}{\mu\epsilon(V_{GS} - V_T)z} \qquad (8\text{-}45)$$

so that

$$I_D = \frac{V_{DS}}{R} = \left(\frac{V_T V_{DS} \mu \epsilon z}{L t_{ox}}\right)\left(\frac{V_{GS}}{V_T} - 1\right) \tag{8-46}$$

The above development has neglected the effect of V_{DS} on the channel. However, the existence of the channel depends on creating an electric field greater than V_T/t_{ox} in the oxide layer. As V_{DS} increases, the voltage across the oxide layer at the drain is given by $(V_{GS} - V_T) - V_{DS}$. If this term becomes zero, a depletion region develops at the drain, producing channel pinch-off and a saturated output characteristic curve.

The boundary between the transition and saturation regions is thus given by the equation

$$V_{DS} = (V_{GS} - V_T) \tag{8-47}$$

The saturation characteristic curve is defined by Eq. (8-46) on the boundary given by Eq. (8-47). Therefore,

$$I_D = I_{D0}\left(\frac{V_{GS}}{V_T} - 1\right)^2 \tag{8-48}$$

with

$$I_{D0} = \frac{V_T^2 \mu \epsilon z}{L t_{ox}} \tag{8-49}$$

approximates the saturated transfer characteristic for the E-MOSFET.

Example 8-6   Enhancement MOSFET Amplifier

The objective for this example is to design a dc-coupled amplifier with a gain $A_V = -4.0$, using an n-channel E-MOSFET with the characteristic curves of Fig. 8-25(A) and the value $V_{GS} = 5.5$ V.

Solution   The schematic to be used is shown in Fig. 8-26. For an n-channel E-MOSFET, the gate must be made positive with respect to the source to produce a current flow.

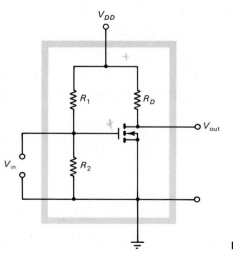

Fig. 8-26   E-MOSFET DC-coupled amplifier.

This is accomplished by the voltage divider $R_1 R_2$. (A similar strategy is used in Chapter 7 to forward bias a BJT for linear operation.)

With the input resistance of the source neglected, the gain of this amplifier is given by

$$A_V \cong -g_m R_D$$

From Fig. 8-20,

$$g_m \cong 4.0 \text{ mA}/2.0 \text{ V} \cong 2.0 \text{ m}\mho$$

Therefore, to obtain $A_V = -4.0$,

$$R_D = 4.0/2.0 \text{ m}\mho = 2.0 \text{ k}\Omega$$

and

$$V_{DD} = V_{DS} + (4.0 \text{ mA})(2.0 \text{ k}\Omega) = V_{DS} + 8.0 \text{ V}. \qquad \blacksquare$$

8-14 FABRICATION AND PACKAGING OF JFETS AND MOSFETS

JFETs are fabricated using the same epitaxial and diffusion process introduced for junction diodes and BJTs. Figure 8-27 shows a cross-sectional profile of the arrangement. The n-type epitaxial layer is like a "bucket," with a cylindrical p-type tube placed inside the bucket so that it almost touches the bottom. The two gate and two drain regions are joined in three dimensions. The channel is formed by the narrow n-type region that passes under the p-type cylinder. Isolation is maintained by reverse biasing the diode between the substrate and epitaxial layer.

Depletion MOSFETs are formed as shown in Fig. 8-28(A). A p-type diffusion is used around the edge of each device to achieve isolation between devices.

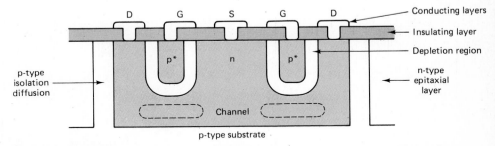

Fig. 8-27 Geometry for an n-channel JFET. The depletion region is used to open and close the channel.

An enhancement MOSFET is shown in Fig. 8-28(B). The gate is made sufficiently positive to form a depletion region, attract electrons into the channel, and allow a current flow to take place.

Figure 8-29 shows the metallization pattern for a discrete JFET whereas Fig. 8-30 illustrates several different types of discrete FET packages.

The tradeoffs involved in FET design and fabrication are discussed in Chapter 18. The relationships derived in this chapter are used in Chapter 18 to further describe the linkages between device operation and performance parameters.

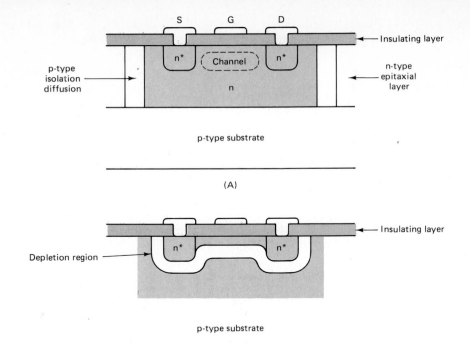

Fig. 8-28 MOSFET geometries. (A) N-channel depletion MOSFET (conducting). (B) N-channel enhancement MOSFET (conducting).

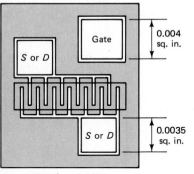

0.015 in. × 0.017 in.

Fig. 8-29 Metallization pattern of a JFET.

Fig. 8-30 Discrete FET packages. (Courtesy of Unitrode Corporation)

In the usual n-type semiconductor, the doping process adds both electrons and ions to the channel of an FET. The ions interact with the electrons and slow them down. However, the electrons and ions can be separated by placing an undoped layer next to the doped layer. The undoped layer has lower energy states available to the electrons. The ions stay locked in place while the electrons migrate to the layer with the lower energy states. The resulting electron-filled layer can then be modified by an electric field from a gate, as shown in Fig. 8-31. The resulting device is called a MODFET (modulation-doped FET) or HEMT (high-electron mobility transistor) and is discussed further in Chapter 15.

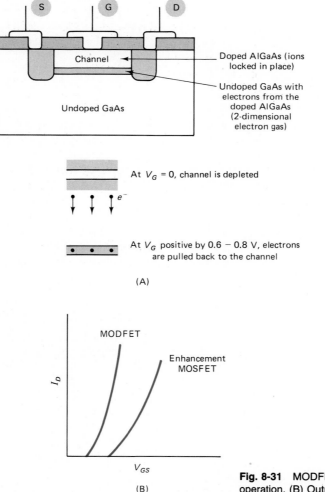

Fig. 8-31 MODFET (HEMT). (A) Functional operation. (B) Output characteristic.

Special MOSFET designs have evolved for high-power applications. The VMOS (vertical MOS) device shown in Fig. 8-32 makes use of a groove cut into the layered structure. Oxide and metal layers are then produced on the surface. An enhancement channel results when a positive voltage is applied to the gate. Due to this

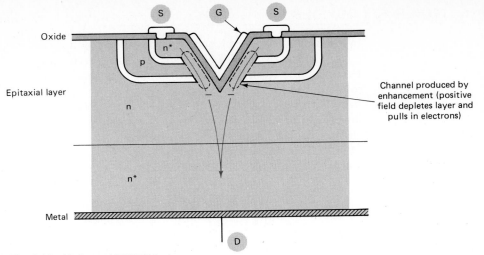

Fig. 8-32 N-channel VMOS device.

geometry, the channel is wide compared with its length, and the power-handling ability of the device is improved.

The VMOS device has been modified to a more U-shaped groove for ease of fabrication and to reduce the electric field at the apex of the groove. It is also possible to create a power FET by using a polycrystalline layer on the surface for a gate and overlaying the entire surface with a layer of metal for the source. This device is called a vertical, double-diffused MOS (VDMOS) and is shown in Fig. 8-33. Power MOS devices can handle over 10 A and 500 to 1000 V from source to drain.

Many other possible FET configurations are possible as materials, fabrication processes, and geometries are changed. Major effort is today being devoted to constructing FET devices from GaAs instead of Si. The advantages of GaAs include a smaller effective mass m^* for electrons (as shown in Fig. 2-6) that results in faster

Fig. 8-33 N-channel VDMOS device.

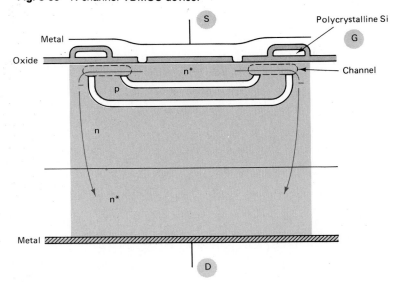

device speed and lower internal power dissipation. Fabrication difficulties have slowed the broad availability of GaAs FET devices, but this situation is improving rapidly. Depletion and enhancement versions of the MESFET are now being produced in quantity and HEMT versions are being pursued. The characteristics of GaAs-based devices are further discussed in Chapter 15.

8-16 OVERVIEW

Field-effect transistors have been explored by using several different dc bias methods and amplifier designs. In each case, the design strategy involves the simultaneous solution of several equations. By making use of the transfer characteristic representation, the design process has been simplified. Figure 8-34 illustrates the basic design

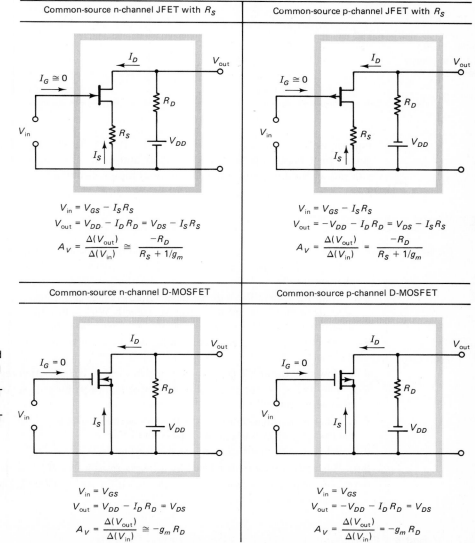

Fig. 8-34 DC-coupled amplifier designs using FETs. (A) Common-source n-channel JFET with R_S. (B) Common-source p-channel JFET with R_S. (C) Common-source n-channel D-MOSFET. (D) Common-source p-channel D-MOSFET. (Continued on p. 316.)

Common-source n-channel JFET with R_S

$$V_{in} = V_{GS} - I_S R_S$$
$$V_{out} = V_{DD} - I_D R_D = V_{DS} - I_S R_S$$
$$A_V = \frac{\Delta(V_{out})}{\Delta(V_{in})} \cong \frac{-R_D}{R_S + 1/g_m}$$

Common-source p-channel JFET with R_S

$$V_{in} = V_{GS} - I_S R_S$$
$$V_{out} = -V_{DD} - I_D R_D = V_{DS} - I_S R_S$$
$$A_V = \frac{\Delta(V_{out})}{\Delta(V_{in})} = \frac{-R_D}{R_S + 1/g_m}$$

Common-source n-channel D-MOSFET

$$V_{in} = V_{GS}$$
$$V_{out} = V_{DD} - I_D R_D = V_{DS}$$
$$A_V = \frac{\Delta(V_{out})}{\Delta(V_{in})} \cong -g_m R_D$$

Common-source p-channel D-MOSFET

$$V_{in} = V_{GS}$$
$$V_{out} = -V_{DD} - I_D R_D = V_{DS}$$
$$A_V = \frac{\Delta(V_{out})}{\Delta(V_{in})} = -g_m R_D$$

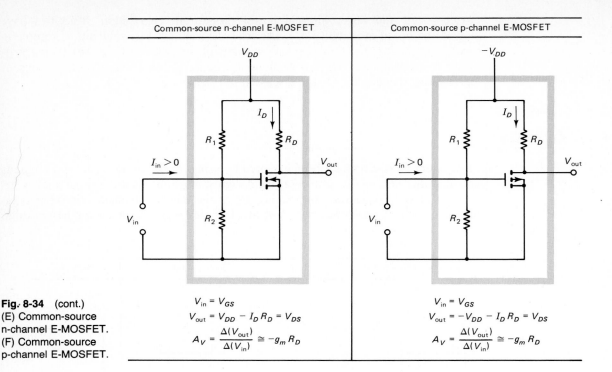

Common-source n-channel E-MOSFET

$V_{in} = V_{GS}$

$V_{out} = V_{DD} - I_D R_D = V_{DS}$

$A_V = \dfrac{\Delta(V_{out})}{\Delta(V_{in})} \cong -g_m R_D$

Common-source p-channel E-MOSFET

$V_{in} = V_{GS}$

$V_{out} = -V_{DD} - I_D R_D = V_{DS}$

$A_V = \dfrac{\Delta(V_{out})}{\Delta(V_{in})} \cong -g_m R_D$

Fig. 8-34 (cont.)
(E) Common-source
n-channel E-MOSFET.
(F) Common-source
p-channel E-MOSFET.

equations for n- and p-channel dc-coupled amplifiers using FETs and MOSFETS. Based on Chapters 6 to 8, both bipolar and field-effect transistors can now be considered as fundamental units for more complex circuits.

QUESTIONS AND ANSWERS

Questions

8-1. What general strategy can be used to find the direction of current flow in an FET?

8-2. Why is the transfer characteristic of an FET more useful for circuit design than the input characteristic?

8-3. In an FET amplifier with an input signal coupled in through a capacitor, a large resistance is often placed from the gate to ground. Why is this an important aspect of circuit design with such devices? How may the resistance value be defined?

8-4. Why are the characteristic curves for the JFET not highly sensitive to the exact form of $A(x)$ in Eq. (8-1)?

8-5. How may the JFET performance be described in the transition region?

8-6. Why is Eq. (8-10) a logical extension of Eq. (8-9) for a depletion region with a more gradual boundary?

8-7. What is the importance of the transconductance of a transistor? What insight does this parameter give into design with BJTs and FETs?

8-8. What is the role of minority-carrier flow in FETs?

8-9. What tradeoffs are introduced by a source resistor in a JFET amplifier?

8-10. Under what conditions do the n-channel JFET and D-MOSFET have the same characteristic curve for $V_{GS} < 0$?

8-11. What is the relationship between the characteristic curves for the JFET and D-MOSFET?

8-12. What is the advantage of using an FET as an active load for an amplifier?

8-13. What is the advantage of using a D-MOSFET for designing a dc-coupled amplifier?

8-14. Under what conditions does the E-MOSFET channel become pinched off?

8-15. What is the advantage of using an E-MOSFET for designing a dc-coupled amplifier?

8-16. What important advantages of FET amplifiers have been introduced in this chapter?

Answers

8-1. Carrier flow in an FET is always from source to drain. For an n-channel device, with electrons as carriers, the resulting current is into the drain and therefore positive. A positive supply voltage is required. For a p-channel device, with holes as carriers, the current is out of the drain and therefore negative. A negative supply voltage is required.

8-2. The input characteristic of a JFET displays a small leakage current as a function of the reverse-bias voltage across the pn junction, and, therefore, is of little use in determining the device operating point. The dc input current to a MOSFET is almost zero, due to the effect of the insulating layer between the gate and channel, and is also of little use in determining the device operating point.

8-3. Due to the small gate currents associated with FETs, an electrostatic charge can build up on the gate unless a dc pathway to ground is provided. If the path through the input voltage is blocked (by a coupling capacitor), a resistor must be added from the gate to ground to prevent erratic device behavior. The resistance should be made as large as possible subject to the constraint that the voltage drop across the resistor due to leakage current should be small with respect to the smallest input voltage variations that are to be amplified. (This consideration is further discussed in Chapter 9.)

8-4. As given in Eq. (8-1), the characteristic curves for the JFET depend on the integral of the function $A(x)$ and are thus not highly sensitive to the exact form of this function.

8-5. In the transition region, the JFET functions as a voltage-controlled variable resistor, with V_{GS} and V_{DS} determining the resistance of the channel.

8-6. For a depletion region with a more gradual boundary, the resistance of the channel decreases more rapidly than the function $1 - (V_{GS}/V_P)^{1/2}$ of Eq. (8-9). The function $1 - V_{GS}/V_P$ used in Eq. (8-10) has the desired property, and thus represents a modification in the expected direction.

8-7. The transconductance g_m describes the ability of a device to convert an input voltage waveform to an output current waveform. To maximize the current change associated with a changing input voltage, g_m should be maximized. For the JFET, the voltage gain is given by the product of the transconductance capability of the device times the value of the output resistor being used. Similar relationships can be formulated for BJTs. As discussed in Chapter 9, g_m is an important parameter for both BJTs and FETs. For many cases, g_m is much larger for bipolar devices than for field-effect devices. Therefore, the BJT often has the inherent capability for producing a higher voltage gain in a dc-coupled amplifier (for a given supply voltage and load resistor).

8-8. As discussed in Chapter 6, BJTs can be regarded as minority-carrier devices. The carriers injected from emitter to base become minority carriers in the base, then again become majority carriers when they enter the collector. Minority carriers also play an important role in reverse current flow in the BJT. As discussed in this chapter, FETs are majority-carrier devices. Current flow through the channel consists of majority carriers. Minority-carrier flow affects the leakage flow in the JFET, but does not enter into basic device operation.

8-9. A source resistor in a JFET amplifier can be used to improve temperature stability and produce the required Q-point bias for the gate without an additional supply voltage. In turn, the introduction of the source resistor reduces the voltage gain of the amplifier.

8-10. The characteristic curves for the JFET and D-MOSFET are the same in the transition region ($V_{GS} < 0$) if the oxide layer in the D-MOSFET is much thinner than the depletion region in the channel.

8-11. The characteristic curves for the JFET and D-MOSFET are approximately the same for $V_{GS} < 0$. For $V_{GS} \geq 0$, the D-MOSFET characteristic represents an extension of the parabolic form

$$I_D = I_{D0}(1 - V_{GS}/V_P)^2$$

8-12. Since the ac resistance of an FET in the saturation region is very high, the gain of the amplifier can be increased without resort to very high supply voltages that are difficult to control and that can result in device breakdown.

8-13. Since a D-MOSFET can be operated around a zero gate-source bias level, such an amplifier requires neither an extra voltage source (needed for a fixed-bias JFET) nor a source resistor (needed for a self-bias JFET).

8-14. The E-MOSFET channel becomes pinched off when the voltage across the oxide layer at the drain becomes zero.

8-15. The required bias for an E-MOSFET amplifier can be obtained by an input voltage divider across the gate, eliminating the need for an extra voltage source.

8-16. FETs require only very small input currents and are therefore used to produce isolation between the input and output of an amplifier. And because design emphasizes the transfer and output characteristics, the equations and procedures are simplified.

EXERCISES AND SOLUTIONS

Exercises
(8-3)

8-1. For a JFET, $I_{D0} = 6.0$ mA and $V_P = -5.5$ V. (a) What voltage V_{GS} must be used to produce $I_D = 4.0$ mA? (b) Find the transconductance for the JFET at this Q point.

(8-3) **8-2.** Find the ac resistance of the output characteristic curve of Fig. 8-4(B) at point D.

(8-3) **8-3.** Refer to Fig. 8-4(B). For $V_{GS} = -1.0$ V, what is the smallest value of V_{DS} that can be used to maintain this JFET in linear operation? What current I_D flows?

(8-3) **8-4.** Show that

$$g_m = -(2/V_P)(I_D I_{D0})^{1/2}$$

(8-3) **8-5.** How can the previous study of semiconductor materials and devices provide insight into the operation of the JFET in the saturated region?

(8-3) **8-6.** An n-channel Si JFET is designed with $L = 5.0$ μm, $z = 20$ μm, $N_{CH} = 1.0 \times 10^{22}/m^3$, and $W_o = 1.0$ μm. (a) Find the resistance for the characteristic curve $V_{GS} = 0$ near the origin. (b) Find the pinch-off voltage.

(8-3) **8-7.** Discuss how a gradual profile in the channel can contribute to the "rounding off" of the JFET output characteristic between the transition and saturation regions.

(8-5) **8-8.** Find the possible range of values for g_m for the J210, J211, J212, J270, and J271 FETs.

(8-5) **8-9.** For the J210 JFET, what is the maximum temperature at which the device can be operated to dissipate 100 mW?

(8-6) **8-10.** You are given the JFET characteristic curve shown in Fig. 8-4 and the Q point at D. Design a common-source fixed-bias dc-coupled amplifier with a gain $A_V = -12$.

(8-6) **8-11.** If a drain resistor $R_D = 2.0$ kΩ is used, find A_V for a common-source JFET dc-coupled amplifier with the transfer characteristic shown in Fig. 8-4, biased at $V_{GS} = -3.0$ V (in the saturation region).

(8-7)   **8-12.** Design a self-biased common-source JFET dc-coupled amplifier to achieve a gain $A_V = -5.0$ using the characteristic curve in Fig. 8-5 with $V_{GS} = 1.0$ V as the bias point.

(8-12) **8-13.** An n-channel D-MOSFET is designed using the values in Exercise 8-6. What current flows in the channel at saturation if $V_{GS} = 0$?

(8-12) **8-14.** A D-MOSFET dc-coupled amplifier is designed to make use of a device with $g_m = 3.5$ m℧ and $V_P = -3.0$ V(at $V_{GS} = 0$). (a) What I_{D0} is required? (b) If I_{D0} for a selected device type can vary by ± 20 percent, what percentage change is observed in g_m?

(8-13) **8-15.** For an n-channel E-MOSFET, $V_T = 2.0$ V, $z = 10$ μm, $L = 5.0$ μm, and $t_{ox} = 1.0$ μm. What current flows in the channel for $V_{GS} = 4.0$ V?

(8-13) **8-16.** Design a dc-coupled amplifier with a gain $A_V = -2.5$ using the p-channel E-MOSFET of Fig. 8-25(B) and the schematic of Fig. 8-26.

(8-13) **8-17.** For the amplifier of Fig. 8-26, R_1 and R_2 are selected so that the E-MOSFET is biased at $V_{GS} = 4.0$ V. When $V_{in} = 5.0$ V, 10 μA flows through R_2. Find R_1 if $V_{DD} = 12$ V.

(8-13) **8-18.** The transfer characteristic of an E-MOSFET is to be approximated by a curve of the form

$$I_D = A + BV_{GS} + CV_{GS}^2$$

(a) Find expressions for A, B, and C in terms of previously established device parameters. (b) Find an expression for g_m.

8-19. A company can produce an amplifier circuit using a JFET or a depletion MOSFET. To produce the JFET circuit, the company has to invest $8,000 in equipment, and each circuit costs $18.00 to fabricate. To produce the MOSFET circuit, the company has to invest $12,000 in equipment, and each circuit costs $12.00 to produce. (a) What is the break-even circuit volume for which costs are equal in either case? (b) If volume is expected to be greater than this level, which device should be selected?

Solutions **8-1.** From Eqs. (8-10) and (8-12),

(a)
$$I_D = I_{D0}(1 - V_{GS}/V_P)^2$$

$$V_{GS} = V_P\left[1 - \left(\frac{I_D}{I_{D0}}\right)^{1/2}\right]$$

$$= (-5.5\text{ V})\left[1 - \left(\frac{4.0\text{ mA}}{6.0\text{ mA}}\right)^{1/2}\right] \cong -1.0\text{ V}$$

(b)
$$g_m = \frac{-2I_{D0}}{V_P}\left(1 - \frac{V_{GS}}{V_P}\right)$$

$$= \frac{(-2)(6.0\text{ mA})}{-5.5\text{ V}}\left(1 - \frac{1.0\text{ V}}{5.5\text{ V}}\right) \cong 1.8\text{ m℧}$$

8-2. As shown in Fig. 8-35,

$$r \cong 10\text{ V}/0.20\text{ mA} \cong 50\text{ k}\Omega$$

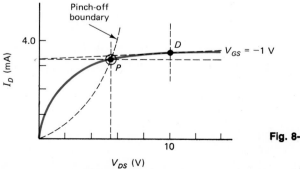

Fig. 8-35 Solution for Exercise 8-2.

8-3. From Fig. 8-35, to maintain $V_{GS} = -1.0$ V and stay in the linear (saturated) region, the smallest value of $V_{DS} \cong 5.0$ V and $I_D \cong 3.2$ mA.

8-4. From Eq. (8-12),

$$g_m = \frac{-2I_{D0}}{V_P}\left(1 - \frac{V_{GS}}{V_P}\right)$$

and from Eq. (8-10),

$$V_{GS}/V_P = [1 - (I_D/I_{D0})^{1/2}]$$

By substitution,

$$g_m = -(2/V_P)(I_D I_{D0})^{1/2}$$

8-5. The pinch-off process occurs when the gate-channel depletion regions almost touch in the middle of the channel. The current I_D flows through a narrow conducting region near the center of the JFET channel. As $V_{DS} + V_{GS}$ increases so that pinch-off begins, a high resistance develops in the pinch-off region and a rapidly growing electric field is created in the region.

For high electric fields, the drift velocity is no longer given by

$$V_d = \mu E$$

from Chapter 2 [Eq. (2-7)]. The drift velocity saturates and becomes approximately independent of E. As this change in mobility occurs, the current flow through the pinch-off region becomes approximately constant. The conducting portions of the pinch-off region have a finite width and length and inside the high electric field produces saturated current flow levels.

An increase in $V_{DS} + V_{GS}$ increases the length of the depletion region along the channel axis and the electric field across the region, but the drift velocity remains approximately the same as does the current I_D. Thus, to the first order, the output characteristic curves for the JFET are approximately independent of V_{DS} in the saturation region. The shift in μ from the transition region value (where it is approximately constant) to the saturation region value (where it is proportional to $1/E$) contributes to the rounded characteristic curve between the two regions.

A more careful examination of the output characteristic curves for the FETs in the active region reveals that they, like the BJTs, show a finite output resistance r. This can be expected because the increasing values of V_{DS} expand the length of the depletion region along the channel axis, thus reducing the effective channel length L. In the saturation (active) region, Eq. (8-2) becomes

$$I_D \cong \frac{(V_{DS})_{\text{boundary}}}{\rho \dfrac{L}{A_{\text{boundary}}}}$$

As the effective channel length L decreases, I_D increases, producing the observed finite output resistance r. It can be shown (Gray and Meyer, 1984) that an equation of the form

$$r \cong V_A/I_D$$

can be used to estimate r, given I_D, in analogy with the Early voltage calculation introduced in Exercise 6-8 of Chapter 6. Typical values are $V_A \cong 30$ to 200 V.

8-6. (a) From Eq. (8-6), with $V_{GS} = 0$,

$$I_D = \frac{V_{DS}zW_o}{\rho L}$$

so that

$$R = \frac{V_{DS}}{I_D} = \frac{\rho L}{z W_o} = \frac{L}{z W_o (N_{CH} q \mu_n)}$$

$$= \frac{5.0 \times 10^{-6} \text{ m}}{(20 \times 10^{-6} \text{ m})(1.0 \times 10^{-6} \text{ m})(1.0 \times 10^{22}/\text{m}^3)(1.6 \times 10^{-19} \text{ C})(0.14 \text{ m}^2/\text{V} \cdot \text{s})}$$

$$\cong 1.1 \text{ k}\Omega$$

(b) From Eq. (8-7),

$$V_P = \frac{W_o^2 q N_{CH}}{8\epsilon}$$

$$= \frac{(1.0 \times 10^{-6} \text{ m})^2 (1.6 \times 10^{-19} \text{ C})(1.0 \times 10^{22}/\text{m}^3)}{8(12 \times 8.85 \times 10^{-12} \text{ F/m})}$$

$$\cong 1.9 \text{ V}$$

8-7. The slopes of the output characteristic curves in the transition region result from a smaller V_{DS}/V_P value for a given I/I_{D0}. The effective resistance of the channel for small V_{DS} is then reduced for this modified case. In Fig. 8-9, the slope of the $V_{GS} = 0$ line in the transition region changes from I_{D0}/V_P to I_{D0}/V_P', so the resistance is reduced by the ratio V_P/V_P'. From Eq. (8-7), the baseline width W_o of the channel must increase by the ratio W_o'/W_o:

$$W_o' = W_o \left(\frac{W_o'}{W_o}\right) = W_o \sqrt{\frac{V_P'}{V_P}}$$

The saturation transfer characteristic does not change because the old and new boundaries intercept the $V_{GS} = $ constant curves at the same values of I_D/I_{D0}.

As might be expected, a more gradual depletion region profile results in the same device current levels only if the baseline width W_o of the channel is increased (providing a larger V_P and I_{D0}) to compensate for the additional channel resistance.

8-8. From Figs. 8-12(A) and (B), the data of Fig. 8-36 are found.

8-9. From Fig. 8-8(A), 350 mW can be dissipated at 25°C ambient temperature and the derating factor is 3.5 mW/°C. Therefore

$$P(T) = P(T_o) - (\text{derating factor})(T - T_o)$$

$$T - T_o = \frac{350 \text{ mW} - 100 \text{ mW}}{3.5 \text{ mW/°C}} \cong 71°C$$

$$T \cong 25°C + 71°C = 96°C$$

FIGURE 8-36
Solution data for Exercise 8-8

JFET Type	g_m Range (m℧)
J210	4–12
J211	7–12
J212	7–12
J270	6–15
J271	8–18

8-10. At point D in Fig. 8-4(B), $I_D \cong 3.2$ mA. At this same current level in Fig. 8-4(A), the slope g_m can be found as shown in Fig. 8-37.

$$g_m \cong 5.5 \text{ mA}/2.3 \text{ V} \cong 2.4 \text{ m}\mho$$

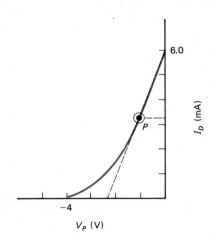

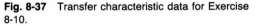

Fig. 8-37   Transfer characteristic data for Exercise 8-10.

From Eq. (8-20),

$$A_V = -g_m R_D$$

so that

$$R_D = -12/2.4 \text{ m}\mho \cong 5.0 \text{ k}\Omega$$

and

$$V_{DD} \cong 10 \text{ V} + I_D R_D$$

$$\cong 10 \text{ V} + (3.2 \text{ mA})(5.0 \text{ k}\Omega) \cong 26 \text{ V}$$

which may well be too large, requiring a further design iteration.

8-11. At $V_{GS} = -3.0$ V, the slope g_m is found as shown in Fig. 8-38.

$$g_m \cong 2.8 \text{ mA}/3.8 \text{ V} \cong 0.74 \text{ m}\mho$$

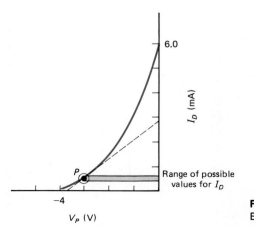

Range of possible
values for I_D

Fig. 8-38   Transfer characteristic data for Exercise 8-11.

Therefore, from Eq. (8-20),

$$A_V \cong -g_m R_D = -(0.74 \text{ m}\mho)(2.0 \text{ k}\Omega) \cong -1.5$$

8-12. From Fig. 8-5,

$$g_m \cong 2.3 \text{ mA}/1.8 \text{ V} \cong 1.3 \text{ m}\mho$$

at the operating point $V_{GS} = 1.0$ V.

Further, to achieve the desired Q point,

$$R_S = -V_{GS}/I_D \cong 1.0 \text{ V}/1.0 \text{ mA} = 1.0 \text{ k}\Omega$$

from Eq. (8-23). Therefore, from Eq. (8-26),

$$A_V = \frac{-R_D}{R_S + 1/g_m}$$

$$R_D = -(-5.0)(1.0 \text{ k}\Omega + 0.8 \text{ k}\Omega) \cong 9.0 \text{ k}\Omega$$

To maximize the linear operating range, let $(I_D)_{max}$ for the load line be -5.0 mA. Then,

$$(I_C)_{max} = \frac{V_{CC}}{R_D + R_S}$$

$$V_{CC} = (5.0 \text{ mA})(9.0 \text{ k}\Omega + 1.0 \text{ k}\Omega) \cong 50 \text{ V}$$

In many cases, such a high value for V_{CC} is not feasible. If this is so, R_D and V_{CC} must be reduced together to keep the same Q point, and the amplifier gain will drop.

8-13. From Eq. (8-40),

$$I_D = I_{D0}(1 - V_{GS}/V_P)^2$$

where, from Eq. (8-8),

$$I_{D0} = \frac{V_P z W_o}{\rho L}$$

$$= \frac{(1.9 \text{ V})(20 \times 10^{-6} \text{ m})(1.0 \times 10^{-6} \text{ m})(1.0 \times 10^{22}/\text{m}^3)(1.6 \times 10^{-19}\text{C})(0.14 \text{ m}^2/\text{V} \cdot \text{s})}{5.0 \times 10^{-6} \text{ m}}$$

$$\cong 1.7 \text{ mA}.$$

so that at $V_{GS} = 0$, $I_D = I_{D0} = 1.7$ mA.

8-14. (a) From Eq. (8-22),

$$g_m \cong -2I_{D0}/V_P$$

$$I_{D0} \cong \frac{-g_m V_P}{2} = \frac{(-3.5 \text{ m}\mho)(-3.0 \text{ V})}{2} \cong 5.3 \text{ mA}$$

(b) Since $g_m \propto I_{D0}$ for bias at $V_{GS} = 0$, a ± 20 percent change is also observed in g_m.

8-15. From Eq. (8-49),

$$I_{D0} = \frac{(2.0 \text{ V})^2(0.14 \text{ m}^2/\text{V} \cdot \text{s})(12 \times 8.85 \times 10^{-12} \text{ F/m})(10 \times 10^{-6} \text{ m})}{(5.0 \times 10^{-6} \text{ m})(1.0 \times 10^{-6}\text{m})}$$

$$\cong 0.12 \text{ mA}$$

and, from Eq. (8-48),

$$I_D = (0.12 \text{ mA})\left(\frac{4.0}{2.0} - 1\right)^2 = 0.12 \text{ mA}$$

8-16. Choose $V_{GS} = -6$ V. Then from Fig. 8-20(B),

$$g_m \cong 6.0 \text{ mA}/1.5 \text{ V} \cong 4.0 \text{ m}\mho$$

To achieve a gain of -2.5,

$$R_D \cong 2.5/4.0 \text{ m}\mho \cong 0.63 \text{ k}\Omega$$

The voltage across R_D at the Q point is then

$$I_D R_D = (6.0 \text{ mA})(0.63 \text{ k}\Omega) \cong 3.8 \text{ V}$$

With $(V_{DS})_Q \cong -6$ V to achieve linear operation,

$$V_{DD} \cong 9.8 \text{ V}$$

With $V_{GS} = -6$ V, the voltage divider $R_1 R_2$ must then satisfy the relationship

$$\left(\frac{R_2}{R_1 + R_2}\right)V_{DD} = 6.0 \text{ V}$$

If $R_2 = 1.0 \text{ M}\Omega$,

$$R_1 = (1.0 \text{ M}\Omega)\left(\frac{9.8 \text{ V}}{6.0 \text{ V}} - 1\right) \cong 0.63 \text{ M}\Omega$$

8-17. For the data given,

$$R_2 = 5.0 \text{ V}/10 \text{ }\mu\text{A} \cong 0.5 \text{ M}\Omega$$

$$\left(\frac{R_2}{R_1 + R_2}\right)12 \text{ V} = 4.0 \text{ V}$$

$$R_1 = (0.5 \text{ M}\Omega)\left(\frac{12 \text{ V}}{4.0 \text{ V}} - 1\right) \cong 1.0 \text{ M}\Omega$$

8-18. From the characteristic curve for the E-MOSFET,

$$I_D = I_{D0}\left(\frac{V_{GS}^2}{V_T^2} - 2\frac{V_{GS}}{V_T} + 1\right)$$

so that

$$A = I_{D0}$$

$$B = \frac{-2I_{D0}}{V_T}$$

$$C = \frac{I_{D0}}{V_T^2}$$

and

$$g_m = \frac{\partial I_D}{\partial V_{GS}} = B + 2CV_{GS}$$

$$= \frac{-2I_{D0}}{V_T} + 2\frac{I_{D0}}{V_T^2}V_{GS} = \frac{2I_{D0}}{V_T}\left(\frac{V_{GS}}{V_T} - 1\right)$$

8-19. (a) Let N = number of circuits produced. Then,

For JFET: $\qquad\qquad\qquad$ Cost = \$8,000 + \$18N

For MOSFET: $\qquad\qquad\quad$ Cost = \$12,000 + \$12N

Break-even occurs when

$$\$8,000 + \$18N = \$12,000 + \$12N$$

$$N = 667$$

(b) For production levels above 667, the MOSFET circuit is preferred.

PROBLEMS

(8-3) **8-1.** For a JFET, $V_{GS} = -2.0$ V, $V_P = -6.0$ V, $V_{CE} = 4.0$ V, $V_{DS} = 4.0$ V, and $I_{D0} = 4.8$ mA. (a) Locate the Q point on the output and transfer characteristic curves. (b) Find g_m at the Q point.

(8-3) **8-2.** Show that

$$I_D \cong -\frac{g_m V_P}{2}\left(1 - \frac{V_{GS}}{V_P}\right)$$

From this equation, show that V_P and I_D (and, therefore, I_{D0}) are always of opposite sign.

(8-3) **8-3.** Given $I_D = -6.0$ mA and $I_{D0} = -8.5$ mA, find V_{GS} if $V_P = 4.0$ V.

(8-3) **8-4.** Using the definition of Eq. (8-12), find g_m for the saturation transfer characteristic of Fig. 8-4 at $V_{GS} = -2.0$ V.

(8-3) **8-5.** Find I_D and the transconductance g_m for a JFET with $V_P = 6.0$ V and $I_{D0} = -12$ mA at the Q point $V_{GS} = 4.5$ V.

(8-4) **8-6.** In Fig. 8-11, find g_m for $T = T_1$ and $T = T_2$ at $V_{GS} = -2.0$ V. What happens to the value of g_m near the $V_{GS} = 0$ axis as the temperature increases?

(8-5) **8-7.** Compare the pn junction leakage currents for the J210–12 and J270–71 devices. From this perspective, which type of JFET more closely approaches having ideal properties?

(8-5) **8-8.** Compare the values of V_P for the J210–12 and J270–71 devices.

(8-5) **8-9.** Using the approximation of Eq. (8-10), plot the possible range of transfer characteristics for the J210–12 and J270–71 JFETs on the same axes. Discuss the relative design uncertainties for each case.

(8-5) **8-10.** Prepare a (saturation region) transfer characteristic for the 2N5457 JFET [Fig. 8-13(A)].

(8-5) **8-11.** Prepare a set of (transition and saturation region) transfer characteristics for the JFET shown in Fig. 8-13(B).

(8-5) **8-12.** The maximum power dissipation for a JFET is 300 mW at 25°C. If the derating factor is 3.0 mW/°C, (a) find the maximum power that can be dissipated at 100°C, and (b) find the maximum temperature that can be used (for which the allowable dissipation is zero).

(8-6) **8-13.** A JFET $I_{D0} = -4.8$ mA, $V_P = 5.5$ V, and $V_{GS}/V_P = 0.60$ is used in an amplifier with $R_D = 1.5$ kΩ and fixed bias. Find the voltage gain A_V of the amplifier.

(8-6) **8-14.** For a specific JFET type, $I_{D0} = -3.5$ mA(± 30 percent) and $V_P = 3.0$ V(± 20 percent). If the bias point $V_{GS}/V_P = 0.40$ is used with $R_D = 1.0$ kΩ and fixed bias, what range of amplifier gain can result?

(8-6) **8-15.** You are given the JFET characteristic curve shown in Fig. 8-13(B). Design a common-source fixed-bias dc-coupled amplifier with a gain $A_V = -4.0$. Bias is in the middle of the linear (saturated) operating region.

(8-7) **8-16.** For a given JFET self-bias dc-coupled amplifier, $R_D = 1.8$ kΩ, $R_S = 0.60$ kΩ, and $g_m = 2.5$ m℧. (a) Find the gain A_V. (b) If g_m varies by ±20 percent among different devices, what percentage variation in A_V is observed?

(8-7) **8-17.** Design a self-biased common-source JFET dc-coupled amplifier using the characteristic curve of Fig. 8-13(B) to achieve the highest gain possible with a 12-V dc power supply. Allow for a maximum input signal amplitude of 1.0 V around the bias point.

(8-7) **8-18.** A self-biased common-source JFET dc-coupled amplifier uses a device for which I_{D0} and V_P are known within ±30 percent. By what percentage can the amplifier gain change for different devices if $R_S = 0.20$ kΩ, $V_{GS}/V_P = 0.50$, and $I_{D0}/V_P = -2.0$ m℧ (nominal)?

(8-7) **8-19.** Design a JFET dc-coupled amplifier with self-bias using $I_{D0} = 10$ mA, $V_P = -3.0$ V, and $(V_{GS})_Q = -1.0$ V to achieve $A_V = -2.0$. Find R_D and V_{DD} if $(V_{DS})_Q > 3.0$ V.

(8-8) **8-20.** A common-source dc-coupled amplifier uses an active load with an ac resistance of 25 kΩ. What range of gain A_V is possible using the J270 JFET?

(8-11) **8-21.** You are given a depletion MOSFET with $I_{D0} = -12$ mA and $V_P = 6.0$ V. Draw the schematic for a dc-coupled amplifier. Find the values of R_D and V_{DD} needed to achieve a gain of $A_V = -8.0$ if the bias point $V_{GS} = 0$ is selected. Assume $|V_{DS}| > 6.0$ V is required to assure saturated operation.

(8-11) **8-22.** Design a dc-coupled amplifier with a gain of $A_V = -4.0$ using a depletion MOSFET with $I_{D0} = -6.5$ mA and $V_P = 3.5$ V (at $V_{GS} = 0$).

(8-11) **8-23.** You are given a fixed-bias depletion MOSFET dc-coupled amplifier with $(V_{GS})_Q = -1.0$ V and $(I_D)_Q = 2.0$ mA. If $V_P = -5.0$ V and $R_D = 2.2$ kΩ, what is the gain A_V of the amplifier?

(8-13) **8-24.** Using the n-channel E-MOSFET of Fig. 8-25(A) and schematic of Fig. 8-26, design an ac amplifier with $V_{GS} = 6.0$ V and $V_{DD} = 10$ V. Bias is near the middle of the linear operating region. Choose a current of 50 μA to flow in R_1 and R_2.

COMPUTER APPLICATIONS

8-1. Figures 8-39 and 8-40 show a SPICE simulation of a common-source dc-coupled JFET amplifier. What is the gain of this amplifier as determined from the graphical results? How does this finding compare with the gain prediction using the methods of analysis of this chapter?

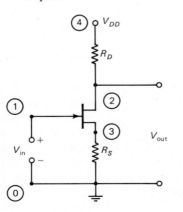

Fig. 8-39 SPICE simulation for a JFET dc-coupled amplifier.

```
     1      JFET D.C. AMPLIFIER
     2      *
     3      * SELF BIASING JFET D.C. AMPLIFIER
     4      *
     5      * RESISTORS
     6      *
     7      RS 3 0 620
     8      RD 4 2 1.2K
     9      *
    10      * N-CHANNEL JFET
    11      *
    12      J1 2 1 3 MOD1
    13      .MODEL MOD1 NJF(VTO=-4.2 BETA=230E-3 LAMBDA=1.0E-4 RD=100
    14      + RS=100 CGS=5PF CGD=1PF PB=0.6)
    15      *
    16      * INPUT VOLTAGE
    17      *
    18      VIN 1 0 PWL(0US 0 0.1US -1 20US -1 20.1US -3 30US -3 30.1US -0.5
    19      + 35US -0.5 35.1US 1 50US 1 50.1US -2 70US -2 70.1US 0)
    20      *
    21      * DC POWER SUPPLY
    22      *
    23      VDD 4 0 DC 12
    24      *
    25      * OUTPUT
    26      *
    27      .TRAN 1US 80US 0 0.1US
    28      .PRINT TRAN V(1) V(2)
    29      .PLOT TRAN V(1) V(2)
    30      .END
```

Note:
VTO = V_P
BETA = I_{D0}/V_P^2

Fig. 8-39 *(cont.)*

Fig. 8-40 Output waveform for a SPICE simulation of a JFET dc-coupled amplifier. Note that the effective capacitances associated with the JFET do not produce an observable effect for the time scale being used.

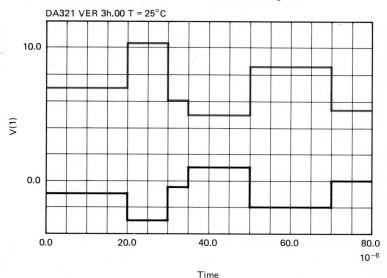

DA321 VER 3h.00 T = 25°C

Time

* 8-2. The purpose of this assignment is to develop a computer program that can assist in the design of self-biased common-source JFET amplifiers. Write a program with the device parameters (I_{D0}, V_P), the Q point [$(V_{DS})_Q$, $(I_D)_Q$], and the power supply voltage (V_{DD}) as input variables. The program output is A_V. Use the program to design an amplifier, selecting typical set of input variables. Turn in your program listing and example run.

* 8-3. Use SPICE to find the output waveform for the D-MOSFET amplifier of Fig. 8-23 using the device of Fig. 8-21, and the same R_D, V_{DD}, and input of Figs. 8-39 and 8-40.

EXPERIMENTAL APPLICATIONS

8-1. The objective of this experiment is to study JFET characteristic curves. Use a curve tracer or equivalent circuit to obtain the output characteristic, and then prepare the transfer characteristic. Determine I_{D0} and V_P for the device. Graphically compare the shape of the saturation transfer curve with a plot of the characteristic curve given by Eq. (8-10). Determine values for g_m at $V_{GS} = 0$ and $V_{GS} = V_P/2$ using both curves, and determine the percentage differences. Discuss and interpret your findings. Mark and retain your JFET for future experiments.

8-2. The objective of this experiment is to design, build, and test a JFET common-source dc-coupled amplifier using fixed bias so that $A_V = -2.0(\pm 10$ percent). Choose the operating point $V_{GS} = V_P/2$ and use standard resistor values. Record dc bias voltages at all nodes. Once these are confirmed as satisfactory, measure $\Delta V_{\text{out}}/\Delta V_{\text{in}}$. Evaluate the design strategy you have used and the relationships between predicted and experimental results.

8-3. The purpose of this experiment is to design, build, and test a JFET common-source dc-coupled amplifier using self-bias so that $A_V = -2.0(\pm 10$ percent). Choose the bias point $V_{GS} = V_P/2$ and $V_{DS} = 6.0$ V. Consider how the gain of your amplifier is affected if V_{DD} is gradually reduced from the design value to zero. Predict the approximate shape of the graph that results if A_V is plotted as a function of V_{DD}. Then, gather data to experimentally determine the same relationship. Compare the predicted and observed results.

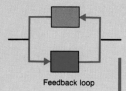

Feedback loop

PART 4
EQUIVALENT CIRCUITS, FREQUENCY RESPONSE, AND DISTORTION

Part 3 emphasizes the biasing of various semiconductor devices for linear operation and the design of dc-coupled amplifiers as a way to understand some of the fundamental properties of devices and circuits. For a dc-coupled amplifier, small changes in the input voltage or current around the input bias point are linearly related to the resultant changes in the output voltage and current. Such waveforms can vary widely over time, ranging from dc (fixed) signals to complex, rapidly changing waveforms. The operational amplifier, which is an important type of dc-coupled amplifier, is discussed in detail in Chapters 12 and 13.

In some applications, dc-coupled amplifiers are not the most effective way to meet performance objectives. When such an amplifier is connected to an input coupling circuit and drives an output coupling circuit, dc currents flow between the amplifier and the input and output circuits. These currents can undesirably affect the input and output circuits and can shift the dc operating point of the amplifier away from its design value. In many applications, there is a need to prevent dc interconnections between an amplifier and the surrounding circuit.

The needed isolation can be obtained by coupling input and output signals to the amplifier through capacitors. The result is an *ac-coupled amplifier,* often referred to simply as an ac amplifier. Since capacitors have been added to the circuit, the performance of the circuit depends on the frequency of operation as well as the dc-bias design. The coupling capacitors affect ac-amplifier performance at low frequencies and intrinsic device and wiring capacitances affect amplifier performance at high frequencies.

A variety of different equivalent circuit models can be developed to support the study of ac amplifiers. Chapter 9 describes methods for the study of such amplifiers in the midrange region, where capacitive effects are not important. Basic equivalent circuit models are developed to improve the understanding of circuit operation in this region.

Chapter 10 expands the discussion to consider the characteristics of ac amplifiers at low and high frequencies, where capacitive effects are important. Methods are developed for producing equivalent circuits that represent the frequency-dependent properties of an ac amplifier over all frequency ranges.

Chapter 11 introduces amplifier distortion that can arise due to frequency-dependent circuit performance, nonlinear amplification, and circuit noise sources. Several design strategies are discussed for use to achieve desired performance characteristics in the presence of realistic circuits in which distortion is a consideration.

Chapters 9 to 11 thus provide more multidimensional viewpoints of transistors and amplifier circuits. The shift from dc- to ac-coupled amplifiers brings with it a concern for many new circuit features, including frequency response and distortion. These chapters also provide important insights into the limitations associated with the devices themselves when they are used in amplifier circuits. Equivalent circuit models are developed to aid in this analysis, and a renewed concern develops over the realistic characteristics of R, C, and L functions that appear in circuits in discrete or distributed form. As will be noted, SPICE simulations can be a significant design aid as these more complex circuits are studied.

CHAPTER 9
MIDRANGE
AC AMPLIFIER DESIGN

Equivalent circuits are an important design aid. This chapter describes equivalent circuits that apply to both transistors and amplifier circuits operated to achieve linear ac amplification at midrange frequencies. These equivalent circuits are useful because, for small signal amplitudes, both devices and circuits can be treated as having approximately constant parameter values. The equivalent circuit is described in terms of fixed resistor values and dependent voltage and current sources.

As discussed in earlier chapters, these models must always be viewed as having constant elements only to a first-order approximation. Higher-order nonlinear terms exist and can be important. The value of these models must always be considered in the context set by the application and performance objectives.

Within the recognized limits, the modeling techniques developed here are quite useful. They enable the designer to quickly create easily remembered interpretive strategies for the operation of ac amplifier circuits. They provide a useful means for building the insights required for the design process. They provide building block concepts for evaluating circuit operation and encourage development of the pattern-recognition skills essential for more advanced applications.

The models described here introduce parameters and relationships that are high in information content, easy to remember, and useful for generalizing the results of previous chapters. Other types of parameters are also in common use, and the designer will naturally accommodate them through experience.

Much of Chapter 9 is concerned with developing additional insights regarding the design process for complex circuits. Important features of such circuits include the required linkages between the dc and ac aspects of circuit operation, the comparative advantages and disadvantages of alternative devices and circuit configurations, and the advantages and disadvantages of negative feedback as a design strategy. The chapter thus serves to bring together many of the concepts introduced previously into the design process.

When a CAD system is used for circuit design, it is essential for the designer to follow the calculations being revealed, understand the implications of the results, and provide directive input to the software system. The ability to conceptualize ac-circuit operation in terms of equivalent circuits and to develop an understanding for the integration patterns that exist among circuit elements results in the most effective computer-aided design procedure.

9-1 INTRODUCTION TO AC AMPLIFIERS

The objective of an ac amplifier is to take an input signal and amplify it to produce a similar output signal with a different amplitude. From Fourier analysis, any time-varying signal can be thought of as a combination of sine waves with different frequencies and amplitudes. Thus, an ac amplifier can be used to linearly amplify a variety of waveforms as long as the amplifier characteristics are independent of frequency and amplitude for the important Fourier components of the waveform.

Figure 9-1(A) shows a dc-coupled amplifier designed according to the procedures that have been developed. Figure 9-1(B) illustrates how an ac amplifier can be created by providing parallel dc and ac circuit inputs and outputs.

Capacitors C_1 and C_2, called coupling capacitors, isolate the dc currents flowing through the BJT and thus simplify the design task. The time-varying ac input signal passes through C_1 and the ac output signal passes through C_2, while the dc currents are blocked.

Capacitor C_E, called a bypass capacitor, can be added to prevent R_E from reducing the amplifier gain through negative feedback. In Chapter 7, Eq. (7-32) illustrates how the gain of a dc amplifier is reduced by the addition of an emitter resistor R_E. The addition of C_E does not affect the dc operating point of the amplifier. However, the ac signal is directed through C_E and is thus not affected by R_E.

The total ac amplifier operation involves a superposition of dc voltages and currents and ac voltages and currents. For small ac signals, these two design aspects can be treated separately, with carefully defined linkages to assure a consistent approach.

In order to understand how ac amplifiers respond to ac signals (modifying operation around the Q point), it is useful to derive appropriate *ac equivalent circuits*. In this chapter, equivalent circuits are derived for applications at *midrange* frequencies.

At such frequencies, all capacitances in the circuit can be replaced by either open circuits or short circuits. The reactances of the coupling capacitors C_1 and C_2, which are in series with the signal, are assumed sufficiently small at the signal frequency to have negligible effect on the circuit. The reactances associated with internal device capacitance and distributed wiring capacitance, which connect the signal to ground, are sufficiently large to have no effect on the circuit of interest. And if a bypass capacitor of the proper value is placed across R_E, as shown in Fig. 9-1(B), the reactance of C_E is much smaller than R_E. In Chapter 10, more complex equivalent circuits are derived for use with the above capacitances determining circuit performance.

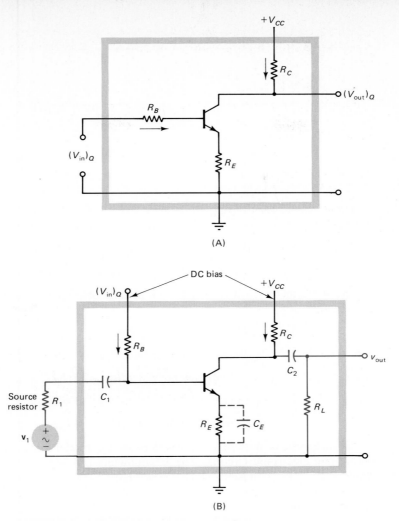

Fig. 9-1 (A) DC- and (B) ac-coupled amplifiers.

9-2 AC EQUIVALENT CIRCUITS

Figure 9-2(A) illustrates a general three-terminal device from the ac perspective. The variables of concern are no longer dc currents and voltages (V_i, I_i, V_o, I_o), but small-amplitude ac voltages and currents that vary the circuit operation around a selected Q point. An ac equivalent circuit provides a useful method for relating the *effective* values of the ac signal components (v_i, i_i, v_o, and i_o). By convention, voltage polarities are measured with respect to the common terminal and positive currents are defined as shown.

The three-terminal device of Fig. 9-2(A) is based on the definition of four ac variables and the device provides two equations relating the input and output. Therefore, there are always two independent variables, and two dependent variables are appropriate to describe device performance.

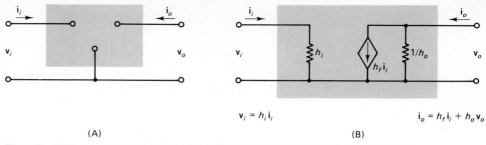

$$v_i = h_i\,i_i \qquad\qquad i_o = h_f\,i_i + h_o\,v_o$$

(A)   (B)

Fig. 9-2 (A) Input and output ac variables for a general three-terminal device or circuit. (B) AC equivalent circuit for the BJT.

Three sets of relationships in general use are (1) the z parameters, which relate the input and output voltages to the input and output currents:

$$\mathbf{v}_i \text{ (linear function of } \mathbf{i}_i \text{ and } \mathbf{i}_o) = Z_{11}\,\mathbf{i}_i + Z_{12}\,\mathbf{i}_o$$

$$\mathbf{v}_o \text{ (linear function of } \mathbf{i}_i \text{ and } \mathbf{i}_o = Z_{21}\,\mathbf{i}_i + Z_{22}\,\mathbf{i}_o$$

(9-1)

(2) the y parameters, which relate the input and output currents to the input and output voltages:

$$\mathbf{i}_i \text{ (linear function of } \mathbf{v}_i \text{ and } \mathbf{v}_o) = y_{11}\,\mathbf{v}_i + y_{12}\,\mathbf{v}_o$$

$$\mathbf{i}_o \text{ (linear function of } \mathbf{v}_i \text{ and } \mathbf{v}_o = y_{21}\,\mathbf{v}_i + y_{22}\,\mathbf{v}_o$$

(9-2)

and (3) the hybrid, or h, parameters, which use a mixed format:

$$\mathbf{v}_i \text{ (linear function of } \mathbf{i}_i \text{ and } \mathbf{v}_o) = h_i\,\mathbf{i}_i + h_r\,\mathbf{v}_o$$

$$\mathbf{i}_o \text{ (linear function of } \mathbf{i}_i \text{ and } \mathbf{v}_o) = h_f\,\mathbf{i}_i + h_o\,\mathbf{v}_o$$

(9-3)

All three sets of parameters are widely used to model the ac performance of transistors and electronic circuits. The h parameters provide a useful starting point for study of the BJT because the BJT characteristic curves are drawn by holding the input current at a constant dc value (I_i = constant, $\mathbf{i}_i = 0$) for the output characteristic and holding the output voltage at a constant dc value (V_o = constant, $\mathbf{v}_o = 0$) for the input characteristic.

9-3   THE h PARAMETERS

The physical significance of the h parameters can be found as follows. First, imagine a situation for which the ac output voltage is set to zero ($\mathbf{v}_o = 0$). Then, $\mathbf{v}_i = h_i\,\mathbf{i}_i$ and $\mathbf{i}_o = h_f\,\mathbf{i}_i$, so that

$$h_i = \left.\frac{\mathbf{v}_i}{\mathbf{i}_i}\right|_{\mathbf{v}_o = 0}$$

$$h_f = \left.\frac{\mathbf{i}_o}{\mathbf{i}_i}\right|_{\mathbf{v}_o = 0}$$

(9-4)

Parameter h_i is the ac input resistance of the device with the output voltage held constant, and h_f is the ac current gain of the device with the output voltage held constant.

Now, consider a situation for which the input current is held constant ($\mathbf{i}_i = 0$). Then, $\mathbf{v}_i = h_r \mathbf{v}_o$ and $\mathbf{i}_o = h_o \mathbf{v}_o$, so that

$$h_r = \left. \frac{\mathbf{v}_i}{\mathbf{v}_o} \right|_{\mathbf{i}_i = 0}$$

$$1/h_o = \left. \frac{\mathbf{v}_o}{\mathbf{i}_o} \right|_{\mathbf{i}_i = 0}$$

$(9\text{-}5)$

Parameter $1/h_o$ is the ac output resistance of the device with the input current held constant, and h_r measures the reverse ac voltage gain of the device. For many practical design applications, h_r can be taken to be approximately zero because a changing output voltage has only a slight effect on the input characteristic.

This assumption has been used in earlier chapters, where multiple input characteristic curves have been replaced by a single average curve. The assumption continues to be reasonable and is often used in this and later chapters. If the approximation does not hold, the ac equivalent circuits described here must be modified. By the above definitions, h_i and $1/h_o$ are always positive and h_f can be positive or negative.

In Chapter 7, an assumption that output characteristic curves were approximately flat resulted in setting $1/h_o$ so large that it could be neglected. As discussed below, this assumption is sometimes appropriate, but at other times, this term affects circuit performance.

The above definitions lead to the BJT ac equivalent circuit shown in Figure 9-2(B). Equation (9-3) (with $h_r = 0$) corresponds to the equivalent circuit arrangement as shown. Therefore, the representation of Fig. 9-2(B) can be used as an equivalent circuit for modeling the ac parameters of the BJT. As expected, h_i is the input resistance of the BJT, h_f is the current gain, and $1/h_o$ is the output resistance.

The positive directions of ac current flow are defined as shown in Fig. 9-2(B). For the common-base configuration, the ratio $\mathbf{i}_o/\mathbf{i}_i$ is negative, and for the common-emitter configuration, the ratio $\mathbf{i}_o/\mathbf{i}_i$ is positive.

The assumption that $h_r = 0$ results in an equivalent circuit consisting of an input portion and an output portion, with the dependent current generator $h_f \mathbf{i}_i$ the only connection between the two portions. The result is a useful approximate circuit that lends itself well to ac circuit design.

The h parameters take on different values depending on the BJT configuration used. A second subscript is added to each h parameter to indicate the configuration, so that h_{ie}, h_{fe}, and h_{oe} relate to a common-emitter configuration, and h_{ib}, h_{fb}, and h_{ob} relate to a common-base configuration.

By comparing the above results with definitions from earlier chapters, the identification $h_{fb} = \alpha_{ac} \cong -1$ can be made for a BJT in the common-base configuration and $h_{fe} = \beta_{ac}$ for a BJT in the common-emitter configuration.[1] Further, h_i can be associated with the ac input resistance r in each case.

The common-base output resistance $1/h_{ob}$ is so large that it is always taken to be negligible here. The magnitude of the common-emitter output resistance $1/h_{oe}$ can sometimes be important, so a method is needed to estimate its magnitude.

Based on a detailed analysis of BJT base narrowing effects (as presented in Exercise 6-8 in Chapter 6), the approximate model of Fig. 9-3 can be developed. The slopes of the common-emitter output curves are asymptotic to straight lines and all

---

[1] Also recall that the representation $\beta_{dc} = h_{FE}$ was introduced in Chapter 6.

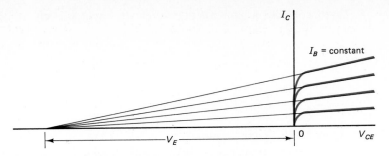

Fig. 9-3 Using the Early voltage to estimate $1/h_{oe}$.

of these lines intercept the V_{CE} axis at approximately the same point. The *Early voltage* V_E is the distance from the origin to this intercept point.

From Fig. 9-3,

$$1/h_{oe} \cong \frac{V_E}{|I_C|} \tag{9-6}$$

for each output characteristic. For typical devices, V_E is 50 to 100 V. Therefore, once a Q point is selected, I_C is known and $1/h_{oe}$ can be estimated. For most applications here, the value $V_E = 100$ V is used.

9-4 THE y PARAMETERS

Equation (9-3) for the h parameters can be rearranged in the format of Eq. (9-2) to describe the BJT in terms of y parameters. Continuing to let $h_r \cong 0$, Eq. (9-3) can be rewritten:

$$\mathbf{i}_i = \frac{1}{h_i}\mathbf{v}_i$$

$$\mathbf{i}_o = \frac{h_f}{h_i}\mathbf{v}_i + h_o\mathbf{v}_o \tag{9-7}$$

and the y parameters can be identified by comparison with Eq. (9-2).

The ratio h_f/h_i is defined as the *transconductance* of the BJT:

$$g_m = h_f/h_i \tag{9-8}$$

For a large output impedance ($h_o \cong 0$), g_m describes how the ac output current is related to the ac input voltage $\mathbf{v}_i$ appearing on the device input. For the common-base configuration, $g_{mb} = h_{fb}/h_{ib} \cong -1/h_{ib}$ and is negative, while for the common-emitter configuration, $g_{me} = h_{fe}/h_{ie}$ and is positive.

It is often possible to take $h_{fe} \cong h_{FE}$. For this case,[2]

$$g_{me} = \frac{h_{fe}}{h_{ie}} \cong \frac{h_{fe}}{0.026 \text{ V}/|I_B|} \cong \frac{|I_C|}{0.026 \text{ V}} \cong \frac{|I_E|}{0.026 \text{ V}} \tag{9-9}$$

---

[2] The same result can be obtained by performing the operation $g_m = (\partial I_E/\partial V_{BE})|_{V_{CB}}$ on the ideal diode characteristic equation.

Therefore, g_{me} can be estimated if $|I_C| \cong |I_E|$ is known. Both $1/h_{oe}$ and g_{me} are functions of the output current of the BJT. To the degree that the amplifier gain is dependent on these two variables, the gain varies with the output Q point.

Example 9-1   **Variation of BJT Parameters with Q Point**

(a) A BJT is biased at a Q point for which $I_C = 1.0$ mA. Estimate $1/h_{oe}$ and g_{me}.
(b) If the Q point shifts to 100 mA, estimate how these parameters change.

Solution   (a) For the given operating point,

$$1/h_{oe} \cong 100\,\text{V}/1.0\,\text{mA} \cong 1.0 \times 10^5\,\Omega$$

$$g_{me} \cong \frac{I_C}{0.026\,\text{V}} = \frac{1.0\,\text{mA}}{0.026\,\text{V}} \cong 38\,\text{m}\mho$$

(b) For the revised operating point,

$$1/h_{oe} \cong 100\,\text{V}/100\,\text{mA} \cong 1.0\,\text{k}\Omega$$

$$g_{me} \cong 100\,\text{mA}/0.026\,\text{V} \cong 3.8\ \mho$$

For circuits that use high impedance levels (such as the 741 IC operational amplifier discussed in Chapter 12), the $1/h_{oe}$ value is important even for $I_C = 1.0$ mA. For $I_C = 100$ mA, $1/h_{oe}$ is of the order of typical resistor values used in dc bias design. Since g_m is so sensitive to I_C, one possible strategy for providing high-g_m circuits is to reduce the dc current levels. These relationships are thus useful in developing circuit design strategies. ∎

9-5 THE Z PARAMETERS

Equation (9-3) can be rearranged in the form of Eq. (9-1) to describe the BJT in terms of the Z parameters. Continuing the above development,

$$\mathbf{v}_i = h_i \mathbf{i}_i$$

$$\mathbf{v}_o = \frac{h_f}{h_o}\mathbf{i}_i + \frac{\mathbf{i}_o}{h_o} \tag{9-10}$$

The results

$$Z_{11} = h_i \qquad \text{and} \qquad Z_{12} \cong 0 \tag{9-11}$$

$$Z_{21} = h_f/h_o \qquad \text{and} \qquad Z_{22} \cong 1/h_o \tag{9-12}$$

are obtained by comparison. Therefore, h_i represents the input impedance of the BJT with $\mathbf{i}_o = 0$, and $1/h_o$ represents the output impedance of the BJT with $\mathbf{i}_i = 0$. These concepts can be extended to entire amplifier circuits. For this use, let Z_i be the input impedance of the amplifier with $\mathbf{i}_o = 0$, and let Z_o be the output impedance of the amplifier with $\mathbf{i}_i = 0$.

When impedance concepts are extended for application to entire circuits, $\mathbf{i}_o = 0$ is interpreted as meaning that the circuit input has been terminated with an effective load R_1 and that no current flows past R_L into a subsequent circuit. Similarly, $\mathbf{i}_i = 0$ is interpreted as meaning that the circuit input has been terminated with the internal resistance R_I of the source in a similar way.

The impedance relationships that exist between a voltage source and the amplifier input, as shown in Fig. 9-4(A), and between the amplifier output and the equivalent circuit load, as shown in Fig. 9-4(B), are an important aspect of ac-amplifier design. The input impedance Z_i of this circuit is the Thevenin equivalent impedance of the circuit looking into the input, with the actual load replaced by equivalent load resistor R_L on the output. The output impedance Z_o of the circuit is the Thevenin equivalent impedance of the circuit looking into the output, with the internal resistance of the source on the input.

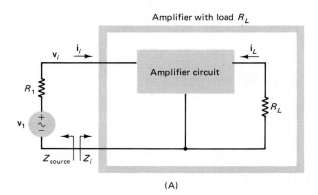

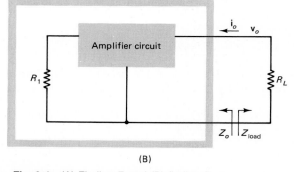

Fig. 9-4 (A) Finding Z_i and (B) finding Z_o.

From the maximum power transfer theorem, the maximum power is transferred from the source to the input of the circuit when $R_1 = Re(Z_i)$. The maximum power is transferred from the circuit to the load when $Re(Z_o) = R_L$. These relationships are important when efforts are made to maximize the power gain of an amplifier.

9-6 CURRENT, VOLTAGE, AND POWER GAIN

It is also important to introduce the current gain A_i, voltage gain A_V, and power gain A_P for an ac amplifier. These can be defined as follows:

A_i is the ac load current divided by the ac source current:

$$A_i = \mathbf{i}_L / \mathbf{i}_1 \qquad (9\text{-}13)$$

A_V is the ac output voltage divided by the ac source voltage:

$$A_V = \mathbf{v}_o/\mathbf{v}_1 \tag{9-14}$$

and A_P is the circuit power gain:

$$A_P = |A_i A_V| \tag{9-15}$$

The above parameters measure the properties of a complete ac amplifier circuit and are functions of the h parameters associated with the BJT used in the circuit.

It is further useful to note that

$$A_V = \frac{\mathbf{v}_o}{\mathbf{v}_1} = \frac{-(\mathbf{i}_L R_L)}{\mathbf{i}_1(R_1 + Z_i)} = -A_i \frac{R_L}{R_1 + Z_i} \tag{9-16}$$

Once A_i and Z_i are known, A_V can be found by applying Eq. (9-16). The above relationships are summarized and compared with the h parameters in Fig. 9-5.

	BJT equivalent circuit parameters	BJT equivalent parameters from previous chapters	Amplifier parameters
Input resistance/impedance	$h_i = \dfrac{\mathbf{v}_i}{\mathbf{i}_i}\bigg|_{\mathbf{v}_o = 0}$	$h_{ib} = r_{(CB)}$ $h_{ie} = r_{(CE)}$	$Z_i = \dfrac{\mathbf{v}_i}{\mathbf{i}_i}$ Load $= R_L$
Current gain	$h_f = \dfrac{\mathbf{i}_o}{\mathbf{i}_i}\bigg|_{\mathbf{v}_o = 0}$	$h_{fb} = \alpha_{ac}$ $h_{fe} = \beta_{ac}$	$A_i = \dfrac{\mathbf{i}_L}{\mathbf{i}_1}$ Load $= R_L$
Reverse voltage gain	$h_r = \dfrac{\mathbf{v}_i}{\mathbf{v}_o}\bigg|_{\mathbf{i}_i = 0} \cong 0$	Assumed $\cong 0$	Depends on circuit configuration
Output resistance/impedance	$\dfrac{1}{h_o} = \dfrac{\mathbf{v}_o}{\mathbf{i}_o}\bigg|_{\mathbf{i}_i = 0}$	Assumed very large	$Z_o = \dfrac{\mathbf{v}_o}{\mathbf{i}_o}$ Input resistance $= R_1$
Voltage gain	$\dfrac{\mathbf{v}_o}{\mathbf{v}_i}\bigg|_{\mathbf{i}_o = 0} = \dfrac{-g_m}{h_o}$	$A_V = \dfrac{\Delta(V_{out})}{\Delta(V_{in})}$	$A_V = \dfrac{\mathbf{v}_o}{\mathbf{v}_1}$ Load $= R_L$

Fig. 9-5 Definition of BJT and circuit parameters (with simplifying assumptions).

The following sections describe methods that can be used to calculate midrange values for the parameters Z_i, Z_o, A_i, and A_V for different ac amplifier circuits as a function of circuit components. The design objectives for these and other parameters can be related to the selection of appropriate circuit elements. The ac and dc designs are linked through those elements that affect both circuit aspects.

It is important to realize that ac equivalent circuits can occur in many different configurations and that it is neither feasible nor helpful to attempt to memorize the various parameter equations. A more useful approach is to learn the methods of analysis and the patterns that can be observed in the equations that are derived. These skills can then be applied to the analysis of both familiar and unfamiliar circuits as they are encountered.

A note of caution should be added. At low and high frequencies, the models described here are no longer adequate. And at all frequencies, these models are approximations. More precise results can be obtained by using computer simulations.

9-7 BJT COMMON-EMITTER AC AMPLIFIER

A BJT common-emitter ac amplifier is shown in Fig. 9-6(A). Resistor R_i is added to control the input current from the source v_1, and resistor R_L represents the load as seen by the amplifier.

The purpose of this circuit is to amplify the ac input signal v_1, to produce an ac output signal v_o that is larger in amplitude. The capacitors act as short circuits for the midrange frequencies of interest in this chapter. Resistors R_{B1}, R_{B2}, R_C, and R_E provide the dc bias so that the BJT operates in the linear region [Fig. 9-6(B)].

In order to calculate the voltage gain A_V of the circuit, refer to the ac equivalent circuit of Fig. 9-6(C). Where the voltages and currents in Fig. 9-6(B) are total dc

Fig. 9-6 (A) BJT common-emitter ac amplifier circuit. (B) DC circuit. (C) Midrange ac equivalent circuit.

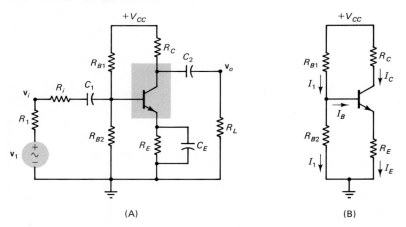

(A)

(B)

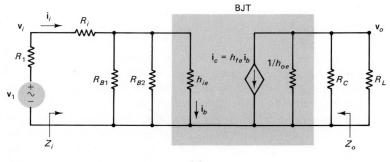

(C)

variables, the voltages and currents in Fig. 9-6(C) are incremental ac variables. Complete circuit operation requires superposition of the effects of Fig. 9-6(B) and 9-6(C).

In order to develop Fig. 9-6(C), the following observations and assumptions are made:

- Any node whose voltage is constant in time is considered to be ac ground. The resistance of all supplies is assumed negligible with respect to circuit parameters, so power supply nodes are ac ground.
- C_1, C_2, and C_E act as short circuits at midrange frequencies. This assumption defines the midrange region.
- Device and wiring capacitances act as open circuits at midrange frequencies. Such effects are considered in Chapter 10.
- The BJT input behaves as a diode having ac resistance h_{ie}. A base current $\mathbf{i}_b$ flows in the device.
- The BJT output behaves as a current generator $h_{fe}\mathbf{i}_b$ with an output resistance $1/h_{oe}$ (applying the above discussion and definitions).

The device model from Fig. 9-2(B) has been used to form Fig. 9-6(C). All capacitors have been shorted out (for midrange consideration). Wherever resistors connect to dc sources, they are routed directly to ground. The result is the effective ac circuit shown. The device equivalent circuit is now integrated into the total amplifier circuit. For small ac signals, the ac amplifier can be represented by the combination of the linear circuit components shown.

The key ac parameters can be found by inspecting Fig. 9-6(C). The input impedance is given by

$$Z_i = R_i + R_{B1} \| R_{B2} \| h_{ie} \tag{9-17}$$

A signal applied on the output cannot directly affect the current through a dependent current generator. Since the output circuit does not include a pathway to the controlling current source (the current through h_{ie}), the dc collector current must remain constant ($\mathbf{i}_c = 0$). No current flows through the dependent generator. Therefore, the output impedance becomes

$$Z_o = \frac{1}{h_{oe}} \| R_C \tag{9-18}$$

Changes in the device currents originate with the controlling current source $\mathbf{i}_b$. Changes in $\mathbf{i}_b$ (through $\mathbf{h}_{ie}$) drive the dependent current generator $\mathbf{i}_c = h_{fe}\mathbf{i}_b$. The current gain of the device is $\mathbf{h}_{fe}$, but the current gain of the circuit is reduced due to the current dividers at the input and output:

$$\frac{\mathbf{i}_b}{\mathbf{i}_i} = \frac{R_{B1} \| R_{B2} \| h_{ie}}{h_{ie}} \tag{9-19}$$

$$\frac{\mathbf{i}_L}{\mathbf{i}_C} = \frac{(1/h_{oe}) \| R_C \| R_L}{R_L} \tag{9-20}$$

so that the current gain of the circuit becomes

$$A_i = h_{fe} \frac{R_{B1} \| R_{B2} \| h_{ie}}{h_{ie}} \frac{(1/h_{oe}) \| R_C \| R_L}{R_L} \tag{9-21}$$

From Eq. (9-16), the voltage gain is given by

$$A_V = -A_i \frac{R_L}{R_1 + Z_i}$$

$$= \frac{-h_{fe}}{h_{ie}} \frac{(R_{B1} \parallel R_{B2} \parallel h_{ie})[(1/h_{oe}) \parallel R_C \parallel R_L]}{(R_1 + R_i + R_{B1} \parallel R_{B2} \parallel h_{ie})} \qquad (9\text{-}22)$$

by substituting from Eqs. (9-17) and (9-21). The equivalence $g_{me} = h_{fe}/h_{ie}$ can also be used in writing this equation.

Resistors R_{B1} and R_{B2} in Fig. 9-6 are added to the ac amplifier to improve the thermal stability of the circuit. The following effects are associated with these resistors:

- The input impedance Z_i of the amplifier is reduced.
- The current gain A_i of the amplifier is reduced.
- The voltage gain A_V of the amplifier is approximately unchanged for the case $(R_1 + R_i) \ll (R_{B1} \parallel R_{B2} \parallel h_{ie})$. Otherwise, A_V also is reduced.

Resistors R_{B1} and R_{B2} thus improve thermal stability but require additional source current, whereas A_V stays about the same for low values of $R_1 + R_i$.

Equation (9-22) can be better understood by taking the special case for which $R_1 \ll R_i$, R_{B1} and $R_{B2} \gg h_{ie}$, and $1/h_{oe}$ and $R_L \gg R_C$:

$$A_V \cong -h_{fe} \frac{R_C}{R_i + h_{ie}} \qquad (9\text{-}23)$$

This expression for the BJT common-emitter ac gain is identical in form with the previously derived equation for the dc-coupled amplifier gain using the same configuration [Eq. (7-21)] if the following associations are made:

$$R_C \longrightarrow R_C \qquad R_i \longrightarrow R_B$$
$$h_{ie} \longrightarrow r \qquad h_{fe} \longrightarrow \beta_{ac}$$

This means that the dc-coupled amplifier gain and the ac amplifier gain (in the midrange region) are equivalent concepts. Comparable analyses have been made for the two cases and have produced similar results.

A possible design strategy to produce an ac amplifier with a specified gain A_V can be seen by examining Eq. (9-22). To start, determination of the dc operating point can be used to specify appropriate values of R_{B1}, R_{B2}, and R_C. For a given device (defining h_{fe} and $1/h_{oe}$), the amplifier gain then depends on R_L and R_i. If R_L is given, the gain can be adjusted by varying R_i. This is one type of design problem that is discussed below.

Example 9-2 **Designing a Common-Emitter AC Amplifier**

Design an ac amplifier with a midrange gain $A_V = -10$ and $R_L = 5.0 \text{ k}\Omega$ using the schematic of Fig. 9-6(A) and the device output characteristic and load line shown in Fig. 9-7. Assume that $R_1 = 0.50 \text{ k}\Omega$. For this example, select the design approach that varies R_i to produce the desired gain.

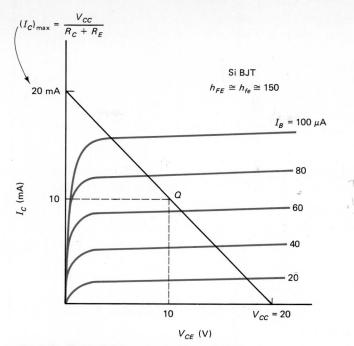

Fig. 9-7 Characteristic curve for Example 9-1.

Solution   **DC Design**

Referring to the load line,

$$(I_C)_Q = 10 \text{ mA}$$

$$(V_{CE})_Q = 10 \text{ V}$$

$$V_{CC} = 20 \text{ V}$$

Given the vertical intercept $V_{CC}/(R_C + R_E)$ as part of the load line definition,

$$R_C + R_E = 20 \text{ V}/20 \text{ mA} = 1.0 \text{ k}\Omega$$

It is now necessary to select a value for R_E. As R_E becomes larger, R_C must be reduced to keep the sum a constant, and the amplifier gain is reduced. However, R_E must be made large enough to stabilize the circuit. If R_E is taken as $R_C/9$, in keeping with approximate design guidelines,[3] then

$$R_C \cong 0.9 \text{ k}\Omega$$

$$R_E \cong 0.1 \text{ k}\Omega$$

To achieve circuit stability, the current in R_{B1} and R_{B2} must be greater than I_B. From Fig. 9-7, this condition is satisfied if $I_1 \cong I_2 = 1.0$ mA. Since the voltage

---

[3] Refer to Chapter 7, Exercise 7-15, for some of the considerations leading to this selection.

across the emitter resistor is $(100 \ \Omega)(10 \ \text{mA}) = 1.0$ V, the voltage divider must create a voltage 1.0 V $+ 0.7$ V $= 1.7$ V at the base of the BJT. Therefore,

$$R_{B2}I_1 \cong 1.7 \ \text{V}$$

$$R_{B2} \cong 1.7 \ \text{V}/1.0 \ \text{mA} = 1.7 \ \text{k}\Omega$$

so that

$$R_{B1} + R_{B2} = 20 \ \text{V}/1.0 \ \text{mA} = 20 \ \text{k}\Omega$$

$$R_{B1} = 18.3 \ \text{k}\Omega$$

All dc values are now determined. The Q point does not drift with T and is independent of $\beta_{dc} = h_{FE}$, which is not needed in the dc calculation.

AC Design

Now proceed to the ac circuit design. In order to find the ac circuit parameters Z_i, Z_o, and A_V, values are needed for h_{ie}, h_{fe}, and h_{oe}. From Fig. 9-7,

$$h_{fe} \cong 150$$

and from Eq. (9-6),

$$1/h_{oe} \cong 100 \ \text{V}/10 \ \text{mA} \cong 10 \ \text{k}\Omega$$

From Fig. 9-7, $h_{FE} \cong 150$ is given so that

$$(I_B)_Q \cong 10 \ \text{mA}/150 \cong 67 \ \mu\text{A}$$

which is also observable directly on the characteristic curve. Therefore, by applying the estimation technique for h_{ie},

$$h_{ie} \cong 0.026 \ \text{V}/67 \ \mu\text{A} = 0.39 \ \text{k}\Omega$$

From the equivalent circuit of Fig. 9-6(C) and Eq. (9-22),

$$A_V = -(150)\frac{(1.7 \ \text{k}\Omega \| 18.3 \ \text{k}\Omega \| 0.39 \ \text{k}\Omega)}{0.39 \ \text{k}\Omega}$$

$$\times \frac{(10 \ \text{k}\Omega \| 0.9 \ \text{k}\Omega \| 5.0 \ \text{k}\Omega)}{R_i + 0.50 \ \text{k}\Omega + (1.7 \ \text{k}\Omega \| 18.3 \ \text{k}\Omega \| 0.39 \ \text{k}\Omega)}$$

$$-10 = -150 \left(\frac{0.32 \ \text{k}\Omega}{0.39 \ \text{k}\Omega}\right)\left(\frac{0.71 \ \text{k}\Omega}{R_i + 0.50 \ \text{k}\Omega + 0.32 \ \text{k}\Omega}\right)$$

$$R_i \cong 7.9 \ \text{k}\Omega$$

Observe that there is a maximum gain that can be obtained with these circuit values. As A_V increases, R_i must decrease. The limiting case occurs when $R_i = 0$:

$$A_V = -150 \left(\frac{0.32 \ \text{k}\Omega}{0.39 \ \text{k}\Omega}\right)\left(\frac{0.71 \ \text{k}\Omega}{0.82 \ \text{k}\Omega}\right) \cong -1.1 \times 10^2$$

The gain can be adjusted to this level by adjusting R_i. Since the objective was for a gain of -10, a design solution is possible. The other ac characteristics of the circuit can be found by inspection:

$$Z_i = R_i + R_{B1} \| R_{B2} \| h_{ie}$$

$$= 7.9 \ \text{k}\Omega + 1.7 \ k\Omega \| 18.3 \ k\Omega \| 0.39 \ k\Omega \cong 8.2 \text{k}\Omega$$

$$Z_o = (1/h_{oe}) \| R_C$$
$$= 10\ k\Omega \| 0.90\ k\Omega \cong 0.83 k\Omega$$

The design of the ac amplifier is thus complete, with determination of all resistors and voltages necessary to bias the device in the linear region. An ac gain of -10 has been produced, as desired. The dc and ac designs are interlinked and meet all design conditions.

If the load-line Q point is changed, the values of R_C and R_E shift, affecting both dc and ac circuit properties. A bias point associated with reduced current values leads to higher values for R_C and R_E and less dc power dissipation. Reduced current values also lead to larger values for $1/h_{oe}$ and h_{ie}. From Eq. (9-21), these increases raise current gain A_i. Amplifier performance can often be improved by designing with reduced current values. On the other hand, the ability to apply this technique is limited by the magnitude of the signal voltages and currents, which must remain small with respect to the dc bias voltage and current levels, and the power output requirements. ∎

9-8 DESIGN WITH NEGATIVE FEEDBACK

The above design strategy has both advantages and disadvantages. As an advantage, the design process can begin with dc considerations to fix the operating point, with the amplifier gain A_V then adjusted over a wide range by varying R_i. As a disadvantage, the amplifier gain is a function of h_{fe}, h_{ie}, and h_{oe}. The parameter h_{fe} varies widely among devices and with temperature. The parameters h_{ie} and h_{oe} are clearly dependent on the dc operating point and thus vary as the Q point shifts.

An alternative design strategy can be developed by returning to Fig. 9-6 and considering the ac performance of this circuit with C_E removed, as shown in Fig. 9-8(A). To simplify the analysis, assume that $1/h_{oe}$ is large enough to be neglected.

In Fig. 9-8(B), the voltage from point A to ground is

$$\mathbf{v}_A \cong \mathbf{i}_b h_{ie} + (h_{fe} + 1)\mathbf{i}_b R_E \cong \mathbf{i}_b(h_{ie} + h_{fe}R_E) \qquad (9\text{-}24)$$

In the following analysis, the assumption that $h_{fc} \gg 1$ is often used. When viewed from the input, R_E acts as a resistor of value $\cong h_{fe}R_E$. This result is equivalent to the effect of R_E derived in Chapter 7, Eq. (7-32).

Fig. 9-8 Common-emitter ac amplifier with $C_E = 0$ to produce negative feedback. (A) Circuit schematic. (B) Equivalent circuit.

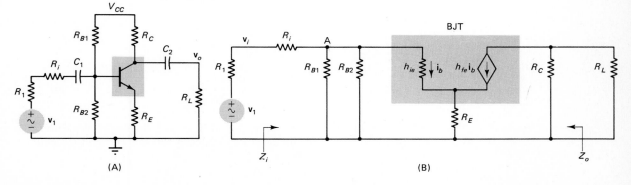

The current $\mathbf{i}_b$ through h_{ie} is amplified by h_{fe}. The total current $\mathbf{i}_b + \mathbf{i}_c$ is then fed through R_E. The voltage across R_E increases by a factor h_{fe}, making R_E seem to have a larger value.

By inspecting Fig. 9-8, the ac circuit properties are found to be

$$Z_i = R_i + R_{B1} \| R_{B2} \| (h_{ie} + h_{fe} R_E) \tag{9-25}$$

$$Z_o = R_c \tag{9-26}$$

$$A_i = h_{fe} \frac{R_{B1} \| R_{B2} \| (h_{ie} + h_{fe} R_E)}{h_{ie} + h_{fe} R_E} \frac{R_C \| R_L}{R_L} \tag{9-27}$$

The negative feedback caused by R_E has led to the replacement of h_{ie} by $h_{ie} + h_{fe}R_E$. If $R_{B1} \| R_{B2}$ is much larger than h_{ie}, one result is to increase Z_i. Negative feedback can thus be used to increase the input impedance of an amplifier.

This technique can be used to minimize loading between the different portions of a circuit. The addition of negative feedback also reduces A_i by shifting additional current into the combination $R_{B1} \| R_{B2}$.

By substitution,

$$A_V = -h_{fe} \frac{R_{B1} \| R_{B2} \| (h_{ie} + h_{fe} R_E)}{(R_1 + R_i) + R_{B1} \| R_{B2} \| (h_{ie} + h_{fe} R_E)} \frac{R_C \| R_L}{h_{ie} + h_{fe} R_E} \tag{9-28}$$

In some circuit applications, it is possible to require

$$R_1 + R_i \ll R_{B1} \| R_{B2} \| (h_{ie} + h_{fe} R_E)$$

so that

$$A_V \cong -\frac{h_{fe} R_C \| R_L}{h_{ie} + h_{fe} R_E} \tag{9-29}$$

For future reference, it is useful to rewrite Eq. (9-28) in a different form. Expanding the representation and letting $R_B = R_{B1} \| R_{B2}$ and $R_1 + R_i = R_A$,

$$A_{Vf} = \frac{-h_{fe} R_B (R_C \| R_L)}{R_A R_B + (R_A + R_B)(h_{ie} + h_{fe} R_E)} \tag{9-30}$$

where A_{Vf} is the gain with negative feedback due to R_E. With $R_E = 0$,

$$A_{Vo} = \frac{-h_{fe} R_B (R_C \| R_L)}{R_A R_B + (R_A + R_B) h_{ie}} \tag{9-31}$$

With a little manipulation, Eq. (9-30) can now be written as

$$A_{Vf} = \frac{A_{Vo}}{1 - A_{Vo} B} \tag{9-32a}$$

with

$$B = \frac{R_E}{R_C \| R_L} \left(\frac{R_A + R_B}{R_B} \right) \tag{9-32b}$$

The resistor R_E is causing part of the output signal to be fed back in series with (and out of phase with) the input, and this negative feedback reduces the voltage gain of the amplifier. A_{Vo} is the voltage gain without feedback, A_{Vf} is the gain with feedback, and B is the *feedback factor*.

The ratio $R_E/(R_C\|R_L)$ in B represents the fraction of the output voltage that is being fed across R_E. The factor $(R_A + R_B)/R_B$ arises because the voltage feedback across R_E is weighted more strongly than v_1.

Equation (9-32a) is a standard form that is used in Chapter 11. Returning to Eq. (9-29) and requiring that $h_{ie} \ll h_{fe}R_E$ (or $1/g_{me} \ll R_E$), the result

$$A_V \cong \frac{-R_C\|R_L}{R_E} \tag{9-33}$$

is obtained. This equation for ac amplifier performance is equivalent to Eq. (7-41) found for dc-coupled amplifier performance under similar conditions.

An ac amplifier can thus be designed with A_V independent of h_{fe} and h_{ie} if C_E is removed. R_C, R_L, and R_E are chosen to give the desired gain, and R_{B1}, R_{B2}, and V_{CC} to give the desired Q point. This design strategy takes advantage of the negative feedback created by R_E to reduce the circuit dependence on the BJT parameters. However, this design approach is only possible for small values of $R_1 + R_i$, as noted.

The use of negative feedback is further discussed in Chapter 11.

Example 9-3   Effect of Feedback on Gain

A BJT biased at 1.0 mA is used in an ac amplifier with $R_C = R_L = 1.0$ kΩ and $(R_1 + R_i) \ll R_{B1}\|R_{B2}\|h_{ie}$. (a) Find the gain of the amplifier without feedback. (b) If a feedback resistor $R_E = 0.1R_C$ is introduced, find the feedback factor B. (c) Find the amplifier gain with feedback.

Solution   From Eq. (9-31), with $R_B \gg R_A$ and $h_{ie} \gg R_A$,

$$A_{Vo} \cong -g_{me}R_C\|R_L$$

$$\cong -(1.0 \text{ mA}/0.026 \text{ V})(1.0 \text{ k}\Omega\|1.0 \text{ k}\Omega) = -19$$

$$B = \frac{0.10 \text{ k}\Omega}{0.50 \text{ k}\Omega} \cong 0.20$$

$$A_{Vf} = \frac{-19}{1 + 19(0.20)} \cong -4.0$$

The gain A_{Vo} without feedback is proportional to the transconductance g_m of the BJT and is highly device dependent. With feedback factor $B = 0.20$, the gain with feedback A_{Vf} is almost independent of the device parameters, so the desired predictable and stable amplifier gain has been obtained. However, a price has been paid. The 20 percent feedback factor reduces the amplifier gain from -19 to -4.0. A tradeoff exists between a voltage gain independent of the device parameters and the obtainable gain of the amplifier. ∎

Many different types of design problems can be encountered in working with ac amplifiers. The objective may be to produce a specific Z_i, Z_o, A_i, A_V, or A_P, or to produce balanced performance that places requirements on a number of different circuit parameters. The following discussion emphasizes obtaining a specific value of

A_V as the primary circuit objective, recognizing that similar strategies follow for other objectives. The problems at the end of the chapter address some alternative design aspects that may be encountered.

9-9 BJT COMMON-BASE AC AMPLIFIER

A BJT common-base circuit, its dc portion, and its ac equivalent are shown in Fig. 9-9. In this configuration, the output resistance $1/h_{ob}$ is usually sufficiently large to neglect its impact on $R_C \| R_L$. By inspection,

$$Z_i = R_i + R_E \| h_{ib} \tag{9-34}$$

$$Z_o = R_C \tag{9-35}$$

Further, for the input current divider,

$$\frac{\mathbf{i}_e}{\mathbf{i}_i} = \frac{R_E \| h_{ib}}{h_{ib}} \tag{9-36}$$

and for the output current divider,

$$\frac{\mathbf{i}_L}{\mathbf{i}_C} = \frac{-R_C \| R_L}{R_L} \tag{9-37}$$

(adding the minus sign by convention) so that

$$A_i = h_{fb} \frac{R_E \| h_{ib}}{h_{ib}} \frac{R_C \| R_L}{R_L} \tag{9-38}$$

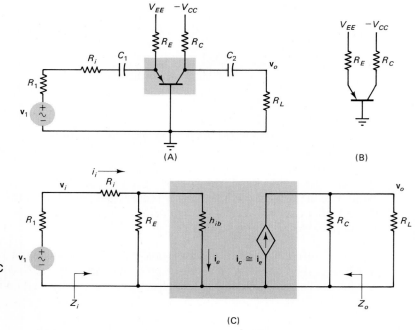

Fig. 9-9 (A) BJT common-base ac amplifier circuit. (B) DC circuit. (C) Equivalent midrange ac circuit (assuming $1/h_{ob} \gg R_C \| R_L$.

The voltage gain is

$$A_V = -A_i \frac{R_L}{Z_i + R_1} = -h_{fb} \frac{R_E \| h_{ib}}{h_{ib}} \frac{R_C \| R_L}{(R_1 + R_i) + R_E \| h_{ib}} \qquad (9\text{-}39)$$

where, from the previous discussion, $h_{fb} \cong -1$.

For the special case $R_1 \ll R_i$, $R_E \gg h_{ib}$, and $R_L \gg R_C$,

$$A_V \cong -h_{fb} \frac{R_C}{R_i + h_{ib}} \qquad (9\text{-}40)$$

This expression for the BJT common-base amplifier gain is identical in form with the previously derived equation [Eq. (7-12)] for the dc-coupled amplifier gain using the same configuration when the following associations are made:

$$R_C \longrightarrow R_C \qquad R_i \longrightarrow R_E$$

$$h_{ib} \longrightarrow r \qquad h_{fb} \longrightarrow \alpha_{ac} \cong -1$$

Another variation of the common-base ac amplifier, requiring only one power supply, is discussed in Exercise 9-15.

Example 9-4   **Designing a Common-Base AC Amplifier**

Suppose that you wish to design a common-base ac amplifier with $A_V = 10$ and $R_L = 5.0$ kΩ using the schematic shown in Fig. 9-9. The dc circuit is shown in Fig. 9-9(B). Assume $R_1 = 50$ Ω, $(I_C)_Q = -1.0$ mA, $V_{CC} = 20$ V, and $V_{EE} = 20$ V.

Solution   **DC Design**

Since $V_{EB} \cong 0.7$ V,

$$R_E \cong \frac{V_{EE} - V_{EB}}{I_E} = \frac{V_{EE} - 0.7 \text{ V}}{1.0 \text{ mA}} \cong 19 \text{ k}\Omega$$

Further, biasing with $(V_{CB})_Q = V_{CC}/2$,

$$R_C = \frac{V_{CC} - (V_{CB})_Q}{(I_C)_Q} = \frac{10 \text{ V}}{1.0 \text{ mA}} = 10 \text{ k}\Omega$$

AC Design

The ac design involves Fig. 9-9(C) and Eq. (9-39). Noting that

$$h_{ib} \cong 0.026 \text{ V}/1.0 \text{ mA} = 26 \text{ }\Omega$$

the value for R_i can be found

$$A_V = \frac{(1)(19 \text{ k}\Omega \| 26 \text{ }\Omega)(10 \text{ k}\Omega \| 5.0 \text{ k}\Omega)}{(26 \text{ }\Omega)[(R_1 + R_i) + (19 \text{ k}\Omega \| 26 \text{ }\Omega)]}$$

$$10 \cong \frac{3.33 \text{ k}\Omega}{(R_1 + R_i) + 26 \text{ }\Omega}$$

$$(50 + R_i + 26) \text{ }\Omega \cong 330 \text{ }\Omega$$

$$R_i \cong 0.25 \text{ k}\Omega$$

There is an upper limit to the gain that can be obtained by varying R_i for the prescribed values of other circuit components. As A_V increases, R_i must decrease. The limiting case is when

$$A_V = \frac{3.33 \text{ k}\Omega}{50 \text{ }\Omega + 26 \text{ }\Omega} \cong 44$$

The given design objective, $A_V = 10$, is safely beneath this maximum. For other circuit characteristics,

$$Z_i = R_i + R_E \| h_{ib} = 0.25 \text{ k}\Omega + 19 \text{ k}\Omega \| 26 \text{ }\Omega \cong 0.28 \text{ k}\Omega$$

$$Z_o = R_C = 10 \text{ k}\Omega$$

The selected values have led to an achievable selection of resistor and voltage values. If R_1 has an appreciably higher value, the maximum possible value of A_V drops, possibly leading to constraints that prevent a circuit solution with the given bias point. Since the voltage gain is proportional to $h_{fb} \cong -1$, a higher potential gain can be obtained only by modifying the voltages and resistors in the circuit. For a given load resistor R_L, the maximum value of A_V can be substantially limited. A comparison between Examples 9-2 and 9-4 illustrates this result. In comparing, also note that Z_i is much smaller for the common-base configuration because Z_i is dominated by $h_{ib} \ll h_{ie}$ for this case. The low input impedance typical of common-base ac amplifiers can present a problem in circuit design. ∎

9-10 BJT COMMON-COLLECTOR (EMITTER-FOLLOWER) AC AMPLIFIER

In ac circuit design, it is sometimes useful to consider a new configuration, the common-collector or emitter-follower, as shown in Fig. 9-10. The collector is attached to the power supply, which is ac ground. The output is taken from the emitter to ground. Figure 9-10(B) gives the dc portion of this circuit and Fig. 9-10(C) gives the ac equivalent. The term $1/h_{oe}$ is assumed to be sufficiently large to neglect its impact on this circuit. As described above, the current through R_E and R_L is about h_{fe} times the current through h_{ie}. Based on Eq. (9-25),

$$Z_i = R_i + R_{B1} \| R_{B2} \| (h_{ie} + h_{fe} R_x) \tag{9-41}$$

where $R_x = R_E \| R_L$. In order to determine Z_o, first observe that a voltage $\mathbf{v}_o$ placed on the output changes the current through R_E. This causes a change in the current $\mathbf{i}_b$ through h_{ie}, which drives the dependent current generator $h_{fe} \mathbf{i}_b$. The output circuit now includes a pathway to the controlling current source $\mathbf{i}_b$, so the output impedance Z_o is a function of the input parameters. Because of the current generator, $\mathbf{i}_e = (h_{fe} + 1)\mathbf{i}_b \cong h_{fe} \mathbf{i}_b$. Current $\mathbf{i}_b$ is smaller than the current through R_E by a factor of h_{fe}.

Thus, the relationship

$$R_E \mathbf{i}_e = [h_{ie} + (R_1 + R_i) \| R_{B1} \| R_{B2}] \mathbf{i}_b$$

$$\cong \frac{\mathbf{i}_e}{h_{fe}} [h_{ie} + (R_1 + R_i) \| R_{B1} \| R_{B2}] \tag{9-42}$$

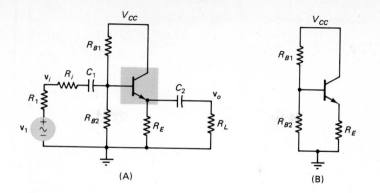

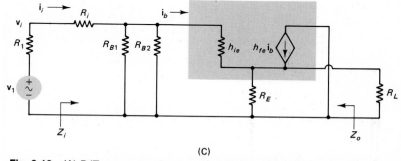

Fig. 9-10 (A) BJT common-collector (emitter-follower) circuit. (B) DC portions of the circuit. (C) Midrange ac equivalent circuit.

is found. When viewed from the output, the effective resistance of the input portion of the circuit is reduced by h_{fe}. The output impedance Z_o therefore becomes

$$Z_o = R_E \left\| \frac{h_{ie} + (R_1 + R_i) \| R_{B1} \| R_{B2}}{h_{fe}} \right. \tag{9-43}$$

By inspection, the current gain A_i is found to be

$$A_i = h_{fe} \frac{R_{B1} \| R_{B2} \| (h_{ie} + h_{fe} R_x)}{h_{ie} + h_{fe} R_x} \frac{R_x}{R_L} \tag{9-44}$$

so that

$$A_V = h_{fe} \frac{R_{B1} \| R_{B2} \| (h_{ie} + h_{fe} R_x)}{(R_1 + R_i) + R_{B1} \| R_{B2} \| (h_{ie} + h_{fe} R_x)} \frac{R_x}{h_{ie} + h_{fe} R_x} \tag{9-45}$$

If $R_1 + R_i$ is negligible,

$$A_V \cong \frac{1}{1 + h_{ie}/(h_{fe} R_x)} = \frac{1}{1 + 1/(g_{me} R_x)} \tag{9-46}$$

For many circuits, $A_V \cong 1$ and the output voltage *follows* the signal on the base. The common-collector amplifier is often called an *emitter follower*.

Example 9-5   Designing an Emitter Follower

Example 9-2 (for a common-emitter amplifier) is to be modified to use the same dc design for an emitter follower using the schematic shown in Fig. 9-10 with $R_L = 50\ \Omega$. Assume that $R_i = 0.50\ \text{k}\Omega$ and $R_i = 0$.

Solution   If R_C and R_E are replaced for the common-emitter circuit with $R_C' = 0$, and $R_E' = R_C + R_E$, then the transformation is complete. The circuits have the same operating points and load lines in each case.

For the emitter-follower circuit,

$$R_E = 1.0\ \text{k}\Omega \qquad h_{ie} = 0.39\ \text{k}\Omega \qquad \textit{mistake}$$

$$R_{B1} = 18.3\ \text{k}\Omega \qquad h_{fe} = 150\ \text{k}\Omega$$

$$R_{B2} = 1.7\ \text{k}\Omega \qquad R_1 + R_i = 0.50\ \text{k}\Omega$$

From Eq. (9-46), the voltage gain of the circuit is

$$A_V = \cfrac{1}{1 + \cfrac{0.39\ \text{k}\Omega}{(150)(50\ \Omega \,\|\, 1.0\ \text{k}\Omega)}} \cong 0.95$$

The input impedance is

$$Z_i = (1.7\ \text{k}\Omega \,\|\, 18.3\ \text{k}\Omega) \,\|\, [0.39\ \text{k}\Omega + 150(50\ \Omega \,\|\, 1.0\ \text{k}\Omega)]$$

$$\cong 1.3\ \text{k}\Omega$$

and the output impedance is

$$Z_o = 1.0\ \text{k}\Omega \,\left\|\, \frac{0.39\ \text{k}\Omega + 1.7\ \text{k}\Omega \,\|\, 18.3\ \text{k}\Omega \,\|\, 0.50\ \text{k}\Omega}{150}\right.$$

$$\cong 5\ \Omega$$

This emitter-follower circuit has a voltage gain of about $+1$ with an input impedance 1.3 kΩ and an output impedance of 5 Ω. This result is obtained because the negative feedback increases the effective input resistance of the amplifier. The circuit can thus be helpful in impedance matching from higher input impedance sources to lower impedance loads and achieving isolation between the stages of more complex circuits.   ∎

9-11   JFET COMMON-SOURCE AC AMPLIFIER

To derive the JFET equivalent circuit, the y parameters are used. From the earlier introduction, these parameters can take the form

$$\mathbf{i}_o \ (\text{linear function of } \mathbf{v}_i, \mathbf{v}_o) = g_m \mathbf{v}_i + \frac{1}{r_d} \mathbf{v}_o$$

$$\mathbf{i}_i = \frac{1}{h_i} \mathbf{v}_i \cong 0 \qquad\qquad (9\text{-}47)$$

where g_m is the slope of the transfer characteristic and $1/r_d$ is the slope of the output characteristic at the operating point.

$$g_m = \left. \frac{\mathbf{i}_o}{\mathbf{v}_i} \right|_{\mathbf{v}_o = 0}$$

$$r_d = \left. \frac{\mathbf{v}_o}{\mathbf{i}_o} \right|_{\mathbf{v}_i = 0} \tag{9-48}$$

In many design applications, $r_d \gg 1/g_m$, so the effect of r_d can be neglected. If this condition does not hold, r_d can be easily accommodated in the ac equivalent circuit.

A JFET common-source amplifier is shown in Fig. 9-11. The source is connected to ac ground, so the common-source JFET and the common-emitter BJT are analogous configurations. The resistor R_G is added to provide a dc path from the gate to ground. The objective is to make R_G as large as possible so as to limit loading of the input signal but keeping it small enough so that $I_o R_G \ll |V_{GS}|$, where I_o is the gate leakage current. Typical values for R_G range from 1 to 10 MΩ. The dc portion of the circuit and its ac equivalent are shown in Figs. 9-11(B) and 9-11(C), respectively.

By inspection,

$$Z_i = R_i + R_G \tag{9-49}$$

and opening the constant-current generator,

$$Z_o = R_D \tag{9-50}$$

Fig. 9-11 (A) JFET common-source ac amplifier. (B) DC circuit. (C) AC equivalent circuit.

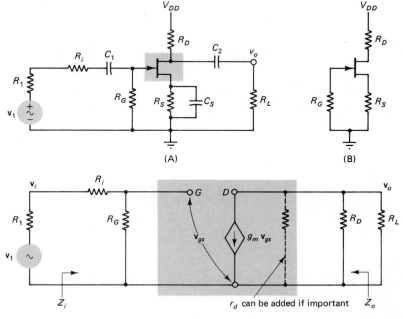

Summing voltages around the input and output loops,

$$\mathbf{v}_{gs} = \frac{R_G}{(R_1 + R_i) + R_G} \mathbf{v}_1$$

$$\mathbf{v}_o = -g_m \mathbf{v}_{gs}(R_D \| R_L)$$

so that

$$\frac{R_G}{(R_1 + R_i) + R_G} \mathbf{v}_1 = \frac{\mathbf{v}_o}{g_m(R_D \| R_L)}$$

and

$$A_V = \frac{\mathbf{v}_o}{\mathbf{v}_i} = \frac{-R_G}{(R_1 + R_i) + R_G} g_m R_D \| R_L \qquad (9\text{-}51)$$

The gain of the common-source amplifier is proportional to the transconductance g_m of the JFET and the combined load $R_D \| R_L$.

Example 9-6 **Designing a Common-Source AC Amplifier**

Design the amplifier in Fig. 9-11 with $A_V = -4.0$ and $R_L = 5.0$ kΩ. Assume for the JFET characteristic the following values and Q point

$$I_{D0} = 10 \text{ mA} \qquad (I_D)_Q = 4.0 \text{ mA}$$

$$V_P = -6.0 \text{ V} \qquad (V_{DS})_Q = 10 \text{ V}$$

Solution **DC Design**

Since

$$(I_D)_Q \cong I_{D0}(1 - V_{GS}/V_P)^2$$

$$\frac{V_{GS}}{V_P} = 1 - \left[\frac{(I_D)_Q}{I_{D0}} \right]^{1/2}$$

the bias point on the transfer characteristic is

$$(V_{GS})_Q = (-6.0)(1 - 0.63)\text{V} = -2.2 \text{ V}$$

and

$$R_S = \frac{-(V_{GS})_Q}{(I_D)_Q} = \frac{2.2 \text{ V}}{4.0 \text{ mA}} = 0.55 \text{ k}\Omega$$

for the needed bias resistor. If $V_{DD} = 20$ V is chosen as a supply, then the voltage across R_D is

$$V_{R_D} = V_{DD} - (V_{DS})_Q + V_{GS}$$

$$= 20 \text{ V} - 10 \text{ V} - 2.2 \text{ V} \cong 7.8 \text{ V}$$

so that

$$R_D = \frac{7.8 \text{ V}}{4.0 \text{ mA}} \cong 2.0 \text{ k}\Omega$$

and, typically,

$$R_G \cong 1\ M\Omega$$

AC Design

The dc design is now complete. For the ac design, g_m can be estimated from the approximate theoretical equation:

$$g_m \cong \left.\frac{\partial I_D}{\partial V_{GS}}\right|_Q = \frac{-2I_{D0}}{V_P}\left(1 - \frac{V_{GS}}{V_P}\right) = \frac{-2(10\ mA)}{-6.0\ V}\left(1 - \frac{2.2}{6.0}\right) \cong 2.1\ m\mho$$

Therefore,

$$-4.0 = -\left(\frac{R_G}{R_i + R_G}\right)(2.1\ m\mho)\ (2.0\ k\Omega\|5.0\ k\Omega)$$

$$-4.0 = (-3.0)\frac{R_G}{R_i + R_G}$$

There is *no solution* for this equation because a maximum A_V of -3.0 can be provided by the circuit. To obtain $A_V = -4.0$ requires a JFET with a larger value of g_m or a redesign of the dc circuit. Assume that a different JFET with $g_m \cong 6.0\ m\mho$ is used in the circuit. Then, the requirement on R_i becomes

$$-4.0 = (-8.6)\frac{R_G}{R_i + R_G}$$

and

$$R_i \cong 1.1R_G = 1.1\ M\Omega$$

This example illustrates that performance objectives are not always achievable for an initially proposed circuit configuration. The low value of g_m for the JFET, in comparison with typical BJT values found above, has limited the maximum voltage gain A_V to a relatively low value. As indicated, more design flexibility is achieved for a higher value of g_m, so this is a figure of merit for transistors. Another way to increase A_V is to choose larger values for R_D and V_{DD}. An increased power dissipation is the price paid for a higher maximum voltage gain. ■

If the capacitor C_S is removed from the amplifier circuit of Fig. 9-11, negative feedback occurs. From the equivalent ac circuit of Fig. 9-12,

Fig. 9-12 AC equivalent circuit for the common-source amplifier with feedback.

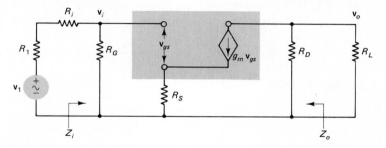

$$\mathbf{v}_o = (-g_m\mathbf{v}_{gs})\,(R_D\|R_L)$$

$$\mathbf{v}_i = \mathbf{v}_{gs} + (g_m\mathbf{v}_{gs})R_S = \mathbf{v}_{gs}(1 + g_m R_S)$$

so that

$$\frac{\mathbf{v}_o}{\mathbf{v}_i} = \frac{-R_D\|R_L}{1/g_m + R_S}$$

If the voltage divider due to $R_1 + R_i$ and R_G is included,

$$A_V = \frac{-R_G}{(R_1 + R_i) + R_G}\frac{R_D\|R_L}{1/g_m + R_S} \tag{9-52}$$

This result is compatible with the gain derivation in Chapter 8. Observe that Z_i and Z_o are unchanged using the model shown. The feedback form of Eq. (9-32a) can again be used:

$$A_{Vf} = \frac{A_{Vo}}{1 - A_{Vo}B} \tag{9-53}$$

with A_{Vo} given by Eq. (9-51), and

$$B = \frac{R_S}{R_D\|R_L}\frac{(R_1 + R_i) + R_G}{R_G} \tag{9-54}$$

The effect of negative feedback is equivalent for the common-emitter and common-source amplifiers.

9-12 JFET COMMON-DRAIN (SOURCE-FOLLOWER) AC AMPLIFIER

A JFET common-drain circuit along with its dc circuit and the ac equivalent are shown in Fig. 9-13. The output is taken from the source to ground. From the figure,

$$Z_i = R_i + R_G \tag{9-55}$$

A voltage applied to the output increases the voltage across R_S and therefore changes the value of $\mathbf{v}_{gs}$. In turn, $\mathbf{v}_{gs}$ drives the dependent current generator $i_d = g_m\mathbf{v}_{gs}$, further increasing the current through R_S and effectively reducing the value of R_S. From Fig. 9-13,

$$\mathbf{v}_o/R_S = \mathbf{i}_o + g_m\mathbf{v}_{gs} \tag{9-56}$$

where $\mathbf{i}_o$ is the current into the circuit as a result of $\mathbf{v}_o$. With $\mathbf{v}_1 = 0$, no current flows in R_G and the gate-to-ground voltage is zero. Therefore, $\mathbf{v}_o = -\mathbf{v}_{gs}$ and

$$\mathbf{i}_o = \frac{\mathbf{v}_o}{R_S} + g_m\mathbf{v}_o \tag{9-57}$$

so that the output impedance is

$$Z_o = \frac{\mathbf{v}_o}{\mathbf{i}_o} = \frac{\mathbf{v}_o}{\mathbf{v}_o/R_S + g_m\mathbf{v}_o} = R_S\|\frac{1}{g_m} \tag{9-58}$$

The voltage gain can be found from Fig. 9-13:

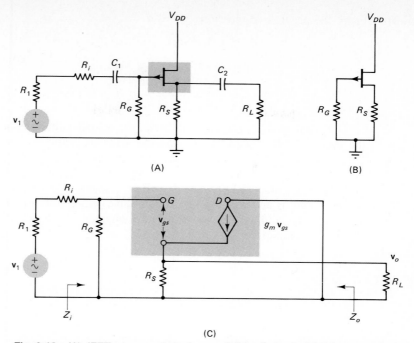

Fig. 9-13 (A) JFET common-drain (source-follower) circuit. (B) DC circuit. (C) AC equivalent circuit.

$$\mathbf{v}_1 \frac{R_G}{(R_1 + R_i) + R_G} = \mathbf{v}_{gs} + \mathbf{v}_o$$

$$\mathbf{v}_o = g_m \mathbf{v}_{gs} (R_S \| R_L) \tag{9-59}$$

so that

$$\mathbf{v}_1 \frac{R_G}{(R_1 + R_i) + R_G} = \frac{\mathbf{v}_o}{g_m(R_S \| R_L)} + \mathbf{v}_o \tag{9-60}$$

and

$$A_V = \frac{\mathbf{v}_o}{\mathbf{v}_1} = \frac{R_G}{(R_1 + R_i) + R_G} \frac{R_S \| R_L}{1/g_m + R_S \| R_L}$$

$$= \frac{R_G}{(R_1 + R_i) + R_G} \frac{1}{1 + \dfrac{1}{g_m(R_S \| R_L)}} \tag{9-61}$$

If $R_1 + R_i \ll R_G$ and $g_m (R_S \| R_L) \gg 1$, then $A_V \cong 1$ and the source follows the gate voltage.

Example 9-7   AC Parameters for a Source Follower

Calculate the ac parameters for the circuit of Fig. 9-13 with

$$R_1 + R_i = 0 \qquad\qquad R_S = 1.0 \text{ k}\Omega \qquad g_m = 1.0 \text{ m}\mho$$

$$R_G = 1.0 \text{ M}\Omega \qquad R_L = 5.0 \text{ k}\Omega$$

Solution   From the previous discussion,

$$Z_i \cong 1.0 \ \text{M}\Omega$$

$$Z_o \cong R_S \| 1/g_m \cong 1.0 \ \text{k}\Omega \| 1.0 \ \text{k}\Omega = 0.50 \ \text{k}\Omega$$

$$A_V = \frac{1}{1 + 1.0 \ \text{k}\Omega/0.83 \ \text{k}\Omega} \cong 0.45$$

The result is a *source follower*, with the output voltage approximately following the input. As illustrated by this and the previous example, the design equations for FET ac amplifiers are often simpler in form than those for BJTs because the current into the device is negligible (in many typical circuits, as assumed here). Both types of devices are essential because device parameters vary significantly, leading to different strengths and weaknesses. Bipolar and field-effect devices are complementary in many ways and, as discussed in future chapters, both are used widely in modern microelectronic circuits. ∎

9-13   D-MOSFET AC AMPLIFIER

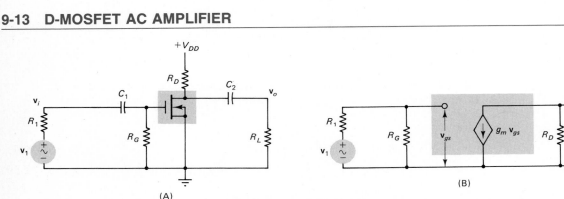

Fig. 9-14   Depletion MOSFET ac amplifier. (A) Schematic. (B) AC equivalent circuit.

The depletion MOSFET is a useful device for the design of ac amplifiers because it can operate with no dc bias on the gate and the input impedance is high. The simple circuit design of Fig. 9-14 is a useful one.

The dc biasing of the depletion MOSFET is discussed in Chapter 8. For midrange frequencies, the ac design equations are the same as those for the common-source JFET. However, there is no need for the bypass capacitor across R_S with the depletion MOSFET design. The ac amplifier in Fig. 9-14 has two fewer components than does the equivalent JFET circuit.

Assuming $R_G \gg R_1$, the voltage gain of this amplifier is

$$A_V = -g_m R_D \| R_L \tag{9-62}$$

Example 9-8   D-MOSFET Amplifier

Given a D-MOSFET amplifier biased at $V_{GS} = 0$, find the value of I_{D0}/V_P required for the MOSFET to result in an amplifier gain $A_V = -6.0$. Assume $R_D = R_L = 0.50 \ \text{k}\Omega$. If $V_P = -2.0 \ \text{V}$, what I_{D0} is required?

Solution From the characteristic equation,

$$(-6.0) = -\left(\frac{-2I_{D0}}{V_P}\right)(5.0 \text{ k}\Omega \| 5.0 \text{ k}\Omega)$$

so that

$$I_{D0}/V_P = -3.0/2.5 \text{ k}\Omega = -1.2 \text{ m}\mho$$

For $V_P = -2.0$ V,

$$I_{D0} = (2.0 \text{ V})(1.2 \text{ m}\mho) = 2.4 \text{ mA}$$

The circuit is appealing because of its simplicity, but it displays the limitations associated with FET device parameters. Increasing the gain A_V may require increasing R_D and V_{DD}, a solution that is limited by the maximum voltage drop V_{DS} allowed across the FET and power supply restrictions. ∎

9-14 E-MOSFET AC AMPLIFIER

One possible form for an E-MOSFET ac amplifier is shown in Fig. 9-15. The voltage divider $R_{B1}R_{B2}$ is used to provide a fixed bias $V_{GS} > V_T$ on the gate of the MOSFET. The circuit can be designed with or without R_S and C_S. Figure 9-16 shows the two bias strategies.

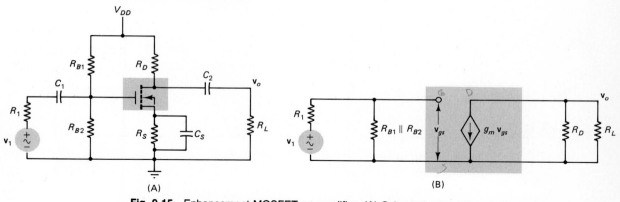

(A)

(B)

Fig. 9-15 Enhancement MOSFET ac amplifier. (A) Schematic. (B) AC equivalent circuit.

Fig. 9-16 DC bias for the E-MOSFET ac amplifier. (A) Bias without R_S. (B) Bias with R_S.

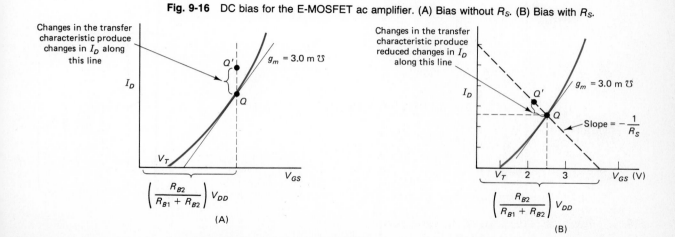

In Fig. 9-16(A), the fixed voltage-divider bias produces the dependence on temperature T that is shown. In Fig. 9-16(B), the variation of I_D with T has been reduced by the addition of R_S. Capacitor C_S can be used to prevent negative feedback or removed to provide a circuit with A_V less dependent on g_m.

The ac equivalent circuit is shown in Fig. 9-15(B) with C_E included. The divider $R_{B1}R_{B2}$ can be made large because (essentially) no dc current flow exists into the gate.

The input resistance to the circuit remains large. Assuming that $R_{B1} \| R_{B2} \gg R_1$, the amplifier voltage gain is again

$$A_V = -g_m R_D \| R_L \qquad (9\text{-}63)$$

Example 9-9   E-MOSFET AC Amplifier

Figure 9-16(B) shows the Q point for an E-MOSFET ac amplifier. If $R_L = 2.0 \text{ k}\Omega$, what value of R_D must be chosen to produce $A_V = -4.0$?

Solution   From the graph,

$$g_m \cong 3.0 \text{ m} \mho$$

so that

$$(R_D \| 2.0 \text{ k}\Omega) = 4.0/3.0 \text{ m}\mho = 1.3 \text{ k}\Omega$$

$$R_D \cong 3.7 \text{ k}\Omega$$

The value of R_D that is chosen in turn fixes the required supply voltage V_{DD}. For a given g_m, as the gain A_V increases, the value of V_{DD} must also increase. As noted previously, the maximum gain that can be achieved is thus limited by dc circuit considerations. ∎

9-15   SUMMARY

Midrange equivalent circuits for ac amplifiers are introduced in this chapter. Several different device and amplifier configurations are explored. The results of this analysis are shown in Figs. 9-17 and 9-18.

Fig. 9-17   Design equations for BJT ac amplifiers.

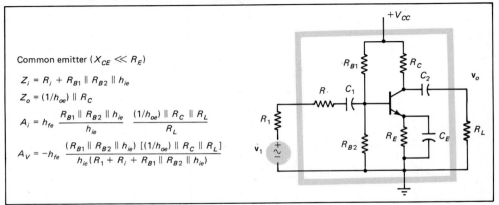

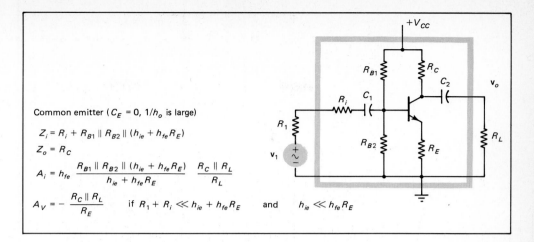

Common emitter ($C_E = 0$, $1/h_o$ is large)

$$Z_i = R_i + R_{B1} \parallel R_{B2} \parallel (h_{ie} + h_{fe}R_E)$$

$$Z_o = R_C$$

$$A_i = h_{fe} \frac{R_{B1} \parallel R_{B2} \parallel (h_{ie} + h_{fe}R_E)}{h_{ie} + h_{fe}R_E} \quad \frac{R_C \parallel R_L}{R_L}$$

$$A_V = -\frac{R_C \parallel R_L}{R_E} \quad \text{if } R_1 + R_i \ll h_{ie} + h_{fe}R_E \quad \text{and} \quad h_{ie} \ll h_{fe}R_E$$

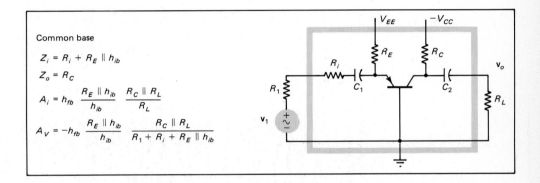

Common base

$$Z_i = R_i + R_E \parallel h_{ib}$$

$$Z_o = R_C$$

$$A_i = h_{fb} \frac{R_E \parallel h_{ib}}{h_{ib}} \quad \frac{R_C \parallel R_L}{R_L}$$

$$A_V = -h_{fb} \frac{R_E \parallel h_{ib}}{h_{ib}} \quad \frac{R_C \parallel R_L}{R_1 + R_i + R_E \parallel h_{ib}}$$

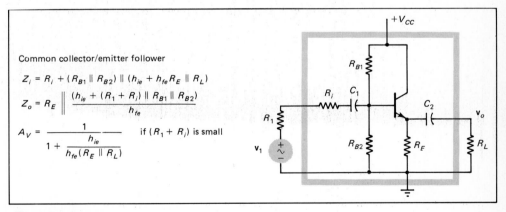

Common collector/emitter follower

$$Z_i = R_i + (R_{B1} \parallel R_{B2}) \parallel (h_{ie} + h_{fe}R_E \parallel R_L)$$

$$Z_o = R_E \parallel \frac{(h_{ie} + (R_1 + R_i) \parallel R_{B1} \parallel R_{B2})}{h_{fe}}$$

$$A_V = \frac{1}{1 + \dfrac{h_{ie}}{h_{fe}(R_E \parallel R_L)}} \quad \text{if } (R_1 + R_i) \text{ is small}$$

Fig. 9-17 *(cont.)*

Numerous design variations for ac amplifiers are possible. It is important to be able to apply the above methods to the treatment of new configurations as they are encountered and to develop new configurations to meet objectives. The approaches developed here can be generalized for a variety of additional circuits.

JFET Common source ($X_{C_S} \ll R_S$)

$$Z_i = R_i + R_G$$

$$Z_o = R_D$$

$$A_V = \frac{-R_G}{(R_1 + R_i) + R_G}\, g_m (R_D \parallel R_L)$$

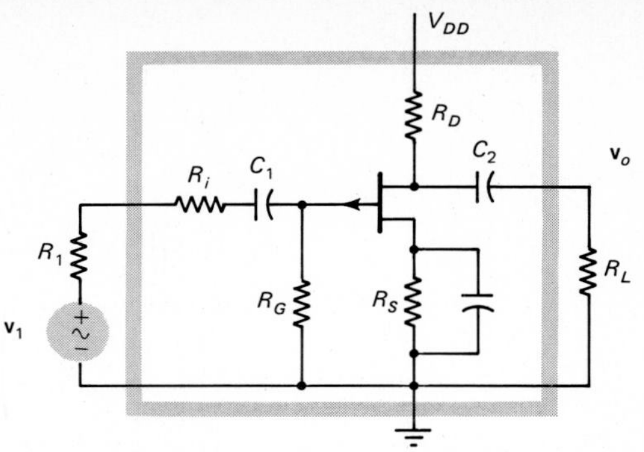

JFET Common source ($C_S = 0$)

$$Z_i = R_i + R_G$$

$$Z_o = R_D$$

$$A_V = \frac{-R_G}{(R_1 + R_i) + R_G} \quad \frac{R_D \parallel R_L}{1/g_m + R_S}$$

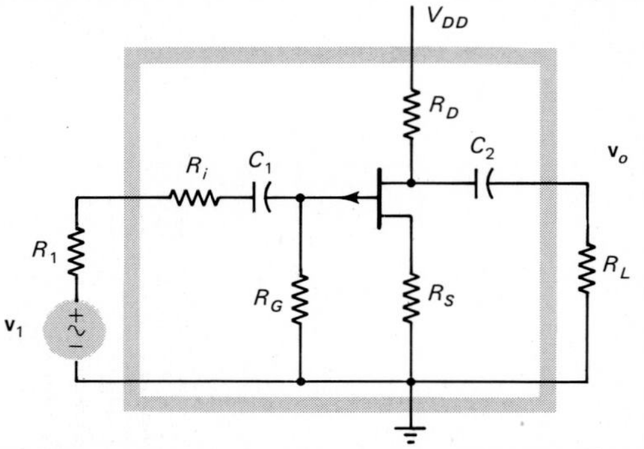

JFET Common drain/source follower

$$Z_i = R_i + R_G$$

$$Z_o = R_S \parallel 1/g_m$$

$$A_V = \frac{R_G}{(R_1 + R_i) + R_G} \quad \frac{1}{1 + \dfrac{1}{g_m (R_S \parallel R_L)}}$$

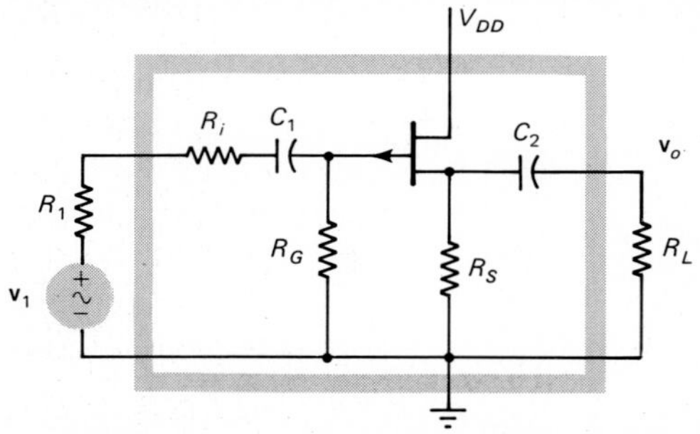

D-MOSFET Common source

$$Z_i = R_i + R_G$$

$$Z_o = R_D$$

$$A_V = -g_m R_D \parallel R_L \qquad \text{if } R_G \gg R_1 + R_i$$

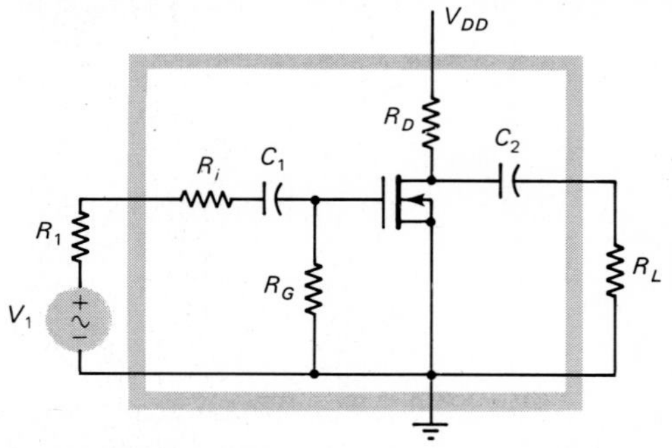

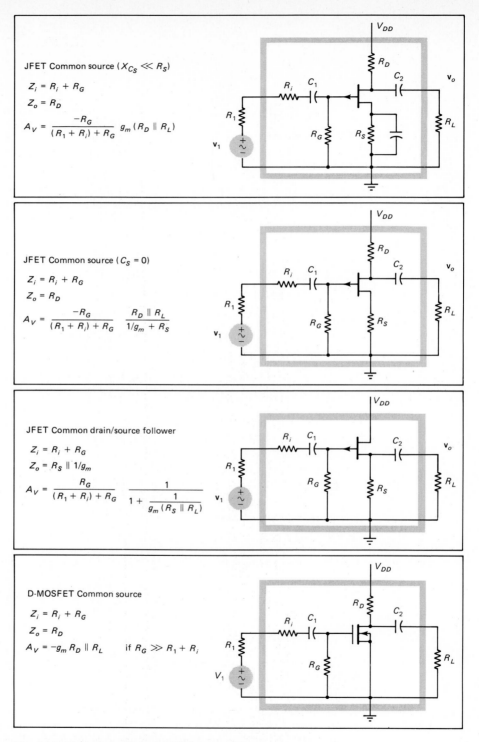

Fig. 9-18 Design equations for FET ac amplifiers.

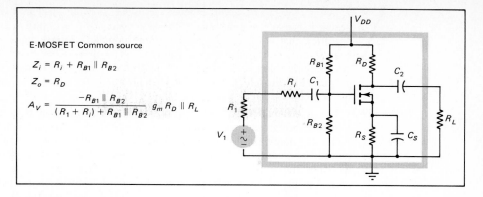

E-MOSFET Common source

$Z_i = R_i + R_{B1} \parallel R_{B2}$

$Z_o = R_D$

$A_V = \dfrac{-R_{B1} \parallel R_{B2}}{(R_1 + R_i) + R_{B1} \parallel R_{B2}} \, g_m R_D \parallel R_L$

Fig. 9-18 *(cont.)*

QUESTIONS AND ANSWERS

Questions

9-1. What are the major differences between dc-coupled and ac amplifiers?

9-2. Why is it useful to develop a family of ac equivalent circuit models for use in ac amplifier design?

9-3. How are dc and ac design constraints linked in an ac amplifier?

9-4. Why do the *h* parameters provide a convenient way to describe the ac operation of BJTs?

9-5. What linkages exist between the BJT ac parameters (h_{ie}, h_{oe}, and g_{me}) and the dc bias point?

9-6. How do typical transconductance values for BJTs and FETs compare?

9-7. What similarities exist between the gain equations of this chapter and the gain equations for the dc-coupled amplifiers derived in Chapters 7 and 8?

9-8. What two ways for designing a common-emitter ac amplifier with a known gain are developed in this chapter? What are the relative advantages and disadvantages of each?

9-9. Describe the tradeoffs involved in the use of common-base, common-emitter, and common-collector ac amplifier configurations to achieve a given A_V design objective.

9-10. How can the value of the collector resistor be modified to increase the gain of an ac amplifier?

9-11. Describe the tradeoffs associated with the use of a common-source FET versus a common-emitter BJT in an ac amplifier circuit.

9-12. What biasing tradeoffs are involved in deciding whether to use a JFET, D-MOSFET, or E-MOSFET in an ac amplifier?

Answers

9-1. A dc-coupled amplifier allows current to flow between the amplifier and the input and output coupling circuits, where the coupling capacitors used in ac amplifiers block this current flow. The ac amplifier thus achieves dc isolation between the amplifier and the input and output coupling circuits. However, the introduction of the coupling capacitors reduces the gain of the ac amplifier at low frequencies (limiting the use of the amplifier), where the midrange gain of dc-coupled amplifiers extends to dc input signals. When these two types of amplifiers are considered, a tradeoff exists between dc isolation and amplifier gain at low frequencies. As discussed in Chapter 10, both types of amplifier experience gain limitations at high frequencies.

9-2. The ac equivalent circuit models are useful to obtain insights that are required for the design process. They encourage the development of an intuitive appreciation for circuit operation that is of assistance when more complex circuits must be understood. The ac equivalent circuit models provide an easily remembered way to develop an "inventory" of circuit types that can be referenced, so that the designer can work with larger, more aggregated building block units rather than always be restricted to conceptualizing a circuit at the component level.

9-3. The dc design constraints are associated with achieving the desired Q-point bias for the transistor being used, whereas the ac design is associated with achieving the desired current and voltage gains and input and output impedances for the amplifier. The dc operating point helps determine the ac device parameters, so a modified operating point couples to the gain and impedance equations. Similarly, the attainment of desired ac properties impacts the dc resistors and voltages selected. For small-signal operation, for which the device parameters can be treated as approximately constant, the dc and ac design tasks are linked through a set of linear equations. If the device parameters vary with input signal, so that nonlinearities exist, computer modeling is required to predict circuit performance. The simple models of this chapter are thus a baseline reference that can be used to initially estimate circuit performance and to guide in the use and evaluation of subsequent computer modeling.

9-4. The h parameters provide a useful description of BJT ac operation because they are closely linked to the input and output characteristic curves for the device. The parameters thus can be associated with graphical equivalents, aiding in application and interpretation.

9-5. The BJT parameters (h_{ie}, h_{oe}, g_{me}) can all be linked to the device dc operating point through the relationships:

$$h_{ie} \cong 0.026 \text{ V}/I_B \qquad 1/h_{oe} \cong V_E/I_C \qquad g_{me} \cong |I_C|/0.026 \text{ V}$$

These equations provide a powerful way to explore how various dc design strategies can be used to affect ac circuit performance. High current levels are associated with low input and output impedances and a high transconductance. Low current levels are associated with high input and output impedance levels and a low transconductance.

9-6. In general, for typical current levels, the transconductance g_m for the BJT is much larger than that for the FET. As a result, high amplifier gain is often easier to obtain using bipolar devices. A tradeoff exists between the high g_m values for BJTs and the high input impedance values associated with FETs.

9-7. The ac gain equations of this chapter represent generalizations of the gain equations of Chapters 7 to 8 because they represent more complex circuits and reduce, in appropriate circumstances, to the equations for dc-coupled amplifiers derived earlier.

9-8. This chapter describes how R_i can be varied to control A_V without use of negative feedback, or how the feedback fraction B can be varied using negative feedback to control A_V. The former strategy allows higher gains, which are dependent on device parameters, accompanied by less Q-point stabilization, whereas the latter strategy produces a lower gain that is independent of device parameters and associated with improved Q-point stability.

9-9. A common-base ac amplifier has a stable Q point and a moderate gain potential because A_V is proportional to $h_{fb} \cong -1$, combined with a low input impedance Z_i and a high output impedance Z_o associated with the device properties. A common-emitter ac amplifier with negative feedback has a reasonably stable Q point and a moderate potential gain, whereas the amplifier without negative feedback has an unstable Q point and a high potential gain (because $h_{fe} \gg 1$). The common-emitter amplifier has a higher input impedance (which can be very high with the use of R_E) and a moderate output impedance Z_o associated with the device properties. A common-collector ac amplifier has a gain $A_V \cong 1$, a moderately high to high input impedance, and a low output impedance; it is therefore useful for driving low-impedance loads and achieving isolation between stages.

9-10. The gain of an ac amplifier can be increased by using a larger value of R_C. However, this solution also requires a larger value of V_{CC}, which can lead to higher power dissipation in the circuit and breakdown of the transistor. An alternative strategy is to use an active load (as introduced in Section 8-8), which has a high ac resistance, but offers only a modest dc voltage drop across itself. Both BJTs and FETs can be used as such active loads. The IC operational amplifier, introduced in Chapter 12, uses this strategy to produce high-gain, high-input-impedance bipolar amplifier stages.

9-11. The FET common-source ac amplifier is characterized by a lower g_m and a higher input impedance. The BJT common-emitter ac amplifier is characterized by a higher g_m and a lower input impedance. In many IC applications, the choice of technology depends on relative power consumption levels and the fabrication constraints associated with each strategy.

9-12. A JFET ac amplifier requires use of a supply voltage or source resistor to bias the gate and maintain reverse bias on the gate-channel junction. A D-MOSFET ac amplifier can be operated without any gate bias, whereas the E-MOSFET ac amplifier requires a voltage divider across the base to maintain a turn-on bias and create a channel.

EXERCISES AND SOLUTIONS

Exercises

(9-3) **9-1.** Graphically determine the Early voltage V_E for the BJT characteristic curves in Fig. 6-14(B).

(9-3) **9-2.** The common-emitter output impedance of a BJT is measured as 10 kΩ, and $V_E = 70$ V is determined from the output characteristic curve for the device. What current level is thus associated with the device as it is being operated?

(9-3) (9-4) **9-3.** A BJT is operated with an emitter current of 50 μA. Estimate h_{oe} and g_{me}.

(9-4) **9-4.** Show that

$$g_{mb} \cong -h_{FE}/h_{ie}$$

for a BJT, so that g_{mb} and g_{me} are usually of the same order of magnitude for a given device.

(9-5) **9-5.** Given $h_{ie} = 2.0$ kΩ, $h_{fe} = 150$ kΩ and $1/h_{oe} = 10$ kΩ, find Z_{11}, Z_{12}, Z_{21}, and Z_{22} for the amplifier model used in this chapter.

(9-6) **9-6.** An ac amplifier has a current gain of 20 and a voltage gain of 10. If the power into the amplifier is 1.0 mW, what is the output power?

(9-6) **9-7.** An ac amplifier has a current gain of 100 and an input impedance of 2.5 kΩ. For a source resistance of 3.0 kΩ and load resistance of 5.0 kΩ, what is the voltage gain of the amplifier?

(9-6) **9-8.** An ac amplifier has a current gain $A_i = 15$. If $R_1 = 1.0$ kΩ, $Z_i = 2.3$ kΩ, and $R_L = 2.2$ kΩ, find the voltage gain of the circuit. Is this a common-base or common-emitter circuit?

(9-7) **9-9.** For the circuit of Fig. 9-6(A), assume that $R_E = 150$ Ω, $R_C = 1.1$ kΩ, $V_{CE} = 8.0$ V, $I_E = -8.0$ mA, and $h_{FE} = 200$. Choose values for R_{B1} and R_{B2} so that a current of about 1.0 mA flows in each resistor. Assume a Ge BJT.

(9-7) **9-10.** For the circuit shown in Fig. 9-19 (see p. 366), (a) choose the value of R_x so that $V_B = -1.7$V (assuming I_B is small), (b) choose the value of R_E to produce $I_E = 5.0$ mA, (c) choose R_C so that $V_{CE} = -5.0$ V, (d) find h_{ie}, and (e) find the midrange gain A_V.

(9-7) **9-11.** Show that for the common-emitter amplifier circuit of Fig. 9-6(A), the maximum possible voltage gain of the amplifier is $A_V \cong g_{me}/h_{oe}$ and is approximately independent of the collector current. What is the approximate value of this maximum gain?

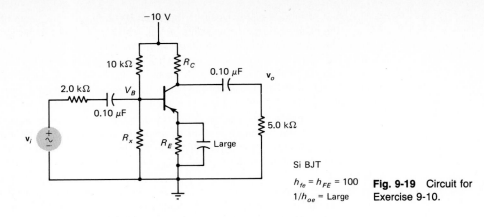

Si BJT

$h_{fe} = h_{FE} = 100$

$1/h_{oe}$ = Large

Fig. 9-19 Circuit for Exercise 9-10.

(9-8) **9-12.** For a common-emitter ac amplifier with feedback, $R_A = 0.50 \text{ k}\Omega$, $R_B = 1.0 \text{ k}\Omega$, $R_C = 0.80 \text{ k}\Omega$, $R_L = 2.4 \text{ k}\Omega$, and $R_E = 0.10 R_C$. Find the feedback factor B.

(9-8) **9-13.** A BJT is biased at $I_C = 500 \ \mu\text{A}$ in a common-emitter ac amplifier with $R_L = 1.0 \text{ k}\Omega$, $R_C = 2.0 \text{ k}\Omega$, $R_1 + R_i$ is negligible, and $R_E = 0.10 \text{ k}\Omega$. (a) Find the gain of the amplifier with R_E shorted by capacitor C_E, (b) find the feedback factor if C_E is removed, and (c) find the amplifier gain with feedback. No voltage divider $R_{B1}R_{B2}$ is used.

(9-9) **9-14.** For the BJT common-base amplifier of Figure 9-9(A), you are given $R_C = 5.0 \text{ k}\Omega$, $R_L = 4.0 \text{ k}\Omega$, $R_1 = 0.20 \text{ k}\Omega$, $R_E = 20 \text{ k}\Omega$, and $h_{ib} \cong 50 \ \Omega$. (a) Design an ac amplifier with a gain $A_V = 5.0$. (b) Find Z_o and Z_i.

(9-9) **9-15.** The common-base circuit shown in Fig. 9-20 requires only one voltage source. Assume that the bypass and coupling capacitors are large enough to produce midrange circuit behavior. (a) Draw the schematic for the midrange ac equivalent circuit, (b) find h_{ib}, and (c) find the value of R_L to produce a gain $A_V = 5.0$.

Si BJT

$h_{FE} = h_{fe} = 150$

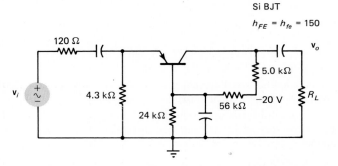
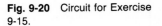

Fig. 9-20 Circuit for Exercise 9-15.

(9-11) **9-16.** You are given a JFET common-source ac amplifier with $R_G = 1.0 \text{ M}\Omega$, $R_D = 2.0 \text{ k}\Omega$, $R_L = 20 \text{ k}\Omega$, and $R_1 = R_i = 0.25 \text{ M}\Omega$. Your design objective is $A_V = -5.0$. What value of g_m is required? Find Z_i and Z_o.

(9-13) **9-17.** A D-MOSFET is biased at $V_{GS} = 0$. Assume that $R_D = 1.0 \text{ k}\Omega$, $R_L = 2.0 \text{ k}\Omega$, $V_P = -3.5 \text{ V}$, and $I_{D0} = 4.5 \text{ mA}$. What amplifier gain A_V results?

9-18. An ac amplifier is manufactured using hybrid microelectronics (discussed in Chapters 1 and 18). A major issue arises over how much testing is to be done on the circuit, and the data of Fig. 9-21 are collected. During testing, most failures occur between 4 to 6 hours. If almost no testing is done, operational failures are high (0.7 percent). After 6 hours of testing, very few operational failures occur. The data indicate that testing is more stressful than actual operations and thus produces a higher percentage of failures. What is the opti-

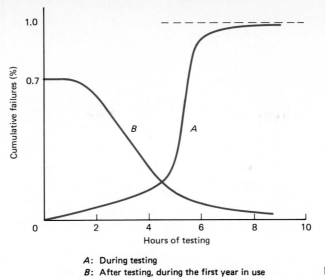

A: During testing
B: After testing, during the first year in use

Fig. 9-21 Data for Exercise 9-18.

mum number of hours of testing to be done on this amplifier to minimize the total percentage of failures during and after testing? What other choice of "hours of testing" might reasonably be made as a sound business decision?

Solutions **9-1.** From Figure 6-14(B), the results of Fig. 9-22 are obtained.

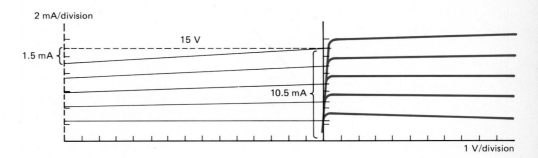

Fig. 9-22 Graphical determination of the Early voltage for Exercise 9-1.

9-2. From Eq. (9-6),

$$|I_C| \cong \frac{V_E}{1/h_{oe}} \cong \frac{70 \text{ V}}{10 \text{ k}\Omega} \cong 7.0 \text{ mA}$$

9-3. Using Eqs. (9-6) and (9-9) and $V_E \cong 1.0 \times 10^2$ V,

$$h_{oe} \cong \frac{|I_C|}{V_E} \cong \frac{50 \ \mu\text{A}}{1.0 \times 10^2 \text{ V}} \cong 0.5 \ \mu\mho$$

and

$$g_{me} \cong 50 \ \mu\text{A}/0.026 \text{ V} \cong 1.9 \ \text{m}\mho$$

9-4. From Eq. (9-8),

$$g_{mb} = h_{fb}/h_{ib} \cong \frac{-1}{h_{ib}}$$

Since $h_{ib} \cong V_o/|I_C|$, then

$$g_{mb} \cong \frac{h_{fb}|I_C|}{V_o} \cong \frac{-h_{FE}|I_B|}{V_o} \cong \frac{-h_{FE}}{h_{ie}}$$

Since $g_{me} \cong h_{fe}/h_{ie}$, then

$$|g_{mb}| \cong |g_{me}|$$

9-5. From Eqs. (9-11) and (9-12),

$$Z_{11} = h_{ie} = 2.0 \text{ k}\Omega$$

$$Z_{12} \cong 0$$

$$Z_{21} \cong 150(10 \text{ k}\Omega) = 1.5 \text{ M}\Omega$$

$$Z_{22} \cong 10 \text{ k}\Omega$$

9-6. From Eq. (9-15),

$$A_P = |A_i A_V| = (20)(10) = 200$$

so that

$$P_{\text{out}} = A_P P_{\text{in}} = 200(1.0 \text{ mW}) = 0.20 \text{ W}$$

9-7. From Eq. (9-16),

$$A_V = -A_i \frac{R_L}{R_i + Z_i} = (-100) \frac{5.0 \text{ k}\Omega}{3.0 \text{ k}\Omega + 2.5 \text{ k}\Omega}$$

$$\cong -91$$

9-8. By convention, the current gain is positive for a common-emitter amplifier and negative for a common-base amplifier. Therefore, with A_i positive, a common-emitter amplifier is considered. From Eq. (9-16),

$$A_V = (-15) \frac{2.2 \text{ k}\Omega}{1.0 \text{ k}\Omega + 2.3 \text{ k}\Omega} \cong -10$$

9-9. Refer to Fig. 9-6(B). Note that

$$I_E R_E = (-8.0 \text{ mA}) (0.15 \text{ k}\Omega) = -1.2 \text{ V}$$

Given a Ge BJT, the required voltage on the base is

$$V_B \cong 1.2 + 0.3 \text{ V} = 1.5 \text{ V}$$

Since $h_{FE} = 200$, $I_B = 8.0 \text{ mA}/200 = 40 \text{ }\mu\text{A} \ll 1.0 \text{ mA}$, so

$$I_B \ll \text{ current in } R_{B1} \text{ and } R_{B2}$$

If 1.0 mA flows in R_{B2},

$$V_R = (1.0 \text{ mA})R_{B2} \cong 1.5 \text{ V}$$

$$R_{B2} \cong 1.5 \text{ k}\Omega$$

The power supply voltage V_{CC} is given by

$$V_{CC} = -I_E R_E + V_{CE} + I_C R_C = 1.2 \text{ V} + 8.0 \text{ V} + (8.0 \text{ mA})(1.1 \text{ k}\Omega) \cong 18.0 \text{ V}$$

Therefore,

$$R_{B1} = \frac{18.0 - 1.5 \text{ V}}{1.0 \text{ mA}} \cong 16.5 \text{ k}\Omega$$

9-10. (a) Assume tentatively that $I_B \ll I_{R_x}$. Then R_x is given by

$$(10 \text{ V}) \frac{R_x}{R_x + 10 \text{ k}\Omega} = 1.7 \text{ V}$$

$$R_x \cong 2.0 \text{ k}\Omega$$

(b) Given $V_B = -1.7 \text{ V}$, $V_E = -1.7 \text{ V} + 0.7 \text{ V} = -1.0 \text{ V}$, so that

$$R_E = 1.0 \text{ V}/5.0 \text{ mA} = 0.20 \text{ k}\Omega$$

Since $I_E = 5.0 \text{ mA}$ and $h_{fe} = 100 \text{ }\Omega$,

$$I_B \cong 5.0 \text{ mA}/100 \text{ }\Omega = 0.05 \text{ mA}$$

The current through the voltage divider on the base is

$$10 \text{ V}/12 \text{ k}\Omega \cong 0.83 \text{ mA}$$

Therefore, I_B is much smaller than the divider current, as originally assumed.

(c) From above, $V_E = -1.0 \text{ V}$. If $V_{CE} = -5.0 \text{ V}$, then 4.0 V must be dropped across R_C and

$$R_C \cong 4.0 \text{ V}/5.0 \text{ mA} = 0.80 \text{ k}\Omega$$

(d) To estimate h_{ie}, use the relationship

$$h_{ie} \cong 0.026 \text{ V}/|I_B| = 0.026 \text{ V}/0.05 \text{ mA} = 0.52 \text{ k}\Omega$$

(e) From Eq. (9-22),

$$A_V = \frac{-100}{0.52 \text{ k}\Omega} \frac{(10 \text{ k}\Omega \| 2.0 \text{ k}\Omega \| 0.52 \text{ k}\Omega)(0.80 \text{ k}\Omega \| 5.0 \text{ k}\Omega)}{2.0 \text{ k}\Omega + 10 \text{ k}\Omega \| 2.0 \text{ k}\Omega \| 0.52 \text{ k}\Omega}$$

$$\cong -22$$

9-11. From Eq. (9-22), let all of the external resistors take on values to maximize A_V. Thus, $R_{B1} \| R_{B2} \gg h_{ie}$, $R_1 + R_i \ll h_{ie}$, and $R_C \| R_L \gg 1/h_{oe}$. Then

$$(A_V)_{max} = \frac{-h_{fe}}{h_{ie} h_{oe}} \cong \frac{-g_{me}}{h_{oe}}$$

Given $1/h_{oe} \cong V_E/|I_C|$ and $g_{me} \cong |I_E|/0.026 \text{ V}$,

$$(A_V)_{max} = \frac{-|I_E|}{0.026 \text{ V}} \frac{V_E}{|I_C|} \cong \frac{-V_E}{0.026 \text{ V}}$$

For $V_E \cong 100 \text{ V}$,

$$(A_V)_{max} \cong -3.8 \times 10^3$$

The attainable values for A_V are much lower for typical circuit designs.

9-12. From Eq. (9-32b),

$$B = \frac{R_E}{R_C \| R_L} \frac{R_A + R_B}{R_B}$$

$$= \frac{0.08 \text{ k}\Omega}{0.80 \text{ k}\Omega \| 2.4 \text{ k}\Omega} \frac{0.50 \text{ k}\Omega + 1.0 \text{ k}\Omega}{1.0 \text{ k}\Omega} \cong 0.20$$

9-13. (a) From Eq. (9-28), with $R_E = 0$, and for the relationships given,

$$A_V = -h_{fe} \frac{R_C \| R_L}{h_{ie}} = -g_{me}(R_C \| R_L)$$

From Eq. (9-9),

$$g_{me} \cong 0.500 \text{ mA}/0.026 \text{ V} \cong 19.2 \text{ m}\mho$$

Therefore,

$$A_V = -(19.2 \times 10^3)(2.0 \text{ k}\Omega \| 1.0 \text{ k}\Omega) = -12.8$$

(b) From Eq. (9-32b),

$$B = \frac{0.10 \text{ k}\Omega}{2.0 \text{ k}\Omega \| 1.0 \text{ k}\Omega} \cong 0.15$$

(c) From Eq. (9-32a),

$$A_{Vf} = \frac{-12.8}{1 + (12.8)(0.15)} \cong -4.4$$

9-14. (a) From Eq. (9-39) and for the values given,

$$A_V = -h_{fb} \frac{R_E \| h_{ib}}{h_{ib}} \frac{R_C \| R_L}{(R_1 + R_i) + R_E \| h_{ib}}$$

$$\cong \frac{5.0 \text{ k}\Omega \| 4.0 \text{ k}\Omega}{R_i + 250 \ \Omega} = 5.0$$

$$R_i \cong 0.19 \text{ k}\Omega$$

(b) From Eq. (9-35),

$$Z_o = R_C = 5.0 \text{ k}\Omega$$

From Eq. (9-34),

$$Z_i = R_i + R_E \| h_{ib} \cong 190 + 50 = 0.24 \text{ k}\Omega$$

9-15. (a) The midrange ac equivalent circuit is shown in Fig. 9-23.

(b)

$$h_{ib} \cong 0.026 \text{ V}/|I_E|$$

Assume that I_B is much smaller than the current through the (24 kΩ, 56 kΩ) voltage divider on the base. Then

$$V_B = \frac{24 \text{ k}\Omega}{24 \text{ k}\Omega + 56 \text{ k}\Omega} (-20 \text{ V}) = -6.0 \text{ V}$$

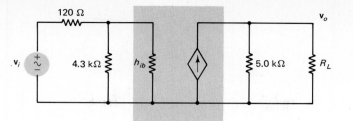

Fig. 9-23 Midrange ac equivalent circuit for Exercise 9-15.

Therefore, $V_E \cong -6.0$ V + 0.7 V = -5.3 V, and

$$I_E \cong 5.3 \text{ V}/4.3 \text{ k}\Omega = 1.2 \text{ mA}$$

and

$$I_B = 1.2 \text{ mA}/150 = 8 \text{ }\mu\text{A}$$

Because the current through the divider is 20 V/80 kΩ = 250 μA, the initial assumption is confirmed. Therefore,

$$h_{ib} \cong 0.026 \text{ V}/1.2 \text{ mA} = 22 \text{ }\Omega$$

(c) By comparing the circuits of part (a) and Fig. 9-9(C), Eq. (9-39) can be applied

$$A_V \cong \frac{5.0 \text{ k}\Omega \| R_L}{120 + 22} = 5.0$$

$$R_L \cong 0.84 \text{ k}\Omega$$

9-16. From Eq. (9-51),

$$A_V = \frac{-1.0 \text{ M}\Omega}{0.50 \text{ M}\Omega + 1.0 \text{ M}\Omega} g_m(2.0 \text{ k}\Omega \| 20 \text{ k}\Omega) = -5.0$$

$$g_m \cong 4.1 \text{ m}\mho$$

9-17. From Eq. (9-62),

$$A_V = -g_m R_D \| R_L$$

To find g_m,

$$g_m = -2I_{D0}/V_P = -2(4.5 \text{ mA})/-3.5 \text{ V} = 2.57 \text{ m}\mho$$

and

$$A_V = (-2.57 \times 10^{-3})(1.0 \text{ k}\Omega \| 2.0 \text{ k}\Omega) = -1.7$$

9-18. For the data given, the solution of Fig. 9-24 results. The preferred solution is about 4.5 hours of testing based on the data shown. However, at this testing level, about equal numbers of failures occur during testing and during operation. A preferred business solution might be to use 6 hours of testing to assure that most failures are caught prior to shipping the product and to accept the higher total failure level.

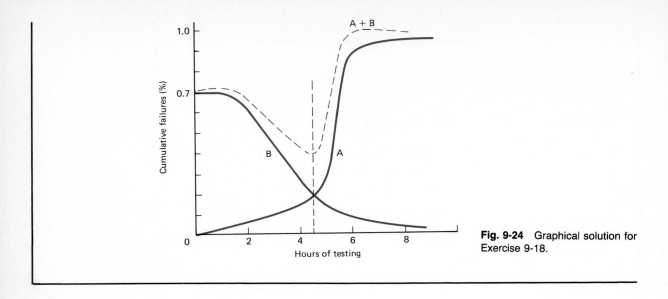

Fig. 9-24 Graphical solution for Exercise 9-18.

PROBLEMS

(9-4) **9-1.** A BJT is operated with $I_C = 100\ \mu A$. Estimate the output resistance $1/h_{oe}$ and g_{me}.

(9-4) (9-5) **9-2.** Find the y and Z parameters as functions of the h parameters for a BJT if $h_r \neq 0$.

(9-5) **9-3.** Show that maximum power transfer takes place between a source and circuit input if $R_1 = R_e(Z_i)$ and between the circuit output and load if $R_e(Z_o) = R_L$.

(9-6) **9-4.** An ac amplifier has an input impedance of 4.8 kΩ and is used with a source having a resistance of 0.8 kΩ. With a 4.0-kΩ load resistor, a voltage gain of -20 is obtained. What is the current gain of the amplifier?

(9-7) **9-5.** A common-emitter ac amplifier is operated with $I_C = 0.50$ mA. If the design objective is $Z_o = 50$ kΩ, what value of R_C should be used?

(9-7) **9-6.** For a common-emitter ac amplifier, R_{B1} and R_{B2} are used with a 12-V power supply to achieve 2.0 V on the BJT base with a divider current of 1.5 mA. If $I_B = (1/20)$ (divider current), what value of R_i should be used to produce $Z_i = 0.5$ kΩ?

(9-7) **9-7.** For a common-emitter ac amplifier, $R_{B1} = 20$ kΩ, $R_{B2} = 2.0$ kΩ, and $I_B = 50\ \mu A$. Find the fractional reduction in amplifier current gain associated with the input voltage divider.

(9-7) **9-8.** For a common-emitter ac amplifier operated with $I_E = 2.5$ mA, $R_C = 1.5$ kΩ, and $R_L = 3.0$ kΩ, find the fractional reduction in amplifier current gain associated with current division at the amplifier output section.

(9-7) **9-9.** For the amplifier in Fig. 9-6(A), let $R_C = 2.0$ kΩ, $R_E = 0.10R_C$, and $V_{CE} = 5.5$ V. Bias at $I_C = 2.5$ mA and choose a Si BJT with $h_{FE} = 150$. Find values for R_{B1} and R_{B2} so that a current of about $20I_B$ flows in each resistor.

(9-7) **9-10.** For the circuit of Fig. 9-6(A), assume that

$$R_{B1} = 3.3\ \text{k}\Omega \qquad h_{FE} \cong h_{fe} = 175 \qquad R_L = 4.7\ \text{k}\Omega$$

$$R_{B2} = 0.47\ \text{k}\Omega \qquad h_{oe} \cong 10^{-5}\ \mho \qquad R_i = 0.10\ \text{k}\Omega$$

$$I_E = -1.5\ \text{mA} \qquad R_C = 3.3\ \text{k}\Omega \qquad R_1 = 0.10\ \text{k}\Omega$$

(a) Find Z_i, Z_o, A_i, A_V, and A_P. (b) What is the maximum gain that can be achieved with this circuit if $R_i = 0$? (c) To achieve a linear operating point, assume that $V_{CE} = 8.0$ V. What value of V_{CC} is required for this circuit if $R_E = 0.1R_C$?

(9-7)   **9-11.** For the BJT common-emitter amplifier of Fig. 9-6(A), you are given

$$R_C = 1.0 \text{ k}\Omega \qquad R_{B1} = 2.0 \text{ k}\Omega \qquad 1/h_o \cong 100 \text{ k}\Omega$$

$$R_L = 5.0 \text{ k}\Omega \qquad R_{B2} = 0.20 \text{ k}\Omega \qquad R_1 = 50 \,\Omega$$

$$h_{ie} = 0.50 \text{ k}\Omega \qquad h_{FE} \cong h_{fe} = 150$$

For a gain $A_V = -8.0$, find the required value for R_i.

(9-7)   **9-12.** For the circuit of Fig. 9-6(A), assume that

$$R_1 + R_i \ll R_{B1} \| R_{B2} \| h_{ie}$$

$$R_L = 4.7 \text{ k}\Omega$$

$$g_{me} = 60 \text{ m}\mho$$

$$1/h_{oe} \cong 200 \text{ k}\Omega$$

What value of R_C is required to achieve $Z_o = 50$ kΩ? What is the amplifier gain?

(9-8)   **9-13.** Show that Eq. (9-32a) follows from Eq. (9-30).

(9-8)   **9-14.** A common-emitter ac amplifier with negative feedback is operated at $I_C = 1.8$ mA and $R_E = 0.20$ kΩ. (a) If $h_{fe} = 200$ for the device, $R_{B1} = 35$ kΩ and $R_{B2} = 5.0$ kΩ for the base voltage divider, and $R_i = 3.8$ kΩ, find the input impedance for the amplifier. (b) If $R_1 \ll R_i$, $R_C = 2.5$ kΩ, and $R_L = 5.0$ kΩ, find the output impedance and the current, voltage, and power gains of the amplifier. (c) By how much has R_E increased the input impedance Z_i of the amplifier?

(9-8)   **9-15.** A common-emitter ac amplifier has a gain of -15 without negative feedback. If a feedback factor $B = 0.35$ is introduced to the circuit, what is the gain of the circuit with feedback?

(9-8)   **9-16.** Assume for a common-emitter ac amplifier that $R_1 + R_i$ is negligible and $h_{ie} \ll h_{fe} R_E$. If $R_E = 0.10R_C$ and $R_L = R_C$, find the voltage gain of the amplifier.

(9-9)   **9-17.** A common-base ac amplifier with $R_i = 0$ has an emitter resistor R_E with a current 15 μA flowing through it. What emitter resistor R_E should be used to produce $Z_i = 1.5$ kΩ?

(9-9)   **9-18.** If $h_{ib} = 0.30R_E$, what fractional reduction in amplifier gain is experienced due to the input current divider?

(9-9)   **9-19.** For the common-base ac amplifier of Fig. 9-9(A), let $R_C = 4.0$ k$\Omega = 9R_E = R_L$ and $I_E = 1.0$ mA, with $R_1 + R_i$ negligible. (a) Find Z_i and Z_o of the amplifier. (b) Find the current, voltage, and power gains of the amplifier.

(9-9)   **9-20.** You wish to design a common-base ac amplifier with $Z_o = 5.0$ kΩ and $R_L = 10$ kΩ using the schematic of Fig. 9-9. Assume $R_1 + R_i = 50 \,\Omega$. What is the voltage gain of the amplifier? (Assume $h_{ib} = 20 \,\Omega \ll R_E$).

(9-10)   **9-21.** An emitter follower operates with $h_{fe} = 150$, $I_B \cong 25 \,\mu$A, and $R_1 + R_i$ small enough to be negligible. What value of R_E produces an output impedance of 5.0 Ω?

(9-10)   **9-22.** A BJT in an emitter follower is operated with $I_C = 0.50$ mA. If $R_E = R_L = 0.10$ kΩ, what is the voltage gain of the amplifier? (Assume $R_1 + R_i$ negligible.)

(9-11)   **9-23.** What is the maximum voltage gain that can be obtained with the amplifier of Fig. 9-11(A) if $R_D = 1.0$ kΩ, $R_L = 5.0$ kΩ, and the device parameters are

$$I_{D0} = 5.0 \text{ mA}$$

$$V_P = -2.0 \text{ V}$$

$$(I_D)_Q = 2.0 \text{ mA}$$

(9-11) **9-24.** A JFET common-source ac amplifier is operated with $R_G \gg (R_1 + R_i)$. (a) If $g_m = 3.0 \text{ m}\mho$, $R_D = 1.8 \text{ k}\Omega$, and $R_L = 2.3 \text{ k}\Omega$, find the voltage gain without feedback (with C_E bypassing R_E). (b) If C_E is removed and $R_S = 0.20 R_D$, find the gain with negative feedback.

(9-12) **9-25.** Calculate A_V for the circuit of Fig. 9-13, with $R_G = 1.5 \text{ M}\Omega$, $R_L = 4.5 \text{ k}\Omega$, and $R_S = 0.85 \text{ k}\Omega$. Assume that $R_1 + R_i = 1.0 \text{ M}\Omega$ and the JFET is characterized by $I_{D0} = -6.0 \text{ mA}$, $V_{GS} = 0.50\text{V}$, $V_P = 2.2 \text{ V}$.

(9-12) **9-26.** You are given a JFET common-drain circuit with the following values:

$$R_G = 2.0 \text{ M}\Omega \qquad R_S = 3.6 \text{ k}\Omega \qquad g_m = 2.0 \text{ m}\mho$$

$$R_1 = 5.0 \text{ M}\Omega \qquad R_L = 18 \text{ k}\Omega$$

$$R_i = 0$$

Find Z_i, Z_o, and A_V.

(9-13) **9-27.** For the D-MOSFET ac amplifier circuit of Fig. 9-14,

$$I_{D0} = -4.0 \text{ mA} \qquad R_D = 1.2 \text{ k}\Omega$$

$$V_P = 2.0 \text{ V} \qquad R_L = 4.8 \text{ k}\Omega$$

$$R_G = 5.1 \text{ M}\Omega \qquad R_1 = 2.0 \text{ M}\Omega$$

Find Z_i, Z_o, and A_V.

(9-14) **9-28.** Find Z_i, Z_o, and A_V for the E-MOSFET ac amplifier of Fig. 9-15(A) with the capacitor C_S removed. Assume that the E-MOSFET of Fig. 9-16(B) is used, biased at the Q point shown. Choose $V_{DD} = 12 \text{ V}$ and $I_{\text{divider}} = 100 \text{ }\mu\text{A}$, and assume that $R_D = 2.0 \text{ k}\Omega$, $R_L = 3.0 \text{ k}\Omega$, and $R_1 = 50 \text{ k}\Omega$.

COMPUTER APPLICATION

* **9-1.** Write a computer program that can be used to calculate Z_i, Z_o, A_i, and A_V for a BJT common-emitter amplifier using the equivalent circuit shown in Fig. 9-6(C). Use the data of Example 9-2 as input, and obtain the output predicted by your program. Compare your answer with the results of the example. Turn in your program listing, output data, and interpretation.

EXPERIMENTAL APPLICATIONS

9-1. The purpose of this experiment is to explore the processes involved in the design of ac equivalent circuit models for midrange frequencies and the test and evaluation of such models. First, design a common-emitter ac amplifier, as shown in Fig. 9-8(A), using the strategy of Section 9-8 (design with negative feedback) to obtain $A_V = -4.0$. Data are to be collected at $f = 10$ kHz, so choose capacitor values that produce $X_{C1} \ll Z_i$ and $X_{C2} \ll Z_o$ at this frequency to obtain midrange operation. Develop an ac equivalent circuit model to represent the amplifier circuit, and calculate theoretical values of Z_i, Z_o, A_i, and A_V using the model.

Now construct your circuit and measure Z_i, Z_o, and A_V at $f = 10$ kHz. Determine $Z_i = \mathbf{v}_i/\mathbf{i}_i$ by attaching a signal generator to the input and load R_L to the output [Fig. 9-4(A)]. Find $Z_o = \mathbf{v}_o/\mathbf{i}_o$ by attaching the signal generator to the output and with a shorted input [Fig. 9-4(B)]. Use resistors in series with the generator to enable current measurements to be made. Compare the measured and predicted values of Z_i, Z_o, and A_V. Assess the accuracy and usefulness of the ac equivalent circuit model.

9-2. The objective of this experiment is to design, build, and test a common-source JFET ac amplifier using self-bias and without a capacitor C_S. For a given JFET, calculate all circuit values to produce $A_V = -2.5$ and a midrange region that includes a frequency of 10 kHz. Check the actual gain of your circuit at this frequency. Compare the predicted and observed results and discuss.

Discussion

A typical output characteristic curve (for the 2N5460 p-channel JFET) is shown in Fig. 9-25. Figure 9-26 shows how a load line can be established using the characteristic curve. Figure 9-27 shows the transfer characteristic and the calculation of g_m and R_S. All parameters are now determined and the amplifier can be constructed to yield the desired gain.

Fig. 9-25 Output characteristic for Experimental Application 9-2.

1 mA/division

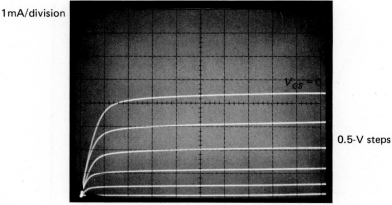

$V_{GS} = 0$

0.5-V steps

2 V/division

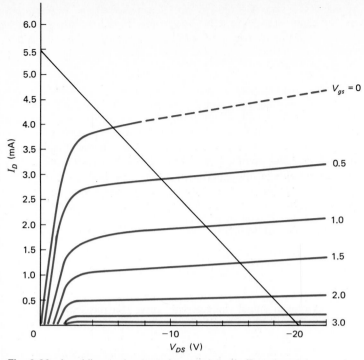

Fig. 9-26 Load line and output characteristic for Experimental Application 9-1.

Fig. 9-27 Transfer characteristic and solution for Experimental Application 9-2.

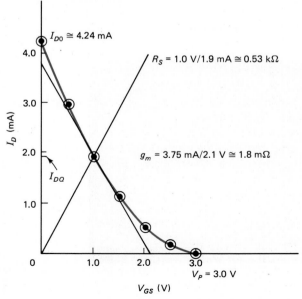

CHAPTER 10
FREQUENCY RESPONSE OF AC AMPLIFIERS

The midrange ac models of Chapter 9 are now extended to include frequency-dependent effects. The models become more complex and, at the same time, higher in information content. As will be seen, all of the previous results become special cases of the models developed here. This progression from more limited to more extended models enables the expansion of modeling skills through incremental growth.

This chapter also describes some new modeling strategies that are powerful design aids. The "poles-and-zeros" approach to understanding frequency response enables the viewing of a complex circuit interaction from a simplified perspective. Once proficient in this technique, the designer can estimate the frequency response of an amplifier circuit by inspection. The models developed in this chapter enhance the ability to interact in a CAD setting and provide a foundation for working with the more complex circuits to follow.

It has been demonstrated that the addition of coupling capacitors can be beneficial because they provide isolation for the dc design task. At midrange frequencies, these capacitors appear as short circuits and do not affect amplifier parameters. However, at low frequencies, the X values of the coupling capacitors increase, affecting circuit performance. If a bypass capacitor is used to prevent signal feedback through an emitter or source resistor, low frequencies also affect the degree of bypass achieved. Equivalent circuits must now be developed for study of low-frequency circuit behavior.

In addition, the circuit is affected at high frequencies by internal device capacitances and distributed wiring capacitances. Equivalent circuits must also be found to model the ways in which these capacitances affect circuit gain.

The hybrid-π equivalent circuit is an important ac representation for the BJT, as shown in Fig. 10-1. The hybrid-π is a useful starting point because of its generality. Several assumptions are then to be made to simplify the model for use in this and the following chapters.

By comparing Figs. 10-1 and 9-3, it can be seen that the hybrid-π model of the BJT differs from the midrange model by the addition of resistors and capacitors. The model shown in Fig. 10-1 recognizes that, in an electrical sense, the effective center of the base region is not directly connected to the base terminal of the device. If b' represents the effective center of the base region and b the base terminal, the ac resistance between these two points can be described as $r_{bb'}$. The effective ac resistances between point b' and the collector and emitter terminals can be written $r_{b'c}$ and $r_{b'e}$.

The origin of internal device capacitances also must be considered. Each of the pn junctions in the BJT is formed by semiconducting regions separated by a deple-

Fig. 10-1 AC equivalent circuits for the BJT common-emitter configuration: (A) Hybrid-π ac equivalent circuit. (B) Approximate ac equivalent circuit.

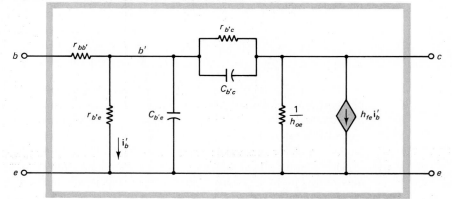

(A)

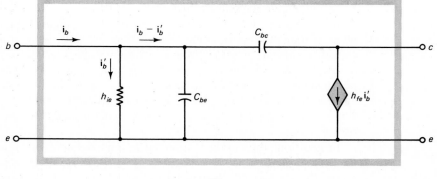

(B)

tion region. Two conducting materials separated by a nonconducting material constitute a capacitor. The junctions thus provide small but important capacitances across the base-emitter junction ($C_{b'e}$) and the base-collector junction ($C_{b'c}$). If the resistances and junction capacitances are added to the midrange ac equivalent circuit model of Fig. 9-3, the hybrid-π equivalent circuit model of Fig. 10-1(A) is obtained.

The value of this model lies in its generality. However, its complexity also limits its usefulness. It is often helpful to simplify the hybrid-π model for a given application, depending on the objectives of interest. This chapter is primarily concerned with amplifier circuits, so the capacitors in the hybrid-π model are important. The resistance $r_{bb'}$ is less important for the particular purposes of this chapter.

Given the stated objectives, now assume that $r_{bb'}$ is very small, so that points b and b' can be treated as approximately the same location. In many cases, $r_{b'c}$ and $1/h_o$ are very large. The result of the above assumptions is the approximate ac equivalent circuit model of Fig. 10-1(B).

The midrange equivalent circuit has been modified by the addition of two junction capacitances. Observe that for ac signals, part of the base current flows through the resistor h_{ie} and part through the device capacitances C_{be} and C_{bc}. Only the portion of the total current that flows through h_{ie} is amplified by the current generator. Also note that the gain of the current generator h_{fe} is determined for midrange frequencies.

The approximate model of Fig. 10-1(B) is still awkward to use in circuit design applications because of the location of the capacitor C_{bc}. It is useful to develop an alternative way to represent the effect of C_{bc} on the input and output portions of this model. First, consider the effect of C_{bc} on the input circuit, as shown in Fig. 10-2(A). The input impedance Z_x looking into the combination of C_{bc} and C_{be} is given

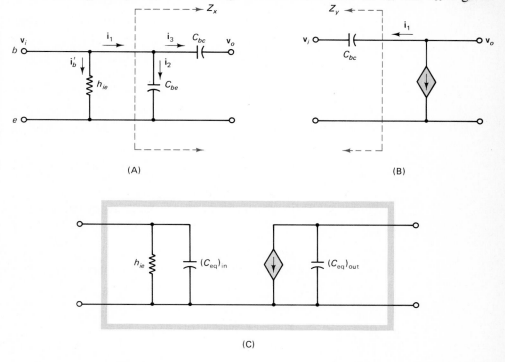

Fig. 10-2 Simplified BJT common-emitter ac equivalent circuit. (A) Equivalent input capacitance. (B) Equivalent output capacitance. (C) BJT model.

by

$$Z_x = \frac{\mathbf{v}_i}{\mathbf{i}_i} = \frac{\mathbf{v}_i}{\mathbf{i}_2 + \mathbf{i}_3} \qquad (10\text{-}1)$$

Since

$$\mathbf{i}_2 = \frac{\mathbf{v}_i}{-jX_{C_{be}}} \qquad \text{and} \qquad \mathbf{i}_3 = \frac{\mathbf{v}_i - \mathbf{v}_o}{-jX_{C_{bc}}} \qquad (10\text{-}2)$$

Eq. (10-1) can be written

$$Z_x = \frac{\mathbf{v}_i}{\dfrac{\mathbf{v}_i}{-jX_{C_{be}}} + \dfrac{\mathbf{v}_i - \mathbf{v}_o}{-jX_{C_{bc}}}} \qquad (10\text{-}3)$$

so that

$$\frac{1}{Z_x} = \frac{1}{-jX_{C_{be}}} + \frac{1 - \mathbf{v}_o/\mathbf{v}_i}{-jX_{C_{bc}}} \qquad (10\text{-}4)$$

$$\frac{1}{Z_x} = j\omega C_{be} + j\omega C_{bc}\left(1 - \frac{\mathbf{v}_o}{\mathbf{v}_i}\right) \qquad (10\text{-}5)$$

The equivalent capacitance C_{eq} as seen looking into C_{be} and C_{bc} is thus

$$(C_{\text{eq}})_{\text{in}} = C_{be} + C_{bc}\left(1 - \frac{\mathbf{v}_o}{\mathbf{v}_i}\right) \qquad (10\text{-}6)$$

From Fig. 10-1, the voltage gain of the BJT can be written

$$\frac{\mathbf{v}_o}{\mathbf{v}_i} = -\frac{h_{fe}\mathbf{i}'_b}{h_{ie}\mathbf{i}'_b} Z_o \| R_L = -g_{me} Z_o \| R_L \qquad (10\text{-}7)$$

where $Z_o \| R_L$ is the impedance of the output circuit as seen by the current generator $h_{fe}\mathbf{i}'_b$. By substitution,

$$(C_{\text{eq}})_{\text{in}} = C_{be} + C_{bc}(1 + g_{me} Z_o \| R_L) = C_{be} + C_M \qquad (10\text{-}8)$$

The second term arises because the voltage across C_{bc} is given by $\mathbf{v}_i - \mathbf{v}_o$. Since $\mathbf{v}_o$ is larger than $\mathbf{v}_i$ by $g_{me} Z_o \| R_L$ and is out of phase by 180 degrees, the total voltage across the capacitor is larger by the factor $(1 + g_{me} Z_o \| R_L)$. The larger effective capacitance arises because the larger voltage swing on the collector side of the capacitor makes it seem, when looking into the capacitor from the base side, that the capacitor has a larger value. This second term, which can cause a large contribution to $(C_{\text{eq}})_{\text{in}}$, is called the *Miller capacitance* C_M.

Now consider the effect of C_{bc} on the output portion of the circuit of Fig. 10-2(B). Looking into the combination of C_{bc} and C_{ce},

$$Z_y = \frac{\mathbf{v}_o}{\mathbf{i}_1} \qquad \text{and} \qquad \frac{1}{j\omega C_{bc}}\mathbf{i}_1 = \mathbf{v}_o - \mathbf{v}_i \qquad (10\text{-}9)$$

so that

$$Z_y = \frac{\mathbf{v}_o}{(\mathbf{v}_o - \mathbf{v}_i)\,j\omega C_{bc}} = \frac{1}{j\omega\,(C_{\text{eq}})_{\text{out}}} \tag{10-10}$$

$$(C_{\text{eq}})_{\text{out}} = C_{bc}\left(1 - \frac{\mathbf{v}_i}{\mathbf{v}_o}\right) \tag{10-11}$$

In most amplifier applications, the output voltage $\mathbf{v}_o$ is much greater than the voltage $\mathbf{v}_i$ on the base of the BJT. For the moment, assume that $\mathbf{v}_o \gg \mathbf{v}_i$. This is a good assumption for most circuits and frequency ranges of interest.

Equation (10-11) thus becomes

$$(C_{\text{eq}})_{\text{out}} \cong C_{bc} \tag{10-12}$$

By looking into C_{bc} from the collector, the input side is seen to be approximately at ground. Therefore, no multiplier effect develops.

Figure 10-2(C) shows the BJT model that results from the above discussion. Note that the input and output portions of the model are now isolated from one another.

For the common-base configuration, the BJT is arranged so that the equivalent Miller capacitance is much smaller, as shown in Fig. 10-3. The emitter and collector are separated by the (grounded) base. The common-base equivalent circuit can thus be described by the capacitors

$$(C_{\text{eq}})_{\text{in}} \cong C_{eb}$$

$$(C_{\text{eq}})_{\text{out}} \cong C_{cb} \tag{10-13}$$

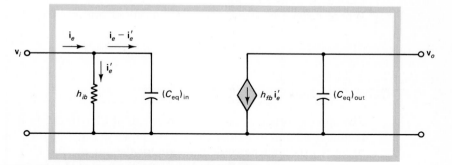

Fig. 10-3 BJT common-base ac equivalent model.

By comparing Eqs. (10-8) and (10-12) for the common-emitter amplifier with Eq. (10-13) for the common-base amplifier, it can be noted that $(C_{\text{eq}})_{\text{in}}$ is much smaller in the latter case. This difference has an important effect on the comparative frequency response characteristics of the two configurations.

The JFET has equivalent capacitors between the gate and source and between the gate and drain, as shown in Fig. 10-4. Therefore, the JFET common-source configuration is similar to the BJT common-emitter configuration. By analogy,

$$(C_{\text{eq}})_{\text{in}} = C_{gs} + C_{gd}(1 + g_{ms}Z_o\|R_L)$$

$$(C_{\text{eq}})_{\text{out}} \cong C_{gd} \tag{10-14}$$

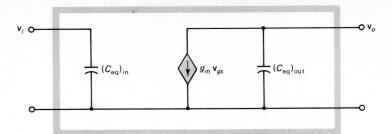

Fig. 10-4 JFET common-source ac equivalent circuit.

For typical amplifiers, g_{ms} for the JFET is less than g_{me} for the BJT. Therefore, the Miller capacitance is usually smaller for the JFET.

By introducing $(C_{eq})_{in}$ and $(C_{eq})_{out}$, BJT and JFET models have been developed with the input and output portions of the transistor isolated from one another. This turns out to be an advantage in developing an understanding of total circuit performance as a function of frequency.

A remaining problem is presented by the Miller capacitance C_M. In general, since $\mathbf{v}_o/\mathbf{v}_i$ and Z_o are functions of frequency, C_M behaves as a frequency-dependent capacitor. However, a treatment based on a frequency-dependent C_M can lead to analytic complexity that prevents an understanding of amplifier operation.

A more workable strategy is to consider the limiting cases for which C_M can be treated as a constant, and then to develop transitions between the limiting cases where necessary. This strategy is followed below. Both $(C_{eq})_{in}$ and $(C_{eq})_{out}$ are regarded as constants for the derivation of basic frequency-response relationships. Where needed in applications, modifications are introduced to cope with transition regions.

10-2 FREQUENCY RESPONSE
FOR THE BJT COMMON-EMITTER AC AMPLIFIER

The approximate device equivalent circuit of Fig. 10-2(C) can be incorporated into a common-emitter amplifier circuit, as shown in Fig. 10-5. C_1 and C_2 are coupling capacitors used to achieve dc isolation for the circuit, and C_E is used to bypass R_E.

If C_E is chosen sufficiently large, it acts as an ac short circuit across R_E for all frequencies of interest. Therefore, the ac gain is not reduced by the presence of R_E. At the same time, the dc bias stabilization is still obtained. For the moment, X_{C_E} is assumed much smaller than R_E over the range of frequencies of primary interest. The impact of the R_E, C_E combination on frequency response is discussed in Section 10-13.

Figure 10-5 introduces the effective distributed wiring capacitances C_{W1} and C_{W2}. These distributed capacitances arise between all conductors that are separated by a nonconducting layer or region. C_{W1} represents the effective capacitance of all input connections from the base to ground and C_{W2} represents the effective capacitance of all output connections from the collector to ground.

Figure 10-5(B) shows the ac equivalent circuit for the entire amplifier, using the device ac model developed above. Capacitor C_i is the parallel combination of $(C_{eq})_{in}$

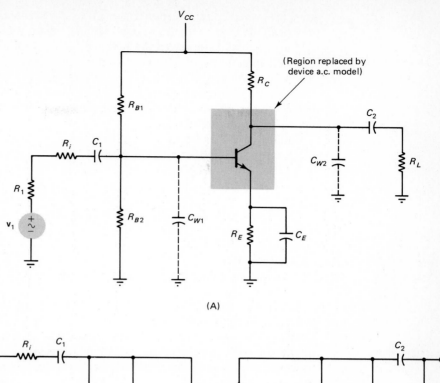

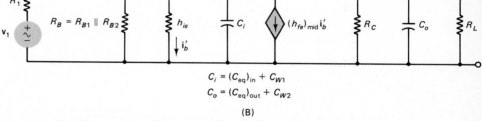

$$C_i = (C_{eq})_{in} + C_{W1}$$
$$C_o = (C_{eq})_{out} + C_{W2}$$

(B)

Fig. 10-5 (A) BJT common-emitter ac amplifier. (B) Composite ac equivalent circuit.

and C_{W1}, and capacitor C_o is the parallel combination of $(C_{eq})_{out}$ and C_{W2}. These effective capacitances thus represent a combination of junction properties inside the device and circuit properties outside the device.

In the following development, it is assumed that the frequencies for which C_1 and C_2 are important are much lower than the frequencies for which C_i and C_o are important. Thus, capacitors C_i and C_o can be placed on either side of C_1 and C_2, since C_1 and C_2 can be regarded as short circuits at frequencies for which C_i and C_o are important.

In the midfrequency range, the composite equivalent circuit for Fig. 10-5(B) reduces to the midrange circuit of Fig. 9-6(C). Therefore, the composite circuit can be used to see how this same common emitter amplifier behaves at low, midrange, and high frequencies.

It is important for the notation used to distinguish between amplifier performance at midrange (covered in Chapter 9) and amplifier performance at low and high frequencies, where circuit capacitances affect performance. The symbol A_V has been previously defined as the (midrange) voltage gain for an amplifier. A new parameter θ_1 is now added to represent the midrange phase shift associated with the amplifier. From Chapter 9, $\theta_1 = 0°$ or $\theta_1 = 180°$, depending on the amplifier configuration that is used. A_V and θ_1 are midrange parameters.

The voltage attenuation due to any frequency-sensitive portion of a circuit is now defined as α_V, and the phase shift associated with this circuit portion as θ_2. Both α_V and θ_2 are functions of frequency.

Finally, let A_V^* and θ^* represent the total gain and phase shift, respectively, associated with an amplifier when frequency-sensitive aspects of the circuit are considered. The above variables are thus related so that

$$|A_V^*| = |A_V||\alpha_V|$$

$$\theta^* = \theta_1 + \theta_2 \qquad (10\text{-}15)$$

10-3 BJT COMMON-EMITTER LOW-FREQUENCY RESPONSE DUE TO OUTPUT COUPLING

The coupling capacitors C_1 and C_2 pass midrange and high frequencies almost unimpeded but significantly attenuate low frequencies. DC voltages are completely blocked. The low-frequency equivalent circuit is shown in Fig. 10-6. The input and output sections can be analyzed separately because they are connected by only a single wire, as shown in Fig. 10-6(A). Each section constitutes a *high-pass filter*.

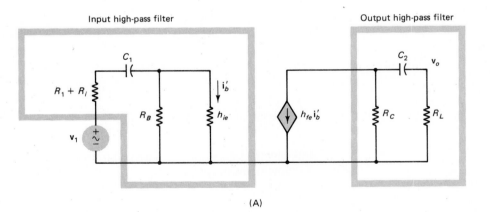

(A)

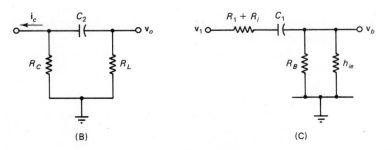

Fig. 10-6 (A) Common-emitter, low-frequency, ac amplifier equivalent circuit. (B) Output portion. (C) Input portion.

(B)   (C)

At midrange, both filters have a gain of unity. As the signal frequency is reduced below midrange, the gain begins to drop. The lower *half-power frequency* (or *point*) is defined as the frequency for which the magnitude of the gain is $1/\sqrt{2}$ times the midrange value.

First, consider the effect of C_2 alone on the low-frequency gain of the amplifier. The circuit can be characterized by finding the frequency f_1 at which the output of Fig. 10-6(B) is reduced by $1/\sqrt{2}$.

With C_2 in place,

$$\mathbf{v}_{o2} = \mathbf{i}_c Z_{\text{eff}} = \frac{R_C(R_L - jX_C)}{R_C + R_L - jX_C} \frac{R_L}{R_L - jX_C} \tag{10-16}$$

and with the capacitor shorted at midrange frequencies ($X_C \to 0$),

$$\mathbf{v}_{o1} = \mathbf{i}_c \frac{R_C R_L}{R_C + R_L} \tag{10-17}$$

Let α_V be the reduction in circuit gain due to this portion of the circuit. Then,

$$\alpha_V = \frac{\mathbf{v}_{o2}}{\mathbf{v}_{o1}} = \frac{R_C + R_L}{R_C + R_L - jX_C} = \frac{1}{1 - j\dfrac{X_C}{R_C + R_L}} \tag{10-18}$$

The half-power point occurs when $R_C + R_L = X_C$, or

$$f_1 = \frac{1}{2\pi C_2(R_C + R_L)} \tag{10-19}$$

so that

$$\alpha_V = \frac{1}{1 - j\dfrac{X_C(f)}{X_C(f_1)}}$$

Since

$$\frac{X_C(f)}{X_C(f_1)} = \frac{2\pi f_1 C_2}{2\pi f C_2} = \frac{f_1}{f} \tag{10-20}$$

the result becomes

$$\alpha_V = \frac{1}{1 - j\dfrac{f_1}{f}} \tag{10-21}$$

In component form,

$$|\alpha_V| = \frac{1}{[1 + (f_1/f)^2]^{1/2}}$$
$$\theta_2 = \tan^{-1}(f_1/f) \tag{10-22}$$

Both positive and negative values for the midrange gain A_V have been encountered in previous chapters, depending on the midrange amplifier phase shift ($0°$ or $180°$). Now consider both the midrange phase shift and phase shifts introduced by the coupling circuits. For the common-emitter amplifier circuit combined with the output coupling circuit,

$$|A_v^*| = |A_v||\alpha_v| = \frac{|A_v|}{[1 + (f_1/f)^2]^{1/2}}$$

$$\theta^* = \theta_1 + \theta_2 = 180° + \tan^{-1}(f_1/f) \qquad (10\text{-}23)$$

where A_v and θ are the midrange gain and phase shift, respectively, for the circuit, and A_v^* and θ^* represent the gain and phase shift, respectively, with the coupling circuit.[1] From the previous discussion, $\theta_1 = 180°$ for BJT amplifiers in the common-emitter configuration and for JFET amplifiers in the common-source configuration. (Also, $\theta_1 = 0°$ for the BJT common-base and common-collector arrangements and for the JFET common-drain arrangement.)

Example 10-1 Low-Frequency Response due to C_2

A BJT common-emitter amplifier has a midrange voltage gain $A_v = -25$. If $C_2 = 1.0\ \mu\text{F}$, $R_C = 1.0\ \text{k}\Omega$, and $R_L = 5.0\ \text{k}\Omega$, find the low-frequency half-power point f_1 due to the output coupling circuit. Plot the total gain $|A_v^*|$ and the total phase shift θ^* for the amplifier for low frequencies (neglecting the effects of the input coupling circuit).

Solution From Eq. (10-19),

$$f_1 = \frac{1}{2\pi(1.0\ \mu\text{F})(6.0\ \text{k}\Omega)} \cong 27\ \text{Hz}$$

and from Eq. (10.23),

$$|A_v^*| = \frac{25}{[1 + (27/f)^2]^{1/2}}$$

$$\theta^* = 180° + \tan^{-1}(27/f)$$

$|A_v^*|$ and θ^* are plotted as a function of f in Fig. 10-7.

Fig. 10-7 Frequency response due to C_2 (Example 10-1).

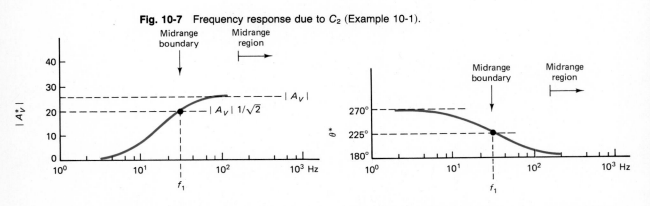

---

[1] The phase angles are, of course, periodic over 360°. Therefore, the phase angles can all be reduced by 360°, so that the midrange value is $-180°$

As expected for a high-pass filter, $|A_v^*|$ is asymptotic to $|A_v| = 25$ for $f \gg f_1$. At the half-power point f_1, $|A_v^*| = \dfrac{1}{\sqrt{2}}|A_v|$. The voltage gain drops off below f_1, asymptotically approaching zero. The phase shift is 180° at midrange, as expected, and increases by 90° for $f \ll f_1$. The total amplifier phase shift is 180° + 45° = 225° at f_1. ∎

10-4 BJT LOW-FREQUENCY RESPONSE DUE TO INPUT COUPLING

Now consider the effect of C_1 alone on the low-frequency gain. Circuit performance is described in terms of the frequency f_1 at which the portion of the circuit shown in Fig. 10-6(C) reduces the gain by $1/\sqrt{2}$.

With C_1 in place

$$\mathbf{v}_{o2} = \frac{\mathbf{v}_1(R_B \parallel h_{ie})}{(R_1 + R_i) + (R_B \parallel h_{ie}) - jX_C} \tag{10-24}$$

and with $X_C \rightarrow 0_1$

$$\mathbf{v}_{o1} = \frac{\mathbf{v}_1(R_B \parallel h_{ie})}{(R_1 + R_i) + (R_B \parallel h_{ie})} \tag{10-25}$$

The reduction in gain is

$$\alpha_V = \frac{\mathbf{v}_{o2}}{\mathbf{v}_{o1}} = \frac{(R_1 + R_i) + (R_B \parallel h_{ie})}{(R_1 + R_i) + (R_B \parallel h_{ie}) - jX_C}$$

$$= \frac{1}{1 - \dfrac{jX_C}{(R_1 + R_i) + (R_B \parallel h_{ie})}} = \frac{1}{1 - j\dfrac{f_1}{f}} \tag{10-26}$$

The half-power point occurs when

$$f_1 = \frac{1}{2\pi C_1(R_1 + R_i + R_B \parallel h_{ie})} \tag{10-27}$$

Since Eqs. (10-19) and (10-27) are of the same form, the low-frequency behavior caused by C_2 is similar to that of C_1. The only difference is in the redefinition f_1 when considering one coupling circuit instead of the other.

Example 10-2 **Low-Frequency Response Due to C_1**

For a BJT common-emitter ac amplifier, $C_1 = 0.10\ \mu F$, $R_i = 1.0\ k\Omega$, $R_B = 50\ k\Omega$, and $h_{ie} = 1.0\ k\Omega$. R_1 is negligible. The midrange gain is -12. Circuit design calls for the magnitude of the gain to be greater than 6.0 at $f = 500$ Hz. What total gain and phase shift does the amplifier exhibit at 500 Hz? Neglect the effects of the output coupling circuit.

Solution   From Eq. (10-27),

$$f_1 = \frac{1}{2\pi \,(0.10\,\mu\text{F})(2.0\;\text{k}\Omega)} \cong 8.0 \times 10^2\,\text{Hz}$$

At 500 Hz,

$$|A_V^*| = \frac{12}{\left[1 + \left(\dfrac{0.80}{0.50}\right)^2\right]^{1/2}} \cong 6.4$$

$$\theta^* = 180° + \tan^{-1}\!\left(\frac{0.80}{0.50}\right) \cong 240°$$

A plot of the low-frequency response of this circuit is shown in Fig. 10-8. The circuit meets the given performance objective.

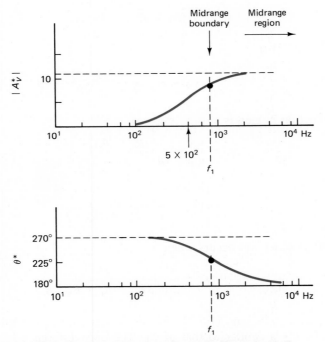

Fig. 10-8   Frequency response due to C_1 (Example 10-2).

The gain $|A_V^*|$ is again asymptotic to $|A_V|$ for $f \gg f_1$. The frequency of interest is 500 Hz, which is less than $f_1 = 800$ Hz. Therefore, $|A_V^*|$ is less than $\dfrac{1}{\sqrt{2}}|A_V|$ for the frequency of interest. The phase shift at 500 Hz is above 225°, as expected. The response characteristics of Figs. 10-7 and 10-8 are similar, differing only in the definitions for f_1. ■

At high frequencies, the gain of the BJT common-emitter circuit is reduced due to device capacitances and distributed wiring capacitances. The high-frequency equivalent circuit is shown in Fig. 10-9. As the signal frequency is increased above midrange, the gain begins to drop.

The upper half-power point f_2 occurs at the frequency for which the gain is $1/\sqrt{2}$ times its midrange value. The values of f_2 that are produced by C_i and C_o are analyzed separately because only one connection exists between the input and output. Each section is a *low-pass filter*.

Consider the effect of C_o alone on the high-frequency gain of the amplifier. What is the frequency f_2 at which the output of Fig. 10-9(B) is reduced by $1/\sqrt{2}$?

With C_o in place

$$\mathbf{v}_{o2} = \mathbf{i}_c \frac{R_C R_L(-jX_C)}{R_C R_L + (R_C + R_L)(-jX_C)} \tag{10-28}$$

and with the capacitor open-circuited at the midrange frequencies ($X_C \to \infty$),

$$\mathbf{v}_{o1} = \mathbf{i}_c \frac{R_C R_L}{R_C + R_L} \tag{10-29}$$

Fig. 10-9 (A) High-frequency ac equivalent circuit. (B) Output portion. (C) Input portion.

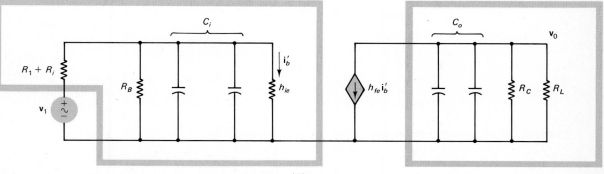

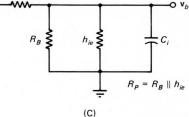

The reduction in gain due to this portion of the circuit is

$$\alpha_V = \frac{\mathbf{V}_{o2}}{\mathbf{V}_{o1}} = \frac{(R_C + R_L)(-jX_C)}{R_C R_L + (R_C + R_L)(-jX_C)} = \frac{1}{1 + j\left(\dfrac{R_C \| R_L}{X_C}\right)} \quad (10\text{-}30)$$

The upper half-power point occurs when $X_C = R_C \| R_L$, or

$$f_2 = \frac{1}{2\pi C_o(R_C \| R_L)} \quad (10\text{-}31)$$

Since

$$\frac{R_C R_L}{R_C + R_L} = X_C(f_2)$$

then

$$\frac{R_C R_L}{R_C + R_L}\frac{1}{X_C} = \frac{X_C(f_2)}{X_C(f)} = \frac{f}{f_2}$$

so that

$$\alpha_V = \frac{1}{1 + j(f/f_2)} \quad (10\text{-}32)$$

Note the similarities and differences between Eq. (10-21) (for low-frequency roll-off) and Eq. (10-32) (for high-frequency roll-off). The denominator of Eq. (10-21) has the term $1 - jf_1/f$, whereas the denominator of Eq. (10-32) has the term $1 + jf/f_2$.

Following the procedure used above,

$$|A_V^*| = |A_V||\alpha_V| = \frac{|A_V|}{[1 + (f/f_2)^2]^{1/2}} \quad (10\text{-}33)$$

$$\theta^* = \theta_1 + \tan^{-1}(-f/f_2) = \theta_1 - \tan^{-1}(f/f_2)$$

for the amplifier combined with the output coupling circuit.

Example 10-3 Amplifier Phase Shift at High Frequencies

An amplifier has a gain $A_V = -30$ at midrange. The circuit objective is to operate at a frequency ratio f/f_2 for which the total θ^* for the amplifier is $125°$. What is the amplifier ratio $|A_V^*/A_V|$ at this frequency?

Solution The midrange phase shift of the amplifier is $\theta_1 = 180°$. At lower frequencies, θ^* increases above θ_1, so there will be no value $\theta^* = 125°$ in the low-frequency region. In the high-frequency region, θ^* decreases from its midrange value of $180°$ to an asymptotic value of $90°$. A solution to the problem can exist in this region.

The results of Fig. 10-10 can be obtained by plotting Eq. (10-33). The value $\theta^* = 125°$ corresponds to $f/f_2 \cong 1.4$.

Observe that $|A_V^*/A_V| \cong 0.57$ at this frequency ratio.

For the low-pass filter, $|A_V^*|$ is asymptotic to $|A_V|$ for $f \ll f_2$ and is asymptotic to zero for $f \gg f_2$. The phase shift associated with this filter stage is $180°$ at midrange $f \ll f_2$, decreases to $180° - 45° = 135°$ at $f = f_2$, and is asymptotic to $90°$ for $f \gg f_2$. ■

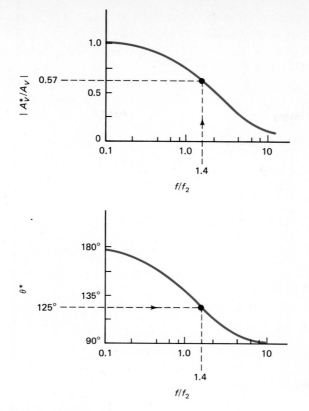

Fig. 10-10 Finding the value $|A_v^*/A_v|$ that corresponds to a given θ^* (Example 10-3).

10-6 BJT COMMON-EMITTER HIGH-FREQUENCY RESPONSE DUE TO INPUT LOADING

Now consider the effect of C_i on the high-frequency gain of the amplifier, as shown in Fig. 10-9(C). With C_i in place,

$$\mathbf{v}_{o2} = \cfrac{\mathbf{v}_1 \cfrac{R_p(-jX_C)}{R_p - jX_C}}{(R_1 + R_i) + \cfrac{R_p(-jX_C)}{R_p - jX_C}} \qquad R_p = R_B \| h_{ie} \qquad (10\text{-}34)$$

and with C_i open-circuited at midrange frequencies ($X_C \rightarrow \infty$)

$$\mathbf{v}_{o1} = \mathbf{v}_1 \frac{R_P}{(R_1 + R_i) + R_P} \qquad (10\text{-}35)$$

The resultant reduction in gain is given by

$$\alpha_V = \frac{\mathbf{v}_{o2}}{\mathbf{v}_{o1}} = \frac{(R_1 + R_i + R_P)(-jX_C)}{(R_1 + R_i)R_P + (R_1 + R_i + R_P)(-jX_C)}$$

$$= \cfrac{1}{1 + j\cfrac{(R_1 + R_i)\|R_P}{X_C}} \qquad (10\text{-}36)$$

Since this equation takes the same form as Eq. (10-30), the above results apply equally well to values of f determined by C_o or C_i.

Example 10-4   High-Frequency Half-Power Point

For a common-emitter amplifier, $R_1 + R_i = 1.0\,\text{k}\Omega$, $R_P = 3.0\,\text{k}\Omega$, and $C_i = 25\,\text{pF}$. Frequency f_2 due to C_i must be greater than 5 MHz. Does this circuit perform as desired?

Solution   From Eqs. (10-31) and (10-32), define

$$\frac{(R_1 + R_i)\,\|\,R_P}{X_C} = \frac{f}{f_2}$$

so that

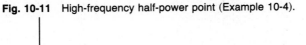

$$f_2 = \frac{1}{2\pi C_i[(R_1 + R_i)\,\|\,R_P]} = \frac{1}{2\pi\,(25\,\text{pF})(1.0\,\text{k}\Omega\,\|\,3.0\,\text{k}\Omega)} \cong 8.5\,\text{MHz}$$

The high-frequency circuit has the response shown in Fig. 10-11 and meets the performance objective.

The low-pass filter is again characterized by $|A_v^*|$ asymptotic to $|A_v|$ for $f \ll f_2$ and to zero for $f \gg f_2$. At $f = f_2$, $|A_v^*| = \dfrac{1}{\sqrt{2}}|A_v|$ and $\theta^* = 180° - 45° = 135°$. ■

Fig. 10-11   High-frequency half-power point (Example 10-4).

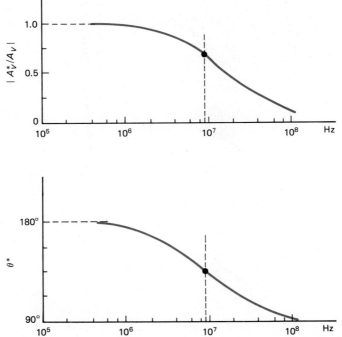

The above results can be adapted to use with common-base ac amplifier. By comparing the midrange transistor equivalent circuits in Figs. 9-6 and 9-9, it can be seen that one maps into the other with the substitution

$$h_{ie} \rightarrow h_{ib}$$

$$h_{fe} \rightarrow h_{fb}$$

Values of $(C_{eq})_{in}$ and $(C_{eq})_{out}$ can be found from Eq. (10-13).

By combining the above changes, the common-base equivalent circuit shown in Fig. 10-12 can be derived. This common-base circuit is similar to the common-

Fig. 10-12 (A) BJT common-base ac amplifier. (B) Composite ac equivalent circuit.

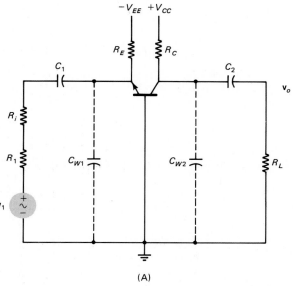

(A)

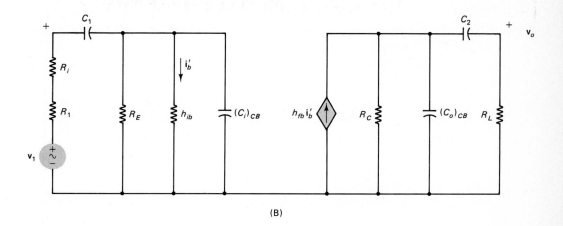

(B)

emitter circuit shown in Fig. 10-5(B). The only differences are in the particular resistor and capacitor values and the gain of the current generator. By following the step-by-step procedures outlined in the sections above, the frequency response parameters for the common-base configuration can be determined. (Section 10-9 summarizes the results of such calculations.)

Example 10-5   Common-Base Frequency Response

You are designing a common-base ac amplifier, as shown in Fig. 10-12, with $C_2 = 1.0\,\mu\text{F}$. $R_C = 2.2\,\text{k}\Omega$, and $R_L = 6.0\,\text{k}\Omega$. Find the low-frequency half-power point f_1 that corresponds to this portion of the amplifier circuit.

Solution   By referring to Fig. 10-5(B), it can be seen that the common-base and common-emitter problems are in a similar form. Therefore, applying Eq. (10-19),

$$f_1 = \frac{1}{2\pi C_2(R_C + R_L)}$$

$$= \frac{1}{2\pi\,(1.0\,\mu\text{F})(2.2\,\text{k}\Omega + 6.0\,\text{k}\Omega)} \cong 19\,\text{Hz}$$

The result is a high-pass filter with the asymptotes: $|A_v^*/A_v| \to 1$ and $\theta^* \to 0°$ for $f \gg f_1$ and $|A_v^*/A_v| \to 0$ and $\theta^* \to 90°$ for $f \ll f_1$. The curve is again uniquely identified in terms of the lower half-power frequency f_1. ∎

10-8   FET LOW- AND HIGH-FREQUENCY RESPONSE

The ac equivalent circuits for FETs are similar to those derived above for BJTs. Figure 10-13 shows a JFET common-source amplifier along with an ac equivalent circuit. At midrange frequencies, Fig. 10-13(B) reduces to the circuit shown earlier in Fig. 9-10. At low frequencies, the circuit is dominated by C_1 and C_2, which are coupling capacitances added to the circuit. The lower half-power frequency f_1 can be calculated directly using the BJT findings because the external circuits are the same in each case.

At high frequencies, C_i and C_o dominate circuit performance. Although the magnitudes of these equivalent capacitances can be quite different for the JFET case, the circuits take on the same form, so the upper half-power frequency f_2 can again be calculated directly using the BJT results.

Example 10-6   JFET Frequency Response

Find f_2 for the output section of a JFET common-source ac amplifier with $R_D = 10\,\text{k}\Omega$, $R_L = 20\,\text{k}\Omega$, and $C_o = 10\,\text{pF}$.

Solution   The frequency f_2 for the BJT common-emitter amplifier is given by Eq. (10-31):

$$f_2 = \frac{1}{2\pi C_o(R_C \| R_L)}$$

By comparing Figs. 10-9 and 10-13, note that the high-frequency characteristics of the two circuits are similar if R_C (the collector resistor) is exchanged for R_D (the

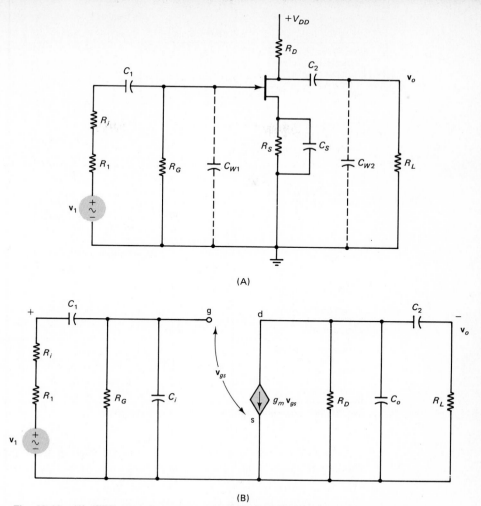

Fig. 10-13 (A) JFET common-source ac amplifier. (B) Composite ac equivalent circuit.

drain resistor). Thus, for the problem given here:

$$f_2 = \frac{1}{2\pi\,(10\,\text{pF})(10\,\text{k}\Omega\,\|\,20\,\text{k}\Omega)} \cong 2.4\,\text{MHz}$$

The upper half-power point for the JFET has been found by analogy. Because the input and output coupling sections of the BJT and JFET amplifiers are of a similar form, the BJT equations can be extended to JFET applications. The low-pass frequency-response curve due to f_2 takes the same shape shown in Fig. 10-10. ∎

10-9 SUMMARY OF FREQUENCY RESPONSES

The low- and high-frequency half-power points for three amplifier configurations are given in Fig. 10-14. At each of these half-power points, the amplifier gain is reduced below the midrange value by a factor of $1/\sqrt{2}$.

Low-frequency half-power points (f_1)

	BJT/CE	BJT/CB	JFET/CS
Determined by the output coupling capacitor C_2	$\frac{1}{2\pi C_2(R_C + R_L)}$	$\frac{1}{2\pi C_2(R_C + R_L)}$	$\frac{1}{2\pi C_2(R_D + R_L)}$
Determined by the input coupling capacitor C_1	$\frac{1}{2\pi C_1(R_1 + R_i + R_B \parallel h_{ie})}$	$\frac{1}{2\pi C_1(R_1 + R_i + R_E \parallel h_{ib})}$	$\frac{1}{2\pi C_1(R_1 + R_i + R_G)}$
Determined by the bypass capacitor	Assume $X_{CE} \ll R_E$	—	Assume $X_{C_x} \ll R_S$

High-frequency half-power points (f_2)

	BJT/CE	BJT/CB	JFET/CS
Determined by the output device capacitance and distributed capacitance	$\frac{1}{2\pi C_o(R_C \parallel R_L)}$ $C_o = C_{W2} + C_{bc}$	$\frac{1}{2\pi C_o(R_C \parallel R_L)}$ $C_o = C_{W2} + C_{cb}$	$\frac{1}{2\pi C_o(R_D \parallel R_L)}$ $C_o = C_{W2} + C_{gd}$
Determined by the input device capacitance and distributed capacitance	$\frac{1}{2\pi C_i[(R_1 + R_i) \parallel R_B \parallel h_{ie}]}$ $C_i = C_M + C_{W1} + C_{be}$	$\frac{1}{2\pi C_i[(R_1 + R_i) \parallel R_E \parallel h_{ib}]}$ $C_i = C_{W1} + C_{be}$	$\frac{1}{2\pi C_i[(R_1 + R_i) \parallel R_G]}$ $C_i = C_M + C_{W1} + C_{gs}$

Fig. 10-14 Low- and high-frequency half-power points.

It is not necessary to attempt to memorize each of these equations. Observe that the low-frequency half-power points f_1 are always in the form

$$f_1 = \frac{1}{2\pi R_{\text{series}} C_{\text{series}}} \tag{10-37}$$

where C_{series} is the series capacitance of interest, and R_{series} is the total circuit resistance in series with C_{series}. Using this approach, f_1 can be obtained by inspection.

Similarly, the high-frequency half-power points f_2 are always in the form

$$f_2 = \frac{1}{2\pi R_{\text{parallel}} C_{\text{parallel}}} \tag{10-38}$$

where C_{parallel} is the parallel capacitance of interest, and R_{parallel} is the total circuit resistance in parallel with C_{parallel}.

10-10 FREQUENCY RESPONSE FOR THE EMITTER FOLLOWER

The frequency response of emitter followers and source followers can be found by applying the above methods. Consider the emitter follower shown in Fig. 10-15. The values of f_1 and f_2 can be found by applying Eq. (10-37). For low frequencies, f_1 due to the input becomes

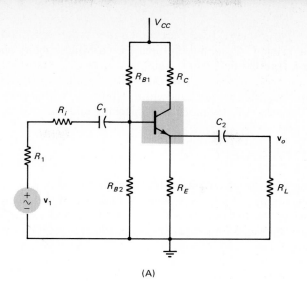

(A)

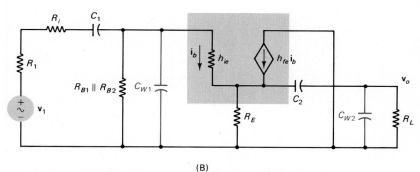

(B)

Fig. 10-15 (A) Emitter-follower circuit. (B) AC equivalent circuit.

$$f_1 = \frac{1}{2\pi C_1[(R_1 + R_i) + R_{B1}\,\|\,R_{B2}\,\|\,(h_{ie} + h_{fe}R_E\,\|\,R_L)]} \tag{10-39}$$

because R_E has an effective value $h_{fe}R_E$ when seen from the input. The f_1 due to the output is

$$f_1 = \frac{1}{2\pi C_2\left\{R_L + R_E\,\|\,\left[\dfrac{h_{ie} + R_{B1}\,\|\,R_{B2}\,\|\,(R_1 + R_i)}{h_{fe}}\right]\right\}} \tag{10-40}$$

because the input portion of the circuit has an effective resistance reduced by h_{fe} when seen from the output. Similar extensions can be used to find values for f_2. For the input and output sections, respectively,

$$f_2 = \frac{1}{2\pi C_i[(R_1 + R_i)\,\|\,R_{B1}\,\|\,R_{B2}\,\|\,(h_{ie} + h_{fe}R_E\,\|\,R_L)]} \tag{10-41}$$

$$f_2 = \frac{1}{2\pi C_o\left\{R_L\,\|\,R_E\,\|\,\left[\dfrac{h_{ie} + R_{B1}\,\|\,R_{B2}\,\|\,(R_1 + R_i)}{h_{fe}}\right]\right\}} \tag{10-42}$$

The value of $(C_{eq})_{in}$ for the emitter follower is modified. Because the collector is ac ground, C_{bc} is not multiplied by a Miller factor, and because $A_V \cong 1$, C_{be} has no effect to the first order. Thus, $(C_{eq})_{in} \cong C_{bc}$. The value for $(C_{eq})_{out}$ is negligible to the first order. As a result of the above calculations, it can be concluded that the emitter follower has a wide bandwidth. However, with a gain $A_V \cong 1$, its application is limited primarily to impedance matching and buffer situations.

10-11 COMPOSITE FREQUENCY RESPONSE

Each amplifier circuit (of the form discussed in this chapter) has two values of f_1 and two values of f_2. The frequency response of the amplifier around midrange is determined by the *interior* frequencies $(f_1)_{max}$ and $(f_2)_{min}$, which determine the edges of the midrange region, as shown in Fig. 10-16. The *bandwidth BW* of the amplifier is given by

$$BW = (f_2)_{min} - (f_1)_{max} \tag{10-43}$$

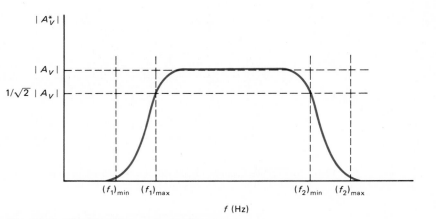

Fig. 10-16 Combined frequency-response curve.

The bandwidth is measured between the interior lower and upper half-power points, for which the gain of the circuit has dropped below the midrange value by $1/\sqrt{2}$. For frequencies between these half-power points, the gain is approximately constant. For frequencies below $(f_1)_{max}$ or above $(f_2)_{min}$, the gain drops off rapidly.

Example 10-7 Circuit Frequency Response

For the circuit of Fig. 10-17(A), find the interior values $(f_1)_{max}$ and $(f_2)_{min}$.

Solution The ac equivalent circuit is shown in Fig. 10-17(B). First, find

$$I_E = \frac{20\,\text{V} - 0.7\,\text{V}}{2.0\,\text{k}\Omega} \cong 9.7\,\text{mA}$$

$$h_{ib} \cong 0.026\,\text{V}/I_E \cong 2.7\,\Omega$$

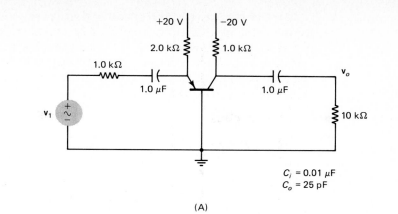

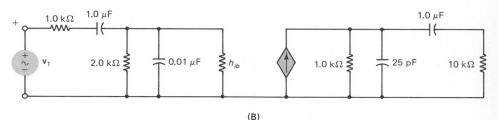

Fig. 10-17 (A) Common-base ac amplifier. (B) Equivalent circuit.

For the input,

$$f_1 \cong \frac{1}{2\pi \, (1.0 \, \mu\text{F})(1.0 \, \text{k}\Omega)} \cong 1.6 \times 10^2 \, \text{Hz}$$

$$f_2 \cong \frac{1}{2\pi \, (0.01 \, \mu\text{F})(2.7 \, \Omega)} \cong 5.9 \, \text{MHz}$$

and for the output,

$$f_1 \cong \frac{1}{2\pi \, (1.0 \, \mu\text{F})(11 \, \text{k}\Omega)} \cong 14 \, \text{Hz}$$

$$f_2 \cong \frac{1}{2\pi \, (10 \, \text{pF})(10 \, \text{k}\Omega \, \| \, 1.0 \, \text{k}\Omega)} \cong 18 \, \text{MHz}$$

The interior values are $(f_1)_{\text{max}} = 1.6 \times 10^2 \, \text{Hz}$ and $(f_2)_{\text{min}} = 5.9 \, \text{MHz}$.

The strategies developed for calculating f_1 and f_2 are now being applied. Once the ac equivalent circuit for an amplifier is drawn, the general forms given by Eqs. (10-37) and (10-38) can be used to find f_1 and f_2 by inspection. The resultant values of the half-power points can be used to draw graphs of the type shown in Fig. 10-16. ∎

In the above example, the values for f_1 and f_2 have been determined independently for the input and output coupling circuits. However, the results are modified when two roll-off curves occur close to one another. For the example, the two values for the upper half-power point are $f_{2a} = 5.9$ MHz and $f_{2b} = 18$ MHz. For this case, the correct (combined) half-power point f_2' is given by setting the product of the individual frequency responses equal to $1/\sqrt{2}$.

$$\left| \left(\frac{1}{1 + j\dfrac{f_2'}{f_{2a}}} \right) \left(\frac{1}{1 + j\dfrac{f_2'}{f_{2b}}} \right) \right| = \frac{1}{\sqrt{2}} \tag{10-44}$$

This equation can be solved to yield a value for f_2' for the case shown (see Problem 10-17). However, given the assumptions made in deriving the basic models in use and parameter variations among devices, it is often reasonable to continue to use the independently derived values of f_1 and f_2 as estimates. If this is done, an additional approximation has been introduced into the prediction of circuit performance.

Example 10-8   Amplifier Design and Frequency Response

Design a JFET ac amplifier to achieve $A_V = -3.0$ using the schematic shown in Fig. 10-13 and the following values:

DC design		AC design
$(I_D)_Q = 5.0$ mA		C_s = Large
$(V_{DS})_Q = 5.0$ V	$C_{gs} = 5.0$ pF	$C_1 = 0.10$ μF
$R_L = 2.5$ kΩ	$C_{gd} = 9.0$ pF	$C_2 = 0.10$ μF
$I_{D0} = 10$ mA	$C_{ds} = 4.0$ pF	$R_1 + R_i = 1.0$ kΩ
$V_p = -4.0$ V	$C_{w1} = 6.0$ pF	
$R_G = 1.0$ MΩ	$C_{w2} = 6.0$ pF	

Find Z_i and Z_o, the limiting values of f_1 and f_2, and bandwidth BW.

Solution   From the transfer characteristic equation,

$$(I_D)_Q = I_{D0}(1 - (V_{GS})_Q/V_P)^2$$

so that

$$(V_{GS})_Q = V_P\{1 - [(I_D)_Q/I_{D0}]^{1/2}\}$$

To find g_m at the Q point,

$$g_{ms} = \left. \frac{\partial I_D}{\partial V_{GS}} \right|_Q = \frac{-2I_{D0}}{V_p}\left(1 - \frac{V_{GS}}{V_P}\right)$$

$$= \frac{-2(10 \text{ mA})}{-4.0 \text{ V}}\left(1 - \frac{1.2 \text{ V}}{4.0 \text{ V}}\right) \cong 3.5 \text{ m}\mho$$

The midrange gain is given by

$$A_V \cong -g_{ms}(R_D \| R_L) = -3.0$$

$$= (-3.5 \text{ m}\mho)(R_D \| 2.5 \text{ k}\Omega) = -3.0$$

so that

$$R_D \cong 1.3 \text{ k}\Omega$$

To find R_S,

$$R_S = \frac{-(V_{GS})_Q}{(I_D)_Q} \cong \frac{1.2 \text{ V}}{10 \text{ mA}} \cong 0.12 \text{ k}\Omega$$

and the value of V_{DD} must be

$$\begin{aligned}
V_{DD} &= I_D R_S + V_{DS} + I_D R_D \\
&= (10 \text{ mA})(0.12 \text{ k}\Omega) + 5.0 \text{ V} + (10 \text{ mA})(1.3 \text{ k}\Omega) \\
&\cong 19 \text{ V}
\end{aligned}$$

By inspection,

$$Z_i \cong 1.0 \text{ M}\Omega$$
$$Z_o \cong R_D = 1.3 \text{ k}\Omega$$

For the input section,

$$f_1 = \frac{1}{2\pi C_1[(R_1 + R_i) + R_G]} = \frac{1}{2\pi (0.1 \ \mu\text{F})(1.0 \text{ M}\Omega)} \cong 1.6 \text{ Hz}$$

$$\begin{aligned}
C_i &= C_{W1} + C_{gd}(1 + g_{ms} R_D \| R_L) + C_{gs} \\
&= 6.0 \text{ pF} + 9.0 \text{ pF}[1 + 3.5 \text{ m}\mho(1.3 \text{ k}\Omega \| 2.5 \text{ k}\Omega)] + 5.0 \text{ pF} \cong 47 \text{ pF}
\end{aligned}$$

$$f_2 = \frac{1}{2\pi C_i[(R_1 + R_i) \| R_G]} = \frac{1}{2\pi (47 \text{ pF})(1.0 \text{ k}\Omega)} \cong 3.4 \text{ MHz}$$

and for the output section,

$$f_1 = \frac{1}{2\pi C_2(R_D + R_L)} = \frac{1}{2\pi (0.1 \ \mu\text{F})(3.8 \text{ k}\Omega)} \cong 4.2 \times 10^2 \text{ Hz}$$

$$\begin{aligned}
C_o &= C_{W2} + C_{gd} + C_{ds} \\
&= 6.0 \text{ pF} + 9.0 \text{ pF} + 4.0 \text{ pF} \cong 19 \text{ pF}
\end{aligned}$$

$$f_2 = \frac{1}{2\pi C_o(R_D \| R_L)} = \frac{1}{2\pi (19 \text{ pF})(0.86 \text{ k}\Omega)} \cong 9.7 \text{ MHz}$$

so that

$$\begin{aligned}
BW &= f_2 - f_1 \\
&= 3.4 \text{ MHz} - 4.2 \times 10^2 \text{ Hz} \cong 3.4 \text{ MHz}
\end{aligned}$$

This example applies many of the design techniques introduced earlier in Chapter 9 and in this chapter. Starting from an amplifier schematic, both dc and ac aspects of circuit operation are considered to determine all resistor and voltage values to achieve a desired voltage gain. The resistor values are then combined with the appropriate capacitor values to find the lower and upper half-power points and bandwidth for the circuit. ∎

Further insights into the frequency response of circuits can be obtained by making a change in the way $|A_v^*|$ is plotted. So far, graphs have been prepared showing the function $|A_v^*|$ (on a linear scale) as a function of the frequency f (on a logarithmic scale). The logarithmic scale for f provides an overview of the behavior of the circuit over a wide frequency range and is maintained in the approach described here. However, it is useful to change the way the gain function is plotted. Subsequently, the magnitude of the gain will be plotted in *decibels* (*dB*).

Appendix 10 introduces the decibel and shows that

$$\left| \frac{A_v^*}{A_v} \right|_{dB} = |\alpha_v|_{dB} = 20 \log |\alpha_v| \tag{10-45}$$

where $|\alpha_v|$ includes the effects of both low- and high-pass filters. By application of Eq. (10-45), the gain ratio $|A_v^*/A_v|$ can be converted into its dB equivalent.

To see the advantage of this approach, consider what happens when $|\alpha_v|$ is plotted in dB for a high-pass filter at frequencies for which $f \ll f_1$. From Eq. (10-22),

$$|\alpha_v|_{dB} = 20 \log \frac{1}{[1 + (f_1/f)^2]^{1/2}} \tag{10-46}$$
$$\cong 20 \log (f/f_1) = 20 \log f - 20 \log f_1$$

This is the equation of a straight line, which can be written

$$y = ax + b \tag{10-47}$$

with

$$y = |\alpha_v|_{dB} \qquad x = \log f$$
$$a = 20 \qquad b = -20 \log f_1$$

If $|\alpha_v|_{dB}$ is plotted (on a linear scale) versus f (on a log scale), a straight line is obtained for $f \ll f_1$. The slope of this line (given by a) is positive. Every time f changes by a factor of 10 (a *decade*), the log f changes by a factor of 1 and $|\alpha_v|_{dB}$ changes by a factor of 20 dB. The slope of the straight line is thus *+20 dB per decade*. For $f \ll f_1$, the frequency response curve is asymptotic to a straight line with a slope of 20 dB/decade.

A similar result can be obtained for a low-pass filter at frequencies f for which $f \gg f_2$. From Eq. (10-32),

$$|\alpha_v|_{dB} = 20 \log \frac{1}{[1 + (f/f_2)^2]^{1/2}} \tag{10-48}$$
$$\cong 20 \log (f_2/f) = 20 \log f_2 - 20 \log f$$

To recognize the straight line for this case, set

$$y = ax + b \tag{10-49}$$

with

$$y = |\alpha_v|_{dB} \qquad x = \log f$$
$$a = -20 \qquad b = 20 \log f_2$$

A plot of $|\alpha_V|_{dB}$ (on a linear scale) versus f (on a logarithmic scale) produces a straight line for $f \gg f_2$. The slope of this line is negative, equal in value to -20 $dB/$ $decade$. For $f \gg f_2$, the frequency response curve is asymptotic to a straight line with a slope of -20 dB/decade.

By drawing on the above results, a simple construction can be used to sketch the frequency-response curve for any amplifier once f_1 and f_2 are known. For a complete description, both amplitude and phase must be shown as a function of frequency. These combined results constitute a *Bode* plot (pronounced Bō-dē).

To prepare an amplitude plot, a horizontal line is drawn to represent the normalized midrange gain $|A_V^*/A_V|$ (at 0 dB). The lower and upper half-power points and the asymptotes for low and high frequencies are then drawn, and a final curve is produced. The key to construction is correctly linking the various frequency regions to produce the final result.

The step-by-step procedure is given below and is illustrated in Fig. 10-18(A). For interior values of f_1 and f_2, the amplitude plot is thus constructed as follows:

1. Locate f_1 and f_2 on the graph.
2. Draw a horizontal line at a value of 0 dB, extending between the values of f_1 and f_2 being used.
3. Locate the lower and upper half-power points. By definition, f_1 and f_2 are associated with a gain level of

$$|\alpha_V| = 1/\sqrt{2}$$

$$|\alpha_V|_{dB} = -3 \text{ dB}$$

The half-power points then occur at f_1 and f_2 and at gain values of -3 dB.

4. For the low-frequency region, draw a straight line with a slope of $+20$ dB/decade and passing through the intersection point ($f = f_1$, $|\alpha_V| = 0$ dB).
5. For the high-frequency region, draw a straight line with a slope of -20 dB/decade and passing through the intersection ($f = f_2$, $|\alpha_V| = 0$ dB.)

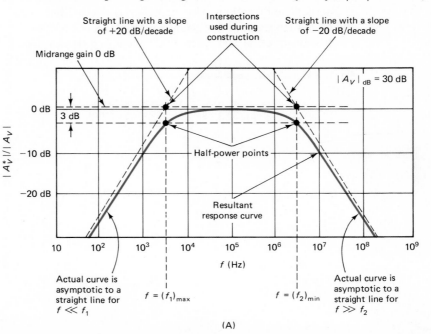

Fig. 10-18 Bode plot construction. (A) Amplitude plot with single values of f_1 and f_2. (Continued on p. 404.)

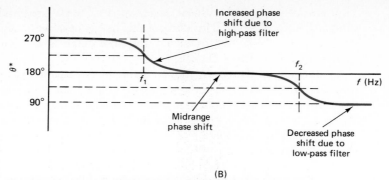

(B)

Fig. 10-18 *(cont.)* (B) Phase plot with single values of f_1 and f_2.

6. At the corners, connect the straight lines with curves passing through the points $(f_1, -3 \text{ dB})$ and $(f_2, -3 \text{ dB})$.

The solid line in Fig. 10-18(A) is the amplitude plot for an amplifier with the interior values of f_1 and f_2 indicated. Since

$$|A_V^*| = |A_V||\alpha_V|$$
$$|A_V^*|_{\text{dB}} = |A_V|_{\text{dB}} + |\alpha_V|_{\text{dB}} \tag{10-50}$$

the results of Fig. 10-18(A) can be adjusted for any value of A_V by moving the entire curve up by $|A_V|_{\text{dB}}$. Alternatively, it is often useful to leave the graph as it is, with a maximum gain of 0 dB, and to add the appropriate value of $|A_V|_{\text{dB}}$ as a separate statement on the graph, as shown.

The phase portions of the Bode plot are plotted to show θ^* (on a linear scale) as a function of f (on a logarithmic scale). By combining the results of previous sections in this chapter, the results of Fig. 10-18(B) are obtained for interior values of f_1 and f_2.

Further complications are encountered if the Bode plots are extended to lower and higher frequencies that include multiple values of f_1 and f_2. For multiple values of f_1, the construction shown in Fig. 10-19 can be developed. Each additional half-

Fig. 10-19   Low-frequency response with two values of f_1.

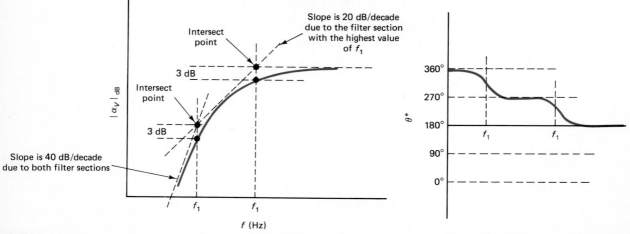

power point makes the slope of the asymptotic line more positive by 20 dB/decade and increases the phase shift by 90°. Increased shifts of 45° are experienced at the half-power frequencies, as shown.

For multiple values of f_2, each additional half-power point makes the slope of the roll-off more negative by 20 dB/decade and decreases the phase shift by 90°, as shown in Fig. 10-20.

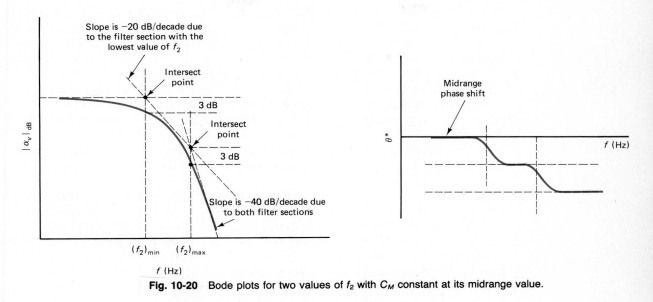

Fig. 10-20 Bode plots for two values of f_2 with C_M constant at its midrange value.

10-13 POLES AND ZEROS

A Bode plot is constructed around specific values of f_1 and f_2 that have been determined for a circuit of interest. In previous sections, methods for calculating f_1 and f_2 were found for simple input and output amplifier sections. It is useful to seek a general method for handling the frequency response of amplifier circuits in order to more efficiently approach different configurations as they are encountered.

The previously derived equations for α_V at low and high frequencies [Eq. (10-21) and (10-32)] take the form:

$$\alpha_V = \frac{1}{1 - j\dfrac{f_1}{f}} \qquad \text{(low-frequency cutoff, high-pass filter)}$$

$$\alpha_V = \frac{1}{1 + j\dfrac{f}{f_2}} \qquad \text{(high-frequency cutoff, low-pass filter)}$$

One useful way to generalize these results is to make a change of variable, defining $s = j2\pi f$, $s_1 = 2\pi f_1$, and $s_2 = 2\pi f_2$.

By direct substitution,

$$\alpha_V = \frac{s}{s + s_1} \qquad \text{(high-pass filter)} \qquad (10\text{-}51a)$$

$$\alpha_V = \frac{s_2}{s + s_2} \qquad \text{(low-pass filter)} \qquad (10\text{-}51b)$$

The variable s is called the *transfer variable* and α_V has been rewritten in a standard *transfer function* format. The meaning of this format is discussed below.

Any linear circuit can be represented by a transfer function that consists of a product of terms of the form $1/(s + s_a)$ and $(s + s_b)$.[2] In the complex s plane $s = \sigma + j\omega$, the former term can be described as producing a *pole*: when $s = -s_a$, the term becomes infinitely large. The latter term can be described as producing a *zero*: when $s = -s_b$, the function becomes zero. For the types of circuits being considered here, a generalized transfer function thus consists of the product of pole and zero terms.

For the circuit described by Eqs. (10-51a) and (10-51b), s_1 and s_2 are real constants. This is a property of steady state circuit analysis. For transient circuit analysis, s_1 and s_2 will be complex constants.

As has been demonstrated, Eqs. (10-51a) and (10-51b) describe high- and low-pass filters with half-power points at $f = f_1$ and $f = f_2$. By linking these equations with the above discussion, it can be concluded that each half-power point is associated with a transfer function term that has a pole in the complex s plane.

This provides an alternative way to find the half-power points for a circuit. If the total transfer function for the circuit is written as a product-of-terms in the form $1/(s + s_a)$ and $(s + s_b)$, the half-power points can be identified by the relationship $2\pi f = |s_a|$ for each term with a pole. The half-power points can be found by observation of the transfer function in this form. The identification of the half-power points is accomplished by finding the poles of the transfer function and linking the value of s_a with f.

The properties of individual high- and low-pass filters, as developed in this chapter, can now be summarized as follows:

- For a high-pass filter, a pole in the transfer function is associated with $|\alpha_V| = 1/\sqrt{2}$ and the asymptotic slope of the amplitude characteristic for $f \gg f_1$ is less positive by 20 dB/decade than the asymptotic slope of $f \ll f_1$. A $-90°$ phase shift is observed in passing from $f \ll f_1$ to $f \gg f_1$.

- For a low-pass filter, a pole in the transfer function is associated with $|\alpha_V| = 1/\sqrt{2}$ and the asymptotic slope of the amplitude characteristic for $f \gg f_2$ is again less positive (or more negative) by 20 dB/decade than the asymptotic slope for $f \ll f_2$. A $-90°$ phase shift is observed in passing from $f \ll f_2$ to $f \gg f_2$.

Because of the product-of-terms transfer function, these results can be extended to circuits with many poles.

In general, then, a pole is always associated with a reduction in the asymptotic slope of the amplitude characteristic by 20 dB/decade and a $-90°$ asymptotic phase shift for the phase characteristic.

---

[2] Refer to a standard text on Laplace transforms.

The properties associated with a zero in the transfer function can be developed by extending the above results. First generalize Eq. (10-51a) (for a low-pass filter) to

$$\alpha_V = \frac{s + s_3}{s + s_1} \qquad (10\text{-}52a)$$

This term includes both a pole (at $s = -s_1$) and a zero (at $s = -s_3$). Refer to Figs. 10-21(A) and (B). For s large with respect to s_1 and s_3, $\alpha_V \rightarrow 1$ (the midrange region). For s small with respect to s_1 and s_3, $\alpha_V \rightarrow s_3/s_1$, which is a real number.

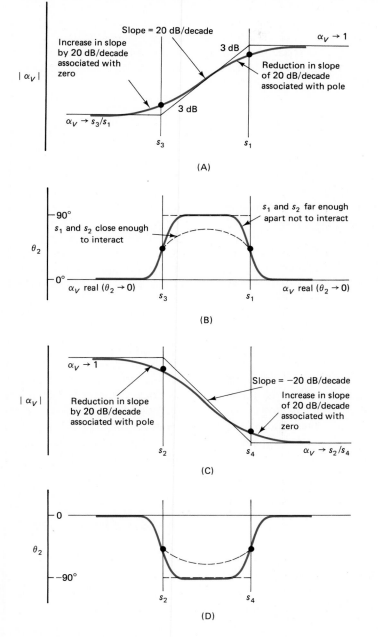

Fig. 10-21 Effects of poles and zeros on Bode plots.

Therefore, $\theta_2 = 0°$ at both high and low frequencies for this function. The $|\alpha_V|$ decreases from unity at high frequencies to s_3/s_1 at low frequencies. In between these two limits, α_V is a complex function with both real and imaginary parts.

Consider the case for which $|s_3| \ll |s| \ll |s_1|$ in the transition region. Then $\alpha_V \rightarrow s/s_1$. Using the methods of Section 10-12, the roll-off in the region between s_3 and s_1 has a slope of 20 dB/decade and $\theta_2 \rightarrow 90°$. The pole associated with s_1 produces the expected reduction in slope of 20 dB/decade and a phase shift of $\Delta\theta_2 = -90°$. The zero due to s_3 increases the slope by 20 dB/decade and has produced a phase shift of $\Delta\theta_2 = 90°$. These results are general and serve to define the significance of poles and zeros. In general, a zero is associated with an increase in the asymptotic slope of the amplitude characteristic by 20 dB/decade and a 90° asymptotic phase shift for the phase characteristic.

At high frequencies, Eq. (10-51b) can be extended to the form

$$\alpha_V = \frac{s_2}{s_4} \frac{s + s_4}{s + s_2} \qquad (10\text{-}52b)$$

A similar result is obtained, as shown in Figs. 10-21(C) and (D). For s small, $\alpha_V \rightarrow 1$, and for s large, $\alpha_V \rightarrow s_2/s_4$. In each limit, α_V is real and $\theta_2 \rightarrow 0$. In between is the transition region shown.

The initial equations for α_V [Eqs. (10-51a) and (10-51b)] correspond to the special cases $s_3 = 0$ and $s_4 \rightarrow \infty$. Therefore, the zeros associated with these functions never produce observable results on the Bode plot (which extends over all finite frequencies).

Because no interaction takes place between the input and output amplifier stages (based on the models used) and the s_n values are assumed sufficiently separated to prevent interaction, the equivalent α_V for amplifiers of the type previously discussed in this chapter (with two values of f_1 and two values of f_2) can be written as

$$\alpha_V = \left(\frac{s}{s + s_{1a}}\right)\left(\frac{s}{s + s_{1b}}\right)\left(\frac{s_{2a}}{s + s_{2a}}\right)\left(\frac{s_{2b}}{s + s_{2b}}\right) \qquad (10\text{-}53)$$

which is the *transfer function* for the amplifier.

Any circuit (for practical circuits of interest here) can be written in the transfer function format. However, considerable effort may be required to convert circuits to a product-of-terms representation. Once the transfer function is known, all lower and upper half-power points can be found by inspection.

The asymptotic behavior of the transfer function at low or high frequencies can be used to start construction of the Bode plot, and the rules discussed above can be applied to produce the desired description of the frequency characteristics of the amplifier circuit.

Example 10-9   Poles and Zeros

Suppose that a given circuit can be described by the transfer function

$$\alpha_V = \frac{s + s_{1a}}{s + s_{1b}} \frac{s + s_{2a}}{s + s_{2b}}$$

with

$$s_{1a} = 2\pi \times 10 \text{ Hz}$$

$$s_{1b} = 2\pi \times 10^2 \text{ Hz}$$

$$s_{2a} = 2\pi \times 10^5 \text{ Hz}$$

$$s_{2b} = 2\pi \times 10^4 \text{ Hz}$$

Draw the Bode plot that results.

Solution As $s \to 0$, $\alpha_v \to \dfrac{s_{1a}}{s_{1b}} \dfrac{s_{2a}}{s_{2b}} = $ constant, which corresponds to a zero slope for the amplitude characteristic and a zero phase shift for the coupling circuit. The poles in the function are associated with half-power points at $2\pi f = |s_{1b}|$ and $|s_{2b}|$, while the zeros in this function are associated with frequencies $2\pi f = |s_{1a}|$ and $|s_{1b}|$. The amplitude and phase plots of Fig. 10-22 result. ■

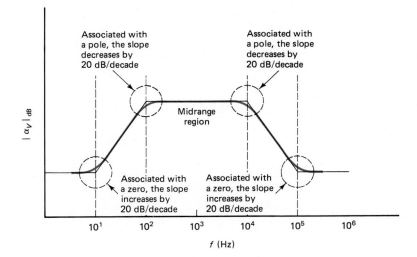

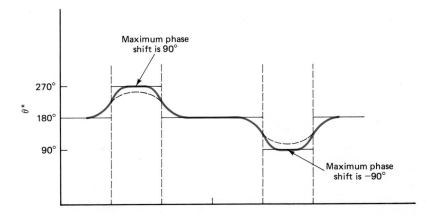

Example 10-10   Roll-Off due to C_E

Assume that the low-frequency roll-off for a common-emitter ac amplifier is dominated by the R_E, C_E combination. Find the Bode plot to be expected.

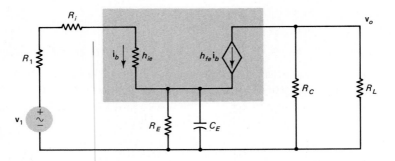

Fig. 10-23   Circuit for Example 10-10.

Solution   The equivalent circuit of Fig. 10-23 is used. By an extension of previous treatments of negative feedback,

$$Z_i = R_i + h_{ie} + h_{fe}\left(R_E \| \frac{1}{j\omega C_E}\right)$$

$$A_i = h_{fe}\frac{R_C \| R_L}{R_L} \tag{10-54}$$

$$A_V^* = \frac{-h_{fe}(R_C \| R_L)}{R_T + h_{fe}\left(R_E \| \dfrac{1}{j\omega C_E}\right)} \tag{10-55}$$

where $R_T = R_1 + R_i + h_{ie}$.

At midrange,

$$A_V = \frac{-h_{fe}(R_C \| R_L)}{R_T} \tag{10-56}$$

Therefore,

$$\alpha_V = \frac{R_T}{R_T + h_{fe}\left(R_E \| \dfrac{1}{j\omega C_E}\right)} \tag{10-57}$$

$$= \frac{s + 1/R_E C_E}{s + 1/R_E C_E + h_{fe}/C_E R_T}$$

This transfer function has a zero at

$$f_{1a} = 1/2\pi R_E C_E \tag{10-58}$$

and a pole at

$$f_{1b} = (1/2\pi)(1/R_E C_E + h_{fe}/C_E R_T) \tag{10-59}$$

The result shown in Fig. 10-24 combines the transfer function of Eq. (10-57) with two additional high-pass coupling circuits with half-power points at f_{1c} and f_{1d} (assumed due to the input and output coupling circuits). First, consider the effect of

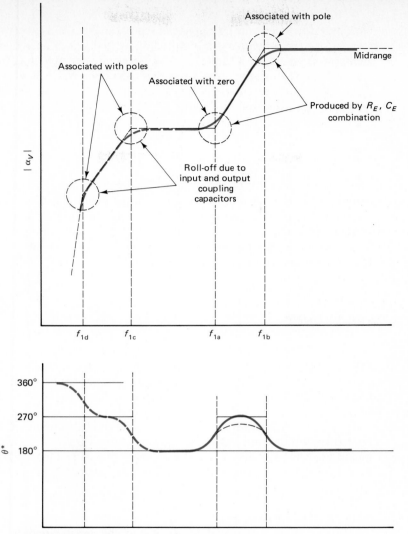

Fig. 10-24 Frequency response due to R_E, C_E.

R_E, C_E as expressed in Eq. (10-57). For $s \rightarrow 0$, $\alpha_v \rightarrow$ real constant so that slope of the amplitude characteristic approaches zero and the phase of the coupling circuit is zero. The R_E, C_E combination has produced a plateau in the low-frequency roll-off region. The two additional high-pass coupling circuits at f_{1c} and f_{1d} will have the form given in Eq. (10-51a), producing the effects shown in the figure. If all three transfer function terms are multiplied together, $\alpha_v \rightarrow s^2$ for $s \rightarrow 0$, producing the expected asymptotic slope of 40 dB/decade in this limit.

The effect of the R_E, C_E combination was postponed in earlier treatments in this chapter. The introduction of poles and zeros has now provided a means for including R_E, C_E in combination with other low-pass and high-pass filter effects.

The roll-off due to R_E, C_E in Fig. 10-24 has a simple explanation. For $f > f_{1b}$, the capacitive reactance is small, shorting R_E, and the midrange gain is obtained. For $f < f_{1a}$, the capacitive reactance is large, leaving R_E to produce negative feedback and a reduced gain. The plateau region in Fig. 10-24, between f_{1d} and f_{1a}, corresponds to the amplifier gain with negative feedback. These two limiting cases are analyzed separately in Chapter 9. The frequency-dependent A_V^* of this section thus bridges the two limiting conditions derived previously, showing the transition that occurs between common-emitter ac amplifier circuits with and without negative feedback. As discussed in Chapter 9, the gains A_{Vf} and A_{Vo} with and without feedback can be related by the expression

$$A_{Vf} = \frac{A_{Vo}}{1 - A_{Vo}B} \tag{10-60}$$

The curve in Fig. 10-24 shows how the effects of the pole and zero due to R_E, C_E can be drawn on the same Bode plot with the additional low-frequency poles due to the input and output coupling capacitors. Observe that the relative positions of the various poles and the single zero depend on the circuit component values. The low-frequency poles due to the coupling capacitors can occur above or below the pole-zero combination produced by R_E, C_E. ■

10-14 VARIATIONS IN THE MILLER CAPACITANCE

When f_2 for a Bode plot is calculated, a decision must be made as to the appropriate values to use for $(C_{eq})_{in}$ and $(C_{eq})_{out}$. If $(f_2)_{min}$ is determined by the input portion of the circuit, the output section capacitances have not yet begun to display shunting effects and the midrange (constant) value $Z_o \| R_L$ can be used to calculate $(C_{eq})_{in}$ and $(f_2)_{min}$.

If $(f_2)_{min}$ is determined by the output portion of the circuit, $(C_{eq})_{out}$ can be approximated by a (constant) value C_{bc}. [An amplifier with $A_V > 1$ usually has $\mathbf{v}_o \gg \mathbf{v}_i$ at midrange, where $\mathbf{v}_o$ and $\mathbf{v}_i$ are the voltages shown in Fig. 10-2. If $(f_2)_{min}$ is determined by the output section, the gain at this half-power frequency is reduced from the midrange maximum by only $\frac{1}{\sqrt{2}}$. In the general vicinity of $(f_2)_{min}$, $\mathbf{v}_o \gg \mathbf{v}_i$ remains a good approximation.]

Values can be calculated for f_2 due to the input circuit with midrange output impedance $Z_o \| R_L$ and for f_2 due to the output circuit with $\mathbf{v}_o \gg \mathbf{v}_i$. The smaller result for f_2 yields the correct $(f_2)_{min}$.

In many typical amplifier circuits, $(f_2)_{min}$ is determined by the input coupling circuit because of the impact of C_M. However, if a large capacitor load is added to the amplifier or if $R_1 + R_i$ is very small, then $(f_2)_{min}$ can be determined by the output coupling circuit.

In order to extend the plots to higher frequencies [above $(f_2)_{min}$], a method must be developed for dealing with variations in C_M with frequency. One approach is to keep C_M at its midrange value even when shunting effects by output capacitance reduce the value. This is a worst-case approach, since the resulting high-frequency circuit roll-off is estimated to occur at frequencies below those that would actually be experienced. The f_2 value for the input circuit is taken as a constant (minimum) value for all frequencies. Due to the simplicity of this approach, it is often used and produces the construction shown in Fig. 10-20.

An alternative strategy is to find the extreme values for C_M that occur in a circuit, establish these as limiting cases, and connect them across a transit ion region. This construction is shown in Fig. 10-25.

The value for f_{2a} is calculated with Z_o equal to its midrange value and corresponds to the input section half-power point. The value for f_{2c} is calculated with $Z_o = 0(C_M \cong C_{bc})$. At frequency f_{2b}, Z_o begins to decrease with frequency. In Fig.

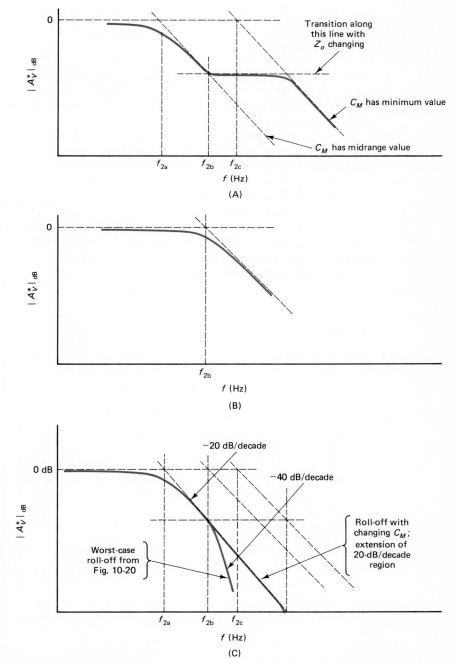

Fig. 10-25 Amplitude plots for two values of f_2 with C_M varying over a transition region. (A) Roll-off due to the input circuit. (B) Roll-off due to the output circuit. (C) Composite.

10-25, f_{2a} and f_{2c} set the extremes on circuit performance. As discussed in Exercise 10-25, the transition between regions is along a horizontal line that intersects the f_{2a} roll-off at the frequency f_{2b} associated with the output circuit. When the output capacitance begins to shunt across $R_C \| R_L$ (identifying f_{2b}), Z_o starts to drop, reducing C_M and beginning the transition to the lower C_M value. As shown in Fig. 10-25(C), the combined effect is to extend the -20-dB/decade roll-off to higher frequencies.

10-15   BODE PLOT APPLICATIONS

Bode plots have a variety of applications. They can be used to predict the performance of a given circuit, in which case the frequency response curves describe predicted behavior. They can be used to place limits on circuit capacitances. And they can also be used to interpret experimental data.

For example, suppose experimental amplitude data are provided, and the reasonableness of the results is to be determined. The procedure illustrated in Fig. 10-26 can be used to aid in interpretation. Draw a smooth curve through the experimental points to create an experimental curve. By measuring 3 dB down from the maximum gain, identify the interior values of f_1 and f_2. Then add lower and upper roll-off curves (asymptotes) with slopes of 20 dB/decade and -20 dB/decade, respectively, passing through the intercepts shown in the figure. The theoretical plot created by the above means is helpful in interpreting the experimental data. Comparison between the experimental curve and theoretical asymptotes can be used to draw conclusions about circuit performance.

Fig. 10-26   Interpreting experimental frequency-response data.

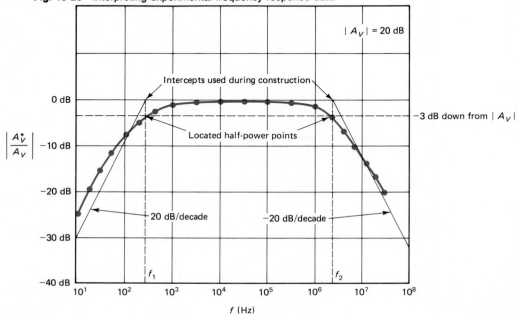

Example 10-11   Amplitude Frequency Response

Find the magnitude of the voltage gain of the circuit shown in Fig. 10-27 as a function of frequency. Draw the amplitude portion of the Bode plot using only the interior half-power points.

Fig. 10-27 Common-emitter ac amplifier.

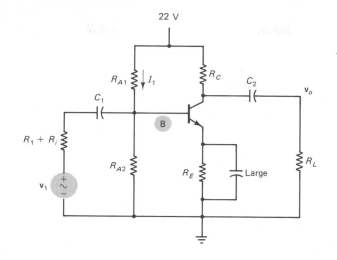

22 V

$$R_{A1} = 47\,\text{k}\Omega \qquad R_C = 5.6\,\text{k}\Omega \qquad C_1 = 1.0\,\mu\text{F} \qquad C_{W1} = 8\,\text{pF}$$
$$R_{A2} = 4.7\,\text{k}\Omega \qquad R_L = 3.3\,\text{k}\Omega \qquad C_2 = 0.5\,\mu\text{F} \qquad C_{W2} = 4\,\text{pF}$$
$$R_1 + R_i = 0.50\,\text{k}\Omega \qquad R_E = 1.0\,\text{k}\Omega \qquad h_{FE} = h_{fe} = 100 \qquad C_{bc} = 2\,\text{pF}$$
$$C_{be} = 6\,\text{pF}$$

Solution   First, solve for the dc operating point of the circuit. From Kirchhoff's Laws,

$$22\,\text{V} - I_1(47\,\text{k}\Omega) = V_B$$

$$I_1 = V_B/4.7\,\text{k}\Omega + I_B$$

and

$$22\,\text{V} - 11V_B = (47\,\text{k}\Omega)(I_B)$$

$$h_{FE}I_B = I_C$$

Also

$$V_E \cong I_C(1.0\,\text{k}\Omega)$$

$$V_B \cong V_E + 0.7\,\text{V}$$

and

$$V_B \cong 0.7\,\text{V} + h_{FE}I_B(1.0\,\text{k}\Omega)$$

From which

$$I_B \cong 12.5\,\mu\text{A} \qquad\qquad I_1 = 0.43\,\text{mA}$$

$$V_B \cong 1.9\,\text{V}$$ (10-61)

$$I_C \cong 1.2\,\text{mA} \qquad\qquad h_{ie} \cong 0.026\,\text{V}/12.5\,\mu\text{A} \cong 2.0\,\text{k}\Omega$$

Given $I_1 = 0.43$ mA and $I_B = 12.5\,\mu$A, $I_1 \gg I_B$ as desired.

For the midrange voltage gain,

$$A_V = -h_{fe}\frac{R_A \| h_{ie}}{h_{ie}}\frac{R_C \| R_L}{R_1 + R_i + R_A \| h_{ie}}$$

where $R_A = R_{A1} \| R_{A2}$.

$$A_V = (-100)\frac{4.3\,\text{k}\Omega \| 2.0\,\text{k}\Omega}{2.0\,\text{k}\Omega}\frac{5.6\,\text{k}\Omega \| 3.3\,\text{k}\Omega}{0.50\,\text{k}\Omega + 4.3\,\text{k}\Omega \| 2.0\,\text{k}\Omega} \cong -76$$

The values of f_1 and f_2 due to the input circuit are

$$f_1 \cong \frac{1}{2\pi C_1(R_1 + R_i + R_A \| h_{ie})}$$

$$= \frac{1}{2\pi(1.0\,\mu\text{F})(1.9\,\text{k}\Omega)} \cong 80\,\text{Hz}$$

$$C_i = C_{W1} + C_{be} + C_{bc}(1 + g_{me}R_C \| R_L)$$

$$= 8.0\,\text{pF} + 6.0\,\text{pF} + 2.0\,\text{pF}\left[1 + 100\frac{(5.6\,\text{k}\Omega \| 3.3\,\text{k}\Omega)}{2.0\,\text{k}\Omega}\right] \cong 0.23\,\text{nF}$$

$$f_2 \cong \frac{1}{2\pi C_i[(R_1 + R_i) \| R_A \| h_{ie}]}$$

$$= \frac{1}{2\pi(0.23\,\text{nF})(0.50\,\text{k}\Omega \| 4.3\,\text{k}\Omega \| 2.0\,\text{k}\Omega)} \cong 1.9\,\text{MHz}$$

and for the output circuit,

$$f_1 = \frac{1}{2\pi C_2(R_C + R_L)} = \frac{1}{2\pi(0.5\,\mu\text{F})(8.9\,\text{k}\Omega)} \cong 36\,\text{Hz}$$

$$C_o = C_{W2} + C_{bc}$$

$$= 4.0\,\text{pF} + 2.0\,\text{pF} = 6.0\,\text{pF}$$

$$f_2 \cong \frac{1}{2\pi C_o(R_C \| R_L)}$$

$$= \frac{1}{2\pi(6.0\,\text{pF})(2.1\,\text{k}\Omega)} \cong 13\,\text{MHz}$$

The resultant plot is shown in Fig. 10-28.

The above analysis combines dc biasing, the construction of an ac equivalent circuit, midrange amplifier design, and the calculation of the half-power frequencies f_1 and f_2 to create the amplitude plot of Fig. 10-28. The result is a description of the gain of the amplifier as a function of frequency in the midrange region and the regions around the interior half-power points.   ■

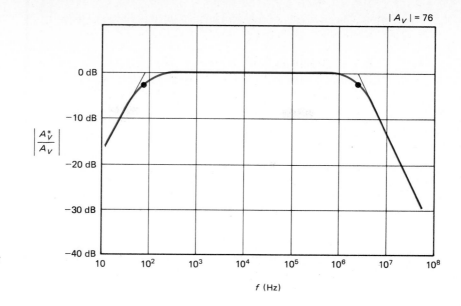

Fig. 10-28 Plot for the circuit of Fig. 10-27.

QUESTIONS AND ANSWERS

Questions

10-1. Why is the equivalent input capacitance for a common-emitter ac amplifier so large?

10-2. Why is the Miller capacitance for the BJT usually larger than the Miller capacitance for the JFET?

10-3. What circuit capacitances dominate amplifier frequency response at low and high frequencies?

10-4. How do the general frequency-dependent models of this chapter relate to the midrange models of Chapter 9?

10-5. What phase shifts are associated with the amplifier half-power points (at frequencies f_1 and f_2)?

10-6. For a typical ac amplifier of the type discussed here, how many values of f_1 and f_2 exist?

10-7. Why is f_2 usually larger for the common-base configuration than for the common-emitter configuration?

10-8. What are the advantages and problems associated with the use of poles and zeros in analyzing the frequency response of a circuit?

Answers

10-1. The equivalent input capacitance to a common-emitter ac amplifier is dominated by the Miller effect. Since voltage changes on the collector terminal of C_{bc} are amplified versions of the voltage on the base, small voltage changes on the base produce the effect of a larger capacitance from base to collector.

10-2. Since g_{me} for the common-emitter BJT is usually larger than g_{ms} for the common-source JFET, the Miller capacitance associated with the former is larger than that associated with the latter.

10-3. The input and output coupling capacitors dominate amplifier performance at low frequencies, whereas device and distributed wiring capacitances dominate amplifier performance at high frequencies.

10-4. The frequency-dependent models of this chapter reduce to the midrange models of Chapter 9 for frequencies for which capacitive effects are not important.

10-5. At frequencies f_1 and f_2, the midrange phase shift of the amplifier (0° or 180°, depending on the configuration) increases or decreases by 45°, respectively.

10-6. For the amplifier models used in this chapter, two values of f_1 and two values of f_2 exist. (The low-frequency performance is also affected by C_E, which can introduce an additional f_1 value.)

10-7. Since the effective Miller capacitance is smaller for the common-base configuration than for the common-emitter configuration, f_2 for the former is typically larger than f_2 for the latter.

10-8. The advantage of a poles-and-zeros approach to finding the frequency response of a circuit is that once the circuit is described in the standard transfer function form, the Bode plot can be immediately constructed. The disadvantage is that the derivation of this standard circuit transfer function can be a tedious task.

EXERCISES AND SOLUTIONS

Exercises
(10-1)

10-1. A common-emitter ac amplifier with $Z_o = 2.5\,\text{k}\Omega$ at midrange uses a BJT with $I_C = 5.0\,\text{mA}$ and a load resistor $R_L = 3.5\,\text{k}\Omega$. If $C_{bc} = 10\,\text{pF}$, find the midrange C_M for the amplifier.

(10-1)
(10-2)

10-2. For a BJT common-emitter amplifier,

$$C_{be} = 15\,\text{pF} \qquad h_{fe} \cong h_{FE} = 150$$

$$C_{bc} = 5.0\,\text{pF} \qquad I_E = 10\,\text{mA}$$

(a) Find the Miller capacitance associated with the amplifier, given $R_C \| R_L = 1.0\,\text{k}\Omega$.
(b) If $C_{W1} = 50\,\text{pF}$ and $C_{W2} = 100\,\text{pF}$, find C_i and C_o.

(10-1)
(10-2)

10-3. Assume that the BJT of Exercise 10-2 is used to create a common-base ac amplifier. If the wiring capacitances are unchanged, find C_i and C_o.

(10-1)
(10-2)

10-4. For a JFET, $C_{gd} = 3.0\,\text{pF}$, $C_{gs} = 3.0\,\text{pF}$, and $g_{ms} = 2.0\,\text{m}\mho$. The JFET is used in an amplifier with $C_{W1} = C_{W2} = 3.0\,\text{pF}$. Find C_i and C_o. Assume a common-source amplifier with $R_D \| R_L = 1.0\,\text{k}\Omega$.

(10-2)

10-5. For a common-emitter ac amplifier with a midrange gain of -10, $\alpha_V = 0.5e^{j\pi/6}$. Find $|A_V^*|$ and θ^*.

(10-3)

10-6. A common-emitter ac amplifier is designed with $R_C = 2.0\,\text{k}\Omega$, $R_L = 1.5\,\text{k}\Omega$, and $C_2 = 0.10\,\mu\text{F}$. Find the lower half-power frequency f_1.

(10-3)

10-7. A common-emitter ac amplifier with $A_V = -18$ is designed with $f_1 = 150\,\text{Hz}$. Find $|A_V^*|$ and θ^* at $f = 200\,\text{Hz}$.

(10-3)
(10-6)

10-8. For a common-emitter ac amplifier, $C_o = 25\,\text{pF}$, $C_2 = 0.47\,\mu\text{F}$, and $R_C = R_L = 1.0\,\text{k}\Omega$. Find the f_1 and f_2 values associated with the output section of this amplifier.

(10-3)
(10-6)

10-9. For a common-emitter ac amplifier, $C_i = 50\,\text{pF}$, $C_1 = 0.10\,\mu\text{F}$, $R_1 + R_i = 0.1\,\text{k}\Omega$, $R_{B1} \| R_{B2} = 22\,\text{k}\Omega$, and $h_{ie} = 0.50\,\text{k}\Omega$. Find the f_1 and f_2 values associated with the input section of the amplifier.

(10-3)
(10-6)

10-10. Assume that Exercises 10-8 and 10-9 describe the same amplifier. Which values of f_1 and f_2 dominate the frequency response of this amplifier?

(10-11)

10-11. For the circuit in Fig. 10-29, a bandwidth of greater than 0.1 MHz is to be obtained. Find the largest value of C_{W2} that can be used.

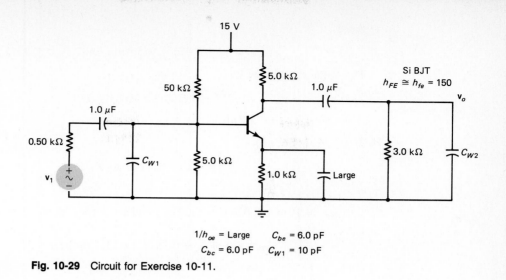

15 V

50 kΩ

5.0 kΩ

1.0 μF

Si BJT
$h_{FE} \cong h_{fe} = 150$

v_o

1.0 μF

0.50 kΩ

C_{W1}

5.0 kΩ

1.0 kΩ

Large

3.0 kΩ

C_{W2}

v_1

$1/h_{oe}$ = Large C_{be} = 6.0 pF
C_{bc} = 6.0 pF C_{W1} = 10 pF

Fig. 10-29 Circuit for Exercise 10-11.

(10-13) **10-12.** The transfer function for an ac amplifier can be written

$$\alpha_V = \frac{s + s_3}{s + s_1} \frac{s_2}{s_4} \frac{s + s_4}{s + s_2}$$

$$s_1 = 2\pi \times 10^6 \, \text{Hz}$$

$$s_3 = 2\pi \times 10^8 \, \text{Hz}$$

$$s_2 = 2\pi \times 10^4 \, \text{Hz}$$

$$s_4 = 2\pi \times 10^2 \, \text{Hz}$$

Prepare a Bode plot for the amplifier, using 0 dB as the midrange gain.

(10-13) **10-13.** Given

$$R_E = 0.10 \, \text{k}\Omega \qquad R_T = 5.0 \, \text{k}\Omega$$

$$C_E = 1.0 \, \mu\text{F} \qquad h_{fe} = 200$$

find f_{1a} and f_{1b} values associated with the roll-off due to R_E, C_E. Plot your results.

(10-13) **10-14.** Show that for a JFET common-source ac amplifier with self-bias, the transfer function term due to $R_S C_S$ is

$$\alpha_V = \frac{s + 1/R_S C_S}{s + 1/R_S C_S + g_{ms}/C_S}$$

Check the limiting cases for $f \gg f_1$ and $f \ll f_1$ to show that the above function is compatible with the earlier discussion of the amplifier with and without feedback.

10-15. An amplifier chain is to be designed so that $f_2 > 1.0 \times 10^6 \, \text{Hz}$ and $A_V > 1.0 \times 10^3$. IC amplifier style No. 1 (with $f_2 = 10 \, \text{MHz}$ and $A_V = 25$) is available for \$5.00, and IC amplifier No. 2 (with $f_2 = 5 \, \text{MHz}$ and $A_V = 50$) is available for \$10.00. How should the amplifier chain be configured to minimize costs?

Solutions **10-1.** From Eq. (10-8),

$$C_M = C_{bc}(1 + g_{me} Z_o \| R_L)$$

Since

$$g_{me} \cong |I_E|/0.026\,\text{V} \cong 5.0\,\text{mA}/0.026\,\text{V} \cong 0.19\,\text{U}$$

The Miller capacitance is

$$C_M \cong 10\,\text{pF}[1 + 0.19\,\text{U}(2.5\,\text{k}\Omega\,\|\,3.5\,\text{k}\Omega)]$$

$$\cong 2.8\,\text{nF}$$

10-2. (a) From Eq. (10-8),

$$C_M = C_{bc}(1 + g_{me}Z_o\,\|\,R_L)$$

and

$$g_{me} \cong 10\,\text{mA}/0.026\,\text{V} \cong 0.38\,\text{U} \qquad \text{and} \qquad 1/h_{oe} \cong 100\,\text{V}/10\,\text{mA} = 10\,\text{k}\Omega$$

So that

$$C_M \cong 5.0\,\text{pF}[1 + 0.38\,\text{U}(1.0\,\text{k}\Omega\,\|\,10\,\text{k}\Omega)]$$

$$\cong 1.7\,\text{nF}$$

(b) By definition,

$$C_i = C_{W1} + C_{be} + C_M$$

$$= 50\,\text{pF} + 15\,\text{pF} + 1.7\,\text{nF} \cong 1.8\,\text{nF}$$

$$C_o = C_{W2} + C_{bc}$$

$$= 100\,\text{pF} + 5.0\,\text{pF} \cong 0.11\,\text{nF}$$

10-3. From Eq. (10-13),

$$C_i = C_{W1} + C_{eb} \cong 50\,\text{pF} + 15\,\text{pF} = 65\,\text{pF}$$

$$C_o = C_{W2} + C_{cb} \cong 100\,\text{pF} + 5.0\,\text{pF} \cong 0.11\,\text{nF}$$

Observe that

$$(C_i)_{CE}/(C_i)_{CB} \cong 1.8\,\text{nF}/0.065\,\text{nF} \cong 28$$

which is approximately the voltage gain of the amplifier.

10-4. From Eq. (10-14),

$$C_M = C_{gd}(1 + g_{ms}R_D\,\|\,R_L)$$

$$= 3.0\,\text{pF}[1 + 2.0\,\text{m}\text{U}(1.0\,\text{k}\Omega)]$$

$$= 9.0\,\text{pF}$$

so that

$$C_i = C_{gs} + C_M + C_{W1}$$

$$= 3.0\,\text{pF} + 9.0\,\text{pF} + 3.0\,\text{pF} = 15\,\text{pF}$$

$$C_o = C_{gd} + C_{W2} = 3.0\,\text{pF} + 3.0\,\text{pF} = 6.0\,\text{pF}$$

10-5. From Eq. (10-15),

$$|A_v^*| = |A_v||\alpha_v| = 10(0.50) = 5.0$$

$$\theta^* = \theta_1 + \theta_2 = 180° + 30° = 210°$$

10-6. From Fig. 10-19,

$$f_1 = \frac{1}{2\pi (0.10 \ \mu\text{F})(2.0 \ \text{k}\Omega + 1.5 \ \text{k}\Omega)}$$

$$\cong 4.5 \times 10^2 \, \text{Hz}$$

10.7. From Eq. (10-23),

$$|A_V^*| = \frac{18}{[1 + (150/200)^2]^{1/2}} \cong 14$$

$$\theta^* = 180° + \tan^{-1}(150/200) = 180° + 37° = 217°$$

10-8. From Eq. (10-19),

$$f_1 = \frac{1}{2\pi (0.47 \ \mu\text{F})(1.0 \ \text{k}\Omega + 1.0 \ \text{k}\Omega)}$$

$$\cong 1.7 \times 10^2 \, \text{Hz}$$

From Eq. (10-31),

$$f_2 = \frac{1}{2\pi (25 \ \text{pF})(1.0 \ \text{k}\Omega \parallel 1.0 \ \text{k}\Omega)}$$

$$\cong 13 \, \text{MHz}$$

10-9. From Eq. (10-27),

$$f_1 = \frac{1}{2\pi C_1(R_1 + R_i + R_B \parallel h_{ie})}$$

$$= \frac{1}{2\pi (0.10 \ \mu\text{F})(0.10 \ \text{k}\Omega + 22 \ \text{k}\Omega \parallel 0.50 \ \text{k}\Omega)} \cong 2.7 \, \text{kHz}$$

From Example 10-4,

$$f_2 = \frac{1}{2\pi C_i[(R_1 + R_i) \parallel R_B \parallel h_{ie}]}$$

$$= \frac{1}{2\pi (50 \ \text{pF})(0.10 \ \text{k}\Omega \parallel 22 \ \text{k}\Omega \parallel 0.50 \ \text{k}\Omega)} \cong 38 \, \text{MHz}$$

10-10. From Exercises 10-8 and 10-9,

$$f_1 = 170 \ \text{Hz}, \ 2.7 \ \text{kHz}$$

$$f_2 = 13 \ \text{MHz}, \ 38 \ \text{MHz}$$

so the bandwidth is limited by

$$(f_1)_{\text{max}} = 2.7 \ \text{kHz}$$

$$(f_2)_{\text{min}} = 13 \ \text{MHz}$$

10-11. (a) First, determine the dc operating point. Assuming that the current through the base voltage divider is much greater than I_B,

$$V_B = 15 \ \text{V} \left(\frac{5.0 \ \text{k}\Omega}{50 \ \text{k}\Omega + 5.0 \ \text{k}\Omega} \right) = 1.4 \ \text{V}$$

For a *Si* BJT,

$$V_E \cong 1.4 \text{ V} - 0.7 \text{ V} = 0.7 \text{ V}$$

and

$$I_E = 0.7 \text{ V}/1.0 \text{ k}\Omega = 0.70 \text{ mA}$$

with

$$I_B \cong 0.70 \text{ mA}/150 \cong 4.7 \ \mu\text{A}$$

Since the divider current is 15 V/55 kΩ = 0.27 mA, the initial assumption regarding the divider current is confirmed.

For the given BJT,

$$g_{me} \cong 0.70 \text{ mA}/0.026 \text{ V} \cong 27 \text{ m}\mho$$

$$h_{ie} \cong 0.026 \text{ V}/4.7 \ \mu\text{A} \cong 5.5 \text{ k}\Omega$$

(b) To find the capacitances C_i and C_o,

$$C_i = C_{be} + C_{W1} + C_{bc}[1 + g_{me}Z_o \| R_L]$$

$$= 6.0 \text{ pF} + 10 \text{ pF} + 6.0 \text{ pF}[1 + 27 \text{ m}\mho(5.0 \text{ k}\Omega \| 3.0 \text{ k}\Omega)]$$

$$\cong 0.32 \text{ nF}$$

$$C_o = C_{bc} + C_{W2} = 6.0 \text{ pF} + C_{W2}$$

(c) Find f_1 and f_2 due to the input section:

$$f_1 = \frac{1}{2\pi(1.0 \ \mu\text{F})(0.50 \text{ k}\Omega + 50 \text{ k}\Omega \| 5.0 \text{ k}\Omega \| 5.5 \text{ k}\Omega)}$$

$$\cong 51 \text{ Hz}$$

$$f_2 = \frac{1}{2\pi(0.32 \text{ nF})(0.50 \text{ k}\Omega \| 50 \text{ k}\Omega \| 5.0 \text{ k}\Omega \| 5.5 \text{ k}\Omega)}$$

$$\cong 1.2 \text{ MHz}$$

and f_1 and f_2 due to the output section:

$$f_1 = \frac{1}{2\pi(1.0 \ \mu\text{F})(5.0 \text{ k}\Omega + 3.0 \text{ k}\Omega)}$$

$$\cong 20 \text{ Hz}$$

$$f_2 = \frac{1}{2\pi C_o(5.0 \text{ k}\Omega \| 3.0 \text{ k}\Omega)} = \frac{8.5 \times 10^{-5}}{C_o} \text{ Hz}$$

(d) Given the small values of f_1, $BW \cong f_2$. Given $f_2 \gg 0.1$ MHz for the input section, the BW of the amplifier is equal to 0.1 MHz if

$$C_o = \frac{8.5 \times 10^{-5}}{0.1 \times 10^6} = 0.85 \text{ nF}$$

10-12. With the data given, the results of Fig. 10-30 are obtained.

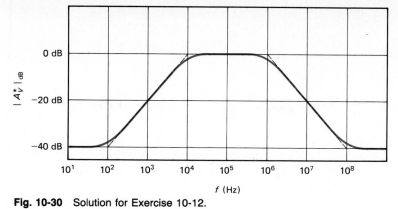

Fig. 10-30 Solution for Exercise 10-12.

10-13. From Eqs. (10-58) and (10-59),

$$f_{1a} = \frac{1}{2\pi R_E C_E} = \frac{1}{2\pi (0.10 \text{ k}\Omega)(1.0 \text{ }\mu\text{F})} = 1.6 \text{ kHz}$$

$$f_{1b} = \frac{1}{2\pi}\left(\frac{1}{R_E C_E} + \frac{h_{fe}}{C_E R_T}\right)$$

$$= \frac{1}{2\pi}\left[\frac{1}{(0.10 \text{ k}\Omega)(1.0 \text{ }\mu\text{F})} + \frac{200}{(1.0 \text{ }\mu\text{F})(5.0 \text{ k}\Omega)}\right] \cong 8.0 \text{ kHz}$$

Therefore, the results of Fig. 10-31 are obtained.

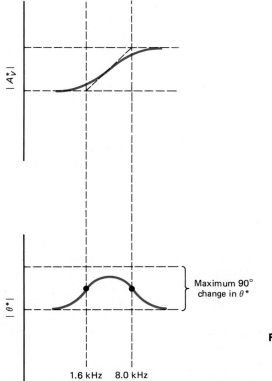

Fig. 10-31 Solution for Exercise 10-13.

423

10-14. The gain of the JFET common-source ac amplifier is given by

$$A_V = \frac{-R_G}{(R_1 + R_i) + R_G}\left(\frac{R_D \| R_L}{1/g_{ms} + R_S \| X_{CS}}\right)$$

where at midrange, $X_{CS} = 0$. Therefore,

$$\alpha_V = \frac{1/g_{ms}}{1/g_{ms} + R_S \| X_{CS}} = \frac{1}{1 + g_{ms} R_S \| X_{CS}}$$

$$= \frac{j\omega + 1/R_S C_S}{j\omega + 1/R_S C_S + g_{ms}/C_S} = \frac{s + 1/R_S C_S}{s + 1/R_S C_S + g_{ms}/C_S}$$

For $f \gg f_1$, $\alpha_V \cong 1$ and $A_V^* \cong 1$, producing midrange amplifier behavior without feedback. For $f \ll f_1$,

$$\alpha_V \cong \frac{1}{1 + g_{ms} R_S}$$

and

$$A_V^* = \frac{-R_G}{(R_1 + R_i) + R_G}\frac{R_D \| R_L}{1/g_{ms} + R_S}$$

which is the midrange amplifier gain with feedback.

10-15. For style No. 1, to achieve $A_V > 1.0 \times 10^3$ requires

$$(25)^n > 1.0 \times 10^3, \text{where } n = 3$$

For this case,

$$f_2' = 10 \text{ MHz}[2^{1/3} - 1]^{1/2} \cong 5.1 \text{ MHz}$$

For style No. 2, to achieve $A_V > 1.0 \times 10^3$ requires

$$(50)^n > 1.0 \times 10^3, \text{where } n = 2$$

For this case,

$$f_2' = 5 \text{ MHz}[2^{1/2} - 1]^{1/2} \cong 3.2 \text{ MHz}$$

Both choices have $f_2 > 1.0$ MHz.

The cost for No. 1 is $3 \times \$5.00 = \$15.00$, and for No. 2 is $2 \times \$10.00 = \$20.00$. Therefore, style No. 1 is the preferred choice.

PROBLEMS

(10-1) **10-1.** A common-emitter ac amplifier makes use of a BJT biased at $I_E = 0.8$ mA. If $Z_o = 2.0$ kΩ, $R_L = 2.5$ kΩ, and $C_{bc} = 15$ pF, find the midrange C_M for the amplifier.

(10-1) **10-2.** For a common-emitter ac amplifier with $v_o/v_i = -4.0$, find the equivalent input capacitance if $C_{cb} = 15$ pF.

(10-1) **10-3.** For a common-source ac amplifier, $C_{gs} = C_{gd} = 10$ pF, $I_{D0} = 6.0$ mA, and $V_P = -2.5$ V. If the JFET is biased at $V_{GS} = -1.5$ V, find the equivalent input capacitance if $Z_o = 1.5$ kΩ and $R_L = 1.0$ kΩ.

(10-1) **10-4.** A common-base ac amplifier has a midrange gain of $+20$ and an attenuation factor $\alpha_V = 2^{-1/2}$ at frequency of operation f_1. Find $|A_V^*|$ and θ^* for the amplifier at f_1.

(10-1)
(10-2) **10-5.** For a common-emitter ac amplifier, $C_{be} \cong C_{bc} = 5.0$ pF, $h_{fe} \cong h_{FE} = 200$, and the Q point is at $I_C = 2.5$ mA. Find the Miller capacitance if $R_C \| R_L = 0.80$ kΩ, and find C_i and C_o if $C_{W1} \cong C_{W2} = 25$ pF.

(10-3) **10-6.** For a common-emitter ac amplifier, the input coupling capacitor $C_2 = 0.01$ μF and $R_C \cong R_L = 2.0$ kΩ. Find the lower half-power frequency f_1.

(10-3) **10-7.** A common-emitter ac amplifier with midrange gain -20 is operated at 100 Hz. If $f_1 = 50$ Hz, find $|A_V^*|$ and θ^* at the frequency of operation.

(10-3) **10-8.** For the circuit of Example 10-1, find $|A_V^*|$ and θ^* at $f = 50$ Hz.

(10-4) **10-9.** For a common-emitter ac amplifier, the coupling capacitor $C_1 = 0.10$ μF. If $R_1 + R_i$ is negligible and $R_{B1} = 25$ kΩ, $R_{B2} = 5.0$ kΩ, what is f_1 for $I_C = 0.5$ mA and $h_{FE} = 150$?

(10-4) **10-10.** For the circuit of Example 10-2, find $|A_V^*|$ and θ^* at $f = 1.0 \times 10^2$ Hz.

(10-5) **10-11.** A common-emitter ac amplifier has an effective value of $C_o = 8.0 \times 10^2$ pF. Find the upper half-power point if $R_C = 3.0$ kΩ and $R_L = 0.8$ kΩ.

(10-5) **10-12.** A common-emitter ac amplifier with a midrange gain $A_V = -8.0$ is operated at 2.0 MHz. If $f_2 = 1.0$ MHz, find $|A_V^*|$ and θ^* at the frequency of operation.

(10-6) **10-13.** A common-emitter ac amplifier has a gain $A_V = -12.5$ at midrange. If $f_2 = 1.0$ MHz, find $|A_V^*|$ and θ^* at $f = 0.50$ MHz.

(10-7) **10-14.** (a) Find $|A_V^*|$ and θ^* for the circuit of Example 10-5 at $f = 10$ Hz if the midrange gain is $A_V = 10$. (b) If $R_E = 10$ kΩ and $I_E = 10$ mA, find the value of $R_1 + R_i$ that should be used to produce the desired midrange gain.

(10-11) **10-15.** An amplifier has the following half-power points: $f_1 = 2.5$ kHz and 8.5 kHz, $f_2 = 180$ kHz and 240 kHz. Find the bandwidth of the amplifier (neglecting coupling).

(10-11) **10-16.** For the circuit in Fig. 10-32, $f_2 > 1.0$ MHz is to be obtained. Find the largest value of C_{W2} that can be used.

Fig. 10-32 Circuit for Problem 10-16.

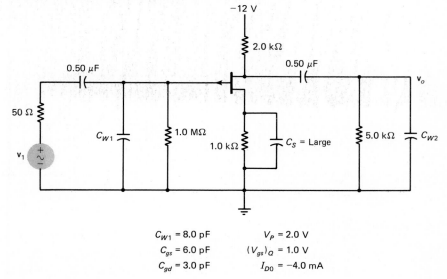

$$C_{W1} = 8.0 \text{ pF} \qquad V_P = 2.0 \text{ V}$$
$$C_{gs} = 6.0 \text{ pF} \qquad (V_{gs})_Q = 1.0 \text{ V}$$
$$C_{gd} = 3.0 \text{ pF} \qquad I_{D0} = -4.0 \text{ mA}$$

(10-11)   **10-17.** For an ac amplifier, the following values are found for f_2

$$(f_2)_{min} = 6.0 \text{ MHz} \qquad (f_2)_{max} = 7.4 \text{ MHz}$$

If coupling between the input and output sections is considered, what is the correct value of (f_2)?

(10-12)   **10-18.** For an ac amplifier, the inner half-power points are $f_1 = 500$ Hz and $f_2 = 5$ MHz. The midrange gain is $A_V = 50$. Draw a Bode plot for this amplifier.

(10-12)   **10-19.** Draw a Bode plot for the circuit of Fig. 10-29, using $C_{W2} = 30$ pF and only interior values of f_1 and f_2.

(10-12)   **10-20.** Draw a Bode plot for the circuit of Fig. 10-29, using $C_{W2} = 1.0$ nF and only interior values of f_1 and f_2.

(10-13)   **10-21.** Given $R_E C_E = 2.5$ ms, $R_T = 20 R_E$, and $h_{fe} = 125$, find the f_{1a} and f_{1b} values associated with the roll-off due to R_E, C_E. Plot your results.

(10-13)   **10-22.** For a JFET common-source ac amplifier with self-bias, $R_S = 0.6$ kΩ, $C_S = 0.001$ μF, $g_m = 4.8$ m℧, and $R_1 + R_i$ is negligible. Plot the low-frequency roll-off due to the source capacitor C_S.

(10-13)   **10-23.** An amplifier is described by the transfer function

$$\alpha_V = \frac{s}{s + 1.5 \times 10^2 \text{rad}/s} \frac{s + 5.5 \times 10^3 \text{rad}/s}{s + 1.5 \times 10^4 \text{rad}/s}$$

Draw the Bode plot that results.

(10-13)   **10-24.** Show that the midrange gain A_{vo} of Eq. (10-56) $(1/\omega C_E \ll R_E)$ and the plateau gain of the amplifier with feedback $(1/\omega C_E \gg R_E)$ are related by

$$A_{Vf} = \frac{A_{vo}}{1 - A_{vo}B}$$

Interpret the expression for B.

(10-14)   **10-25.** Show that at high frequencies, the changing value of C_M results in $|A_V^*|$ approximately constant until C_M becomes small, producing a plateau effect in the frequency-response curve due to the input coupling circuit.

(10-15)   **10-26.** For an ac amplifier, the data points of Fig. 10-33 are recorded. Construct an interpretive amplitude plot around these points. Find the inner half-power points f_1 and f_2.

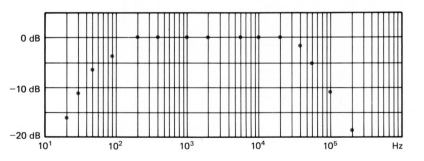

Fig. 10-33 Data points for Problem 10-26.

COMPUTER APPLICATIONS

10-1. A midrange version of the Ebers-Moll model that is suitable for use in simulation programs such as SPICE was introduced in Chapter 5. Figure 10-34 shows the further evolution of this model to include internal capacitors.

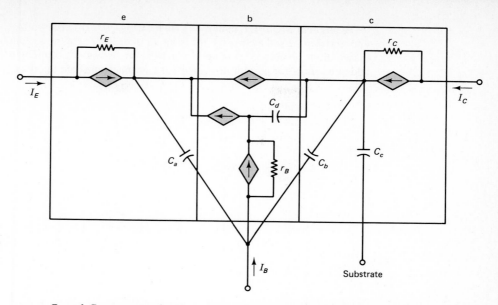

Fig. 10-34 SPICE BJT representation using a variation of the Ebers-Moll model, showing effective capacitances.

C_a and C_b represent the base-emitter and base-collector junction capacitances, respectively, that can divert current away from the amplification process. C_c represents the coupling between the device and the substrate. At high frequencies, this capacitance can divert part of I_C to ground, reducing the device current gain. C_d allows a capacitive coupling between the interior base and collector currents.

Figure 10-35 shows a comparable JFET model. The capacitive coupling C_a and C_b is between the gate and the channel. Normally I_{G1} and I_{G2} are very small, corresponding to the input diode leakage current.

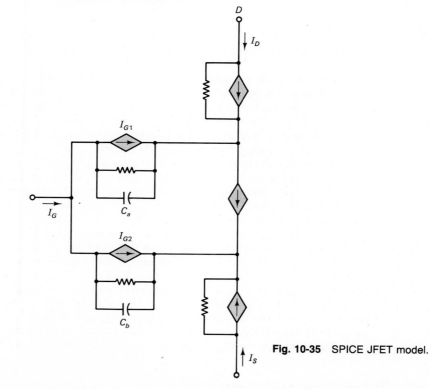

Fig. 10-35 SPICE JFET model.

427

A possible MOSFET model is shown in Fig. 10-36. Observe that the current generators and resistors associated with the gate of the JFET are now gone because the reverse-biased diode is replaced with an insulating layer. However, coupling to the substrate has been added to allow for current flow across the channel-substrate interface.

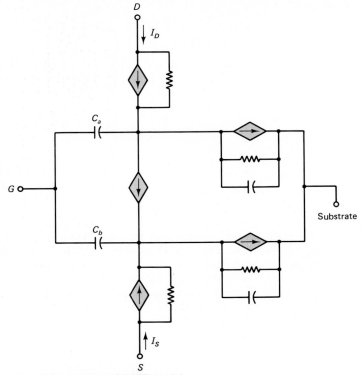

Fig. 10-36 SPICE MOSFET model.

The above models and others like them are suitable for use in computer simulations. More complexity is often introduced into the models in an effort to accurately describe device performance. Unfortunately, more complex models require much more input data, which are often not available. (Up to 40 to 50 data elements can be defined for each device.) In most applications, typical or default parameters are used in these models, except for a few parameters that are of specific interest.

Bode plots can be produced directly from simulations such as SPICE. Consider the amplifier schematic shown in Fig. 10-37 and the input file in Fig. 10-38. There are six resistors and three capacitors. The simulation makes use of a BJT model named MOD1, with a base-emitter capacitor (C_{JE}) of 25 pF, a base-collector capacitor (C_{JC}) of 8 pF, and a current gain (BF) of 150. The power supply is $V_{CC} = 18.7$ V and the input signal is 1.0 V in amplitude. The output to be plotted is the ac amplitude (V_{DB}) and the phase (V_P) at node 6.

The results of the SPICE run are shown in Fig. 10-39. As discussed in the next chapter, Section 11-2, negative feedback can change the frequency-response characteristics of a circuit. However, for the amplifier circuit in Figure 10-37 (for large values of $1/h_{oe}$), the negative feedback due to R_E and the frequency-response characteristics of the circuit due to the complex load do not couple together. Therefore, for the case shown, the gain with feedback can be predicted from Eq. (11-9), and the upper half-power point f_2 can be predicted from the relationship $f_2 = 1/2\pi(R_L \| R_C)C_{CG}$. Configurations for which negative feedback will change the circuit frequency response are discussed in Chapter 11.

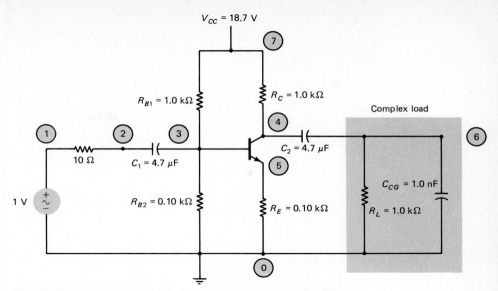

Fig. 10-37 BJT ac amplifier with nodes used for a SPICE simulation.

Fig. 10-38 SPICE data entries (CJE = C_{BE},

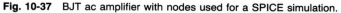

Format:	Component name	Node connections	Component value

```
.OPTIONS NOMOD
*
* RESISTORS
*
RIN 1 2 10
RB1 3 7 1K
RB2 3 0 100
RC 4 7 1K
RE 5 0 100
RL 6 0 1K
*
* CAPACITORS
*
C1 2 3 4.7U
C2 4 6 4.7U
CCG 4 0 1000P
*
* TRANSISTOR
*
Q1 4 3 5 MOD1
.MODEL MOD1 NPN CJE=25P CJC=8P BF=150
*
* SOURCES
*
VCC 7 0 DC 18.7
VIN 1 0 AC 1.0
*
* OUTPUT SECTION
*
.OP
.AC DEC 4 1 10G
.PLOT AC VDB(6)
.PLOT AC VP(6)
.END
```

$$R_L \| R_C = \frac{1 \cdot 10^6}{2K} = .5 \cdot 10^3$$

$$.5 \cdot 10^3 (1000 \cdot 10^{-12})$$

$$= .5 \cdot 10^3 \cdot 10^{-9}$$

$$= \frac{1}{500 \cdot 10^{-9}}$$

$$= \frac{1}{3.1455 \cdot 10^{-6}}$$

$$= 318,309$$

$$f_2 = \frac{1}{2\pi C}$$

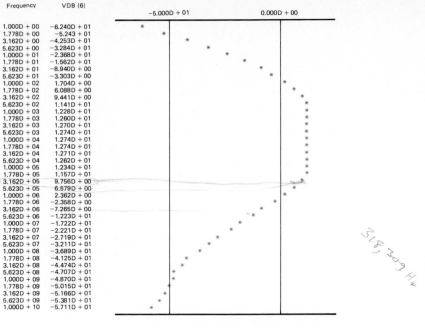

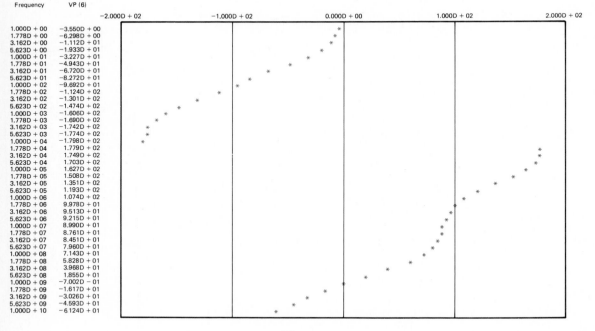

Fig. 10-39 SPICE output using a line printer. (A) Amplitude plot. (B) Phase plot. Note that values of θ^* are defined to fall between $-180°$ and $180°$.

10-2. The JFET ac amplifier of Fig. 10-40 and the input data of Fig. 10-41 have been used in a SPICE simulation to produce the Bode plots of Fig. 10-42. Compare these simulation results with the performance results produced by the analysis methods of this chapter.

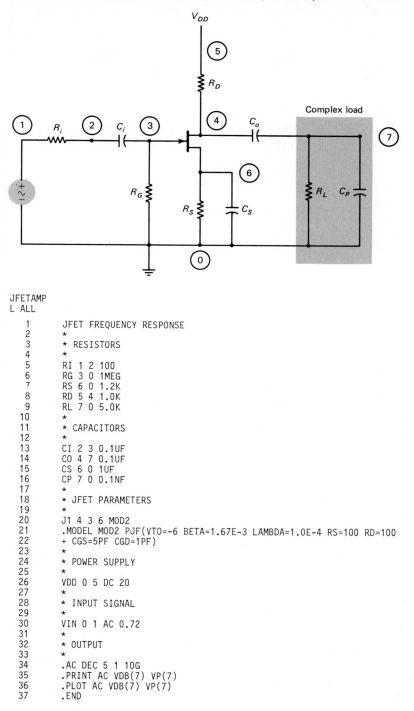

Fig. 10-40 JFET ac amplifier with nodes used for a SPICE simulation.

Fig. 10-41 SPICE data entries $(VTO = V_P,$ $BETA = I_{D0}/V_P^2)$.

```
JFETAMP
L ALL
   1        JFET FREQUENCY RESPONSE
   2        *
   3        * RESISTORS
   4        *
   5        RI 1 2 100
   6        RG 3 0 1MEG
   7        RS 6 0 1.2K
   8        RD 5 4 1.0K
   9        RL 7 0 5.0K
  10        *
  11        * CAPACITORS
  12        *
  13        CI 2 3 0.1UF
  14        CO 4 7 0.1UF
  15        CS 6 0 1UF
  16        CP 7 0 0.1NF
  17        *
  18        * JFET PARAMETERS
  19        *
  20        J1 4 3 6 MOD2
  21        .MODEL MOD2 PJF(VTO=-6 BETA=1.67E-3 LAMBDA=1.0E-4 RS=100 RD=100
  22        + CGS=5PF CGD=1PF)
  23        *
  24        * POWER SUPPLY
  25        *
  26        VDD 0 5 DC 20
  27        *
  28        * INPUT SIGNAL
  29        *
  30        VIN 0 1 AC 0.72
  31        *
  32        * OUTPUT
  33        *
  34        .AC DEC 5 1 10G
  35        .PRINT AC VDB(7) VP(7)
  36        .PLOT AC VDB(7) VP(7)
  37        .END
```

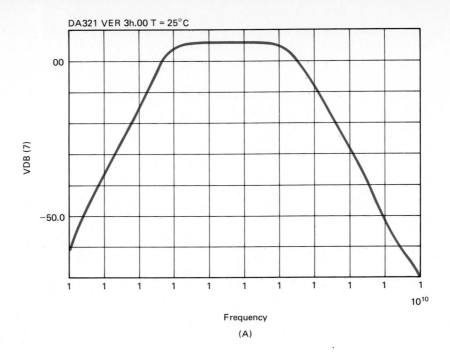

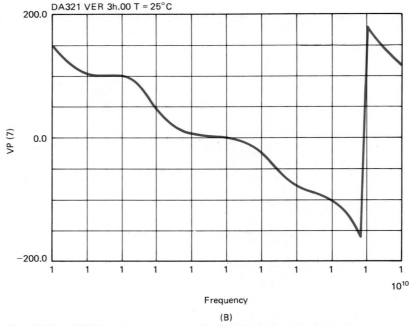

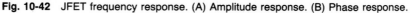

Fig. 10-42 JFET frequency response. (A) Amplitude response. (B) Phase response.

*   **10-3.** Write a computer program that produces Bode plot (amplitude and phase) values for an
         amplifier with a known midrange gain A_V and half-power points f_1 and f_2. Choose values
         for A_V, f_1, and f_2, and show the output that results. Plot the resultant Bode plot on the com-
         puter or by hand.

* **10-4.** Use a SPICE simulation to prepare a Bode plot for the amplifier of Problem 10-19.

* **10-5.** Use a SPICE simulation to show how two high-pass filters can interact if the f_1 values are sufficiently close. Compare with the predictions of Eq. (10-44). What criterion might be used for the separation between f_1 values to neglect any interactions?

* **10-6.** Appendix 11 describes the shift in θ at the half-power points produced by multiple, identical amplifier stages. Write a computer program that can be used to find θ for an arbitrary number of stages n, and apply to calculate θ for $n = 1$ to 10. Compare with the results of Appendix 11.

EXPERIMENTAL APPLICATIONS

10-1. The objective of this experiment is to determine the Bode plot for the BJT ac amplifier used in the SPICE simulation of Fig. 10-37. Refer to the data sheet for your BJT to obtain the necessary estimates for the device capacitances. Use the methods of analysis described in this chapter to find the expected midrange gain A_V and the values for f_1 and f_2 (poles and zeros) for the circuit. Compare your predictions with the SPICE simulation and discuss the likely reasons for any differences.

Now build your circuit and measure $|A_V^*|$ and θ^* as a function of f_1 using a range that extends at least one decade below the maximum f_1 and one decade above the minimum f_2. Take at least four data points per decade. Plot your data on semilog graph paper and prepare an interpretive Bode plot. Measure the values of f_1 and f_2 and compare with both theoretical and SPICE predictions. What conclusions can you draw? Turn in your data, analysis, findings, and discussion.

10-2. The purpose of the experiment is to determine the Bode plot for a JFET ac amplifier, using the schematic shown in Fig. 10-40. Give the low- and high-frequency ac equivalent circuit models for the amplifier and predict values for f_1 and f_2. Then build your circuit, collect data, and prepare an interpretive Bode plot (follow the data guidelines suggested in Experimental Application 10-1). Measure the expected values of f_1 and f_2 and compare with theoretical predictions. Turn in your data, analysis, findings, and discussion.

CHAPTER 11
DISTORTION
AND AMPLIFIER
PERFORMANCE

Several different types of model development are included in this chapter, extending the core concepts of Chapter 10. Fourier analysis (introduced in Chapter 5) is used again as a major technique for appreciating the significance of nonsinusoidal waveforms. This powerful method for conceptualizing arbitrary waveforms is a necessary addition to the designer's array of resources and is applied to understand how several types of nonsinusoidal waveforms interact with linear amplifier circuits.

The emphasis on linear circuits that has prevailed in previous chapters evolves to consideration of other ways for utilizing basic amplifier circuits. Nonlinear amplifiers are shown to have both advantages and disadvantages, depending on the application.

A brief introduction to the concept of circuit noise is included. Several key parameters are introduced to provide initial contact with this important limitation to circuit performance. The concept of circuit noise is further considered in Chapter 14, in relationship to sine-wave oscillators.

11-1 DISTORTION DUE TO FREQUENCY-DEPENDENT GAIN

Chapter 10 developed the Bode plot as a way to describe the frequency-dependent properties of an amplifier. Based on the techniques of Fourier analysis (Appendix 9), an arbitrary input signal can be represented as a superposition of sine waves with different amplitudes and phases. The distortion due to a frequency-dependent gain can thus be determined by reducing the input to its Fourier components, finding the effect of the amplifier on each component, then reassembling the components at the output to produce the modified signal.

This method of study is possible only because linear amplifiers are being considered. The rectifier and filter circuit of Section 5-6, which is nonlinear because of the diodes, could not be analyzed the same way. (In the latter case, the actual nonlinear circuit could be treated as two linear circuits, with boundary conditions used to match the solutions together.)The Bode plot can be used to evaluate how each Fourier component is modified in amplitude and phase as it passes through the amplifier. Frequency components in the midrange region, where capacitive effects are unimportant, are amplified with the same gain and experience the same phase shifts. Frequencies outside of this region experience different gains and phase shifts. The bandwidth of the amplifier determines which Fourier components are in the roll-off regions where distortion occurs. An amplifier with a wider bandwidth results in less distortion.

Distortion produced by the limited frequency response of an amplifier can be demonstrated using a square-wave input, as shown in Fig. 11-1(A). From Appendix 9, the Fourier series for this square wave is

$$v = \frac{4V_m}{\pi}[\sin(\omega t) + \tfrac{1}{3}\sin(3\omega t) + \tfrac{1}{5}\sin(5\omega t) + \cdots] \qquad (11\text{-}1)$$

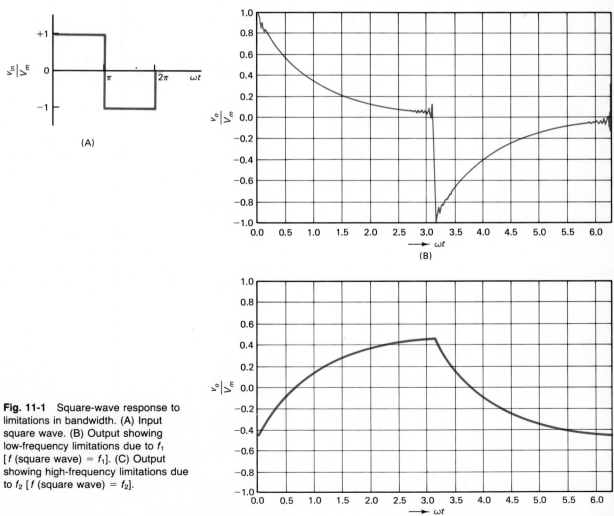

Fig. 11-1 Square-wave response to limitations in bandwidth. (A) Input square wave. (B) Output showing low-frequency limitations due to f_1 [f (square wave) $= f_1$]. (C) Output showing high-frequency limitations due to f_2 [f (square wave) $= f_2$].

First consider what happens if ω is selected so that the low-frequency amplifier roll-off affects the Fourier components. Those components at lower frequencies are reduced in amplitude and shifted to larger phase angles. The maximum change affects the component in $\sin(\omega t)$ (the fundamental). Each term in the series of Eq. (11-1) must be modified according to the appropriate gain reduction and phase shift given by Eq. (10-22). By combining Eqs. (10-22) and (11-1), the output can be written in the form

$$v_o(t) = \frac{4V_m}{\pi}\left\{\frac{\sin[2\pi ft + \tan^{-1}(f_1/f)]}{[1 + (f_1/f)^2]^{1/2}} + \frac{\sin(3 \cdot 2\pi ft + \tan^{-1}(f_1/3f)]}{3[1 + (f_1/3f)^2]^{1/2}} + \cdots\right\}$$

(11-2)

If the square-wave frequency f is selected equal to f_1,

$$v_o(t) = \frac{4V_m A_V}{\pi}\left[\frac{\sin(2\pi ft + 45°)}{(2)^{1/2}} + \frac{\sin(6\pi ft + 18°)}{(10)^{1/2}} + \cdots\right]$$

(11-3)

Now consider what happens if ω is selected so that the high-frequency amplifier roll-off affects the Fourier components. Those components at higher frequencies are reduced in amplitude and shifted to smaller phase angles. The maximum change affects the highest frequency components. By combining Eqs. (10-33) and (11-1), the output can be written in the form

$$v_o(t) = \frac{4V_m}{\pi}\left\{\frac{\sin[2\pi ft - \tan^{-1}(f/f_2)]}{[1 + (f/f_2)^2]^{1/2}} + \frac{\sin[3 \cdot 2\pi ft - \tan^{-1}(3f/f_2)]}{3[1 + (3f/f_2)^2]^{1/2}} + \cdots\right\}$$

(11-4)

If the square-wave frequency f is selected equal to f_2,

$$v_o(t) = \frac{4V_m A_V}{\pi}\left[\frac{\sin(2\pi ft - 45°)}{(2)^{1/2}} + \frac{\sin(6\pi ft - 72°)}{(90)^{1/2}} + \cdots\right]$$

(11-5)

Figures 11-1(B) and 11-1(C) show the output waveforms that result. (Methods for calculating these results are discussed in a computer application at the end of this chapter.)

To perfectly reproduce the horizontal portion of the square wave requires a dc amplifier ($f_1 = 0$) because the horizontal line can be interpreted as a dc signal. The effect of removing the low frequencies from a square wave is to produce a "sag," as shown in Fig. 11-1(B). From Section 10-9, the lower half-point f_1 can be found from

$$f_1 = \frac{1}{2\pi R_{\text{series}} C_{\text{series}}}$$

(11-6)

The product $R_{\text{series}} C_{\text{series}}$ is the time constant τ_1 associated with the high-pass filter response and can be measured by finding the time it takes for the waveform to fall to $1/e$ of its maximum value. The duration of the input square wave must be of the order of τ_1 for an accurate measurement.

Figure 11-1(C) illustrates the effect of the removal of the higher-frequency components of the square wave due to the roll-off in gain beyond the upper half-power point. The leading and trailing edges of the waveform show the effects of exponential charging and discharging.

From Section 10-9, the upper half-power point f_2 can be written in the form

$$f_2 = \frac{1}{2\pi R_{\text{parallel}} C_{\text{parallel}}} \tag{11-7}$$

The product $R_{\text{parallel}} C_{\text{parallel}}$ is the time constant τ_2 associated with the low-pass filter response and can be measured from the waveform. Clearly, the rise time of the input square wave must be much smaller than τ_2 for an accurate measurement.

Example 11-1   Square-Wave Testing

Show that the time constant of the exponential decay in Fig. 11-1(B) can be used to correctly predict the lower half-power frequency f_1.

Solution   From Fig. 11-1(B), which was found by computer simulation, the waveform has dropped to $1/e$ of its final value by $\omega\tau = 1.0$. Therefore,

$$f_1 = 1/2\pi\tau$$

as given in Eq. (11-6).

As discussed in Computer Application 11-2 at the end of this chapter, the results of Figs. 11-1(B) and 11-1(C) were found by computer simulation, using Eqs. (11-3) and (11-5). These results thus serve to demonstrate how the Bode plot and Fourier analysis techniques, when used as input to a computer simulation of a linear circuit, can be used to predict circuit output. The utility of square-wave pulse testing in this context is also established. ∎

11-2   FREQUENCY RESPONSE OF A COMMON-EMITTER AMPLIFIER WITH NEGATIVE FEEDBACK

The distortion of an ac amplifier can be reduced by increasing the bandwidth. Thus, strategies for expanding the middle range of an amplifier are sometimes useful in the design process. One way to improve amplifier bandwidth is to use negative feedback.

The frequency response of the common-emitter amplifier is discussed in detail in Chapter 10. An emitter capacitor is used to prevent negative feedback through R_E at high frequencies. As illustrated in Section 10-13, feedback reduces the amplifier gain at low frequencies because C_E no longer acts as a short circuit across R_E.

If C_E is removed from the amplifier, as shown in Fig. 11-2, the midrange amplifier gain is reduced and the frequency-response characteristics of the circuit

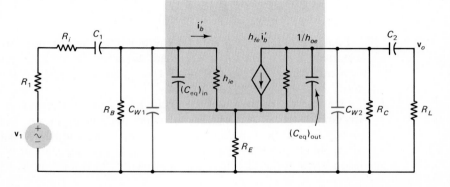

Fig. 11-2 Equivalent circuit of a common-emitter amplifier with negative feedback.

change. Determining these characteristics through analysis is a formidable task due to circuit complexity. Fortunately, there is another approach that can be used.

Figure 11-3 shows a block diagram with a (general) frequency-dependent feedback factor B^*. This diagram applies to a BJT ac amplifier operated in the linear region with negative feedback due to R_E. If $1/h_{oe}$ is sufficiently large to be neglected, then the amplifier transconductance g_m^* is load independent. Both g_m^* and the load Z_o^* can vary with frequency.

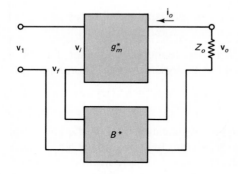

Fig. 11-3 Representation of a transconductance amplifier with feedback.

The gain of the amplifier can be found as follows:

$$\mathbf{v}_1 + \mathbf{v}_f = \mathbf{v}_i$$

$$\mathbf{v}_i g_m^* = \mathbf{i}_o$$

where g_m^* applies to the amplifier circuit,

$$\mathbf{i}_o(B^*Z_o^*) = \mathbf{v}_f$$

so that

$$\mathbf{v}_1 + \mathbf{i}_o B^*Z_o^* = \frac{\mathbf{i}_o}{g_m^*}$$

$$\mathbf{v}_1 = \mathbf{i}_o\left(\frac{1}{g_m^*} - B^*Z_o^*\right) = \frac{\mathbf{v}_o}{Z_o^*}\left(\frac{1}{g_m^*} - B^*Z_o\right)$$

$$A_{Vf}^* = \frac{\mathbf{v}_o}{\mathbf{v}_1} = \frac{g_m^*Z_o^*}{1 - g_m^*Z_o^*B^*} \tag{11-8a}$$

$$= \frac{A_V^*}{1 - A_V^*B^*} \tag{11-8b}$$

Equation (11-8a) is useful for the amplifier under consideration since g_m^* is approximately independent of the load. Equation (11-8b) illustrates a form that has been previously used in Chapter 9 to represent the effect of feedback on amplifier gain.

Consider the operation of the circuit of Figure 11-3 applied to a BJT amplifier under circumstances for which the output section produces the dominant high-frequency roll-off near midrange (such a circuit was studied in Chapter 10, Experimental Application 10-1). For this case, $g_m^* = g_m$ is independent of f and the frequency dependence of the circuit is produced by Z_o^*. Given $B^* = R_E/Z_o^*$,

$$A_{Vf}^* = \frac{g_m Z_o^*}{1 - g_m R_E} \tag{11-9}$$

The gain of the circuit is reduced by negative feedback according to Eq. (11-9) and f_2 is independent of the feedback factor. This type of performance is obtained because the frequency-dependence of the circuit is produced external[1] to the feedback loop [Fig. 11-4(A)].

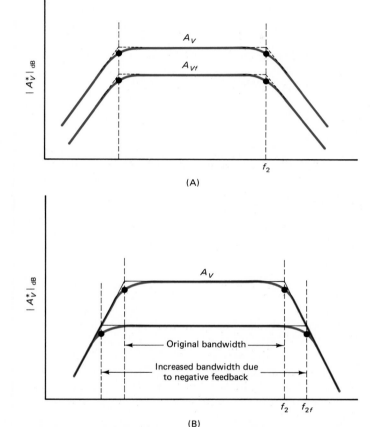

(A)

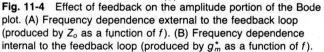

(B)

Fig. 11-4 Effect of feedback on the amplitude portion of the Bode plot. (A) Frequency dependence external to the feedback loop (produced by Z_o as a function of f). (B) Frequency dependence internal to the feedback loop (produced by g_m^* as a function of f).

Now consider the operation of this circuit under circumstances for which the input section produces the dominant circuit roll-off near midrange. For this case, Z_o^* is equal to $R_C \| R_L$ (a real constant) but g_m^* becomes a function of f. For this configuration, the negative feedback will affect both the gain and the upper half-power point f_2 of the circuit.

---

[1] Used in this context, a frequency-dependent circuit that is *external* to the feedback loop cannot produce a change in the voltage across R_E, where such a circuit that is *internal* to the feedback loop can produce a change in the voltage across R_E.

The frequency response of a low-pass coupling filter can be written

$$g_m^* = \frac{g_m}{1 + j\dfrac{f}{f_2}}$$

and by substituting into Eq. (11-8-a)

$$A_{Rf}^* = \frac{g_m^* R_C \| R_L}{1 - g_m^* R_E} = \frac{g_m R_C \| R_L}{\left(1 + j\dfrac{f}{f_2}\right) - g_m R_E}$$

for the present case. Dividing numerator and denominator by $(1 - g_m R_E)$,

$$A_{Vf}^* = \frac{\left(\dfrac{g_m R_C \| R_L}{1 - g_m R_E}\right)}{1 + j\dfrac{f}{f_2(1 - g_m R_E)}} = \frac{A_{Rf}}{1 + j\dfrac{f_2}{f_{2f}}} \qquad (11\text{-}10a)$$

with

$$f_{2f} = f_2(1 - g_m R_E) \qquad (11\text{-}10b)$$

At this point, a useful design parameter can be found by multiplying the absolute value of the midrange amplifier gain by the amplifier bandwidth BW. This parameter is called the *Gain Bandwidth Product (GBWP)*. An important result can be obtained if the *GBWP* is evaluated for the amplifier with feedback described above.

$$GBWP \cong |A_{Vf}|\, f_{2f} = \frac{|A_V|\, f_2(1 - g_m R_E)}{(1 - g_m R_E)} = |A_V|\, f_2 \qquad (11\text{-}11)$$

The gain bandwidth product is a constant,[2] independent of the feedback factor.

This result is obtained because the frequency dependence of the circuit is internal to the feedback loop. The current divider action produced by the interior low-pass filter reduces the base current drive for the amplifier and thus reduces the feedback voltage in series with v_1. Therefore, an increasing R_E both reduces the amplifier gain and reduces the magnitude of the feedback signal, causing the frequency-response characteristic to roll off more slowly than without R_E. As a result, f_2 increases in such a way that the *GBWP* remains approximately constant.

In summary, if the frequency-dependent portions of an amplifier are internal to the feedback loop, the *GBWP* of the op amp will be approximately constant. The gain of the amplifier can be set precisely by resistors, and the *GBWP* will hold constant as $B = R_E / R_C \| R_L$ is varied. The result is a useful, flexible building block circuit.

The use of negative feedback with interior frequency-dependent circuits is an important approach to producing a tradeoff between amplifier gain and bandwidth. This strategy is widely used in operational amplifier applications, as discussed in Chapter 12.

For the interior coupling circuit, since the product $|A_{Vf}|\, f_{2f}$ remains constant, the roll-off curves will coincide at high frequencies.

---

[2] In contrast, the *GBWP* for an amplifier without feedback is not a constant.

For $f \gg f_2$,

$$A_V^* \cong \frac{A_V}{j\dfrac{f}{f_2}} = \frac{A_V f_2}{jf} \qquad (11\text{-}12)$$

and for $f \gg f_2'$

$$A_{Vf}^* = \frac{A_{Vf}}{j\dfrac{f}{f_2'}} = \frac{A_{Vf} f_2'}{jf} \qquad (11\text{-}13)$$

If the *GBWP* is constant, the two curves will coincide.

A similar analysis shows that the curves will also coincide at low frequencies (see Problem 10-9). The result is a simple construction shown in Fig. 11-4 to illustrate the effect of R_E on circuit frequency response. The introduction of negative feedback has resulted in a reduced gain and increased bandwidth. The circuit is more limited in the amplitude of the output signal but distortion has been reduced.

Example 11-2 Effect of Feedback on the Gain and Bandwidth

An ac common-emitter amplifier has a midrange voltage gain of 20 dB. The design objective is to produce an increase in the value of f_2 by a factor of 10 through an internal frequency-dependent coupling circuit. Will $B = 0.50$ be satisfactory?

Solution From Eq. (11-8b) (taken at midrange),

$$A_{Vf} = \frac{-10}{1 + (10)(0.50)} = \frac{-10}{6.0} = -1.7$$

and with the *GBWP* constant,

$$f_2(10) = f_2'(1.7)$$

$$f_2'/f_2 = 6.0$$

The design objective is not met. A larger feedback factor B is required.

When the *GBWP* of an amplifier is constant, the gain and bandwidth are inversely proportional. By adjusting the feedback fraction B, a range of gain-bandwidth combinations can be achieved. A tradeoff exists between amplifier gain and bandwidth, and negative feedback can be used to achieve the desired performance objectives within circuit constraints. ∎

11-3 MULTISTAGE AND CASCADED AC AMPLIFIERS

The bandwidth of an amplifier chain is different from that of an individual amplifier stage. To see how this effect occurs, consider an ac amplifier unit that consists of two identical cascaded stages, as shown in Fig. 11-5. How will the gain of these combined circuits vary with frequency?

First, consider the midrange gain of the combination. The overall gain is given by the product of the individual stage gains:

$$A_V = A_{V1} A_{V2} \qquad (11\text{-}14)$$

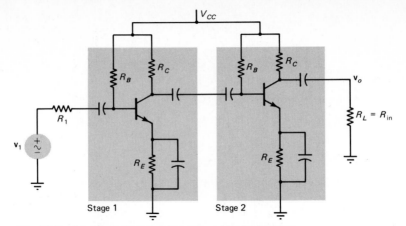

Fig. 11-5 Amplifier unit consisting of two identical stages.

The equations from Chapter 9 can be used. When calculating A_{V1} and A_{V2}, note that the equivalent output loading of the first stage is given by R_C in parallel with R_{in} of the second stage.

Both f_1 and f_2 are affected by the combination of stages. It is shown in Appendix 11 that for n identical stages,

$$f_1' = \frac{f_1}{(2^{1/n} - 1)^{1/2}} \tag{11-15}$$

$$f_2' = (2^{1/n} - 1)^{1/2} f_2 \tag{11-16}$$

where f_1 and f_2 are the single-stage (interior) half-power points and f_1' and f_2' are the half-power points for n identical stages. For $n = 2$, $f_1' = 1.55 f_1$, and $f_2' = 0.64 f_2$. The bandwidth of the multistage amplifier is less than the bandwidth of an individual stage. From previous analysis, the roll-off at low and high frequencies is now (20 dB)(n) and (-20 dB) (n) in the vicinity of the interior half-power points.

Example 11-3   Two-Stage Amplifier

Consider a single-stage amplifier with $A_V = 10$, $f_1 = 100$ Hz, and $f_2 = 100$ kHz. Draw the amplitude plot for an amplifier consisting of two identical stages having these characteristics.

Solution   From Eq. (11-14), the midrange gain of the combined stages is $A_V = 100$, 20 dB above the gain of a single stage. From Eqs. (11-15) and (11-16), the interior half-power points are given by

$$f_1' = \frac{100}{(2^{1/2} - 1)^{1/2}} \cong 1.6 \times 10^2 \text{ Hz}$$

$$f_2' = (100 \times 10^3)(2^{1/2} - 1)^{1/2} \cong 0.64 \times 10^5 \text{ Hz}$$

See the solution in Fig. 11-6. Note that the overall gain of the amplifier is increased at midrange, whereas the bandwidth is reduced.

The two curves in Fig. 11-6 reveal how the bandwidth is decreased as the number of stages increases. For two stages, the midrange gain is larger, but the low- and

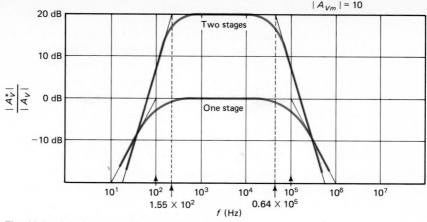

Fig. 11-6 Amplitude plot for two identical stages.

high-frequency roll-off is increased. Due to the more rapid roll-off, the half-power points (3 dB down from the peak gain $|A_V|$) occur nearer the midrange region for the case of two stages. By extrapolating this result, it can be seen that many stages produce a higher midrange gain and narrower midrange region, more rapid roll-off, and a frequency-response curve that more closely resembles the shape of a narrow pulse. ∎

Example 11-4 **Cascaded Amplifiers**

Find the phase shift associated with the half-power frequency f_2 for a chain of three identical amplifier circuits.

Solution Refer to Appendix 11. By definition, the half-power point occurs for

$$\left\{ \frac{1}{[1 + (f_2'/f_2)^2]^{1/2}} \right\}^3 = \frac{1}{(2)^{1/2}}$$

and

$$f_2'/f_2 = (2^{1/3} - 1)^{1/2} = 0.51$$

The phase angle is given by

$$\tan \theta^* = \frac{Im(1 - j0.51)^3}{Re(1 - j0.51)^3} = \frac{-1.40}{0.22} = -6.4$$

$$\theta^* = -80°$$

The upper half-power point drops by 49 percent and a phase shift of $-80°$ is observed at the half-power point. For three identical stages, the maximum total phase shift associated with the low-frequency roll-off is $3 \times 90° = 270°$, and the total phase shift associated with the high-frequency roll-off is $3 \times -90° = -270°$. Only a small fraction of the total phase shift ($80°$ or $-80°$ for f_1' and f_2') is observed at the half-power points. This can be contrasted with the $45°$ or $-45°$ phase shifts observed for f_1 and f_2 for a single stage, with maximum phase shifts of $90°$ and $-90°$. ∎

As shown in Fig. 11-7(A), the *cascode amplifier* consists of a common-emitter stage followed by a common-base stage. Figure 11-7(B) shows the circuit redrawn to indicate the nature of each stage. The dc operation of this circuit can be found as shown in Fig. 11-7(C). First, operating points are selected for Q_1 and Q_2 to obtain linear operation. Because I_C is about the same in Q_1 and Q_2,

$$V_{CC} \cong I_C R_E + (V_{CE})_Q + (V_{CE})_Q + I_C R_C \tag{11-17}$$

Typically, let $R_E \cong 0.1 R_C$. Then, for a selected value of V_{CC}, R_C is determined. The collector and emitter voltages for Q_1 and Q_2 are fixed. The base voltages are then about 0.7 V above the emitter voltages, and $I_{B1} \cong I_{B2}$. If R_1, R_2, and R_3 are chosen so that the current in each resistor is much larger than I_B, the appropriate base voltages can be obtained through the resistor ratios:

$$V_{CC} \frac{R_3}{R_1 + R_2 + R_3} \cong I_C R_E + 0.7 \text{ V}$$

$$\tag{11-18}$$

$$V_{CC} \frac{R_2 + R_3}{R_1 + R_2 + R_3} \cong I_C R_E + V_{CE} + 0.7 \text{ V}$$

To understand the ac circuit operation, refer to the ac equivalent circuit of Fig. 11-7(D), where C_1 is assumed large enough to force the base of Q_2 to ac ground, and C_E is sufficient to short R_E for frequencies of interest.

Fig. 11-7 (A) Cascode amplifier schematic. (B) Circuit redrawn to indicate the nature of each stage. (C) DC circuit. (D) AC equivalent circuit.

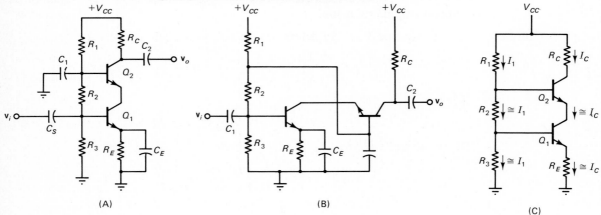

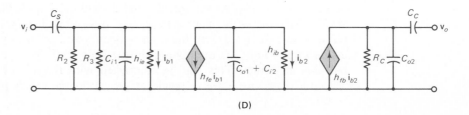

For the voltage gain of the common-emitter stage,

$$A_{V1} = -A_i \frac{Z_o}{Z_i}$$

$$= -h_{fe} \frac{R_2 \| R_3 \| h_{ie}}{h_{ie}} \frac{h_{ib}}{R_2 \| R_3 \| h_{ie}}$$

$$= -h_{fe} \frac{h_{ib}}{h_{ie}} \qquad (11\text{-}19)$$

Since $h_{ie} \cong V_0/|I_B|$ and $h_{ib} \cong V_0/|I_E|$ and if $h_{fe} \cong h_{FE}$, the gain becomes

$$A_{V1} \cong -1 \qquad (11\text{-}20)$$

For the voltage gain of the common-base stage,

$$A_{V2} = -h_{fb} \frac{R_C}{h_{ib}} \cong \frac{R_C}{h_{ib}} \qquad (11\text{-}21)$$

which can be much greater than 1.

What are the advantages of this circuit? The bandwidth of the common-emitter stage is often limited by the effective Miller capacitance; having a low gain ($A_V = -1$) for this stage helps assure a maximum value for f_2 in this configuration. At the same time, the circuit results in an input impedance h_{ie} that is approximately h_{fe} times the input impedance h_{ib} of the common-base stage.

The result is a compromise. The gain responsibility of the amplifier is shifted to the common-base stage and the input impedance responsibility of the circuit is shifted to the common-emitter stage, sacrificing gain to maximum bandwidth. The result is a circuit with both increased bandwidth and a reasonable input impedance.

11-5   NYQUIST PLOTS

At this point, it is useful to introduce Nyquist plots as a way to describe the frequency-dependent properties of amplifiers. These plots can provide a useful design and evaluation aid in describing the performance of amplitude and phase relationships for multiple stages.

For a single ac-coupled amplifier stage, θ^* varies from $180° + 90°$ to $180° - 90°$, whereas $|A_v^*|$ is constant across a middle range and drops off at low and high frequencies. Figure 11-8(A) displays the Bode plot for a single amplifier stage. The information shown can be plotted in the polar or Nyquist plot, as shown in Fig. 11-8(B). The midrange amplifier response becomes a single point on the Nyquist plot (corresponding to a maximum value of $|A_v^*|$ and $\theta^* = 180°$).

For $f \ll f_1$, the curve starts out with $\theta^* = 180° + 90°$ and with $|A_v^*|$ very small. The point $\theta^* = 180° + 45°$ corresponds to $f = f_1$. The high-frequency roll-off begins past $\theta^* = 180°$ and reaches $\theta^* = 180° - 45°$ at $f = f_2$. Finally, as $\theta^* \to 180° - 90°$, $|A_v^*|$ again becomes very small.

A similar process can be followed to draw Nyquist plots for two- and three-stage (and n-stage) amplifiers. For the two-stage amplifier of Fig. 11-9, the midrange phase shift is $\theta^* = 0$. The extreme phase shifts at very low and high frequencies are

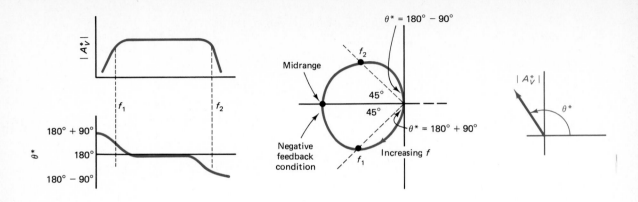

Fig. 11-8 Frequency response of a one-stage ac amplifier. (A) Bode plot. (B) Nyquist plot.

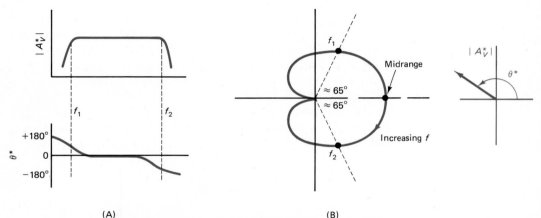

Fig. 11-9 Frequency response of a two-stage ac amplifier. (A) Bode plot. (B) Nyquist plot.

$-180°$ and $+180°$, respectively, and the gain $|A_V^*|$ becomes very small as these limits are approached. At the negative feedback condition, $\theta^* = 180°$ and the gain approaches zero. As shown in Appendix 11, the half-power frequencies f_1 and f_2 occur at about a $\pm 65°$ phase shift with respect to midrange.

For the three-stage amplifier of Fig. 11-10, the midrange phase shift is back to $\theta^* = 180°$ (the negative feedback condition). At low and high frequencies, the phase shift rotates through $+270°$ and $-270°$, producing the illustrated pattern. As shown in Fig. 11-10(B), the half-power frequencies f_1 and f_2 occur at about a $\pm 80°$ phase shift with respect to midrange. Nyquist plots provide another way to conceptualize amplifier performance and are of use in the study of sine-wave oscillators (Chapter 14).

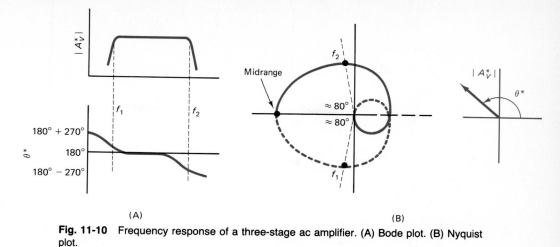

Fig. 11-10 Frequency response of a three-stage ac amplifier. (A) Bode plot. (B) Nyquist plot.

11-6   CLASSES OF AMPLIFIERS

Amplifier distortion can result if large input signals cause the use of a wide region of the characteristic curve. In many applications, this distortion is undesirable. Linear operation is needed, and nonlinear distortion can be a limiting consideration on circuit performance.

In some other applications, however, it is useful to design amplifiers that take advantage of characteristic curve distortion. The various modes for operation are defined by introducing a set of *classes* that describe the relationships between the amplifier and the signal being amplified.

To explain the various classes of operation, it is helpful to describe amplifiers in terms of plots showing v_{out} as a function of v_{in}. Plots of this type for common-base and common-emitter BJT amplifiers are discussed in Chapter 7. Similar plots can be prepared for FET amplifiers.

A plot of v_{out} as a function of v_{in} for a BJT amplifier demonstrates how the linear operating region is bounded by a cutoff region for low values of v_{in} and by the saturation region for large values of v_{in}. Figure 11-11 contrasts small-signal linear operation with large-signal nonlinear operation. The wave shape is distorted in the latter case as the transistor nears cutoff and saturation. Figure 11-12 shows the regions used for each class of operation.

Class A operation corresponds to the use of bias points and input signals so that transistor operation stays within the linear operating region. For large input signals, the output signal is no a faithful reproduction of the input because the gain (slope of the plot) varies somewhat over the region. For small signals, the amplifier provides approximately linear operation. The small-signal linear amplifiers discussed in this and previous chapters thus constitute a special case of class A operation. Figure 11-11 shows small- and large-signal class A operation for a BJT amplifier.

If an input signal ranges from 0 V to a forward-bias value, class B operation results. The output is no longer a linear replica of the input. A common-collector

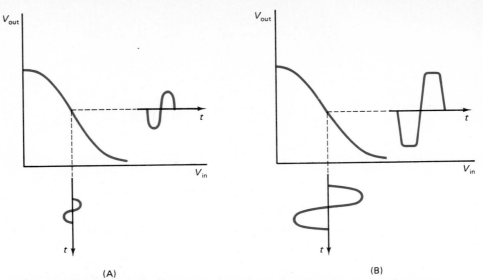

Fig. 11-11 Class A operation. (A) Small-signal linear operation (a special case of class A). (B) Large-signal class A operation.

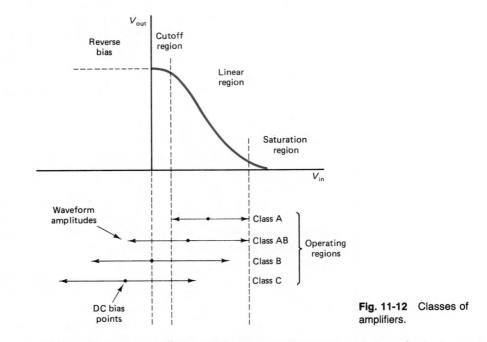

Fig. 11-12 Classes of amplifiers.

(emitter-follower) amplifier configuration often results in class B operation due to the base-emitter characteristic curve, as shown in Fig. 11-13. Voltage v_{out} is thus proportional to v_{in} except for the offset v_{BE}. The dc operating point is located on the boundary between the linear and cutoff regions. Only half of the input signal appears at the output and the other half is clipped off. (The function of the class B amplifier is thus similar to that of a half-wave rectifier.)

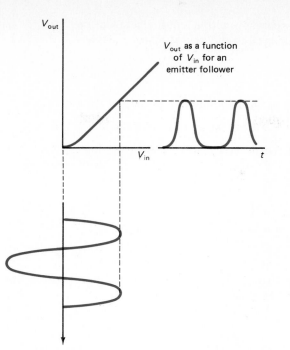

Fig. 11-13
Class B operation.

Two emitter followers in class B operation can be arranged to function as shown in Fig. 11-14. The output is somewhat distorted, but is still an approximate replica of the input signal. (Similar effects are noted in Chapter 5 for full-wave diode rectifiers.) This configuration is called a *push-pull amplifier*.

To see the advantage of this class of operation, consider the efficiency of both class A and class B amplifiers. Amplifier efficiency η is defined as

$$\eta = \frac{\text{Average ac output power delivered to the load}}{\text{Total dc power supplied to the transistor and load}} \times 100\% \qquad (11\text{-}22)$$

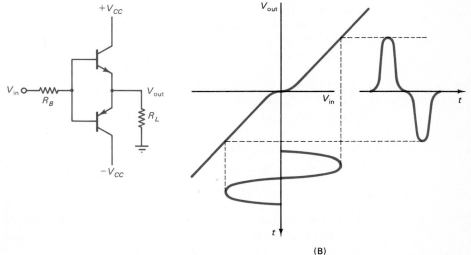

Fig. 11-14 Push-pull class B amplifier. (A) Schematic. (B) Input and output signals.

(B)

For class A operation with a sinusoidal signal that uses the entire linear region, the ac power delivered to the load is approximately

$$\left(\frac{1}{\sqrt{2}}\frac{V_{CC}}{2}\right)^2 \bigg/ R_L \tag{11-23}$$

where $V_{CC}/2$ is the maximum amplitude of the signal, and the effective value of a sine wave is $\dfrac{1}{\sqrt{2}}$ times the amplitude of the wave. The dc power delivered to the BJT is $V_Q I_Q$ and the dc power delivered to the load is $I_Q^2 R_L$. From Fig. 11-12, $V_Q = V_{CC}/2$ and $I_Q = \frac{1}{2} V_{CC}/R_L$ if the Q point is in the middle of the linear operating region. Therefore,

$$\eta = \frac{\left(\frac{1}{\sqrt{2}}\frac{V_{CC}}{2}\right)^2 \bigg/ R_L}{V_Q I_Q + I_Q^2 R_L} \times 100\% = \frac{\left(\frac{1}{\sqrt{2}}\frac{V_{CC}}{2}\right)^2 \bigg/ R_L}{\frac{1}{4}\frac{V_{CC}^2}{R_L} + \frac{1}{4}\frac{V_{CC}^2}{R_L}} = 25\% \tag{11-24}$$

If the input signal does not use the entire linear region, the average ac power delivered to the load is

$$\left(\frac{1}{\sqrt{2}} V_m\right)^2 \bigg/ R_L$$

The dc (bias) power is unchanged, so

$$\eta = \left(\frac{V_m}{V_{CC}}\right)^2 \times 100\% < 25\% \tag{11-25}$$

For class B push-pull operation, the maximum ac power delivered to the load is

$$\left(\frac{1}{\sqrt{2}} V_{CC}\right)^2 / R_L \tag{11-26}$$

because V_{CC} is now the output amplitude. For this class of amplifier, no dc power is supplied to the transistor unless a signal is present. Therefore, the denominator of the expression for the amplifier efficiency is given by $I_{av} V_{CC}$, where I_{av} is the dc component of the Fourier representation for a full-wave rectified output. From Eq. (5-16), this term is

$$\frac{2}{\pi}\frac{V_{CC}}{R_L}$$

and the efficiency for class B operation is

$$\eta = \frac{\left(\frac{1}{\sqrt{2}} V_{CC}\right)^2 / R_L}{\left(\frac{2}{\pi}\frac{V_{CC}}{R_L}\right) V_{CC}} \times 100\% = \frac{\pi}{4} \times 100\% \cong 78\% \tag{11-27}$$

If the input signal does not use the entire active region, the average ac power delivered to the load is

$$\left(\frac{1}{\sqrt{2}}V_m\right)^2 / R_L$$

The dc (bias) power is

$$\left(\frac{2}{\pi}\frac{V_m}{R_L}\right)V_{CC}$$

so that

$$\eta = \frac{\pi}{4}\left(\frac{V_m}{V_{CC}}\right) \times 100\% < 78\% \tag{11-28}$$

The net effect of class B operation is to reduce dc power consumption (and heating) in the circuit for a given ac output power level, at the expense of some amplifier distortion. As might be expected, class B operation is most often concerned with power amplifiers.

Class AB operation provides an intermediate case, with the input signal extending partially, but not completely, into the cutoff region. This class of operation is used to reduce the signal distortion that takes place as signals cross the boundary between the linear and cutoff regions. Push-pull power amplifiers can also be designed using class AB operation. Class AB amplifier configurations are further discussed in Chapter 12 as part of operational amplifier circuits.

Class C operation takes place when the transistor is biased in the cutoff region and an output signal is obtained only when the input signal brings transistor operation into the linear region. The resulting output is a series of pulses, one per cycle, occurring near the peak of the input signal, as shown in Fig. 11-15. The efficiency of the amplifier is increased even further at an additional sacrifice in linearity.

Nonlinear amplifier operation results in the creation of new frequencies that are

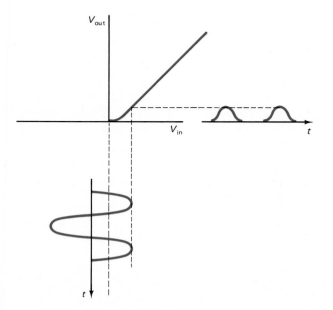

Fig. 11-15
Class C operation.

not present in the input waveform. The distortion in the output waveform is due to these new frequency components. Figures 11-11(B), 11-13, 11-14, and 11-15 show waveforms that have been subjected to such distortion.

The rectifier circuits discussed in Chapter 5 are also designed to introduce new frequency components in a waveform through distortion. With the ideal diode model, the outputs of full-wave and half-wave rectifiers using the ideal diode model are shown in Figs. 5-10 and 5-13. In each case, the nonlinearities in the diode have produced new frequency components in the output. Equations (5-15) and (5-16) give the specific frequency components that are generated.

If a distorted waveform is passed through a band-pass filter, a particular sine-wave component can be extracted. This is an approach to producing high-frequency sine waves from lower-frequency ones. A potential application for class C operation can be seen even though the output is no longer a recognizable version of the input. Class C operation can be used in tuned amplifiers, with a resonant circuit replacing the load resistor, as shown in Fig. 11-16. The resonant circuit selects a narrow band of frequencies from the distorted waveform and thus provides a known amplifier output. The efficiency of tuned amplifiers can approach 100 percent and they are used in transmitters and other high-power applications. (The tuned amplifier can also reduce the effects of circuit noise, as discussed in the next section.)

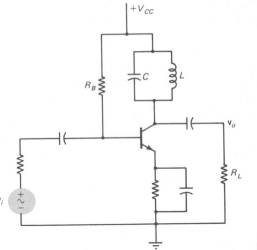

Fig. 11-16   Tuned amplifier.

Example 11-5   Amplifier Efficiency

A class A amplifier makes use of a waveform that is $V_{CC}/4$ in amplitude. Estimate the efficiency of this amplifier. What will be the efficiency for a class B amplifier using the same amplitude?

Solution   For a class A amplifier,

$$\eta = (V_m/V_{CC})^2 \times 100\%$$

Given $V_m = V_{CC}/4$,

$$\eta = \frac{100}{16}\% \cong 6\%$$

For a class B amplifier,

$$\eta = \frac{\pi}{4}\left(\frac{V_m}{V_{CC}}\right) \times 100\%$$

$$= \frac{\pi}{4}\left(\frac{100}{4}\right)\% \cong 20\%$$

The maximum efficiency for a class A amplifier (25 percent) is reduced to 6 percent by a waveform that is one-half of its maximum value. The maximum efficiency for a class B amplifier (78 percent) is reduced to 20 percent by the same waveform. Thus, for amplifiers with waveforms significantly less than the maximum values that can be accommodated, the efficiency drops rapidly. Observe that for class A small-signal operation, with $V_m \ll V_{CC}$, the efficiency is of the order of a few percent or less. ■

11-7 NOISE LIMITATIONS IN AMPLIFIERS

Every amplifier has an output, even with no signal input. Such an output is due to *noise*, which places a limit on circuit performance. Noise sets a lower boundary on the signal that can be successfully amplified because the signal-to-noise ratio must be high enough to prevent masking of the signal. Since the signal is smallest at the input to the first stage of an amplifier, noise developed at this input is likely to be the central factor in a multistage circuit.

Any unwanted current or voltage in a circuit can be referred to as noise in that such effects interfere with the intended performance of the circuit. Noise can arise from distributed capacitance and inductance in a circuit, which allows unwanted coupling from one part of the circuit to another. Extenal sources, including power supply variations, can also couple to the input and affect performance. Both of these types of noise can be reduced by careful design and shielding.

Other types of noise are more fundamental in that they arise from within the functional elements used to create a circuit. Noise of this type is related to the physical behavior of the elements and represents an ultimate limit on amplifier performance.

Every resistor develops noise across its terminals. Inside the resistor, electrons are in constant motion due to their thermal energy. The number of electrons moving in a given direction fluctuates with time. Therefore, a small net current can develop between the resistor terminals. On the average, these small currents must average to zero if no external voltage is applied. However, the variations that exist around this average value produce small voltages across the resistor terminals and thus produce unwanted noise.

This *thermal noise* (also called Nyquist or Johnson noise) can be described by the mean-square variation in the voltage produced across the device

$$\overline{v_{nt}^2} = 4kTR(f_2 - f_1) \tag{11-29}$$

where k is Boltzmann's constant, T is the temperature of the resistor in °K, R is the resistor value, and $f_2 - f_1$ is the bandwidth of the amplifier. The thermal noise is thus dominated by the temperature T, the effective R at the input to the amplifier, and the bandwidth of the amplifier. Careful amplifier design calls for minimizing R

and limiting the bandwidth to only the needed range. Also, low-noise amplifiers can be produced by cooling the front-end circuitry.

The tuned amplifier of Fig. 11-16 provides one way to reduce the effect of amplifier noise. Because the thermal noise is proportional to the bandwidth, the narrow bandwidth of the tuned amplifier reduces noise effects while amplifying the signal.

Additional *excess* or *flicker* noise develops when a dc current flows through a resistor. For a bandwidth Δf around frequency f, the mean-square variation in voltage due to this effect can be written

$$\overline{v_{ne}^2} = \frac{C}{f}\Delta f \tag{11-30}$$

where C is a constant dependent on resistor materials, construction, and on the magnitude of the current in the resistor, and f is the frequency of interest. This increase in noise voltage above thermal noise becomes more important at low frequencies.

To maximize circuit performance, the type of resistor fabrication can be matched to the circuit application. A resistor with a large excess noise level is more suitable for use at higher frequencies.

Over a bandwidth $f_2 - f_1$, the noise contribution from $\overline{v_{ne}^2}$ is

$$\overline{v_{ne}^2} = C\int_{f_1}^{f_2} \frac{df}{f} = C \ln\frac{f_2}{f_1} \tag{11-31}$$

The total *noise power* due to both thermal and excess noise terms is

$$N = \frac{\overline{v_{nt}^2}}{R} + \frac{\overline{v_{ne}^2}}{R} = 4kT(f_2 - f_1) + \frac{C}{R}\ln\frac{f_2}{f_1} \tag{11-32}$$

The noise properties of many amplifiers are described in terms of a *corner frequency* f_c at which the two noise terms are equal, as shown in Fig. 11-17.

Fig. 11-17 Noise power density as a function of frequency.

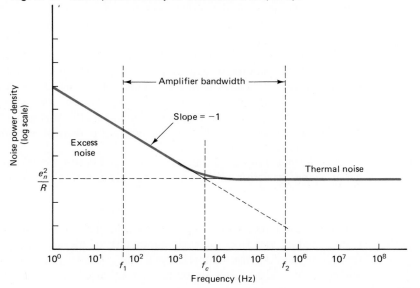

The constant C can be found in terms of f_c, providing a means for describing the performance of specific circuits. Equation (11-32) can be rewritten in the form

$$\frac{N}{f_2 - f_1} = 4kT + \frac{C}{R(f_2 - f_1)} \ln\left(\frac{f_2}{f_1}\right) \tag{11-33}$$

The *noise power density* e_n^2/R can be defined as equal to the ratio $N/(f_2 - f_1)$ for $f > f_c$. From Fig. 11-17, thermal noise dominates in this region, so that

$$\frac{e_n^2}{R} = \left.\frac{N}{f_2 - f_1}\right|_{f \gg f_c} \cong 4kT \tag{11-34}$$

By definition, the thermal and excess noise density components are equal to f_c. From Eq. (11-30),

$$\left.\frac{\overline{v_{nt}^2}}{R(f_2 - f_1)}\right|_{f_c} = \left.\frac{\overline{v_{ne}^2}}{R(f_2 - f_1)}\right|_{f_c} \tag{11-35}$$

so that

$$4kT = C/f_c R \quad\text{and}\quad C = 4kTRf_c$$

Or, from Eq. (11-34),

$$C = e_n^2 f_c \tag{11-36}$$

By substituting into Eq. (11-32),

$$N = \frac{e_n^2}{R}(f_2 - f_1) + \frac{e_n^2 f_c}{R} \ln\left(\frac{f_2}{f_1}\right) \tag{11-37}$$

and the *total noise voltage* takes the form

$$(\overline{v_n^2})^{1/2} = e_n[(f_2 - f_1) + f_c \ln(f_2/f_1)]^{1/2} \tag{11-38}$$

Example 11-6 Thermal Noise in a Signal Source

An amplifier uses a 100-Ω input resistor. Find the thermal noise voltage developed across the resistor, as seen by the amplifier, if the amplifier has $f_2 = 1.0$ MHz and $f_1 = 100$ Hz at 300°K.

Solution

$$(\overline{v_{nt}^2})^{1/2} = [4(1.38 \times 10^{-23} \text{ J/°K})(300°\text{K})(100 \ \Omega)10^6 \text{ Hz}]^{1/2}$$
$$\cong 1.3 \ \mu\text{V}$$

If a random *rms* signal is multiplied by 6, the peak-to-peak value that results is larger than the actual signal over 99 percent of the time. For the above case, this boundary is

$$v_{\text{pk-pk}} = 6 \times 1.3 \ \mu\text{V} = 7.8 \ \mu\text{V}$$

As indicated here, noise voltages are often measured in μV. As can be seen from the above development, the amplifier of Example 11-6 is not able to successfully amplify input signals of this magnitude (unless sufficient signal information is available to apply various types of filtering techniques). ∎

Example 11-7 Effect of Excess Noise

Assume that the amplifier of Example 11-6 has a corner frequency $f_c = 50$ kHz. Find the total noise voltage at the input.

Solution From above,

$$1.3 \ \mu\text{V} = e_n (10^6 \text{ Hz})^{1/2}$$

$$e_n = 1.3 \ \text{nV}/(\text{Hz})^{1/2}$$

For the excess noise,

$$f_c \ln (f_2/f_1) = (5.0 \times 10^4) \ln (10^6/10^2) = 0.46 \text{ MHz}$$

so that

$$(\overline{v_n^2})^{1/2} = 1.3 \frac{\text{nV}}{(\text{Hz})^{1/2}} [(1.0 \times 10^6) + (0.46 \times 10^6)]^{1/2}$$

$$= 1.6 \ \mu\text{V}$$

In this case, the thermal and excess noise contributions are of the same magnitude. If f_c is much smaller, then the thermal noise contribution is negligible, whereas if f_c is much larger, this term dominates. Observe that the thermal noise term is proportional to $f_2 - f_1$, where the excess noise term is proportional to $\ln (f_2/f_1)$. Therefore, changes in f_2 and f_1 affect the two terms in different ways. ∎

Other types of noise can also be experienced in electronic circuits. Some noise sources are better represented by effective current variations and can be analyzed by extending the above results. *Shot noise* arises because current flow consists of the motion of discrete charges. The *rms* variation in a current I is given by

$$\overline{i_{ns}^2} = 2qI (f_2 - f_1) \tag{11-39}$$

The same approach to analysis given above can be used to calculate the relative importance of different current noise sources and to combine their composite effects. Voltage and current noise sources can be combined by converting both to noise powers.

$$N = \overline{v_{nt}^2}/R + \overline{v_{ne}^2}/R + \overline{i_{ns}^2} R \tag{11-40}$$

The performance of an amplifier depends on the ratio of the signal power S to the noise power N. The *signal-to-noise ratio SNR* is defined as

$$SNR (\text{dB}) = 10 \log_{10} (S/N) \tag{11-41}$$

Based on communications criteria, acceptable levels for S/N can be established for a given application. Levels of 15 to 20 dB may be adequate for a digital communications system, whereas 40 dB is often required for telephone communications.

Real amplifiers always display a noise level that is greater than the ideal prediction of Eq. (11-29). It is usual to compare the actual noise level in an amplifier with its ideal value by the *noise figure* (*NF*), which is often expressed in dB. Alternatively, an *effective noise temperature* (T_{eff}) can be defined to accommodate the increase in noise above the ideal level. By definition,

$$NF = (1 + T_{\text{eff}}/T_O) \qquad T_O = 300°\text{K} \tag{11-42}$$

so the two representations are equivalent.

The effective *SNR* is reduced below the ideal value associated with Eq. (11-29) by the *NF:*

$$SNR \text{ (dB)} = 10 \log \left(\frac{v_S}{R}\right)^2 /4kT\,(f_2 - f_1) - NF \text{ (dB)} \qquad (11\text{-}43)$$

Example 11-8 **Amplifier Selection Based on Noise Performance**

An input signal with $v_S = 40 \ \mu V$ and a bandwidth $\Delta f = 1.0$ kHz is to be amplified. Given a 1.0-kΩ input resistance for the source and a matched amplifier input, what is the maximum *NF* that is acceptable to achieve $(S/N) = 40$ dB if only thermal noise is considered?

Solution From Eq. (11-43),

$$NF \text{ (dB)} \le 10 \log \frac{(40 \ \mu V)^2}{(1.0 \text{ k}\Omega)(4)(1.38 \times 10^{-23} \text{ J/°K})(300°\text{K})(1.0 \times 10^3 \text{ Hz})} - 40 \text{ dB}$$

$$\le 10 \text{ dB}$$

For this case, the noise level of the amplifier can be no more than ten times the ideal level. Realistic operational amplifier IC circuits can be rated with noise figures ranging from 3 dB (a device designed to have a very low NF) to 10 dB (which is relatively easy to obtain). ∎

QUESTIONS AND ANSWERS

Questions **11-1.** What are some of the ways in which distortion can affect waveform amplification?

11-2. How can Fourier analysis assist in the study of amplifier distortion?

11-3. How can square-wave testing be used to find f_1 and f_2 for an amplifier?

11-4. What effect does negative feedback have on the frequency distortion produced by an amplifier?

11-5. How can the half-power points of a multistage amplifier be calculated?

11-6. What are some of the advantages and disadvantages of the cascode amplifier?

11-7. What is the relationship between a Bode plot and a Nyquist plot?

11-8. What are some of the tradeoffs involved in deciding to use a linear or nonlinear amplifier for a given application?

11-9. What is the significance of the corner frequency f_c?

Answers **11-1.** As shown in Fig. 11-18, an amplifier can produce distortion due to: (a) nonlinear amplification that changes the frequency components of the input waveform, (b) frequency-

Fig. 11-18 Types of distortion in an amplifier circuit.

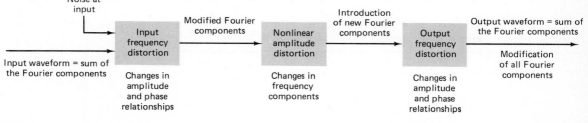

dependent amplification that changes the amplitude and phase relationships among the Fourier components of the input waveform, and (c) noise that adds unwanted variations to the input signal.

11-2. Again, refer to Fig. 11-18. By applying Fourier analysis, an arbitrary input waveform to an amplifier can be reduced to its harmonic (sine-wave) components. The effects of frequency-dependent distortion can then be considered by analyzing the changes in amplitude and phase for each component, and the output can be constructed from the superposition of these component terms. If nonlinear amplitude distortion is important, Fourier analysis can be used to represent the new waveform, and the effects of the output frequency distortion on the new Fourier components can also be established.

11-3. A low-frequency square wave can be used to find f_1 by measuring the time constant associated with the exponential "droop" in the output waveform, as shown in Fig. 11-1(B). A high-frequency square wave can be used to find f_2 by measuring the rise and fall time constants for the output waveform, as shown in Fig. 11-1(C).

11-4. Negative feedback can be used to reduce f_1 and increase f_2, thereby increasing the amplifier bandwidth and reducing the frequency distortion associated with Fourier components within this frequency range. A reduction in amplifier gain is associated with the introduction of negative feedback into the amplifier.

11-5. To find the half-power points associated with a multistage amplifier, the absolute value of the complex gain functions of all amplifier stages multiplied together $\left| A_{V1}^* A_{V2}^* A_{V3}^* \cdots \right|$ is set equal to $\dfrac{1}{\sqrt{2}}$. If all stages are identical, this statement simplifies to $\left| A_v^* \right|^n = \dfrac{1}{\sqrt{2}}$. This equation can be used to find the new values f_1' and f_2', which in turn can be used to find the phase shifts associated with the half-power points.

11-6. The cascode amplifier is designed so that the same amplifier circuit can produce the high gain and input impedance associated with a common-emitter amplifier and the large bandwidth associated with a common-base amplifier. A tradeoff must be made for these circuit advantages against the added complexity of a two-stage amplifier circuit with associated devices and components.

11-7. The two (amplitude and phase) plots that constitute a Bode plot contain the same information as the polar Nyquist plot. The two different ways for presenting the amplitude and phase relationships associated with an amplifier provide complementary insights into frequency-dependent amplifier properties.

11-8. Nonlinear amplifiers result in higher efficiency operation than linear amplifiers, but they introduce distortion into the waveform being amplified. If linearity is important, this distortion is a significant amplifier characteristic and leads to tradeoff decisions. On the other hand, nonlinear amplifiers can create new frequency components in a waveform, where linear amplifiers cannot. If this frequency-generating process is important, then the nonlinear aspect is an advantage to be utilized.

11-9. At the corner frequency f_c, the noise contributions due to thermal and excess noise are equal. At lower frequencies, excess noise dominates, and at higher frequencies, thermal noise dominates. The selection of a resistor technology for a given application is influenced by the value f_c associated with the component.

EXERCISES AND SOLUTIONS

Exercises
(11-1)

11-1. From Fig. 11-1(B), show that

$$\frac{T_{\text{sq. wave}}}{2} = \pi \tau_1$$

(11-1) **11-2.** From Fig. 11-1(C), show that

$$\frac{T_{\text{sq. wave}}}{2} = \pi \tau_2$$

(11-1) **11-3.** A square wave is used to determine f_1 and f_2 for an ac amplifier. If the values $f_{\text{sq. wave}} = 2\pi f_1$ and $f_{\text{sq. wave}} = f_2/2\pi$ are selected, sketch the expected output waveforms.

(11-1) **11-4.** For an ac amplifier, the output waveform of Fig. 11-1(B) is observed. (a) If the time constant of the output is 1.0 ms, what is the fundamental frequency for the circuit? (b) If $R_{\text{series}} = 1.5\ \text{k}\Omega$, what is the effective coupling series capacitance for the circuit?

(11-1) **11-5.** For an ac amplifier, the output waveform of Fig. 11-1(C) is observed. (a) If the upper half-power point $f_2 = 1.0\ \text{MHz}$, what is the (high-frequency) time constant for the circuit? (b) If $R_{\text{parallel}} = 5.0\ \text{k}\Omega$, what is the effective parallel capacitance for the circuit?

(11-2) **11-6.** For the circuit representation of Fig. 11-19, feedback is obtained by placing a fraction $B*$ of the output voltage v_o in series with v_i. Show that, for this case, the results of Eq. (11-8) are also obtained. [There are four types of feedback possible because the output current or voltage can be in series or parallel with the input. The relationship of Eq. (11-8) holds true for all four types, given the correct definition of $B*$.]

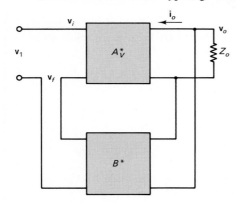

Fig. 11-19 Feedback method for Exercise 11-6.

(11-2) **11-7.** A common-emitter ac amplifier has a midrange voltage gain $|A_{vo}| = f_1 = 1.0 \times 10^2$ Hz and $f_2 = 1.0 \times 10^6$ Hz without feedback. With the addition of feedback, the gain drops to 5.0 dB. (a) Find the feedback factor B (assumed to be real). (b) Find the bandwidth with feedback. (c) Draw an amplitude plot for the circuit without and with feedback.

(11-2) **11-8.** An ac amplifier with $A_V = -15$ is modified to give a real feedback factor of $B = 0.3$. (a) What will be the resultant midrange voltage gain (in dB)? (b) What fractional change is observed in f_2?

(11-2) **11-9.** For an inverting ac amplifier with $R_C \| R_L = 3.0\ \text{k}\Omega$, the $GBWP$ is approximately constant at 10 MHz. The amplifier midrange gain is -35 without feedback. If the circuit is modified by the addition of $R_E = 1.0\ \text{k}\Omega$, find the voltage gain and bandwidth for the amplifier with feedback.

(11-3) **11-10.** An ac amplifier has $A_V = -10$, $f_1 = 20$ Hz, and $f_2 = 20$ kHz. Find the midrange gain and upper and lower half-power frequencies if three such amplifiers are cascaded together.

(11-5) **11-11.** Draw Nyquist plots corresponding to the amplifier plots of Fig. 11-20 (p. 460).

(11-6) **11-12.** An ac amplifier is operated class A (with a maximum amplitude) using a 1.0-kΩ load and a 20-V power supply. (a) Find the average ac power delivered to the load. (b) Find the total dc power delivered to the BJT and load.

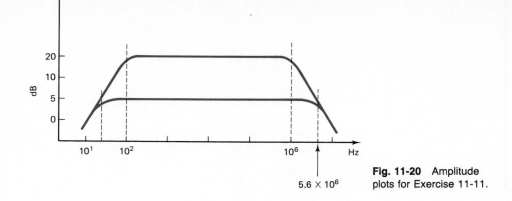

Fig. 11-20 Amplitude plots for Exercise 11-11.

5.6×10^6

(11-6) **11-13.** An amplifier is operated class B (with a maximum amplitude) using a 1.0-kΩ load and a 15-V power supply. (a) Find the ac power delivered to the load. (b) Find the dc power delivered to the BJT and load.

(11-7) **11-14.** An amplifier circuit has a corner frequency $f_c = 200$ Hz. If $f_2 = 1.0$ kHz and the contribution to noise is neglected for frequencies below 10 Hz, find $(\overline{v_n^2})^{1/2}$. Assume that $e_n = 20$ μV/$\sqrt{\text{Hz}}$.

Solutions **11-1.** Since $f_{\text{sq. wave}} = f_1$,

$$T_{\text{sq. wave}} \equiv 1/f_{\text{sq. wave}} = 1/f_1 = 2\pi/2\pi f_1 = 2\pi \tau_1$$

and

$$T_{\text{sq. wave}}/2 = \pi \tau_1$$

11-2. Since $f_{\text{sq. wave}} = f_2$,

$$T_{\text{sq. wave}} \equiv 1/f_{\text{sq. wave}} = \frac{1}{f_2} = \frac{2\pi}{2\pi f_2} = 2\pi \tau_2$$

and

$$T_{\text{sq. wave}}/2 = \pi \tau_2$$

11-3. Since $f_{\text{sq. wave}} = 2\pi f_1$ for the first case,

$$T_{\text{sq. wave}} \equiv 1/f_{\text{sq. wave}} = 1/2\pi f_1 = \tau_1$$

and

$$T_{\text{sq. wave}}/2 = \tau_1/2$$

The results are shown in Fig. 11-21, which can be compared with Fig. 11-1(B). As expected, the higher square-wave frequency reduces the effect of the low-frequency amplifier roll-off.

Since $f_{\text{sq. wave}} = f_2/2\pi$ for the second case,

$$T_{\text{sq. wave}} \equiv 1/f_{\text{sq. wave}} = 2\pi/f_2 = (2\pi)^2/2\pi f_2 = (2\pi)^2 \tau_2$$

and

$$T_{\text{sq. wave}}/2 = 2\pi^2 \tau_2$$

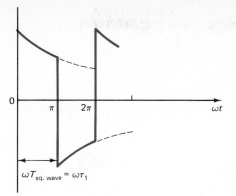

Fig. 11-21 Waveform for the first case in Exercise 11-3.

The results are shown in Figure 11-22, which can be compared with Fig. 11-1(C). As expected, the lower square-wave frequency reduces the effect of the high-frequency amplifier roll-off.

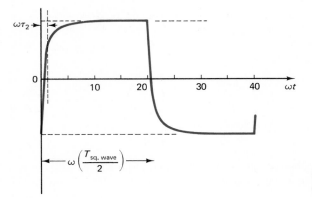

Fig. 11-22 Waveform for the second case in Exercise 11-3.

11-4. (a) From Fig. 11-1(B), $\omega\tau = 1.0$. Therefore,

$$f_1 = \frac{1.0}{2\pi\tau_1} = \frac{1.0}{2\pi(1.0 \text{ ms})} = 0.16 \text{ kHz}$$

The fundamental frequency is the lowest frequency term in the Fourier expansion. Therefore, given $f_1 = f_{\text{sq. wave}}$, the fundamental frequency for the circuit is 0.16 kHz.
(b) From Eq. (11-6),

$$C_{\text{series}} = \frac{1}{2\pi f_1 R_{\text{series}}}$$

Since $f = f_1$,

$$C_{\text{series}} = \frac{1}{2\pi(0.16 \text{ kHz})(1.5 \text{ k}\Omega)} \cong 0.67 \ \mu\text{F}$$

11-5. (a) From Fig. 11-1(C), $\omega\tau = 1.0$. Therefore,

$$\tau_2 = \frac{1.0}{2\pi(1.0 \text{ MHz})} \cong 0.16 \ \mu\text{s}$$

(b) From Eq. (11-7),

$$C_{\text{parallel}} = \frac{1}{2\pi f_2 R_{\text{parallel}}}$$

Since $f_2 = f_{\text{sq. wave}}$,

$$C_{\text{parallel}} = \frac{1}{2\pi (1.0 \text{ MHz}) 5.0 \text{ k}\Omega} \cong 32 \text{ pF}$$

11-6. From Fig. 11-19,

$$\mathbf{v}_1 + \mathbf{v}_f = \mathbf{v}_i$$

$$\mathbf{v}_i g_m = \mathbf{i}_o$$

$$\mathbf{i}_o R_L = \mathbf{v}_o$$

$$\mathbf{v}_o B^* = \mathbf{v}_f$$

Therefore,

$$\mathbf{v}_1 + \mathbf{v}_o B^* = \frac{\mathbf{i}_o}{g_m} = \frac{\mathbf{v}_o}{g_m R_L}$$

$$\mathbf{v}_1 = \mathbf{v}_o(1/g_m R_L - B^*)$$

and

$$A_{Vf}^* = \frac{\mathbf{v}_o}{\mathbf{v}_1} = \frac{1}{1/g_m R_L - B^*} = \frac{g_m R_L}{1 - g_m R_L B^*}$$

$$= \frac{A_V^*}{1 - A_V^* B}$$

11-7. (a) Observe that 20 dB corresponds to a voltage gain of $|A_V| = 10$, and 5 dB corresponds to a voltage gain of $|A_V| = 1.8$. Therefore, from Eq. (11-8),

$$A_{Vf} = \frac{A_{Vo}}{1 - A_{Vo}B}$$

$$-1.8 = \frac{-10}{1 + 10B}$$

so that

$$B = 0.46$$

(b) Given the *GBWP* = constant,

$$10(1.0 \text{ MHz}) = 1.8 f_2'$$

$$f_2' = 5.6 \text{ MHz}$$

(c) The resultant amplitude plot is shown in Fig. 11-20.

11-8. (a) From Eq. (11-8), $A_{Vf} = \dfrac{-15}{1 + 15(0.30)} \cong -2.7$

The matching voltage gain in dB is

$$20 \log 2.7 = 8.6 \text{ dB}$$

(b) Given a constant *GBWP*,

$$15f_2 = 2.7f_2'$$

$$\frac{f_2' - f_2}{f_2} \cong 4.6$$

11-9. From Section 11-2,

$$B = \frac{R_E}{R_C \| R_L} = \frac{1.0 \text{ k}\Omega}{3.0 \text{ k}\Omega} \cong 0.33$$

So, from Eq. (11-8),

$$A_{Vf} = \frac{A_{Vo}}{1 - A_{Vo}B} = \frac{-35}{1 + 35(0.33)} \cong -2.8$$

and

$$GBWP = 10 \text{ MHz} = (35)f_2 \quad \text{and} \quad f_2 = 0.29 \text{ MHz}$$

$$10 \text{ MHz} = (2.8)f_2' \quad \text{and} \quad f_2' = 3.6 \text{ MHz}$$

11-10. From Eq. (11-14),

$$A_V = (-10)^3 = -10^3$$

Then, from Eqs. (11-15) and (11-16),

$$f_1' = \frac{f_1}{(2^{1/3} - 1)^{1/2}} = \frac{20 \text{ Hz}}{0.51} \cong 39 \text{ Hz}$$

$$f_2' = f_2(2^{1/3} - 1)^{1/2} = (20 \text{ kHz})(0.51) \cong 10 \text{ kHz}$$

11-11. The phase shifts associated with the initial and modified half-power points are the same. Therefore, the results of Fig. 11-23 are obtained.

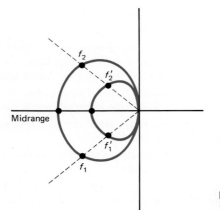

Fig. 11-23 Nyquist plot for the amplitude plots in Fig. 11-20.

11-12. (a) From the data given, the average ac power delivered to the load is

$$P = \left(\frac{1}{\sqrt{2}}\frac{V_{CC}}{2}\right)^2 / R_L = \left(\frac{10}{\sqrt{2}}\right)^2 / 1.0 \text{ k}\Omega \cong 50 \text{ mW}$$

(b) The total dc power delivered to the BJT and load is

$$P = \frac{1}{2}\frac{V_{CC}^2}{R_L} = \frac{1}{2}\frac{(20\text{ V})^2}{1.0\text{ k}\Omega} = 200\text{ mW}$$

The amplifier efficiency is

$$\eta = \frac{50\text{ mW}}{200\text{ mW}} = 0.25$$

as expected.

11-13. (a) The average ac power delivered to the load is

$$P = \left(\frac{1}{\sqrt{2}}15\text{ V}\right)^2/1.0\text{ k}\Omega = 112\text{ mW}$$

(b) The total dc power delivered to the BJT and load is

$$P = \frac{2}{\pi}\left(\frac{15\text{ V}}{1.0\text{ k}\Omega}\right)(15\text{ V}) = 143\text{ mW}$$

The amplifier efficiency is

$$\eta = 112\text{ mW}/143\text{ mW} = 0.78$$

as expected.

11-14. From Eq. (11-38),

$$(\overline{v_n^2})^{1/2} = e_n[(f_2 - f_1) + f_1\ln(f_2/f_1)]^{1/2}$$

$$= 20\ \mu\text{V}/(\text{Hz})^{1/2}\left[(1.0\times10^3) + 200\ln\left(\frac{1.0\times10^3}{10}\right)\right]^{1/2}$$

$$\cong 8.7\times10^2\ \mu\text{V}$$

PROBLEMS

(11-1) **11-1.** A square-wave test is run to determine f_1 for an ac amplifier. If $f_{\text{sq. wave}} = 4\pi f_1$, find the relationship between $T_{\text{sq. wave}}$ and τ_1. Sketch the expected output waveform.

(11-1) **11-2.** A square-wave test is run to determine f_2 for an ac amplifier. If $f_{\text{sq. wave}} = f_2/2$, find the relationship between $T_{\text{sq. wave}}$ and τ_2. Sketch the expected output waveform.

(11-1) **11-3.** For an ac amplifier, the effective series resistance and capacitance are 5.0 kΩ and 1.0 μF, respectively. What value of f_1 results?

(11-1) **11-4.** For an ac amplifier, the effective parallel resistance and capacitance are 2.4 kΩ and 50 pF, respectively. What value of f_2 results?

(11-2) **11-5.** A common-emitter ac amplifier has a midrange voltage gain of -15 with $f_1 = 50$ Hz and $f_2 = 5.0\times10^5$ Hz without feedback. With the addition of negative feedback, the voltage gain drops to -5. What (real) feedback factor B is required? What is the *GBWP* of the amplifier with and without feedback?

(11-2) **11-6.** A common-source amplifier with a midrange gain $A_V = -5.0$ is modified by the addition of negative feedback providing (a real) $B^* = 0.40$. What fractional increase in f_2 results?

(11-2)   **11-7.** For Exercise 10-11 in Chapter 10, capacitor C_E is removed. What is the midrange gain of the amplifier with negative feedback?

(11-2)   **11-8.** For Problem 10-16 in Chapter 10, the capacitor C_S is removed. What is the midrange gain of this amplifier with negative feedback?

(11-2)   **11-9.** For a given amplifier, show that the low-frequency response curves all approach the same asymptote, independent of the (real) feedback fraction B.

(11-2)   **11-10.** For an inverting ac amplifier, $B = 0.60$ is desired. If $R_D \| R_L = 2.0 \text{ k}\Omega$, what source resistor value R_S is required?

(11-3)   **11-11.** Four amplifiers are cascaded together to produce $f_2 = 400 \text{ kHz}$. Find the upper half-power point for each individual amplifier.

(11-3)   **11-12.** You desire to use three identical amplifier stages to produce an effective $BW \geq 1 \text{ MHz}$ and $A_V = -5.0 \times 10^2$. (a) What must be the gain of each stage? (b) What is the minimum value of f_2 for each stage alone?

(11-3)   **11-13.** Find the frequencies f_1 and f_2 associated with a four-stage cascaded amplifier if $f_1 = 1.0 \times 10^2 \text{ Hz}$ and $f_2 = 1.0 \times 10^6 \text{ Hz}$ for each amplifier alone. Find the phase shift associated with each (composite) half-power point.

(11-3)   **11-14.** Show that the f_2 values determined by the input coupling circuits for common-emitter and common-base amplifiers are such that the condition $f_{2CB} \gg f_{2CE}$ usually holds.

(11-4)   **11-15.** Find the midrange input impedance and gain for the cascode amplifier of Fig. 11-7 with

$$I_{\text{divider}} = 1.0 \text{ mA} \qquad h_{FE} = h_{fe} = 200$$

$$R_E = 0.1R_C \qquad I_C = 1.0 \text{ mA}$$

$$R_C = 4.0 \text{ k}\Omega \qquad V_{CC} = 12 \text{ V}$$

(11-5)   **11-16.** Draw Nyquist plots corresponding to the Bode plots in Figs. 10-22 and 10-24.

(11-6)   **11-17.** A class A amplifier is operated with $V_{CC} = 12 \text{ V}$ and an input signal with an amplitude of 4.0 V. Estimate the efficiency of the amplifier.

(11-6)   **11-18.** A class B push-pull amplifier is operated with $V_{CC} = 12 \text{ V}$ and an input signal with an amplitude of 8.0 V. Estimate the efficiency of the amplifier.

(11-6)   **11-19.** Assume that the output of a class C amplifier consists of portions of (undistorted) sine waves, as shown in Fig. 11-24. Find the Fourier components for this output, and contrast the frequencies present in the input and output waveforms.

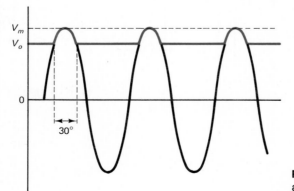

Fig. 11-24 Waveform for a class C amplifier.

(11-6)   **11-20.** A tuned amplifier is to be designed with a resonant frequency of 5.0 MHz and a Q of 100. (a) What is the bandwidth of the amplifier? (b) If $L = 10\ \mu H$, what value of C should be used? (c) What is the approximate resistance of the coil?

(11-7)   **11-21.** For an ac amplifier, $f_2 = 1.0 \times 10^5$ Hz, $f_1 = 2.0 \times 10^2$ Hz, $e_n = 5.8\ nV/\sqrt{Hz}$, and $\sqrt{\overline{v_n^2}} = 6.0\ \mu V$. (a) Find the value of f_c for the amplifier. (b) Find the voltage range outside of which a noise signal results less than 1 percent of the time.

(11-7)   **11-22.** For Exercise 11-14, assume that the input resistor to the amplifier is cooled to the temperature of liquid nitrogen (77°K). Find the new value of $\sqrt{\overline{v_n^2}}$.

(11-7)   **11-23.** Find the effective noise temperature for an amplifier with a NF of 5 dB.

(11-7)   **11-24.** Given $S/N = 40$ dB and $NF = 6$ dB, find the SNR.

(11-7)   **11-25.** Find the maximum NF that is allowed for Example 11-8 if excess noise is considered with $f_c = 10$ kHz and the center frequency of operation is 5.0 kHz.

COMPUTER APPLICATIONS

11-1. A SPICE simulation of the class B amplifier of Fig. 11-25 is shown in Figs. 11-26 to 11-28. Note how the crossover distortion is more important at lower voltage levels. Graphically determine the amplifier efficiency for each case shown.

*   **11-2.** A FORTRAN computer program has been used to produce the output waveforms of Fig. 11-1 with $f_{sq.\ wave} = f_1$ and $f_{sq.\ wave} = f_2$. Write your own computer program and plot the expected output waveforms that result for $f_{sq.\ wave} = 4f_1$ and $f_{sq.\ wave} = f_2/4$. Discuss your findings.

Fig. 11-25 Class B push-pull amplifier schematic used for a SPICE simulation.

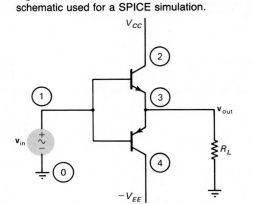

Fig. 11-26 Input file data for a SPICE simulation of the class B push-pull amplifier.

```
/T CLSB
/L ALL
    1        CLASS B PUSH PULL AMPLIFIER
    2    *
    3    * RESISTORS
    4    *
    5    RL 3 0 10K
    6    *
    7    * TRANSISTORS
    8    *
    9    Q1 2 1 3 MOD1
   10    Q2 4 1 3 MOD2
   11    .MODEL MOD1 NPN(BF=180)
   12    .MODEL MOD2 PNP(BF=180)
   13    *
   14    * DC SOURCES
   15    *
   16    VCC 2 0 DC 15
   17    VEE 0 4 DC 15
   18    *
   19    * INPUT SIGNAL
   20    *
   21    VI 1 0 SIN(0 2 1K)
   22    .TRAN .01MS 2MS
   23    *
   24    * OUTPUT
   25    *
   26    .PRINT TRAN V(3) V(1)
   27    .PLOT TRAN V(3) V(1)
   28    .END
```

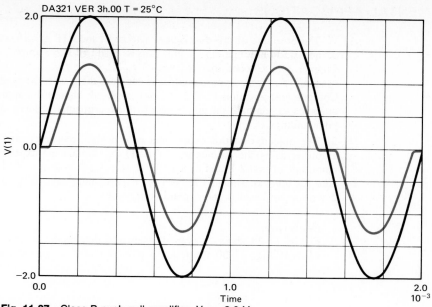

Fig. 11-27 Class B push-pull amplifier, V_m = 2.0 V.

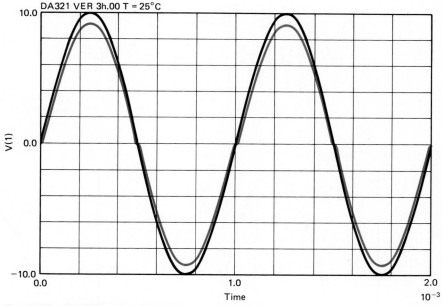

Fig. 11-28 Class B push-pull amplifier, V_m = 10.0 V.

* **11-3.** Assume that an arbitrary square wave is passed through an amplifier dominated by a single high-pass and low-pass filter response. Use the transient capability of SPICE to predict the output waveform and the Fourier components of the output.

EXPERIMENTAL APPLICATIONS

11-1. The objective of this experiment is to apply the BJT amplifier that was used in Fig. 10-37 and study how the circuit functions as a square-wave amplifier (see Problem 10-27 in Chapter 10). Choose the frequencies $f_{sq.\,wave} = f_1$; $f_{sq.\,wave} = $ a midrange frequency; and $f_{sq.\,wave} = f_2$. Record your results and compare them with the predictions of Fig. 10-39. If Computer Application 11-2 has been completed, choose the appropriate square-wave frequencies to compare the simulation with experimental observation. Discuss and interpret your findings.

11-2. The objective of this experiment is to design, construct, and test a class B push-pull amplifier using the schematic of Fig. 11-29. Predict the expected output using an input amplitude of first 2 V and then 10 V. (Computer Application 11-1 can be used to help prepare for this experiment.) Record the experimental data for your circuit, and graphically determine η. Discuss and explain your results.

Discussion
Figure 11-29 shows a typical schematic that might be used for this purpose, whereas Fig. 11-30 illustrates a plot of typical collected data that have been

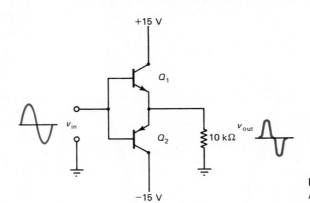

Fig. 11-29 Schematic for Experimental Application 11-2.

Fig. 11-30 Graphical calculation of η for Experimental Application 11-2. (A) Input. (B) Output.

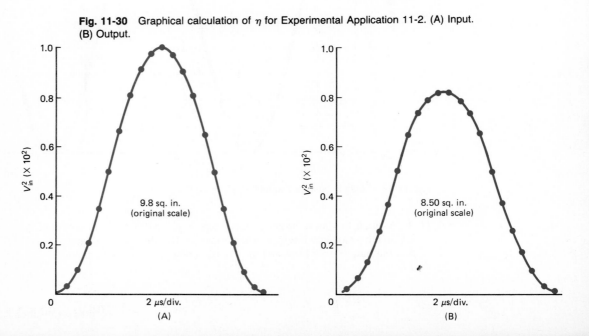

squared on a point-by-point basis. The calculation of η for the amplifier can be done graphically by counting squares (refer to Example 5-9). Based on the output plot shown, an efficiency of about $8.5/9.8 \times 100 = 87$ percent was determined for a 10 percent measurement error with respect to theoretical predictions.

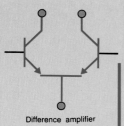

Difference amplifier

PART 5
LINEAR APPLICATIONS

Chapters 12 to 14 are concerned with linear ICs and their applications. Chapter 12 draws upon the modeling and circuit design efforts of previous chapters to demonstrate the operation of IC *operational amplifiers* (or *op amps*). An evolutionary approach illustrates how the results of previous amplifier design efforts can be used to understand these more complex circuits. A simplified version of an op amp that draws directly on dc- and ac-coupled amplifier concepts is used as a learning transition. This simple op amp then leads into a discussion of IC op amps and their properties. Chapter 12 includes an important section on the frequency-response characteristics of IC op amps, extending the concepts presented in Chapters 10 and 11.

Chapters 13 and 14 apply the IC op amps introduced in Chapter 12, and give additional insights that are required to effectively use these building block devices in circuit and system design. Chapter 13 emphasizes fundamental subsystem circuits (consisting of ICs and supporting circuitry) that can themselves be used as building blocks and incorporated into larger systems. The flexibility of the op amp and the usefulness of the subsystem circuits built around this device are demonstrated by example. This strategy of combining IC devices with supplementary circuitry and integrating the entire unit into a new device is a widely used approach to system design today.

Chapter 14 considers the operation and application of sine-wave oscillators built with either ICs or discrete devices. The discussion of alternative oscillator designs emphasizes the plurality of approaches that can be used to achieve a given performance objective and thereby the importance of recognizing all material, device, and circuit parameters that affect the selection process. The chapter concludes with a discussion of oscillator applications, with an emphasis on communication systems.

CHAPTER 12
OPERATIONAL AMPLIFIERS

It is difficult to design amplifiers that have a fixed, predictable voltage gain. Variations in transistor parameters, resistor tolerances, and thermal effects can all affect amplifier gain and prevent expected performance.

However, Sections 9-8, 9-11, and 11-2 show how negative feedback can be used to produce amplifiers with overall gain that is independent of transistor parameters. The *operational amplifier* (*op amp*) is a high-gain amplifier that is used with a large negative feedback factor to produce several attractive performance features.

The op amp is a specific type of dc-coupled amplifier characterized by a high voltage gain and built-in protection against certain types of noise. By using negative feedback, the gain of an op amp can be set with accuracy and high input impedance and a low output impedance can be obtained. The applications of op amp are numerous and limited only by the ingenuity of the designer.

For many years, most op amps were fabricated using hybrid microcircuit technology. However, once design techniques were available for the fabrication of integrated circuit (IC) operational amplifiers, the reduced costs resulted in a rapid shift in technology. (Both hybrid microcircuits and IC technologies are discussed in Chapter 18.)

There are hundreds of different types of IC op amps currently available on the commercial market with prices ranging from less than $1.00 per unit for high-volume, standard devices to much more expensive special-purpose devices. These versatile, inexpensive circuits are widely used in linear circuit applications.

The op amp, which is the most common linear IC, represents the logical extension of the amplifier circuits from the previous chapters. Linear ICs are quite complex, and many of the skills developed during Parts 1 to 4 of this text are drawn upon to obtain the necessary insights.

The strategy followed in this chapter is first to introduce a simple (non-IC) op-amp design and to apply the methods of previous chapters to understand the design objectives and implementation techniques for this circuit. (This simple op amp is also useful as an experimental introduction to op amps, as described in Experimental Application 12-1. Two IC op amps are then discussed in detail.

The 741 op amp is manufactured by a number of semiconductor companies and the schematic for this device is more or less a starting point for many IC op-amp studies. As discussed in previous chapters, schematics provide circuit representations that are limited in accuracy and completeness. There are always many aspects of circuit operation that are not defined by a schematic. This is particularly true regarding the schematics of IC operational amplifiers, where the physical dimensions associated with circuit devices are an important aspect of circuit design.

A comparison of the 741 op-amp schematics of several manufacturers usually reveals small areas of difference among the products. Further, an interpretation of any of these schematics requires that assumptions be made regarding circuit characteristics that are not described through the schematics. In the following sections, assumptions have been made regarding the design and operation of the IC op amp under discussion. Minor circuit changes have also been incorporated to reflect aspects of several different products. Therefore, the following discussion applies to a generic view of a typical 741 op amp, making use of the methods of circuit study of previous chapters to explore both dc- and ac-circuit functions. Because of the importance of IC op amps and the representative nature of the 741, this discussion is worth careful study and thought.

The NE 5230 IC op amp is a recent Signetics entry on the market that is designed to achieve specific performance objectives and obtain commercial use because of its unique features. Much of the 741 analysis can be carried forward, but several new factors and methods for understanding circuit operation are also introduced.

The 741 and the NE 5230 provide an "end point" for the discussion of linear amplifiers that began in Chapter 7. The material and device concepts of earlier chapters are used to produce steadily more complex amplifier circuits and introduce a wide diversity of circuit models leading to the IC op amp. This step-by-step procedure builds an understanding of the IC op amp from several perspectives and serves to encourage a hierarchical viewpoint of the many parameters that are involved in linear circuit design.

This chapter ends with a review of several different IC op amps that are commercially available, in terms of the external parameters used to predict circuit performance. These parameters are defined by the internal materials, circuit configuration, and integration technologies used. The final circuit performance depends on the external IC parameters in combination with all other circuit elements and the composite interconnection and packaging methods. As the options increase in all of these areas, a systems approach to circuit design requires that tradeoffs be conducted among important factors. CAD support is required in such cases, and the designer needs an understanding of the factors that are driving the tradeoffs.

A central component of an operational amplifier is a difference amplifier, consisting of two interlocked common-emitter amplifier stages, as shown in Fig. 12-1. The two stages are linked by having both emitters connected to a constant-current generator; as the current through one emitter increases, the current through the other decreases.

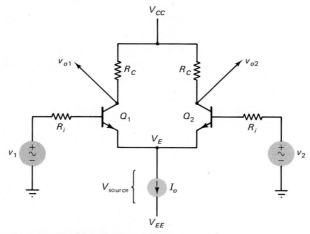

Fig. 12-1 Basic difference amplifier.

Consider the operation of the circuit shown in Fig. 12-1 with no input signals. With $v_1 = v_2 = 0$, an emitter current $I_0/2$ flows in each BJT. Therefore, $I_C \cong I_0/2$ and

$$V_{01} = V_{02} = V_{CC} - (I_0/2)R_C \tag{12-1}$$

For the base current,

$$I_B \cong I_0/2h_{FE} \tag{12-2}$$

and

$$V_E \cong -I_B R_i - 0.7 \text{ V} \tag{12-3}$$

where $V_{BE} = 0.7$ V is assumed for each BJT. V_{CE} must be chosen large enough to bias each BJT in the center of its linear operating region. Once V_{CE} is selected,

$$V_{CC} - V_{EE} = V_{\text{source}} + V_{CE} + I_C R_C \tag{12-4}$$

The ac equivalent circuit for the difference amplifier is shown in Fig. 12-2. By remembering that current generators have a very high resistance and noting that the inputs and outputs of each stage are connected only through the current source,[1]

---

[1] Midrange operation is being considered, so that Z_i and Z_o have no reactive component and can be interpreted as input and output resistances. The parameter h_{ie} is equal to the ac input resistance r introduced in Chapters 7 and 8.

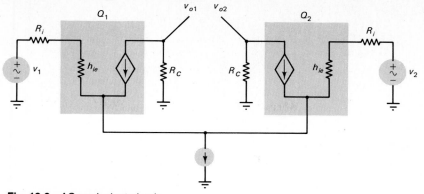

Fig. 12-2 AC equivalent circuit.

$$Z_i \cong 2(R_i + h_{ie})$$

$$Z_o \cong R_C$$

and

$$i_{b1} = \frac{v_1 - v_2}{2(R_i + h_{ie})} \quad \text{and} \quad i_{b2} = \frac{v_2 - v_1}{2(R_i + h_{ie})}$$

(12-5)

$$v_{o1} = -i_{C1}R_C \quad \text{and} \quad v_{o2} = -i_{C2}R_C$$

$$v_{o1} = \frac{-h_{fe}R_C}{2(R_i + h_{ie})}(v_1 - v_2) = -A_{Vd}(v_1 - v_2)$$

(12-6)

$$v_{o2} = \frac{-h_{fe}R_C}{2(R_i + h_{ie})}(v_2 - v_1) = -A_{Vd}(v_2 - v_1)$$

(12-7)

where A_{Vd} is the *differential gain* and is defined as positive.

To better understand these results, consider the case for which $v_2 = 0$. Then

$$v_{o1} = -A_{Vd}v_1$$

$$v_{o2} = +A_{Vd}v_1$$

(12-8)

The output voltage v_{o1} is an amplified, inverted version of v_1, whereas v_{o2} is an amplified, noninverted version of v_1, as shown in Fig. 12-3(A). A similar result can be obtained for $v_1 = 0$:

$$v_{o1} = +A_{Vd}v_2$$

$$v_{o2} = -A_{Vd}v_2$$

(12-9)

Superposition can be used to find v_{o1} and v_{o2} when signals exist at both v_1 and v_2, as shown in Fig. 12-3(B).

The difference amplifier is thus a useful circuit. It provides both inverted and noninverted output signals that have been multiplied by the same gain factor. And it has another advantage. If the same signal appears on both inputs, there is no contribution to v_{o1} or v_{o2} [for an ideal current generator, from Eqs. (12-6) and (12-7)].

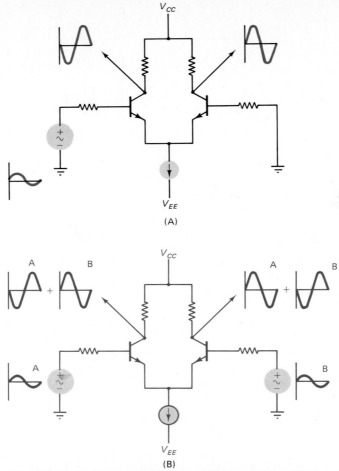

Fig. 12-3 Input and output waveforms. (A) Input on one side of the difference amplifier. (B) Input on both sides of the difference amplifier.

This provides a valuable degree of protection from noise that is common to both inputs. However, this feature does not protect against noise that affects only one of the inputs.

12-2 CONSTANT-CURRENT GENERATORS AND THE COMMON-MODE REJECTION RATIO

As is discussed in Chapter 8, a JFET with its gate tied to its source can be used to approximate a constant-current generator, as shown in Fig. 12-4(A). BJTs can be used in a similar way, with voltage-divider bias, as shown in Fig. 12-4(B). In both cases, the output current is approximately constant as long as the voltage across the device is sufficient.

In order to analyze the noise-rejection features of the difference amplifier, con-

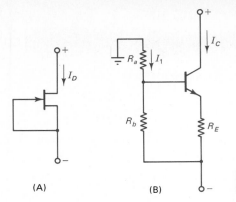

Fig. 12-4 Current sources. (A) JFET current source. (B) BJT current source.

(A)

(B)

sider the actual ac output resistance of the current source. Simply saying that r_o is "large" is not adequate for this purpose. The resistance r_o can be measured from the slope of the output characteristic, as shown in Fig. 12-5(A).

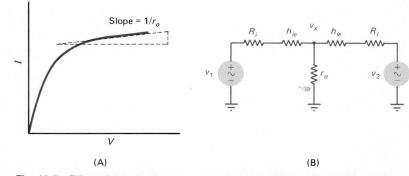

(A)

(B)

Fig. 12-5 Effect of nonconstant current source. (A) JFET output characteristic with $V_{GS} = 0$. (B) AC equivalent circuit showing dependence on r_o.

The finite value of r_o prevents the difference amplifier from completely rejecting noise that appears on both inputs. Refer to Fig. 12-5(B). When v_1 and v_2 change in the same way,

$$\frac{v_i - v_x}{R_i + h_{ie}} + \frac{v_i - v_x}{R_i + h_{ie}} = \frac{v_x}{r_o} \tag{12-10}$$

so that

$$v_x = \frac{v_i}{1 + \dfrac{R_i + h_{ie}}{2r_o}} \cong v_i\left(1 - \frac{R_i + h_{ie}}{2r_o}\right)$$

and

$$i_c \cong i_e = \frac{v_i - v_x}{R_i + h_{ie}} \cong \frac{v_i}{2r_o}$$

so that

$$v_{o1} = v_{o2} = R_C i_C \cong \frac{R_C}{2r_o} v_i \tag{12-11}$$

The *common-mode gain* is defined to be

$$A_{Vc} \cong \frac{R_C}{2r_o} \tag{12-12}$$

The noise-rejection capability of the difference amplifier is described by the *common-mode rejection ratio* (*CMRR*):

$$CMRR = \frac{\text{Output with signal applied to } v_1 \text{ or } v_2}{\text{Output with the same signal applied to } v_1 \text{ and } v_2} = \frac{A_{Vd}}{A_{Vc}}$$

$$= \frac{h_{fe} R_C}{2(R_i + h_{ie})} \bigg/ \frac{R_C}{2r_o} = h_{fe} \frac{r_o}{R_i + h_{ie}} \tag{12-13}$$

Typical *CMRR* values can range from 50 to 100 dB.

Example 12-1 Difference Amplifier

Consider the difference amplifier shown in Fig. 12-6. (a) With no input signals ($v_1 = v_2 = 0$), find I_E, I_B, and V_B for Q_1 and Q_2. (b) Given

$$v_1 = (40 \text{ mV}) \sin(\omega t) + (20 \text{ mV}) \cos(\omega t)$$

$$v_2 = (80 \text{ mV}) \sin(\omega t) + (20 \text{ mV}) \cos(\omega t)$$

find $\mathbf{v}_{o1}$ and $\mathbf{v}_{o2}$. (c) Find the *CMRR*.

Fig. 12-6 Difference amplifier for Example 12-1.

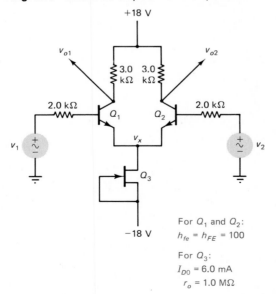

For Q_1 and Q_2:
$h_{fe} = h_{FE} = 100$

For Q_3:
$I_{DO} = 6.0 \text{ mA}$
$r_o = 1.0 \text{ M}\Omega$

Solution (a) From the above discussion,

$$I_{E1} \cong I_{E2} = 3.0 \text{ mA}$$

$$I_{B1} \cong I_{B2} = 30 \text{ }\mu\text{A}$$

$$V_{B1} = V_{B2} = -(2.0 \text{ k}\Omega)(30 \text{ }\mu\text{A}) = -0.06 \text{ V}$$

$$V_{01} = V_{02} = 18 \text{ V} - (3.0 \text{ k}\Omega)(3.0 \text{ mA}) = 9.0 \text{ V}$$

(b) For the given input signals,

$$h_{ie} \cong \frac{0.026 \text{ V}}{30 \text{ }\mu\text{A}} \cong 0.87 \text{ k}\Omega$$

$$A_{Vd} = \frac{(100)(3.0 \text{ k}\Omega)}{2(2.0 \text{ k}\Omega + 0.87 \text{ k}\Omega)} \cong 52$$

$$\mathbf{v}_{o1} = \frac{(52)(40 \text{ mV})}{\sqrt{2}} \cong 1.5 \text{ V}$$

$$\mathbf{v}_{o2} = \frac{(52)(40 \text{ mV})}{\sqrt{2}} \cong 1.5 \text{ V}$$

Notice that the terms (20 mV) cos (ωt) appearing on both inputs cancel each other and do not appear on the output (due to the noise-rejection characteristics of the circuit).

(c) From part (b), $A_{Vd} = 52$ with a differential input. For the common-mode gain,

$$A_{Vc} = \frac{3.0 \text{ k}\Omega}{2(1.0 \text{ M}\Omega)} = 1.5 \times 10^{-3}$$

and

$$CMRR = \frac{52}{1.5} \times 10^3 \cong 34.7 \times 10^3 \cong 91 \text{ dB}$$

Common input signals are reduced at the output by this ratio. ∎

12-3 DESIGN OF A SIMPLE OPERATIONAL AMPLIFIER

Figure 12-7 shows a simple operational amplifier consisting of a difference amplifier, a voltage-level shifter, and an emitter follower. Although this circuit does not meet the specifications for a high-quality operational amplifier, it can be used to illustrate the basic operation of such circuits.

First, consider the dc circuit operation. Begin with the difference amplifier and with both inputs set to zero. Transistor Q_3 serves as a constant-current source, controlled by the divider resistors R_a and R_b and emitter resistor R_{E3}. Assume that I_{B3} is much smaller than the divider current (by choice of R_a and R_b). Then

$$I_{E3} = \frac{(V_{CC} - V_{EE})\dfrac{R_b}{R_a + R_b} - 0.7 \text{ V}}{R_{E3}} \tag{12-14}$$

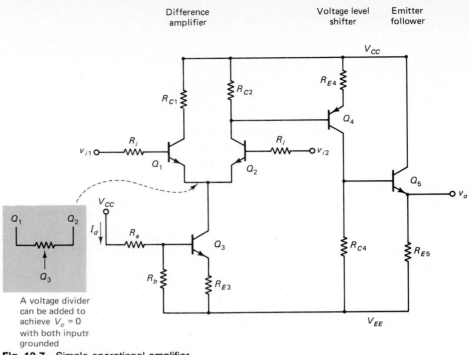

Fig. 12-7 Simple operational amplifier.

The voltage on the emitters of Q_1 and Q_2 is approximately

$$V_{E1} = V_{E2} \cong -0.7 \text{ V} \tag{12-15}$$

neglecting the small drop across R_i. To maximize the linear operating range of the difference amplifier, select

$$V_{CE1} = V_{CE2} = V_{CC}/2$$

Assume that all input resistors and transistors are perfectly matched. Then R_{C1} and R_{C2} must be chosen to satisfy the relationship

$$R_{C1} = \frac{V_{CC} - V_{CE1} + 0.7 \text{ V}}{I_{C3}/2} = \frac{V_{CC}/2 + 0.7 \text{ V}}{I_{C3}/2} = R_{C2} \tag{12-16}$$

and

$$\begin{aligned} V_{C2} &= V_{CC} - I_{C2}R_{C2} \\ &= V_{CC} - (I_{C3}/2)\left(\frac{V_{CC}/2 + 0.7 \text{ V}}{I_{C3}/2}\right) = \frac{V_{CC}}{2} - 0.7 \text{ V} \end{aligned} \tag{12-17}$$

Transistor Q_4 shifts the dc voltage level. The level V_{E4} is

$$V_{E4} = V_{02} + 0.7 \text{ V} = V_{CC}/2 \tag{12-18}$$

The current through R_{E4} is

$$I_{B4} = \frac{V_{CC}/2}{R_{E4}} \tag{12-19}$$

and the base current of Q_4 is

$$I_{B4} = \frac{V_{CC}}{2R_{E4}h_{FE}} \tag{12-20}$$

To prevent excessive dc loading of Q_2, the condition

$$I_{C2} = I_{C3}/2 \gg I_{B4}$$

must be placed. By substitution,

$$\frac{I_{C3}}{2} = \frac{V_{CC}/2 + 0.7 \text{ V}}{R_{C2}} \gg \frac{V_{CC}}{2R_{E4}h_{FE}} \tag{12-21}$$

This condition is satisfied if R_{C2} and R_{E4} are chosen with the same order of magnitude.

Since $I_{C4} \cong I_{E4}$,

$$V_{C4} = V_{EE} + I_{C4}R_{C4} = V_{EE} + \frac{V_{CC}/2}{R_{E4}} R_{C4} \tag{12-22}$$

The voltage V_0 is then

$$V_0 = V_{EE} + \frac{V_{CC}}{2} \frac{R_{C4}}{R_{E4}} - 0.7 \text{ V} \tag{12-23}$$

The base current I_{B5} is much less than I_{C4} if $R_{E5}h_{FE} \gg R_{C4}$. R_{E5} is limited in size by the need to keep $V_{BE5} \gtrsim 0.7$ V for linear operation.

For an operational amplifier, it is desired to have $V_0 = 0$ when $v_{i1} - v_{i2} = 0$. Therefore, values can be selected so that

$$V_{EE} + \frac{V_{CC}}{2} \frac{R_{C4}}{R_{E4}} - 0.7 \text{ V} = 0 \tag{12-24}$$

Due to parameter mismatches (particularly in V_{BE}) between Q_1 and Q_2, the circuit of Fig. 12-7 does not produce a zero output even when the above condition is met. This is a problem with all high-gain amplifiers. A small mismatch in device properties at the input is amplified at the output. As discussed in the remainder of this chapter, most operational amplifiers are used with negative feedback that reduces the effective circuit gain. Under this circumstance, output offset can be significantly reduced.

One possible approach for trimming the input offset of the operational amplifier of Fig. 12-7 is to place a voltage divider between the emitters of Q_1 and Q_2, with the slide contact connected to the collector of Q_3. The divider can be adjusted to produce $V_0 = 0$ when $v_i = 0$.

A voltage drop of a few tenths of a volt is usually adequate, so a total 1.0-V drop across the balancing resistance allows for adequate compensation. The value of the divider should thus be chosen so that

$$\frac{I_{C3}}{2} \frac{R_{\text{divider}}}{2} \cong 0.5 \text{ V} \tag{12-25}$$

This completes the dc operating-point analysis of the circuit. The gain analysis starts with the difference amplifier. From previous discussion,

$$A_{Vd} \cong \frac{h_{fe}R_{C2}}{2(h_{ie1} + R_i)}$$

$$Z_i \cong 2(h_{ie1} + R_i) \tag{12-26}$$

$$Z_o \cong R_C$$

Since the input impedance to Q_4 is approximately

$$Z_{i4} \cong h_{fe} R_{E4} \tag{12-27}$$

and R_{E4} and R_{C2} are chosen with the same order of magnitude, Q_4 is not expected to load down the difference amplifier.

The gain of the level shifter Q_4 is

$$|A_{V4}| = h_{fe}\frac{R_{C4}}{h_{ie4} + h_{fe}R_{E4}} \cong \frac{R_{C4}}{R_{E4}} \tag{12-28}$$

due to the large amount of negative feedback present.

The gain of the emitter follower is about unity, so the overall gain of the operational amplifier is

$$A_V \cong \frac{h_{fe}R_{C2}}{2(h_{ie1} + R_i)} \frac{R_{C4}}{R_{E4}} \tag{12-29}$$

The output impedance of the operational amplifier is

$$Z_o \cong R_{E5} \left\| \frac{R_{C4} + h_{ie5}}{h_{fe}} \cong \frac{R_{C4} + h_{ie5}}{h_{fe}} \right. \tag{12-30}$$

For an ideal operational amplifier, Z_i is very large, Z_o very small, and A_V very large. The circuit shown in Fig. 12-7 shows reasonable amplifier characteristics over some operating conditions, but is still limited in performance. ■

Example 12-2   **A Simple Operational Amplifier**

Design an op amp using the circuit of Fig. 12-7 with $V_{CC} = 12$ V and $V_{EE} = -12$ V. Assume that $h_{FE} = h_{fe} = 200$ for all BJTs and bias Q_3 for a current of 1.0 mA. Choose $R_i = 1.0$ kΩ.

Solution   To achieve $I_{E3} = 1.0$ mA, $I_{B3} = 1.0$ mA$/200 = 5.0$ μA. Choose a divider current I_d of 100 μA to ensure the stability of the circuit. Then

$$I_d = \frac{V_{CC} - V_{EE}}{R_a + R_b} = \frac{24 \text{ V}}{R_a + R_b} = 0.10 \text{ mA}$$

$$R_a + R_b = 0.24 \text{ M}\Omega$$

From Eq. (12-14),

$$I_{E3}R_{E3} = (V_{CC} - V_{EE})\frac{R_b}{R_a + R_b} - 0.7 \text{ V}$$

$$(1.0 \text{ mA})R_{E3} = (0.10 \text{ mA})R_b - 0.7 \text{ V}$$

If R_{E3} is chosen to be 1.3 kΩ, then

$$R_b = 2.0 \text{ V}/0.10 \text{ mA} \cong 20 \text{ k}\Omega$$

$$R_a = 0.22 \text{ M}\Omega$$

From Eq. (12-16), the collector resistors for the difference amplifier then become

$$R_{C1} = R_{C2} = \frac{6.0 \text{ V} + 0.7 \text{ V}}{0.50 \text{ mA}} = \frac{6.7 \text{ V}}{0.50 \text{ mA}} \cong 13 \text{ k}\Omega$$

and from Eqs. (12-17) and (12-18),

$$V_{C2} = 6.0 \text{ V} - 0.7 \text{ V} = 5.3 \text{ V}$$

$$V_{E4} = 6.0 \text{ V}$$

To prevent loading, R_{C2} and R_{E4} are to be of the same order of magnitude. Select $R_{E4} = 10$ kΩ. Then

$$I_{E4} = 6.0 \text{ V}/10 \text{ k}\Omega = 0.60 \text{ mA}$$

$$I_{B4} = 0.60 \text{ mA}/200 = 3.0 \text{ }\mu\text{A}$$

To obtain $V_o = 0$, Eq. (12-24) is applied:

$$(-12 \text{ V}) + (6.0 \text{ V})R_{C4}/10 \text{ k}\Omega - 0.7 \text{ V} = 0$$

$$R_{C4} = (12.7 \text{ V}/6.0 \text{ V})10 \text{ k}\Omega = 21 \text{ k}\Omega$$

If $I_{C5} = 1.0$ mA assures linear operation,

$$R_{E5} \cong 12 \text{ V}/1.0 \text{ mA} = 12 \text{ k}\Omega$$

and $R_{E5}h_{FE} \gg R_{C4}$ as required.

From Eq. (12-29), the gain of the op amp is

$$A_{Vd} = \frac{(200)(13 \text{ k}\Omega)}{2\left[\left(\dfrac{0.026 \text{ V}}{0.50 \text{ mA}}\right)200 + 1.0 \text{ k}\Omega\right]}\left(\frac{21 \text{ k}\Omega}{10 \text{ k}\Omega}\right) \cong 2.4 \times 10^2$$

The input impedance to the amplifier is

$$Z_i \cong 2(13 \text{ k}\Omega) = 26 \text{ k}\Omega$$

and the output impedance is

$$Z_o = \frac{21 \text{ k}\Omega + (0.026 \text{ V}/1.0 \text{ mA})(200)}{200} \cong 0.13 \text{ k}\Omega$$

The result is an op amp that can be easily constructed in the lab (see Experimental Application 12-1). All of the essential operational characteristics of more complex IC op amps can be demonstrated with this circuit. Since the values of A_{Vd}, Z_i, and Z_o are sufficiently reasonable to produce the expected type of performance, but are still significantly different from the ideal case ($A_{Vd} \to \infty$, $Z_i \to \infty$, $Z_o \to 0$), the nonideal characteristics of the op amp can be explored experimentally. ∎

Assume that the operational amplifier of Section 12-3 is used with a feedback resistor R_f from the output to the input and an input resistor R_1. The result is negative feedback from v_o to v_i, using the type of feedback discussed in Exercise 11-6. In order to find the overall gain for this amplifier configuration, refer to the circuit of Fig. 12-8 showing the Thevenin equivalent.

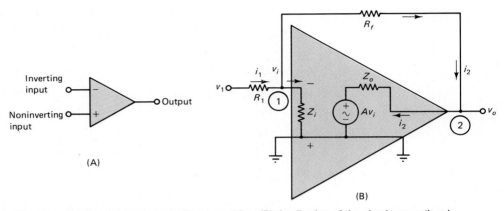

(A)

(B)

Fig. 12-8 (A) Symbol for an operational amplifier. (B) Application of the simple operational amplifier using negative feedback.

Summing currents at nodes 1 and 2,

$$\frac{v_1 - v_i}{R_1} = \frac{v_i}{Z_i} + \frac{v_i - v_o}{R_f} \tag{12-31}$$

and for v_o

$$v_o = A_V v_i + i_2 Z_o \qquad \text{and} \qquad i_2 = \frac{v_i - v_o}{R_f} \tag{12-32}$$

Eliminating i_2,

$$v_o = A_V v_i + \frac{Z_o}{R_f}(v_i - v_o) \tag{12-33}$$

and rewriting,

$$v_i(A_V R_f + Z_o) = v_o(R_f + Z_o) \tag{12-34}$$

Substituting for v_i in Eq. (12-31),

$$\frac{v_1}{R_1} = \frac{v_o(R_f + Z_o)}{A_V R_f + Z_o}\left(\frac{1}{Z_i} + \frac{1}{R_1} + \frac{1}{R_f}\right) - \frac{v_o}{R_f}$$

$$A_{Vf} = \frac{v_o}{v_1} = \frac{1}{\left(\dfrac{R_f + Z_o}{A_V R_f + Z_o}\right)\left(1 + \dfrac{R_1}{Z_i} + \dfrac{R_1}{R_f}\right) - \dfrac{R_1}{R_f}} \tag{12-35}$$

where A_{Vf} is the gain with feedback. For an ideal operational amplifier, the open-loop gain is very large ($A_V \to \infty$), so that

$$A_{Vf} \cong \frac{-R_f}{R_1} \qquad (12\text{-}36)$$

which is dependent only on the external resistors R_f and R_1. For these conditions, the operational amplifier provides an accurate, known gain that is independent of transistor parameters and variations among op amps.

The significance of this result can be appreciated by considering the operation of the circuit of Fig. 12-8 for the idealized case. With Z_i very large, the current flowing into the input of the operational amplifier can be neglected. The current i_1 through R_1 must then be approximately equal to the current through R_f. The resistors R_1 and R_f form a voltage divider across the input. Voltage v_1 on one end of the divider is an independent variable. Voltage v_o is determined by the amplifier and feedback circuit.

Since A_V is very large, v_i must be very small, approaching but not equal to zero. The circuit adjusts the output voltage until the divider formed by R_1 and R_f produces a voltage of approximately zero at v_i. When this equilibrium state is reached, the result

$$\frac{v_o}{v_1} = \frac{-i_1 R_f}{i_1 R_1} = \frac{-R_f}{R_1}$$

is obtained as shown in Eq. (12-36). The negative feedback from v_o to v_i enables the circuit to remain in an operating state with a well-defined overall circuit gain.

Example 12-3   Gain with Feedback

Suppose that an op amp is to be used in the configuration shown in Fig. 12-8 with $R_f = 2.0$ kΩ and $R_1 = 1.0$ kΩ. What is the resultant circuit gain for the amplifier of Example 12-2?

Solution The gain with a high-quality operational amplifier is

$$(A_{Rf})_{\text{ideal}} = -R_f/R_1 = -2.00$$

For the simple operational amplifier, the gain is

$$A_{Vf} = \cfrac{1}{\cfrac{2.0 \text{ k}\Omega + 0.13 \text{ k}\Omega}{-(2.5 \times 10^2)(2.0 \text{ k}\Omega) + 0.13 \text{ k}\Omega}\left(1 + \cfrac{1.0 \text{ k}\Omega}{26 \text{ k}\Omega} + \cfrac{1.0 \text{ k}\Omega}{2.0 \text{ k}\Omega}\right) - \cfrac{1.0 \text{ k}\Omega}{2.0 \text{ k}\Omega}}$$

$$= -1.97$$

The simple op amp of Example 12-2 thus has a predictable voltage gain A_{Vf} that is less than 2 percent from the ideal value of -2.00. Since the gain is dominated by the ratio R_f/R_1, it is almost independent of the amplifier parameters (Z_i, Z_o, and A_V) as well as temperature. As is discussed below, the parameters associated with IC op amps represent a substantial additional improvement over this simple op amp. ∎

This circuit does not function in the same way if the operational amplifier is used with a noninverting input. If v_1 and v_o have the same sign, v_i cannot be brought near to ground. The currents created by v_1 and v_o combine, forcing v_i to increase and the output is as close to the positive or negative supply as the circuit allows. Whereas negative feedback produces stable states, positive feedback results in unstable circuits.

12-5 FREQUENCY RESPONSE

Since the operational amplifier is dc-coupled, the Bode plot follows the form shown in Fig. 12-9(A), with no low-frequency roll-off. The lowest value of f_2 (f_{2a}) determines the initial high-frequency roll-off, and there is, in general, a number of higher-frequency poles. Assuming that the signal is placed on the negative input, θ^* begins at 180° and decreases by 90° as each additional pole is passed. Figure 12-9(B) shows the Nyquist plot.

Observe that for $f > f_{2b}$, the input and output of the amplifier are in phase ($\theta^* \to 0°$). A feedback path from the output to the input can result in an in-phase signal returned to the input. If the loop gain (through the amplifier and feedback path) has a total gain greater than unity, the returned signal is greater than the initial one placed on the input. As discussed in Chapter 14, the result is that the circuit can *oscillate*, providing an output with no input.

This is an undesirable result because the circuit can no longer function as a reliable amplifier. In order to prevent such oscillation, the loop gain must be less than unity for $f > f_{2b}$. Compensation capacitors are used internally or externally to pro-

Fig. 12-9 (A) Bode plot showing two poles and (B) Nyquist plot showing two poles for an operational amplifier (open loop).

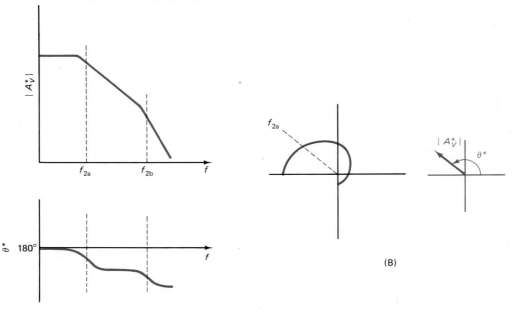

(A)

(B)

duce a sufficiently low value of f_{2a} to achieve the desired result. For this case, the useful frequency range of the amplifier is almost totally determined by the roll-off that results from the pole at f_{2a}. This pole is, therefore, called the *dominant pole*. Depending on the value of the compensation capacitor used, this pole can have a very low value (for example, $f_{2a} \cong 5\,\text{Hz}$ for the 741 op amp discussed in the next section). The resultant amplitude plot takes the form shown in Fig. 12-10(A).

The parameter f_{unity}, called the *unity-gain bandwidth,* defines the frequency for which $|A_V^*| = 1$ (or 0 dB). The second pole f_{2b} is above f_{unity} to prevent any possibility of oscillation. As discussed in Chapter 11, the *GBWP* of an amplifier with negative feedback is approximately constant as long as the frequency-dependent portion of the circuit is interior to the feedback loop, and the effective amplifier load is real. This finding provides the key to use of the op amp. If negative feedback is applied to the open-loop op amp shown in Figs. 12-9 and 12-10, using external resistors to achieve a real output load, then the *GBWP* will remain constant as the feedback factor is varied. Increasing feedback will reduce the amplifier gain and increase the bandwith *BW*, providing the family of possible frequency response curves shown in Fig. 12-10.

For the type of feedback circuit shown in Fig. 12-8, the frequency response curve that is selected depends on the ratio R_f/R_1. By adjusting this ratio, the midrange gain of the op amp is reduced while the midrange bandwidth increases. If $R_f/R_1 = 1.0$, the unity-gain bandwidth is obtained. For larger values of R_f/R_1, the bandwidth is found by holding the *GBWP* constant. If R_f is removed, the open-loop frequency response is obtained. ∎

Example 12-4   Op-Amp Bandwidth

Figure 12-10(B) shows an operational amplifier with $f_{\text{unity}} = 1.0\,\text{MHz}$. If the op amp is operated with $A_{Vf} = -10$, the midrange gain is 20 dB. What is the bandwidth for the amplifier?

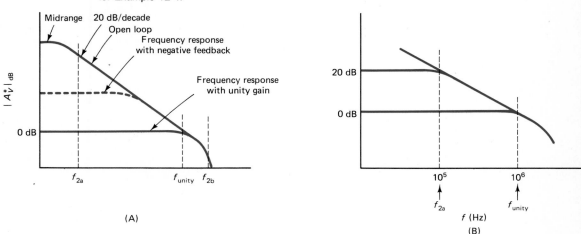

Fig. 12-10   Effect of negative feedback on the amplitude plot. (A) Amplitude plots. (B) Data for Example 12-4.

Solution From above, the bandwidth for the amplifier is

$$BW = \frac{GBWP}{|A_{Vf}|} = \frac{1.0 \times 10^6}{10} = 0.10 \text{ MHz}$$

The negative feedback introduced by $R_f/R_1 = 10$ results in an amplifier with 0.10-MHz bandwidth. Reducing R_f/R_1 further increases the available bandwidth. With $R_f/R_1 = 1.0$, the unity-gain bandwidth results. ∎

Square-wave testing of the type introduced in Chapter 11 can be used with operational amplifiers to measure f_{2f} as a function of the feedback ratio R_f/R_1. As the gain is adjusted upward, the output waveform shows distortion at lower frequencies.

As shown in Fig. 12-11, the rise time t_R of the output waveform is defined as the time it takes for the waveform to shift from 10 to 90 percent of its final value. The time constant of the transient between these levels is related to the rise time through the relationship:

$$t_R = t_B - t_A = -\tau (\ln 0.1 - \ln 0.9)$$
$$= 2.2\tau$$

(12-37)

From Chapter 11,

$$f_2 = 1/2\pi\tau_2$$

so

$$f_2 = 2.2/2\pi t_R \cong 0.35/t_R$$

(12-38)

Fig. 12-11 Square-wave testing to determine f_{2f} for an operational amplifier.

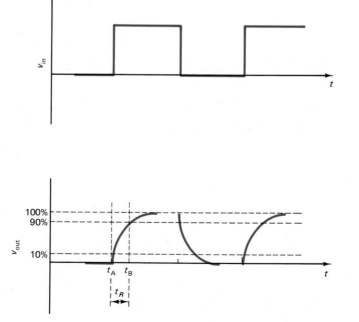

The rise time and upper half-power frequency of an amplifier are thus related in this simple way. As the amplifier gain rises above the 0-dB level, f_{2f} decreases and the rise time of the output waveform increases.

The *slew rate* is defined as the maximum rate at which the output can change (in V/μs). From Fig. 12-11,

$$\frac{dv}{dt} = \frac{d}{dt} V_m (1 - e^{-t/\tau}) = \frac{V_m}{\tau} (1 - e^{-t/\tau})$$

$$\cong V_m/\tau \text{ for } t \ll \tau \tag{12-39}$$

which is an estimate of the slew rate.

12-6 THE 741 IC OPERATIONAL AMPLIFIER: DC OPERATING-POINT BIAS

The 741 operational amplifier, which is emphasized in this section, is included in the product lines of many different manufacturers and has been widely applied. Figure 12-12 shows the schematic of the Raytheon version of this circuit and Fig. 12-13 shows the metallization pattern.

This is a complicated circuit, with 24 transistors. However, it is possible to analyze the circuit so as to realize the basic simplicity of the design and its similarity to the operational amplifier of Section 12-3. In addition, a detailed analysis of the 741 illustrates many design techniques that are widely used in other IC operational amplifiers.

Fig. 12-12 Schematic of the RC741 IC operational amplifier. All resistance and capacitance values are nominal. (Courtesy of Raytheon Corp.)

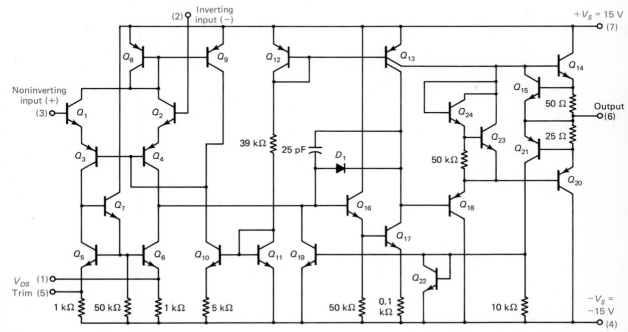

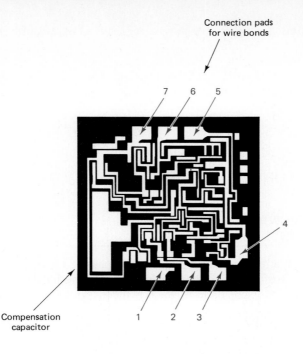

Die size: 55 × 55 mils
Min. pad dimension: 4 × 4 mils

Fig. 12-13 Metallization pattern for the RC741 IC operational amplifier. (Courtesy of Raytheon Corp.)

The 741 version discussed here is shown in Fig. 12-14, for the input portion of the circuit, and in Fig. 12-15, for the output portion of the circuit.

Transistors Q_1 and Q_2 form emitter-follower inputs to the operational amplifier. The difference amplifier is formed by Q_3 and Q_4, whereas Q_8 is the constant-current source, and Q_5 and Q_6 create an active load. The output of the difference amplifier is taken from the collector of Q_4.

A useful starting point for discussion is to consider the combination formed Q_{11} and Q_{12}. For each of these BJTs, the base and collector terminals are connected. As a result, $V_{BC} = 0$ and the transistors behave as diodes with voltage drop V_{BE} across them. The combination Q_{11}, R_4, and Q_{12} thus can be considered as two diodes and a resistor in series between voltages 15 V and -15 V. The current through the 40-kΩ resistor must be

$$I_{C11} = I_{C12} = \frac{30 \text{ V} - 2V_{BE}}{40 \text{ k}\Omega} \cong 0.7 \text{ mA}$$

($V_{BE} \cong 0.7$ V is appropriate because of the current level.) The combination Q_{10} and Q_{11} converts the reference current through Q_{11} to a reduced current level through Q_{10}. From Figure 12-14,

$$I_{E10}R_5 + V_{BE10} = V_{BE11}$$

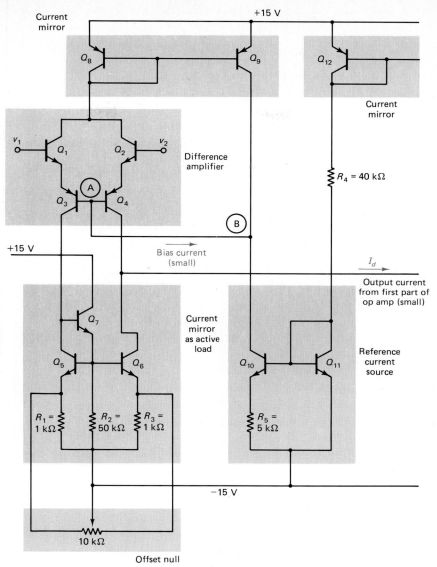

Fig. 12-14 Input portion of a generic 741 operational amplifier drawn to aid in analysis.

If $I_E \cong I_0 \exp (V_{BE}/V_0)$ for each diode, then

$$I_{E10} R_5 + V_0 \ln (I_{E10}/I_0) \cong V_0 \ln (I_{E11}/I_0)$$

$$I_{E10} R_5 \cong V_0 \ln (I_{E11}/I_{E10}) \tag{12-40}$$

For the circuit shown, $R_5 = 5.0 \text{ k}\Omega$ and $I_{E11} \cong 0.7$ mA. If $V_0 = 0.026$ V (at 300°K), it can be confirmed by substitution that

$$I_{E10} \cong 2 \times 10^{-5} \text{A} \cong I_{C10}$$

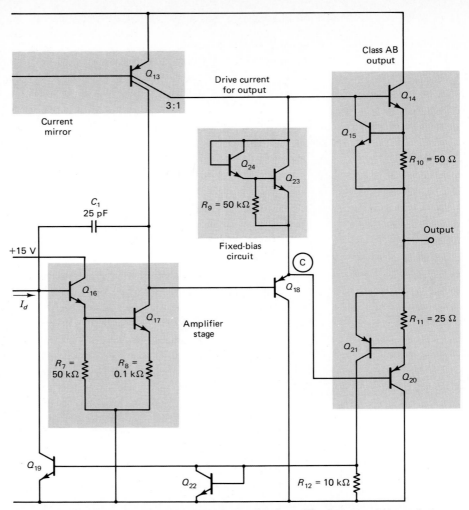

Fig. 12-15 Output portion of a generic 741 operational amplifier drawn to aid in analysis.

because $I_{B10} \ll I_{C10}$. The 0.7-mA reference current has been transformed into a 2×10^{-5}-A current source.

Transistors Q_8 and Q_9 form a *current mirror,* so named because the currents I_{C8} and I_{C9} must be almost the same (assuming identical geometries). Current I_{C9} follows current I_{C8}. Q_8 and Q_9 are assumed matched (as a result of the IC fabrication process) and must have equal values of V_{BE}. Therefore, $I_{B8} = I_{B9}$, and the total current from the combined base-collector output of Q_8 is approximately equal to the collector current I_{C9}.

The current source formed by Q_{10} and Q_{11} and the current mirror formed by Q_8 and Q_9 are linked by the difference amplifier and the bias connection A to B. Through negative feedback, this loop forces currents I_{C8} and I_{C9} to equal I_{C10}. Consider what happens if $I_{C10} > I_{C9}$. A large current must then flow from A to B, forcing increased emitter currents in Q_3 and Q_4, and hence an increase in I_{C8}. Due to the current-mirror action, I_{C9} must then increase. On the other hand, if $I_{C10} < I_{C9}$, then the bias current from A to B is reduced. The condition $I_{C10} \cong I_{C9} \cong I_{C8} \cong 2 \times 10^{-5}$A must thus be satisfied.

Transistors Q_5 and Q_6 form another variation of a current mirror, used as an active load for the difference amplifier. If I_{C5} is greater than I_{C6}, then $V_{E5} > V_{E6}$. Since the bases of Q_5 and Q_6 are tied together, the requirement that $V_{BE5} < V_{BE6}$ is obtained. But if $I_{C5} > I_{C6}$, this result cannot be obtained. The only possible solution is for $I_{C5} = I_{C6}$. For $v_1 - v_2 = 0$, the collector currents for Q_3 and Q_4 are identical and the collector currents into Q_5 and Q_6 are identical.

The emitters are at approximately -15V because $(1.0 \text{ k}\Omega)(2 \times 10^{-5}\text{A}) = 0.02$ V.

To adequately turn on Q_5 and Q_6 for a 1×10^{-5}-A collector current, the bases of Q_5 and Q_6 must be at about

$$V_0 \ln (I/I_0) = V_0 \ln (10^{-5} \text{ A}/10^{-14} \text{ A}) \cong 0.5 \text{ V}$$

Therefore, $I_{E7} \cong 0.5 \text{ V}/50 \text{ k}\Omega \cong 1 \times 10^{-5}$A.

Transistors Q_7 and Q_{16} (Figs. 12-14 and 12-15) are connected as emitter followers to the collectors of Q_5 and Q_6. Transistor Q_7 provides base current for Q_5 and Q_6.

If $v_1 - v_2 > 0$ so that $I_{C1} \cong I_{C3}$ increases, the constant-current source Q_8 results in a decrease in $I_{C2} \cong I_{C4}$ of the same magnitude. When used as a load, the current mirror Q_5 and Q_6 transfers the increase in I_{C3} to an increase in I_{C6}. The decrease in I_{C4} and the increase in I_{C6} combine to produce an increase in I_d given by $\Delta I_d = 2(\Delta I_{C3})$.

The emitter follower formed by Q_{16} isolates the output portion of the circuit from the difference amplifier. The combinations Q_7 and R_2 and Q_{16} and R_7 are identical, so from symmetry $I_{E16} = I_{E7} = 1 \times 10^{-5}$ A. The base currents and quiescent bias voltages of Q_7 and Q_{16} are thus identical and maintain balance in the input stage.

The offset null connections are used to externally zero out any mismatches in transistors or resistors that produce a nonzero output voltage from the operational amplifier with $v_1 - v_2 = 0$. A 10-kΩ voltage divider is used between the null connections, with the sliding contact connected to -15 V. The additional current flow through the divider serves to adjust I_{C3} and I_{C4} to obtain the desired Q point for the circuit output.

Since $I_{C12} \cong 0.7$ mA, then $I_{C13} \cong 0.7$ mA for the combined collectors of Q_{13} (if the total collector area is the same for Q_{12} and Q_{13}). Given a 3:1 dimensional ratio between the two collectors of Q_{13} (as indicated in Fig. 12-15, based on circuit fabrication), $I_{C17} \cong 0.5$ mA.

Q_{18} is biased *on* by the current flowing from Q_{13} through Q_{23}. The second emitter of Q_{13} is designed so that $I_{E23} \cong 0.2$ mA to provide adequate current drive for the output. Q_{18} is an emitter follower with a gain $\cong 1$. Q_{23} and Q_{24} provide a fixed bias voltage across V_{CE23}.

The collector current through Q_{23} is about 0.2 mA, so $V_{BE23} \cong 0.6$ V. The current through R_9 is

$$0.6 \text{ V}/50 \text{ k}\Omega \cong 1 \times 10^{-5} \text{ A}$$

The total voltage across Q_{23} is approximately

$$V_{BE24} + V_{BE23} \cong 1.2 \text{ V}$$

The output is driven by emitter followers Q_{14} and Q_{20} in push-pull class AB operation. Either Q_{14} or Q_{20} is *on*, depending on the current drive to Q_{18}. Changes in I_d due to input signals are amplified through Q_{16} and Q_{17} and produce corresponding changes in I_{C17}. As a result, I_{B18} and I_{E18} increase or decrease. Decreases in I_{E18} produce more drive for Q_{14} and less for Q_{20}. Increases in I_{E18} produce more drive for Q_{20} and less for Q_{14}. The function of Q_{18} is thus to guide current changes in I_d to the output transistors Q_{14} and Q_{20}.

The voltage at C is determined by the load on the output and the current through Q_{14} or Q_{20}. The voltage across Q_{23} (1.2 V) appears between the bases of Q_{14} and Q_{20} and is adequate to cut off Q_{14} as V_{out} becomes negative and to cut off Q_{20} as V_{out} becomes positive.

With zero input, the bias circuit Q_{23} and Q_{24} provides the needed voltage difference across the bases of Q_{14} and Q_{20} to maintain a low bias current in both transistors and to keep them in a region of linear operation. In this way, *crossover distortion* is reduced. (As discussed in Section 11-6, because of the nonlinear nature of the output transistors, such distortion arises when the output does not follow the input until sufficient forward bias has been developed across the emitter-base junctions of the output transistors.)

Bypass transistors Q_{15} and Q_{21} protect Q_{14} and Q_{20} from excessive current flow. When enough current flows through R_{10} or R_{11} to forward-bias the base-emitter junction of the bypass transistor,

$$I_C \cong I_E \cong 0.7 \text{ V}/50 \text{ } \Omega \cong 0.02 \text{ A}$$

and base current is diverted from the output stage, preventing it from overloading.

Q_{19} and Q_{22} provide short-circuit protection for the output. If Q_{21} is *on*, then Q_{19} is turned *on*, diverting current from the base of Q_{16}. Capacitor C_1 is the compensation capacitor used to prevent oscillation of the operational amplifier at high frequencies.

12-7 THE 741 IC OPERATIONAL AMPLIFIER: GAIN ANALYSIS

The gain analysis of the circuit can be performed with previously developed techniques. The difference amplifier is formed by Q_1 and Q_2 and Q_3 and Q_4; in each case, an emitter-follower input is followed by a common-base amplifier stage. The current gain of the emitter-follower stage is h_{fe}, and the current gain of the common-base stage is $\cong 1$. The combinations Q_1 and Q_3 and Q_2 and Q_4 each function as equivalent common-emitter input stages. The reason for the configuration shown is to provide a way to link the bases of Q_3 and Q_4 and create the feedback loop A to B.

From Section 12-1, the gain of the difference amplifier is given by

$$A_{Vd} \cong h_{fe}\frac{R_C}{2h_{ie}} \times 2 \cong g_{me}R_C \tag{12-41}$$

remembering that the current mirror Q_5 and Q_6 doubles the effective current and voltage gain of the difference amplifier stage.

The effective R_C is produced by the parallel combination of the output impedance of Q_4 and Q_6 and the input impedance of Q_{16}. The Early voltage of the pnp Q_4 is much lower than that of the npn Q_6, so that the assumption $1/h_{oe6} \gg 1/h_{oe4}$ can be made. Because $I_{C8} \cong 1 \times 10^{-5}$A, $I_{E6} \cong 1 \times 10^{-5}$A. Therefore, taking the Early voltage $V_E \cong 50$V,

$$\frac{1}{h_{oe4}} \cong \frac{50 \text{ V}}{1 \times 10^{-5} \text{ A}} \cong 5 \text{ M}\Omega$$

If $h_{fe} = 200$, the input impedance of Q_{16} is given by

$$Z_{i16} = h_{ie} + h_{fe}[R_7 \| (h_{ie} + h_{fe}R_8)]$$

$$= \left(\frac{0.026 \text{ V}}{1 \times 10^{-5} \text{ A}} \times 200 \right)$$

$$+ 200 \left\{ 50 \text{ k}\Omega \| \left[\left(\frac{0.026 \text{ V}}{0.5 \text{ mA}} \times 200 \right) \right. \right.$$

$$\left. \left. + (200 \times 0.1 \text{ k}\Omega) \right] \right\}$$

$$\cong 4.3 \text{ M}\Omega$$

(12-42)

The effective R_L is thus $5 \text{ M}\Omega \| 4.3 \text{ M}\Omega \cong 2.3 \text{ M}\Omega$.

Since $I_{E1} \cong I_{E2} \cong 1 \times 10^{-5}$ A,

$$g_{me} = h_{fe}/h_{ie} \cong I_E/0.026 \text{ V}$$

$$= 1 \times 10^{-5} \text{ A}/0.026 \text{ V} \cong 0.38 \text{ m}\mho$$

for the combinations Q_1 and Q_3 and Q_2 and Q_4. The gain of the input stage is thus

$$A_V = 0.38 \text{ m}\mho \times 2.3 \text{ M}\Omega \cong 8.7 \times 10^2$$

The active load Q_5 and Q_6 is designed so that Q_7 and Q_{16} are approximately symmetric. The effective resistance on the emitter of Q_7 is

$$R_2 \| (h_{ie5} + h_{fe}R_1) \| (h_{ie6} + h_{fe}R_1) \cong 50 \text{ k}\Omega \| 200 \text{ k}\Omega \| 200 \text{ k}\Omega$$

$$\cong 30 \text{ k}\Omega$$

(12-44)

The effective resistance on the emitter of Q_{16} is

$$R_7 \| (h_{ie} + h_{fe}R_8) \cong 50 \text{ k}\Omega \| 20 \text{ k}\Omega$$

$$\cong 15 \text{ k}\Omega$$

(12-45)

The magnitude of the gain of the amplifier stage Q_{16} (an emitter follower) and Q_{17} (a common-emitter amplifier) is given by

$$A_V = \frac{h_{fe}R_C}{h_{ie} + h_{fe}R_E}$$

$$= \frac{R_C}{1/g_{me} + R_E}$$

(12-46)

Since $I_{C17} \cong 0.5$ mA, $I_{B17} \cong 0.5$ mA$/200 = 2.5$ μA and

$$h_{ie17} \cong 0.026 \text{ V}/2.5 \text{ } \mu\text{A}$$

$$\cong 10 \text{ k}\Omega$$

With $I_C = 0.5$ mA, the output impedances $1/h_{oe}$ for Q_{13} and Q_{17} are approximately

$$1/h_{oe} \cong 50 \text{ V}/0.5 \text{ mA}$$

$$\cong 100 \text{ k}\Omega$$

The effective resistance looking into Q_{18} is approximately given by

$$Z_{i18} \cong h_{fe}\left[\frac{1}{h_{oe13B}}\|(h_{ie14} + h_{fe}R_L)\|(h_{ie20} + h_{fe}R_L)\right] \qquad (12\text{-}47)$$

Since $R_L > 1$ kΩ in typical applications to prevent excess output current, Z_{i18} is large with respect to the output resistances of Q_{13} and Q_{17}.

The voltage gain of the amplifier stage is thus

$$A_V \cong 200\frac{100 \text{ k}\Omega\|100 \text{ k}\Omega}{10 \text{ k}\Omega + 200(0.1 \text{ k}\Omega)} \cong 3.3 \times 10^2$$

Since the output consists of two emitter-follower stages, the voltage gain of these stages is about unity. Therefore, based on the above approximate estimates, the total circuit gain is about

$$(8.7 \times 10^2)(3.3 \times 10^2) \cong 2.9 \times 10^5$$

The effective impedance looking into Q_{18} from point C to ground is of the order:

$$\frac{1}{h_{fe18}}\left(h_{ie18} + \frac{1}{h_{oe13}}\right) \cong \frac{1}{200}\left(\frac{0.026 \text{ V}}{0.5 \text{ mA}}200 + \frac{50 \text{ V}}{0.5 \text{ mA}}\right)$$

$$\cong 0.5 \text{ k}\Omega$$

When Q_{14} is conducting heavily (up to 10 mA), the effective output impedance of the operational amplifier is

$$Z_o = \left[R_{10} + \frac{h_{ie14}}{h_{fe14}} + \frac{0.5 \text{ k}\Omega}{h_{fe14}}\right] \qquad (12\text{-}48)$$

where the drops across the fixed-bias circuit Q_{23} and Q_{24} are neglected. Therefore,

$$Z_o \cong 50 \text{ }\Omega + 0.026 \text{ V}/10 \text{ mA} + 0.5 \text{ k}\Omega/200 \cong 55 \text{ }\Omega$$

When Q_{20} conducts heavily,

$$Z_o \cong \left[R_{11} + \frac{h_{ie20}}{h_{fe20}} + \frac{0.5 \text{ k}\Omega}{h_{fe20}}\right] \cong 30 \text{ }\Omega$$

The larger emitter and base resistance of pnp Q_{20} produces a closer balance between these two values than is indicated above.

The above discussion of the 741 operational amplifier provides only an order-of-magnitude approach to calculations. Detailed analysis of each amplifier stage produces more accurate estimates of circuit performance (see Gray and Meyer, 1984). A computer simulation of the circuit provides the most exact results.

The objective of the type of analysis shown here is to illustrate the dominant factors in amplifier performance and to show how simple methods can be used to estimate the performance of complex circuits.

Many variations in the basic operational amplifier configuration are used to improve performance for specific objectives. The NE 5230 op amp, introduced by Signetics in 1985, is designed to allow use of very low supply voltages and to utilize the largest possible part of the supply-voltage range for signal operations.

The two supply voltages V_{CC} and V_{EE} are known as the amplifier *rails* because of the ways in which schematics are drawn, with parallel voltage buses across the top and bottom of the op amp. The maximum output voltage swing is thus from rail to rail. The NE 5230 op amp is designed to achieve almost rail-to-rail operation on the output.

The input stage is shown in Fig. 12-16. The analysis can usefully begin with the reference voltage generator V_{B1}. Since V_{B5} is held at 0.8 V, Q_5 is *on* if $V_a >$ $0.8 + 0.7 = 1.5$ V. This condition holds if Q_1 and Q_2 are *off*. Q_1 and Q_2 are *off* as long as v_1 and v_2 remain above 0.8 V.

Under these conditions, the input difference amplifier Q_3 and Q_4 is in operation. Current generator I_{B1} feeds through Q_5 into the current mirror Q_6 and Q_7, providing a constant current I_{C7} to drive Q_3 and Q_4. The collector currents for Q_3 and Q_4 are each equal to $I_{B1}/2$ at balance. R_{10} and R_{11} are selected so that $(I_{B1}/2)R_{10} = (I_{B2}/2)R_{11} < V_{B2}$. Therefore, forward biasing is applied to the emitter-base junctions of Q_{10} and Q_{11}. With Q_1 and Q_2 *off*, matched currents $I_{C8} = I_{C9}$ develop until $V_{E10} = V_{E11} = V_{B2} + 0.7$ V.

If input voltage v_1 increases, current I_{C3} increases and, because of the constant current through Q_7, I_{C4} decreases. To maintain the voltage $V_{B2} + 0.7$ V on V_{E11}, I_{E11} must increase by an equal amount and I_{E10} must decrease by an equal amount. Since I_{C9} is forced to equal I_{E10} by the current mirror Q_8 and Q_9, I_{out} increases by an amount equal to twice the initial change in I_{C4}.

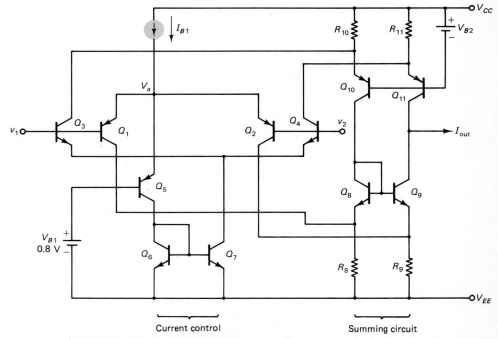

Fig. 12-16 Input section for the NE5230 IC operational amplifier. (Courtesy of Signetics Corporation)

If input voltage v_1 decreases, I_{C3} decreases and I_{C4} increases. As a result, I_{out} decreases. Similar changes in I_{out} are observed for changes in the input voltage v_2. Transistors Q_{10} and Q_{11} thus serve to guide changes in I_{C3} and I_{C4} to the output of the stage.

The above analysis applies to the case for which Q_3 and Q_4 are *on*, whereas Q_1 and Q_2 are *off*. This condition results if v_1 and $v_2 > 0.8$ V. Note that the input voltage can increase slightly beyond the value V_{CC} without allowing Q_3 and Q_4 to become saturated (with the base-collector junction forward biased).

Now consider what happens when v_1 and v_2 drop below about 0.8 V. The base-emitter junction of Q_5 is turned *off*, reducing I_{C6} and I_{C7} to zero. As a result, Q_3 and Q_4 turn *off*. The current generator I_{B1} flows directly into Q_1 and Q_2, providing the required constant-current source.

With Q_3 and Q_4 off, fixed currents flow through R_{10} and R_{11} to produce quiescent currents through R_8 and R_9. If v_1 decreases toward V_{EE}, I_{C1} increases and I_{C2} decreases, causing I_{E8} to decrease and I_{E9} to increase. I_{out} must then decrease by twice the change in I_{C1}. If v_1 increases, I_{out} must increase. Similar changes in I_{out} are observed for input changes in v_2.

Note that the input voltage can decrease below V_{EE} without bringing Q_1 and Q_2 into saturation. Therefore, the input common-mode voltage range extends beyond the power supply voltages (or rails) by small amounts. Input voltages below 0.8 V are handled by the pnp input stage, whereas input voltages above 0.8 V are handled by the npn input stage.

The input is followed by a common-emitter stage to increase amplifier gain, applying the general method discussed for the 741. The output stage is then designed as shown in Fig. 12-17. The output transistors Q_3 and Q_1 are connected in a com-

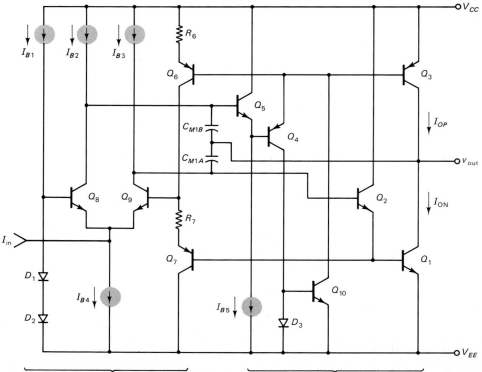

Fig. 12-17 Output section for the NE5230 IC operational amplifier. (Courtesy of Signetics Corporation)

mon-emitter configuration in order to allow v_{out} to almost reach V_{CC} and V_{EE}. The collector-emitter voltage drops across Q_3 and Q_1 can become about 0.1 V before saturation occurs.

The input current i_{in} is applied to a difference pair Q_8 and Q_9. Current changes in I_{C8} are fed through the current amplifier Q_4 and Q_5 to drive Q_3, and current changes in I_{C9} are fed through Q_2 to drive Q_1. As the current in Q_3 increases, the current in Q_1 decreases, and vice versa. The combination D_3 and Q_{10} increases the base current of Q_3 and therefore increases the potential maximum value for I_{OP}.

Class AB operation is obtained by balancing a reference voltage across D_1 and D_2 with the voltage on the base of Q_9. The emitter-base drop across Q_3 fixes V_{E6} and causes a fixed current to flow in R_6 and Q_6, and the emitter-base drop across Q_1 fixes V_{E7}. As a result, the voltage drop across R_7 is determined, as is V_{B9}.

The difference stage Q_8 and Q_9 restrains V_{B9} to a range for which neither Q_1 nor Q_3 can cut off. Class AB operation is thus maintained.

The NE 5230 indicates how many different improvements can be made in op-amp design to produce new products with enhanced properties.

This state-of-the-art op amp can handle inputs that go 0.25 V beyond the power-supply voltages and the ouput can swing to within 0.10 V of the power supply. Power-supply current drain is programmable over a range of 0.1 to 0.6 mA, yielding a unity-gain bandwidth of 0.10 to 0.60 MHz, depending on the supply current, and specified performance is achieved with a total supply voltage as low as 1.8 V (or ±0.9 V).

The next section introduces many of the different parameters used to describe IC op amps and compares several products available on the market.

12-9 PERFORMANCE SPECIFICATIONS FOR IC OPERATIONAL AMPLIFIERS

The operational amplifier is the most common linear IC and is manufactured in many different variations. In order to understand the strengths and weaknesses of alternative op amp designs, it is important to gain an understanding of the performance specifications that are in common use.

The earlier discussion of the op amp introduced the differential voltage gain A_{Vd}, the common-mode rejection ratio CMRR, and the input and output impedances Z_i and Z_o, respectively. Many other parameters are also used to specify operational-amplifier performance.

The *input bias current* is the average of the currents into the two input terminals with the output forced to zero volts. For an ideal operational amplifier, the input bias current is zero. The *input offset voltage* is the dc voltage that must be applied between the input terminals to force the output voltage to zero. For an ideal operational amplifier, the input offset voltage is zero.

The *differential input-voltage range* is the maximum differential signal that is allowed between the input terminals. The *common-mode input-voltage range* describes the range of input voltages that can be placed simultaneously on both input terminals. Restrictions on these ranges serve to limit the potential applications. The *maximum peak output voltage* is the maximum positive or negative peak output voltage that can be obtained with the operational amplifier still providing linear performance (the dc-output Q point is assumed equal to zero volts).

The *supply voltage* and *supply current* define the power supply requirements of the op amp. The *standby power dissipation* describes how much power is required to operate the circuit with zero output current. The power dissipation increases as the output current increases. The *supply-voltage sensitivity* is the magnitude of the change in supply voltage divided by the change in input offset voltage producing it and describes how sensitive the IC is to variations in the power supply.

The *short-circuit output current* is the maximum current that is available from the amplifier with the output shorted to ground. Many IC operational amplifiers are designed with *output short-circuit protection* to prevent damage to the amplifier. The *equivalent input-noise voltage* and the *equivalent input-noise current* are different ways for specifying the effect of the input-noise level as it is observed at the output for a given amplifier gain.

IC op amps are designed both with and without *frequency compensation* to prevent unwanted amplifier oscillation. *Internal compensation* means that a capacitor has been internally fabricated as part of the IC to assure that the criteria for oscillation cannot be satisfied for the device. *External compensation* means that the stability of the operational amplifier must be assured by externally adding the necessary restrictions on bandwidth.

The *unity-gain bandwidth* describes the frequency response of the op amp, giving the range of frequencies for which the open-loop voltage amplification is greater than unity. Frequency roll-off clearly begins far below the unity-gain bandwidth.

The *slew rate* denotes the maximum rate of change of the closed-loop amplifier output voltage for a step function input and the *rise time* is the time for the output to transition from 10 to 90 percent of its final value with a step input signal.

The allowable *temperature range* describes the minimum and maximum temperature over which the IC can be used. A typical commercial range is from 0 to 70°C, and military products (which are more expensive) are often rated at -55 to $+125$°C.

The above specifications can be used to distinguish between the many IC op amps that are now on the market. Figure 12-18 compares several examples on the basis of some of the parameters defined above.

The prefixes that appear on these devices are associated with specific manufacturers. For example, μA is used by Fairchild, LF/LM and LH by National, OP by Precision Monolithics, NE/SA/SE by Signetics, MC by Motorola, RC by Raytheon, and TL/TLC by Texas Instruments. Once an IC of a particular type is introduced to the market, it is common practice for other companies to begin to produce an equivalent product to provide multiple sourcing.

A quick review of Fig. 12-18 can serve to establish typical parameter ranges. The open-loop gains A_V range from 6000 to 200,000, with the upper end much more typical. *CMRR* ratings are from 86 to 123 dB. Input bias currents are typically from 10^{-11} to 10^{-7}A, depending on the way in which the circuit has been optimized. Input offset voltages usually range from 1 to 3 mV except for those op amps designed for ultra-low values. The frequency response can be indicated by the unity-gain bandwidth or the rise time for a step input (refer to the discussion of Section 12-5).[2] Typical supply voltages are ±15 V and standby supply currents are of the general magnitude of 2 mA.

The bipolar operational amplifier is the fundamental linear IC building block.

---

[2] For comparison, state-of-the-art GaAs op amps can achieve unity gain bandwidths of 150 MHz (Goodenough, 1986).

FIGURE 12-18
Comparison of Typical Values for IC Operational Amplifiers

Device Code	Description	A_{vd} ($\times 10^3$)	CMMR (db)	Input Bias Current (nA)	Input Offset Voltage (mV)	Frequency Response Unity GBW (MHz)	Rise Time (μs)	Slew Rate (V/μs)	Supply Voltage (V)	No-load Current (mA)
RC741	General purpose	200	90	80	1		0.3	0.5	±15	1.7
NE5230	Low supply voltage	300	97	40	0.3	0.6		0.25	±0.75 min.	0.5
LM124	Quad op amp/single supply	100	85	45	2	1		0.3	±15 30,0	1.5
NE5512	High input impedance	200	100	6	1	3		1	±15	3.4
NE530	High slew rate	200	90	65	2		0.06	35	±18	2
NE5535	Dual op amp/high slew rate	500	90	65	2		0.25	15	±18	1.8
LM218	High speed	200	100	120	2	15		50	±15	5
OP-07E	Ultra-low offset voltage	400	123	±1.2	0.03	0.6		0.3		
TL060M	Low-power JFET input	6	86	0.03	3	1		3.5	±15	0.2
LF155A	JFET	200	100	0.03	1	5		5	±15	2
MC1539	Uncompensated	120	110	200	1		0.08–0.19	6–34	±15	3
MC1776	Programmable[a]	200	86	2	2		0.6	0.03–0.35	±3	0.013–0.13

[a]Q-point currents set by an external resistor.

FIGURE 12-19
Packaging and Thermal Properties (Courtesy of Texas Instruments Incorporated)

| Package | PNS | Thermal Resistance | |
		Junction-to-Case Thermal Resistance, $R_{\theta JC}(°C/W)$	Junction-to-Ambient Thermal Resistance, $R_{\theta JA}(°C/W)$
D plastic dual-in-line	8	51	172
	14, 16	33	131
FH ceramic chip carrier	20, 28	35	104
FK ceramic chip carrier	20	35	91
FN plastic chip carrier	20	37	114
J ceramic dual-in-line (glass-mounted chips)	14 thru 20	60	122
J ceramic dual-in-line[a] (alloy-mounted chips)	14 thru 20	29[a]	91[a]
JG ceramic dual-in-line (glass-mounted chips)	8	58	151
JG ceramic dual-in-line[a] (alloy-mounted chips)	8	26[a]	119[a]
LP plastic plug-in	3	40	160
LU plastic plug-in	3	40	178
N plastic dual-in-line	14 thru 20	72	143
	28	45	100
	40	40	100
P plastic dual-in-line	8	79	172
U ceramic flat	10, 14	55	185
W ceramic flat	14, 16	60	125

[a] In addition to those products so designated on their data sheets, all devices having a type number prefix of "SNC" or "SNM," or a suffice of "/883B" have alloy-mounted chips.

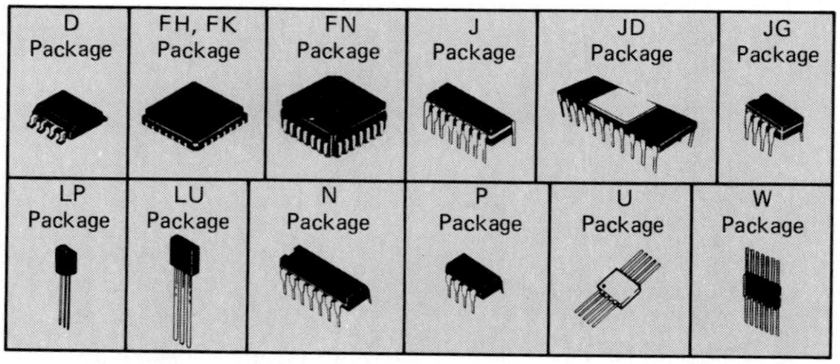

Bipolar technology has long dominated the analog segment of the IC market and the thousands of different op-amp configurations that dominate the available circuitry. However, the situation is rapidly changing today. As discussed in Section 13-11, many other types of specialized analog ICs are available on the market, and the number continues to increase rapidly. Companies are developing a range of CAD tools and methods for semiconductor design that is making analog ICs more accessible and widely used. At the same time, fabrication methods continue to evolve. There is a growing availability of MOS and combined bipolar/MOS technology available for linear applications.

There is also a major expansion occurring in the growth of both analog and digital circuitry on the same chip. This topic is further discussed in Chapter 17.

12-10 PACKAGING

Several different standard IC packages are shown in Fig. 12-19. The DIP configurations are available in plastic or ceramic packages, with leads prepared for insertion into a printed circuit board or in a leadless format prepared for surface mounting. The thermal characteristics of these various packages are also given. The particular terminal connections for each IC are given on the data sheets provided by the manufacturers. (The package designations and thermal resistance values can vary among manufacturers.)

QUESTIONS AND ANSWERS

Questions

12-1. What is a linear IC?

12-2. Do schematics usually provide a sufficiently complete description of IC op-amp performance to predict all details of circuit operation?

12-3. What type of noise protection is provided by a difference amplifier?

12-4. What determines the degree of common-mode noise protection in a difference amplifier?

12-5. What limits the maximum difference-mode input signal that can be used with a BJT difference amplifier?

12-6. What are the functions of each BJT in the simple op amp of Fig. 12-7?

12-7. Under what conditions can a feedback loop result in oscillation of an op amp?

12-8. How can the ratio R_f/R_1 be used to modify the properties of an op-amp inverting amplifier circuit?

12-9. What alternatives can be used to describe the frequency-response characteristics of an IC op amp?

12-10. For the current mirror Q_8 and Q_9 in Fig. 12-14, does $I_{C8} \cong I_{C9}$ always hold true?

12-11. In order to analyze the 741 op amp using the schematics shown in Figs. 12-14 and 12-15, what supplementary assumptions must be made?

12-12. What is the advantage of rail-to-rail operation for an op amp?

12-13. Why is the op amp such a useful building block element for analog circuits?

Answers

12-1. A linear IC is one that performs linear operations on the input waveform.

12-2. Schematics of IC op amps usually are not complete in that they do not define the geometries of the devices in use, basic device parameters (such as h_{FE}, h_{fe}, and V_E), and the distributed R, C, and L effects. Without this information, any circuit analysis is approximate and incomplete. Many of the circuit details not released by manufacturers are proprietary.

12-3. An op amp provides protection against identical noise signals that appear on the two inputs. Protection thus exists with respect to externally caused noise sources that create the same noise on both inputs, and this is called common-mode protection.

12-4. The common-mode noise protection in a difference amplifier is determined by the ac resistance of the constant-current source at its bias point. Since this value is always finite, the noise protection is never perfect.

12-5. The maximum difference-mode input signal (representing the difference between the signals on the two inputs) is limited by the desire to maintain linear operation. Since the emitters are connected, the input base-emitter diodes limit the available voltage excursions. The base-emitter voltages on each BJT must remain in the range 0.3 to 0.7 V (for Si). Therefore, for this case, the maximum voltage difference that can appear on the BJT bases is of this same order of magnitude.

12-6. In Fig. 12-7, Q_1 and Q_2 form an input difference amplifier and Q_3 provides the constant-current source required for difference-amplifier operation. Q_4 is used to shift the dc voltage level between the collector of Q_2 and the output (emitter-follower) stage Q_5 to obtain $V_0 = 0$ with grounded inputs.

12-7. An op amp oscillates if the output is connected to the input (through a feedback loop) in such a way that the total loop gain is a multiple of 360° and the loop gain is greater than unity. Oscillators of this type are discussed in detail in Chapter 14.

12-8. The ratio R_f/R_1 can be used to determine the gain of an op-amp inverting amplifier and to thereby determine the bandwidth of the amplifier. Given a constant *GBWP*, adjustment of the ratio R_f/R_1 directly controls the tradeoff between gain and bandwidth.

12-9. The frequency-response characteristics of an op amp can be described in terms of the amplifier bandwidth or rise time. As shown in Eq. (12-38), a reciprocal relationship exists between these two parameters.

12-10. For the current mirror Q_8 and Q_9 in Fig. 12-14, $I_{C8} \cong I_{C9}$ only if the device geometries are identical. If the collector areas are not the same, the output currents are in proportion to the areas.

12-11. In order to analyze the 741 op amp shown in the schematic of Figs. 12-14 and 12-15, additional assumptions must be made regarding the input and output characteristic curves for the devices used (defining h_{FE}, h_{fe}, $1/h_{oe}$, and g_{me}), the effective device capacitances, the distributed R, C, and L effects in the circuit, and the tolerance levels associated with these parameters as a result of the fabrication process.

12-12. Rail-to-rail operation for an op amp provides a circuit that can accommodate the maximum amplitude input and output signals for a given power-supply voltage. This is of particular advantage when very low power-supply voltages are being used.

12-13. With its very high open-loop gain, the op amp can be used with many types of feedback loops to produce circuit properties that are precisely predicted and controlled by external component parameters. The op amp can thus be used in any circuit application that requires the use of a high-performance dc-coupled amplifier.

Exercises
(12-1)

12-1. Find the dc output levels V_{o1} and V_{o2} for the circuit in Fig. 12-20 with $v_i = 0$.

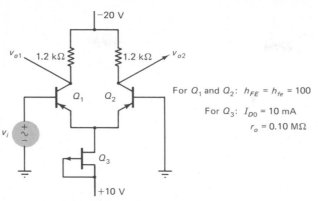

For Q_1 and Q_2: $h_{FE} = h_{fe} = 100$

For Q_3: $I_{DO} = 10$ mA
$r_o = 0.10$ MΩ

Fig. 12-20 Circuit for Exercises 12-1 to 12-3.

(12-1)   **12-2.** Find the differential gain A_{vd} associated with each half of the difference amplifier of Fig. 12-20.

(12-2)   **12-3.** (a) Find the common-mode gain A_{vc} associated with the difference amplifier of Fig. 12-20. (b) Find the *CMRR* for the amplifier.

(12-1)   **12-4.** (a) Find A_{vd} for the difference amplifier circuit of Fig. 12-21. (b) What is the *CMRR*
(12-2)   in dB?

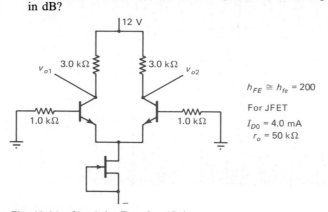

$h_{FE} \cong h_{fe} = 200$

For JFET

$I_{DO} = 4.0$ mA
$r_o = 50$ kΩ

Fig. 12-21 Circuit for Exercise 12-4.

(12-2)   **12-5.** A difference amplifier has a 10-V peak-to-peak ac output when a given source is applied to one input and the other input is grounded. If the same source is applied to both inputs, a 3.0-mV peak-to-peak output is observed. Find the *CMRR* in dB.

(12-2)   **12-6.** The BJT of Fig. 12-22 (p. 506) is used as a constant-current source in a difference amplifier. Given $R_i = 0$, $I_E = 2.0$ mA, and $h_{FE} \cong h_{fe} = 150$, estimate the *CMRR* in dB.

(12-3)   **12-7.** For the simple op amp of Fig. 12-7, Q_3 is to be biased so that $r_o = 100$ kΩ. Given $V_{B3} = 4.0$ V, $V_{CC} = 8.0$ V, $V_{EE} = -8.0$ V, and $V_E = 1.0 \times 10^2$ V, select values for R_a, R_b, and R_{E3}. Assume $h_{FE} > 100$ and let $I_{div} > 20I_B$.

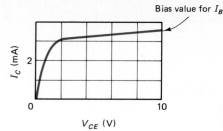

Bias value for I_B

I_C (mA)

2

0   10

V_{CE} (V)

Fig. 12-22 BJT characteristic used for a constant-current generator (Exercise 12-6).

(12-3) **12-8.** For the simple op amp of Fig. 12-7, $I_{E3} = 2.0\,\text{mA}$. Choose R_{C1} and R_{C2} so that $V_{02} = 8.0\,\text{V}$ with $v_{i1} = v_{i2} = 0$. Assume that $V_{CC} = 12\,\text{V}$.

(12-3) **12-9.** For the simple op amp of Fig. 12-7, $I_{E4} = 2.0\,\text{mA}$ is selected with $V_{CC} = 12\,\text{V}$ and $V_{EE} = -12\,\text{V}$. (a) Select values for R_{C4} and R_{E4} if $V_{CE4} = -3.0\,\text{V}$. (b) What value for V_{C2} is required for consistency?

(12-4) **12-10.** For an amplifier with negative feedback,

$$A_{Vf} = \frac{A_{Vo}}{1 - A_{Vo}B}$$

From Eq. (12-36), $A_{Vf} = -R_f/R_1$. Assuming $|A_{Vo}B| >>> 1$, find B.

(12-5) **12-11.** Given the unity-gain bandwidth for an op amp equal to 5.0 MHz, find A_{Vf} if $f_{2f} = 500$ kHz.

(12-5) **12-12.** Given the unity-gain bandwidth equal to 1.0 MHz and $f_2 = 5.0$ Hz for a 741 op amp (with internal compensation), estimate the open-loop gain A_{Vo}.

(12-6) **12-13.** Assume that the 741 op amp of Figs. 12-14 and 12-15 is operated with ± 12-V power supplies. Find the reference current I_{R4}. Assume that V_{BE} is $0.6\,\text{V} \pm 0.1\,\text{V}$ at this current level.

(12-6) **12-14.** Assume that the 741 op amp of Figs. 12-14 and 12-15 is modified so that $R_5 = 10\,\text{k}\Omega$. What change is observed in I_{C10}?

(12-7) **12-15.** Assume that a reduction in the power-supply voltages for the 741 op amp reduces all current levels by a factor of two. What happens to the gain of the op amp?

(12-9) **12-16.** Fill in the missing values of unity-gain bandwidth and t_R in Fig. 12-18.

(12-9) **12-17.** Estimate A_{Vc} for the operational amplifiers in Fig. 12-18.

(12-10) **12-18.** (a) If 50 mW is dissipated in an 8-pin type P DIP package, at an ambient temperature of 30°C, what is the (same) temperature of the transistor junctions? (b) If the temperature of the case is held at 30°C using a heat sink, what is the temperature of the junctions?

Solutions **12-1.** Given $I_{D0} = 10$ mA, the emitter current in each BJT is about 5.0 mA. Therefore, the voltage drop across each collector resistor is

$$V_{R_c} = (5.0\,\text{mA})(1.2\,\text{k}\Omega) = 6.0\,\text{V}$$

With balanced inputs ($v_i = 0$),

$$V_{01} = V_{02} = -20\,\text{V} + 6.0\,\text{V} = -14\,\text{V}$$

12-2. From Eqs. (12-6) or (12-7),

$$A_{Vd} = \frac{h_{fe}R_C}{2(h_{ie} + R_i)}$$

With $R_i = 0$ and

$$h_{ie} \cong 0.026 \text{ V}/I_B \cong 0.52 \text{ k}\Omega$$

A_{Vd} can be determined:

$$A_{Vd} \cong \frac{(100)(1.2 \text{ k}\Omega)}{2(0.52 \text{ k}\Omega)} \cong 1.2 \times 10^2$$

12-3. (a) From Eq. (12-12),

$$A_{Vc} \cong R_C/2r_o = 1.2 \text{ k}\Omega/2(0.10 \text{ M}\Omega) \cong 6 \times 10^{-3}$$

(b) From Eq. (12-13),

$$CMRR = \frac{A_{Vd}}{A_{Vc}} = \frac{1.2 \times 10^2}{6 \times 10^{-3}} \cong 2 \times 10^4 \cong 86 \text{ dB}$$

12-4. (a) From Eq. (12-6) or (12-7),

$$A_{Vd} = \frac{h_{fe} R_C}{2(h_{ie} + R_i)}$$

From Fig. 12-21, $R_C = 3.0 \text{ k}\Omega$, $R_i = 1.0 \text{ k}\Omega$, $h_{fe} = 200$, and

$$h_{ie} \cong \frac{0.026 \text{ V}}{\dfrac{4.0 \text{ mA}}{2(200)}} \cong 2.6 \text{ k}\Omega$$

so that

$$A_{Vd} \cong \frac{(200)(3.0 \text{ k}\Omega)}{2(2.6 \text{ k}\Omega + 1.0 \text{ k}\Omega)} \cong 83$$

(b) From Eq. (12-12),

$$A_{Vc} \cong R_C/2r_o = 3.0 \text{ k}\Omega/2(50 \text{ k}\Omega) \cong 0.03$$

and from Eq. (12-13),

$$CMRR = 83/0.03 \cong 2.8 \times 10^3 \cong 69 \text{ dB}$$

12-5. From the exercise,

$$A_{Vd} = 10 \text{ V}/2v_i$$

$$A_{Vc} = \frac{3.0 \times 10^{-3} \text{ V}}{2v_i}$$

so that from Eq. (12-13),

$$CMRR = \frac{A_{Vd}}{A_{Vc}} = \frac{10 \text{ V}}{3.0 \times 10^{-3} \text{ V}} \cong 3.3 \times 10^3 \cong 70 \text{ dB}$$

12-6. From Fig. 12-22, the slope of the BJT output characteristic is about $0.5 \text{ mA}/8.0 \text{ V}$, so

$$r_o \cong 8.0 \text{ V}/0.5 \text{ mA} = 16 \text{ k}\Omega$$

From Eq. (12-13),

$$CMRR = h_{fe} \frac{r_o}{R_i + h_{ie}}$$

$$= \frac{150(16 \text{ k}\Omega)}{\dfrac{0.026 \text{ V}}{2.0 \text{ mA}/150}} \cong 1.2 \times 10^3 \cong 62 \text{ dB}$$

12-7. From Chapter 9, for a BJT,

$$r_o = \frac{1}{h_{oe}} \cong \frac{V_E}{|I_E|} = \frac{1.0 \times 10^2}{|I_E|}$$

To achieve $r_o = 100 \text{ k}\Omega$,

$$I_E = \frac{1.0 \times 10^2}{1.00 \times 10^5} = 1.0 \times 10^{-3} = 1.0 \text{ mA}$$

Given $V_B = 4.0 \text{ V}$, $V_E \cong 3.3 \text{ V}$, so that

$$R_E \cong \frac{3.3 \text{ V} + 8.0 \text{ V}}{1.0 \text{ mA}} \cong 11 \text{ k}\Omega$$

To achieve a stable divider, $I_{\text{div}} > 20I_B$. Then choose

$$I_{\text{div}} = 20(1.0 \text{ mA}/100) = 0.2 \text{ mA}$$

$$R_b(0.2 \text{ mA}) = 12 \text{ V} \qquad \text{and} \qquad R_b = 60 \text{ k}\Omega$$

$$R_a(0.2 \text{ mA}) = 4.0 \text{ V} \qquad \text{and} \qquad R_a = 20 \text{ k}\Omega$$

12-8. With $v_{i1} = v_{i2} = 0$,

$$I_{E1} \cong I_{E2} = 1.0 \text{ mA}$$

and

$$V_{02} = 12 \text{ V} - (R_{C2})(1.0 \text{ mA}) = 8.0 \text{ V}$$

so that

$$R_{C2} = 4.0 \text{ k}\Omega$$

and to maintain symmetry,

$$R_{C1} = 4.0 \text{ k}\Omega$$

12-9. (a) The circuit design requires $V_0 = 0$ when $v_{i1} = v_{i2} = 0$. Therefore, $V_{B5} = 0.7 \text{ V}$ and

$$R_{C4} = \frac{12 \text{ V} + 0.7 \text{ V}}{2.0 \text{ mA}} \cong 6.4 \text{ k}\Omega$$

The voltage on the emitter of Q_4 must be

$$0.7 \text{ V} + 3.0 \text{ V} = 3.7 \text{ V}$$

Therefore,

$$R_{E4} = \frac{12 \text{ V} - 3.7 \text{ V}}{2.0 \text{ mA}} \cong 4.2 \text{ k}\Omega$$

(b) Given $V_{E4} = 3.7$ V, the voltage on the collector of Q_2 must be

$$V_{C2} = 3.7 \text{ V} - 0.7 \text{ V} = 3.0 \text{ V}$$

12-10. Given

$$A_{Vf} = \frac{A_{Vo}}{1 - A_{Vo}B}$$

and $|A_{Vo}B| \gg 1$,

$$A_{Vf} \cong A_{Vo}/-A_{Vo}B = -1/B$$

Therefore,

$$B = -1/A_{Vf} = R_1/R_f$$

The feedback fraction B is given by the ratio of the input resistance R_1 to the feedback resistance R_f.

12-11. Given the unity-gain bandwidth of 5.0 MHz, the *GBWP* for the amplifier is

$$(1)(5.0 \text{ MHz}) = 5.0 \text{ MHz}$$

If the amplifier bandwidth with feedback is 500 kHz, then

$$|A_{Vf}|f_{2f} \cong 5.0 \text{ MHz}$$
$$|A_{Vf}| \cong 5.0 \text{ MHz}/500 \text{ kHz} = 10$$

12-12. Assuming the *GBWP* $\cong$ constant,

$$(1)(1.0 \text{ MHz}) = (A_{Vo})(5.0 \text{ Hz})$$
$$A_{Vo} \cong 1.0 \text{ MHz}/5.0 \text{ Hz} = 2.0 \times 10^5$$

12-13. From the discussion of Section 12-6,

$$(I_{R4})_{min} \cong \frac{24 \text{ V} - 2(0.7 \text{ V})}{40 \text{ k}\Omega} \cong 0.57 \text{ mA}$$
$$(I_{R4})_{max} \cong \frac{24 \text{ V} - 2(0.5 \text{ V})}{40 \text{ k}\Omega} \cong 0.58 \text{ mA}$$

For large power-supply voltages, the reference current is not very sensitive to V_{BE}.

12-14. From Eq. (12-40),

$$I_{E10}R_5 \cong V_0 \ln (I_{E11}/I_{E10})$$

Given $R_5 = 10$ kΩ, $V_0 = 0.026$ V, and $I_{E11} = 0.7$ mA,

$$I_{E10}(10 \text{ k}\Omega) \cong (0.026 \text{ V}) \ln (0.7 \text{ mA}/I_{E10})$$

The left- and right-hand sides of this equation can be plotted as shown in Fig. 12-23. The solution is

$$I_{E10} \cong 1 \times 10^{-5} \text{ A}$$

The change in R_5 from 5.0 to 10 kΩ results in a decrease in I_{E10} from 2×10^{-5} to 1×10^{-5} A.

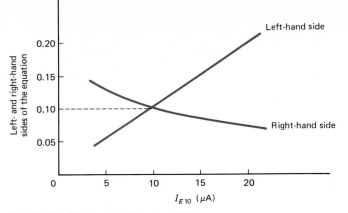

Fig. 12-23   Solution for Exercise 12-14.

12-15. From Eq. (12-41), the gain of the difference amplifier is proportional to $g_{me}R_C$. The effective R_C is dominated by $1/h_{oe4} \cong V_E/I_{E4}$ and the input impedance to Q_{16}. If I_{E4} is reduced by a factor of two, $1/h_{oe4}$ will double to 10 MΩ. If all currents are reduced by a factor of two, Eq. (12-42) can be used to find the input impedance to Q_{16}:

$$Z_{i16} = \left[\left(\frac{0.026 \text{ V}}{0.5 \times 10^{-5} \text{ A}} \right) 200 \right] + 200 \left\{ 50 \text{ k}\Omega \left\| \left[\left(\frac{0.026 \text{ V}}{0.25 \text{ mA}} \right) 200 + 200(0.1 \text{ k}\Omega) \right] \right\}$$
$$\cong 5 \text{ M}\Omega$$

Therefore, the effective load increases from 2 MΩ to 10 MΩ∥5 MΩ $\cong$ 3.3 MΩ, a factor of 1.7. The gain of the difference amplifier thus increases by a factor of

$$(2.0)(1.7) = 3.4$$

The gain of the amplifier stage is given by Eq. (12-46). As the current is reduced by a factor of two, $1/h_{oe13} \cong 1/h_{oe17}$ and h_{ie17} each increases by a factor of two. Therefore,

$$A_V \cong 200 \, \frac{200 \text{ k}\Omega \| 200 \text{ k}\Omega}{20 \text{ k}\Omega + 200(0.1 \text{ k}\Omega)} \cong 5 \times 10^2$$

an increase by a factor of 1.7.

The total op-amp gain thus increases by a factor of

$$(3.4)(1.7) \cong 5.8$$

12-16. From Eq. (12-38), the bandwidth and rise time for an op amp are related by

$$f_2 \cong 0.35/t_R$$

Therefore, the values of Fig. 12-24 can be found for Fig. 12-18.

12-17. From Fig. 12-18, A_{vd} and the *CMRR* (in dB) are known. Values for A_{vc} can then be found:

$$A_{vc} \cong A_{vd}/CMRR$$

The results are shown in Fig. 12-25.

FIGURE 12-24
Solution for Exercise 12-16

Device Code	Unity-Gain Bandwidth (MHz)	Rise Time (μs)
RC741	1.2	0.3
NE5230	0.6	0.6
LM124	1.0	0.4
NE5512	3.0	0.1
NE530	5.8	0.06
NE5535	1.4	0.25
LM218	1.5	0.02
OP-07E	0.6	0.6
TL060M	1.0	0.4
LF155A	5.0	0.07
MC1539	1.8–4.4	0.08–0.19
MC1776	0.6	0.6

FIGURE 12-25
Solution for Exercise 12-17

Device Code	A_{vd} ($\times$ 10^5)	CMRR ($\times$ 10^4)	A_{Vc}
RC741	2.0	3.2	6.3
NE5230	3.0	7.1	4.2
LM124	1.0	1.8	5.6
NE5512	2.0	10	2.0
NE530	2.0	3.2	6.3
NE5535	5.0	3.2	16
LM218	2.0	10	2.0
OP-07E	4.0	1.4×10^2	0.3
TL060M	0.06	2.0	0.3
LF155A	2.0	10	2.0
MC1539	1.2	32	0.4
MC1776	2.0	2.0	10

12-18. (a) Using data from Fig. 12-19,

$$\Delta T = P_D\, \theta_{JA}$$

$$= (50 \text{ mW})(172°\text{C/W}) = 8.6°\text{C}$$

$$T_J = 8.6°\text{C} + 30°\text{C} = 38.6°\text{C}$$

(b) Again, using data from Fig. 12-19,

$$\Delta T = P_D\, \theta_{JC}$$

$$= (50 \text{ mW})(79°\text{C/W}) = 4.0°\text{C}$$

$$T_J = 4°\text{C} + 30°\text{C} = 34°\text{C}$$

(12-1)   **12-1.** A difference amplifier operates with a 2.4-mA constant-current source and $V_{CC} = 6$ V, $V_{EE} = -6$ V power supplies. If $R_C = 1.2$ kΩ, find the dc voltages on the collectors of the BJTs.

(12-1)   **12-2.** A difference amplifier is designed with $R_C = 2.3$ kΩ and $R_i = 1.5$ kΩ. If $h_{fe} = 150$ and $h_{ie} = 2.5$ kΩ for the BJTs being used, find the differential gain A_{Vd}.

(12-2)   **12-3.** A difference amplifier operates with $R_C = 2.3$ kΩ and a BJT constant-current source with $1/h_{oe} = 35$ kΩ. Find the common-mode gain A_{Vc} in dB.

(12-2)   **12-4.** A BJT difference amplifier makes use of a JFET constant-current source with $I_E = 1.0$ mA. If $R_C = 2.3$ kΩ, find A_{Vc} in dB.

(12-1) (12-2)   **12-5.** Find A_{Vd} and the *CMRR* in dB for the difference amplifier of Fig. 12-26.

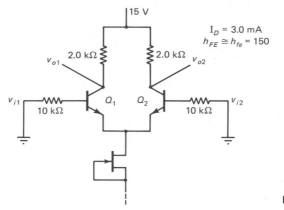

$$I_D = 3.0 \text{ mA}$$
$$h_{FE} \cong h_{fe} = 150$$

Fig. 12-26   Circuit for Problem 12-5.

(12-2)   **12-6.** Draw the schematic for a difference amplifier using pnp transistors and a JFET constant-current source. Label inputs and outputs and note whether dc voltages are positive or negative. Assume that there are available power sources at $+10$ V and -10 V. Using 5-kΩ collector resistors and no input resistor ($R_i = 0$), find A_{Vd}, A_{Vc}, and the *CMRR* (in dB) for your circuit using the device characteristics given below:

BJT: $h_{FE} \cong h_{fe} = 150$

JFET: $r_o = 0.5$ MΩ, $I_D = 2.0$ mA for $V_{GS} = 0$ and $V_{DS} > 1.0$ V

(12-2)   **12-7.** (a) A difference amplifier uses a BJT constant-current source with $I_C = 100$ μA. Estimate $r_o = 1/h_{oe}$ by using the Early voltage $V_E \cong 50$ V. (b) Find r_o if I_C is reduced to 10 μA (c) What effect does a reduced I_C have on the *CMRR*?

(12-3)   **12-8.** For the simple op amp of Fig. 12-7, assume $V_{CC} = 6$ V, $V_{EE} = -6$ V, and $h_{FE} \cong h_{fe} = 100$ for all BJTs. Design to maximize the linear operating region. Both inputs are grounded. (a) Find V_{E1}, V_{E2}, V_{C1}, and V_{C2}. (b) If $I_{C3} = 100$ μA, find R_{C1} and R_{C2}. (c) If $I_{C4} = 100$ μA, find R_{E4}. (d) Find R_{C4}.

(12-3)   **12-9.** For the simple op amp of Fig. 12-7, assume that Q_1 and Q_2 have values $n = 1.5$ and 1.6, respectively, for the ideality factor. What ratio of values for I_E is observed if $v_{i1} = v_{i2} = 0$?

(12-3)   **12-10.** Estimate the *CMRR* for the simple op amp of Example 12-2.

(12-4)   **12-11.** For a simple op amp, $Z_o = 0.20$ kΩ, $A_V = 1.0 \times 10^3$, and $Z_i = 2.0$ kΩ. For an application with $R_1 = 5.0$ kΩ and $R_f = 15$ kΩ, find the ideal circuit gain and the circuit gain that is observed considering the actual amplifier characteristics.

(12-5) **12-12.** An op amp with $A_{vf} = 25$ and $BW = 1.0$ MHz is to be used. What must be the open-loop midrange gain of the op amp if the open-loop bandwidth is 50 Hz?

(12-5) **12-13.** An op amp has an input resistance of 10 kΩ. An output noise voltage greater than 10 V causes the amplifier to saturate. Assuming that thermal noise dominates, what is the maximum bandwidth that can be used to assure that saturation due to noise will occur less than 1 percent of the time? Let $A_v = 1.0 \times 10^3$ and $T = 300°$K.

(12-5) **12-14.** An op amp is tested with a square wave and $t_R = 0.10$ μs is observed. Find the bandwidth of the op amp.

(12-5) **12-15.** An op amp is switched from an output of 5.0 V to an output of 0 V and a time constant $\tau = 1.0$ μs is observed. Find the slew rate for the op amp.

(12-6) **12-16.** The 741 op amp functions with ±10-V power supplies. Estimate the reference current I_{R4}.

(12-6) **12-17.** For the circuit of Fig. 12-27, find I_1 and I_2.

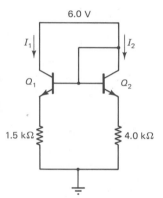

Fig. 12-27 Circuit for Problem 12-17.

(12-6) **12-18.** Assume that the 741 op amp is modified so that $R_5 = 20$ kΩ. Find the I_{C10} value that results.

(12-7) **12-19.** Assume that all current levels in the 741 op amp increase by a factor of three. What happens to the gain of the op amp?

(12-7) **12-20.** What tolerance in the nominal current levels for the 741 op amp is acceptable to maintain the open-loop gain within ±50 percent? What range of supply voltages is acceptable given this constraint?

(12-8) **12-21.** The NE 5230 op amp is used with ±3.0-V power-supply voltages. What are the maximum input and output signal amplitudes that can be accommodated?

(12-9) **12-22.** You require an op amp with an open-loop gain of 1.0×10^2 concurrent with a bandwidth of ≥50 kHz. Which ICs listed in Fig. 12-18 can be used?

(12-9) **12-23.** An op amp with a rise time of less than 0.2 μs is required. Which ICs listed in Fig. 12-18 can be used?

(12-9) **12-24.** An op-amp application requires that the output change from 1.0 to −1.0 V in less than 4.0 μs. Which ICs listed in Fig. 12-18 can be used?

(12-10) **12-25.** Which type of package in Fig. 12-19 provides the maximum protection against heat rise due to internal power dissipation?

(12-10) **12-26.** For a given IC application, a dissipation of 100 mW must lead to less than a 10°C increase in junction temperature if the ambient air is maintained at a constant temperature. Which packages in Fig. 12-19 can be used?

COMPUTER APPLICATIONS

* **12-1.** Use SPICE to stimulate the difference amplifier of Fig. 12-6 (with one input grounded). Use the simulation to plot V_{out} as a function of V_{in} for the circuit. What can you conclude about the limitations of the amplitude of the voltage placed across the amplifier input?

* **12-2.** Use SPICE to simulate the simple operational amplifier of Fig. 12-7. Compare the prediction of voltage gain A_V with that given by the discussion of Example 12-2.

* **12-3.** Write a computer program that can be used to solve Eq. (12-37) for I_{E10}, given I_{E11}, R_5, and V_0. Plot I_{E10} as a function of 10 μA $\leq I_{E11} \leq$ 10 mA for $R_5 =$ 1 kΩ, 5 kΩ, and 10 kΩ. Use $V_0 =$ 0.026 V. Turn in your program listing and graph.

EXPERIMENTAL APPLICATIONS

12-1. The objective of this experiment is to construct the simple operational amplifier described in Example 12-2 and to study its properties. Once the circuit is complete, check and record all dc bias voltages and compare them with prediction. Test the ability to zero the output with 1.0-kΩ input resistors connected to ground. Experimentally determine the open-loop gain at 1.0 kHz. Measure the *CMRR* by placing the same ac signal at both inputs. Choose $R_f =$ 20 kΩ and $R_1 =$ 10 kΩ; compare theoretical and experimental values of the gain with feedback. Discuss and interpret your findings.

12-2. The purpose of this experiment is to explore the operation of the 741 op amp, with an emphasis on the frequency-response characteristics. (a) Set up your 741 with $R_f =$ 20 kΩ, $R_1 =$ 10 kΩ, and a voltage divider to use for zeroing the output. Check the operation with an input sine wave. Collect data and draw a Bode plot for your op amp. (b) Change R_f to 10 kΩ and measure the unity-gain bandwidth of the op amp. Confirm by measuring the distortion produced by a high-frequency square-wave input at this frequency. (c) Change the gain to 20 and again measure the bandwidth of the op amp. (d) Plot the f_{2f} values for parts (a), (b), and (c) to show how they relate on the amplitude portion of a Bode plot. (e) If f_2 (open loop) = 5 Hz, what is the open-loop gain of the op amp (by prediction)? (f) Measure the dc standby power consumption of the 741 under the condition of part (a) with the input resistor R_1 grounded.

CHAPTER 13
DESIGN WITH IC OP AMPS

IC op-amp applications that form a reference base for the designer are introduced in this chapter. These applications develop proficiency in the techniques for building and interpreting op-amp circuits. With this experience base, the designer is able to draw on IC op amps as building block elements in the future.

Several different op-amp types are discussed in these applications to illustrate how ICs with desired parameters can be selected for optimum circuit performance. To the extent that these parameters define the desired performance objectives, IC selection becomes a matter of external evaluation. When more complex issues are raised regarding device performance, the internal construction of the IC and the materials used must also be considered. Further, the integration technology can significantly affect circuit performance. The familiarity with basic IC parameters is a starting point for a more complete consideration of tradeoffs and system design.

13-1 VIRTUAL GROUND

Operational amplifiers are typically used with feedback loops that connect the output to the input through a resistive or reactive component. A typical configuration is shown in Fig. 13-1(A). The triangular symbol represents the operational amplifier. The negative ($-$) input produces an inverted output, and the positive ($+$) input produces a noninverted output. The dc supply voltages and external balance connections are often suppressed (not shown) as part of the representation. Pin connections for a typical DIP configuration are shown in Fig. 13-1(B).

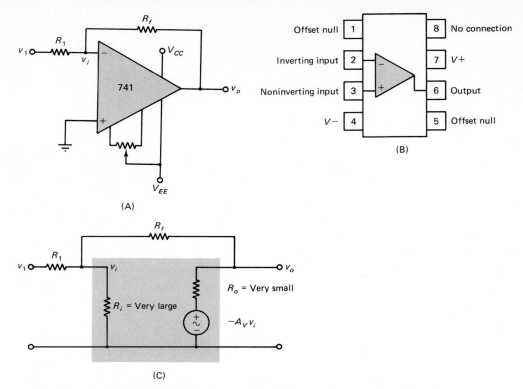

Fig. 13-1 Inverting amplifier using the 741 operational amplifier. (A) Operational amplifier circuit. (B) DIP carrier pin configuration. (C) Equivalent circuit.

In order to understand the operation of this circuit, assume that the operational amplifier has near-ideal features, with R_i very large, R_o very small, and the open-circuit gain (without R_f) very large. In addition, when $v_i = 0$, the output $V_0 = 0$. For the near-ideal operational amplifier, the equivalent circuit of Fig. 13-1(C) can be drawn.

First, note that $v_o \cong -A_V v_i$. Since the open-loop gain A_V is very large by design, $v_i \ll v_o$. If v_o is limited to a typical range of ± 10 V, then v_i is measured in mV or μV. It is possible to treat v_i as approximately equal to ground; v_i is spoken of as being at *virtual ground*. Despite the high input impedance from the input to ground, the feedback loop forces v_i to a value close to (but not exactly) 0 V.

The virtual-ground concept can be further explored by recalling that the output of an operational amplifier depends on both inputs. In Fig. 13-2, let v_1 be an unknown voltage and v_x an externally fixed reference voltage. Since

$$v_o = A_V(v_x - v_y)$$

and

$$\frac{v_o - v_y}{R_f} = \frac{v_y - v_1}{R_1}$$

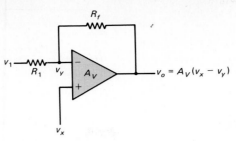

Fig. 13-2 Relationship between v_x and v_y.

substitution produces

$$\frac{A_V(v_x - v_y) - v_y}{R_f} = \frac{v_y - v_1}{R_1}$$

or, rearranging terms

$$v_y = v_x \frac{A_V}{A_V + 1 + \dfrac{R_f}{R_1}} + \frac{(R_f/R_1)v_1}{A_V + 1 + \dfrac{R_f}{R_1}} \qquad (13\text{-}1)$$

For large A_V,

$$v_y \cong v_x \qquad (13\text{-}2)$$

Therefore, $v_x - v_y \cong 0$ when a feedback resistor is present. If $v_x = 0$, then v_y becomes virtual ground. If $v_x \neq 0$, then $v_x - v_y$ is very small and v_y can be effectively taken equal to the reference voltage v_x. This general observation is of use in analyzing the behavior of more complex circuits that use operational amplifiers.

13-2 INVERTING AMPLIFIERS

Referring to the equivalent circuit of Fig. 13-1(C) (in the limiting case), the relationship

$$\frac{v_1 - v_i}{R_1} \cong \frac{v_i - v_o}{R_f}$$

can be found, and with $v_i = 0$,

$$\frac{v_o}{v_1} \cong -\frac{R_f}{R_1} \qquad (13\text{-}3)$$

as previously found from a detailed analysis [Eq. (12-36)]. The gain of the operational amplifier with the feedback loop is given by $-R_f/R_1$ and is determined only by the resistor values, which can be set with accuracy. The circuit of Fig. 13-1 is called an inverting amplifier.

As shown, the circuit requires the use of two power supplies (V_{CC} and V_{EE}). An alternative configuration is shown in Fig. 13-3. The LM 124 is a quad op amp, meaning that four different op amps are included on a single chip. The LM 124 amplifier is designed specifically to operate from a single power supply over a wide range of voltages.

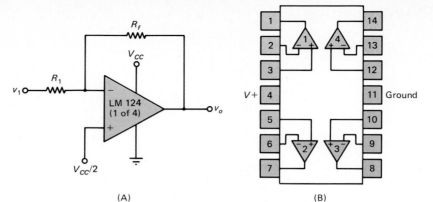

Fig. 13-3 Single-supply inverting amplifier using one op amp on the LM 124 chip. (A) Amplifier circuit. (B) LM 124 pin configuration showing quad op-amp construction.

(A)

(B)

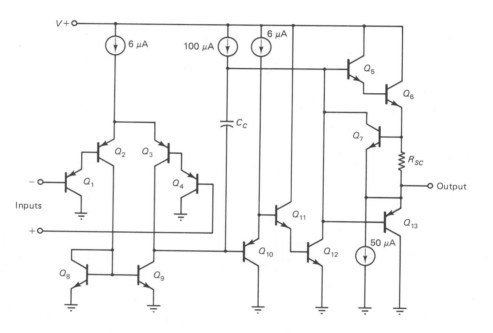

Fig. 13-4 Equivalent circuit for one op amp contained in the LM 124 quad op amp. (Courtesy of Signetics Corporation)

Figure 13-4 shows an equivalent circuit for one of the op amps. Note the similarity with the 741 discussed in Chapter 12. The input difference amplifier Q_1 and Q_4, active load Q_8 and Q_9, gain stage Q_{10} and Q_{11}, current-steering transistor Q_{12}, and output drivers Q_6 and Q_{13} are readily identifiable.

Example 13-1   **Tracking between v_x and v_y in an Inverting Amplifier**

For the inverter amplifier of Fig. 13-2, assume that

$$A_V = 1.0 \times 10^4 \qquad v_1 < 1.0 \text{ V}$$

$$\frac{R_f}{R_1} = 10 \qquad\qquad v_x = 0 \text{ V}$$

What percentage accuracy is associated with the assumption $v_y \cong v_x$?

Solution From Eq. (13-1),

$$v_y = v_x \frac{1.0 \times 10^4}{1.0 \times 10^4 + 1 + 10} + \frac{10v_1}{1.0 \times 10^4 + 1 + 10}$$

Given $v_1 < 1.0$ V, the maximum difference is

$$v_y \cong 1.0 \times 10^{-3}v_1 = 1.0 \text{ mV}$$

The assumption $v_y = v_x$ is thus valid within about 0.1 percent of v_1. ■

13-3 NONINVERTING AMPLIFIERS

Figure 13-5(A) shows a noninverting amplifier configuration. The virtual ground concept can be used to show [Fig. 13-5(B)] that

$$v_o \frac{R_1}{R_1 + R_f} \cong v_1$$

$$\frac{v_o}{v_1} \cong 1 + \frac{R_f}{R_1} \tag{13-4}$$

Note that the gain of this amplifier is $+2$ when $R_f = R_1$.

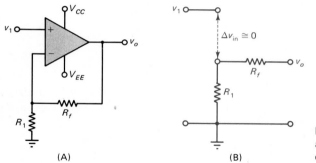

Fig. 13-5 (A) Noninverting amplifier circuit. (B) Equivalent circuit.

Figure 13-6 shows how the LM 124 can be used to produce the same function with a single power supply, assuming an ac application. From the dc perspective, the noninverting input is at $V_{CC}/2$ and the inverting input seeks the same level. The capacitor C prevents a dc gain from developing. From an ac perspective, the capacitor is assumed a short circuit.

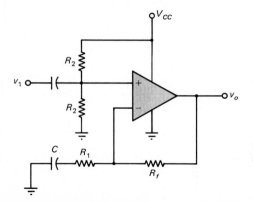

Fig. 13-6 Noninverting amplifier with a single supply voltage.

Figure 13-7 shows the equivalent circuit for a unity follower, which is a special case of the noninverting amplifier with $R_f = 0$. The result

$$v_o \cong v_1 \qquad (13\text{-}5)$$

is obtained.

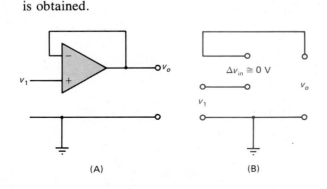

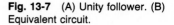

Fig. 13-7 (A) Unity follower. (B) Equivalent circuit.

(A)   (B)

For the summing amplifier of Fig. 13-8, nodal currents can be summed:

$$\frac{v_1}{R_1} + \frac{v_2}{R_2} + \frac{v_3}{R_3} \cong -\frac{v_o}{R_f}$$

$$v_o = -\left(\frac{R_f}{R_1}v_1 + \frac{R_f}{R_2}v_2 + \frac{R_f}{R_3}v_3\right) \qquad (13\text{-}6)$$

and the output is a weighted linear combination of the inputs.

Resistor R_a is added to help cancel the effects of input bias currents, with $R_a = R_1 \| R_2 \| R_3 \| R_f$ for balance.

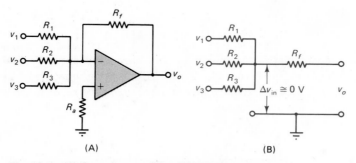

(A)   (B)

Fig. 13-8 (A) Summing amplifier. (B) Equivalent circuit.

A subtracting amplifier can be formed as shown in Fig. 13-9 to produce

$$\frac{v_1 - v_2/2}{R_1} = \frac{v_2/2 - v_o}{R_1}$$

$$v_o \cong v_2 - v_1 \qquad (13\text{-}7)$$

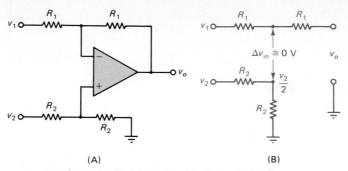

(A)   (B)

Fig. 13-9 (A) Subtracting amplifier. (B) Equivalent circuit.

Example 13-2 **DC Calculation**

For the circuit of Fig. 13-10, find v_{o2}.

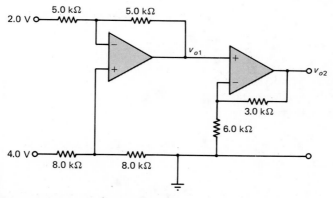

Fig. 13-10 Circuit for Example 13-2.

Solution The first op amp is connected as a subtracting amplifier. From Eq. (13-7),

$$v_{o1} = (4.0 - 2.0) \text{ V} = 2.0 \text{ V}$$

The second op amp is connected as a noninverting amplifier, with v_{o1} as the input. From Eq. (13-4),

$$v_{o2} = (2.0 \text{ V})(1 + 3.0 \text{ k}\Omega/6.0 \text{ k}\Omega) = 3.0 \text{ V}$$ ■

13-5 DIFFERENTIATORS AND INTEGRATORS

If a capacitor is used in the operational-amplifier feedback loop, an integrator results, as shown in Fig. 13-11. If (the small value of) v_i is included,

$$i = C\frac{d}{dt}(v_i - v_o) = \frac{v_1 - v_i}{R}$$

$$v_1 = v_i + RC\frac{dv_i}{dt} - RC\frac{dv_o}{dt} \tag{13-8}$$

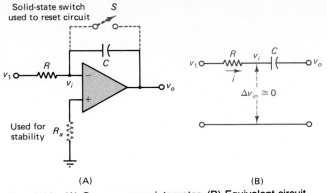

Fig. 13-11 (A) Op amp as an integrator. (B) Equivalent circuit.

Since $v_i \cong 0$,

$$v_1 \cong -RC\ dv_o/dt$$

$$v_o \cong -\frac{1}{RC}\int_0^t v_1\ dt \qquad (13\text{-}9)$$

given $v_o(0) = 0$.

   If the feedback capacitor and the input resistor are exchanged, a differentiator results, as shown in Fig. 13-12, with

$$i = C\frac{d}{dt}(v_1 - v_i) = \frac{v_i - v_o}{R}$$

$$v_o = v_i + RC\ dv_i/dt - RC\ dv_1/dt$$

and with $v_i \cong 0$,

$$v_o \cong -RC\ dv_1/dt \qquad (13\text{-}10)$$

For the integrator and differentiator circuits, R_a is selected approximately equal to R.

Fig. 13-12 (A) Op amp as a differentiator. (B) Equivalent circuit.

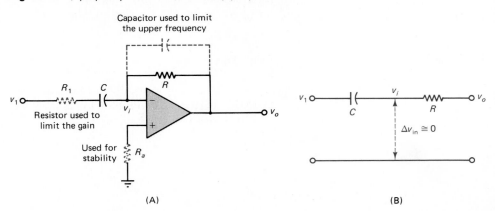

(A)

(B)

Example 13-3   Differentiator

Figure 13-13 shows an active differentiator using the NE 5512, which has a very high input impedance. The second stage is a buffer with a gain of 10X. Given the v_1 shown in Fig. 13-14(A), find v_{o2}.

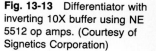

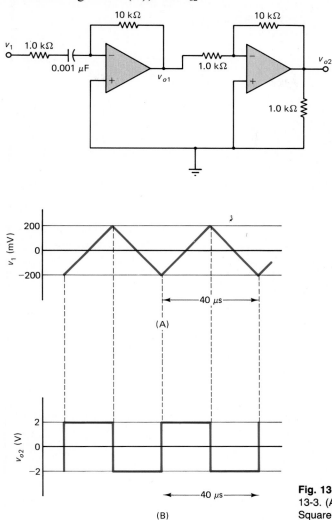

Fig. 13-13   Differentiator with inverting 10X buffer using NE 5512 op amps. (Courtesy of Signetics Corporation)

Fig. 13-14   Waveforms for Example 13-3. (A) Triangle-wave input v_1. (B) Square-wave output v_{o2}.

Solution   The input is a triangle wave, so the expected output is a square wave. From Eq. (13-10),

$$v_{o1} = (\mp 10 \ \mu s)\frac{400 \text{ mV}}{20 \ \mu s} = \mp 200 \text{ mV}$$

for the first stage. With the 10X buffer,

$$v_{o2} = \pm 2.00 \text{ V}$$

The result is shown in Fig. 13-14(B).

By straightforward analysis (see Exercise 13-9), it can be shown that the effect of R_i is to modify Eq. (13-10) to read

$$v_o = -RC \, dv_1/dt - R_1C \, dv_o/dt \qquad (13\text{-}11)$$

Since the output is a square wave, dv_o/dt is small except during transients and the correction term is negligible for the example given as far as the square-wave amplitude is concerned. ∎

13-6 ACTIVE FILTERS

Simple RC filters are encountered in Chapter 10 in the input and output sections of single-stage ac amplifiers. As is shown, these filters produce low- and high-frequency roll-offs for the amplifier and define the amplifier bandwidth. As is discussed in Chapters 10 and 11, filters can control the bandwidth and produce desired circuit properties. A wider bandwidth results in more faithful reproduction of complex input signals, whereas a narrower bandwidth can reduce the effective amplifier noise and increase the signal-to-noise ratio. The choice of bandwidth thus depends on the application.

Active filters make use of op amps in combination with resistors, capacitors, and inductors to produce frequency responses for many uses. An op amp can isolate stages of a filter from one another, buffer the output of a filter stage, increase circuit gain, and provide noise rejection. The op amp simplifies filter design and improves performance.

Figure 13-15 shows an ideal low-pass filter and a single-pole, active low-pass filter. As can be seen, the actual filter response only approximately resembles the idealized response. The op amp of Fig. 13-15(C) provides a high input impedance as

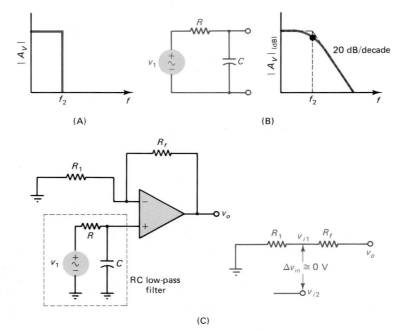

Fig. 13-15 Low-pass active filter. (A) Ideal low-pass filter. (B) RC low-pass filter with one pole. (C) Active filter with equivalent circuit.

seen by the RC filter, preventing any subsequent circuit load from affecting the performance of the filter. The inverting input of the op amp must follow the voltage on the noninverting input, so that

$$\frac{0 - v_{i2}}{R_{o1}} = \frac{v_{i2} - v_o}{R_{of}} \quad \text{and} \quad v_i \cong v_1$$

$$v_o \cong v_{i2}(1 + R_f/R_1) \tag{13-12}$$

The result is a noninverting amplifier of the type shown in Fig. 13-5, with v_{i2} a function of time.

The arrangements of Figs. 13-5 and 13-15 provide an advantageous way to apply the op amp because the input impedance of the op amp is so high. When used as an inverting amplifier, as shown in Fig. 13-1, the input impedance to the op amp circuit is about R_1 (because v_i is virtual ground).

Figure 13-16 indicates how the response of the above filter can be improved with a three-pole, active low-pass filter (called a Butterworth maximally flat filter). The actual response is much closer to the desired ideal. The derivation of the capacitor values used in this circuit is covered in courses on filter design.

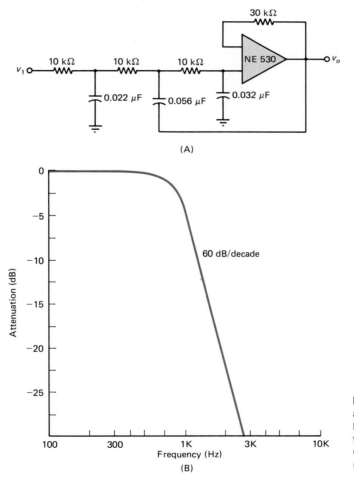

Fig. 13-16 Three-pole, active, low-pass filter with Butterworth maximally flat frequency response. (A) Circuit. (B) Frequency response.

The NE 530 is a high-slew-rate op amp that can directly replace general-purpose amplifiers such as the 741. As can be seen from Fig. 12-18, the slew rate of the NE 530 is 35 V/μs compared with 0.5 V/μs for the 741. As a result, the response of the op amp is improved.

Figures 13-17 and 13-18, respectively, show high-pass and band-pass ideal and active filters. Note that isolation is achieved between the two stages for the band-pass filter (Fig. 13-18) by use of the op amps. The performance of these circuits can be improved in each case by using multipole circuit configurations.

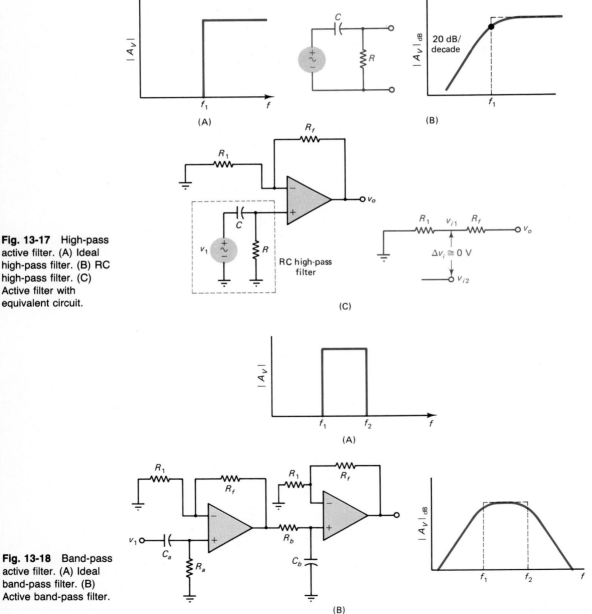

Fig. 13-17 High-pass active filter. (A) Ideal high-pass filter. (B) RC high-pass filter. (C) Active filter with equivalent circuit.

Fig. 13-18 Band-pass active filter. (A) Ideal band-pass filter. (B) Active band-pass filter.

Example 13-4   **High-Pass Active Filter**

For the high-pass active filter of Figure 13-17(C), discuss the relationship between v_1 and v_{i2} for $f \leqslant f_1$ and $f \gg f_1$.

Discussion   For $f \leqslant f_1$ (around or below cutoff),

$$v_{i2} = v_1 \frac{R}{R - jX_C}$$

The RC voltage divider provides signal attenuation and phase shift as expected. As $f \rightarrow 0$, $v_{i2} \rightarrow 0$. For $f \gg f_1$ (well above cutoff), $X_C \ll R$ and

$$v_{i2} \cong v_1$$

∎

Example 13-5   **Band-Pass Filter**

A band-pass filter of the type shown in Fig. 13-18(B) is available with

$$R_1 = 1.0 \text{ k}\Omega \qquad C_a = 0.10 \text{ }\mu\text{F} \qquad R_a = 40 \text{ k}\Omega$$

$$R_f = 10 \text{ k}\Omega \qquad C_b = 1.0 \text{ nF} \qquad R_b = 8.0 \text{ k}\Omega$$

For a given application, a high-pass filter with $f_1 < 100$ Hz and $f_2 > 10$ kHz is required. Can the available filter be used?

Solution   From the above discussion,

$$f_1 = \frac{1}{2\pi R_1 C_a} = \frac{1}{2\pi (40 \text{ k}\Omega)(0.10 \text{ }\mu\text{F})} \cong 40 \text{ Hz}$$

$$f_2 = \frac{1}{2\pi R_2 C_b} = \frac{1}{2\pi (8.0 \text{ k}\Omega)(1.0 \text{ nF})} \cong 20 \text{ kHz}$$

The available filter is acceptable.

∎

13-7   PRECISION RECTIFIERS

Op amps can improve the precision of rectifiers discussed in Chapter 5. Figures 13-19 and 13-20, respectively, show half-wave and full-wave precision rectifiers that use the high slew rate of the NE 5535 dual op amp.

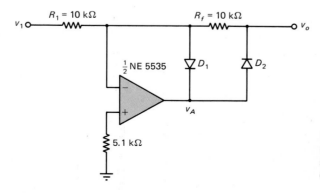

Fig. 13-19 Precision half-wave rectifier using the NE 5535 high-slew-rate dual op amp. (Courtesy of Signetics Corporation)

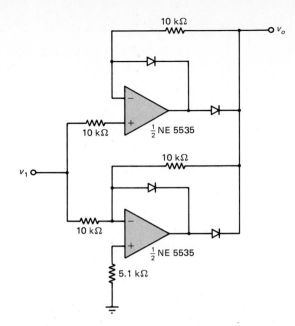

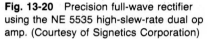

Fig. 13-20 Precision full-wave rectifier using the NE 5535 high-slew-rate dual op amp. (Courtesy of Signetics Corporation)

First consider the half-wave rectifier of Fig. 13-19. For $v_1 > 0$, $v_A < 0$. Diode D_1 is *on* and diode D_2 is *off*. No current flows through R_f, so that $v_o \cong v_{in} \cong 0$ V. The input current all flows through D_1.

For $v_1 < 0$, $v_A > 0$. Diode D_1 is *off* and diode D_2 is *on*. The currents through R_1 and R_f must be approximately the same. Given $v_{in} \cong 0$, $v_o \cong -v_1$. The result is a half-wave rectified output, with no diode drops to reduce the signal.

Two half-wave rectifier circuits can be joined, as shown in Fig. 13-20, to produce a precision full-wave rectifier. Note that the input connections are reversed for the second op amp.

13-8 VOLTAGE COMPARATOR

Figure 13-21 shows a simple voltage comparator that can use almost any operational amplifier. The two input voltages are compared and the difference produces a large v_o. For $v_o > v_z$, the zener saturates at the desired output level. For $v_o < v_z$, the zener is forward-biased so v_o is small. The voltage comparator is discussed further in Section 17-6, with reference to analog-to-digital converters.

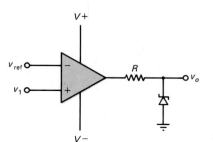

Fig. 13-21 Voltage comparator.

A useful application of the op amp is shown in Fig. 13-22. The objective is to observe how ΔR changes with an external variable (such as temperature, mechanical stress, or light intensity). Changes of the order of 1 percent in $\Delta R/R$ are typically of interest in an environment in which there is a large common-mode (noise) signal.

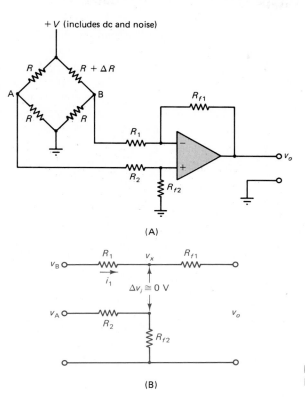

(A)

(B)

Fig. 13-22 (A) Bridge amplifier. (B) Equivalent circuit.

The circuit of Fig. 13-22 can provide the needed combination of gain and noise protection. From the figure,

$$v_A = V/2 \quad \text{and} \quad v_B = V\frac{R}{2R + \Delta R}$$

so

$$v_B - v_A = \frac{R}{2R + \Delta R} - \frac{1}{2} \cong -\frac{V}{4}\frac{\Delta R}{R} \tag{13-13}$$

From the equivalent circuit of Fig. 13-22(B),

$$\frac{v_A - v_x}{R_2} = \frac{v_x}{R_{f2}}$$

$$\frac{v_B - v_x}{R_1} = \frac{v_x - v_o}{R_{f1}}$$

so that

$$v_o = -\frac{R_{f1}}{R_1}(v_B - v_A) \tag{13-14}$$

Combining Eqs. (13-13) and (13-14),

$$v_o \cong \frac{V}{4} \frac{R_{f1}}{R_1} \frac{\Delta R}{R} \tag{13-15}$$

which is the desired result. To the degree that the same noise signal appears at A and B, the noise does not appear on the output of the op amp.

A variation of the bridge amplifier can create a tracking circuit, as shown in Fig. 13-23. D_1 and D_2 are two photosensitive resistors. If the antenna is pointed directly at the source, $v_A = v_B$, and $v_o = 0$. If the antenna is not pointed directly at the source, D_1 and D_2 are not the same; an output signal proportional to $v_B - v_A$ is developed and can be used to drive the antenna motor.

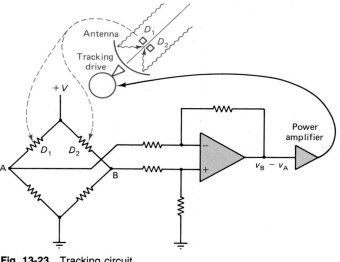

Fig. 13-23 Tracking circuit.

Example 13-6   **Bridge Amplifier**

For a bridge amplifier, $\Delta R = 0.1\ \Omega$ is observed in a 50-Ω resistor. The desired amplifier output is 10 mV. What ratio R_{f1}/R_1 should be used to set the amplifier gain for a 1.0-V supply?

Solution   From Eq. (13-15),

$$10 \times 10^{-3}\text{V} = \frac{1}{4}\frac{R_{f1}}{R_1}\frac{0.10}{50}(1.0\ \text{V})$$

$$R_{f1}/R_1 = 20$$ ∎

The various configurations introduced above can be used to create analog computers to solve a variety of problems. Figure 13-24 indicates how several operational amplifiers can be used to solve for $i(t)$ in a series resonant circuit, given an arbitrary driving function $v(t)$.

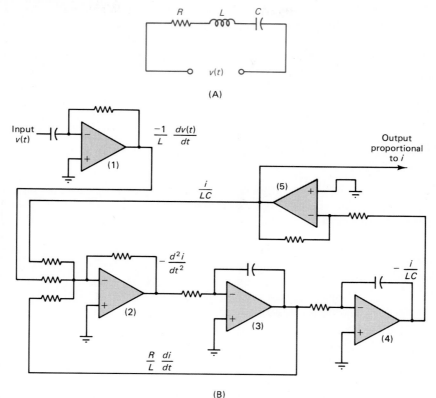

(A)

(B)

Fig. 13-24 Analog computer solution to find $i(t)$. (A) Circuit to be analyzed. (B) Analog computer.

The circuit of Fig. 13-24(A) must satisfy the differential equation

$$\frac{d}{dt}(v_R + v_L + v_C) = \frac{d}{dt}v(t)$$

$$R\frac{di}{dt} + L\frac{d^2i}{dt^2} + \frac{i}{C} = \frac{dv(t)}{dt}$$

or

$$\frac{d^2i}{dt^2} + \frac{R}{L}\frac{di}{dt} + \frac{i}{LC} = \frac{1}{L}\frac{dv(t)}{dt} \tag{13-16}$$

Solving for the second-order term,

$$-\frac{d^2i}{dt^2} = \frac{R}{L}\frac{di}{dt} + \frac{i}{LC} - \frac{1}{L}\frac{dv(t)}{dt} \tag{13-17}$$

Operational amplifier (2) performs this summation if the three terms can be placed on the inputs. The term

$$-\frac{1}{L}\frac{dv(t)}{dt}$$

is obtained by differentiating the signal $v(t)$ [operational amplifier (1)]. To obtain the other two terms, the output of operational amplifier (2) is integrated once [via amplifier (3)] to obtain the term in di/dt and again [amplifier (4)] to obtain the term in i. By feeding back the outputs of amplifiers (3) and (4), the function is satisfied.

Analog computers were widely used in electronics in the 1950s and 1960s, then began to decline in general use as they were replaced by digital systems (discussed in Chapters 15 to 17). For some specialized applications, however, analog computers can still provide a preferred design alternative (Fitzgerald, 1987).

13-11 LINEAR IC APPLICATIONS

The IC operational amplifier is perhaps the most widely used analog circuit in electronics, with millions sold each year. The operational amplifier makes use of the negative feedback concept traced through earlier chapters, while adding new features to produce a highly versatile circuit. Years of intense design activity have resulted in a variety of different IC configurations for a range of potential applications. Specialized features have evolved to produce a highly adaptable family of circuits. Only a few potential applications have been presented here because a full understanding of IC op-amp design and uses constitutes an entire text.

There are also many other types of linear ICs available. Typically, these address common linear-circuit applications and they are a major resource for the electronics designer.

Many of the linear ICs available are designed for use in communication systems. Figure 13-25 shows the TDA 1522 IC intended for use as a playback amplifier for

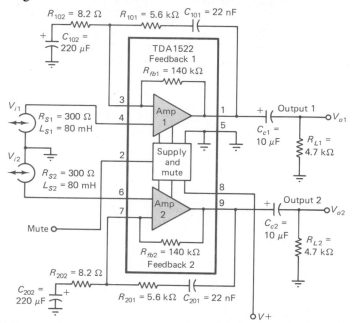

Fig. 13-25 TDA 1522 playback amplifier. (Courtesy of Signetics Corporation)

car radio and cassette players. Note that the IC includes two amplifiers, a supply and mute circuit, and two feedback resistors. Figure 13-26 illustrates the two sides of the printed circuit board that can be used with the IC. The IC can complete the preamplifier circuit with a minimum number of external components.

This circuit is representative of the many linear ICs that are available today. These units can be treated as composite devices and used in producing complex systems.

However, extreme care must be observed in developing IC-based circuits. Data sheets can be complex and a complete understanding of specifications and their implications is required for satisfactory application.

Fig. 13-26 Printed circuit board for the TDA 1522 playback amplifier (75 mm × 65 mm). (A) Component side showing layout. (B) Interconnection side. (Courtesy of Signetics Corporation)

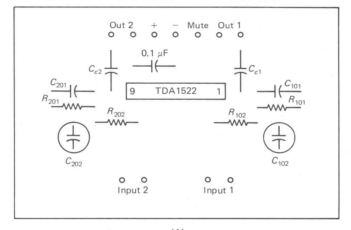

(A)

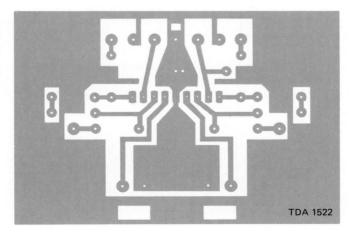

(B)

Questions
13-1. Why is the concept of virtual ground useful in studying op-amp applications?

13-2. What are the advantages of a quad IC op amp?

13-3. In what ways can the inverting and noninverting amplifier applications of the op amp be contrasted?

13-4. Under what circumstances might the unity follower be a useful circuit?

13-5. How can you obtain a summing amplifier without inversion?

13-6. What are the functions of the switch S and resistor R_a in Fig. 13-11?

13-7. What are the functions of the capacitor C_f, resistor R_i, and resistor R_a in Fig. 13-12?

13-8. What is the purpose of the op amp in the low-pass active filter of Fig. 13-15?

13-9. How is the number of poles in a low-pass filter related to the slope of the high-frequency roll-off?

13-10. How can the precision rectifier of Figs. 13-19 and 13-20 be useful in instrumentation design?

Answers
13-1. The virtual ground concept enables the circuit to be analyzed by assuming that with negative feedback present the voltage across the positive (+) and negative (−) terminals of the op amp is very small. If the positive terminal is ground, the negative terminal becomes virtual ground. If the positive terminal is set at a fixed voltage, the negative terminal takes on approximately the same value. For these cases, the ability to initially assume a value for the voltage present on the negative terminal of the op amp simplifies circuit analysis and understanding.

13-2. For a quad IC op amp, the integrated circuit is fabricated with four op amps on the same chip, producing a single high-density package with matched op-amp parameters and (almost) identical responses to external environmental effects.

13-3. For the inverting amplifier (Fig. 13-1), the gain is given by $-R_f/R_1$ and the input resistance to the circuit is approximately R_1. For the noninverting amplifier (Fig. 13-5), the gain is $1 + R_f/R_1$ and the input resistance to the circuit is very high (given by the input resistance of the op amp itself.) The choice of circuit configuration depends on whether signal inversion is desired, on whether the magnitude of the gain must be less than unity, and on constraints regarding the loading of the prior stage or input signal source.

13-4. The unity follower provides a method to achieve isolation between an input signal and an output load without changing the characteristics of the input waveform (within the bandwidth of the circuit.) The high input impedance of the op amp prevents loading of the input source, whereas the low output impedance prevents the load from affecting output-signal properties (for load resistances much greater than the output resistance of the op amp).

13-5. A summing amplifier without inversion can be obtained by combining the summing circuit of Fig. 13-8 with a unity follower of Fig. 13-7, as shown in Fig. 13-27. The use of the unity follower in series with one or more input terminals enables the summing am-

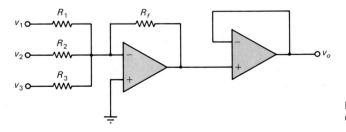

Fig. 13-27 Solution for Question 13-5.

plifier to be modified into a combined summing/subtracting amplifier, providing any desired combination of input signs.

13-6. The output v_o of the integrator is dependent on the initial voltage on the capacitor C. If switch S is closed, then the initial condition $v_o(0) = 0$ is obtained. Resistor R_a is added to correct for differing input bias currents.

13-7. Capacitor C_f of the differentiator is used to limit the upper-frequency components present in the output (by introducing integrator effects) to prevent oscillation in the output. Resistor R_i limits the gain of the circuit, so that with a rapidly changing input, the op amp does not saturate. R_a is again used to balance the effects of the input bias current.

13-8. The op amp is used in the low-pass filter to obtain isolation between the output of the RC filter itself and the subsequent load. The gain of this buffer stage can be set as desired by adjusting the ratio R_f/R_1.

13-9. From Chapter 10, each pole contributes a 20-dB/decade roll-off at high frequencies for a low-pass filter. The circuit of Fig. 13-15 thus rolls off at 20 dB/decade, whereas the circuit of Fig. 13-16 rolls off at 60 dB/decade.

13-10. In instrumentation design, it is often necessary to measure very accurately the effective value of an ac waveform. An ac voltmeter requires such a measurement. Precision rectifiers in combination with an integrating circuit can be used to provide the desired output.

EXERCISES AND SOLUTIONS

Exercises
(13-2)

13-1. For an inverting amplifier, $R_f = 10$ kΩ and $R_1 = 5.0$ kΩ. Find the relationship between v_o and v_1.

(13-2)

13-2. As discussed in Chapter 12, the maximum input current that can flow into an op amp is called the *input bias current*. To prevent significant error, this current must be small with respect to the currents through the external circuitry. For an inverting amplifier configuration, $v_1 = 1.0$ V. (a) For the 741 op amp (from Fig. 12-18), what is the maximum value of R_1 that can be used so that $I_{bias}R_1 < (1.0 \times 10^{-3})v_1$? The input bias current thus limits the maximum resistance that can be used in the external feedback circuit. (b) What design factor enters in to limit the lower value of R_1?

(13-2)

13-3. Find an expression for Z_i for the circuit of Fig. 13-28, using a nonideal operational amplifier.

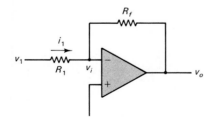

Fig. 13-28 Circuit for Exercise 13-3.

(13-3)

13-4. A noninverting amplifier is to be applied with a gain of 1.5. If $R_1 = 3.0$ kΩ, what value of R_f should be used?

(13-4)

13-5. Find v_o for the circuit of Fig. 13-29 (p. 536).

(13-4)

13-6. Find v_o for the circuit of Fig. 13-30 (p. 536).

(13-4)

13-7. Find V_0 for the circuits of Fig. 13-31 (p. 536). What problems might arise with the circuit of Fig. 13-31(A)?

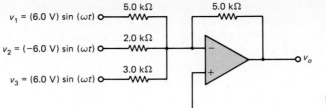

$v_1 = (6.0\ V)\sin(\omega t)$

$5.0\ k\Omega$

$5.0\ k\Omega$

$v_2 = (-6.0\ V)\sin(\omega t)$

$2.0\ k\Omega$

$v_3 = (6.0\ V)\sin(\omega t)$

$3.0\ k\Omega$

v_o

Fig. 13-29 Circuit for Exercise 13-5.

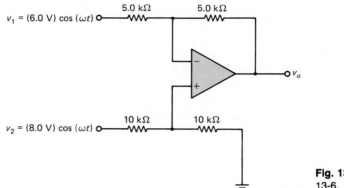

$v_1 = (6.0\ V)\cos(\omega t)$

$5.0\ k\Omega$

$5.0\ k\Omega$

v_o

$v_2 = (8.0\ V)\cos(\omega t)$

$10\ k\Omega$

$10\ k\Omega$

Fig. 13-30 Circuit for Exercise 13-6.

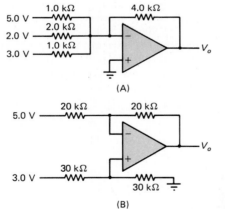

5.0 V

$1.0\ k\Omega$

$4.0\ k\Omega$

2.0 V

$2.0\ k\Omega$

3.0 V

$1.0\ k\Omega$

V_o

(A)

5.0 V

$20\ k\Omega$

$20\ k\Omega$

V_o

3.0 V

$30\ k\Omega$

$30\ k\Omega$

(B)

Fig. 13-31 Circuits for Exercise 13-7.

(13-2) (13-3)
(13-4)

13-8. Draw the schematics for operational amplifier circuits with the following characteristics: (a) $v_o = -2v_1$ and (b) $v_o = -6v_1 + 8v_2 - 3v_3$.

(13-5)

13-9. Show that R_1 in Fig. 13-12 produces the result of Eq. (13-11).

(13-6)

13-10. The low-pass active filter in Fig. 13-15 is designed with $R_f = 5.0\ k\Omega$, $R_1 = 2.0\ k\Omega$, $R = 5.0\ k\Omega$, and $C = 0.001\ \mu F$. If $v_1 = 0.10\ V$, find v_o at frequencies for which $f \ll f_2$.

(13-6)

13-11. Redraw the frequency-response characteristic for the 3-pole low-pass active filter of Fig. 13-16 using semilog scales to compare it with the ideal response for such a filter.

(13-7)

13-12. What factors limit the accuracy of the precision rectifier, preventing ideal operation?

(13-9) **13-13.** For a bridge amplifier, $\Delta R/R = 0.1$ percent. If $(v_o)_{max}/V = 0.10$ and $R_1 > 10$ kΩ to limit current flow from the source, what is the largest feedback resistor R_{f1} that can be used?

13-14. Figure 13-32 shows the schematic for a logarithmic amplifier. Prove that

$$v_o \cong v_T \ln(v_1/V_{ref})$$
$$V_{ref} = 1/I_0 R_i$$

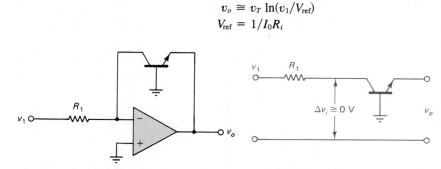

Fig. 13-32 Logarithmic amplifier with equivalent circuit for Exercise 13-14.

13-15. Voltage regulators, introduced in Chapter 5, are a necessary part of a high-performance power supply. Discuss how an IC op amp can be used to produce a simple circuit that functions as a voltage regulator.

Solutions **13-1.** From Eq. (13-3),

$$v_o/v_1 = -R_f/R_1 = -10 \text{ k}\Omega/5.0 \text{ k}\Omega = -2.0$$

13-2.(a) Accurate op-amp performance requires that

$$I_{bias}R_1 < (1.0 \times 10^{-3})v_1$$

From Fig. 12-18, the input-bias current for the 741 op amp is 80 nA. Therefore, for $v_1 = 1.0$ V,

$$R_1 < \frac{1.0 \times 10^{-3} \text{ V}}{80 \times 10^{-9} \text{ A}} \cong 1.3 \times 10^4 \text{ } \Omega$$

(b) As R_1 is reduced in value, the output current of the op amp must increase. The magnitude of the voltage drop across Z_o with respect to v_o and limitations on the output current place a lower limit on R_1.

13-3. For the circuit of Fig. 13-28,

$$\frac{v_1 - v_i}{R_1} = \frac{v_i}{Z_i} + \frac{v_i - v_o}{R_f}$$

so that

$$\frac{v_1}{R_1} = \left(\frac{1}{R_1} + \frac{1}{Z_i} + \frac{1}{R_f}\right)v_i - \frac{v_o}{R_f}$$

From Eq. (12-34),

$$v_o = v_i \frac{Z_o + A_V R_f}{Z_o + R_f}$$

so that

$$\frac{v_1}{R_1} = \left(\frac{1}{R_1} + \frac{1}{Z_i} + \frac{1}{R_f}\right)v_i - \frac{v_i}{R_f}\frac{Z_o + A_V R_f}{Z_o + R_f}$$

and

$$\frac{v_i}{v_1} = \frac{1}{1 + \dfrac{R_1}{Z_i} + \dfrac{R_1}{R_f} - \dfrac{R_1}{R_f}\dfrac{Z_o + A_V R_f}{Z_o + R_f}}$$

Given

$$i_1 = \frac{v_1 - v_i}{R_1}$$

$$Z_{in} = \frac{v_1}{i_1} = \frac{v_1}{v_1 - v_i}R_1$$

$$= \frac{R_1}{1 - v_i/v_1}$$

$$= \frac{R_1}{1 - \dfrac{1}{1 + \dfrac{R_1}{Z_i} + \dfrac{R_1}{R_f} - \dfrac{R_1}{R_f}\dfrac{Z_o + A_V R_f}{Z_o + R_f}}}$$

Observe that as A_V becomes very large, Z_{in} approaches R_1.

13-4. From Eq. (13-4),

$$\frac{v_o}{v_1} = 1 + \frac{R_f}{R_1}$$

so that

$$R_f = R_1(v_o/v_1 - 1)$$

$$= 3.0 \text{ k}\Omega(1.5 - 1) = 1.5 \text{ k}\Omega$$

13-5. From Eq. (13-6),

$$v_o = -\left(\frac{R_f}{R_1}v_1 + \frac{R_f}{R_2}v_2 + \frac{R_f}{R_3}v_3\right)$$

Using the values from Fig. 13-29,

$$v_o = -\left[\frac{5 \text{ k}\Omega}{5 \text{ k}\Omega}(6 \text{ V}) + \frac{5 \text{ k}\Omega}{2 \text{ k}\Omega}(-6 \text{ V}) + \frac{5 \text{ k}\Omega}{3 \text{ k}\Omega}(3 \text{ V})\right]\sin(\omega t)$$

$$= (4 \text{ V})\sin(\omega t)$$

13-6. By comparing Figs. 13-30 and 13-9, Eq. (13-7) can be used:

$$v_o = v_2 - v_1$$

$$= (8 \text{ V})\cos(\omega t) - (6 \text{ V})\cos(\omega t)$$

$$= (2 \text{ V})\cos(\omega t)$$

13-7. (a) For the summing amplifier shown in Fig. 13-31(A),

$$v_o = -\left[\frac{4.0 \text{ k}\Omega}{1.0 \text{ k}\Omega}(5.0 \text{ V}) + \frac{4.0 \text{ k}\Omega}{2.0 \text{ k}\Omega}(2.0 \text{ V}) + \frac{4.0 \text{ k}\Omega}{1.0 \text{ k}\Omega}(3.0 \text{ V})\right]$$

$$= -36 \text{ V}$$

This circuit will probably not operate as desired because most op amps saturate for voltages much less than 36 V. The output is close to the negative rail voltage.

(b) For the subtracting amplifier shown in Fig. 13-31(B),

$$v_o = 3 \text{ V} - 5 \text{ V} = -2 \text{ V}$$

13-8. Possible solutions for the desired circuits are shown in Fig. 13-33.

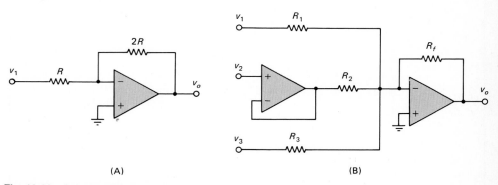

(A)   (B)

Fig. 13-33 Solutions for Exercise 13-8.

(a) For the circuit of Fig. 13-33(A), by choosing $R_f = 2R$ and $R_1 = R$, the relationship

$$v_o = -\frac{R_f}{R_1}v_1 = -2v_1$$

is obtained.

(b) For the circuit of Fig. 13-33(B) and Eq. (13-6),

$$v_o = -\left[\frac{R_f}{R_1}(v_1) - \frac{R_f}{R_2}(-v_2) - \frac{R_f}{R_3}(v_3)\right]$$

The desired result is obtained if

$$R_f/R_1 = 6, \qquad R_f/R_2 = 8 \qquad \text{and} \qquad R_f/R_1 = 3$$

13-9. From Fig. 13-34,

$$i = C\frac{dv_C}{dt} = \frac{0 - v_o}{R}$$

$$v_C = v_1 - iR_1$$

By substitution,

$$C\frac{d}{dt}(v_1 - iR_1) = -\frac{v_0}{R}$$

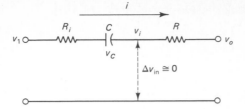

Fig. 13-34 Equivalent circuit for Exercise 13-9.

and

$$v_o = -RC\frac{dv_1}{dt} - R_1C\,dv_o/dt$$

as given in Eq. (13-11).

13-10. From Fig. 13-15,

$$f_2 = \frac{1}{2\pi RC} = \frac{1}{2\pi\,(5.0\ \text{k}\Omega)(0.001\ \mu\text{F})} \cong 32\ \text{kHz}$$

and the gain of the inverting amplifier is

$$v_o/v_1 = -\frac{R_f}{R_1} = -5.0\ \text{k}\Omega/2.0\ \text{k}\Omega = -2.5$$

If $\mathbf{v}_1 = 0.10$ V,

$$\mathbf{v}_o = (2.5)(0.10\ \text{V}) = 0.25\ \text{V}$$

in the pass band with $f \ll f_2$.

13-11. The results are shown in Fig. 13-35. Note that the dB values in Fig. 13-16 must be converted to ratios before the graph is prepared.

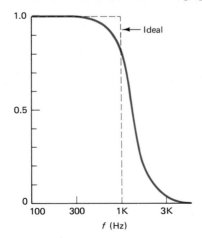

Fig. 13-35 Solution for Exercise 13-11.

13-12. The precision rectifier operates in an ideal way if $v_{in} = 0$ V, $R_{in} = \infty$, the wire from diode D_1 to input v_{in} has zero resistance, and diode D_2 has no leakage current. Actual values of all these parameters prevent the precision rectifier from operating in an ideal way.

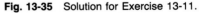

13-13. From Eq. (13-15),

$$v_o \cong \frac{V}{4} \frac{R_{f1}}{R_1} \frac{\Delta R}{R}$$

For the condition given,

$$0.10 > \frac{1}{4} \frac{R_{f1}}{10 \text{ k}\Omega} (0.001)$$

so that

$$R_{f1} < \frac{4(0.10)(10 \text{ k}\Omega)}{0.001} = 4 \text{ M}\Omega$$

13-14. For the BJT in Fig. 13-32,

$$I_E \cong I_0 \exp (V_{BE}/V_0)$$

With $v_i \cong 0$ V and $V_{BE} = v_o$,

$$\frac{v_1 - 0}{R_1} \cong I_0 \exp (v_o/V_0)$$

and

$$v_o \cong V_0 \ln \left(\frac{v_1}{V_{\text{ref}}}\right) \qquad \text{and} \qquad V_{\text{ref}} = I_0 R_1$$

The output is thus proportional to the natural logarithm of the input.

13-15. The simple circuit of Fig. 13-36 functions as a voltage regulator. The op amp is used as a noninverting amplifier comparing voltage V_A with the (temperature-compensated) voltage drop across a zener diode. From the figure,

$$(V_{\text{ref}} - V_A)A_V - V_{BE} = V_{\text{out}}$$

$$V_A = V_{\text{out}} \frac{R_1}{R_1 + R_f}$$

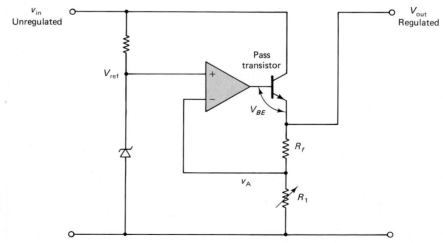

Fig. 13-36 Op amp used as a voltage regulator, solution for Exercise 13-15.

so that

$$V_{\text{out}} \cong \frac{V_{\text{ref}} A_V - V_{BE}}{1 + \dfrac{R_1}{R_1 + R_f} A_V}$$

For $A_V \gg 1$,

$$V_{\text{out}} \cong V_{\text{ref}} (1 + R_f/R_1)$$

which has the same form as Eq. (5-24) obtained for the UC150 op amp. The regulated output voltage is independent of v_{in} and the load resistor and can be adjusted by varying R_1.

PROBLEMS

(13-2) **13-1.** An inverting amplifier is to be used with a gain of -2.5. Given $R_1 = 5.6$ kΩ, find R_f.

(13-2) **13-2.** The circuit of Fig. 13-28 is applied with $R_f = 5.000$ kΩ and $R_1 = 2.000$ kΩ. If $A_V = -1.000 \times 10^4$, and the real input and output impedance are $Z_i = 100.0$ kΩ and $Z_o = 50.00$ kΩ, respectively, find Z_{in} to four significant figures. Compare this with the approximate value $Z_{\text{in}} \cong R_1$.

(13-2) **13-3.** For the inverting amplifier of Fig. 13-2, $A_V = 1000$, $R_f/R_1 = 5.00$, $v_x = 5.000$ V, and $v_i < 1.0$ V. What percentage accuracy is associated with the assumption $v_y \cong v_x$?

(13-3) **13-4.** A noninverting amplifier is applied with a gain of 5.0. If $R_f = 10$ kΩ, what value of R_1 should be used?

(13-3) **13-5.** An inverting and noninverting amplifier are arranged so that the output of the inverting amplifier is the input to the noninverting amplifier. If $R_f = 25$ kΩ and $R_1 = 10$ kΩ for each portion of the circuit, what is the overall system gain?

(13-4) **13-6.** A summing amplifier is operated with $V_1 = 4.0$ V, $V_2 = -6.0$ V, and $V_3 = 2.5$ V. If $R_f = 8.5$ kΩ and $R_1 = R_2 = R_3 = 2.5$ kΩ, find V_0.

(13-4) **13-7.** Show two different op-amp arrangements that satisfy the equation:

$$v_o = -v_1 + 6v_2$$

(13-4) **13-8.** Find v_o for the circuit of Fig. 13-37.

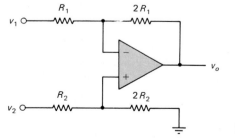

Fig. 13-37 Circuit for Problem 13-8.

(13-4) **13-9.** Find v_o for the circuit of Fig. 13-38.

(13-5) **13-10.** For an integrator circuit, $v_1 = (1 \text{ V}) \sin (\omega t)$. If $R = 5.0$ kΩ and $C = 1.0 \times 10^3$ pF, find v_o at $\omega t = \pi/2$ if $v_o(0) = 0$, $\omega = 1.0$ MHz.

(13-5) **13-11.** For a differentiator circuit, $v_1 = (1 \text{V}) \sin (\omega t)$. If $R = 5.0$ kΩ and $C = 1.0 \times 10^3$ pF, find v_o at $\omega t = \pi/2$.

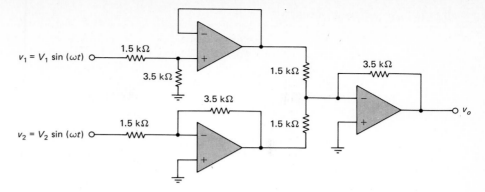

$v_1 = V_1 \sin (\omega t)$

1.5 kΩ

3.5 kΩ

3.5 kΩ

1.5 kΩ

3.5 kΩ

3.5 kΩ

1.5 kΩ

1.5 kΩ

$v_2 = V_2 \sin (\omega t)$

v_o

Fig. 13-38 Circuit for Problem 13-9.

(13-6) **13-12.** The low-pass active filter of Fig. 13-15 is designed with $f_2 = 100$ kHz. If $R_f = 5.0$ kΩ and $R_1 = 10$ kΩ, find the gain v_o/v_1 of the circuit at 200 kHz.

(13-6) **13-13.** Plot the amplitude frequency-response characteristics for one-, two-, and three-pole low-pass active filters (consisting of identical stages of the type shown in Fig. 13-15) using semilog scales and compare with the ideal characteristics for a low-pass filter.

(13-6) **13-14.** The high-pass active filter of Fig. 13-17 is designed with $R_f = 5.0$ kΩ, $R_1 = 2.0$ kΩ, $R = 50$ kΩ, and $C = 1.0 \times 10^3$ pF. If $v_1 = 0.10$ V, find f_1 and v_o at $f = 0.5f_1$.

(13-6) **13-15.** Plot the frequency-response characteristics for one-, two-, and three-pole high-pass active filters (consisting of identical stages of the type shown in Fig. 13-17) using semilog scales and compare with the ideal characteristics for a high-pass filter.

(13-7) **13-16.** (a) Plot the frequency-response characteristics for the circuit of Fig. 13-18 using semilog scales. (b) Compare with the ideal shape for a band-pass filter.

(13-9) **13-17.** A bridge amplifier is designed with $R = 100$ Ω for all four values at $T = 300°$K. If the *TCR* for one resistor is 1.00×10^3 ppm/°K and 1.00×10^2 ppm/°K for the others, what output is observed when the temperature increases to $400°$K? Assume $R_{f1}/R_1 = 10$ and $V = 1.0$ V.

(13-10) **13-18.** An analog computer circuit is to be designed to solve the differential equation

$$A \, d^2v/dt^2 + B \, dv/dt = 0$$

Individual component costs are

$$\text{IC operational amplifier} = \$1.00$$

$$\text{resistor} = \$.20$$

$$\text{capacitor} = \$.50$$

Select a possible configuration and determine the cost for the circuit.

COMPUTER APPLICATIONS

* **13-1.** Figure 13-16(A) shows the schematic for a three-pole, active low-pass filter. Write a computer program that calculates v_o as a function of v_i for an arbitrary frequency f. Then use this program to produce the expected amplitude and phase responses for the circuit. Compare the amplitude response with Fig. 13-16(B). Turn in your program listing, graphs, and a discussion of your findings.

* **13-2.** Use SPICE to simulate the circuit of Fig. 13-32 (represent the op amp by an idealized equivalent circuit). Does the circuit have the expected properties?

13-1. The objective of this experiment is to use the 741 op amp to demonstrate the operation of several types of circuits, including inverting and noninverting amplifiers, a unity follower, a summing amplifier, a subtracting amplifier, a differentiator, and an integrator. Design an experimental procedure to use in implementing and evaluating each type of application. Construct your circuits and evaluate their performance. Are the outputs as expected for each case? Discuss your findings.

Discussion
When used with the input of Fig. 13-39(A), the circuit of Fig. 13-39(B) produces the waveform shown in Fig. 13-39(C). Figure 13-40(A) shows a differentiating amplifier and Figs. 13-40(B) to (D) show input and output waveforms obtained from the amplifier. From the above results, discuss the conclusions that can be drawn regarding the op amp being used.

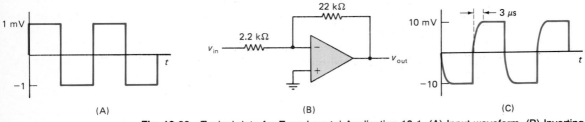

Fig. 13-39 Typical data for Experimental Application 13-1. (A) Input waveform. (B) Inverting amplifier. (C) Output waveform.

Fig. 13-40 Typical data for a differentiating amplifier. (A) Differentiator. (B) Triangle-wave input and resulting output.

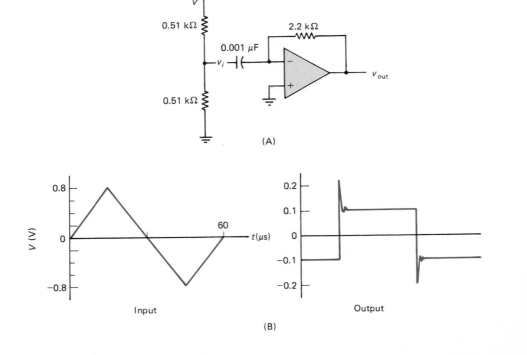

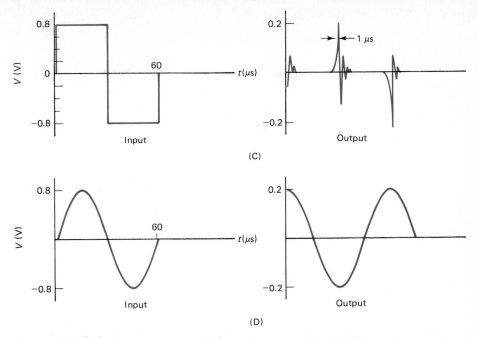

Fig. 13-40 (cont.) (C) Square-wave input and resulting output. (D) Sine-wave input and resulting output.

13-2. The objective of this experiment is to construct and test the three-pole active filter shown in Fig. 13-16 and discussed in Computer Application 13-1. Determine the amplitude and phase responses for the circuit and construct a Bode plot. Compare the theoretical and experimental results and discuss your findings.

13-3. Figure 13-41 shows an op-amp circuit that may be used as a capacitor multiplier. By drawing on the properties of a series resonant circuit, design and conduct an experiment to confirm the properties of this circuit. (Refer to Problem 3-41 in Chapter 3 and Appendix 3 for helpful information.)

Fig. 13-41 Use of an op amp circuit as a capacitor multiplier. (A) Schematic. (B) Equivalent circuit. (C) Use of a JFET in place of R_3.

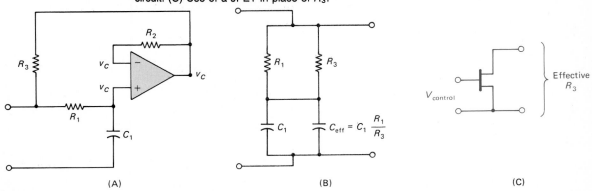

Figure 13-41(A) shows how an op-amp circuit can be used to create a capacitor multiplier. The output voltage of the op amp is forced to be equal to the voltage across the capacitor. Therefore, when a voltage v_{in} is applied to the circuit, the current through R_1 is

$$i_1 = \frac{v_{in} - v_C}{R_1}$$

and the current through R_3 is

$$i_3 = \frac{v_{in} - v_C}{R_3}$$

By definition

$$i_1 = C_1 \frac{dv_C}{dt}$$

$$i_3 = C_{eff} \frac{dv_C}{dt}$$

and by substitution and division,

$$C_{eff} = C_1 \frac{R_1}{R_3}$$

The equivalent circuit of Fig. 13-41(B) represents the circuit of Fig. 13-41(A).

This circuit is useful in situations where a large capacitance is required but circuit design constraints prevent the use of an appropriate component. Observe that by replacing R_3 by a (symmetric) JFET as shown in Fig. 13-41(C), a control voltage can be used to adjust R_3 and thus to vary C_{eff}. The result is a *voltage-controlled capacitance* that is similar in function to the varactor introduced in Section 4-6.

13-4. Figure 13-42 shows an op-amp circuit that can be used as an effective inductance. By drawing on the properties of a series resonant circuit, design and conduct an experiment to confirm the properties of this circuit.

Fig. 13-42 Use of an op amp circuit as an effective inductor. (A) Schematic. (B) Equivalent circuit.

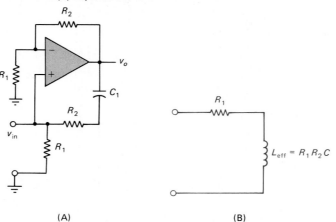

(A)   (B)

Discussion

Figure 13-42(A) shows how an op-amp circuit using resistors and a capacitor can be used to simulate an inductor. To understand the circuit shown, consider the transient response that is observed when a step-function input voltage v_{in} is applied. A current v_{in}/R_1 flows through R_1. The op amp produces an output voltage

$$v_o = v_{in}\left(1 + \frac{R_2}{R_1}\right)$$

so that a current

$$\frac{v_o - v_{in}}{R_2} = \frac{v_{in}\left(1 + \frac{R_2}{R_1}\right) - v_{in}}{R_2} = \frac{v_{in}}{R_1}$$

flows through R_2 at time $t = 0^+$ (immediately after application of the step function). Since the currents through R_1 and R_2 are equal, no input current flows into the circuit ($i_{in} = 0$).

As the capacitor charges up, the current through R_2 decreases, allowing the current i_{in} to increase until the current through R_2 tends to zero and i_{in} approaches v_{in}/R_1. The input current i_{in} is thus given by

$$i_{in} = \frac{v_{in}}{R_1}(1 - e^{-t/R_2 C})$$

The equivalent circuit of Fig. 13-42(B) responds to a step-function v_{in} with the transient

$$i_{in} = \frac{v_{in}}{R_1}(1 - e^{-tR_1/L})$$

The two equations are identical if

$$L = R_1 R_2 C$$

Thus the circuit of Fig. 13-42(A) has the equivalent circuit of Fig. 13-42(B) with the effective inductance given above.

This circuit is useful in situations where an inductive circuit component is needed, but an actual inductor is difficult to include in the circuit. Also, note that by replacing R_1 with a (symmetric) JFET, a control voltage can be used to adjust R_1 and thus to vary L. The result is a *voltage-controlled inductor*.

CHAPTER 14
SINE-WAVE OSCILLATORS AND APPLICATIONS

The fundamental properties of sine-wave oscillators are discussed in this chapter. These oscillators are then applied to create communication systems. Oscillator circuits extend the feedback concepts that have been discussed throughout the text. Positive feedback, introduced here, leads to new ways for understanding the interactions among various parts of complex circuits. Several different types of oscillators are analyzed in detail as a way to expand on the use of modeling techniques and to illustrate how many different options are available to the designer. As will be seen, the tradeoffs among oscillator types depend on both external device parameters and the nonideal properties of the elements used in the system.

Feedback has been previously encountered in several different circuits. Chapter 7 introduces negative feedback through an emitter resistor R_E used to provide temperature stabilization for the BJT common-emitter dc-coupled amplifier. Part of the amplifier gain is sacrificed to produce the desired temperature behavior.

In Chapter 9, the study of ac amplifiers is pursued. With a bypass capacitor C_E placed in parallel with R_E, the (midrange) ac negative feedback is eliminated while the dc negative feedback is maintained for temperature stabilization. However, circuit gain remains sensitive to the h_{fe} of the BJT. With C_E removed, ac negative feedback returns, reducing both amplifier gain and circuit sensitivity to BJT parameters. Chapter 11 introduces a more general model of feedback and shows that negative feedback is associated with an increase in amplifier bandwidth.

In Chapters 12 to 13, negative feedback is used for applications associated with operational amplifiers. The feedback circuits used with the operational amplifiers depend on current summing at the input node. All of the above discussions are concerned with negative feedback, having the feedback signal out of phase with the input signal.

In this chapter, these studies are extended in order to consider the applications of positive feedback. Positive feedback concepts are applied to the design of several different sine-wave oscillator circuits.

The chapter includes an application of the frequency-response methods of Chapter 10 to probe how distributed capacitance can result in oscillation when it is not expected. The materials used in a circuit and the integration technology determine whether spurious oscillations can develop in an amplifier. The problems associated with unwanted oscillation again illustrate how circuit design must extend beyond idealized devices and circuit configurations.

14-1 FEEDBACK MODELS

Two possible forms of amplifier feedback are shown in Fig. 14-1. Figure 14-1(A) illustrates the *current sample–series feedback* approach that can be implemented through the use of an emitter resistor R_E (as discussed in Section 9-8). Figure 14-1(B) illustrates the *voltage sample–parallel feedback* approach that is used with operational amplifiers (as discussed in Chapter 11). Two other forms of feedback can be developed using *current sample–parallel feedback* and *voltage sample–series feedback*.

For all of these cases, the amplifier gain with feedback can be written in the general form:

$$A_{Vf}^* = \frac{A_V^*}{1 - A_V^* B} \tag{14-1}$$

where definition on the feedback factor B varies with the type of feedback present.

Positive feedback is obtained when $Re[A_V^* B]$ is positive. For the critical case for which $A_V^* B$ approaches the value $+1$, A_{Vf}^* goes to infinity, according to Eq. (14-1). As is always the case when nonphysical results are obtained, the model described by this equation is no longer valid. As discussed in Chapter 1, care must be taken to be aware at all times of model limitations. As $Re[A_V^* B]$ tends to $+1$, the performance of any real circuit becomes nonlinear, preventing the possibility of an infinite gain. The behavior of positive feedback circuits is studied below for this limiting case.

Fig. 14-1 Feedback models. (A) Current sample-series feedback. (B) Voltage sample-parallel feedback.

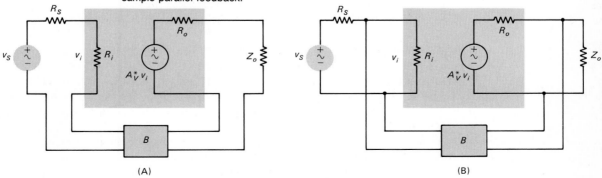

(A)   (B)

To understand what happens to a circuit with positive feedback as $A_v^* B \rightarrow 1$, remember that noise is present at the input of every amplifier circuit (see Section 11-7). This noise appears in amplified form at the output of the amplifier. The feedback circuit sends back a fraction of this output to the input.

Assume for the moment that positive feedback is obtained over the midrange frequencies of an amplifier, with the frequency-independent phase shift of the amplifier over this range combining with the phase shift of the feedback circuit to produce a 2π phase shift around the loop for all midrange frequencies. Over this frequency band, the portion of the output fed back to the input is in phase with the original noise signal. If the total loop gain is greater than unity, a new, larger signal is now present at the input, with a matching phase. By repeating the amplification feedback process, the noise continues to grow until nonlinearities in the circuit prevent further increase.

If B is independent of frequency, the frequency response of the loop is determined by the response of the amplifier circuit. All midrange frequencies can produce oscillation because they satisfy the required gain and phase relationships. Output v_o is noise, with a bandwidth corresponding to that of the amplifier.

Now consider what happens if a frequency-dependent B is introduced into the feedback loop. Only frequencies for which the total phase shift around the loop is $2n\pi$ ($n = 1, 2, 3 \ldots$) reinforce the signal present at the input and so produce oscillation. All other frequencies return to the input with different total phase angles. The net result is a circuit that oscillates at those frequencies determined by the characteristics of the feedback circuit, represented by B.

An output wave form is observed, consisting of the desired frequencies, without the presence of an input signal. The circuit converts dc energy (from the power supplies) to ac energy (in the desired signal), using the noise always present to start the oscillation process.

In many cases, a sine-wave oscillator is desired to produce a single output frequency. The feedback loop is thus designed to satisfy the required phase relationships at only a single frequency. Different ways for designing feedback circuits that oscillate at a single frequency are explained in this chapter.

It is useful to extend the Nyquist-plot concept to plot the characteristics of the combined amplifier building block and feedback circuit. In general, both A_v^* and B are represented by complex numbers, so that the notation

$$A_v^* B = \left| A_v^* B \right| e^{j\theta*} \qquad (14\text{-}2)$$

can be used.

The angle $\theta*$ is the total phase shift associated with the amplifier and feedback loop. Oscillation occurs if $\left| A_v^* B \right| > 1$ and $\theta* = 2n\pi$. The Nyquist plot can be extended by plotting $\left| A_v^* B \right|$ as a function of $\theta*$. The condition for oscillation on such a plot is that the function pass outside of the point $\left| A_v^* B \right| = 1$, $\theta* = 0$ (on the x-axis). Inspection of such a plot reveals if oscillation will occur in a given circuit.

A further note should be added about implementation of all the oscillator circuits to be discussed. In each case, the loop-gain requirement must actually be somewhat greater than the idealized factor to allow for circuit losses. Resistive losses absorb part of the circuit energy. In addition, any oscillator circuit satisfies the oscillation condition for a narrow band of frequencies around a center frequency. Energy is thus spread among the frequencies present, requiring additional gain.

Circuit oscillation begins at a critical value of the loop gain $|A_v^* B|$. Below this critical value, the amplitude of the signal v_o is zero, and above this value, oscillation is observed. The critical gain value is given by $|A_v^* B| = 1$ if circuit losses are neglected, but by a somewhat larger value (perhaps 1.05 to 1.20), if losses are considered.

Based on the use of a linear amplifier model alone, the amplitude of the oscillation is unbounded once the critical loop gain is attained. Circuit nonlinearities must be considered to understand the actual limits on the amplitude of the output.

As the loop gain reaches the critical value, the signal with frequency f_o selected by the feedback circuit response is amplified sufficiently around the loop to return to the input with an amplitude greater than that with which it began its way through the loop. As a result, the signal feeds on itself and grows in amplitude.

This process continues until the linear region of the circuit is exceeded. As soon as the signal is large enough to produce nonlinear operation, the gain drops and distortion occurs. Energy is transferred from the desired signal frequency to other frequencies, reducing the amplitude of the signal and providing an effective drop in amplifier gain. This tends to bring the circuit out of oscillation. Yet, a drop in output voltage brings device operation back into the linear circuit region, increasing the gain to its critical value. As a result of the behavior on the boundary of the linear region, a signal with stable amplitude can be obtained with a little distortion (see point 1 of Fig. 14-2). If the loop gain is increased further, the signal penetrates more deeply into the nonlinear region and significant signal distortion is noted (point 2).

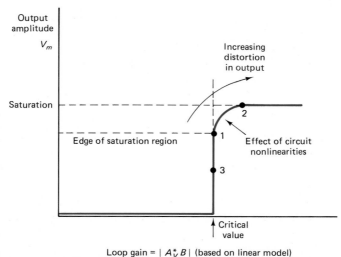

Fig. 14-2 Nonlinear limitation of signal amplitude.

When distortion occurs, a signal at the selected frequency of operation passes through the amplifier and feedback loop, maintaining oscillation. The sine-wave input to the amplifier produces a complex output waveform at the output. The feedback circuit selects the desired frequency from this output and sends it back to the input, maintaining oscillation.

The difference in loop gain between points 1 and 2 can be very small (of the order of 1 percent), so adjustment of $|A_v^* B|$ to achieve a low-distortion output is difficult. To operate a practical version of the circuits in this chapter based on the above nonlinear effect, the gain of the amplifier must be adjustable with a voltage divider to allow the circuit operation to be established just on the edge of the linear region. Due to thermal variations in device gain, the desired setting on this divider changes as the ambient temperature of the circuit changes.

The nonlinear effect introduced above provides a method of *automatic gain control* (*AGC*). The circuit achieves internal balance and maintains a linear output. Other improved approaches are also available.

In general, an AGC circuit requires a nonlinear device to limit the amplitude of oscillation and prevent a distorted output. As shown in Fig. 14-3, a thermistor (see Section 3-1) can be used as such a nonlinear device to control the gain of an oscillator.

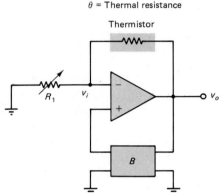

Fig. 14-3 AGC using a thermistor.

As the amplitude of the output voltage V_m increases, more power is dissipated in R_T (remembering that v_i remains at virtual ground). The temperature of the thermistor increases. Following prior development, the increase in temperature can be found from equation 5-12:

$$T - T_o = \theta P_{av}$$
$$= \theta \left(\frac{V_m}{\sqrt{2}}\right)^2 \frac{1}{R_T} = \frac{1}{2} \theta \frac{V_m^2}{R_T} \tag{14-3}$$

where θ is the thermal resistance between the thermistor and its environment, and R_T is the resistance of the thermistor at temperature T.

An increase in temperature reduces R_T according to Eq. (2-11):

$$\frac{R_T}{R_o} = \exp\left[\frac{\mathcal{E}_g}{2k}\left(\frac{1}{T} - \frac{1}{T_o}\right)\right] \tag{14-4}$$

therefore reducing circuit gain A_V and v_o. Less power is dissipated in the thermistor and a balance point is obtained. An input variable resistor sets the initial operating conditions, allowing for circuit losses and variations in the ambient temperature.

This method of automatic gain control is sensitive to external temperature changes, so the divider has to be reset if the environmental temperature increases. Operating point 3 of Fig. 14-2 can be obtained by using such a nonlinear feedback resistor.

Example 14-1   **Thermistor AGC Control**

For a given oscillator circuit, an output amplitude $V_m = 4.0\,V$ is desired. If a thermistor with $\theta = 100°C/W$ is used as a method of AGC, as shown in Fig. 14-3, what resistance value is reasonable for the thermistor?

Solution   Assume that the temperature change $T - T_o$ is sufficiently small so that $R_o \cong R_T$ is a first approximation when using Eq. (14-3). To prevent the circuit from being excessively sensitive to external temperature variations, require that $T - T_o$ be no smaller than $0.10°C$. Thus, from Eq. (14-3),

$$R_o \cong R_T = \frac{\theta}{2}\frac{V_m^2}{T - T_o} = \frac{100°C/W}{2}\frac{(4.0\ V)^2}{0.10°C} \cong 8.0\ k\Omega$$

To check the initial assumption, recall that the variation in R_T due to temperature change can be found from Eq. (14-4). Assuming $T - T_o = \Delta T \ll T_o$ and $\Delta R = R_T - R_o$,

$$\frac{\Delta R}{R_o} = \frac{R_T}{R_o} - 1 \cong \exp\left(\frac{-\mathcal{E}_g}{2kT_o}\frac{\Delta T}{T_o}\right) - 1 = \frac{-\mathcal{E}_g}{2kT_o}\frac{\Delta T}{T_o}$$

For a Si thermistor and $T_o \cong 300°K$,

$$\Delta R = -(8.0\ k\Omega)\frac{1.11\ eV}{2(8.62 \times 10^{-5}\ eV/°K)(300°K)}\left(\frac{0.10°C}{300°K}\right) \cong -57\ \Omega$$

For an 8.0-kΩ thermistor, a temperature increase of $0.10°C$ is associated with a drop in resistance of 57 Ω. The assumption $R_T \cong R_o$ was thus a reasonable one for use in Eq. (14-3). An iterative solution can be used if further accuracy is required.

To produce a stable operating point, $\Delta R = -57\ \Omega$ must change the gain of the op amp sufficiently to prevent saturation. For the values given above,

$$\Delta A_V \cong \Delta R/R_1 = -57\ \Omega/R_1$$

To achieve $\Delta A_V \cong 1$ percent, $R_1 = 5.7\ k\Omega$ ∎

Another approach to AGC is shown in Fig. 14-4(A). The diodes in the feedback loop conduct only when the output signal exceeds ± 0.7 V. When they conduct, the

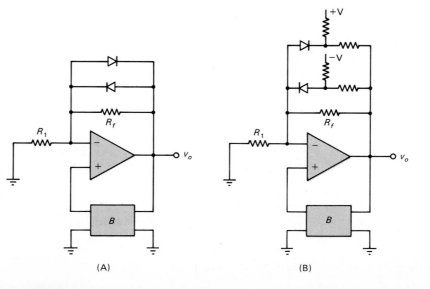

Fig. 14-4 Use of diodes for AGC. (A) Limiting to 0.7-V amplitude. (B) Limiting to larger amplitudes.

(A)   (B)

effective R_f is reduced, bringing down the loop again $|A_v^* B|$ and tending to bring the circuit out of oscillation. However, the diodes then cut off, bringing the loop gain back up. As a result of this balancing, the output stabilizes so that the diodes are just barely conducting with a maximum output signal. The nonlinear characteristic curve of the diodes is being used to stabilize the signal amplitude. To adjust the amplitude of the output, voltage dividers (or other reference voltages) can be used, as shown in Fig. 14-4(B).

Many other AGC strategies are possible, depending on the application and objectives. The following oscillator circuits do not include specific provisions for AGC, except as inherent in device properties. The circuits can be modified by one of the above methods or by other techniques to achieve the desired amplitude stability.

14-4 PHASE-SHIFT OSCILLATOR

Figure 14-5(A) shows a phase-shift oscillator using an operational-amplifier circuit, and Fig. 14-5(B) shows the same circuit using a JFET amplifier. For this oscillator, the frequency-dependent feedback network is obtained with a combination of resistors and capacitors. The Bode and Nyquist plots are shown in Figs. 14-5(C) and (D), respectively. Observe that the criterion for oscillation ($|A_v^* B| > 1$) is met if the Nyquist plots pass outside of the point $|A_v^* B| = 1$, $\theta^* = 0$.

Consider the oscillator circuit of Fig. 14-5(A). The gain of the operational amplifier is determined by v_o/v_1, and the feedback factor B is determined by v_f/v_o. For oscillation, $A_v^* B = +1$. When this relationship holds, the feedback voltage v_f is in phase with and equal to the input voltage v_1. When v_f is connected to v_1, oscillation occurs.

At midrange, the operational amplifier has a gain of R_f/R_1 and a phase shift of $\theta^* = 180°$. There is no low-frequency roll-off because the operational amplifier is dc-coupled. At high frequencies, the gain rolls off at 20 dB per decade and the phase shift decreases toward $\theta^* = 90°$ (assuming a single-pole low-pass filter) over the operational range.

Fig. 14-5 Phase-shift oscillator. (A) Operational-amplifier phase-shift oscillator circuit. (B) JFET phase-shift oscillator.

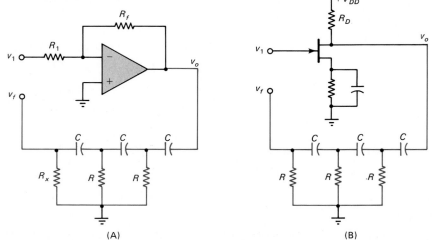

(A)   (B)

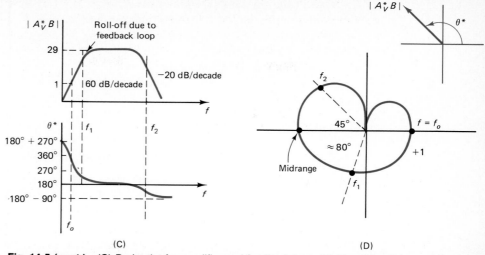

(C)   (D)

Fig. 14-5 (cont.)   (C) Bode plot for amplifier and feedback loop. (D) Nyquist plot.

The effect of the feedback must now be determined. An expression for v_f/v_o is written using complex notation to carry both amplitude and phase information. From Fig. 14-6[1],

$$B = \frac{v_f}{v_o} = \frac{R}{R - jX_C} \frac{Z_a}{Z_a - jX_C} \frac{Z_b}{Z_b - jX_C} \tag{14-5}$$

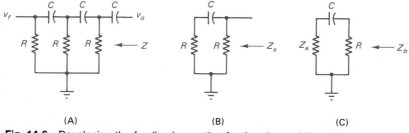

(A)   (B)   (C)

Fig. 14-6   Developing the feedback equation for the phase-shift oscillator.

An oscillator results if R and C are chosen so that the feedback loop provides a $180°$ phase shift and adequate loop gain. The imaginary part of v_f/v_o must vanish. From Appendix 13,

$$f_o = 1/(2\pi\sqrt{6}RC) \tag{14-6}$$

satisfies this requirement. Also for $f = f_o$,

$$v_o/v_f = -1/29 \tag{14-7}$$

---

[1] This analysis assumes that $R_1 \gg R_x$ and $R_x = R$. Otherwise R_x must be adjusted so that $R_1 \| R_x = R$.

To meet the oscillation requirement $A_v^* B = 1$, $|A_v^*|$ must be greater than 29. If this condition holds, the circuit oscillates at frequency f_o.

The combined frequency performance of the operational amplifier and the feedback loop may be viewed together to produce Bode and Nyquist plots, as shown in Figs. 14-5(C) and (D). Note that as $f \to 0$,

$$Z_a \longrightarrow R$$

$$Z_b \longrightarrow R$$

so that

$$B \longrightarrow (j\omega RC)(j\omega RC)(j\omega RC) \qquad (14\text{-}8)$$

Since each $+j$ term represents a $+90°$ phase shift, the total phase shift around the loop is $\theta^* = 180° + 3(90)°$. At midrange, $\theta^* = 180°$, and at high frequencies, θ^* tends toward $90°$. At frequency f_o (below midrange), $\theta^* = 360°$, which defines the oscillator frequency f_o. Because of the three identical high-pass filter terms, the amplitude plot rolls off at the lower end at about 60 dB per decade. It is because of this rapid roll-off that B is so small at f_o ($B = -1/29$).

The conditions for oscillation are shown in Fig. 14-5(D). The total circuit gain is unity, and the voltage v_f is in phase with v_1 for the selected value of operation.

Observe from Fig. 14-5(C) that the loop phase shift is changing slowly as a function of frequency at the operating frequency f_o. The phase-shift oscillator is not highly stable because changes in circuit parameters can result in an appreciable shift in the output frequency. In order to tune the oscillator, three variable resistors or capacitors must be used.

Example 14-2   Phase-Shift Oscillator

A phase-shift oscillator is to be built with $f_o = 1.0$ kHz. If $C = 0.010\,\mu\text{F}$ is selected, then what resistor value must be used?

Solution   From Eq. (14-6),

$$R = \frac{1}{2\pi\sqrt{6}f_o C} = \frac{1}{2\pi\sqrt{6}(1.0 \times 10^3)(1.0 \times 10^{-8})} = 6.5\,\text{k}\Omega \qquad \blacksquare$$

14-5   WIEN BRIDGE OSCILLATOR

Figure 14-7(A) shows a Wien Bridge oscillator, which also achieves frequency selection through a combination of resistors and capacitors. The Bode and Nyquist plots are shown in Figs. 14-7(B) and (C), respectively. As can be seen, the Wien Bridge is frequency selective because $|A_v^* B|$ peaks only near f_o and θ^* is changing rapidly in the vicinity of f_o. Thus, this oscillator is more stable than the phase-shift oscillator.

To find the circuit values for oscillation, set the complex $A_v^* B = 1$, $R_a = R_b = R$ and $C_a = C_b = C$:

$$A_v^* B = \frac{R\|(-jX_C)}{R\|(-jX_C) + (R - jX_C)}\left(1 + \frac{R_2}{R_1}\right) = 1 \qquad (14\text{-}9)$$

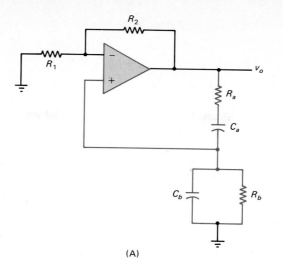

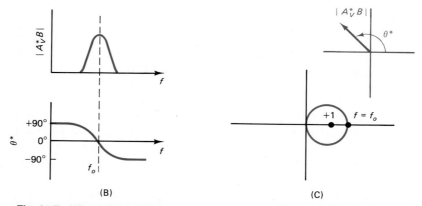

Fig. 14-7 Wien bridge oscillator (with f_2 of op amp $\gg f_o$. (A) Wien bridge oscillator using an operational amplifier. (B) Bode plot. (C) Nyquist plot.

Substitution in Eq. (14-9) shows that

$$A_v^* B = \left(1 + \frac{R_2}{R_1}\right) \frac{1}{3 + j(\omega RC - 1/\omega RC)} \tag{14-10}$$

Oscillation is defined by

$$\omega_0 RC = 1/\omega_o RC \quad \text{and} \quad f_o = 1/2\pi RC \tag{14-11}$$

When this condition holds, $B = 1/3$, so that oscillation can take place if

$$1 + R_2/R_1 > 3 \tag{14-12}$$

The feedback circuit produces a phase shift of $+j$ for $f \ll f_o$ and $-j$ for $f \gg f_o$. There is a peak in $|A_v^* B|$ at f_o, with the magnitude dropping off for higher and lower frequencies. The resultant Bode and Nyquist plots are shown in Figs. 14-7(B) and (C) (assuming that f_2 for the operational amplifier $\gg f_o$).

Example 14-3 Wien Bridge Oscillator

How is Eq. (14-11) modified if the R and C values in a Wien bridge oscillator are not the same?

Solution The equation for B becomes

$$B = \frac{R_b \| (-jX_{cb})}{R_b \| (-jX_{cb}) + (R_a - jX_{ca})} = \frac{1}{1 + \dfrac{R_a - jX_{ca}}{R_b \| (-jX_{cb})}}$$

For resonance, the imaginary part of B equals zero:

$$\text{Im}\left[\frac{R_a + X_{ca}}{R_b \| X_{cb}}\right] = R_a R_b + X_{ca} X_{cb} = 0$$

so that

$$f_o = \frac{1}{2\pi (R_a R_b C_a C_b)^{1/2}}$$ ∎

14-6 LC RESONANT OSCILLATOR

Oscillators can also be developed using combinations of inductors and capacitors to provide the desired frequency characteristics. Figure 14-8(A) shows an LC resonant oscillator using a JFET. The frequency-sensitive feedback loop is provided by the parallel LC combination. Figure 14-8(B) shows the ac equivalent circuit, and Figs. 14-8(C) and (D) show the appropriate Bode and Nyquist plots, respectively.

To find an expression for $A_v^* B$, refer to Fig. 14-8(B). The gain of the amplifier is given by

$$A_v^* = v_o/v_1 = -g_m(r_d \| R_{\text{eff}} \| jX_L \| -jX_C)$$ (14-13)

where R_{eff} is the effective series resistance of the coil converted to a parallel representation (see Exercise 14-8). The secondary of the transformer does not affect v_o because $i_2 \cong 0$.

The feedback fraction B is obtained through the secondary of the transformer. From Fig. 14-8(B),

$$\begin{aligned}
v_o &= -L\,di_1/dt + M\,di_2/dt \cong -L\,di_1/dt \\
v_f &= -L\,di_2/dt + M\,di_1/dt \cong M\,di_1/dt
\end{aligned}$$ (14-14)

where M is the mutual inductance between the two windings.

Therefore,

$$B = v_f/v_o = -M/L$$ (14-15)

Combining the above relationships,

$$\begin{aligned}
A_v^* B &= \frac{M}{L} g_m(r_d \| R_{\text{eff}} \| jX_L \| -jX_C) \\
&= \frac{M}{L} g_m \frac{r_d \| R_{\text{eff}}}{1 + \dfrac{r_d \| R_{\text{eff}}}{jX_L \| -jX_C}}
\end{aligned}$$ (14-16)

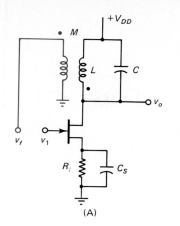

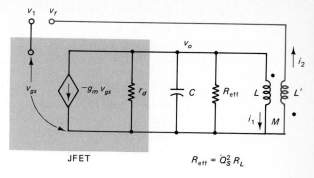

(A)

(B)

$R_{\text{eff}} = Q_S^2 R_L$

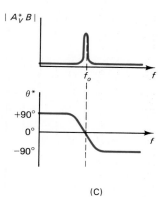

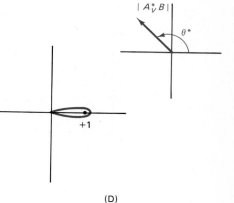

(C)

(D)

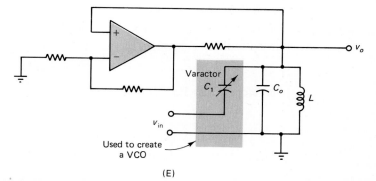

Used to create
a VCO

(E)

Fig. 14-8 LC resonant oscillator. (A) *LC* tuned-oscillator circuit. (B) AC equivalent circuit. (C) Bode plot. (D) Nyquist plot. (E) Alternate *LC* resonant-oscillator circuit.

The total loop phase shift is 360° only when the imaginary part of $A_V^* B$ is zero. Therefore,

$$f_o = \frac{1}{2\pi\sqrt{LC}} \tag{14-17}$$

Oscillation occurs if

$$A_V^* B = \frac{M}{L} g_m r_d \| R_{\text{eff}} = 1 \qquad (14\text{-}18)$$

To draw the Bode plot, observe from Eq. (14-16) that $|A_V^* B|$ is a maximum at $f = f_o$ and smaller for both lower and higher frequencies. At low frequencies, the capacitive term dominates and an additional $+90°$ phase shift results. At high frequencies, the inductive term dominates and an additional $-90°$ phase shift results. (At very low frequencies, there is an additional positive phase shift due to C_S, and at very high frequencies, there is an additional $-90°$ phase shift due to the JFET capacitive roll-off.) The Nyquist plot takes the form shown in Fig. 14-8(D). The LC resonant oscillator can have a narrow bandwidth and a stable f_o due to the high Q of the LC circuit.

However, note that there is a problem associated with the oscillator circuit of Fig. 14-8(A). In order to maintain the value of g_m at a reasonable value, current I_D must not be allowed to drop below a minimum level. This requires that the impedance of the LC combination be limited at resonance. As a result, restrictions occur on the sharpness of resonance that can be obtained.

An alternate LC resonant circuit is shown in Fig. 14-8(E). The complex voltage divider provides the feedback path, and there is no longer any upper limit on the impedance of the LC combination. The gain of the op amp must be just slightly greater than unity to produce oscillation at the resonant frequency. For all other frequencies, the reduced gain and loop phase shift not equal to $360°$ prevent oscillation.

Example 14-4   LC Resonant Oscillator

An LC resonant oscillator is fabricated with a transformer primary that has a series resistance of 5.0 Ω and a Q factor of 100. If $r_d = 50$ kΩ and $g_m = 2.0$ m℧, what value of M/L is required for the transformer to produce oscillation? Assume that the g_m value is valid at resonance.

Solution   As noted in Exercise 14-8, for a high-Q inductor the series resistor R_S can be replaced with an equivalent parallel resistor:

$$R_{\text{eff}} = R_S Q^2 = 5.0(100)^2 = 50 \text{ k}\Omega$$

Therefore, from Eq. (14-18), the value

$$\frac{M}{L} = \frac{1}{g_m r_d \| R_{\text{eff}}} = \frac{1}{(2.0 \times 10^{-3})(50 \text{ k}\Omega \| 50 \text{ k}\Omega)} = 0.02$$

is required for oscillation.   ■

The LC resonant oscillator can be used as a *voltage-controlled oscillator* (VCO), as shown in Fig. 14-8(E). A varactor (introduced in Section 4-6) is placed in parallel with the fixed LC combination and used to change the frequency of oscillation. At the low frequencies associated with v_{in}, the inductor acts as a short circuit. Therefore, the voltage across the varactor is $-v_{\text{in}}$. If the capacitance associated with the source v_{in} is much greater than C_1, then the source capacitance does not affect the resonant frequency. With the varactor, the resonant frequency of oscillation becomes

$$f_o = \frac{1}{2\pi\sqrt{L(C_o + C_1)}}$$

If $C_1 \ll C_o$ as is often the case,

$$f_o \cong \frac{1}{2\pi\sqrt{LC_o(1 + C_1/C_o)}} \cong \frac{1}{2\pi\sqrt{LC_o}}(1 - \tfrac{1}{2}C_1/C_o) \qquad (14\text{-}19)$$

To the first order, changes in frequency Δf are thus linearly related to changes in the capacitance ΔC_1.

14-7 COLPITTS OSCILLATOR

Figure 14-9 shows a Colpitts oscillator, which is based on another LC configuration. To understand how this circuit works, refer to the ac equivalent circuit in Fig. 14-9(B) (valid at midrange frequencies for the JFET version). The product $A_V^* B$ can be written in the complex form:

$$A_V^* B = -g_m[R_D \| -jX_C \| (jX_L - jX_C)]\frac{-jX_C}{jX_L - jX_C}$$

$$= -g_m\frac{R_D X_C^2}{R_D(jX_L - 2jX_C) + (-jX_C)(jX_L - jX_C)} \qquad (14\text{-}20)$$

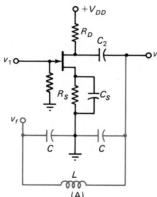

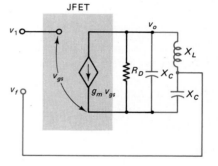

(A)

(B)

Fig. 14-9 Colpitts oscillator. (A) Colpitts oscillator with JFET. (B) Colpitts oscillator with operational amplifier. (C) Midrange ac equivalent circuit for the JFET Colpitts oscillator. (D) AC equivalent circuit for the feedback loop for the Colpitts oscillator using an operational amplifier.

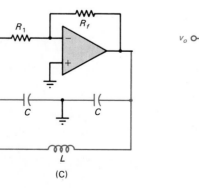

(C)

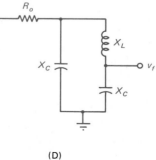

(D)

For resonance, the imaginary part of $A_v^* B$ must vanish, requiring

$$f_o = \frac{\sqrt{2}}{2\pi\sqrt{LC}} \qquad (14\text{-}21)$$

and

$$A_v^* B = g_m R_D \qquad (14\text{-}22)$$

at $f = f_o$. This is a real, positive number. If values are chosen so that $g_m R_D > 1$, the circuit oscillates.

For the Colpitts oscillator using an operational amplifier, as shown in Figs. 14-9(C) and (D), $A_v^* = -R_f/R_1$ at midrange. For the feedback fraction B,

$$B = \frac{(-jX_C)\|(jX_L - jX_C)}{(-jX_C)\|(jX_L - jX_C) + R_o} \frac{-jX_C}{jX_L - jX_C} \qquad (14\text{-}23)$$

$$= \frac{-X_C^2}{R_o(jX_L - 2jX_C) + (-jX_C)(jX_L - jX_C)} \qquad (14\text{-}24)$$

Oscillation occurs when

$$f_o = \frac{\sqrt{2}}{2\pi\sqrt{LC}} \qquad (14\text{-}25)$$

as before. At f_o,

$$B = -1 \qquad (14\text{-}26)$$

and by substitution

$$A_v^* B = R_f/R_1 \qquad (14\text{-}27)$$

Oscillation occurs if $R_f/R_1 > 1$. The Bode and Nyquist plots for the Colpitts oscillator are referenced in Problem 14-13.

Example 14-5   Colpitts Oscillator

For a Colpitts oscillator, $L = 1.0\,\mu\text{H}$ and $C = 1.0 \times 10^3\,\text{pF}$. (a) Find f_o. (b) Find the value of B at $f = 0.9 f_o$ if $R_o = 1.0 \times 10^2\,\Omega$.

Solution   From Eq. (14-25),

$$f_o = \frac{\sqrt{2}}{2\pi\sqrt{LC}} = \frac{\sqrt{2}}{2\pi\sqrt{(10^{-6})(10^{-9})}} = 7.12\,\text{MHz}$$

At $f = 0.9 f_o = 6.41\,\text{MHz}$,

$$X_C = \frac{1}{2\pi(6.41 \times 10^6)(10^{-9})} = 24.8\,\Omega$$

$$X_L = 2\pi(6.41 \times 10^6)(10^{-6}) = 40.3\,\Omega$$

From Eq. (14-24),

$$B = \frac{-(24.8)^2}{100[40.3\,j - 2\,j(24.8)] + 24.8[40.3 - 24.8]}$$

$$= \frac{-1}{-1.5\,j + 0.62} = -(0.24 + 0.58\,j)$$

and

$$|B| = [0.24^2 + 0.58^2]^{1/2} = 0.6$$

For comparison, $|B| = 1.0$ at resonance. ■

14-8   HARTLEY OSCILLATOR

The Hartley oscillator has a design similar to that of the Colpitts oscillator, as shown in Fig. 14-10. Comparing these two oscillators, one can be mapped into the other by exchanging jX_L and $-jX_C$.

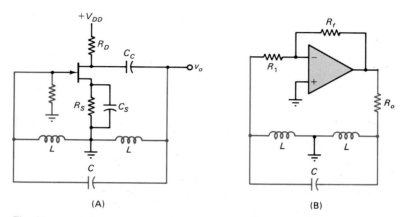

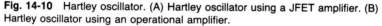

(A)   (B)

Fig. 14-10   Hartley oscillator. (A) Hartley oscillator using a JFET amplifier. (B) Hartley oscillator using an operational amplifier.

For a Hartley oscillator using a JFET amplifier,

$$A_v^* B = \frac{g_m R_D X_L^2}{R_D(-jX_C + 2jX_L) + jX_L(jX_L - jX_C)} \tag{14-28}$$

The frequency of oscillation is given by

$$f_o = \frac{1}{2\pi\sqrt{2LC}} \tag{14-29}$$

and oscillation occurs if $g_m R_D > 1$. A similar result can be obtained using an operational amplifier.

14-9   CRYSTAL OSCILLATOR

In addition to the use of RC and LC circuits, frequency-dependent circuits can be obtained through the use of crystals. A small crystal, usually of quartz, is mounted under tension in a mechanical frame. Stress is applied to create a natural mechanical resonant frequency of vibration, which is a function of both crystal and mount properties. Quartz is a piezoelectric crystal, so that a mechanical-electrical interaction can take place if the crystal is placed in a circuit.

The crystal behaves like the equivalent circuit shown in Fig. 14-11(A). This circuit has both series and parallel resonant frequencies. Typical values for the equivalent circuit result in a very-high-Q resonant circuit, with a narrow bandwidth and very-high-frequency stability.

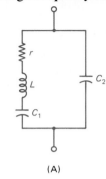

(A)

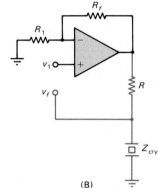

(B)

Fig. 14-11 Crystal oscillator. (A) Electrical equivalent circuit of a mounted quartz crystal. (B) Crystal oscillator using an operational amplifier.

By using an operational amplifier, an oscillator can be created, as shown in Fig. 14-11(B), with

$$A_V^* B = \left(1 + \frac{R_f}{R_1}\right) \frac{Z_{\text{cry}}}{Z_{\text{cry}} + R} \tag{14-30}$$

Near parallel resonance, Z_{cry} is large and $\theta^* = 0°$ for the total circuit. At other frequencies, Z_{cry} is small and the phase-shift requirement for oscillation is no longer met.

Miniaturized crystal oscillators are widely used today in electronics, both in the application shown in Fig. 14-11(B) and by combining the crystal with other circuit configurations. Performance of the oscillator circuits described in the above sections can often be improved by including a crystal in the feedback loop. The Colpitts oscillator is well suited to such an application (Guio, 1987).

14-10 NEGATIVE-RESISTANCE OSCILLATORS

By using a different design approach, it is possible to create an oscillator with a two-terminal device having a negative ac resistance over part of its characteristic curve. Figure 14-12 shows the tunnel diode, for which a negative resistance region exists.

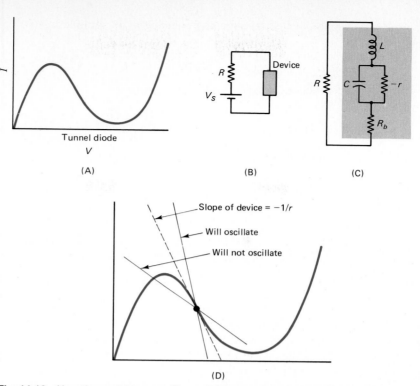

Fig. 14-12 Negative resistance oscillator. (A) Tunnel diode characteristic curve. (B) Schematic. (C) AC equivalent circuit. (D) Criterion for oscillation.

(Other examples of negative-resistance devices, such as the Gunn oscillator, can be found in electronics.)

The schematic and ac equivalent circuit of Figs. 14-12(B) and (C), respectively, can be used to study such devices. The source V_S and resistor R are selected to give a load line that biases the device in the negative-resistance region, as shown in Fig. 14-12(D). Inductance L arises from the wire leads, C is the effective capacitance of the device, and R_b is the bulk resistance of the device.

For this circuit, energy is supplied from the dc source and converted into ac oscillation. When the imaginary part of the total impedance Z of the circuit is zero and the real part is zero or negative, noise present in the circuit reinforces a given frequency and builds up until the magnitude of the oscillation exceeds the extremes of the negative-resistance region.

The ac equivalent circuit can thus be analyzed by finding

$$Z = j\omega L + \left(\frac{1}{j\omega C}\right) \| (-r) + (R_b + R)$$

$$= j\omega L - \frac{j(r/\omega C)(r + 1/j\omega C)}{r^2 + 1/\omega^2 C^2} + (R_b + R) \tag{14-31}$$

In order to have the imaginary part of $Z = 0$,

$$j\left[\omega L - \frac{(r^2/\omega C)^2}{r^2 + (1/\omega C)^2} \right] = 0 \tag{14-32}$$

which defines the angular frequency ω_o of oscillation,

$$\omega_o^2 = \frac{1}{LC}\left(1 - \frac{L}{r^2 C}\right) \qquad (14\text{-}33)$$

To produce a negative real component,

$$\frac{-(r/\omega_o C)(1/\omega_o C)}{r^2 + 1/\omega_o^2 C^2} + (R_b + R) < 0 \qquad (14\text{-}34)$$

and combining with Eq. (14-33), the condition

$$\frac{R_b + R}{r} < \frac{L}{r^2 C} \qquad (14\text{-}35)$$

is obtained.

From Eq. (14-33), the circuit has a real frequency of oscillation ω_o if $L/r^2 C < 1$. Combining this condition and Eq. (14-35) produces the requirement

$$\frac{R_b + R}{r} < \frac{L}{r^2 C} < 1 \qquad (14\text{-}36)$$

for circuit oscillation.

The requirement that $(R_b + R/r) < 1$ is equivalent to stating that the negative slope of the circuit load line must be greater than the slope of the negative-resistance curve, as shown in Fig. 14-12(D).

Example 14-6 **Negative-Resistance Oscillator**

For a tunnel diode, $L = 0.01$ μH and $C = 5.0$ pF. If the negative-resistance region has a negative slope $r = 100$ Ω and $R_b = 50$ Ω, does the circuit of Fig. 14-12(C) produce oscillation?

Solution Since

$$\frac{L}{r^2 C} = \frac{0.01\ \mu H}{(100)^2(5.0\ \text{pF})} = 0.2 < 1$$

this part of the requirement for oscillation is met. However, using the test of Eq. (14-35),

$$R_b + R \overset{?}{<} \frac{L}{rC} = \frac{0.01\ \mu H}{(100\ \Omega)(5.0\ \text{pF})} = 20\ \Omega$$

Since $R_b = 50$ $\Omega > 20$ Ω, no solution exists and the circuit does not oscillate. ∎

14-11 SUMMARY OF OSCILLATOR CHARACTERISTICS

Figure 14-13 summarizes the resonant frequencies and conditions for oscillation for the seven oscillators discussed in this chapter. Because of differences in the Nyquist plots for the various configurations, the output signals of the oscillators differ in the range of frequencies around f_o that can exist. Also, due to losses in the feedback

FIGURE 14-13
Summary of Oscillator Characteristics

Type	Resonant Frequency	Condition for Oscillation
Phase-shift oscillator	$f_0 = \dfrac{1}{2\pi RC \sqrt{6}}$	$\lvert A_v^* \rvert > 29$
Wien bridge oscillator	$f_0 = \dfrac{1}{2\pi RC}$	$1 + R_2/R_1 > 3$
LC-tuned oscillator	$f_0 = \dfrac{1}{2\pi \sqrt{LC}}$ $(M/L)g_m r_d \Vert R_{\text{eff}} > 1$	
Colpitts oscillator	$f_0 = \dfrac{1}{2\pi}\sqrt{\dfrac{2}{LC}}$	$g_m R_D > 1$
Hartley oscillator	$f_0 = \dfrac{1}{2\pi}\left(\dfrac{1}{\sqrt{2LC}}\right)$	$g_m R_D > 1$
Crystal oscillator	f_0 fixed by mechanical variables	$1 + R_f/R_1 > 1$
Negative-resistance oscillator	$f_0 = \dfrac{1}{2\pi}\left[\dfrac{1 - (R_b + R)/r}{LC}\right]^{1/2}$	$\left(\dfrac{R_b + R}{r}\right) < \dfrac{L}{r^2 C} < 1$

loops, it is usual to exceed the minimum requirements for oscillation by a small amount. It is desirable not to exceed this gain requirement by a large amount because distortions can be introduced into the output.

14-12   SPURIOUS OSCILLATIONS

As seen in this chapter, circuits can be designed to oscillate at desired frequencies. The criterion for oscillation is that the Nyquist plot produce a curve that passes outside the point $\lvert A_v^* B \rvert = 1$ for $\theta^* = 2\pi n$.

Based on these findings, it is therefore disconcerting to learn that many circuits not designed as oscillators do indeed oscillate nonetheless. This spurious oscillation is particularly true of complex amplifier circuits. For example, the operational amplifier in Section 12-3 can likely turn out to be an oscillator despite a design that seems to make this impossible.

How can a circuit with negative feedback $\theta^* = 180°$ and with no positive feedback loop break into oscillation? The answer is to recall the limitation of models, as described in Chapter 1, and the discussion of capacitance and inductance in real circuits from Chapter 3. The ideal, discrete model of circuit components is incomplete. Any circuit includes distributed capacitances and inductances due to linkages between wires and conducting surfaces. Every circuit that is drawn is incomplete unless the effects of these distributed capacitances are considered.

For the operational amplifier of Fig. 12-7, some of the more important distributed capacitances can link the feedback loop to the signals at each stage. On examining the circuit, it can be noted that the outputs of Q_1, Q_4, and Q_5 are all 180° out of phase with the input signal to Q_1. Distributed capacitances between these outputs and the input do not stimulate oscillations. However, the output of Q_2 is in phase with the input signal to Q_1, and a distributed capacitance C_d from the collector of Q_2 to the base of Q_1 can possibly set off circuit oscillation.

Consider the loop consisting of the amplifier formed by Q_2 and the feedback formed by C_d. The Bode and Nyquist plots for this loop are shown in Figs. 14-14 and 14-15, respectively.

In Fig. 14-14, two different response curves are shown for $|A_v^*B|$. The first curve is for the dc-coupled amplifier formed by Q_2, showing a midrange band extending from the origin up to a frequency f_2, where high-frequency roll-off begins. The second curve shows the characteristics of the distributed high-pass filter created by C_d and the effective resistance R_{eff} in series with C_d. The gain of the filter is 0 dB

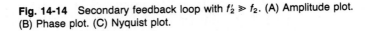

Fig. 14-14 Secondary feedback loop with $f_2' \gg f_2$. (A) Amplitude plot. (B) Phase plot. (C) Nyquist plot.

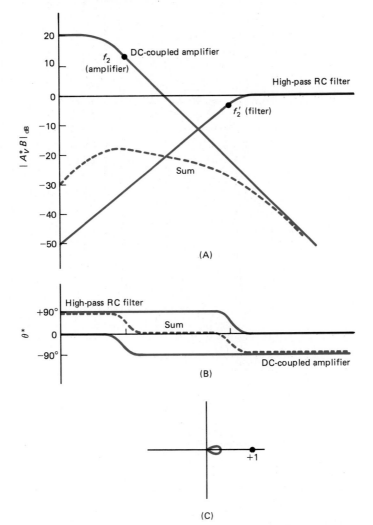

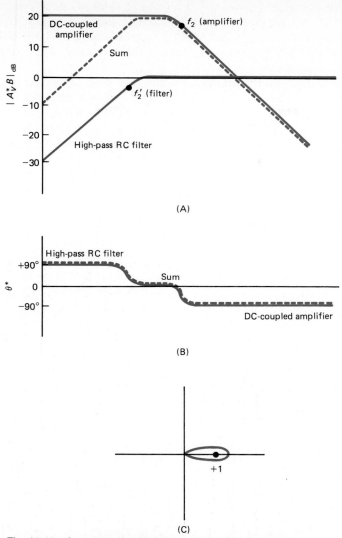

(A)

(B)

(C)

Fig. 14-15 Secondary feedback loop with $f_2' < f_2$. (A) Amplitude plot. (B) Phase plot. (C) Nyquist plot.

at very high frequencies and rolls off rapidly below the half-power point f_2' associated with the filter. If $f_2' \gg f_2$, where

$$f_2' = \frac{1}{2\pi R_{eff} C_d} \tag{14-37}$$

the combined positive feedback loop through C_d always has $\left| A_V^* B \right| < 1$ (0 dB).

The matching phase response of the feedback loop through C_d is shown in Fig. 14-14(B). The two component phase shifts are associated with the amplifier and distributed filter. The combined phase shift does pass across the axis $\theta^* = 0$. However, because $\left| A_V^* B \right| < 1$, oscillation does not occur.

Now consider Fig. 14-15, with a large value of C_d. For this case, the relationship $f_2' < f_2$ can be produced. By combining the gain responses of the amplifier and filter, the peak amplitude $|A_V^* B| > 1$ is obtained. From the phase diagram, the condition $\theta^* = 0°$ is met by the sum of the individual phase shifts. Oscillation occurs for this condition.

If spurious oscillation occurs in this operational amplifier, an attempt can be made to eliminate the problem by increasing f_2' (through careful layout of conductors to reduce C_d), or by decreasing f_2 (by placing a capacitor from the amplifier signal to ground). This combination shifts f_2' for the filter to higher frequencies and f_2 to lower frequencies, causing circuit performance to shift from the condition of Fig. 14-15 to that of Fig. 14-14. The method outlined above can be used to study spurious oscillations in any circuit and to make appropriate design changes when required. The advantages of short leads and careful physical design for amplifiers are obvious.

14-13 OSCILLATOR APPLICATIONS

Two major applications for oscillators involve (1) the experimental determination of the frequency-response characteristics of a circuit and (2) the operation of communication systems. These applications apply many of the concepts and circuits discussed in previous chapters and provide insight into important topics in electronics.

As introduced in Chapter 5 and expanded upon in Chapter 11, an arbitrary periodic waveform can be represented by a Fourier series. If the performance of a linear circuit is known for all (sine-wave) frequencies of interest, the performance in response to an arbitrary waveform can be found by superposition. Thus, the general frequency-response characteristics of linear circuits can be experimentally determined if the response of the circuit to an input sine wave is found over a sufficiently wide range of frequencies, as shown in Fig. 14-16. High-quality oscillators and vector voltmeters or oscilloscopes are essential for experimental studies of circuit performance, which are an important aspect of electronics system design.

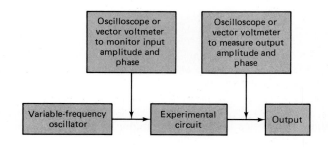

Fig. 14-16 Experimental determination of the frequency response of a circuit, using a variable-frequency oscillator. The oscilloscope or vector voltmeter must have a bandwidth greater than the bandwidth of the experimental circuit for which the frequency response is to be determined.

Another important application area involves the operation of communication systems. Although such systems are complex and require extensive study for in-depth understanding, a brief overview is useful to introduce new concepts that link to the nonlinear properties of diodes (Chapters 4 and 5), earlier discussions of Fourier analysis (Chapter 5), filters (Chapters 10 and 11), and classes of amplifiers (Chapter 11).

A fundamental amplitude-modulation (AM) communication system is shown in Fig. 14-17. The objective is to transfer a specific signal from one physical location to another. The process involves a *transmitter*, a *transmission channel,* and a *receiver*. This text emphasizes circuits and systems that are physically interconnected by conducting materials that allow a current flow to take place between locations. These conductors form the physical channel and require no specific transmitter and receiver.

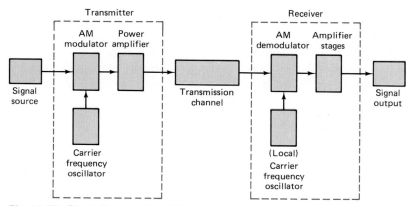

Fig. 14-17   Block diagram of an AM communication system.

However, communication involving high-frequency analog circuits ($f \gtrsim$ 10 MHz) and high-speed digital circuits (with important Fourier components in this region) places additional requirements on system design. In this high-frequency region, wires begin to act as antennas and radiate energy away from the circuit. *Transmission lines* are required to confine the energy to the desired communication channel. (At optical frequencies, optical fibers are used to confine the channel energy, as discussed in Chapter 4.)

These high frequencies are also used to create communication systems that do not require a physical link. Radio, microwave, and laser communication systems all provide the ability to transfer a signal from one location to another through electromagnetic wave propagation.

For these systems, a transmitter is required to modify the signal for transmission through the channel and a receiver is required to return the signal to its original form. Such a communication system can be represented as shown in Fig. 14-17.

An oscillator creates a *carrier* sine wave at the frequency desired for propagation through the channel. The signal then *modulates* the carrier to produce a complex waveform that contains the desired signal information to be passed through the channel. At the receiver, another oscillator *demodulates* the waveform and extracts the desired signal. The ability of the communication system to use electromagnetic wave propagation through the channel requires oscillators that generate carriers of the appropriate frequencies and the ability to modulate and demodulate the oscillator outputs to transfer the desired information from the source to the output at another location.

Information can be transferred by modulating the amplitude or phase angle of the carrier. The following discussion applies to amplitude modulation (AM). Frequency modulation (FM) and phase modulation (PM), which both produce changes in the phase angle, are discussed in the next section, and the coverage of digital communications in Chapter 16 extends the analysis to consider modulation strategies for digitized information.

This chapter describes some of the fundamentals of oscillator design and operation. It is useful to see how previous studies also provide the necessary insight to understand the operation of modulators and demodulators.

Figure 14-18 shows a simple (idealized) modulator circuit. The BJT uses class C operation (as discussed in Chapter 11). The oscillator output, which is the carrier sine wave at frequency f_o, turns *on* the BJT during a series of partial cycles at the carrier frequency. When the BJT is *on,* it saturates, so the output voltage is close to ground. When the BJT is *off,* the collector rises to a voltage determined by the transformer secondary voltage created by the signal source in series with V_{CC}. The isolation inductor, called a *radio-frequency (RF) choke,* prevents the carrier signal from coupling into the signal source.

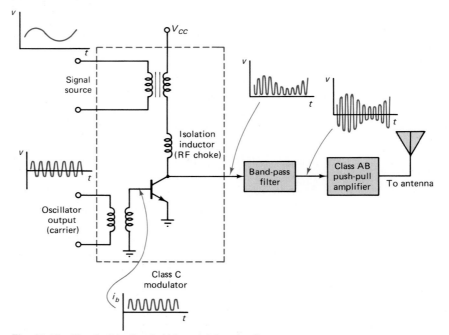

Fig. 14-18 Simple (idealized) AM modulator circuit.

The result is that the amplitude of the carrier becomes modified by the signal amplitude. The frequency components of the total waveform include a carrier component at f_o and a narrow band of frequencies Δf around f_o that represent the information contained in the carrier. A band-pass filter removes the unwanted harmonic frequencies in the output waveform, leaving the amplitude-modulated (AM) carrier. A class AB push-pull amplifier provides linear amplification of the modulated carrier, which is then directed to the antenna and into the channel.

Figure 14-19 shows a simple (idealized) demodulator circuit. The original signal can be extracted from the AM carrier by using the nonlinear characteristic curve of the junction diode (or any other nonlinear device). If the diode is biased at the knee of the characteristic curve, then the actual characteristic can be approximated by a parabola. Thus, around this bias point, the parabolic term results in

$$i \cong kv^2 \tag{14-38}$$

where k is a constant determined by the slope of the characteristic curve and the bias point.

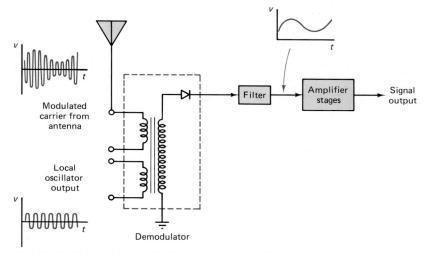

Fig. 14-19 Simple (idealized) AM demodulator circuit.

For the circuit of Fig. 14-19, the small signal voltage v across the diode is given by the sum of the modulated carrier received from the antenna and a second local oscillator at frequency f_o. Therefore, for each angular frequency component $\Delta \omega$,

$$i = k\{V_1 \cos (\omega_o t) + V_2 \cos [(\omega_o + \Delta \omega)t]\}^2 \tag{14-39}$$

Filters are used to remove all terms except for the cross product, which yields

$$2kV_1 V_2 \cos (\omega_o t) \cos [(\omega_o + \Delta \omega)t] \tag{14-40}$$

By using the trigonometric identity,

$$\cos A \cos B = \tfrac{1}{2}[\cos (A + B) + \cos (A - B)] \tag{14-41}$$

Eq. (14 = 40) can be written as

$$kV_1 V_2\{\cos [(2\omega_o + \Delta \omega)t] + \cos (\Delta \omega t)\} \tag{14-42}$$

The term at frequency $\Delta \omega$ represents a Fourier component of the original signal. By superimposing all component terms resulting from the function, the original signal can be created.

By using an alternative approach, it is often convenient to apply a *local oscillator* frequency that is slightly different from the original oscillator frequency. As a result, the output from the demodulator is centered around an intermediate frequency (IF) that is easy to amplify. A second demodulator can then extract the desired signal from the amplified IF waveform.

The demodulator shown in Fig. 14-19 illustrates how a nonlinear device or circuit can be used to produce an output wave with frequency components that are not present in the inputs. This same property is used for the rectifier and wave-shaping circuits of Chapter 5. In general, a nonlinear device can be represented by a power series and will produce many different new frequencies, or harmonics. Filters can be used to select out the desired frequencies and remove those that are unwanted. For the demodulator of Fig. 14-19, the quadratic term is emphasized since it produces the desired difference term. Other terms are assumed removed from the output.

By choosing f_{o1} and f_{o2} for two sine-wave sources, a new difference frequency $f_{o1} - f_{o2}$ is produced. The result is a *beat-frequency oscillator* (BFO) that provides another approach to obtaining a desired sine-wave frequency. An advantage of the BFO is that small percentage changes in f_{o1} and f_{o2} can produce large percentage changes in $f_{o1} - f_{o2}$.

The above discussion of AM communication systems presents a simplified view of the subject. The intent is to illustrate the ways in which previous concepts can be combined to a single purpose, and to provide a bridge to more detailed courses in communication system design.

14-15   FM AND PM COMMUNICATION SYSTEMS

Another way to transfer information through a channel is to modulate the phase angle of the carrier wave instead of the amplitude. If the frequency of the carrier is varied in proportion to a modulating signal, *frequency modulation* (FM) results. If the phase of the carrier is varied in proportion to a modulating signal, then *phase modulation* (PM) results. Since the instantaneous frequency can be obtained by differentiating the instantaneous phase angle, the two types of modulation are closely related. The following discussion emphasizes FM, although many of the concepts can be directly applied to PM.

An FM system uses a constant-amplitude signal and can thus produce efficient communication through a lossy or nonlinear channel. In addition, FM allows a trade-off to be made between the S/N ratio (discussed in Section 11-7) and the available bandwidth of the channel. A wide bandwidth allows smaller values of S/N to be used to achieve a specified performance level (usually specified as a detector error rate) while a narrow bandwidth requires larger values of S/N to achieve the same performance level.

For an AM communication system, the only way to increase the S/N ratio is to increase the amplitude of the signal. For an FM communication system, the S/N ratio can be increased by causing the modulation frequency to vary more rapidly in time. The noise becomes relatively less significant with respect to the signal, but a wider bandwidth is required. Much of the design of FM systems is concerned with ways to achieve the maximum S/N ratio for a given bandwidth.

An FM system requires that the frequency of the carrier be varied in proportion a modulating signal. One possible circuit design for an FM transmitter is shown in Fig. 14-20(A). The transmitter makes use of the voltage-controlled oscillator

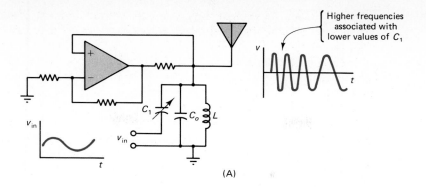

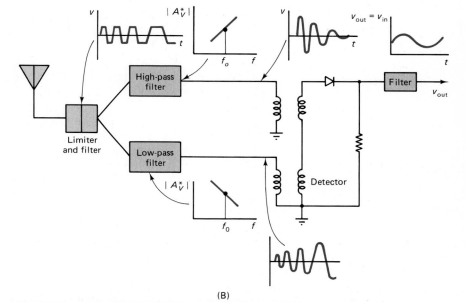

Fig. 14-20 Simple (idealized) FM communication system. (A) Transmitter. (B) Discriminator using high-pass and low-pass filters. (Continued on p. 576.)

(VCO) introduced in Section 14-16.[2] The input voltage v_{in} changes the bias across the varactor, thus shifting the frequency of oscillation. From Eq. (14-19), the frequency of oscillation is given by

$$f_o \cong \frac{1}{2\pi\sqrt{LC_o}}\left(1 - \frac{1}{2}\frac{C_1}{C_o}\right)$$

where the value of C_1 can be found from Eq. (4-41). To the first order,

$$\Delta f = \frac{1}{2\pi\sqrt{LC_o}}\left(-\frac{1}{2}\frac{\Delta C_1}{C_o}\right) = -\frac{1}{4\pi\sqrt{LC_o}}\left[\frac{1}{C_o}\left(\frac{\partial C_1}{\partial V}\right)\right]_{f=f_o}\Delta v_{in}$$

---

[2] The voltage-controlled capacitors and inductors of Experimental Applications 13-3 and 13-4 might also be used.

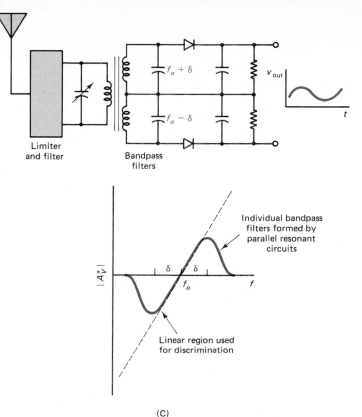

(C)

Fig. 14-20 (cont.)   (C) Discriminator using bandpass filters (ω_o is the carrier frequency).

For small values of Δv_{in}, the shift in frequency Δf around f_o is proportional to the change Δv_{in}. The output from the transmitter will be a carrier that is modulated in frequency, and is (typically) of constant amplitude.

An FM receiver must reverse the above process, converting a frequency-modulated, constant-amplitude carrier to an amplitude-modulated waveform that reproduces v_{in}. This requires the use of a circuit that is able to convert frequency changes to amplitude changes.

The high-pass and low-pass simple filter circuits introduced in earlier chapters (and discussed in Appendix 2) have this property if applied in the regions where roll-off occurs. The amplitude of the signal at the output of such filters is dependent on the frequency of the input. Figure 14-20(B) shows how an FM receiver, or discriminator, can be created by making use of these filter circuits.

The input to the receiver includes a frequency modulated carrier and noise added to the signal during propagation through the channel. Much of the amplitude noise can be removed by passing the signal through a saturated amplifier or limiter that produces an output of pulses. A filter can then be used to remove the unwanted harmonics and reconstruct the desired sine-wave frequency components.

The received sine wave of frequency f can be used as input to parallel high-pass and low-pass filters, as shown in Fig. 14-20(B), which function as a discriminator.

These filters produce an amplitude modulation for each signal, with changes in the signal amplitude linearly related to changes in the frequency.

The envelope detector circuit is similar to the one shown in Fig. 14-19. The output produces a difference signal for which the frequency-modulated carrier terms cancel out, leaving only the frequency components that are associated with the slowly varying output $v_{\text{out}} = v_{\text{in}}$.

Many other types of discriminators are also possible. Figure 14-20(C) illustrates how two bandpass filters (formed by parallel resonant circuits, discussed in Appendix 3) can be used to produce a discriminator that is effective in meeting design requirements. Typical values might be $f_o \cong 10$ MHz and $\delta \cong .01 f_o$, with $Q \cong 100$ for the parallel resonant circuits.

Many types of standard ICs are available to implement both AM and FM communication systems. High-level building block components are available (similar to those discussed in Section 13-11 for a playback amplifier) and are able to satisfy a wide variety of performance objectives. The linear circuit discussions of previous chapters, particularly with respect to the discussions of op amps, oscillators, and communication systems (Chapters 12 through 14) provide the background that is required to understand how these building block circuits function and how to appreciate the tradeoffs involved in system design, to interpret system performance, and to satisfy design objectives.

QUESTIONS AND ANSWERS

Questions

14-1. If $A_V^* B > 1$, what happens to the operation of an amplifier circuit with feedback?

14-2. If $A_V^* B > 1$ for all midrange circuit frequencies, which frequency components are present in the output of the oscillator?

14-3. What types of AGC are discussed in this chapter?

14-4. Why does the frequency f_1 for the phase-shift oscillator of Fig. 14-5 occur at a phase angle of approximately 80° with respect to midrange, as shown in Fig. 14-5(D)?

14-5. Compare the characteristics of the phase-shift and Wien bridge oscillators.

14-6. What are the relative advantages and disadvantages of the LC resonant oscillator?

14-7. What factors might affect the selection of a Colpitts or Hartley oscillator for a given application?

14-8. Why can every mounted quartz crystal be used to create two different frequencies of oscillation?

14-9. Why is it advantageous to have a tunnel diode with a minimum R_b when producing a negative-resistance oscillator?

14-10. If an amplifier circuit is observed to oscillate due to distributed effects, what steps can be taken to suppress these spurious oscillations?

Answers

14-1. If $A_V^* B > 1$, positive feedback produces a feedback signal that is in phase with and greater in amplitude than the initial (noise-generated) input signal. The result is growth in signal amplitude until circuit nonlinearities result in a stable operating condition.

14-2. If $A_V^* B > 1$ for all midrange circuit frequencies, all frequencies in the circuit bandwidth are present at the output of the oscillator. The various frequency components have random phase relationships with respect to one another.

14-3. Several types of automatic gain control (AGC) are discussed in this chapter. The amplitude of the output signal from an oscillator can be stabilized by using inherent amplifier nonlinearities, as shown in Fig. 14-2 (with operation between points 1 and 2). Alternatively, the temperature-dependent resistance of a thermistor can be used to prevent the amplifier from saturating (with operation at point 3 in the same figure). The nonlinear characteristic curve of the junction diode can also be used for AGC, as shown in Fig. 14-4.

14-4. The half-power point for three identical high-pass filters is discussed in Section 11-3 and Appendix 11. The phase shift of approximately 80° results at the half-power point for the three-pole configuration.

14-5. Both the phase-shift and Wien bridge oscillators use RC networks to produce the desired frequency-dependent B. The feedback network for the phase-shift oscillator produces a 180° phase shift at f_o, so the op amp is used as an inverting amplifier with gain $-R_f/R_1$. The feedback network for the Wien bridge oscillator produces 0° phase shift at f_o, so the op amp is used as a noninverting amplifier with gain $1 + R_f/R_1$. The phase-shift oscillator requires three RC combinations and an amplifier gain of 29, whereas the Wien bridge oscillator requires only two RC combinations and an amplifier gain of 3. The Wien bridge oscillator is more frequency-selective due to the rapidly changing phase shift around f_o, and it is used more often in system applications than is the phase-shift oscillator.

14-6. The LC resonant oscillator can provide a high-Q output (including only a narrow band of frequencies around f_o), which is an attractive feature for an oscillator. Only a single component value (L or C) has to be tuned to change f_o, in comparison with two or three components for the Wien bridge and phase-shift oscillators. On the other hand, transformers are often difficult to use efficiently in high-density circuits, which is a disadvantage for some applications.

14-7. Since the Colpitts and Hartley oscillators differ in the use of inductors and capacitors, a choice between oscillator types is based on the most desirable properties of the components. The frequency of operation for the oscillator, the desired Q for the inductors, and the required values and acceptable loss factors for the capacitors all affect the selection process.

14-8. As shown in the equivalent circuit of Fig. 14-11(A), the mounted quartz crystal has both series and parallel resonances. The application of Fig. 14-11(B) makes use of the parallel resonance when Z_{cry} is large. If the divider formed by the crystal and R is reversed, the series resonance can determine the oscillation frequency. These two resonant frequencies are usually close for typical crystal values.

14-9. In order to achieve oscillation, R must be adjusted so that the condition of Eq. (14-35) holds. By minimizing R_b, a maximum range of oscillator frequencies can be determined by varying R.

14-10. In general, spurious oscillations can be reduced or eliminated by reducing the bandwidth of the amplifier or reducing the magnitude of the distributed capacitances between critical circuit points. Both solutions can require a significant circuit redesign if such oscillation occurs. It is difficult to predict whether such distributed effects will be important in a high-frequency amplifier without constructing a breadboard.

EXERCISES AND SOLUTIONS

Exercises
(14-2)

14-1. (a) You are given a two-stage ac amplifier. Assume that $A_V > 1$ for each stage. If the output of the second stage is connected directly to the input of the first stage, does the circuit oscillate? Draw the Bode and Nyquist plots. (b) Determine whether the amplifier versions of Figs. 11-8 through 11-10 produce oscillation if the output is connected to the input through a frequency-independent loop with $|A_v^* B| > 1$.

(14-2)   **14-2.** For a selected circuit, the Bode plot of Figure 14-21 is obtained. Draw a Nyquist plot. Does the circuit oscillate?

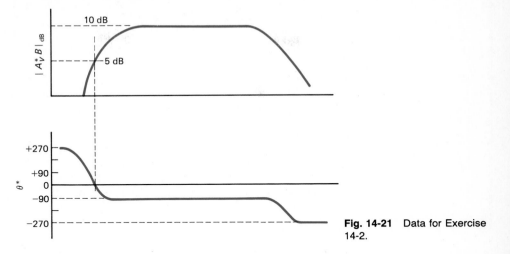

Fig. 14-21   Data for Exercise 14-2.

(14-3)   **14-3.** A Si thermistor has a resistance of 10 kΩ at 300°K and is mounted in a package for which $\theta = 100°C/W$. What power must be dissipated in the thermistor to produce a 100-Ω change in resistance?

(14-4)   **14-4.** A phase-shift oscillator is to be designed with $f_o = 10$ kHz using an op amp with $R_1 = 10$ kΩ. If $C = 0.001$ μF, what values for R and R_f should be selected?

(14-5)   **14-5.** What is the frequency of oscillation for the Wien bridge oscillator circuit of Fig. 14-22? Find R_f to assure oscillation.

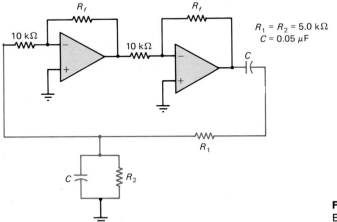

$R_1 = R_2 = 5.0$ kΩ
$C = 0.05$ μF

Fig. 14-22   Circuit for Exercise 14-5.

(14-5)   **14-6.** For a Wien bridge oscillator, the resistor and capacitor tolerances are ±10 percent and ±20 percent, respectively. Find the percentage range for the frequency of oscillation f_o with respect to the nominal value.

(14-5)   **14-7.** A Wien bridge oscillator is designed with $R_1 = 5.0$ kΩ, $R_2 = 10$ kΩ, $C_1 = 0.1$ μF, and $C_2 = 0.01$ μF. What is the frequency of oscillation?

(14-6)   **14-8.** Show that a parallel LC-tuned circuit consisting of an ideal capacitor and a high-Q inductor with resistance R_S can be modeled by an ideal L and C in parallel with an effective resistance $R_{eff} = R_S Q^2$.

(14-6)   **14-9.** Consider an LC resonant oscillator circuit using a JFET, with $L = 100 \ \mu H$ and $C = 0.10 \ \mu F$. Find the frequency of oscillation for the circuit.

(14-6)   **14-10.** An LC-tuned oscillator using a JFET is designed with $M/L = 0.10$, a 10-mH inductor with a 1.0-Ω resistance, and $C = 1.0 \ \mu F$. A given JFET has $g_m = 5.0 \ m\mho$ and $r_d = 50 \ k\Omega$. Show whether this JFET can be used in the oscillator.

(14-7)   **14-11.** A Colpitts oscillator uses a JFET with $I_{D0} = 3.5$ mA and $V_P = -3.5$ V. If $R_D = 1.0 \ k\Omega$, what bias voltage V_{GS} just produces the condition for oscillation?

(14-9)   **14-12.** A crystal has an impedance of 10 kΩ at the parallel resonant frequency. Given $R_f/R_1 = 2.0$, find the values of R in the circuit of Fig. 14-11(B) that result in oscillation.

(14-10)   **14-13.** For a given tunnel diode,

$$R_b = 25 \ \Omega \qquad L = 1.0 \ \mu H$$

$$r = 0.20 \ k\Omega \qquad C = 40 \ pF$$

If the tunnel diode is placed in a circuit with $R = 10 \ \Omega$, does the circuit oscillate?

14-14. Assume that the six oscillators of Fig. 14-23 are available (all fabricated using hybrid microelectronics) to produce a 1.0-MHz sine wave. The lowest-cost oscillator having $Q \geq 10^3$ and (BW) (vol.) < 10 kHz·cm³ is to be selected. Which circuit should be picked?

FIGURE 14-23
Data for Exercise 14

Oscillator	BW (kHz)	Cost ($)	Circuit volume (cm³)
A	1	8	2
B	10	10	2
C	5	12	1
D	10	14	1.5
E	0.1	10	4
F	0.5	6	3

Solutions   **14-1.** (a) As shown in Fig. 14-24, the circuit oscillates for all midrange frequencies. (b) The single-stage ac amplifier of Fig. 11-8 cannot oscillate because a $\theta^* = 0°$ condition does not exist. As noted in part (a), the two-stage ac amplifier of Fig. 11-9 oscillates for all midrange frequencies. The three-stage ac amplifier of Fig. 11-10(A) oscillates if the $\theta^* = 0°$ condition in Fig. 11-10(B) is associated with $|A_v^* B| > 1$. From the discussion of the phase-shift oscillator, this condition requires that the total gain of the three amplifiers be greater than 29, or that each oscillator have a gain of $29^{1/3} \cong 3.1$.

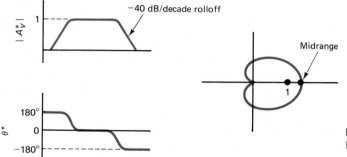

Fig. 14-24 Solution for Exercise 14-1.

14-2. The Nyquist plot corresponding to Fig. 14-21 is shown in Fig. 14-25. Because the curve passes outside the $A_v^* B = 1$ point, the circuit oscillates.

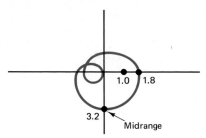

Fig. 14-25 Solution for Exercise 14-2.

14-3. From Example 14-1,

$$\frac{\Delta R}{R_o} \cong \frac{-\mathcal{E}_g}{2kT_o}\frac{\Delta T}{T_o}$$

For the values given,

$$\frac{\Delta T}{T_o} = -\left(\frac{100 \ \Omega}{10 \ k\Omega}\right)\left(\frac{2(0.026 \ \text{eV})}{1.11 \ \text{eV}}\right) \cong -4.7 \times 10^{-4}$$

$$\Delta T = -(300°\text{K})(4.7 \times 10^{-4}) \cong 0.14°\text{C}$$

Therefore, from Eq. (14-3),

$$P_D = \Delta T/\theta \cong 0.14°\text{C}/100°\text{C/W} = 1.4 \ \text{mW}$$

14-4. From Eq. (14-6),

$$R_{\text{eff}} C = \frac{1}{2\pi\sqrt{6} \ f_o} = \frac{1}{2\pi\sqrt{6} \ (10 \ \text{kHz})}$$
$$\cong 0.065 \times 10^{-4} \ \text{s}$$

so that

$$R_{\text{eff}} = \frac{0.065 \times 10^{-4} \ \text{s}}{0.001 \ \mu\text{F}} = 6.5 \ \text{k}\Omega$$

In Fig. 14-5(A), the equivalency $R = R_{\text{eff}} = 6.5 \ \text{k}\Omega$ can be made. However, R_x must be selected so that $R_x \| R_1 = 6.5 \ \text{k}\Omega$ and

$$\frac{10 \ \text{k}\Omega \, R_x}{10 \ \text{k}\Omega + R_x} = 6.5 \ \text{k}\Omega$$

$$R_x \cong 18.6 \ \text{k}\Omega$$

The gain of the amplifier must be greater than 29. Therefore,

$$R_f/R_1 = 29 \quad \text{and} \quad R_f = 29(10 \ \text{k}\Omega) = 0.29 \ \text{M}\Omega$$

As noted in Exercise 13-2 in Chapter 13, the input bias current limits the largest values of R_1 and R_f that can be used for a given application. In a realistic design situation, the objective for the above circuit may have to be reassessed and an iterative strategy used to find an acceptable solution with a lower value of R_f.

14-5. As shown in Fig. 14-22, the effective input resistance to the first op amp (10 kΩ) is in parallel with R_2. Therefore, from Example 14-3,

$$f_o = \frac{1}{2\pi (R_a R_b C_a C_b)^{1/2}}$$

$$= \frac{1}{2\pi [(5.0 \text{ k}\Omega)(5.0 \text{ k}\Omega \| 10 \text{ k}\Omega)(0.05 \ \mu\text{F})(0.05 \ \mu\text{F})]^{1/2}}$$

$$\cong 0.78 \text{ kHz}$$

To achieve the required gain of 3,

$$(R_f/R_1)^3 = 3$$

$$R_f = 3^{1/3}(10 \ k\Omega) \cong 14 \ k\Omega$$

14-6. From Eq. (14-11),

$$f_o = 1/2\pi RC$$

so that

$$f_{\min} = \frac{1}{2\pi (1.1R)(1.2C)} = 0.76 f_{\text{nom}}$$

$$f_{\max} = \frac{1}{2\pi (0.9R)(0.8C)} = 1.39 f_{\text{nom}}$$

and

$$0.76 \leq f_o/f_{\text{nom}} \leq 1.39$$

14-7. From Example 14-3,

$$f_o = \frac{1}{2\pi (R_a R_b C_a C_b)^{1/2}}$$

$$= \frac{1}{2\pi [(5.0 \text{ k}\Omega)(10 \text{ k}\Omega)(0.1 \ \mu\text{F})(0.01 \ \mu\text{F})]^{1/2}}$$

$$\cong 0.71 \text{ kHz}$$

14-8. The objective is to establish the relationship between R_s and R_{eff}, as shown in Fig. 14-26. Equivalency is shown if $X_L + R_s = R_{\text{eff}} \| X_L$. Expanding this expression,

$$jX_L + R_s = jX_L \frac{R_{\text{eff}}}{R_{\text{eff}} + jX_L} = jX_L \frac{1}{1 + jX_L/R_{\text{eff}}}$$

By definition, $Q = X_L/R_s$ at resonance, so by substitution,

$$1 + \frac{1}{jQ} = \frac{1}{1 + jX_L/R_{\text{eff}}}$$

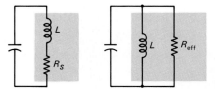

Fig. 14-26 Equivalent circuits for Exercise 14-8.

If $Q \gg 1$ and $X_L \ll R_{eff}$,

$$1 - jQ \cong 1 - jX_L/R_{eff}$$

or

$$1/Q = X_L/R_{eff}$$

which means that the two inequalities are self-consistent. Combining the above relationships,

$$1/Q = QR_s/R_{eff}$$

so that

$$R_{eff} = R_s Q^2$$

which is the desired result.

14-9. From Eq. (14-17),

$$f_o = \frac{1}{2\pi\sqrt{LC}}$$

$$= \frac{1}{2\pi[(100 \ \mu H)(0.10 \ \mu F)]^{1/2}}$$

$$\cong 50 \text{ kHz}$$

14-10. The resonant frequency is

$$f_o = \frac{1}{2\pi\sqrt{LC}} = \frac{1}{2\pi[(10 \text{ mH})(1.0 \ \mu F)]^{1/2}}$$

$$\cong 1.6 \text{ kHz}$$

so that

$$Q = \frac{\omega_o L}{R_s} = \frac{(10 \times 10^3)(10 \text{ mH})}{1.0 \ \Omega} = 100$$

Therefore,

$$R_{eff} = Q^2 R_s = 10^4(1 \ \Omega) = 10 \text{ k}\Omega$$

From Eq. (14-18), oscillation occurs if

$$M/L \ g_m r_d \| R_{eff} \geq 1$$

By substitution,

$$(0.10)(5.0 \text{ m}\mho)(50 \text{ k}\Omega \| 10 \text{ k}\Omega) \stackrel{?}{\geq} 1$$

$$4.2 \geq 1$$

and the circuit oscillates.

14-11. From Eq. (14-22), oscillation will just occur if

$$g_m R_D = 1$$

or

$$g_m = 1/1.0 \text{ k}\Omega = 1.0 \text{ m}\mho$$

Since

$$g_m = -\frac{2I_{D0}}{V_P}\left(1 - \frac{V_{GS}}{V_P}\right)$$

the desired bias point is

$$1 - \frac{V_{GS}}{V_P} = -\frac{g_m V_P}{2I_{D0}} = \frac{(-1.0\ \text{m}\mho)(-3.5\ \text{V})}{2(3.5\ \text{mA})} = 0.50$$

so that

$$V_{GS} = 0.5V_P \cong -1.8\ \text{V}$$

14-12. From Eq. (14-30), the oscillation condition is

$$\left(1 + \frac{R_f}{R_1}\right)\frac{Z_{\text{cry}}}{Z_{\text{cry}} + R} \geq 1$$

or

$$R \leq Z_{\text{cry}}(1 + R_f/R_1) - Z_{\text{cry}} = Z_{\text{cry}}(R_f/R_1)$$

By substitution,

$$R < (10\ \text{k}\Omega)(2.0) = 20\ \text{k}\Omega$$

results in oscillation.

14-13. From Eq. (14-36), oscillation occurs if

$$\frac{R_b + R}{r} < \frac{L}{r^2 C} < 1$$

By substitution,

$$\frac{25\ \Omega + R}{0.20\ \text{k}\Omega} \overset{?}{\lessgtr} \frac{1.0\ \mu\text{H}}{(0.20\ \text{k}\Omega)^2(40\ \text{pF})} \overset{?}{\lessgtr} 1$$

$$\frac{25\ \Omega + R}{0.20\ \text{k}\Omega} \overset{?}{\lessgtr} 0.63 < 1$$

so the circuit oscillates if $R < 100\ \Omega$. From Eq. (14-33), the frequency of oscillation will be

$$\omega_o = \left[\frac{1}{(1.0\ \mu\text{H})(40\ \text{pF})}(1 - 0.63)\right]^{1/2}$$

$$\cong 1.0 \times 10^8\ \text{Hz}$$

14-14. Using the relationship $Q = f_o/BW = 1.0\ \text{MHz}/BW$, the calculations of Fig. 14-27 can be made. The preferred choice is oscillator A.

FIGURE 14-27
Solution for Exercise 14-14

Oscillator	Q (10³)	(BW)(Volume)kHz · cm³	Accept?
A	1.0	2.0	Yes
B	0.10	20	No
C	0.20	5.0	No
D	0.10	15	No
E	10	0.40	Yes
F	0.20	1.5	No

(14-2) **14-1.** For a given amplifier and feedback loop combination, $|A_v^* B| = 1.2$ and $\theta^* = 270°$. Does the circuit oscillate?

(14-3) **14-2.** For a given oscillator circuit, an output $v = (2\ \text{V})\sin(\omega t)$ is to be obtained. What resistance value should be used for a thermistor to provide AGC, given $\theta = 50°\text{C/W}$? Select $\Delta T = 0.2°\text{C}$.

(14-3) **14-3.** A Ge thermistor at 300°K has a resistance of 12 kΩ and is mounted in a package with $\theta = 50°\text{C/W}$. A 150-Ω change in resistance is to be associated with AGC stabilization. (a) What ΔT is required? (b) What average power is required to produce this temperature change?

(14-4) **14-4.** A phase-shift oscillator is made with $R = 10$ kΩ and $C = 0.01\ \mu\text{F}$. What is the resonant frequency? Assume $R_1 \gg R$.

(14-4) **14-5.** A phase-shift oscillator is to be evaluated. By observation, $R_1 = 8.0$ kΩ, $C = 0.001\ \mu\text{F}$, and $R = 5.0$ kΩ. The value for R_x cannot be determined by observation. By implication, what value for R_x should be expected by measurement?

(14-5) **14-6.** For a Wien bridge oscillator, both R values are 25 kΩ and both C values are 0.036 μF. What is the frequency of operation?

(14-5) **14-7.** For the Wien bridge oscillator of Problem 14-6, find $|B|$ for frequencies 20 percent above and below the resonant frequency.

(14-5) **14-8.** For the Wien bridge oscillator of Fig. 14-7(A), $R_a = R_b = 6.6$ kΩ, and $C_a = C_b = 0.026\ \mu\text{F}$. Find the frequency of oscillation for the circuit.

(14-5) **14-9.** For a Wien bridge oscillator, what tolerance in C is necessary to produce an uncertainty of 10 percent in f_o if the R values are known within 2 percent?

(14-6) **14-10.** For an LC resonant oscillator, $L = 100\ \mu\text{H}$ and $C = 2.4 \times 10^2$ pF. Find the resonant frequency.

(14-6) **14-11.** An LC resonant oscillator is designed to oscillate at 10 kHz with a JFET having $r_d = 50$ kΩ and $g_m = 4.0$ m$\mho$. If $M/L = 0.10$ for the transformer being used and $L = 5.0\ \mu\text{H}$, what must be the Q of the primary transformer coil to produce oscillation?

(14-7) **14-12.** Find the resonant frequencies for Colpitts and Hartley oscillators using the component
(14-8) values of Problem 14-10.

(14-7) **14-13.** Sketch the expected Bode and Nyquist plots for a Colpitts oscillator.

(14-7) **14.14.** For a Colpitts oscillator, $L = 5.0\ \mu\text{H}$ and $C = 1.0 \times 10^4$ pF. Find the values of B at frequencies 20 percent above and below the resonant frequency.

(14-9) **14-15.** Draw the schematic for a crystal oscillator using the series resonance properties of the crystal.

(14-10) **14-16.** For a given tunnel diode, $r = 150\ \Omega$ and $L = 2.5\ \mu\text{H}$. What value of capacitance must the device have to allow oscillation to take place? What is the maximum allowable value for $R_b + R$?

(14-12) **14-17.** For a complex amplifier circuit, a distributed high-pass filter has the potential for producing a positive feedback loop. If $C_d = 5.0$ pF and $R_{\text{eff}} = 100$ kΩ, what value of f_2 should be selected for the amplifier to assure that oscillation does not occur from this coupling?

(14-12) **14-18.** Prepare Bode and Nyquist plots for the amplifier and distributed filter coupling described in Figs. 14-14 and 14-15 for the case $f_2 = f_2'$.

COMPUTER APPLICATIONS

* **14-1.** Write a computer program to solve Exercise 14-14 for any number of oscillators N. Then, solve Exercise 14-14 using this program.

14-2. Use SPICE to simulate the open-loop amplitude and phase-shift frequency-response properties of the oscillators of Figs. 14-5 and 14-7. Compare the range of frequencies that might be observed in the oscillator output in each case.

EXPERIMENTAL APPLICATION

14-1. The objective of this experiment is to construct a Wien bridge oscillator using the 741 op amp and prepare a Nyquist plot based on data collected. Choose values for the feedback circuit so that $f_o \cong 10$ kHz and $R \geq 10$ kΩ.

(a) Use a variable resistor R_f to allow the loop gain to be adjusted. Construct the oscillator and see if it oscillates at the expected frequency. Adjust R_f to obtain the most linear output possible.

(b) Use the method of AGC shown in Fig. 14-4. Compare the output properties for these two approaches to amplitude stabilization.

(c) Disconnect the feedback loop at the input. Connect a sine-wave source to the input and record data for $|A_V^* B|$ and θ^* as a function of frequency for the amplifier and feedback circuit together. Use the largest frequency range that can be managed. Prepare an interpretive Bode plot and Nyquist plot. Label the point $|A_V^* B| = 1$. Interpret your results.

Discussion
A typical Wein bridge oscillator can be constructed with $R_a = R_b = 1.5$ kΩ and $C_a = C_b = 0.01$ μF, so that

$$f_o = 1/2\pi RC \cong 11 \text{ kHz}$$

Figure 14-28 shows typical data collected for part (c). The amplitudes of the input and output sine waves and their phase differences are recorded as a function of frequency. Figure 14-29 shows how $|A_V^* B|$ can be calculated from the experimental data. Figure 14-30 presents a Bode plot of the experimental data. The experimental frequency of operation is approximately 9.9 kHz. Figure 14-31 shows the resultant Nyquist plot for this particular experiment. Discuss and interpret these findings in preparation for the experiment.

FIGURE 14-28
Data Collected for the Frequency Response of a Wien Bridge Oscillator

Frequency (Hz)	$v_{in}(V)$	$v_f(V)$	$\theta^*(°)$	$v_{out}(V)$
3.0×10^5	2.0	0.02	-1.6×10^2	0.60
1.5×10^5	2.0	0.08	-1.5×10^2	1.3
8.3×10^4	2.0	0.22	-1.4×10^2	2.2
5.0×10^4	2.0	0.58	-1.1×10^2	3.7
2.0×10^4	2.0	1.9	-28	6.0
10×10^3	2.0	2.1	0	6.0
7.0×10^3	2.0	2.0	15	6.0
4.0×10^3	2.0	1.7	36	6.0
1.5×10^3	2.0	0.85	64	6.0
8.0×10^2	2.0	0.44	76	6.0
5.0×10^2	2.0	0.30	80	6.0
1.5×10^2	2.0	0.10	88	6.0

FIGURE 14-29
Data Used to Prepare Bode and Nyquist Plots

Frequency (Hz)	$v_f/v_{in} = \lvert A_v^* B \rvert$	$\lvert A_v^* B \rvert_{dB}$
3.0×10^5	0.01	-40
1.5×10^5	0.04	-29
8.3×10^4	0.11	-19
5.0×10^4	0.29	-11
2.0×10^4	0.95	-0.5
10×10^3	1.1	0.4
7.0×10^3	1.0	0
4.0×10^3	0.83	-1.7
1.5×10^3	0.43	-7.4
6.0×10^2	0.22	-13
5.0×10^2	0.15	-17
1.5×10^2	0.05	-26

Fig. 14-30 Bode plot of the oscillator circuit.

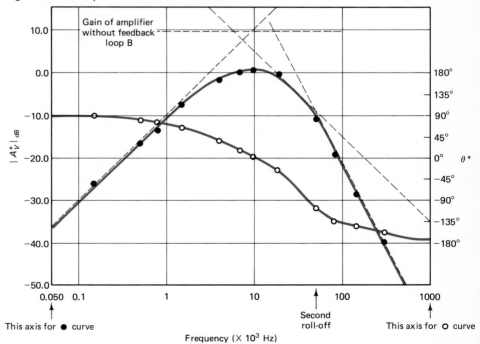

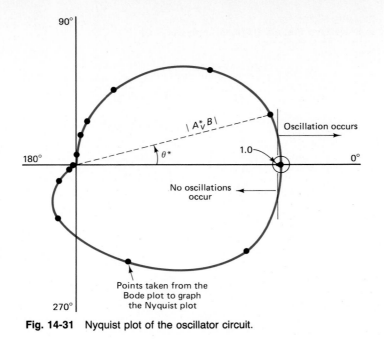

Fig. 14-31 Nyquist plot of the oscillator circuit.

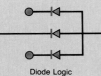

Diode Logic

PART 6
DIGITAL
APPLICATIONS

The previous chapters introduced semiconductor devices, the design process in electronics, and many different types of analog circuit configurations. In Chapters 15 to 17, the results of this previous development are used to gain an understanding of digital electronics. The same types of transistors are used, but circuit design takes on new forms to achieve the operational requirements and performance objectives associated with digital systems. Many of the same system design concepts hold as this new application area is pursued.

Nonlinear applications of transistors have been introduced in Section 11-6, with a discussion of classes of amplifiers. Several important types of circuits have been found to depend on device operation that extends beyond the small-signal linear region.

This part introduces another type of nonlinear operation, with transistors used as switches in order to create digital circuits and systems. Chapter 15 describes how both bipolar and field-effect transistors can be used in a switching mode to achieve the logic functions that are required for digital systems. The evolution of digital technology is traced through several families of circuits and devices, starting with configurations that lend themselves to discrete fabrication methods and progressing to those that are used for integrated-circuit fabrication. The emphasis in Chapter 15 is on the building block circuits that are used in systems and the constraints that enter into such system design.

Whereas the operational amplifier is the most important linear IC circuit and many different types of IC op amps are available, one of the most important digital circuits is the *gate,* and many different types and arrangements of IC gates are produced. However, whereas analog ICs typically contain only one or a few op amps, digital ICs can contain thousands of gates. Chapter 15 provides an introduction to the most important digital gates that are in use today and describes some of the strengths and weaknesses of these various circuits. As will be noted, distributed R, C, and L functions and the required high-density packaging techniques are important in designing high-performance digital circuits.

Chapter 16 introduces some of the basic design features of digital computers in order to provide a systems context for application of the circuits of Chapter 15. The functional design requirements for a digital computer are related to the requirements placed on individual gate circuits, and circuit design must reflect the ultimate system objectives. Chapter 16 provides the means for illustrating how systems and circuit-level performance objectives are linked. The integration strategy used is also an essential determinant of system performance. The concepts of AM and FM communication are applied to digital communication systems in order to consider the operation of networks of computers.

Chapter 17 bridges the circuits of Chapter 15 and the overall systems design of Chapter 16. It is useful to obtain a working knowledge of the types of IC subsystem building block elements that are generally available to the designer to form this bridge. The building block elements (in conjunction with discrete and analog IC circuitry) are the "raw materials" from which more complex digital systems are created.

Chapter 17 considers the strategies that are used in creating these building block elements, the types of external parameters that are used to describe them, the limitations associated with the use of these parameters, how they form the bridge between individual circuits and subsystems, and some of the broader tradeoff issues.

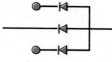

CHAPTER 15
DIGITAL CIRCUITS

Digital applications make use of nonlinear circuits that have only two stable operating points. Changes in circuit operation take place when circuits switch between these two points. This chapter shows how such circuits can be designed to meet application objectives.

Both bipolar and MOSFET devices are widely used for the fabrication of digital IC circuits, and many tradeoffs exist among the alternative circuit types. There is no single "best solution" for all applications. This chapter draws on the earlier semiconductor material descriptions in order to understand some of the important parameters that affect circuit characteristics and on the discussions of device operation to describe the many ways in which transistors are used in digital circuits. Since gate circuits are used as building blocks for large systems, such circuits are designed by considering how they will function within the larger network setting. This requirement leads to some new and useful design methods. The nature of the tradeoffs change, as do the models that are found to be most relevant.

15-1 DIGITAL DESIGN REQUIREMENTS

Since they have only two stable operating states, digital circuits can be best analyzed using a bivalued form of mathematics called *Boolean algebra*. The two operating states are represented by a **0** and a **1**. These symbols can represent positive or negative voltages of any value.

The analysis of digital circuits uses building block units called *gates* that can perform the logical operations that arise in Boolean algebra. The most important logical operations are given names that symbolize their function. The following sections cover digital circuits that function as NOT, AND, OR, NAND, and NOR gates.

The characteristics of NOT, AND, and OR gates are shown in Figs. 15-1(A) through (C). A NOT circuit produces a **1** output with a **0** input and a **0** output with a **1** input. This relationship is summarized in the *truth table* shown in Fig. 15-1(A). An AND circuit produces a **1** output from a multiinput circuit only when all inputs have the value **1**; otherwise, the output is **0**. An OR circuit produces a **1** output if any input has the value **1**. The AND and OR circuit symbols and truth tables are shown in Figs. 15-1(B) and 15-1(C), respectively.

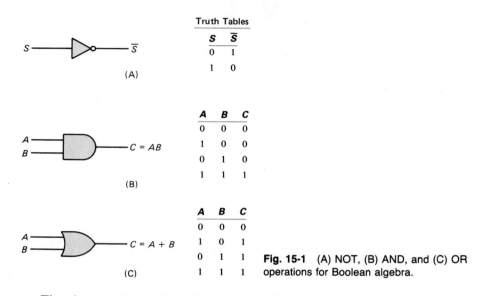

Truth Tables

S	$\bar{S}$
0	1
1	0

(A)

A	B	C
0	0	0
1	0	0
0	1	0
1	1	1

(B)

A	B	C
0	0	0
1	0	1
0	1	1
1	1	1

(C)

Fig. 15-1 (A) NOT, (B) AND, and (C) OR operations for Boolean algebra.

The characteristics of the NAND and NOR logical units are shown in Fig. 15-2. A NAND circuit produces a **0** output only when all inputs have the value **1**. A NOR circuit produces a **0** output if any input has the value **1**.

Fig. 15-2 (A) NAND operation and (B) NOR operation.

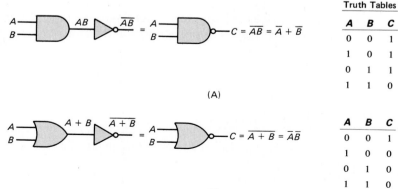

Truth Tables

A	B	C
0	0	1
1	0	1
0	1	1
1	1	0

(A)

A	B	C
0	0	1
1	0	0
0	1	0
1	1	0

(B)

As illustrated in Figs. 15-3 and 15-4, an entire Boolean logic system can be built by using only NAND gates or NOR gates. Combinations of NAND gates can produce the NOT, AND, OR, and NOR functions. Combinations of NOR gates can produce the NOT, AND, OR, and NAND functions. Either type of gate can thus be used as a building block unit for a digital logic system.

Fig. 15-3 Building digital logic functions using NAND gates.

Truth Tables

A	B	C
0	0	1
1	0	1
0	1	1
1	1	0

NAND

S	$\overline{S}$
0	1
1	0

NOT

A	B	C	D
0	0	1	0
1	0	1	0
0	1	1	0
1	1	0	1

AND

A	B	C	D	E
0	0	1	1	0
1	0	0	1	1
0	1	1	0	1
1	1	0	0	1

OR

A	B	C	D	E	F
0	0	1	1	0	1
1	0	0	1	1	0
0	1	1	0	1	0
1	1	0	0	1	0

NOR

Truth Tables

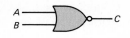

A	B	C
0	0	1
1	0	0
0	1	0
1	1	0

NOR

S	S̄
0	1
1	0

NOT

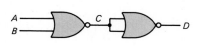

A	B	C	D
0	0	1	0
1	0	0	1
0	1	0	1
1	1	0	1

OR

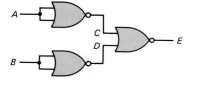

A	B	C	D	E
0	0	1	1	0
1	0	0	1	0
0	1	1	0	0
1	1	0	0	1

AND

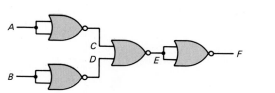

A	B	C	D	E	F
0	0	1	1	0	1
1	0	0	1	0	1
0	1	1	0	0	1
1	1	0	0	1	0

NAND

Fig. 15-4 Building digital logic functions using NOR gates.

15-2 DIODE LOGIC

Circuits that perform the OR and AND functions can be created by using combinations of diodes and resistors, as shown in Fig. 15-5. For now assume that the voltage drops across the diodes can be neglected.

Let the voltage levels for these circuits be 0 V and $-V_1$, or 0 V and $+V_1$, as shown. *Negative logic* is obtained when the most negative voltage level is defined as a **1**, and for *positive logic,* the most positive voltage is represented by a **1**.

In Fig. 15-5(A), consider the case for which both A and B are at 0 V. Then C is

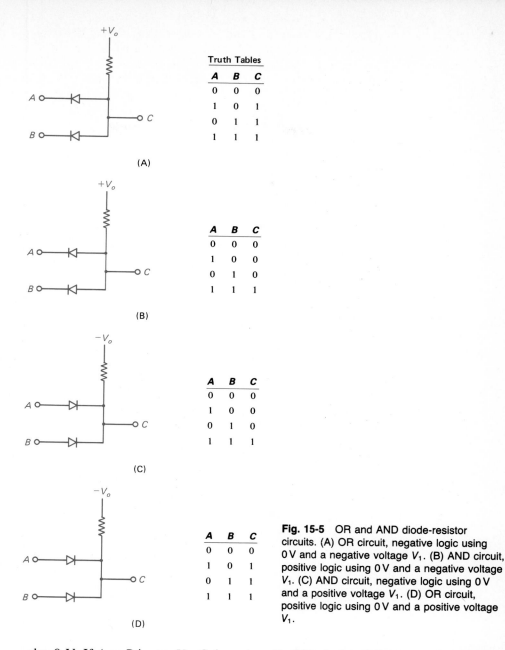

Truth Tables

A	B	C
0	0	0
1	0	1
0	1	1
1	1	1

(A)

A	B	C
0	0	0
1	0	0
0	1	0
1	1	1

(B)

A	B	C
0	0	0
1	0	0
0	1	0
1	1	1

(C)

A	B	C
0	0	0
1	0	1
0	1	1
1	1	1

(D)

Fig. 15-5 OR and AND diode-resistor circuits. (A) OR circuit, negative logic using 0 V and a negative voltage V_1. (B) AND circuit, positive logic using 0 V and a negative voltage V_1. (C) AND circuit, negative logic using 0 V and a positive voltage V_1. (D) OR circuit, positive logic using 0 V and a positive voltage V_1.

also 0 V. If A or B is at $-V_1$, C drops to $-V_1$. If both A and B are at $-V_1$, then C is also at $-V_1$. Referring to Fig. 15-1 reveals that the OR operation is produced.

If 0 V is redefined to be a **1** and $-V_1$ to be a **0**, positive logic is obtained, as shown in Fig. 15-5(B). The truth table symbols in Fig. 15-5(A) now change to those in Fig. 15-5(B), resulting in an AND circuit.

Whether a given diode-resistor circuit performs the OR or AND function depends on the definition of logic levels. Figures 15-5(A) and (B) show a circuit that performs the OR logic function for negative logic and the AND function for positive logic. Figures 15-5(C) and (D) show a circuit that performs the AND function for negative logic and the OR function for positive logic.

In a practical sense, the use of these circuits by themselves is limited because of the voltage drops across the diodes. Digital computer circuits must provide a method for maintaining the **0** and **1** voltage levels at specific values through long chains of logical operations. A way must also be found to produce the NOT function so as to provide a complete set of operations.

The BJT provides a natural solution. When operated in saturation or cutoff, it provides two well-defined output voltage states. The BJT in a common-emitter configuration provides a natural 180° phase shift, giving the needed NOT function. And the current gain of the device can be used to maintain voltage levels through a long chain of circuits.

15-3 THE BJT AS A SWITCH

The use of the BJT as a switch is illustrated in Fig. 15-6. Consider that a current I_B flows in the base of the device.

The transistor can be biased in the linear region if $I_C = h_{FE}I_B$. If I_B is then held fixed and I_C is reduced by increasing R_C, the BJT moves out of the linear region into *saturation*. The voltage V_{CE} becomes small and approximately independent of I_C. The more I_C is reduced, the deeper the BJT moves into saturation.

If a value $(I_C)_{max}$ is defined to represent an upper limit to the collector current

Fig. 15-6 The BJT as a switch.

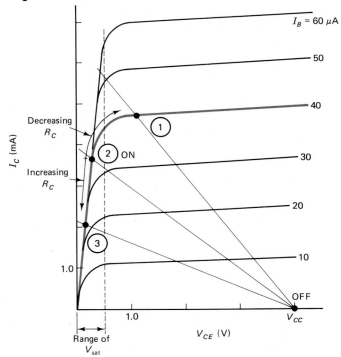

through the BJT, saturation can be assured if I_B is greater than a certain saturation value. For $I_B > (I_B)_{\text{sat}}$, V_{CE} is approximately constant independent of I_C. The BJT begins to enter saturation when $I_B > I_C/h_{FE}$ and is deeply in saturation when $I_B \gg I_C/h_{FE}$. A parameter $h_{\text{sat}} \ll h_{FE}$ is often defined so that if $I_B > I_C/h_{\text{sat}}$, the BJT is sure to be saturated.

The needed NOT (or *inverter*) circuit can be developed by causing the BJT to move between an *off* state and a saturated state. When the BJT is in saturation, $V_{CE} < 0.7$ V (typically 0.2 to 0.5 V for a Si device). When the BJT is *off*, the value of V_{CE} is set by the external circuit. This application of the BJT produces noise protection for a chain of circuits because V_{CE} does not fluctuate appreciably with small variations in I_B and I_C.

15-4 DESIGN OF A DISCRETE-COMPONENT GATE CIRCUIT USING DTL TECHNOLOGY

There are several important design objectives for a gate circuit. It is desirable to use as few circuits as possible to achieve the desired system performance (to minimize size, cost, and power requirements) and to maximize the potential for complex logical configurations. Related design objectives are thus to maximize the ability of each circuit to have its output drive many additional circuits (*fan-out*) and to maximize the ability of the circuit to have its input connected to many different circuits (*fan-in*). Further, acceptable power-dissipation levels, operation within device specifications, and the maximum circuit speed possible are needed. A number of design tradeoffs arise from these conflicting objectives.

The design task can be approached by considering each gate circuit to be one of a long chain. Design values must be selected so that the composite system functions as desired. This can complicate the design because individual circuits cannot be analyzed in isolation.

This section considers a basic gate circuit that operates between the two voltage levels: V_{sat} (the value of V_{CE} with the BJT saturated) and a positive voltage set by the external circuit. The following design uses Si npn devices, but can be extended to Ge and pnp devices.

Figure 15-7 shows a simple diode-transistor logic (DTL) inverter circuit. Consider what happens when the $(N - 1)$th stage is *on* (with the npn conducting). The output of this stage is V_{sat}. Voltage V_1 for the Nth stage is given by $V_{\text{sat}} + V_{D1}$, where V_{D1} is the drop across the input diode to the Nth stage.

The base current through Q_N is very small because V_1 appears across a combination of three diodes (resulting in a voltage $V_{D2} + V_{D3} + V_{BE}$). If $V_{\text{sat}} = 0.5$ V and $V_{D1} = 0.7$ V, then the total voltage across these three diodes is 1.2 V. If 1.2 V$/3 = 0.4$ V appears across the base-emitter junction of Q_N, only a very small current flows. The current I_1 is thus

$$I_1 \cong \frac{V_{CC} - V_{\text{sat}} - V_D}{R_1} \tag{15-1}$$

The BJT characteristic curve (shown in Fig. 15-6) indicates that a small I_B produces $V_{CE} \cong V_{CC}$. Transistor Q_N is essentially *off*. The inverter function results because the *on* $(N - 1)$th stage forces the Nth stage *off*.

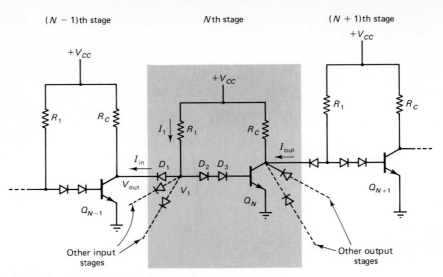

Fig. 15-7 A chain of DTL gate circuits.

What happens if the $(N − 1)$th stage turns *off*? As the current flow through R_C decreases, V_{out} rises toward V_{CC}. The collector resistor R_C is called the *pull-up resistor* because it provides the mechanism to raise V_{out} to V_{CC} when the BJT turns *off*. The voltage V_1 rises toward V_{CC}, but when it reaches $3V_D$, Q_N turns *on*, clamping the voltage at this level. The base current is given by

$$I_B \cong \frac{V_{CC} − 3V_D}{R_1} \tag{15-2}$$

The coupling diode D_1 is reverse-biased and *off*.

The collector current to Q_N flows through R_C and from the $(N + 1)$th stages (to keep them turned *off*). The current through R_C is

$$I_{R_C} = \frac{V_{CC} − V_{sat}}{R_C} \tag{15-3}$$

The current flow I_1 into the collector from each following circuit is given by Eq. (15-1). The maximum fan-out m of the circuit can be determined by the equation

$$I_{R_C} + mI_1 < h_{sat}I_B \tag{15-4}$$

which is the condition requiring that Q_N remain in saturation. A simple DTL inverter circuit can be quickly designed by combining the above equations.

Example 15-1   DTL Gate Circuit

What is the maximum fan-out for a DTL gate circuit with $V_{CC} = 6.0$ V, $R_1 = 5.0$ kΩ, and $R_C = 5.0$ kΩ? Assume that $V_D = 0.7$ V for all diodes, $V_{sat} = 0.5$ V, and $h_{sat} = 20$.

Solution   When an inverter stage is *off*,

$$I_1 = \frac{(6.0 − 0.5 − 0.7) \text{ V}}{5.0 \text{ k}\Omega} = \frac{4.8 \text{ V}}{5.0 \text{ k}\Omega} = 0.96 \text{ mA}$$

When a stage is *on*, the base current into Q is

$$I_B = \frac{[6.0 - 3(0.7)] \text{ V}}{5.0 \text{ k}\Omega} = \frac{3.9 \text{ V}}{5.0 \text{ k}\Omega} = 0.78 \text{ mA}$$

The total collector current is

$$I_C = I_{RC} + mI_1 = \frac{(6.0 - 0.5) \text{ V}}{5.0 \text{ k}\Omega} + m(0.96 \text{ mA}) = 20(0.78 \text{ mA}) = 15.6 \text{ mA}$$

so that

$$m < \frac{(15.6 - 1.1) \text{ mA}}{0.96 \text{ mA}} \cong 15$$ ∎

Example 15-2   Power Dissipation

How much power is dissipated in the above gate circuit when the circuit is (a) *on* and (b) *off*.

Solution   (a) When the gate circuit is *on*,

$$P_{D1} = 0 \text{ V}$$

$$P_{D2} = P_{D3} = (0.7 \text{ V})(0.78 \text{ mA}) \cong 0.6 \text{ mW}$$

$$P_{R1} = (0.78 \text{ mA})^2(5 \text{ k}\Omega) \cong 3.0 \text{ mW}$$

$$P_{RC} = \frac{(6 \text{ V} - 0.5 \text{ V})^2}{5 \text{ k}\Omega} \cong 6.1 \text{ mW}$$

$$P_Q = V_{\text{sat}} I_C + I_B V_{BE} = (0.5 \text{ V})(20)(0.78 \text{ mA}) + (0.78 \text{ mA})(0.7 \text{ V})$$
$$= 8.4 \text{ mW}$$

$$P_{\text{tot}} = P_{D1} + P_{D2} + P_{D3} + P_{R1} + P_{RC} + P_Q$$
$$\cong 18.7 \text{ mW}$$

(b) When the inverter circuit is *off*, the power levels become

$$P_{D1} = (0.7 \text{ V})(0.96 \text{ mA}) = 0.7 \text{ mW}$$

$$P_{D2} = P_{D3} = 0$$

$$P_{R1} = (0.96 \text{ mA})^2 \, 5 \text{ k}\Omega = 5 \text{ mW}$$

$$P_{RC} \cong 0$$

$$P_Q \cong 0$$

$$P_{\text{tot}} = 5.7 \text{ mW}$$

For positive logic, the gate circuit of Fig. 15-8 provides the AND + NOT = NAND function [compare with Fig. 15-5(B)]. For negative logic, the gate provides the OR + NOT = NOR function [compare with Fig. 15-5(A)]. As shown in Figs. 15-3 and 15-4, this gate can become a building block unit for a complex digital system. ∎

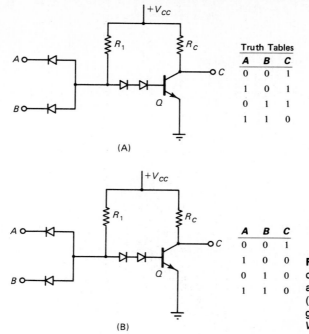

(A)

| Truth Tables | | |
A	B	C
0	0	1
1	0	1
0	1	1
1	1	0

(B)

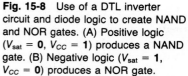

A	B	C
0	0	1
1	0	0
0	1	0
1	1	0

Fig. 15-8 Use of a DTL inverter circuit and diode logic to create NAND and NOR gates. (A) Positive logic ($V_{sat} = 0$, $V_{CC} = 1$) produces a NAND gate. (B) Negative logic ($V_{sat} = 1$, $V_{CC} = 0$) produces a NOR gate.

15-5 SWITCHING SPEED

In the design of digital circuits, circuit speed is always important since high-speed computers require high-speed gate circuits. The DTL circuits of Fig. 15-8 work as designed in terms of their two stable states. However, the switching speed between states leaves much to be desired. To see why this is the case, consider the transient processes that take place between stable states.

Since every circuit includes device and distributed capacitances, changes between states involve delays that relate to charging or discharging currents. Figure 15-9 illustrates how the turn-off path followed by the BJT is affected by the load capacitance C. The load line describes the relationship between V_{CE} and I_C that exists due to the current flow through the load resistor R. An additional current flow takes place due to capacitance effects. During turn-on, V_{CE} is slow to drop until the charge on C is reduced. During turn-off, V_{CE} begins to rise only after internal excess charge levels clear out and a charging current flows into C through R. The turn-on and turn-off times can be quite different due to the time constants and charging mechanisms involved.

In general,

$$\Delta t = \int_{v_1}^{v_2} C \frac{dv}{i_{cap}} \tag{15-5a}$$

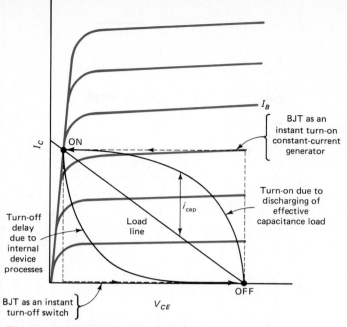

Fig. 15-9 Switching between stable states.

gives the delay associated with a shift in v from v_1 to v_2. If the current through R_L is small compared with the current through C over most of the switching period of interest, then $i_{cap} \cong I_C$. If the BJT behaves as a constant-current generator, then

$$\Delta t \cong C \frac{v_2 - v_1}{I_C} \cong \frac{C}{g_m}\left(\frac{v_2 - v_1}{V_0}\right) \tag{15-5b}$$

From this approximate result, the ratio C/g_m can be regarded as a figure of merit for a BJT during the turn-on switching transient.

Example 15-3   **Estimating the Turn-On Delay Time for a BJT**

A switching circuit is operated between switching levels of 2.5 and 0.5 V. If g_m is typically $\cong 400\,\mathrm{m\mho}$ for the BJT in use and the load capacitance is about $C \cong 50\,\mathrm{pF}$, estimate the turn-on switching time.

Solution   From Eq. (15-5b),

$$\Delta t \cong \frac{50 \times 10^{-12}\,\mathrm{F}}{400 \times 10^{-3}\mho} \frac{2.5\,\mathrm{V} - 0.5\,\mathrm{V}}{0.026\,\mathrm{V}}$$

$$\cong 9.6\,\mathrm{ns}$$

The turn-off delay time is determined by the time required for excess carriers in the BJT to recombine and by the external circuit configuration. ∎

It is useful to trace the switching process in terms of the waveforms that are observed at the input and output to a typical DTL circuit. Figure 15-10(A) shows

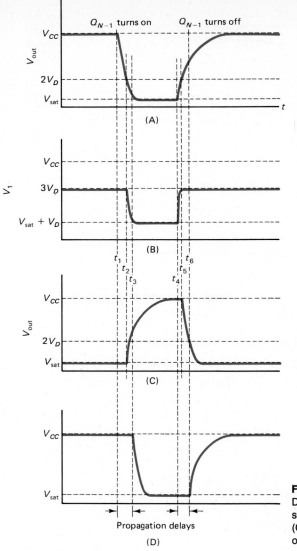

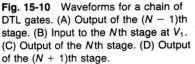

Fig. 15-10 Waveforms for a chain of DTL gates. (A) Output of the $(N - 1)$th stage. (B) Input to the Nth stage at V_1. (C) Output of the Nth stage. (D) Output of the $(N + 1)$th stage.

a waveform that is observed on the output of the $(N - 1)$th stage as it turns *on* and *off*.

The transients between stable states are dominated by the effective capacitances in the circuit. When Q_{N-1} turns *on*, V_{out} drops from V_{CC} toward V_{sat} (refer to Fig. 15-7). The coupling diode D_1 comes *on* when V_{out} drops below $2V_D$, as shown in Fig. 15-10(B). The voltage at V_1 then begins to drop until it reaches $V_{sat} + V_D$. Transistor Q_N begins to turn *off*, so that its output rises toward V_{CC} as shown in Fig. 15-10(C). The $(N + 1)$th stage starts to turn *on* when V_{out} of the Nth stage rises above $2V_D$, as shown in Fig. 15-10(D).

Observe that the output of the $(N + 1)$th stage has been delayed in time with respect to the output of the $(N - 1)$th stage. The average propagation delay time through each circuit is an important parameter. Note that the delays shown in Fig. 15-10(D) are associated with two such gate circuits.

The delays associated with the charging and discharging of circuit capacitances require a minimum pulse width to assure switching. Both the propagation delays and minimum pulse width serve to limit system performance.

In describing the waveforms during turn-on and turn-off, reference can be made to the rise and fall times of each transient. These times are usually measured with respect to the 10 and 90 percent voltage levels, as shown in Fig. 15-11. Rise time refers to an increasing voltage and fall time to a decreasing voltage.

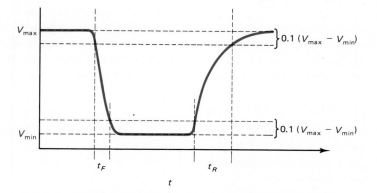

Fig. 15-11 Definitions of rise time t_R and fall time t_F for a pulse.

It is possible to speed up the switching times of the DTL circuit of Fig. 15-7 by making the changes shown in Fig. 15-12. Diode D_4 can be added to limit the maximum value taken by V_{out} to the level $V_{cl} + V_D$, where V_{cl} is a clamp voltage less than V_{CC}. By limiting the voltage extremes of the stable states, the times associated with switching transients can be reduced. The limit to this strategy is reached when noise in the circuit prevents the stable voltages from being closer to one another.

The result obtained by adding D_4 is not the same as reducing V_{CC}. With or without D_4, the collector rises from V_{sat} toward V_{CC} during turn-off. By adding D_4, the output is clamped to make use of the most rapidly changing part of the turn-off time constant.

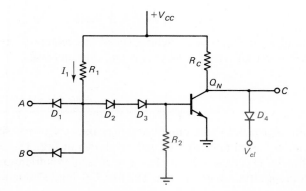

Fig. 15-12 DTL circuit with R_2 and D_4 added to improve switching speed.

The addition of R_2 can also improve circuit performance. The switching times associated with Q_N are limited by the effective capacitance between the base and emitter. This capacitance must be charged and discharged during each switching cycle. By adding R_2, the charging and discharging time constants are reduced. However, R_1 must also be reduced to provide adequate current flow through the parallel combination of the base-emitter junction and R_2.

Finally, note that when Q_N turns *off*, V_{out} rises toward V_{CC} with a time constant $\tau = C_{eff}R_C$. A small value of R_C limits this switching time. However, smaller values of R_C result in higher current levels through R_C when Q_N is *on*, reducing available circuit fan-out.

In all three circuit performance improvement measures suggested above (addition of D_4 and R_2, reduction in R_C), a tradeoff exists between power consumption in the stable states and circuit speed. Higher speed is associated with a higher power-dissipation level. This is one of the important tradeoffs that is encountered throughout digital circuit design. Increased power dissipation requires larger power supplies and circuits and complicates heat dissipation problems.

Example 15-4   **Improving DTL Switching Speed**

Modify the circuit of Example 15-1 by adding resistor R_2, as shown in Fig. 15-12. Assume that $C_{be} = 5.0\,\text{pF}$ and a time constant $\tau = 10\,\text{ns}$ is to exist through R_2. Select a value for R_2 and determine how much R_1 must be reduced to maintain the same fan-out ($m = 15$).

Solution   Given $C_{be}\,R_2 = \tau$,

$$R_2 = 10\ \text{ns}/5.0\ \text{pF} = 2.0\ \text{k}\Omega$$

When Q_N is *on*, $V_{BE} = 0.70$ V, so the current through R_2 is

$$I_{R2} = 0.70\ \text{V}/2.0\ \text{k}\Omega = 0.35\ \text{mA}$$

The design of Example 15-1 requires a base current of 0.78 mA, so the total current through R_1 must be

$$I_{R1} = 0.35\ \text{mA} + 0.78\ \text{mA} = 1.13\ \text{mA}$$

From Fig. 15-12,

$$I_{R1} = \frac{V_{CC} - 3V_D}{R_1} = \frac{6.0\ \text{V} - 2.1\ \text{V}}{R_1} = 1.13\ \text{mA}$$

$$R_1 = 3.9\ \text{V}/1.13\ \text{mA} = 3.5\ \text{k}\Omega$$

The addition of R_2 to improve switching speed has required a reduction in R_1 from 5.0 to 3.5 kΩ and total circuit power consumption increases. ∎

15-6   TTL GATE CIRCUITS

An important approach to gate-circuit design makes use of transistor-transistor-logic (TTL) technology. A simplified TTL gate is shown in Fig. 15-13. The logic and input diodes for the DTL gates are replaced by a single multiemitter transistor Q_1.

The multiple emitters perform the AND or OR logical function. Each emitter is

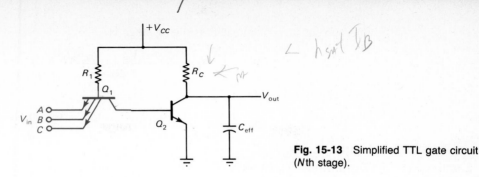

Fig. 15-13 Simplified TTL gate circuit (Nth stage).

connected to a base-emitter diode, with these diodes joined at the base. By making use of the multiemitter configuration, individual logic diodes are no longer required.

Consider what happens when the Nth TTL stage is driven by the previous $(N - 1)$th stage. Assume that the $(N - 1)$th stage is *on*, so that $V_{in} = V_{sat}$ for the Nth stage. The coupling transistor Q_1 comes *on* and saturates because there is a large base current and only a very small (leakage) current can flow out of the base of Q_2. Since there is no driving base current for Q_2, it is *off*. The output is $V_{out} = V_{CC}$.

When the $(N - 1)$th stage turns *off*, no emitter current flows in Q_1. The base-collector diode of Q_1 and the base-emitter diode of Q_2 are forward-biased and current flows out of the collector of Q_1 and into the base of Q_2, turning Q_2 *on*. Transistor Q_1 operates in the *inverse active mode*. The major current flow is across the forward-biased base-collector junction. The flow of carriers across this junction builds up an excess negative-charge density in the base, creating an electric field from the emitter to the base and resulting in a small minority-carrier flow from emitter to base. As might be expected, $I_B \cong I_C$ and $I_E \ll I_B$ for this mode of operation. R_1 and R_C are chosen so that Q_2 saturates and $V_{out} = V_{sat}$ when Q_2 is *on*.

The TTL circuit is configured to achieve rapid turn-on and turn-off for Q_2. When Q_1 turns *on* (for normal saturation operation), a transient current flows into the collector of Q_1 from the base of Q_2, helping clear out charge storage in the base region and turn off Q_2. When Q_1 shifts to inverse active operation, its common-base configuration allows driving current to quickly develop out of the collector of Q_1 and into the base of Q_2.

However, there is a significant delay-time problem associated with the simplified TTL gate. When Q_2 turns off, the rise in V_{out} is dominated by the time constant $\tau = R_C C_{eff}$. The input transistor Q_1 of the next stage does not react until a substantial portion of the time τ has passed. This delay in reaction between stages increases the average gate delay time despite the rapid turn-on and turn-off of Q_2.

The basic operating equations of the simplified TTL gate circuit can be obtained by referring to Fig. 15-13. When the $(N - 1)$th stage is *on*, Q_2 is *off* because the base current in Q_1 prevents any current flow out of the collector of Q_1. Therefore,

$$V_{out} = V_{CC} \tag{15-6}$$

When the $(N - 1)$th stage is *off*, there is no emitter current in Q_1, so that the base current in Q_2 becomes

$$I_B = \frac{V_{CC} - V_{BC}(Q_1) - V_{BE}(Q_2)}{R_1} \tag{15-7}$$

The current flow from each following stage is

$$I_1 = \frac{V_{CC} - V_{sat} - V_{BE}(Q_1)}{R_1} \qquad (15\text{-}8)$$

Q_2 is kept in saturation if

$$\frac{V_{CC} - V_{sat}}{R_C} + m\frac{V_{CC} - V_{sat} - V_{BE}(Q_1)}{R_1} < h_{sat}\frac{V_{CC} - V_{BC}(Q_1) - V_{BE}(Q_2)}{R_1} \qquad (15\text{-}9)$$

Example 15-5 **Simplified TTL Inverter Circuit**

A TTL inverter circuit of the type shown in Fig. 15-13 with a fan-out of 12 is to be designed. You have available a supply voltage $V_{CC} = 6.0$ V and transistors for which $h_{sat} = 20$, $V_{sat} = 0.3$ V, and $V_{BE} = 0.7$ V. Given $R_C = 1.0$ kΩ, what is the largest value of R_1 that can be used?

Solution From Eq. (15-9),

$$\frac{(6.0 - 0.3)\ \text{V}}{1.0\ \text{k}\Omega} + 12\frac{(6.0 - 0.3 - 0.7)\ \text{V}}{R_1} < 20\frac{(6.0 - 0.7 - 0.7)\ \text{V}}{R_1}$$

$$5.7\ \text{mA} + 60\ \text{V}/R_1 < 92\ \text{V}/R_1$$

so that

$$R_1 < 5.6\ \text{k}\Omega$$

For this example, assume that $C_{eff} = 20$ pF. Then V_{out} rises from V_{sat} toward V_{CC} with a time constant $\tau = (1.0\ \text{k}\Omega)(20\ \text{pF}) \cong 20\ \mu\text{s}$, which leads to an unacceptable limit on system operation. ∎

The fan-out capabilities and speed of the TTL gate circuit can be improved with the addition of a *totem pole* output Q_3 and Q_4, as shown in Fig. 15-14(A). The resulting circuit configuration is widely used for fabrication of digital ICs (as discussed in Chapter 17). The resistor values shown in this circuit are typical.

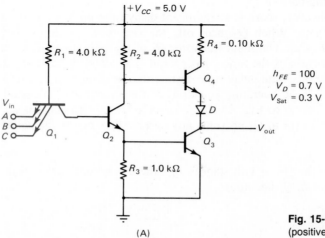

Fig. 15-14 TTL gates. (A) NAND gate (positive logic).

(A)

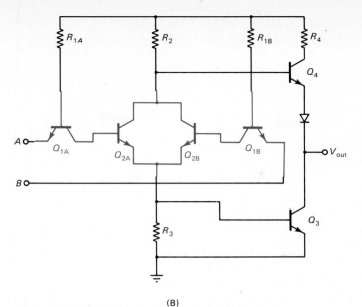

(B)

Fig. 15-14 (cont.) (B) NOR gate (positive logic).

When any input to Q_1 is *low* (at V_{sat}), Q_1 operates in the normal saturation mode and both Q_2 and Q_3 are *off* because no base current is available for either transistor. To understand the state of Q_4, begin by calculating the collector-base voltage for Q_4 in terms of the collector current I_{C4}. By inspection,

$$V_{C4} = V_{CC} - I_{C4}R_4$$

$$V_{B4} = V_{CC} - \frac{I_{C4}}{h_{FE}} R_2$$

(since Q_2 is *off*), so that

$$V_{CB4} = V_{C4} - V_{B4} = \frac{I_{C4}}{h_{FE}}R_2 - I_{C4} R_4 \qquad (15\text{-}10)$$

Transistor Q_4 is saturated if the base-collector diode is forward-biased ($V_{CB} < 0$). Therefore, Q_4 is saturated if

$$R_4 > R_2/h_{FE}$$

Substituting from Fig. 15-14(A), this condition becomes

$$100 > \frac{4.0 \times 10^3}{h_{FE}}$$

$$h_{FE} > 40$$

Assuming that this inequality holds, Q_4 is driven into saturation. The output voltage is given by

$$V_{out} = V_{CC} - (V_D + V_{BE4} + V_{R2}) \qquad (15\text{-}11)$$

Assume for the moment that $I_{E4} < 1.0$ mA. (The reasonableness of this assumption is discussed below.) The voltage drop across R_4 is

$$(1.0 \text{ mA})(0.10 \text{ k}\Omega) = 0.10 \text{ V}$$

or less, and the voltage drop across R_2 is

$$\frac{1.0 \text{ mA}}{100}(4.0 \text{ k}\Omega) = 0.04 \text{ V}$$

or less, choosing $h_{FE} = 100$. The base-collector junction is forward-biased, as expected. In addition, V_{R2} can be seen as negligible when calculating V_{out}. Thus,

$$V_{\text{out}} \cong 5.0 \text{ V} - 1.4 \text{ V} = 3.6 \text{ V}$$

for this operational state.

Since V_{out} is approximately independent of the exact output-current requirements, it quickly seeks the steady-state value and remains there. Q_4 functions as an emitter follower, driving from a low output impedance, and can greatly reduce the time constant τ associated with C_{eff} to increase the average gate-switching speed. A principal disadvantage of the simple TTL circuit is thus addressed.

When all inputs to Q_1 are *high* (at 3.6 V), the base of Q_1 starts to rise toward $V_{CC} = 5.0$ V. However, Q_2 and Q_3 turn *on*, clamping the base at $3V_D = 2.1$ V. The input emitter-base diode of Q_1 is reverse-biased. Therefore,

$$I_{B1} = \frac{5.0 \text{ V} - 2.1 \text{ V}}{4.0 \text{ k}\Omega} \cong 0.7 \text{ mA}$$

The maximum collector current that can flow through Q_2 is

$$\frac{5.0 \text{ V} - 0.7 \text{ V} - 0.3 \text{ V}}{4.0 \text{ k}\Omega} = 1.0 \text{ mA}$$

Therefore, Q_2 is saturated.

Given this current level, Q_3 also saturates. Because $V_{B4} = 0.7 + 0.3 = 1.0$ V and $V_{E4} = 0.7 + 0.3 = 1.0$ V, Q_4 is *off*. Diode D is included in the circuit to assure that Q_4 turns *off* in this state. The current through R_3 is

$$0.7 \text{ V}/1.0 \text{ k}\Omega = 0.7 \text{ mA}$$

and the current through R_2 is

$$4.0 \text{ V}/4.0 \text{ k}\Omega = 1.0 \text{ mA}$$

The base drive for Q_3 is 0.3 mA.

Given $h_{\text{sat}} \cong 30$, up to $(30)(0.3) = 9$ mA can be sunk by Q_3 when $V_{\text{out}} = 0.3$ V. When Q_1 is *on* (in normal saturation), the emitter current is $(5.0 \text{ V} - 0.7 \text{ V} - 0.3 \text{ V})/4.0 \text{ k}\Omega = 1.0$ mA. Therefore, the fan-out for this circuit is about 9. The addition of Q_3 and Q_4 enables obtaining a satisfactory fan-out without sacrificing switching speed.

In the above analysis, it is assumed that $I_{E4} < 1.0$ mA when Q_4 is *on*. When Q_4 is *on*, Q_1 of the next stage is in the inverse active mode, for which $I_{E1} \ll I_{B1} = 0.7$ mA. If the fan-out of the circuit is $m = 10$ and $I_{E1} < 0.1 I_{B1}$ for each circuit, then $I_{E1} < 0.7$ mA, consistent with the assumption that was made.

The TTL gate shown in Fig. 15-14(A) is usually used with positive logic, providing the NAND operation. As shown in Fig. 15-3, multiple NAND gates can provide the NOR operation. However, this approach can reduce system speed and increase power dissipation.

For this reason, the TTL positive-logic NOR gate of Fig. 15-14(B) is also available as part of this family of circuits. Transistors Q_{1A} and Q_{1B} replace Q_1 and Q_{2A} and Q_{2B} replace Q_2. If either Q_{2A} or Q_{2B} conducts, Q_3 is *on* and Q_4 *off*. If both Q_{2A} and Q_{2B} are *off*, $V_{out} = 3.6\,V$ as before.

TTL gate circuits are widely available in IC form, as discussed in Chapter 17. Many other variations on the basic NAND and NOR gates are included to allow for a wide range of digital operations.

15-7   STTL GATE CIRCUIT

The more deeply a BJT moves into saturation during its *on* phase, the more slowly it emerges from saturation to switch *off*. Circuits thus have reduced delay times if the BJT can be maintained on the edge of saturation.

A variation on the TTL gate circuit makes use of a Schottky diode (see Section 4-10) to produce Schottky TTL (STTL) gate circuits to limit the depth of saturation. Consider the circuit shown in Fig. 15-15. When the transistor is *off*, the base-collector junction is reverse-biased so the Schottky diode is also *off*. When the BJT comes *on*, V_{out} drops toward its usual saturation value (typically 0.3 to 0.4 V). However, V_{out} cannot drop below the voltage $V_{BE} - V_D$ because the Schottky diode turns *on* and prevents V_{CE} from falling any further. By fabricating the Schottky diode to produce the desired value of V_D, operation can be obtained on the edge of saturation.

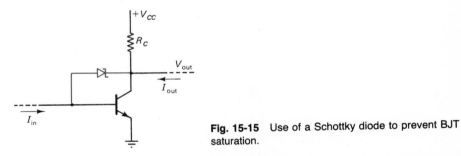

Fig. 15-15   Use of a Schottky diode to prevent BJT saturation.

The Schottky diode is useful in this circuit because it is characterized by a low charge storage that leads to fast switching. A junction diode has its own charge-limited switching time and would prevent the circuit from reaching its potential.

15-8   EMITTER-COUPLED LOGIC

Another family of digital circuits uses the difference amplifier, which is introduced in Chapter 12. The basic emitter-coupled logic (ECL) gate circuit is shown in Fig. 15-16(A). The name describes the sharing of Q_1 and Q_2 emitter currents through the constant-current generator Q_5.

As noted previously, switching speed is slowed down when a BJT moves in and out of saturation. The ECL circuit can achieve rapid switching because all transistors

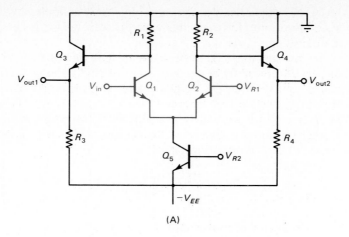

(A)

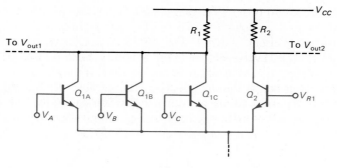

(B)

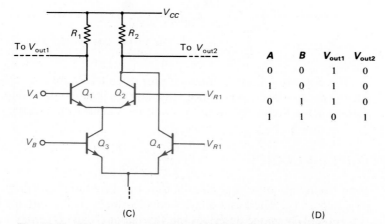

(C)

A	B	V_out1	V_out2
0	0	1	0
1	0	1	0
0	1	1	0
1	1	0	1

(D)

Fig. 15-16 ECL gate circuits. (A) Basic ECL gate circuit. (B) Multiple inputs V_A, V_B, and V_C for OR/NOR gate. (C) Multiple inputs V_A and V_B for AND/NAND gate. (D) Truth table for the gate of part (C). (Courtesy of Fairchild Semiconductor Corporation)

remain in the active region. The current steering through the difference amplifier creates the switching of states. Since the ECL configuration is sensitive to the input voltage levels, the speed also increases because the voltage levels can be set only about 1 V apart. However, the power consumption increases because significant current levels always flow in the circuit.

The reference voltage V_{R1} is chosen to be sufficiently positive to bring Q_2 into the active region when no current flows through Q_1. When Q_2 conducts, its emitter voltage is 0.7 V below V_{R1}. If V_{in} is chosen below this value (the lower switching value), Q_1 is biased *off*. When Q_2 is conducting and Q_1 is *off*, the output of emitter follower Q_3 rises, bringing V_{out1} to the upper switching level. Since Q_2 is *on*, V_{out2} is at the lower switching level.

If V_{in} is now raised to the upper switching level (above V_{R1}), Q_1 obtains active operation and Q_2 is biased *off*. The output of Q_3 is pulled down to the lower switching level while Q_4 rises to the upper level. Observe that Q_3 provides an inverted output and Q_4 provides a noninverted output.

A NOR gate configuration is obtained as shown in Fig. 15-16(B). If any input (A, B, or C) is greater than V_{R1}, the associated input transistor conducts and Q_2 is turned *off*. The output V_{out1} drops, providing a circuit that functions as a NOR gate. If an OR function is desired, the output V_{out2} can be used.

The AND and NAND functions can be produced as shown in Fig. 15-16(C). If V_B is equal to the lower logic level, Q_3 is *off*. Q_1 and Q_2 are then *off*. Q_4 is *on* and V_{out2} is low. This condition holds for V_A high or low.

If V_A is low, Q_1 is *off*. If V_B is high, Q_2 and Q_3 are *on* and Q_4 is *off*. As a result, V_{out2} is low. If V_A and V_B are high, Q_1 and Q_3 are *on* and Q_2 and Q_4 are *off*. V_{out2} is high. The resultant truth table is shown in Fig. 15-16(D). The output V_{out2} provides the AND function and V_{out1} the NAND function.

15-9 FET GATE CIRCUITS

The DTL, TTL, and ECL gate circuits introduced above use bipolar devices (BJTs) to achieve desired circuit properties. It is also possible to design gate circuits that make use of field-effect devices (FETs). Both types of applications are in wide use today in digital systems.

Figure 15-17(A) illustrates a gate circuit using the unique properties of the enhancement MOSFET. The pull-up resistor is created by a diffusion process onto the same substrate material. Assume an input voltage V_{in} that varies between a small positive voltage V_S and V_{DD}. When $V_{GS} = V_{DD}$, a large current I_D flows in the MOSFET and $V_{out} = V_{DS} = V_S$. When $V_{GS} = V_S < V_T$, then $I_D = 0$ and $V_{out} = V_{DD}$. The circuit then performs as an inverter so long as $V_S < V_T$.

Figure 15-17(B) shows how a NAND gate can be formed using this technology. The two transistors Q_1 and Q_2 provide a conducting path only if all the inputs are above V_T (close to V_{DD}). If both Q_1 and Q_2 are *on*, output C is close to ground. If either A or B is below V_T, C rises toward V_{DD}, providing the desired logical result.

Figure 15-17(C) illustrates a related NOR gate. If any input is above V_T (close to V_{DD}), the corresponding MOSFET conducts, bringing output C close to ground. If all inputs are below V_T, C rises toward V_{DD}. The result provides the desired logical function.

Figure 15-17(D) illustrates how the transfer and output characteristics of the FET

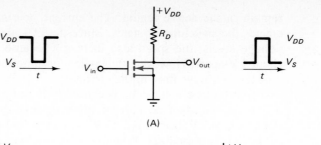

(A)

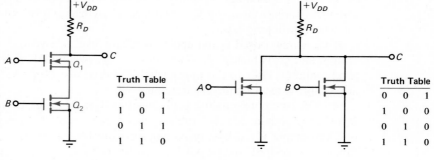

Truth Table

0	0	1
1	0	1
0	1	1
1	1	0

(B)

Truth Table

0	0	1
1	0	0
0	1	0
1	1	0

(C)

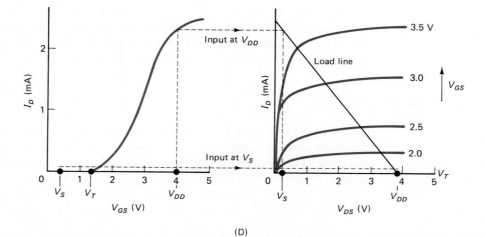

(D)

Fig. 15-17 NMOS gate circuit using pull-up resistor. (A) NMOS gate circuit. (B) NAND gate. (C) NOR gate. (D) Operating levels for the gate.

can be used to determine the two operating levels for the gate. The two logic levels V_S and V_{DD} can be located on both the transfer and output characteristics of the MOSFET. If $V_{in} = V_S$, $I_D \cong 0$ on the transfer characteristic, so $V_{DS} \cong V_{DD}$ on the output characteristic. For $V_{in} \cong V_{DD}$, the FET is *on*, producing a sufficient current to cause V_{DS} to drop to V_S. The graphs clearly indicate how these two levels must bracket the value of V_T. If both V_S and V_{DD} are below V_T, the FET never turns *on*. And if both V_S and V_{DD} are above V_T, the FET is always at least partially *on* and V_{DS} remains below V_{DD}.

Example 15-6   NMOS Gate

An NMOS gate circuit is designed using the device characteristics of Figure 15-17(D). (a) What values of V_{DD} and R_D are used? (b) What are the two logic levels? (c) Assume that for noise protection, V_T must be at least 0.5 V from each logic level. What range of values for V_T is allowed?

Solution   (a) From the load line shown, $V_{DD} = 4.0$ V and $R_D = 4.0$ V/2.5 mA $= 1.6$ kΩ.
(b) From the transfer characteristic, $V_S = 0.3$ V and $V_{DD} = 4.0$ V.
(c) V_T can range from 0.3 V + 0.5 V = 0.8 V to 4.0 V − 0.5 V = 3.5 V. From the transfer characteristic, $V_T = 1.3$ V for the specific device described.   ∎

Since resistors can take up more space than the MOSFET devices themselves, it is sometimes useful to replace the pull-up resistor R with another transistor, as shown in Fig. 15-18(A). The passive load presented by R is replaced by an active load Q_2.

In order to determine how this circuit operates, it is necessary to develop a char-

Fig. 15-18 Active load for the NMOS gate circuit. (A) Schematic. (B) Transfer characteristic for Q_2 (load). (C) Output characteristic for Q_2 (load). (D) Dotted line is the superposition of (B) and (C), creating the curve $V_{DS} = V_{GS}$ as the load due to Q_2.

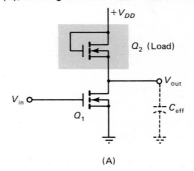

(A)

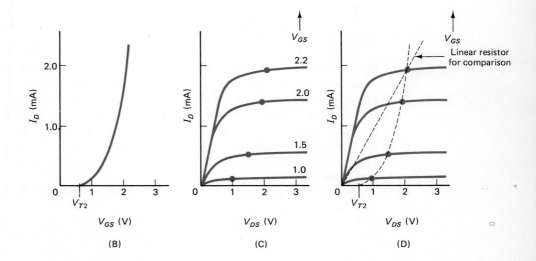

(B)

(C)

(D)

acteristic curve for the load Q_2. This can be done by drawing the transfer and output characteristics for Q_2 [Figs. 15-18(B) and (C), respectively] and combining them. Since $V_{GS} = V_{DS}$ and I_D is the same for both characteristics, the effective $I = f(V)$ curve for Q_2 can be found by superimposing the two characteristics and connecting the common solution points. The result is shown in Fig. 15-18(D). For comparison, the $I = f(V)$ characteristic curve for a linear resistor is also shown. As noted, the two curves are similar except near the origin.

The characteristic of Fig. 15-18(D) describes the load for Q_1. Therefore, using the general relationship developed in Chapter 5, $V_{DD} = V_{\text{load}} + V_Q$ and the nonlinear load line can be drawn on the characteristic for Q_1, as shown in Fig. 15-19. Operating points for Q_1 lie along the load line produced by Q_2. The two stable operating levels for this gate circuit are shown in Fig. 15-19.

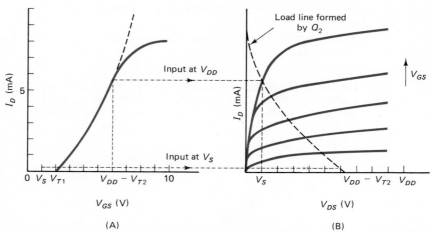

Fig. 15-19 NMOS gate circuit using an enhanced-MOSFET load. (A) Transfer characteristic for Q_1 showing stable states. (B) Output characteristic for Q_1 showing stable states.

Although the two operating points are shifted slightly, the gate circuit still works effectively. It may be necessary to increase V_{DD} to maintain the same upper voltage level for the circuit output. To produce multiple inputs, the transistors can be arranged as indicated in Fig. 15-17.

To see how Q_2 switches between stable states, consider the following stage to be represented by a capacitance C_{eff}. The Q_1 is *on,* a small voltage exists across this capacitance and the next stage is *off* (no channel exists). When Q_1 turns *off,* current through Q_2 charges C_{eff} until V_{DS} across Q_2 reaches V_{T2}. When this condition occurs, Q_2 becomes a high resistance (formed by an intrinsic semiconductor) and C_{eff} remains charged at approximately the value $V_{DD} - V_{T2}$. The next stage turns *on* during the charging transient for C_{eff}. When Q_1 turns *on* again, C_{eff} discharges through Q_1. Observe that for the circuit to function as an inverter, the condition $V_{DD} - V_{T2} > V_{T1}$ must hold.

Another load option is shown in Fig. 15-20. If a depletion MOSFET is used as load Q_2, with the gate and source connected, the characteristic curve seen by the switching MOSFET is the usual $V_{GS} = 0$ curve for the output characteristic. When this load is plotted on the output characteristic of Q_1, the results of Figs. 15-20(B)

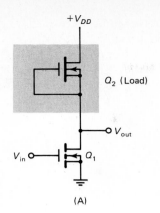

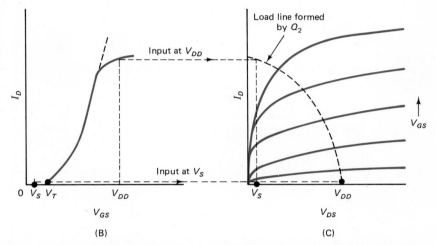

Fig. 15-20 NMOS gate circuit with a depletion-MOSFET load. (A) Schematic. (B) Transfer characteristic for Q_1. (C) Output characteristic for Q_1.

and (C) are obtained. The advantage of this load is that the current through Q_1 remains at a high level during more of the switching transient. This results in more current available to charge and discharge device capacitances and therefore faster circuits.

15-10 CMOS GATE CIRCUITS

A *complementary* enhancement-MOSFET (CMOS) gate circuit is shown in Fig. 15-21(A). It is not helpful to analyze this circuit using the characteristic-curve approach. Rather, an understanding of the physical nature of the devices must be drawn upon to understand circuit operation.

Note that if $V_{in} \cong 0$, the NMOS device has $V_{GS} \cong 0$ and no channel forms, as shown in Fig. 15-21(B). Therefore, the resistance R_n measured between the source and drain is very large. At the same time, a potential $V_{GS} \cong -V_{DD}$ appears across the

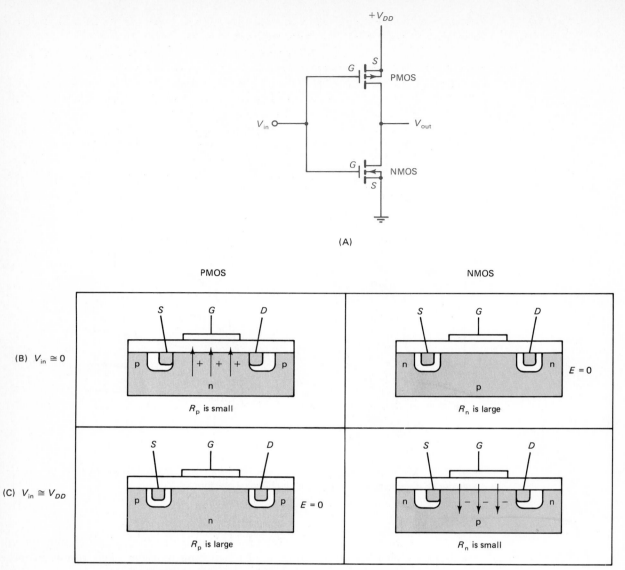

Fig. 15-21 CMOS gate circuit.

PMOS device, creating a channel and resulting in the resistance R_p between the source and drain being very small. Therefore,

$$V_{out} = V_{DD}\frac{R_n}{R_n + R_p} \cong V_{DD} \qquad (15\text{-}12)$$

If $V_{in} = V_{DD}$, as shown in Fig. 15-21(C), then R_p becomes very large, and R_n very small, and

$$V_{out} \cong V_{DD}\frac{R_n}{R_n + R_p} \cong 0 \qquad (15\text{-}13)$$

The minimum value of V_{GS} required to form a channel in each device accommodates the slight voltage shifts away from ideal performance.

For stable states, the current through both devices is very small, limited to leakage-current levels. Therefore, power dissipation is negligible except when the circuit is switching between states. This is a substantial advantage. (A further discussion of CMOS power dissipation is provided in Section 17-4.) As observed above, since $I_D \cong 0$ for both stable states, the characteristic curves are not useful in the analysis.

CMOS gate circuits use various types of series and parallel arrangements of devices. The CMOS circuit of Fig. 15-22 operates with the PMOS devices in series and the NMOS devices in parallel. If both A and B are near 0 V, the n-channel devices (Q_1 and Q_2) are *off* and the p-channel devices (Q_3 and Q_4) are *on*. The output C is thus near V_{DD}.

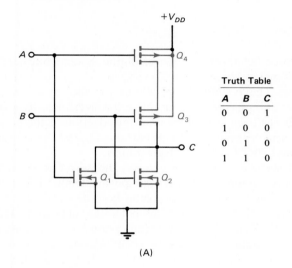

Truth Table

A	B	C
0	0	1
1	0	0
0	1	0
1	1	0

(A)

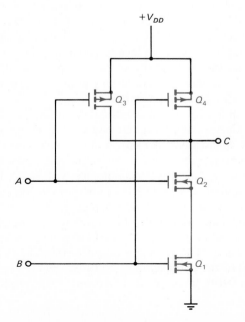

(B)

Fig. 15-22 CMOS logic circuits. (A) CMOS NOR gate (positive logic). (B) CMOS NAND gate (positive logic).

If A increases to the upper switching value, Q_4 turns *off* and Q_1 turns *on*. If B remains near 0 V, Q_2 is *off* and Q_3 is *on*. With Q_4 *off*, the output remains isolated from V_{DD}, whereas with Q_1 *on*, the output is tied to ground. Thus, C is close to 0 V.

If both A and B increase to the upper switching value, both Q_3 and Q_4 are *off*. Q_1 and Q_2 are *on* and the output is again close to 0 V.

If a voltage level near V_{DD} is taken to be a **1** and a voltage level near 0 V is taken to be a **0** (positive logic), the circuit of Fig. 15-22(A) produces the NOR logical function. A similar analysis shows that for these definitions the circuit of Fig. 15-22(B) produces a NAND logical function.

As is discussed in Chapter 17, CMOS technology is now widely applied to create digital ICs. Many of the configurations originally available in TTL or NMOS are now also available in compatible CMOS.

15-11 COMPARISON OF DIGITAL TECHNOLOGIES

The driving force behind the design of digital systems has been the desire to solve increasingly complex computational problems in the shortest time possible and at minimum cost. The potential applications for such systems have grown rapidly, providing a continuing economic incentive for investment to advance the state of the art. The result has been the commercial introduction of the major technologies discussed in this chapter, along with several others, and the continuing exploration of potential technologies for future product lines.

In order to achieve a high gate density, the power consumption associated with each gate must be minimized. Otherwise, power supplies become out of scale for the processing unit and heat removal problems become excessive. And, as noted above, the speed of processing determines the potential applications that are possible for a family of circuits. Since the power-consumption levels and delay times for building block gates are so important in digital system design, one common way to compare technologies is to rate each technology along these two dimensions. Figure 15-23 shows typical ratings for the gate technologies used in this chapter. Such "quick-look" comparisons must be used with caution because the ratings along each axis may depend on many assumptions regarding device fabrication, design constraints and circuit configurations, voltage levels, fan-in and fan-out, and methods of data measurement. Exact values can vary among comparisons. On the other hand, there is value in even such an approximate display because it provides a quick overview of all of the technologies discussed above.

Most of the technologies shown are based on Si device fabrication. However, the rapid evolution of GaAs-based FETs (introduced in Chapter 8) suggests a shift in this direction for many applications. As shown in Fig. 15-23, GaAs devices offer the potential for significant improvements in digital-circuit performance.

Figure 15-23 also includes two technologies with limited availability at present. The MODFET (modulation-doped FET), sometimes called an HEMT (high-electron-mobility transistor), is introduced in Chapter 8. This device uses the ability to change doping properties within an FET structure to produce semiconductor layers with different electrical properties. A doped layer of AlGaAs is fabricated next to an undoped layer of GaAs. The electrons added to the AlGaAs by doping flow out of the doped region into the undoped layer of GaAs, where conduction takes place. As a result, carrier mobility increases and switching time improves. MODFETs can be

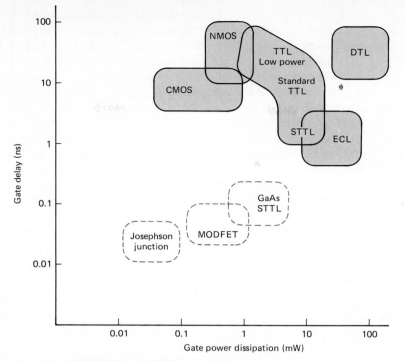

Fig. 15-23 Comparison of digital technologies.

operated at room temperature (about 300°K) or at the temperature of liquid nitrogen (77°K) for improved performance.

The Josephson junction is a device that is based on superconducting phenomena. For many years, such phenomena were observed only for very low temperatures, near the temperature of liquid helium (4°K). However, new discoveries are raising the temperatures at which superconductivity can be observed in custom-designed ceramic materials, holding out the potential for new families of digital circuits with advanced properties.

The commercial production of gate circuits that can achieve competitive operating parameters has depended on the evolution of integrated circuit fabrication techniques. The number of gates on a single integrated circuit, or chip, has grown from a few thousand to almost a million today. Such high-density packaging has resulted in small digital systems that minimize the delays associated with gate interconnections, and thus maximize circuit speed. Chapter 16 discusses further the ways in which digital gates are used to configure systems and Chapter 17 expands on the design of electronic systems using digital ICs.

To see the importance of interconnection distances, recall that the upper limit on the speed of signal propagation is given by the speed of light $c = 3 \times 10^8$m/s. For a conducting line buried in a dielectric medium of dielectric constant ϵ, which is typical of integrated circuit configurations,

$$v = c/\epsilon_r^{1/2} \qquad (15\text{-}14)$$

For the dielectric $SiO_2(\epsilon = 4\epsilon_o)$ used in common ICs, the propagation velocity is $v \cong 1.5$ cm/ns.

As noted in Fig. 15-23, the fastest circuit-switching times range from 0.1 to 1 ns. Therefore, to prevent interconnection propagation times from dominating system performance, typical distances must be of the order of 1 mm. This is about the on-chip interconnect distance associated with current technology. It is also clear that off-chip connections (between ICs) become very important at these speeds. Both digital and analog technologies have reached the point where packaging and interconnect strategies can dominate system performance. A further discussion of this topic, considered in the area of hybrid microelectronics, is included in Chapter 18.

As discussed in Chapter 1, design involves an ongoing iteration among the setting of systems performance specifications, circuit parameters, and the basic materials and fabrication aspects of design. Figure 15-24 illustrates how this interaction takes place in the production of digital systems. Electronics design requires the ability to understand developments and potential implications of improvements in all areas. Digital systems are produced by trading off considerations within and among the different areas shown.

FIGURE 15-24
Aspects of Digital Circuit Design

Digital System Performance	Circuit Parameters	Basic Materials and Fabrication Aspects of Design
Architecture (organization of logical functions and operations)	Combinations of functional elements used to create gate circuits	Materials (Si, GaAs, dielectric, metals, etc.)
Topography (distribution of gates and other features on the surfaces of IC chips)	Device size	Lithography (size of smallest fabrication features)
	Intrinsic device capacitance and carrier mobility	Device geometry and fabrication methods
System size	Wiring length and intrinsic R, L, C	Scale of integration
Power dissipation	Propagation delays	
Average calculation speed	Voltage levels	
Interconnections and packaging	Operating temperature	
	Interconnections and packaging	

QUESTIONS AND ANSWERS

Questions

15-1. What new factor (not encountered in previous chapters) must be considered in the design of gate circuits as digital building blocks in a larger system?

15-2. What are the relationships between the OR and NOR gate functions, and the AND and NAND gate functions?

15-3. Define the terms *positive logic* and *negative logic*.

15-4. Why are chains of diodes and resistors limited in terms of the digital operations they can perform?

15-5. Why does saturated BJT operation produce noise protection in a network of digital circuits?

15-6. What device and circuit properties limit the turn-on and turn-off switching times for a gate circuit?

15-7. What factors affect the current gain of a BJT operated in the inverse active mode?

15-8. What is the purpose of the totem pole output used for TTL gate circuits?

15-9. What power-supply problem might be encountered with TTL gate circuits?

15-10. Why is a Schottky diode used to prevent transistor saturation in STTL circuits instead of a junction diode?

15-11. What unique consideration enters into ECL digital-system logic design?

15-12. Why is the E-MOSFET particularly suited to digital gate-circuit design, for example, when compared with the D-MOSFET?

15-13. What are the various advantages and disadvantages associated with alternative loads for NMOS gate circuits?

15-14. Why are the transfer and output characteristic curves not useful for the study of CMOS gate circuits?

15-15. Based on Fig. 15-23, which of the gate types has (a) the lowest power consumption, (b) the highest speed, (c) the most flexibility, and (d) the least attractive operating characteristics? Neglect the technologies shown within dotted lines.

Answers   **15-1.** Gate circuits are typically applied as building block elements in large digital networks. Therefore, circuit design must consider the design requirements associated with a variable number of input and output connections. Gate circuits must thus be designed in a network context.

15-2. To convert an OR gate to the NOR function, or an AND gate to the NAND function, the NOT function is added. Therefore, the outputs associated with the gate types are reversed ($0 \rightarrow 1$, and $1 \rightarrow 0$) to produce the outputs associated with the latter gate types.

15-3. *Positive logic* is obtained when the most positive voltage level is defined as **1**. *Negative logic* results when the most negative voltage level is defined as **1**.

15-4. Chains of diodes and resistors do not provide the ability to maintain constant **0** and **1** voltage levels through a chain of circuits, nor can they provide the NOT function.

15-5. Noise protection is obtained for digital circuits using saturated BJTs because the voltage V_{CE} across the BJT in the *on* state is approximately independent of the base current I_B. As I_B fluctuates, V_{CE} remains approximately constant, so noise present at the input does not appear at the output.

15-6. The switching times for a BJT gate circuit are limited by (a) the time required for excess charge densities to recombine during turn-off; (b) the times required for excess charge densities to build up during turn-on; and (c) the time required for charging and discharging equivalent capacitances.

15-7. For a BJT operated in the inverse active mode, the effective injection efficiency γ is much smaller than that experienced in the normal active mode. Therefore, the current gain is much reduced. However, since the current gain of the coupling BJT in a TTL gate circuit is not an essential design consideration, this reduction does not have a negative effect on circuit performance.

15-8. The totem pole output is used to speed up the TTL gate circuit. In Fig. 15-14(A), Q_4 acts as an emitter follower to drive V_{out} toward its upper value (3.6 V) when Q_3 turns *off*.

15-9. When a TTL gate circuit is switching between states, both Q_3 and Q_4 are *on* for a brief transient. A temporary low impedance across the 5.0-V power supply can result in "spikes" in the power supply bus, creating noise that interferes with system operation.

Significant design effort must be expended to deal with this problem in practical TTL applications.

15-10. The excess-charge storage associated with a junction diode would add its own charging/discharging delay times to the circuit of Fig. 15-15. The Schottky diode has a very low excess-charge storage, so this disadvantage does not exist.

15-11. ECL digital-system logic design can make use of the two outputs (OR and NOR, or AND and NAND) that are available for each ECL gate. Thus, system-design rules can be expected to be quite different for this case.

15-12. The E-MOSFET forms a channel and conducts only if $V_{GS} > V_T$. The existence of this threshold voltage enables the design of a simple inverter circuit, as shown in Fig. 15-17(A). For the D-MOSFET, any $V_{GS} > 0$ would be adequate to turn the transistor partially *on*, preventing the clear definition of two operating states.

15-13. A diffused resistor as load for an NMOS circuit is a straightforward solution but requires more space than is desirable. An E-MOSFET load allows the use of the same fabrication steps to produce both the gate transistor and load, and approximates the same performance. A D-MOSFET load has the advantage of the E-MOSFET and additionally increases switching performance by making more current available to charge and discharge capacitors.

15-14. The transistors in CMOS gate circuits conduct only very small currents in steady-state conditions. Given $I_D \cong 0$, the transfer and output characteristic curves do not provide useful design information.

15-15. For the gate types shown, (a) the CMOS has the lowest power dissipation, (b) the ECL has the highest speed, (c) the TTL/STTL families provide the most flexibility, and (d) the DTL gates, which are not used in ICs, are the least attractive option.

EXERCISES AND SOLUTIONS

Exercises (15-1)   **15-1.** Prepare a truth table for the combinations of gates in Figs. 15-25(A) and 15-25(B).

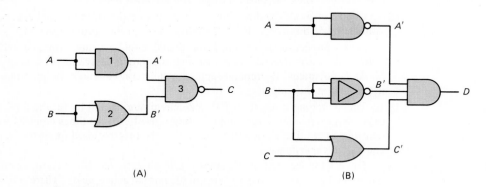

(A)   (B)

Fig. 15-25 Logic circuits for Exercise 15-1.

(15-3)   **15-2.** For the BJT described in Fig. 15-6, find h_{FE} for the three load lines shown. What value might be selected for h_{sat}?

(15-3)   **15-3.** For a given BJT, $h_{FE} = 150$ and $h_{sat} = 30$. If $I_B = 10\,\mu A$, describe the state of the device for (a) $I_C = 0.10\,\text{mA}$, (b) $I_C = 1.0\,\text{mA}$, and (c) $I_C = 1.5\,\text{mA}$. (d) What happens to V_{CE} as the current I_C increases?

15-4. For the DTL gate circuit in Fig. 15-7, $V_{CC} = 5.0$ V, $V_{sat} = 0.2$ V, and $V_D = 0.6$ V. (a) If $R_1 = 1.0$ kΩ, what current must be sunk by the previous stage to keep Q_N *off*? (b) To achieve a maximum fan-out of 10, what current must be sunk by Q_N when it is *on* if $R_C = 1.0$ kΩ? (c) If $h_{sat} = 25$, what I_B is required for this circuit? (d) If I_B drops below the value found in part (c), what happens to V_0 for that stage?

15-5. Develop a truth table for the circuit of Fig. 15-26, using positive logic. Show from your analysis whether this is a NOR or NAND circuit.

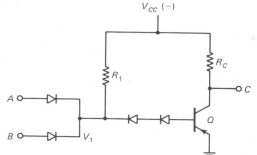

Fig. 15-26 Circuit for Exercise 15-5.

15-6. For the DTL gate circuit of Fig. 15-27, assume that $V_D = 0.7$ V and $V_{sat} = 0.3$ V. (a) If $C_{eff} = 100$ pF, choose R_C so that the turn-off time constant of the output is 0.10 μs. (b) As the preliminary design value, choose $R_1 = 10$ kΩ. What will be the base current through Q when the previous gate is off? (c) If you require $h_{sat} = 20$ to keep Q in saturation, what maximum value of I_C results? (d) How much output current is available to drive the following stages? (e) How much input current flows when Q is off? (f) What is the maximum fan-out for this circuit?

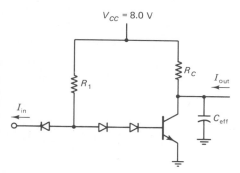

Fig. 15-27 Circuit for Exercise 15-6.

15-7. You are given the input and output waveforms for a gate circuit as shown in Fig. 15-28(A) (p. 624). (a) Find the rise, fall, and delay times for each of the pulses shown. (b) Find the average delay time per gate.

15-8. For the simple TTL gate circuit of Fig. 15-29 (p. 624), answer the following: (a) If $R_1 = 10$ kΩ and $R_C = 1.0$ kΩ, what base current flows in Q_2 when all inputs are at 0.5 V? (b) Using the definition of positive logic, what type of gate is this? (c) If $h_{sat} = 30$, find $(I_C)_{max}$. (d) How much current is available to drive subsequent stages? (e) If Q_1 is *on*, how much current flows in each emitter? (f) What maximum fan-out is possible?

15-9. You are given the simple TTL converter circuit of Fig. 15-30 (p. 624). (a) When $V_{in} = 6.0$ V, what base current flows in Q_2? (b) What current flows in the 0.50-kΩ resistor? (c) What minimum value of h_{sat} is required to ensure that Q_1 stays in saturation if $m = 10$?

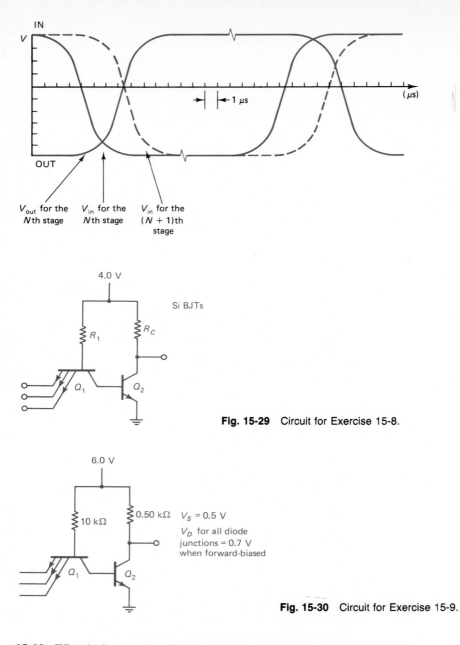

Fig. 15-29   Circuit for Exercise 15-8.

Fig. 15-30   Circuit for Exercise 15-9.

(15-6)   **15-10.** What is the purpose of diode D in the TTL circuit of Fig. 15-14(A)?

(15-9)   **15-11.** What determines the limitation on the fan-out m for the NMOS circuit of Fig. 15-18?

(15-9)   **15-12.** An NMOS gate circuit of the type shown in Fig. 15-18 uses MOSFETS that can be described by the following transfer equations:

$$I_{D1} = I_{D01}(V_{GS1}/V_T - 1)^2 \qquad\qquad I_{D2} = I_{D02}(V_{GS2}/V_T - 1)^2$$

assuming $V_{T1} = V_{T2}$. What design constraints can be placed on the circuit using these relationships?

(15-9) **15-13.** Select V_{DD} = 5.0 V and V_T = 2.0 V for the NMOS circuit of Fig. 15-18(A). Can a viable gate circuit be designed?

(15-9) **15-14.** Can a viable solution to Exercise 15-12 be obtained with the operating voltage levels of 1.0 and 4.0 V with V_T = 2.0 V?

15-10) **15-15.** For a CMOS gate circuit, $R \cong 100\ \Omega$ when either MOSFET is *on* and $R \cong 10.0\ k\Omega$ when either MOSFET is *off*. (a) If V_{DD} = 5.00 V, find the output switching levels. (b) What range of values is acceptable for V_T if a 0.20-V "pad" is required for noise protection?

15-11) **15-16.** (a) Show that the velocity of signal propagation in SiO_2 is $v \cong 1.5$ cm/ns. (b) What typical time delay is required if IC features are 5 mm apart on the average?

15-17. Figure 15-31 describes the relationship between the delay times and unit costs for a family of discrete BJTs. If system design requires a shift from 2.5 to 0.5 ns for typical delay times, what is the incremental increase in system cost per thousand transistors? What other factors might be important in the selection of the preferred BJT?

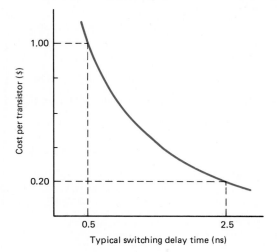

Fig. 15-31   Data for Exercise 15-17.

Solutions **15-1.** (a) For Fig. 15-25(A), gates 1 and 2 are AND gates with the inputs tied together, so the truth table for each gate is shown in Fig. 15-32(A). Gate 3 is a NOR gate, so the truth table for the combination is as shown in Fig. 15-32(B). The three gates shown provide the NOR function. (b) For Fig. 15-25(B), A' is the NOT of A; B' is the NOT of B; C' is the OR of B and C; and D is the AND of A', B', and C'. The output D is **0** unless A', B', and C' are **1**. This happens only if $A = 0$, $B = 0$, and $C = 1$. For this case, $D = 1$. The truth table takes the form shown in Fig. 15-32(C).

FIGURE 15-32
Solution for Exercise 15-1

A	A'	B	B
0	**0**	**0**	**0**
1	**1**	**1**	**1**

(A)

A	A'	B	B'	C
0	**0**	**0**	**0**	**1**
0	**0**	**1**	**1**	**0**
1	**1**	**0**	**0**	**0**
1	**1**	**1**	**1**	**0**

(B)

A	A'	B	B'	C	C'	D
0	**1**	**0**	**1**	**1**	**1**	**1**
All other combinations						**0**

(C)

15-2. At point 1 in Fig. 15-6,

$$h_{FE} = I_C/I_B \cong 4.7 \text{ mA}/40 \text{ } \mu\text{A} \cong 1.2 \times 10^2$$

At point 2,

$$h_{FE} \cong 3.7 \text{ mA}/40 \text{ } \mu\text{A} \cong 90$$

At point 3,

$$h_{FE} \cong 2.0 \text{ mA}/40 \text{ } \mu\text{A} \cong 50$$

A value of $h_{sat} = 50$ would be adequate to assure device saturation for a range of base currents around 40 μA and the range of load lines shown. If I_B can take on values down to 10 μA, then saturation requires

$$h_{sat} \cong 1.0 \text{ mA}/40 \text{ } \mu\text{A} \cong 25$$

15-3. (a) For the condition given,

$$I_C/I_B = 0.10 \text{ mA}/10 \text{ } \mu\text{A} = 10 < h_{sat}$$

Therefore, the BJT is saturated.

(b) Given

$$I_C/I_B = 1.0 \text{ mA}/10 \text{ } \mu\text{A} = 100$$

which is greater than h_{sat} and less than h_{FE}, the BJT is in the transition region between saturated and active operation.

(c) Given

$$I_C/I_B = 1.5 \text{ mA}/10 \text{ } \mu\text{A} = 150 = h_{FE}$$

the BJT is in the active region.

(d) As the current I_C increases from 10 μA to 1.5 mA, the BJT emerges from saturation and V_{CE} increases from V_{sat} to a value in the active operating region.

15-4. (a) From Eq. (15-1),

$$I_1 \cong \frac{5.0 \text{ V} - 0.2 \text{ V} - 0.6 \text{ V}}{1.0 \text{ k}\Omega} = 4.2 \text{ mA}$$

(b) From Eq. (15-3),

$$I_{RC} = \frac{5.0 \text{ V} - 0.2 \text{ V}}{1.0 \text{ k}\Omega} = 4.8 \text{ mA}$$

The total current sunk by the BJT when it is *on* is given by

$$I_{RC} + mI_1 = 4.8 \text{ mA} + 10(4.2 \text{ mA}) \cong 47 \text{ mA}$$

(c) From Eq. (15-4),

$$I_B > 47 \text{ mA}/h_{sat} = 47 \text{ mA}/25 \cong 1.9 \text{ mA}$$

From Eq. (15-2),

$$I_B = \frac{5.0 \text{ V} - 3(0.6 \text{ V})}{1.0 \text{ k}\Omega} = 3.2 \text{ mA}$$

which is more than adequate.

(d) If I_B drops below 1.9 mA, the BJT begins to leave saturation and $V_{CE} = V_{out}$ increases toward V_{CC}.

15-5. From Fig. 15-26, if either A or B is at V_{sat}, V_1 is given by $V_{sat} - V_D$ and Q is *off*. If both A and B are at V_{sat}, Q is also *off*. If both A and B are at V_{CC}, Q is *on*. For positive logic, $V_{sat} = \mathbf{1}$ and $V_{CC} = \mathbf{0}$. Therefore, the truth table of Fig. 15-33 results, and the gate shown fulfills the NOR function.

FIGURE 15-33
Solution for Exercise 15-5

A	B	C
0	0	1
0	1	0
1	0	0
1	1	0

15-6. (a) Given $\tau = R_C C_{eff}$,

$$R_C = 0.10 \ \mu s/100 \ pF = 1.0 \ k\Omega$$

(b) From Eq. (15-2),

$$I_B \cong \frac{V_{CC} - 3V_D}{R_1} = \frac{8.0 \ V - 3(0.7 \ V)}{10 \ k\Omega} \cong 0.6 \ mA$$

(c) To maintain saturation,

$$I_C < h_{sat} \ I_B = (20)(0.6 \ mA) = 12 \ mA$$

(d) From Eq. (15-3),

$$I_{RC} = \frac{8.0 \ V - 0.3 \ V}{1.0 \ k\Omega} = 7.7 \ mA$$

The current available to drive the following stages is

$$12 \ mA - 7.7 \ mA = 4.3 \ mA$$

(e) From Eq. (15-1),

$$I_1 \cong \frac{8.0 \ V - 0.3 \ V - 0.7 \ V}{10 \ k\Omega} = 0.7 \ mA$$

(f) The maximum fan-out is thus

$$m = 4.3 \ mA/0.7 \ mA \cong 6$$

15-7. From Fig. 15-28, the fall times (measured between the 10- and 90-percent points) are about 2.0 μs and the rise and fall times are about 4.0 μs. The average delay time per gate is about 3.5 $\mu s/2 \cong 1.8 \ \mu s$.

15-8. (a) If all inputs are at 0.5 V, Q_1 is in normal saturated operation, so that $I_B \cong 0$ for Q_2.
(b) The logic arrangement is shown in Fig. 15-34. If all inputs are at 0.5 V (a **0**), $D = \mathbf{0}$. If any input rises to 4.0 V (a **1**), $D = \mathbf{0}$ continues. If all inputs rise to 4.0 V, $D = \mathbf{1}$. The truth table is thus as shown in Fig. 15-34(B). The input diodes perform the AND function, and a NAND gate results.
(c) From Eq. (15-7),

$$I_B = \frac{4.0 \ V - 0.7 \ V - 0.7 \ V}{10 \ k\Omega} = 0.26 \ mA$$

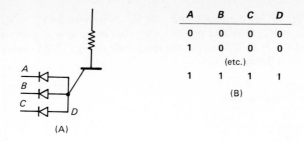

A	B	C	D
0	0	0	0
1	0	0	0
	(etc.)		
1	1	1	1

(B)

(A)

Fig. 15-34 Solution for Exercise 15-8.

Given $h_{sat} = 30$,

$$(I_C)_{max} = (30)(0.26 \text{ mA}) = 7.8 \text{ mA}$$

(d) With Q_2 *on*, the collector current is

$$I_C = \frac{4.0 \text{ V} - 0.5 \text{ V}}{1.0 \text{ k}\Omega} = 3.5 \text{ mA}$$

and the current available to drive subsequent stages is

$$7.8 \text{ mA} - 3.5 \text{ mA} = 4.3 \text{ mA}$$

(e) From Eq. (15-8),

$$I_1 = \frac{4.0 \text{ V} - 0.5 \text{ V} - 0.7 \text{ V}}{10 \text{ k}\Omega} = 0.28 \text{ mA}$$

(f) The maximum possible fan-out is

$$4.3 \text{ mA}/0.28 \text{ mA} \cong 15$$

15-9. (a) When $V_{in} = 6.0$ V, Q_1 is *off*. From Eq. (15-7),

$$I_B = \frac{6.0 \text{ V} - 0.7 \text{ V} - 0.7 \text{ V}}{10 \text{ k}\Omega} = 0.46 \text{ mA}$$

(b) The current in R_C is

$$\frac{6.0 \text{ V} - 0.5 \text{ V}}{0.50 \text{ k}\Omega} = 11 \text{ mA}$$

(c) From Eq. (15-8),

$$I_1 = \frac{6.0 \text{ V} - 0.5 \text{ V} - 0.7 \text{ V}}{10 \text{ k}\Omega} = 0.48 \text{ mA}$$

Q_2 must sink the total current:

$$11 \text{ mA} + 10(0.48 \text{ mA}) \cong 16 \text{ mA}$$

and

$$h_{sat} > 16 \text{ mA}/0.46 \text{ mA} \cong 35$$

15-10. When Q_2 and Q_3 are *on*, it is desired to have Q_4 *off*. The voltages on the base and emitter of Q_4 are given by

$$V_{B4} = V_{BE3} + V_{CE2}$$

$$V_{E4} = V_{CE3} + V_D$$

With the diode in place,

$$V_{B4} \cong 1.0 \text{ V}$$

$$V_{E4} \cong 1.0 \text{ V}$$

and Q_4 is *off*. Without the diode,

$$V_{B4} \cong 1.0 \text{ V}$$

$$V_{E4} \cong 0.3 \text{ V}$$

and Q_4 turns *on*, a condition that must be prevented.

15-11. The turn-on and turn-off switching times given by the circuit of Fig. 15-18(A) are dominated by the time constant

$$\tau = R_{\text{eff}} C_{\text{eff}}$$

where R_{eff} is given by the resistance of Q_2 when Q_1 turns *off*, and by the resistance of Q_1 when Q_1 turns *on*. The effective load capacitance increases in proportion to the number of following gate circuits being driven. Thus, if C_{eff} applies to a single gate circuit and m is the fan-out,

$$\tau = R_{\text{eff}}(mC_{\text{eff}})$$

The fan-out is thus limited by the desired switching time. To achieve smaller values of τ, m must be reduced.

15-12. Refer to Figs. 15-18 and 15-19. The load line due to Q_2 can be written as

$$V_{DS1} = V_{DD} - V_{DS2}$$

and since $V_{DS2} = V_{GS2}$,

$$V_{DS1} = V_{DD} - V_{GS2}$$

Given

$$I_{D2} = I_{D02}(V_{GS2}/V_T - 1)^2$$

$$V_{GS2}/V_T = (I_{D2}/I_{D02})^{1/2} + 1$$

Then

$$V_{DS1} = V_{DD} - V_T\left[(I_{D2}/I_{D02})^{1/2} + 1 \right]$$

or, rearranging terms,

$$I_{D2} = I_{D02}\left(\frac{V_{DD} - V_{DS1}}{V_T} - 1 \right)^2$$

This equation describes the load line shown in Fig. 15-19(B). The transfer characteristic for Q_1 is

$$I_{D1} = I_{D01}(V_{GS1}/V_T - 1)^2$$

As shown in Fig. 15-19(A).

The circuit operating point is determined by setting $I_{D1} = I_{D2}$ when Q_1 is *on*, or

$$I_{D01}\left(\frac{V_{GS1}}{V_T} - 1\right)^2 = I_{D02}\left(\frac{V_{DD} - V_{DS1}}{V_T} - 1\right)^2$$

where $V_{GS1} = V_{high}$ and $V_{DS1} = V_{low}$. The two operating voltage levels must be selected so that $V_{low} < V_T$ (to keep Q_1 *off* when $V_{in} = V_{low}$) and $V_{high} = (V_{DD} - V_T) > V_T$ (to turn Q_1 *on* when $V_{in} = V_{high}$). Exercises 15-13 and 15-14 provide applications of these criteria.

15-13. Choose the values

$$V_{low} = 1.0 \text{ V} < 2.0 \text{ V}$$

$$V_{high} = 5.0 \text{ V} - 2.0 \text{ V} = 3.0 \text{ V} > 2.0 \text{ V}$$

Then, from Exercise 15-12,

$$I_{D01}\left(\frac{3.0 \text{ V}}{2.0 \text{ V}} - 1\right)^2 = I_{D02}\left(\frac{5.0 \text{ V} - 1.0 \text{ V}}{2.0 \text{ V}} - 1\right)^2$$

and

$$I_{D01} = 4I_{D02}$$

The result is an acceptable solution if this relationship is required to hold.

15-14. From Exercise 15-12,

$$I_{D01}\left(\frac{4.0 \text{ V}}{2.0 \text{ V}} - 1\right)^2 = I_{D02}\left(\frac{5.0 \text{ V} - 1.0 \text{ V}}{2.0 \text{ V}} - 1\right)^2$$

$$I_{D01} = I_{D02}$$

The operating voltages must be

$$V_{low} = 1.0 \text{ V} < 2.0 \text{ V}$$

$$V_{high} = 4.0 \text{ V} = V_{DD} - 2.0 \text{ V} > 2.0 \text{ V}$$

so that

$$V_{high} = 6.0 \text{ V}$$

15-15. (a) From Eqs. (15-12) and (15-13),

$$V_{out} = V_{DD}\frac{R_n}{R_n + R_p}$$

Therefore, the two voltage levels are

$$V_{low} = (5.00 \text{ V})\frac{100 \text{ }\Omega}{10 \text{ k}\Omega} \cong 0.050 \text{ V}$$

$$V_{high} = (5.00 \text{ V})\frac{10 \text{ k}\Omega}{10 \text{ k}\Omega + 0.1 \text{ k}\Omega} \cong 4.95 \text{ V}$$

(b) The maximum allowable value of V_T is

$$4.95 \text{ V} - 0.20 \text{ V} = 4.75 \text{ V}$$

and the minimum allowable value of V_T is

$$0.050 \text{ V} + 0.20 \text{ V} = 0.25 \text{ V}$$

15-16. (a) From Eq. (15-14),

$$v = \frac{C}{\epsilon_r^{1/2}} = \frac{3 \times 10^8 \text{ m/s}}{2} = 1.5 \times 10^8 \text{ m/s}$$

$$= 1.5 \text{ cm/ns}$$

(b) If features are 5 mm apart,

$$t_D = 0.5 \text{ cm}/1.5 \text{ cm/ns} \cong 0.3 \text{ ns}$$

15-17. From Fig. 15-31, the incremental increase in cost per transistor is $1.00 - $0.20 = $0.80. For a thousand transistors, the incremental increase is $8.00.

For a realistic cost comparison, the costs of design, fabrication, and heat removal must also be considered as part of any such tradeoff.

PROBLEMS

(15-1) **15-1.** Show that the truth tables in Figs. 15-3 and 15-4 are correct for the given circuits.

(15-2) **15-2.** A gate circuit uses the steady-state voltages of -0.8 and -2.8 V. How should **1** and **0** be defined for positive logic?

(15-2) **15-3.** What conditions must be placed on voltages V_o and V_1 in Fig. 15-5 to achieve the desired logic operations?

(15-3) **15-4.** Assume that the BJT of Fig. 15-6 is operated with $I_B = 40 \ \mu A$ and $V_{CC} = 4.0$ V. What approximate resistor values are associated with (a) operation on the edge of the saturation region and (b) operation so that $I_C = 1.5$ mA?

(15-3) **15-5.** To achieve saturated operation, a BJT must have $I_B \geq 0.10$ mA for $I_C = 2.5$ mA. What is the required value for h_{sat}?

(15-3) **15-6.** A given BJT is operated with $I_C = 1.0$ mA. If $h_{sat} = 30$, describe the state of the device for (a) $I_B = 0.10$ mA and (b) $I_B = 0.01$ mA.

(15-4) **15-7.** What is the maximum fan-out m for a DTL gate circuit with $V_{CC} = 4.5$ V, $R_1 = 3.5$ kΩ, and $R_C = 2.5$ kΩ. Let $V_D = 0.6$ V, $V_{sat} = 0.3$ V and $h_{sat} = 25$.

(15-4) **15-8.** A DTL gate circuit has a maximum fan-out of 20 with $V_{CC} = 6.0$ V and $R_C = 3.0$ kΩ. Given $V_{sat} = 0.3$ V, $h_{sat} = 30$, and $V_D = 0.7$ V, what value of R_1 is being used?

(15-4) **15-9.** For a DTL gate circuit, 3.5 mA flows through R_C and a current of 0.10 mA is required to flow between stages to assure correct inverter operation. If $h_{sat} = 30$ and $I_B = 0.20$ mA, what is the maximum possible fan-out for the circuit?

(15-4) **15-10.** How much power is dissipated in the gate circuit of Problem 15-7 during each state?

(15-4) **15-11.** How much average power is being dissipated in the gate circuit of Problem 15-8 if, on the average, half of all gates in a system are *on* and half are *off*?

(15-4) **15-12.** A DTL gate circuit uses an effective load capacitance of 50 pF. If $V_{CC} = 5.0$ V, $V_{sat} = 0.5$ V, and a collector current of 2.5 mA is used, what is the time constant associated with the output turn-off time?

(15-5) **15-13.** A DTL gate circuit is operated between 0.5 and 3.0 V. (a) Estimate the turn-on time if $C/g_m = 1.0$ ns. (b) If $C = 25$ pF, estimate the dc bias current for the BJT. (c) If $R_C = 1.5$ kΩ, estimate the turn-off time for the circuit.

(15-5) **15-14.** A DTL gate circuit has an output time constant of 0.50 μs. If $V_{sat} = 0.50$ V and $V_{CC} = 4.5$ V, what is the turn-off switching time if a clamping diode with $V_{cl} = 2.0$ V is used?

(15-5) **15-15.** For the circuit of Figure 15-35, (a) find I_{out} that is available to drive the following stages when Q is on, (b) find I_{in} required when Q is off, and (c) find the maximum fan-out m.

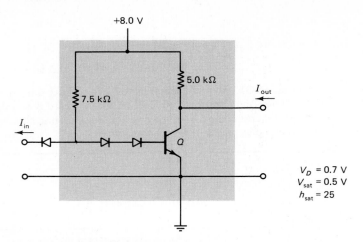

Fig. 15-35 Circuit for Problem 15-15.

(15-5) **15-16.** For the circuit of Figure 15-36, find the maximum collector current that can flow in Q and still have it in saturation.

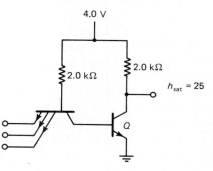

Fig. 15-36 Circuit for Problem 15-16.

(15-6) **15-17.** A simple TTL gate circuit is fabricated in Si. If $V_{CC} = 6.0$ V and $R_1 = 1.5$ kΩ, (a) find the base current that drives Q_2 in Fig. 15-13 (in the *on* condition) and (b) find the current that flows from each following stage (with the following stage in the *off* condition). (c) If $h_{sat} = 25$, find the maximum possible fan-out for the circuit. Take $V_{BE} = 0.7$ V and $V_{sat} = 0.5$ V.

(15-6) **15-18.** Draw a graph showing the maximum fan-out m as a function of R_1 for the simplified inverter circuit of Example 15-5.

(15-6) **15-19.** Calculate all resistor and BJT current levels for the TTL gate circuit of Fig. 15-14(A).

(15-6) **15-20.** Modify the TTL circuit in Fig. 15-14(A) so that (a) $R_1 = 3.0$ kΩ and (b) $R_1 = 5.0$ kΩ. Using the device characteristics given in the text discussion of this circuit, how does the operation of the circuit change as a result of the changes in R_1?

(15-6) **15-21.** Draw the schematic for a three-input TTL NOR gate of the type shown in Fig. 15-14(B).

(15-6) **15-22.** Develop truth tables for the circuits of Figure 15-14.

15-23. For the STTL circuit of Fig. 15-15, assume that $I_{R_C} = 2.0$ mA, $I_{out} = 1.0$ mA per load circuit, the fan-out $m = 10$, and $I_{in} = 0.4$ mA. If $I_z = 0.5$ mA, find h_{FE} for the operating point selected.

(15-8)   **15-24.** For the ECL gate circuit of Fig. 15-16(A), V_{R2} is selected so that $I_{C5} = 2.0$ mA. Find V_{out1} and V_{out2} when Q_1 is *on* and Q_2 is *off*, given $R_1 = R_2 = R_4 = 1.0$ kΩ. What possible range of values for V_{R1} can be selected?

(15-8)   **15-25.** Develop a truth table for the ECL circuit of Fig. 15-16(B).

(15-9)   **15-26.** Select $V_{DD} = 3.5$ V and $V_T = 2.0$ V for the NMOS circuit of Fig. 15-18(A). Can a viable gate circuit be designed? (Use the results of Exercises 15-12 and 15-13.)

(15-9)   **15-27.** Can a viable solution to Exercise 15-13 be obtained with the operating levels of 0.5 and 3.5 V, with $V_T = 1.5$ V?

(15-11)   **15-28.** Suppose that a signal is propagating along a conductor that is embedded in Si($\epsilon_r \cong 12$). What is the velocity of propagation?

COMPUTER APPLICATIONS

* **15-1.** Write a computer program that calculates the smallest supply voltage V_{CC} that can be used to design the DTL logic circuit of Fig. 15-7 as a function of other circuit variables. Assume the parameters of Example 15-1 and $m = 10$, and use the program to find V_{CC}.

* **15-2.** By using a step-function input and transient analysis, SPICE can be used to determine the delay time through an inverter circuit. Explore the properties of the DTL inverter circuit of Example 15-1 by applying SPICE to determine both the gate delay time and the average power dissipation in the gate. Compare your findings with the predictions of Fig. 15-23.

EXPERIMENTAL APPLICATION

15-1. The objective of this experiment is to design and build a DTL inverter circuit and test its performance. (a) Start with the simplest circuit version, without a clamping diode and without R_2. Use Example 15-1 as your model. Construct the inverter and use a square-wave generator to produce an appropriate input waveform (compatible with design values). Observe and record the waveforms at each node and measure the rise, fall, and delay times. Use different scale settings on the oscilloscope as necessary. Compare the predicted and observed waveforms and interpret. Determine the average power dissipation for the gate. (b) Modify the inverter by first adding a clamping diode and then adding R_2. Observe the effects of these changes on circuit performance. (c) For each case, where does your circuit appear on the comparison of digital technologies shown in Fig. 15-23?

CHAPTER 16
DIGITAL SYSTEMS

Chapter 15 describes the fundamental concepts of Boolean algebra and introduces ditigal gates as building block circuits able to perform many important logical functions. Several different technologies are explored for use in fabricating these gate circuits. TTL and ECL are introduced as bipolar applications, and NMOS and CMOS are introduced as MOS applications.

In this chapter, the emphasis turns to digital computer systems, with an emphasis on understanding the ways in which building block circuits provide the means for system implementation. The chapter begins with an introduction to some of the important concepts that underlie computer architecture, then proceeds to demonstrate how the needed subsystems and the complete system can be produced from ICs of different types in combination with input-output peripherals. A complete system can be assembled around a *chip set* that provides the necessary functions through compatible integrated circuits. The chapter concludes with an introduction to computer networks that can link multiple computer locations.

Digital computers are complex systems that can be developed following a variety of different strategies. The discussion here is intended to provide a general introduction and overview of digital systems with an emphasis on key concepts that are required for an understanding of system principles. The discussion is somewhat simpified in places and often centers around one way of achieving a desired system function, where other methods are also available. This chapter should be approached in this light, as introductory material that is intended to lead into more detailed studies of the subject.

It is not feasible to perform digital-circuit tradeoff studies without an appreciation for the larger system in which the circuit will be applied. The many different requirements placed on the circuit relate directly to the necessary interrelationships. At the same time, materials properties and interconnection and packaging techniques provide additional design constraints. The design of a digital system is thus a highly complex task that demands a maximum breadth on behalf of the designer.

Computer-aided design is essential to the development of all digital circuits and to the configuring of IC building block subsystems and systems. Since competitive pressures in the field are intense, every tradeoff and potential advantage must be considered to optimize system performance. Extensive CAD networks are used at every level to achieve the performance objectives, and CAM is required to produce the desired products with sufficient quality and reliability. CAD/CAM strategies are thus applied to all aspects of the digital design task.

16-1 BINARY WORDS

Chapter 15 introduces Boolean algebra, consisting of two symbols: **0** and **1**. The OR, AND, NOR, and NAND operations are defined in terms of these symbols. In order to apply this algebra to the design of a computer system, the concept of a *binary word* is now introduced. The shortest binary word consists of one symbol or *bit*. Only two digital words are possible in a one-bit system: the word **0** and the word **1**.

A longer word can be created by using two bits. For this case, four words are possible:

$$\mathbf{0\ 0}$$

$$\mathbf{0\ 1}$$

$$\mathbf{1\ 0}$$

$$\mathbf{1\ 1}$$

These binary words can be associated with numbers, and the sequence above becomes a way of counting. Let the most right-hand bit be defined as being in the 2^0 place, the next bit on the left being in the 2^1 place, the next bit as the 2^2 place, and so forth for as many bits as exist in the word. The four binary words above then correspond to

$$\mathbf{0\ 0} \rightarrow (0 \times 2^1) + (0 \times 2^0) = 0$$

$$\mathbf{0\ 1} \rightarrow (0 \times 2^1) + (1 \times 2^0) = 1$$

$$\mathbf{1\ 0} \rightarrow (1 \times 2^1) + (0 \times 2^0) = 2$$

$$\mathbf{1\ 1} \rightarrow (1 \times 2^1) + (1 \times 2^0) = 3$$

The total number of different words is 2^n, where n is the number of bits in each word. A word of bit length n can then be used to count from zero to $2^n - 1$.

The binary words can also be associated with many other meanings. For example, they can be related to various computer system operations or various memory storage locations. When used in computer systems, a binary word can have various lengths. Word lengths of 8, 16, 32, and 64 bits are in common use in computer systems. Words are often divided into eight-bit segments called *bytes*.

Example 16-1 Binary Words

How many different eight-bit binary words exist? How high can you count with these words?

Solution (a) The number of eight-bit binary words is

$$2^n = 2^8 = 256$$

(b) The highest number is

$$2^n - 1 = 255$$

The number of available words increases rapidly with the number of bits. For example, there are

$$2^{16} = 256^2 = 65,536$$

16-bit binary words, and

$$2^{32} = 256^4 = 4.295 \times 10^9$$

32-bit binary words. A word with more bits provides more significant figures for representing data and provides more flexibility in operational instructions for a processing system. ■

16-2 COMPUTER ARCHITECTURE

A simple model of a computer system is shown in Fig. 16-1. Each subsystem is discussed below.

The timing control for the system is provided by a *clock*. The various tasks as-

Fig. 16-1 Model of a digital computer system.

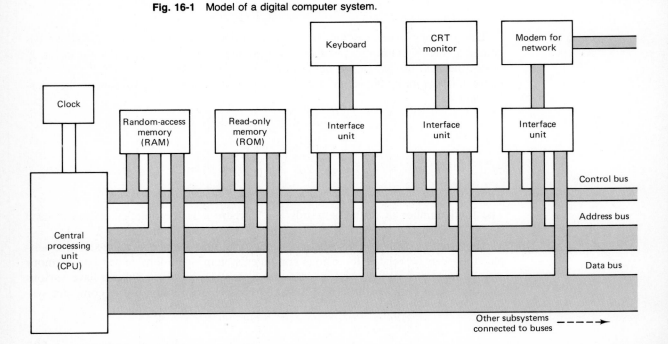

signed a computer system are accomplished through a series of execution cycles. Each cycle consists of a number of component steps, and, in combination, these steps produce a basic operation such as an *add* or *subtract*. The clock sends out pulses of a fixed length and separation to define a fixed time interval that is allocated for each step in the execution cycle.

The clock interval is determined by the delay times inherent in the building block circuits. Adequate time must be allowed during each step for all changes of state to be completed and for the system to stabilize for the next operation.

It is often necessary to have two clock phases in order to alternately enable gates in a chain and control the sequence of logical operations through the chain. Otherwise, pulses would move through the system in an uncontrolled way, become out of phase, and disrupt the planned sequence of operations.

The system clock drives the computer through its set of operations on a step-by-step basis. A typical clock circuit is quite complicated because it must provide a precise time base for the system and must drive many other circuits.

The central processing unit (CPU) consists of a control unit, a set of registers, and an arithmetic logic unit (ALU). The control unit produces the desired sequence of operations to obtain a program objective.

System control is exerted by the application of voltage levels (**1**s and **0**s) to various circuits throughout the system during each clock cycle. In order to obtain a given operation, the appropriate set of signals (voltage levels and locations) must be remembered and then applied. However, it is inefficient to store a long string of one-bit words to produce the control signals. Rather, it is useful to convert a combination of one-bit signals into a single (longer) instruction word. This is the process of *encoding*, which is done during computer system design. During system operation, the coded words are *decoded* in order to apply the desired signals throughout the system. The CPU thus typically includes several decoders.

Registers are CPU memory units that hold and store bytes between execution cycles of the computer. Typically, a word read out of memory is placed in a register and held there until it is needed to implement a specific operation.

The calculating functions are achieved by providing the CPU with an arithmetic logic unit (ALU), which includes combinations of circuits to produce desired logical operations. An important part of system design is the selection of logical operations that allow the types of calculations needed for system application.

An internal *counter* is used to sequence the CPU operations. As the counter sequences through the provided set of instructions (the *program*), control and data words are accessed from various memory locations and the required calculations are performed.

Each program step is treated as an address at which an instruction code can be found. The address is decoded and used to access the instruction word, which is placed in a register. The instruction word is then decoded and used to produce a desired set of signals throughout the CPU to achieve the desired operation.

The ALU consists of logic (gates) that can perform addition, subtraction, multiplication, division, and other functions. When instructions that require arithmetic operations are decoded, the correct signals are applied to the ALU to produce the intended result.

The addition of bytes is one of the most important tasks of the ALU. Suppose that the binary addition

$$00101$$
$$+ \underline{10001}$$

is to be performed, with each number stored in a register. The rules of binary addition are that

$$0 + 0 = 0$$
$$1 + 0 = 1$$
$$0 + 1 = 1$$
$$1 + 1 = 0 \text{ plus a carry of } 1$$

Therefore,

$$00101 \, [\text{decimal} \, (1 \times 2^0) + (1 \times 2^2) = 5]$$
$$+ \underline{10001} \, [\text{decimal} \, (1 \times 2^0) + (1 \times 2^4) = 17]$$
$$10110 \, [\text{decimal} \, (1 \times 2^1) + (1 \times 2^2) + (1 \times 2^4) = 22]$$

Adders are designed to combine the results of two registers according to the logic rules above.

Example 16-2 **Binary Addition**

Add the bytes **10010011** and **10010010** using the rules of binary addition. Check your work by converting each binary number to its decimal equivalent.

Solution Applying the given rules,

Byte	Decimal Equivalent
10010011	$(1 \times 2^0) + (1 \times 2^1) + (1 \times 2^4) + (1 \times 2^7) = 147$
+ 10010010	$(1 \times 2^1) + (1 \times 2^4) + (1 \times 2^7) = \underline{146}$
100100101	293

The decimal equivalent is much more compact because it uses a higher base (10). However, the binary version is more appropriate for use in bivalued computer systems where *off* and *on* symbols are required. ■

Figure 16-2 illustrates how the MC 68000 microprocessor implements the functions of the CPU. The properties of microprocessors are further discussed in Chapter 17.

The *random-access memory* (*RAM*) provides a high-speed capability to write, store, and read data and control words under command of the CPU. The words stored in RAM are constantly changing, and all information in RAM is lost when the system loses power (so that the memory is described as being *volatile*). RAM thus functions as a working short-term memory for the system.

The *read-only memory* (*ROM*) is used to provide long-term memory that can only be read, not written upon, by the CPU. The basic operating codes for the computer are encoded and stored in ROM. When the system receives power, the CPU ac-

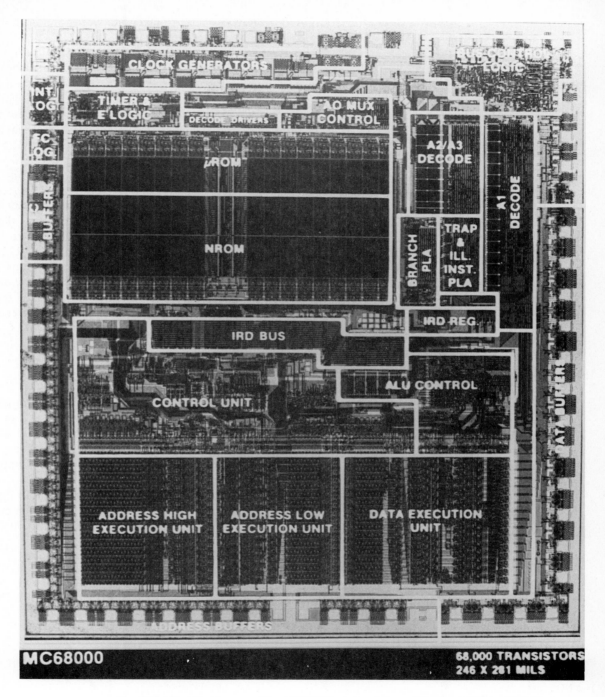

Fig. 16-2 Functional areas of the MC68000 microprocessor. (Courtesy of Motorola, Inc.)

cesses the ROM to provide the capability to initiate and control system functions. ROM remains unchanged when power is lost, and is thus *nonvolatile*.

The RAM and ROM are used to store both control and data words and to make these words available in response to appropriate requests from the CPU.

These subsystems all connect to the three buses than run from the CPU. Each bus consists of a package of wires run side by side, allowing for the parallel transfer of data. When the CPU requires interaction with a subsystem, binary information is placed on the address and control buses to turn *on* the subsystem of interest while turning *off* other subsystems. The address bus also defines where in the subsystem of interest a data word is to be found or placed. The data bus is then used to exchange data words between the CPU and other subsystems. During each execution cycle for which an exchange of information takes place among subsystems, a digital word must be placed on each of the three buses.

In a typical computer system, other subsystems are also attached to the system bus. A *keyboard* can be used to allow the user to provide instructions to the control unit and to input data. A *CRT monitor* is often used to display instructions as they are typed into the keyboard and data results from calculations. Information to be displayed is addressed to a particular screen location and passed from the CPU by way of the address and data buses. A control byte is used to activate the monitor so that the address and data words are received. Serial and parallel ports are provided for communication interface with the outside world. Many different types of networks can be developed to allow the individual computer station to interact efficiently with remote sensors, remote equipment, or other computers.

16-3 SR FLIP-FLOPS AND MEMORY CELLS

One of the building block units for a digital computer is the *set-reset (SR) flip-flop*, which provides a high-speed memory capability. A flip-flop is a circuit with only two stable states. The state is set by an input signal and the circuit retains its state until it is reset by a new input. If the flip-flop is set in the **0** state, the output remains at **0** until the circuit is reset. Similarly, if the flip-flop is set in the **1** state, the output remains at **1** until the circuit is reset.

The SR flip-flop can be constructed from a pair of NOR gates, as shown in Fig. 16-3(A). The output of gate A ($\overline{Q}$) is connected to one input of gate B, and the output of gate B (Q) is connected to one input of gate A. Assume initially that both in-

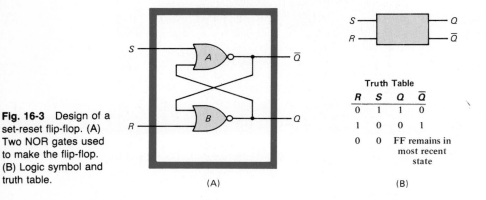

Fig. 16-3 Design of a set-reset flip-flop. (A) Two NOR gates used to make the flip-flop. (B) Logic symbol and truth table.

(A)

Truth Table

R	S	Q	$\overline{Q}$
0	1	1	0
1	0	0	1
0	0	FF remains in most recent state	

(B)

puts S and R are set at the binary $\mathbf{0}$. If gate A has an output $\overline{Q} = \mathbf{0}$, then both inputs to B are $\mathbf{0}$ and the output of gate B must be $Q = \mathbf{1}$. Since $S = \mathbf{0}$ and $Q = \mathbf{1}$, the output of gate A must be $\overline{Q} = \mathbf{0}$. The circuit is stable and remains with $\overline{Q} = \mathbf{0}$ and $Q = \mathbf{1}$ until a change is made in the inputs.

If a $\mathbf{1}$ is placed on the S input, the state of the circuit does not change. The flip-flop is already in the memory state corresponding to an $S = \mathbf{1}$ set.

If a $\mathbf{1}$ is placed on the R input, the output of gate B switches to $Q = \mathbf{0}$. Gate A must then switch so that $\overline{Q} = \mathbf{1}$. The flip-flop has switched to its second stable state. As long as both R and S stay at $\mathbf{0}$, the state is unchanged. The SR flip-flop is a memory element that can remember whether the last input $\mathbf{1}$ appeared on the S or R input and provides outputs (Q and $\overline{Q}$) that can test the condition of the memory. The logic symbol in Fig. 16-3(B) is used to represent this circuit.

The SR flip-flop can be used to create a basic memory cell. As shown in Fig. 16-4, the SR flip-flop is combined with an inverter and three AND gates to produce the desired cell structure. This configuration is often called a D-type flip-flop, or latch.

The select 1 and select 2 lines (S_1 and S_2, respectively) are used to address this specific memory cell. If both select signals are at $\mathbf{1}$, this cell is activated and can respond to a write signal to store new information in the cell. Otherwise, it remains unresponsive to changes in input. It is also necessary to have $S_1 = \mathbf{1}$ and $S_2 = \mathbf{1}$ to read the bit stored in the cell. Unless these two lines are activated, the output of the cell will stay at $\mathbf{0}$. The select signals are obtained by taking the address byte from the memory and decoding it to produce a $\mathbf{1}$ on each of the desired lines.

The input AND gates allow a change of state in the SR flip-flop only when $S_1 = \mathbf{1}$ and $S_2 = \mathbf{1}$. Under these conditions, suppose that the input is a $\mathbf{1}$. This bit is to be stored in the flip-flop, so set write (W) = $\mathbf{1}$. Since all inputs to the S-side AND gate are at $\mathbf{1}$, then the input to $S = \mathbf{1}$. The inverter requires that $R = \mathbf{0}$. The outputs are $Q = \mathbf{1}$ and $\overline{Q} = \mathbf{0}$. Because S_1 and $S_2 = \mathbf{1}$ and $\overline{Q} = \mathbf{1}$, the output of the cell is a $\mathbf{1}$. The input bit has been successfully stored in the cell.

Now set the input bit to $\mathbf{0}$ and let S_1, S_2, and W be activated (set equal to $\mathbf{1}$). Then $R = \mathbf{1}$ and $S = \mathbf{0}$. As a result, $Q = \mathbf{0}$ and the output of the cell is $\mathbf{0}$. The new input bit has been stored.

If $S_1 = \mathbf{1}$ and $S_2 = \mathbf{1}$, but $W = \mathbf{0}$, the bit stored in the cell can be read but not changed. If $Q = \mathbf{1}$, a $\mathbf{1}$ output appears, whereas if $Q = \mathbf{0}$, a $\mathbf{0}$ output appears. All

Fig. 16-4 D-type flip-flop memory cell.

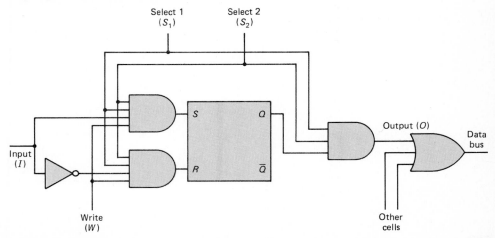

other addresses (memory cells) in the memory have outputs equal to **0** because both select lines are not activated in these other cases.

In a realistic application, the flip-flop must be linked to the system clock. Access commands appear as pulses and the cell can be accessed only when the clock pulse has a value **1**. However, phasing problems can occur for this configuration. The cell remains responsive to the input data line as long as the clock control pulse is present, and an error can occur if the input pulse begins to change state before the clock pulse returns to **0**.

To avoid this type of error, more complex combinations of flip-flops can be used to form memory elements. For the *JK master-slave flip-flop,* two SR flip-flops are interconnected so that the master flip-flop "listens" to the input during the initial portion of the clock cycle and the slave flip-flop changes the output of the memory cell when the clock cycle returns to zero. In a similar manner, edge-triggered flip-flops listen to the input only while the clock is changing state and store the input determined during this period.

Example 16-3   **SR Memory Cell**

The chart of Fig. 16-5 describes the operation of the SR memory cell of Fig. 16-4 with the cell previously set at $S = 1$. Discuss the meaning of the output results O that are obtained.

Discussion   It requires $S_1 = S_2 = 1$ to address a specific memory cell. For control words 01 to 06, this cell is not being accessed, so output O is always **0**. For word 07, the cell is being addressed, but $W = 0$, so the state of the cell cannot be changed. The output O is determined by the previous value set in the cell. For $S = 1$, $Q = O = 1$. For word number 08, the input I can write a new value in the cell. After a switching delay time, the output is determined by the input. If $I = 1$, then Q remains at **1** and $O = 1$. If $I = 0$, then the cell switches, providing $Q = O = 0$.   ∎

FIGURE 16-5
Operation of the SR Memory Cell

Address and Control Word Number	S_1	S_2	W	O	Comments
01	0	0	0	0	Cell not being accessed
02	0	0	1	0	
03	1	0	0	0	
04	1	0	1	0	
05	0	1	0	0	
06	0	1	1	0	
07	1	1	0	1	(read)
08	1	1	1	*	(write/read)

* $I = 1$, memory cell remains set at $S = 1$
  $I = 0$, memory cell sets at $S = 0$

A register can be made of SR flip-flops, as shown in Fig. 16-6. When the write control line is enabled with a **1**, the input word can be read into the register and stored. Figure 16-6 shows a parallel register that might be used as a memory storage device for a CPU. (*Shift registers* are discussed in Exercise 16-18.)

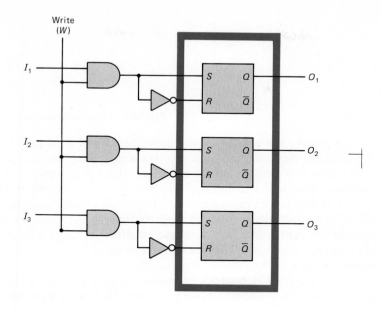

16-4 RANDOM-ACCESS MEMORIES

The RAM is used to write, store, and read data and control words under the command of the CPU. Bits of information can be written into and read from any cell in an array by using the correct address and control signals. In computer design, the objective is usually to store bytes. Therefore, the cells are grouped so that a string of bits can be stored at a single address. The decoder makes sure that each bit is entered into the correct cell within the word. A RAM can be thought of as a place to store bytes of information. An address byte is associated with each data byte, giving the location of the particular word.

In early computer systems, flip-flop-based memory cells of the type described in Section 16-3 were used to create RAMs. Such RAMs are called *static* because the information can be stored indefinitely in the cell as long as power is maintained. However, except in special circumstances, the static RAMs are not well matched to the task because they are large (having too many components) and dissipate too much power.

The evolution of modern computer systems has depended on the development of *dynamic RAMs* (*DRAMs*) to provide high-density, low-power, low-cost memory. A DRAM, as shown in Fig. 16-7, uses the charge stored in a capacitor to represent a **1**, and the lack of charge on the same capacitor to represent a **0**. The E-MOSFET is assumed symmetric. The capacitor C can be the distributed capacitance between the gate and channel or an additional capacitor manufactured on the surface of the IC. Figure 16-7 shows a simplified version of a one-transistor DRAM memory cell. A **1**

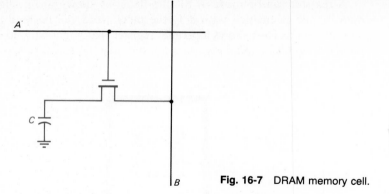

Fig. 16-7 DRAM memory cell.

is written into the cell by raising both lines A and B to potential V. If there is no prior charge on C, the capacitor receives a charge representing a **1**. To read the cell, A is raised to V while B is held at ground. If a **1** is present, the capacitor discharges onto line B, where a *sense amplifier* detects the current and produces a **1** output.

The charge on the capacitor leaks off through the intrinsic channel of the E-MOS-FET, so periodically the charge must be "refreshed." Complex circuitry is required to read the stored data, use the sense amplifier to detect **0**s and **1**s, and replace the data in the capacitor. This type of random-access memory is considered to be *dynamic* because of the refresh requirement.

Clearly, a DRAM requires extensive circuitry to control access to the memory cells and ensure the refresh process. Figure 16-8 provides a functional indication of the required operations. Multiple-DRAM ICs can be combined in a computer to produce a large memory size. This type of volatile memory meets a broad range of performance requirements and is a widely used digital IC.

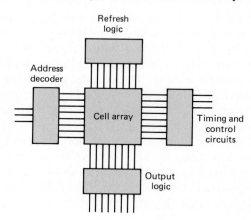
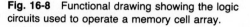

Fig. 16-8 Functional drawing showing the logic circuits used to operate a memory cell array.

16-5 READ-ONLY MEMORIES

Read-only memories (ROMs) are used when the information contained in a memory is not to be changed during computer operations. Various types of semiconductor arrays are commonly used as ROMs. These can be as simple as an array with diodes or transistors present at each address.

Consider the array shown in Fig. 16-9. Each input line corresponds to a single address bit. In order to read from the memory, one of the input address lines is set at **1**. The output data word can then be read. A word is stored in the array by placing a diode at each intersection on the address line where a **1** is to be read. When the input address line is set at **1**, all output lines linked by diodes are also set at **1**.

For Fig. 16-9, input line 3 produces the output word **110**. Input line 5 results in **001**. A three-bit data bit word can be read for each address line, producing the desired memory property.

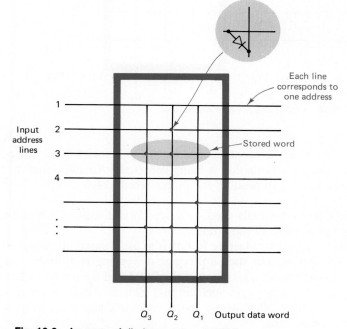

Input address lines

Each line corresponds to one address

Stored word

Q_3 Q_2 Q_1 Output data word

Fig. 16-9 An array of diodes used as a ROM.

Example 16-4   ROM Design

Using an array of diodes, design a ROM that satisfies the following logic relationships:

Address Line (A)	Output Data Word (Q)
1	**001**
2	**010**
3	**011**
4	**100**
5	**101**
6	**110**

Solution   From the discussion above, the solution of Fig. 16-10 is obtained. Each intersection with a diode (dot) provides a connection between the input line and crossing output line. The number of required output lines depends on the number of address lines used. For three output lines, as shown in Fig. 16-10, a maximum of $2^3 - 1 = 7$ ad-

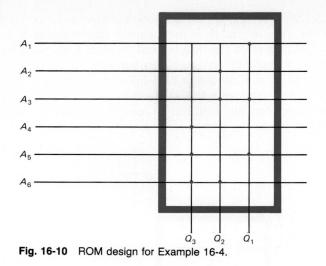

Fig. 16-10 ROM design for Example 16-4.

dress lines can be accommodated. (If $A = \mathbf{0}$ for all lines, the output word is a string of **0**s and no information is being accessed.) An additional A_7 address line could be accommodated to produce the output word **111**. ∎

ROMs are typically used with decoders (discussed in Section 16-9) to reduce the required number of input lines.

One way to manufacture ROMs is to use a *mask*[1] to determine which intersections receive diodes. For such a process, a working relationship must exist between the computer designer who is to use the ROM and the manufacturer who is to fabricate it to order. In large-scale applications, such a relationship may be appropriate. However, in many other cases, more flexibility and independence are desired by the computer designer.

The development of *programmable ROMs* (*PROMs*) helped address this problem. For a PROM, each intersection in Fig. 16-9 can initially include a diode in series with a metal fuse (for example, consisting of nichrome). The user can program the PROM by raising the address lines to a sufficiently high level to "blow" the selected fuses and produce the desired memory pattern.

PROMs are still limiting because they cannot be modified once the programming process has been implemented. The development of the *erasable programmable ROM* (*EPROM*) allowed a new and attractive degree of freedom for the designer. These memory devices can be programmed, erased, and reprogrammed by the user.

The EPROM makes use of the floating-gate structure shown in Fig. 16-11(A). If the upper gate is raised to a potential V with respect to the channel, the electric field lines E extend as shown. The equipotential ($V =$ constant) surfaces are perpendicular to the E field lines. If a floating gate is inserted as shown, it conforms to one of the equipotential surfaces and is thus raised to the potential associated with this location. The potential of the floating gate is, therefore, a function of the potential applied to the upper gate and the relative spacings.

---

[1] Semiconductor fabrication is discussed in Chapter 18.

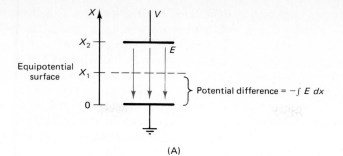

(A)

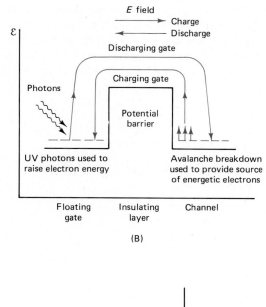

(B)

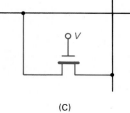

(C)

Fig. 16-11 EPROM operation.
(A) Capacitive coupling to apply a potential to a floating gate.
(B) Method of charge storage.
(C) Memory cell application.

The EPROM stores information by placing charge on the floating gate. This is done by causing the channel to avalanche (as discussed in Chapter 8) and raising the floating gate to a positive potential. The electrons develop sufficient energy to enter the conduction band in the insulator and are attracted to the floating gate, as shown in Fig. 16-11(B). The EPROM is erased by using ultraviolet light to give the stored electrons sufficient energy to again return to the conduction band in the insulator and applying a negative potential to the floating gate to direct the electrons back to the channel.

The presence of a charge on the floating gate is a **1**, and the lack of a charge is a **0**. The storage of a **1** or **0** determines whether a channel forms in the E-MOSFET. Once the PROM is programmed as desired, each intersection becomes a conducting or nonconducting connection, as shown in Fig. 16-11(C). The stored charge is isolated by the high insulator resistance and remains on the floating gate for long periods of time (typically, more than 10 years) unless it is erased.

An additional step in sophistication became available in 1980 with the introduction of the *electrically erasable programmable ROM* (*EEPROM*). This device can be electrically programmed, erased, and reprogrammed without removal from the system.

The EEPROM also makes use of a floating gate. However, the insulating layer between the channel and floating gate is made very thin (of the order of 100 to 200 Å), so energetic electrons from the channel can tunnel to the gate. (Tunneling is also discussed in Chapter 4 with respect to the tunnel diode.) An applied voltage V on the upper gate creates a sufficiently high electric field (about 10 MV/cm) in the insulating layer to produce an adequate probability for tunneling, as shown in Fig. 16-12. This level is close to the breakdown level for the insulating layer and must be chosen carefully.

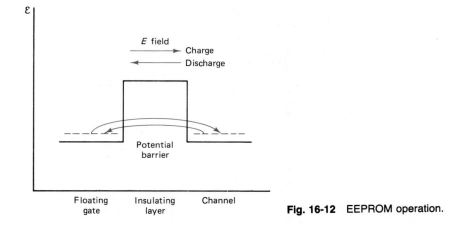

Fig. 16-12 EEPROM operation.

The EEPROM can be programmed and reprogrammed by placing sufficiently large positive and negative voltages (10 to 20 V) on the upper gate to cause electron tunneling in either direction. The EEPROM can thus be programmed and reprogrammed in place (without removal from the circuit) and provides a flexible approach to ROM for the designer.

16-6 MAGNETIC AND CCD MEMORIES

There are many different types of magnetic memory devices based on the hysteresis loop introduced in Chapter 3. Floppy disks, hard disks, and tapes are included in this category. These memories function by associating **0**s and **1**s with different directions of magnetization for small cells in a magnetic material. Two directions of magnetization are defined, corresponding to domain directions, and used to represent **0**s and **1**s. The location of a given domain is its address and the data word is stored by arranging the desired sequence of domain orientations. On disks, these individual cells can be within a few thousandths of an inch of one another.

Reading and writing a word on a disk is accomplished as shown in Fig. 16-13. A loop of magnetic material (the read/write head) is wrapped with a winding. When a current flows through the winding, a B field is created in the magnetic material. There is a gap in the magnetic loop just above the disk surface. A large H field is created in this gap, with field lines extending outside of the loop itself and into the magnetic material on the surface of the disk. A bit is recorded on the disk surface by aligning the domains in a forward or reverse direction, as shown in Fig. 16-14. These directions reverse when the current in the write loop is reversed.

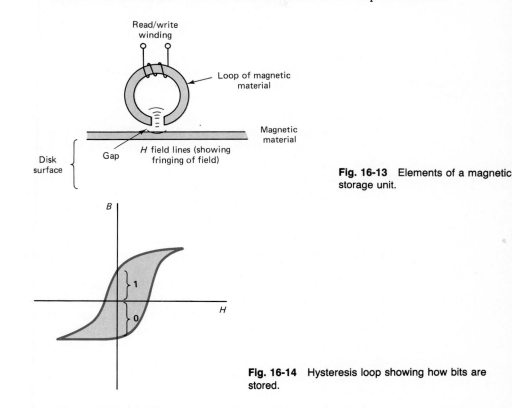

Fig. 16-13 Elements of a magnetic storage unit.

Fig. 16-14 Hysteresis loop showing how bits are stored.

Reading a bit is accomplished by passing the magnetic loop over the address and measuring the B field produced in the loop by the surface domain. A changing B field produces a current pulse in the read/write winding.

Other types of memory devices have been introduced, based on different technologies. Magnetic bubble memories form domains in a thin-film single crystal of magnetic material, and magnetic fields are used to move the bubbles through the film. There are no moving parts to this memory because the bubbles move in response to an external field, providing a memory with many attractive properties.

The operation of a charge control device (CCD) memory is illustrated in Fig. 16-15. The CCD memory functions as a nonequilibrium device, so the following discussion focuses on time-dependent effects.

Shown are three MOS structures, which may be regarded as small E-MOSFETs.

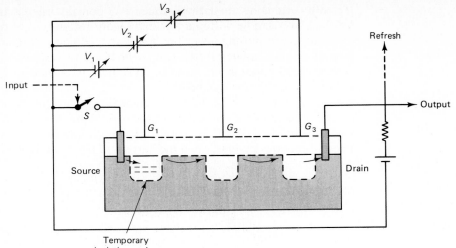

Fig. 16-15 Illustration of a CCD memory.

If a positive charge is placed on gate 1 (G_1), a temporary depletion region is generated in the semiconductor region below. While this depletion region is present, switch S can be briefly closed to allow electrons to flow into the depletion region, where they are stored.

If voltage V_2 is then increased while V_1 is decreased, a similar depletion region is formed under gate 2 (G_2), and the fringing electric fields associated with G_2 attract the electrons from G_1 to G_2, where they are stored. A similar sequence moves the stored charge from G_2 to G_3 and then out of the CCD memory. The existence of the charge at any location in the structure can be associated with the storage of a **1**. Extensions of this technique can thus be used to store digital sequences, as required for a memory.

At each step in the storage sequence, the charge can remain in the depletion region only until the thermally generated carriers begin to fill up the depletion region so that the individual nature of the stored charge is lost. The charge must move from location to location rapidly enough to avoid this problem.

Leakage losses gradually reduce the amount of charge stored at any location, so the charge must periodically be removed from the device, amplified (or "refreshed"), and returned back to circulating storage. The CCD is thus dependent on transient operation.

CCD arrays can also be used to create imaging devices. If depletion regions are formed simultaneously under each gate location and optical energy is allowed to fall on the depletion regions, charge carriers are created in proportion to the light intensity at each location. The sequencing process can then be used to withdraw the charge from the array, read the locations of the charge, and produce a corresponding digital image.

16-7 LOGIC CIRCUITS

Computer design requires that the various registers, memories, and other subsystems described above be linked with many different types of logic circuits. The encoders and decoders introduced in Section 16-1 are required to develop the control words that determine system operation and to decode these words as they are addressed by the CPU. All memories require logic interfaces to manage memory use, as do the other peripherals. The logic circuits provide the linkages that provide functional coherence to the computer system. The next two sections describe encoders and decoders in further detail. The discussion then turns to alternative strategies for implementing logic circuits and systems.

16-8 ENCODERS

A digital system operates by arranging for complex combinations of signals to appear throughout the system during each execution cycle. Each signal corresponds to a bit of information, and to store the needed information of each cycle, these bits are combined into bytes.

The encoding process transforms the desired signals into binary words to be stored in memory. For many internal applications, necessary encoding is accomplished by the system designers when the instruction set for the system is developed. In other cases, involving data input from keyboards and complex networks with multiple external input data lines, encoders must be included as part of the system design.

One type of encoder is shown in Fig. 16-16(A). Seven input signals are encoded to produce three-bit binary words. The truth table for this logic arrangement is shown in Fig. 16-16(B). When all inputs are at **0**, the output word is **000**. If I_1 then becomes a **1**, the output word is **001**. Other results are obtained as shown. A total of $2^n - 1$ input lines can be represented by an n-bit word. For the case shown, $n = 3$ and there are seven input lines. A total of $2^8 - 1 = 255$ input lines can be coded to an eight-bit byte.

This type of logic arrangement can be used to take many different input lines and code them into a set of binary words, where the word length (number of bits) is much less than the number of lines. The word can then be stored in memory. When needed for system operation, the word is read from memory, decoded, and used to provide the desired signals on specified lines.

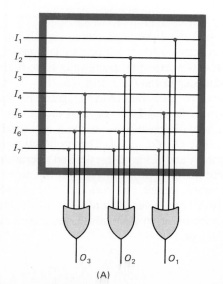

	O_3	O_2	O_1
No input	0	0	0
I_1	0	0	1
I_2	0	1	0
I_3	0	1	1
I_4	1	0	0
I_5	1	0	1
I_6	1	1	0
I_7	1	1	1

(B)

Fig. 16-16 Example of an encoder. (A) Logic diagram. (B) Truth table.

(A)

651

Example 16-5   Encoder Design

You are given an IC chip containing two OR gates. What is the maximum number of input lines that can be encoded? Draw the logic diagram and present the truth table for the encoder.

Solution   The maximum number of input lines is

$$2^2 - 1 = 3$$

The logic diagram and truth table are shown in Fig. 16-17.

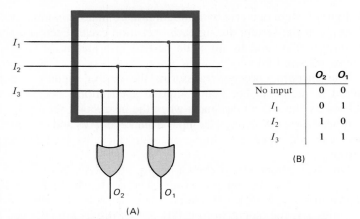

	O_2	O_1
No input	0	0
I_1	0	1
I_2	1	0
I_3	1	1

(B)

(A)

Fig. 16-17   Encoder design for Example 16-5. (A) Logic diagram. (B) Truth table.

For encoders of the type shown in Fig. 16-16 and 16-17, the arrangement of connections can become quite complex for large words. Consider that an eight-bit word can be used to encode $2^8 - 1 = 255$ input lines, and a 16-bit word can be used to encode over 65,000 input lines. The production of encoders is thus a task that is well suited to computer-aided design and manufacturing techniques.   ■

16-9   DECODERS

A decoder takes a binary word and produces output signals on addressed lines. Two decoders could be used to take two select address bytes and decode them to identify a particular memory address (for example, by placing 1s on both select 1 and select 2 in Fig. 16-4). A simple decoder is shown in Fig. 16-18.

Each input line connects to each output AND gate. By using the inverters as shown, a particular pattern of 0s and 1s is required to activate each gate. The input word on I_1, I_2, and I_3 can be withdrawn from memory and used to create the desired signals on the output (O_0, O_1, O_2, O_3, . . .). These outputs activate gates and other circuits throughout the system and achieve the desired operation.

Logic arrangements like the one shown in Fig. 16-18 can be used to "unpack" an instruction that has been coded, stored, and recalled. The encoder-decoder combination of Figs. 16-16 and 16-18 can be used to identify which input lines are set at **1**, create a byte carrying this information (for storage), and then decode this byte to reset the initial lines to values of **1**.

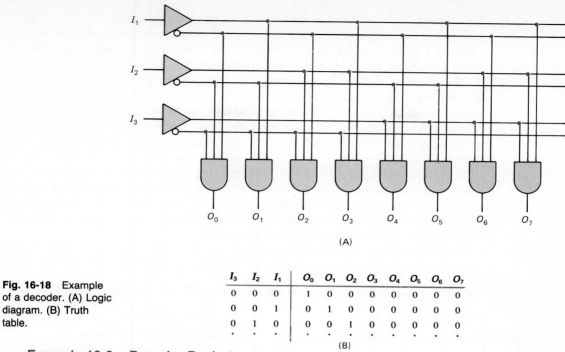

Fig. 16-18 Example of a decoder. (A) Logic diagram. (B) Truth table.

I_3	I_2	I_1	O_0	O_1	O_2	O_3	O_4	O_5	O_6	O_7
0	0	0	1	0	0	0	0	0	0	0
0	0	1	0	1	0	0	0	0	0	0
0	1	0	0	0	1	0	0	0	0	0
·	·	·	·	·	·	·	·	·	·	·

(B)

Example 16-6   **Decoder Design**

Draw the logic diagram and provide the truth table for a decoder that can be used in conjunction with the encoder of Example 16-3, given three AND gates.

Solution   A logic diagram and truth table are shown in Figure 16-19. As can be noted, the advantages of encoding and decoding grow rapidly with the number of bits being considered. For a few bits, the advantage is small. But for large numbers of bits, the packing of data provides a flexible approach to information storage. ■

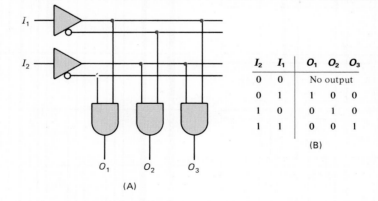

Fig. 16-19 Decoder design for Example 16-6. (A) Logic diagram. (B) Truth table.

I_2	I_1	O_1	O_2	O_3
0	0	No output		
0	1	1	0	0
1	0	0	1	0
1	1	0	0	1

(B)

16-10   STRATEGIES FOR IMPLEMENTING LOGIC CIRCUITS AND SYSTEMS

Digital logic circuits and systems can be designed using standard, programmable, or customized building block elements. As discussed in Chapter 17, many of the most basic logic configurations (for example, AND and OR gates, encoders and decoders, and registers) are available as standard IC chips, and application objectives can be satisfied by interconnecting these chip sets and using a suitable integration strategy (such as a printed circuit board). On the other hand, design applications often require a degree of flexibility that is not allowed by such standard products.

The standard-element approach to system implementation is rapidly changing. The PROM, EPROM, and EEPROM memory technologies of Section 16-5 are now being applied to create programmable logic building block elements. The respective technologies are known as programmable logic devices (PLDs), erasable programmable logic devices (EPLDs), and electrically erasable programmable logic devices (EEPLDs).

Programmable logic is achieved by fabricating arrays of standard gates in conjunction with interconnection matrices that can be programmed (and erased and reprogrammed) using the methods previously applied to memories. From the system designer's perspective, the application of PLDs depends on the types of logic arrangements that are available, performance, and cost.

The PROM introduced above can also be considered as a type of programmable logic. Figure 16-20 shows how a decoder (to make most effective use of input lines)

Fig. 16-20 Combining an input decoder with a programmable array to create a PROM.

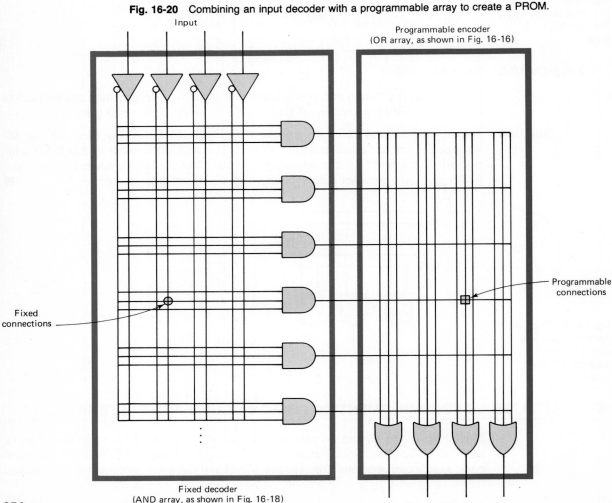

can be combined with a programmable array to produce such a logic unit. Two other forms of PLDs are the programmable logic array (PLA) device and programmable array logic (PAL) device (Coli, 1987). The PLA is arranged as shown in Fig 16-21(A). The decoder portion of the PROM in Fig. 16-20 is now programmable. The entire device allows the programming of AND and OR functions to produce a desired logic form. The PAL is shown in Fig. 16-21(B). The AND gate array is programmable with a fixed OR array.

These and other PLDs are rapidly changing the way in which systems are configured. Programmable memory and logic ICs are forcing computer system designers to have a greater concern with program logic than with the selection of specific hardware units that are appropriate for a given logic task.

Alternatively, sophisticated CAD and CAM systems are enabling the designer to obtain a custom IC that is specifically intended for a given application. The designer prepares a logic diagram for the desired system, and a software program, called a

Fig. 16-21 Programmable logic devices (PLDs). (A) Programmable logic array (PLA). (Continued on p. 656.)

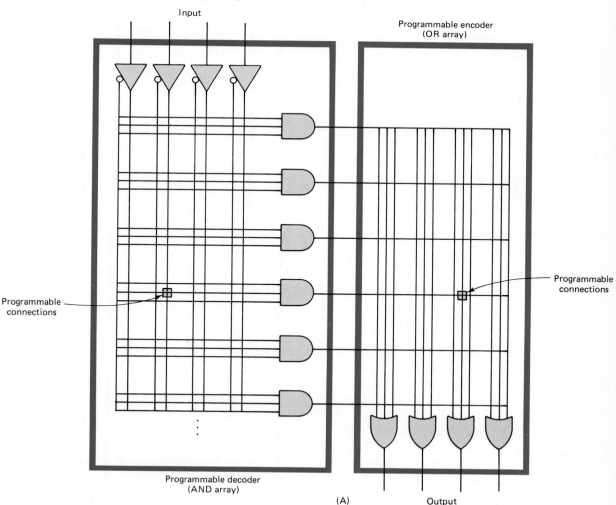

(A)

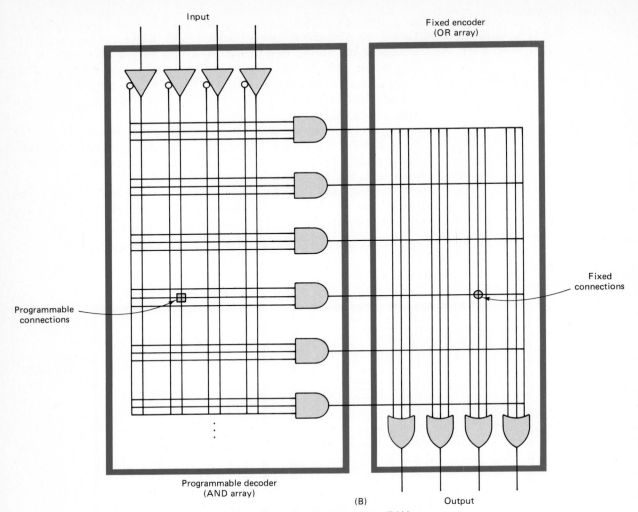

Input   Fixed encoder (OR array)

Fixed connections

Programmable connections

Programmable decoder (AND array)

(B)   Output

Fig. 16-21 (cont.)   (B) Programmable array logic (PAL).

silicon compiler, translates this logic diagram into the input to a manufacturing facility that fabricates the desired IC. It is often feasible to combine several different special-purpose IC designs on a single chip to minimize production costs. An extensive custom IC capability is developing around such application-specific ICs (ASICs).

The microprocessor provides a means of logic implementation that is completely dependent on software. By changing the program codes, a wide range of system operations can be obtained. The development of the microprocessor has provided another dimension of functional flexibility obtained through software instead of hardware. Given the range of microprocessors available (as discussed in Chapter 17), a hardware/software tradeoff always exists in meeting the needs of any logic configuration. Digital logic can thus be programmable or customized by drawing on a variety of hardware and software options. The decision as to how to optimize a given strategy to meet performance objectives can be a complex one that requires an understanding of all the options and tradeoffs.

Many different logic operations are used to produce a functional computer system, as shown in Fig. 16-22. Data are entered from a keyboard and encoded. The coded words pass along the data bus to the CPU, where they are placed in registers, and memory addresses are assigned. The words are then placed in RAM.

Assume that a computer program has been written with the first step requiring the addition of the data in locations **0001** and **0010**. The add execution, represented here by **0001**, calls for a series of logical operations stored in a *stack*, or sequence, starting at **0001**. As these coded words are read one at a time (as the clock cycles), they are placed in a register and then decoded. The decoded information is a sequence of bits (**0**s and **1**s) that act throughout the system to produce the desired execution. By using the sequence of bits, the words at locations **0001** and **0010** are read into the CPU registers and added. An address is assigned to the sum, which is then placed in RAM. The last operation in the **0001** (add) execution returns the system to the next program step.

16-12 EFFECTS OF CIRCUITS ON SYSTEM PERFORMANCE

The performance of a digital system is closely dependent on the reliability of gate design and operation. The importance of fan-in and fan-out restrictions was illustrated in this chapter because these limits affect the logical operations that can be performed. Special driver circuits are often used to connect subsystems to the bus to meet input and output requirements. *Tristate drivers* are able to provide both **0** and **1** outputs with an enable control or to be electrically disconnected from a bus when the enable signal is not present.

With so many devices packed into a small volume, heat dissipation problems can be critical. Efforts are made to reduce power requirements at every stage of design. As noted in Chapter 15, CMOS technology is advantageous because of its low power consumption. However, TTL and ECL technologies provide faster switching times and can thus make use of shorter clock cycles to produce a faster computer.

Distributed capacitance and inductance can severely degrade system performance, even to the point of preventing necessary operations from being performed during the allocated clock cycle. Limits must be placed on coupling between conductors and conductor length to minimize these distributed effects.

The dc supply voltages distributed throughout the system must maintain their levels as the load varies substantially, providing demands on the design of the supply itself and the distribution network. Large supply currents flowing through the system can produce magnetic fields that interfere with subsystem performance.

Internal and external noise can be a major problem. Capacitive coupling between bus lines and other parts of the circuit can induce pulses where none should appear, resulting in a processing error. External electric and magnetic fields can induce voltages and currents that also interfere with performance.

Since system performance depends on both the building block units and on packaging arrangements, system reliability requirements place constraints on the design of individual gate circuits. These circuits must not only reflect the interconnect needs of the logic designer, but must also use a technology matched to system speed objectives and demonstrate insensitivity to variations in supply voltage and noise. The electronics and logic design tasks are thus closely intertwined.

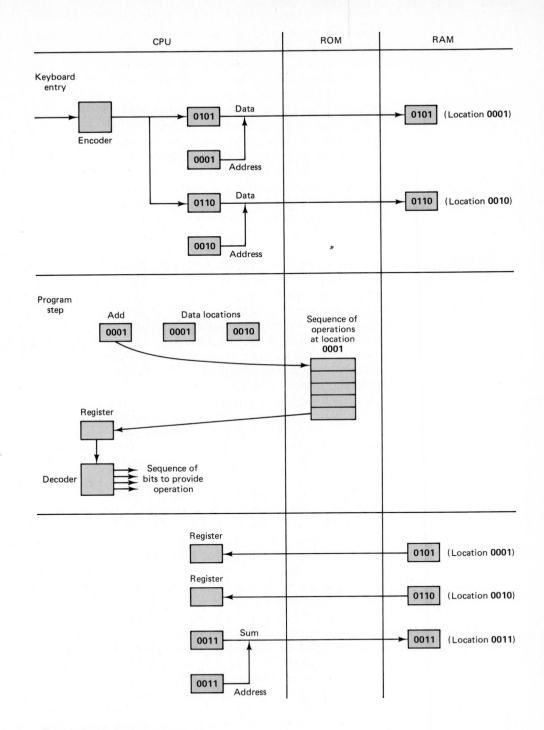

Fig. 16-22 Subsystem function.

In many application settings, performance objectives cannot be met by an individual computer but require many stations, the development of a computer *network*. The network may consist of a single computer with satellite sensors and equipment (such as the programmable power supply introduced in Chapter 5 and further discussed in Chapter 13), numerous computer stations interacting with each other, or computers and factory equipment linked in a production process. Network applications often require a mixture of analog and digital operations, leading to the need for analog-to-digital and digital-to-analog converters (discussed in Chapter 17).

Many different types of computer networks are in operation today. The telephone network is one of the largest, involving a variety of different types of channels and multiple users. Businesses routinely link different computer stations on a nationwide basis, using the telephone channels for interconnection. Other types of networks are used to link different computer stations in an office, or to provide data communications between automated machinery in a manufacturing center.

This text emphasizes *local area networks* (*LANs*) that are useful for computer-integrated design and manufacturing (CID/CIM) in localized facilities. The concepts for this application provide insight into other types of networks.

A basic overview of such networks is provided from three perspectives. This chapter discusses the functional operation of such a system, Chapter 17 describes some of the IC-based hardware that is available to implement such a system, and Chapter 19 discusses how such a system can be applied to the electronics design and manufacturing processes.

Figure 16-23 shows a LAN that includes both computer work stations (for design and management) and manufacturing equipment stations. In an electronics product-

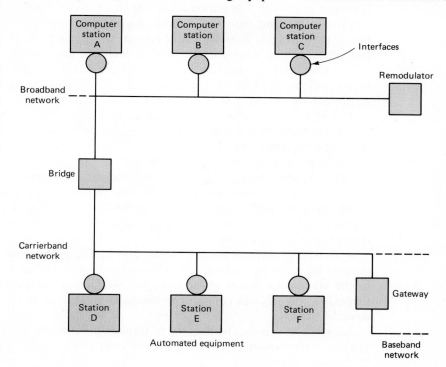

Fig. 16-23 Typical LAN computer network.

development setting, computers A to C might be used for computer-aided design, with the manufacturing center fabricating the desired product.

In order to enable the different computer stations to communicate with each other, an interface must be developed between each station and the network channel, and in order to link the manufacturing stations together, interfaces must also be developed between each equipment item and the network channel. As might be expected, many complex problems can arise in linking different types of computers and a range of production equipment.

One approach to solving this problem is to develop detailed specifications for the network operations and to then produce interface units that enable each station to adapt to the network requirements. The complete specification of the network interface requirement is often called a *protocol*.

Several types of protocols have been developed for office settings in which business activities require the linking of individual stations, and computer vendors are routinely able to achieve a high level of communications over a channel consisting of a pair of twisted wires or a coaxial cable.

One obvious problem is how to "direct traffic" over the network. Some protocols use a statistical access strategy, which has each station "listen" to see if the channel is empty, then send its message. If a collision between messages occurs, the sending stations each wait for a (randomly determined) length of time and try again. This approach works remarkably well and can provide for efficient network operations, but does not guarantee a maximum waiting access time for any one station.

The situation can be different when a manufacturing center is involved. Manufacturing operations often require that each station be provided with a guaranteed maximum waiting time before channel access is provided. The manufacturing automation protocol (MAP) has been developed for this setting. A "token" (an enabling word) is passed from station to station according to a predetermined route, and a station is allowed to transmit onto the channel only if the station has the token. The token moves progressively from station to station, providing a known maximum time before access can be obtained. The result is real-time deterministic access.

Manufacturing operations can also require a faster rate of information flow between stations. A different type of network and channel interface is often appropriate to meet this need.

The MAP strategy has been defined using a layered approach, as shown in Fig. 16-24. Each station is connected to the network by means of a well-defined sequence of functional operations. The advantage of this approach is that modular hardware and software can be developed.

Layer 1, the physical layer, includes the specifications for the channel itself (usually a coaxial cable) and the transmitter and receiver (or *modem*). The digital circuits that are used to implement the higher levels interface with the encoding, decoding, and modulating functions of the modem to produce signals that propagate as desired through the channel.

A *baseband* network is obtained by placing the pulses from the digital circuits directly on a communication link without modulation (so no modem is used). Due to the attenuation losses in the link, this solution is usually limited to short distances. A *gateway* can connect the baseband and the other types of networks discussed below.

Information can also be transferred through the channel by using frequency-modulation (FM) techniques, which are introduced in Section 14-15, applied to digital

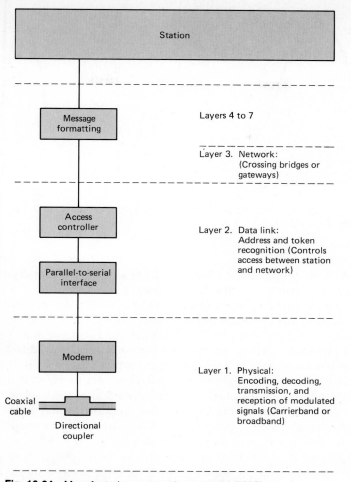

Station

Message
formatting

Layers 4 to 7

Layer 3. Network:
(Crossing bridges or
gateways)

Access
controller

Layer 2. Data link:
Address and token
recognition (Controls
access between station
and network)

Parallel-to-serial
interface

Modem

Layer 1. Physical:
Encoding, decoding,
transmission, and
reception of modulated
signals (Carrierband or
broadband)

Coaxial
cable

Directional
coupler

Fig. 16-24 Manufacturing automation protocol (MAP).

signals. A *carrierband* MAP network uses two different frequencies (with one twice the other) to represent **1**s and **0**s, as shown in Fig. 16-25. Typical data rates are 5 and 10 megabits/second. This type of network can only provide a single communication channel among the stations.

A *broadband* MAP network uses FM to share many channels over the same physical cable. For each channel, two different frequencies are used, one for transmitting and the other for receiving. A *remodulator,* or *repeater,* is used on the network, as shown in Fig. 16-23, to receive all the signals from stations and retransmit them at the appropriate frequency for reception.

It is often preferable to use a carrierband network for a localized manufacturing work cell, where a single channel suffices, and to use a broadband network where multiple channels are required. The two types of channels can communicate using a *bridge,* as shown in Fig. 16-23.

The physical layer of MAP involves specifications of the channel, transmitter, and receiver to enable multiple vendors to produce compatible products. Further details

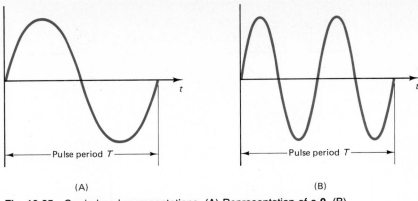

Fig. 16-25 Carrierband representations. (A) Representation of a **0**. (B) Representation of a **1**.

about the properties of FM transmitters and receivers for digital communication systems are provided in the next section.

Layer 2, the data link layer of MAP, provides an interface between the serial bit stream in and out of the physical layer and the special-purpose digital subsystems required for the higher levels of the interface. This layer must have the necessary logic to recognize whether an incoming message is addressed to a particular station and whether a token has been received. Further, this layer is required to route the received message to higher levels if appropriate, to transmit if the token is received, and to place the token back on the network after the transmission is complete.

Layer 2 provides the logic control to allow station nodes to communicate with each other. Layer 3, the network layer, provides the capability for messages to cross over bridges or gateways where required.

Layers 4 to 7 provide the necessary message formatting to allow a specific station to interface with the standard MAP message configurations on the network. These layers thus have a special-purpose computer capability to convert between the standard message format and any other local message format and can be implemented in hardware or software, depending on the application.

16-14   DIGITAL COMMUNICATION SYSTEMS

Digital communication systems are important because of (1) the advantages often associated with the transmission and reception of digitized information and (2) the widespread growth of computer technology that is leading to a growing emphasis on computer networks, which require digital communications between nodes. This important area of digital electronics can be developed by extending the AM and FM communication concepts described in Chapter 14 and considering some of the unique aspects of binary representations.

Digitized information consists of a sequence of discrete symbols. A digital communication system requires the transmission of such representations over time. Digital communication systems are thus concerned with different types of binary representations and with alternative strategies for passing these representations from a transmitter, through a channel, to a receiver. Many types of modulation techniques have been developed for such systems.

The amplitude-modulation (AM) method of Section 14-14 can be applied in a modulation technique called *amplitude-shift keying (ASK)*. Two different amplitude levels are used to represent the binary **1** and **0** symbols. A form of ASK is *on-off keying (OOK)*, which involves turning *on* the carrier to produce a **1** during a bit interval and turning off the carrier to produce a **0** during a bit interval. The two amplitude levels are thus the maximum signal amplitude $v = A_m$ and $v = 0$.

With OOK, a received message consists of a series of amplitude pulses plus noise, as shown in Fig. 16-26. The receiver must distinguish between the existence of a **1** and a **0** in each defined bit interval. This requires that the receiver establish a threshold detector level. If the incoming signal amplitude is greater than the detector level, a **1** is associated with the bit interval. If the incoming signal is less than the detector level, a **0** is associated with the bit interval. Obviously, the likelihood of errors increases as the noise level increases. The threshold is selected so that the total probability of detection error is minimized.

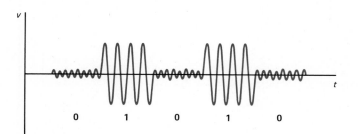

Fig. 16-26   OOK modulation (received signal and noise).

Because the incoming signal amplitude associated with a **1** depends on the channel losses, the optimum threshold is a function of these losses. To overcome this problem, the receiver must incorporate automatic gain control (AGC), as discussed in Chapter 14, that keeps the average power into the detector at approximately a constant level. Because of the implementation complexities associated with the AGC requirement, OOK is rarely used for data communication.

Another approach to digital communication that avoids the detection threshold problem is called *frequency-shift keying (FSK)*. This method of modulation applies the frequency-modulation (FM) concepts developed in Section 14-15.

The voltager-controlled oscillator (VCO) of Fig. 14-20 transmits two different frequencies corresponding to a **0** and **1**, as shown in Fig. 16-27. There is always a

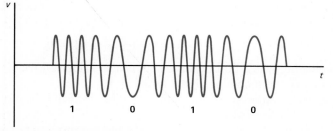

Fig. 16-27   FSK modulation.

signal present during each bit interval, giving a positive message.

As noted for such a modulation strategy, a tradeoff exists between the S/N ratio obtained and the bandwidth required for the communication channel. Since the amplitude of the transmitted signal is constant, the threshold of the receiver can be held constant, avoiding the AGC problem associated with OOK.

The receiver used with FSK modulation must be able to discriminate between the two frequencies that are being transmitted. The objective is to decide over each bit period whether frequency f_1 or f_2 is present while minimizing detection errors. One possible strategy is shown in Fig. 16-28. Over each bit interval, the incoming signal is fed into two mixers or demodulators that use nonlinear detector characteristics to

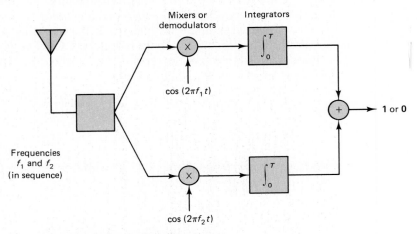

Fig. 16-28 Receiver for FSK modulation.

multiply together the incoming signals and fixed-frequency reference signals from oscillators that are phase-coherent with the input carrier.

As discussed in Chapter 14, the output of such a mixer includes a difference frequency between the two inputs. Neglecting noise, the output of each mixer is then proportional to

Mixer A: $\qquad\qquad\qquad\qquad\qquad \cos[2\pi(f_1 - f_1)] = 1$

or $\qquad\qquad\qquad\qquad\qquad\qquad \cos[2\pi(f_1 - f_2)t]$

Mixer B: $\qquad\qquad\qquad\qquad\qquad \cos[2\pi(f_2 - f_1)t]$

or $\qquad\qquad\qquad\qquad\qquad\qquad \cos[2\pi(f_2 - f_2)] = 1$

The detector output signals are integrated over a bit interval to maximize the S/N ratio. The frequencies f_1 and f_2 are chosen to be orthogonal, so that

$$\int \cos[2\pi(f_2 - f_1)t] = 0$$

By indicating which bit interval output is a dc value and which is a zero, the receiver can identify the incoming signal as a **1** or **0**.

It is often important to maximize the data that can be transmitted through a given channel that is described by a limited bandwidth and noise power N. The waveforms to be transmitted and the types of filters and detectors to be used in the receiver can be modified to meet the requirements associated with a specific channel. Complex coding techniques can be applied to approach channel capacity with a reasonable error rate. In general, the design of optimal communication systems is a complex task covered in courses on this subject and can be implemented in hardware or software, depending on the application.

QUESTIONS AND ANSWERS

Questions

16-1. What is a chip set?

16-2. Why is Boolean algebra appropriate to the study of digital circuits?

16-3. How many bits form a byte?

16-4. What is a *bus*?

16-5. What is the difference between a RAM and a ROM?

16-6. Why aren't RAMs generally used for the ROM function?

16-7. What are the principal features of a dynamic RAM (DRAM)?

16-8. What happens to a **1** stored in a CCD memory if the circulating time between locations is longer than the time required for the thermal generation of carriers in the semiconductor material?

16-9. Why is encoding a basic part of computer design?

16-10. Why is decoding used during computer operation?

16-11. What is a protocol as applied to computer network design?

16-12. What are the advantages and disadvantages of a statistical access strategy for a computer network?

16-13. What are the advantages and disadvantages of a token-controlled computer network?

16-14. What are the characteristics of baseband, carrierband, and broadband computer networks?

16-15. What are the relative advantages and disadvantages of FSK?

Answers

16-1. A chip set is a set of integrated circuits that can operate in a compatible manner to produce a desired system function.

16-2. Because digital circuits have only two stable states (**0** and **1**), a bivalued form of algebra is appropriate for the study of logical operations.

16-3. A word length of eight bits is used to form a byte. A four-bit word is sometimes called a *nibble*.

16-4. A bus is a set of wires that allows for the parallel transfer of data (voltages) from one location in the computer to another.

16-5. A RAM writes, stores, and reads data and control words and functions as a short-term memory. The ROM reads words that have been stored permanently (subject to reprogramming) and functions as a long-term memory.

16-6. Since production costs per bit are typically higher for a RAM than a diode-based ROM, the latter technology is used wherever possible. ROM technology also assures a reliable, unchanging set of operational instructions for the computer.

16-7. A DRAM provides a high-density, low-power, low-cost memory and requires external, extensive control and refresh circuitry.

16-8. If bits stored in a CCD memory are circulated too slowly, the thermal generation of carriers causes the depletion layer to be filled with unwanted carriers, and the information is lost.

16-9. The computer engineer must arrange for each instruction (word) to produce the desired voltage levels throughout the computer to implement the desired logical operation. The required operations are encoded into a word that is available when the memory is accessed.

16-10. When the computer is operated, instruction words are called from memory to produce the desired voltages throughout the system. Decoding transforms the instruction word into the individual voltage levels required.

16-11. A protocol is a set of rules or instructions that defines how various stations on a computer network relate to the communications structure of the network.

16-12. A statistical access strategy has the advantage of simplicity and therefore low cost. On the other hand, there is no guarantee of a maximum access waiting time for any one station, which can prevent satisfactory network operation under heavy loads.

16-13. A token-controlled access strategy provides real-time deterministic access to the network for all stations. The disadvantage of such a system is the resultant complexity and cost.

16-14. Baseband networks directly connect the digital circuits in a computer to the communication link. Carrierband networks use two different frequencies to represent **0**s and **1**s and thus represent a particular type of FM. Only one carrierband channel can be implemented for a given physical link. Broadband networks use FM techniques to frequency-share a given physical link so that several channels can be formed. The **0**s and **1**s are transmitted by modulated carrier frequencies.

16-15. FSK uses a constant-amplitude signal and thus can be used effectively in lossy and nonlinear channels. FSK also allows a design tradeoff between the required S/N ratio and the required signal bandwidth for a given error rate.

EXERCISES AND SOLUTIONS

Exercises
(16-1)

16-1. Find the decimal equivalents of the following binary words:

$$001$$

$$101$$

$$011$$

How many different words are available with three bits?

(16-1) **16-2.** This chapter discusses the use of binary algebra, in which **0**s and **1**s represent powers of 2. Other representations are also commonly used in digital computer design. The hexagonal (hex) representation allows counting to base 6 and octagonal (octal) is to base 8. Numbers in these representations can be converted to decimal equivalents using simple rules. For example, addition in base 6 is obtained as shown in Fig. 16-29. When two digits are included in a square, the most significant digit can be regarded as a carry.

Find the decimal equivalents of the following words using
(a) base 6:

$$21550$$

(b) base 8:

$$6745$$

FIGURE 16-29
Representation for Base 6 (See Exercise 16-2)

	0	1	2	3	4	5
0	0	1	2	3	4	5
1	1	2	3	4	5	10
2	2	3	4	5	10	11
3	3	4	5	10	11	12
4	4	5	10	11	12	13
5	5	10	11	12	13	14

(16-1) **16-3.** Add the following numbers:

$$3321_6 + 4453_6$$

(16-1) **16-4.** Add the following numbers:

$$6432_8 + 5571_8$$

(16-3) **16-5.** A static RAM is designed, as shown in Fig. 16-30, to include 250×250 memory cells.

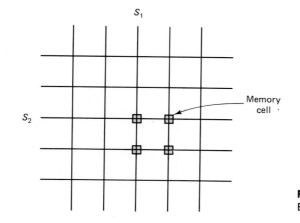

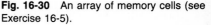

Fig. 16-30 An array of memory cells (see Exercise 16-5).

(a) What is the total number of memory cells?
(b) If the addresses to S_1 and S_2 are to be encoded before use, how many bits are required for the encoded words?
(c) How many gates and transistors are required for the RAM, neglecting the output gates required for interfacing with the data bus, based on Fig. 16-4, and assuming the TTL circuit configuration of Fig. 15-14(B) is representative of all gates used?

(16-3) **16-6.** Assume that the output AND gate is connected to $\overline{Q}$ instead of Q in Fig. 16-4. What form does the truth table take for the memory cell? How could this affect system operation?

(16-3) **16-7.** Modify the chart of Fig. 16-5 for a memory cell that is initially in the $S = 0$ state. Describe the outputs that now correspond to each case and explain the results.

(16-3) **16-8.** For the parallel-input register of Fig. 16-6, $O_1 O_2 O_3 = 101$.
(a) If $I_1 I_2 I_3 = 110$ and $W = 0$, find the resultant output $O_1 O_2 O_3$.
(b) If $I_1 I_2 I_3 = 011$ and $W = 1$, find the resultant output.

(16-5) **16-9.** Develop a truth table for the ROM of Fig. 16-31.

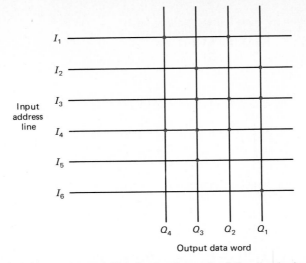

Input address line

I_1
I_2
I_3
I_4
I_5
I_6

Q_4 Q_3 Q_2 Q_1

Output data word

Fig. 16-31   ROM for Exercise 16-9.

(16-5)   **16-10.** Design a ROM that produces the following logical operations:

I_1	I_2	I_3	I_4	Q_3	Q_2	Q_1
1	0	0	0	0	0	1
0	1	0	0	0	1	0
0	0	1	0	0	1	1
0	0	0	1	1	0	0

(16-5)   **16-11.** Assume that voltages in the range of 10 to 20 V are required for programming and erasing an EEPROM. If an adequate probability of tunneling requires 10 MV/cm, what thickness is required for the insulating layer betweeen the channel and gate? Assume that the floating gate is half way between the upper gate and the channel.

(16-6)   **16-12.** Assume that a floppy disk measures 3.5 inches in diameter and that 50 percent of its surface area is usable for data storage. If individual cells require a surface area of 2 square mils, what maximum number of eight-bit bytes can be stored on the disk?

(16-8)   **16-13.** Develop a truth table for the encoder of Fig. 16-32.

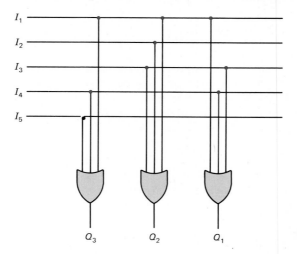

I_1
I_2
I_3
I_4
I_5

Q_3 Q_2 Q_1

Fig. 16-32   Encoder for Exercise 16-13.

(16-8) **16-14.** Given an IC containing four OR gates, design an encoder with the maximum number of input lines.

(16-8) **16-15.** Show how a simple encoder might be used with a keyboard.

(16-9) **16-16.** Develop a truth table for the decoder of Fig. 16-33.

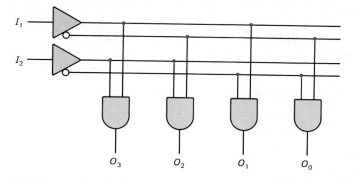

Fig. 16-33 Decoder for Exercise 16-16.

(16-13)
(16-14) **16-17.** A carrierband computer network is designed to accommodate a modem data rate of 5 Mbits/s. Show that the transmission and reception frequencies of 5 and 10 MHz are reasonable selections if a high-efficiency coding strategy is applied.

(16-13) **16-18.** MAP requires that serial messages be converted to a parallel format for input on an internal station bus. Show how such a function might be accomplished.

(16-13) **16-19.** MAP requires that each station recognize its own address and recognize the token when it is received. How might such a recognition function be achieved to allow the user to set the codes?

(16-14) **16-20.** Given a carrierband network with a data rate of 2 Mbits/s, what are the lowest oscillator frequencies that could be used if the transmitter and receiver are to operate only for $f > 1.0$ MHz?

(16-14) **16-21.** A given broadband coaxial cable can be used to transmit frequencies in the range of 50 to 150 MHz. How many 10-Mbit/s data channels can be used according to MAP, choosing $BW \cong 1/T$?

16-22. Figure 16-34 shows how the production cost per gate decreases and the capital invest-

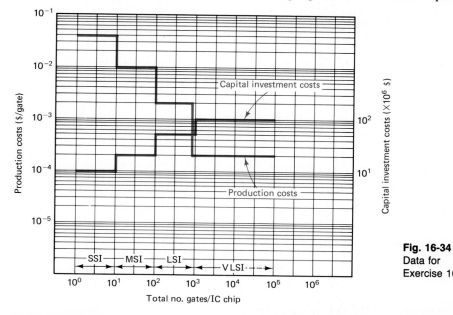

Fig. 16-34 Data for Exercise 16-22.

ment costs increase as a function of the total number of gates per chip for a particular company. If a total market is estimated at 10^9 gates, what gate density is optimum for factory design? How would this change if the market increases to 10^{10} gates?

Solutions

16-1. By definition, the decimal equivalents are given by

$$001 \rightarrow \qquad 1 \times 10^0 = 1$$

$$101 \rightarrow (1 \times 2^2) + (1 \times 2^0) = 5$$

$$011 \rightarrow (1 \times 2^1) + (1 \times 2^0) = 3$$

16-2. By definition, the decimal equivalents are given by

(a)
$$21550_6 = (2 \times 6^4) + (1 \times 6^3) + (5 \times 6^2) + (5 \times 6^1) + (0 \times 6^0)$$
$$= 3,018$$

(b)
$$6745_8 = (6 \times 8^3) + (7 \times 8^2) + (4 \times 8^1) + (5 \times 8^0)$$
$$= 3,557$$

16-3. From Fig. 16-29,

$$
\begin{array}{ll}
\overline{11} & \text{Carries} \\
3321 & \\
4453 & \\
\hline
12214 &
\end{array}
$$

16-4. By using the technique illustrated in Fig. 16-29, the add matrix of Fig. 16-35 can be prepared. Therefore,

$$
\begin{array}{ll}
\overline{11} & \text{Carries} \\
6432 & \\
5571 & \\
\hline
14223 &
\end{array}
$$

FIGURE 16-35
Add Matrix for Base 8

	0	1	2	3	4	5	6	7
0	0	1	2	3	4	5	6	7
1	1	2	3	4	5	6	7	10
2	2	3	4	5	6	7	10	11
3	3	4	5	6	7	10	11	12
4	4	5	6	7	10	11	12	13
5	5	6	7	10	11	12	13	14
6	6	7	10	11	12	13	14	15
7	7	10	11	12	13	14	15	16

16-5. (a) The total number of cells is

$$250 \times 250 = 62,500$$

(b) A word of bit length n can be used to count from zero to $2^n - 1$. For $n = 7$, $2^n - 1 = 127$; and for $n = 8$, $2^n - 1 = 255$. Therefore, an eight-bit word is required.

(c) Based on Fig. 16-4, each memory cell requires three AND gates and two NOR gates. The total number of gates in the memory is thus approximately

$$5(250 \times 250) \cong 312K$$

From Fig. 15-14(B), the total number of transistors for the selected TTL implementation is six per gate, so that the total number of transistors is about

$$312K \times 6 \cong 1.9M$$

16-6. If the output from the memory cell is taken from $\overline{Q}$, the output logic symbol of the cell is the NOT of the stored logic symbol. The cell, therefore, does not function as required.

16-7. The truth table is unchanged for lines 01 to 06 because the cell is still not accessed. The remaining lines are shown in Fig. 16-36.

FIGURE 16-36
Solution for Exercise 16-7

	S_1	S_2	W	O	Comments
07	1	1	0	0	(read)
08	1	1	1	*	(write/read)

*$I = 1$, memory cell set at $S = 1$
$I = 0$, memory cell remains at $S = 0$

16-8. (a) Since $W = 0$, the output remains unchanged at **101**.

(b) Since $W = 1$, the register stores the incoming word **011** and the output becomes **011**.

16-9. The results of Fig. 16-37 can be obtained from Fig. 16-31.

FIGURE 16-37
Solution for Exercise 16-9

Input Address Lines						Output Data Word			
I_1	I_2	I_3	I_4	I_5	I_6	Q_4	Q_3	Q_2	Q_1
1	0	0	0	0	0	1	0	1	0
0	1	0	0	0	0	0	1	0	1
0	0	1	0	0	0	0	1	1	1
0	0	0	1	0	0	1	1	1	0
0	0	0	0	1	0	0	1	0	0
0	0	0	0	0	1	0	0	0	1

Following the model of Example 16-4 and the chapter discussion, the results of Fig. 16-38 are found.

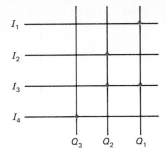

Fig. 16-38   Solution for Exercise 16-10.

16-11. From Fig. 16-11(A), the potential on the floating gate is

$$V\frac{x_1}{x_2} = \frac{V}{2}$$

Therefore,

$$t = \frac{V/2}{10\,\text{MV/cm}}$$

providing a range of values of 50 to 100 Å.

16-12. The surface area of the disk is

$$A = \pi\,(3.5\,\text{in.}/2)^2 \cong 9.6\,\text{in.}^2$$

If 50 percent are usable, the number of individual cells is

$$\frac{1/2 \times 9.6\,\text{in.}^2}{2\ \text{mil}^2(1\ \text{in.}/1000\ \text{mil})^2} = 2.4 \times 10^6$$

And, using eight bits per byte,

$$\text{number of bytes} = (2.4 \times 10^6)/8 = 0.3 \times 10^6$$

16-13. From Fig. 16-32, the results of Fig. 16-39 are obtained.

FIGURE 16-39
Solution for Exercise 16-13

I_1	I_2	I_3	I_4	I_5	Q_3	Q_2	Q_1
1	0	0	0	0	1	1	1
0	1	0	0	0	0	1	0
0	0	1	0	0	0	1	1
0	0	0	1	0	1	0	1
0	0	0	0	1	1	0	0

16-14. An encoder with four OR gates can accommodate $2^4 - 1 = 15$ input lines. The resultant design is shown in Fig. 16-40.

16-15. The signals from a keyboard must be encoded before they can be placed onto the bus and received by the CPU. One simple encoding strategy is shown in Fig. 16-41. Each switch produces a different encoded byte to be placed on the data bus. The keyboard system also places a byte on the address bus, defining the origin of the data byte.

16-16. From Fig. 16-33, the results of Fig. 16-42 are obtained.

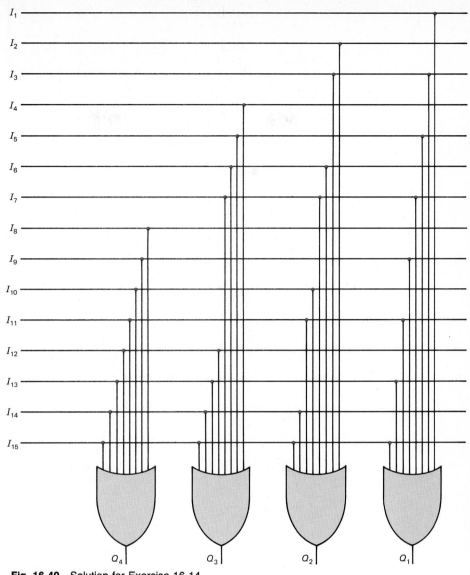

Fig. 16-40 Solution for Exercise 16-14.

16-17. A continuous-phase FSK modulation technique makes efficient use of the available bandwidth in a channel. The most efficient version of this code is called *minimum-shift keying* (*MSK*). For this type of code, the channel bandwidth can be taken as approximately $1/T$, where T is the bit rate. [Based on the data of Stremler (1982), the 99-percent power bandwidth is about $1.17/T$ and the 50-dB bandwidth is about $1.9/T$. The most appropriate relationship depends on the definition of bandwidth and the acceptable error rate.] A modem data rate of 5 Mbit/s thus requires a bandwidth of about 5 MHz. The selection of 5 and 10 MHz as the transmit and receive frequencies, respectively, is illustrated in Fig. 16-43. The minimum required bandwidth is provided and operation of the communication system is maintained in the frequency region of 2.5 to 12.5 MHz.

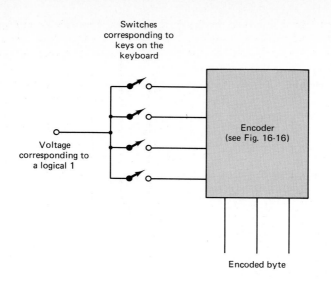

Switches
corresponding to
keys on the
keyboard

Encoder
(see Fig. 16-16)

Voltage
corresponding to
a logical 1

Fig. 16-41 Keyboard interface to bus (Exercise 16-15).

Encoded byte

FIGURE 16-42
Solution for Exercise 16-16

I_1	I_2	Q_3	Q_2	Q_1	Q_0
0	0	0	0	0	1
0	1	0	1	0	0
1	0	0	0	1	0
1	1	1	0	0	0

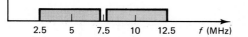

2.5   5   7.5   10   12.5 f (MHz)

Fig. 16-43 Bandwidths for a pulse communication system (Exercise 16-17).

16-18. The desired conversion can take place by using a *shift register*. As shown in Fig. 16-44, a string of JK flip-flops can be used to implement the logic operations. During clock cycle a, flip-flop 1 is set at **1**, corresponding to the first input signal received. During clock

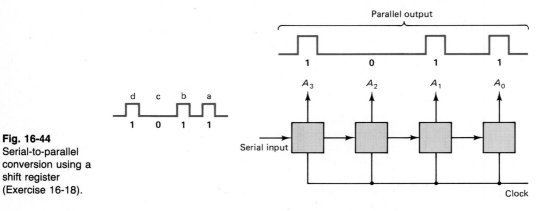

Parallel output

1   0   1   1

A_3 A_2 A_1 A_0

Serial input

Clock

d   c   b   a

1   0   1   1

Fig. 16-44
Serial-to-parallel
conversion using a
shift register
(Exercise 16-18).

cycle b, this **1** moves to flip-flop 2, and flip-flop 1 is set at **0**, corresponding to the second input signal received. Following this process over four clock cycles, the incoming serial word **1011** is read into flip-flops 1 to 4. The parallel output word **1011** can then be obtained.

16-19. Figure 16-45 illustrates how an EPROM can be used to create the desired address and token words, and compare these stored words with an incoming word to determine if a match is obtained.

16-20. Assuming MSK coding as discussed in Exercise 16-17, the results of Fig. 16-46 are obtained, with the frequencies 2.0 and 4.0 MHz selected.

16-21. By choosing $BW \cong 1/T$, each channel requires 20 MHz of bandwidth. Therefore, the maximum number of channels is

$$\frac{(150 - 50)\,\text{MHz}}{20\,\text{MHz}} = 5$$

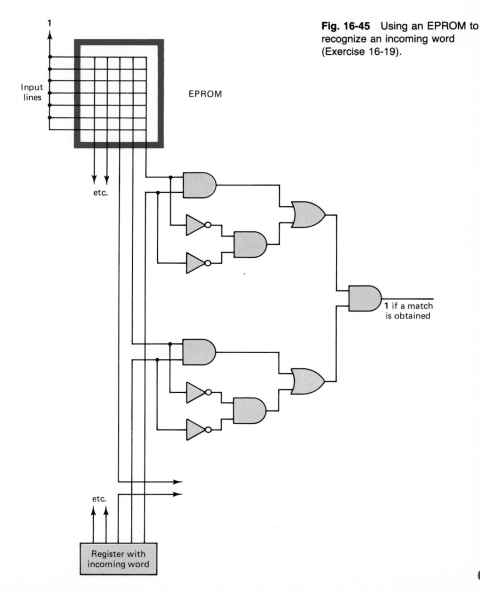

Fig. 16-45 Using an EPROM to recognize an incoming word (Exercise 16-19).

675

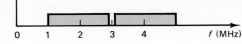

Fig. 16-46 Frequency use (Exercise 16-20).

16-22. The data for this exercise are shown in Fig. 16-47. The preferred technology shifts from MSI to LSI as the market size increases.

FIGURE 16-47
Solution for Exercise 16-22

Total Cost of Production = (Total no. of gates) (Production cost/gate) + Capital Investment Cost

Technology	Production Cost ($)	Capital Investment Cost (M$)	Total Cost (M$)
SSI	$10^9(4 \times 10^{-2})$	10	50
MSI	$10^9(1 \times 10^{-2})$	20	30
LSI	$10^9(2 \times 10^{-3})$	50	52
VLSI	$10^9(2 \times 10^{-4})$	100	100

(A)

Technology	Production Cost ($)	Capital Investment Cost (M$)	Total Cost (M$)
SSI	$10^{10}(4 \times 10^{-2})$	10	410
MSI	$10^{10}(1 \times 10^{-2})$	20	120
LSI	$10^{10}(2 \times 10^{-3})$	50	70
VLSI	$10^{10}(2 \times 10^{-4})$	100	102

(B)

PROBLEMS

(16-1)   **16-1.** Find the decimal equivalents of the following binary words:

001100

110010

111000

(16-1)   **16-2.** Find the decimal equivalents of the following:

23030_6

51617_8

(16-1)   **16-3.** Add the following numbers: $05431_6 + 22503_6$

(16-1)   **16-4.** Add the following numbers: $05431_8 + 22503_8$

(16-3)   **16-5.** Figure 16-48(A) shows two TTL (positive-logic) NOR gates connected to form an SR flip-flop. [Refer to Fig. 15-14(B).] Given the input pulses shown in Fig. 16-48(B), sketch the waveforms that result at A, B, C, D, Q, and $\overline{Q}$.

(16-3)   **16-6.** For Fig. 16-48, assume the values shown in Fig. 15-14(A). Estimate the power consumption in the flip-flop during each portion of the input waveform shown in Fig. 16-48(B).

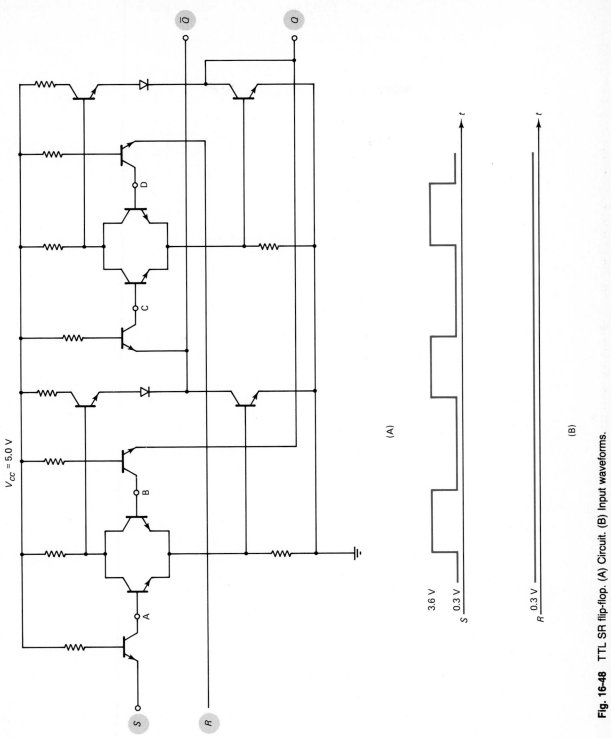

Fig. 16-48 TTL SR flip-flop. (A) Circuit. (B) Input waveforms.

$V_{CC} = 5.0$ V

(A)

(B)

677

(16-3) **16-7.** Figure 16-49 shows two NMOS NOR gates connected to form an SR flip-flop. [Refer to Figs. 15-17(C) and 15-20.] Given the input pulses shown in Fig. 16-48(B), sketch the waveforms that result at Q and $\overline{Q}$.

(16-3) **16-8.** Figure 16-50 shows two CMOS NOR gates connected to form an SR flip-flop. (Refer to Fig. 15-22.) Given the input pulses shown in Fig. 16-48(B), sketch the waveforms that

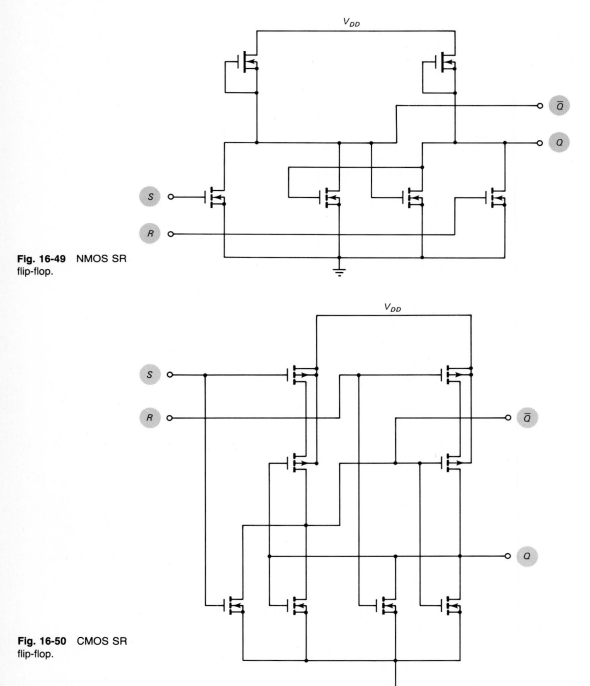

Fig. 16-49 NMOS SR flip-flop.

Fig. 16-50 CMOS SR flip-flop.

result at Q and $\bar{Q}$. Draw a sketch of the current flow from the $+V_{DD}$ source that might be expected.

(16-3) **16-9.** By using the digital logic books available in your library, develop an understanding of the JK master-slave flip-flop and edge-triggered flip-flops. Discuss the requirements that these units place on memory cell design.

(16-3) **16-10.** What problem would be encountered in trying to make a shift register from SR flip-flops?

(16-6) **16-11.** Assume that a positive pulse applied to a CCD gate in Fig. 16-51(A) produces the carrier density profile (resulting in a depletion region) shown in Fig. 16-51(B). If $N < N_o$ is required to prevent loss of signal, what maximum time can the charge remain stored in this location?

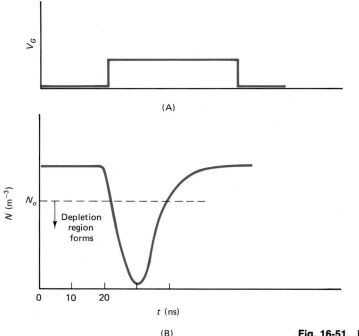

(A)

(B)

Fig. 16-51 Data for Problem 16-11.

Figure 16-52 shows a simplified data processor that illustrates some of the logic functions described in this chapter. Problems 16-12 to 16-16 apply to this figure.

(16-11) **16-12.** Assume that addresses **0001, 0010, 0011**, . . . are to access memory words **0001, 0010, 0011**, ... Design the required decoder unit.

(16-11) **16-13.** The words shown in Fig. 16-53(A) are stored in memory. What sequence of address words should be used to place the sequence of words shown in Fig. 16-53(B) in the data output register?

(16-11) **16-14.** The accumulator performs binary addition. What sequences of two address words could be used to produce the sum **1100** in the accumulator?

(16-11) **16-15.** A control stack in the ALU produces the addresses:

0010   **0100**

0011   **0101**

If all data words are summed in the accumulator, what output sum is obtained?

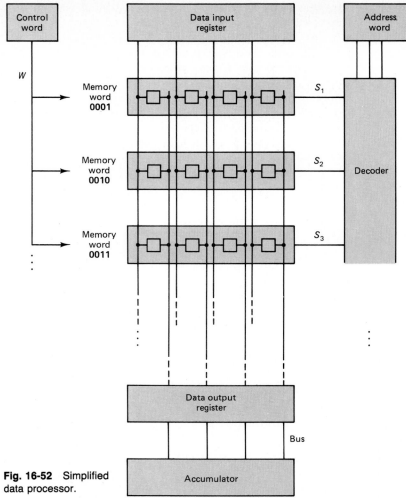

Fig. 16-52 Simplified data processor.

FIGURE 16-53
Data for Problem 16-13

Memory Address Word	Stored Data Word	
0001	0100	
0010	0010	
0011	1000	0010
0100	0010	1100
0101	1100	0001
0110	1010	1000
0111	0001	0100
	(A)	(B)

(16-11) **16-16.** The words shown in Fig. 16-54 are to be read into memory, then read in the order given. Complete the figure to produce these operations.

FIGURE 16-54
Data for Problem 16-16

Operation Step	W	Data Input Word	Address Word	Output Word
1	1	0001		
2	1	0011		
3	1	1010		
4	1	0100		
5	1	0110		
6	0			0100
7	0			1010
8	0			0001
9	0			0011
10	0			0110

(16-14) **16-17.** A carrierband computer network is designed to accommodate a data rate of 10 Mbits/s. What transmission and receiving frequencies are appropriate? What portions of the frequency spectrum would the signals use?

(16-14) **16-18.** Discuss how a parallel-to-serial converter might be achieved for a MAP network using JK flip-flops.

(16-14) **16-19.** What bandwidth is required for five 1.0-Mbit/s data channels for a MAP network?

(16-14) **16-20.** A digital system operates with a pulse rate of 8 Mbits/s. What bandwidth around carrier frequency f_o is required to transmit this pulse rate over a channel?

(16-14) **16-21.** A LAN consists of 10 stations on a carrierband MAP network. If messages transmitted by each station cannot exceed 6 ms in length, what is the maximum time a station can have to wait to transmit on the network?

(16-14) **16-22.** A baseband network uses 5 ± 1 V and 0 ± 1 V signal detection levels. What network resistance produces a detection error, given $I = 10$ mA associated with each data link?

(16-14) **16-23.** A carrierband network has a pulse repetition rate of 1.0 Mbit/s. What bandwidth is required for the channel?

(16-14) **16-24.** For an FSK system, the two incoming signals at the receiver are $f_1 = A_1 \cos (\omega_1 t)$ and $f_2 = A_2 \cos (\omega_2 t)$. Find expressions for the integrator outputs V_{out1} and V_{out2}.

COMPUTER APPLICATIONS

* **16-1.** A set of input lines $(I_1, I_2, I_3, \ldots)$ is to be encoded to a set of output words $(O_1, O_2, O_3, \ldots)$, as shown in Fig. 16-16. Write a program that prints a chart with the appropriate interconnection points, assuming an eight-bit output word size.

* **16-2.** Write a program that can be used to solve Exercise 16-22, assuming arbitrary input data for capital investment costs and production costs, as a function of market-determined gate density.

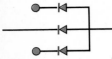

CHAPTER 17
LOGIC AND INTERFACE IMPLEMENTATION

As described in Chapter 16, digital logic can be developed using combinations of standard digital ICs, programmable logic devices (PLDs), custom ICs, and microprocessors. This chapter gives a brief introduction to some of the more often used standard IC logic products and microprocessors. The standard IC applications illustrate a widely used approach for digital logic design, and provide a useful background for programmable devices. Microprocessors are discussed to indicate the utility and flexibility of these software-controlled devices.

The chapter concludes with a discussion of combined digital/analog interface ICs. An introduction is provided to analog-to-digital and digital-to-analog converters, which provide the linkages between analog and digital signals, and to ICs for implementation of MAP computer networks. Many sophisticated ICs are available to fulfill these functions. The information of Chapter 15 provides a foundation for the types of digital circuits that will be discussed; the systems description of Chapter 16 provides the context in which different digital functions can be understood. Chapters 12 to 14 provide the necessary understanding of analog devices.

The following discussion is concerned with the types of IC building block devices available for the design process. An understanding of the technologies, the packaging strategies, and the logical functions of these available devices allows the designer to form the bridge between the four levels of decision making: (1) materials, processes, and device fabrication; (2) the formation of individual gate circuits; (3) the fabrication of large numbers of gates onto a single IC; and (4) the fabrication of systems using multiple ICs.

As discussed in Chapter 16, a digital system consists of different types of logical operations and subsystems. Digital ICs are thus developed in *families, series,* or *chip sets* that provide all of the commonly needed elements to perform a wide variety of digital functions.

Figure 17-1 gives an overview of some of the more common digital IC families. The families are arranged in terms of the *scale of integration* and the *type of fabrication process.* The levels of integration include *small-scale integration (SSI)* with about 10^0 to 10^1 gates/chip, *midscale integration (MSI)* with 10^1 to 10^2 gates/chip, *large-scale integration (LSI)* with 10^2 to 10^3 gates/chip, and very-large-scale integra-

Fig. 17-1 Examples of digital ICs. Many other companies produce similar products to those shown below.

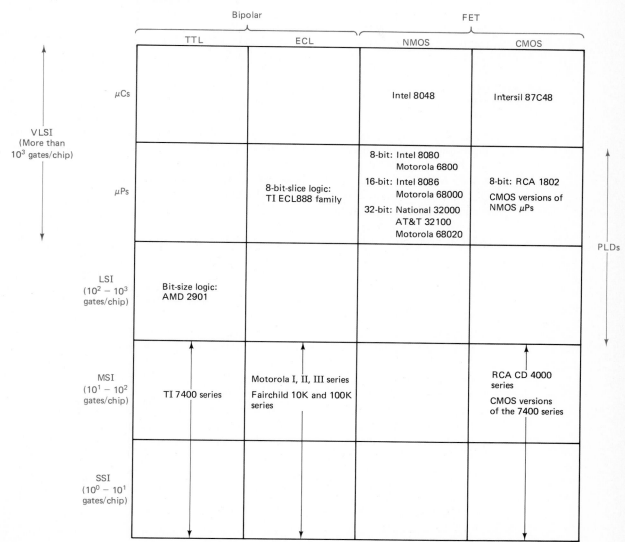

tion (VLSI) with more than 10^3 gates/chip. Two bipolar processes (TTL and ECL) are shown, along with two FET fabrication processes (NMOS and CMOS).

The 7400 family, which is manufactured by Texas Instruments and other semiconductor companies, uses TTL technology and MSI/SSI levels of density. This family was introduced in the mid-1960s and rapidly expanded to a prominent market position. The 7400 series has evolved to include four different variations, including the standard, low-power, high-speed, and Schottky clamped versions. The 7400 series remains an important resource for the electronics designer and is discussed in detail in Section 17-2.

Several families of ECL MSI/SSI ICs have also been developed, including the Motorola ECL I, II, and III series, and the Fairchild 10K and 100K series. The Fairchild 10K family is discussed in Section 17-3.

The evolution of CMOS has made it possible to fabricate digital ICs with reduced power demands, as discussed in Chapter 15. The RCA 4000 family is one of the earliest MSI/SSI CMOS series. In addition, many of the TTL families are now being fabricated using CMOS technology to produce (generally) interchangeable ICs with reduced power requirements. The RCA 4000 series is discussed in Section 17-4.

In the mid-1970s, a large jump in digital IC capability was experienced with the introduction of the microprocessor (μP). As discussed in Chapter 16, a processor unit typically consists of several registers and decoder combinations. Using VLSI/NMOS technology, a complete μP can be fabricated on an IC chip. [Figure 1-11(A) shows the topography for such a chip and Fig. 16-2 shows the functional organization.] The early IC μPs were designed to accommodate eight data bits. The Intel 8080 and the Motorola MC6800 are major products in this category. IC technology has been upgraded to include 16-bit μPs (for example, the Intel 8086 and the Motorola MC68000) and, by the mid-1980s, 32-bit μPs (for example, the National Semiconductor 32000, AT&T 32100, and Motorola 68020). The designer now has available many different families of ICs with wide-ranging capabilities.

CMOS μPs (such as the RCA 1802) were introduced competitively with the NMOS products. CMOS versions of the NMOS μPs are also becoming available. The μPs require memories and input/output subsystems to produce a complete digital system. A complete family of ICs can include both VLSI and LSI modules.

The major reasons for using a microprocessor are to reduce the time required to design a digital system and to replace many MSI/SSI packages with fewer VLSI/LSI devices. One of the major challenges to the industry has been to keep software development moving sufficiently rapidly to take advantage of the μP technology. It is a major undertaking for a semiconductor company to provide for the market a complete family of ICs for μP-based digital systems and to include all of the necessary instructions and software support. Sections 17-5 and 17-6 provide a more detailed look at μPs.

An alternative approach to digital system design uses bit-slice logic with ICs providing LSI and VLSI building block components. Each bit-slice chip contains multiple functions that are able to handle a variety of processing tasks for a computer for a few (two to eight) bits. The chip is thus used in a building-block strategy that involves placing multiple functions on a chip and interconnecting the chips to achieve large-bit words. The bit-slice logic is widely used to create high-speed computers. Available bit-slice logic uses TTL and, more recently, ECL (for example, the Advanced Micro Devices 2901 and the Texas Instruments ECL 888 family).

VLSI integration has gone a step further to the production of a computer-on-a-chip, or microcomputer (μC). A μP, memory, and input/output logic are all fabricated on the same chip. The μCs are useful for applications where a limited memory size is acceptable. The Intel 8048 (NMOS) and Intersil 87C48 (CMOS) are examples of μCs. Where larger computing capability is required, a μP is often mounted on a printed circuit board with ROMs, RAMs, and various input/output devices. The entire unit (a μC board or module) is then sold for wide-ranging applications.

In many applications, digital signal processing must be combined with analog data signals. The necessity exists for analog-to-digital converters (ADCs). Similarly, digital-to-analog converters (DACs) are required to transfer digital outputs to analog signals. Sections 17-7 and 17-8 provide an introduction to ADCs and DACs, respectively, and some of the ICs available to meet needs in this area.

17-2 THE TEXAS INSTRUMENTS 7400 TTL FAMILY

The Texas Instruments TTL 7400 family and similar series offered by other IC companies include hundreds of different MSI/SSI ICs. The result is a set of circuits that produces flexibility for the designer.[1]

One of the simplest members of the 7400 series, the 7400 device, is shown in Fig. 17-2. It consists of four 2-input NAND gates; it is an SSI IC. Figure 17-2(A) identifies the inputs and outputs for each gate and the corresponding pin numbers for the package. Figure 17-2(B) shows an alternative representation that can be found in the TTL data books. Figure 17-2(C) shows the 14-pin type J dual in-line (DIP) container that is used to package the device.

Fig. 17-2 Quadruple 2-input positive NAND gates (7400 IC). (A) Logic circuits. (B) Alternative representation. (Continued on p. 686.)

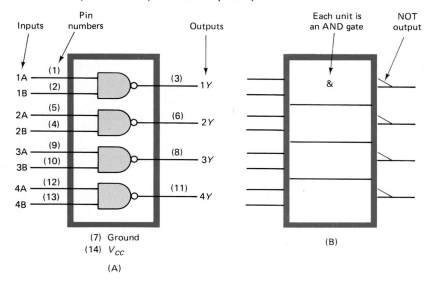

(A)

(B)

---

[1] Texas Instruments also manufactures the 5400 military specification series, which corresponds to the 7400 series.

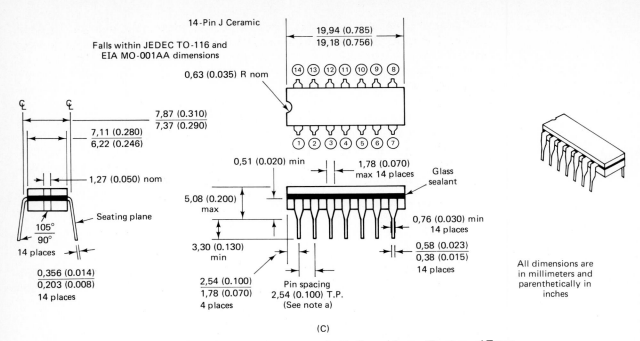

14-Pin J Ceramic

Falls within JEDEC TO-116 and
EIA MO-001AA dimensions

0,63 (0.035) R nom

$\frac{19,94\ (0.785)}{19,18\ (0.756)}$

$\frac{7,87\ (0.310)}{7,37\ (0.290)}$

$\frac{7,11\ (0.280)}{6,22\ (0.246)}$

1,27 (0.050) nom

Seating plane

105°
90°

14 places

$\frac{0,356\ (0.014)}{0,203\ (0.008)}$
14 places

0,51 (0.020) min

1,78 (0.070)
max 14 places

Glass
sealant

5,08 (0.200)
max

3,30 (0.130)
min

0,76 (0.030) min
14 places

$\frac{0,58\ (0.023)}{0,38\ (0.015)}$
14 places

$\frac{2,54\ (0.100)}{1,78\ (0.070)}$
4 places

Pin spacing
2,54 (0.100) T.P.
(See note a)

All dimensions are
in millimeters and
parenthetically in
inches

(C)

Fig. 17-2 (cont.) (C) 14-pin type J dual in-line package. (Courtesy of Texas Instruments Incorporated)

Figure 17-3 shows that 7402 IC, which consists of four 2-input OR gates. Figure 17-3(A) identifies the inputs and outputs for each gate and the corresponding pin numbers for the DIP package. Figure 17-2(B) shows an alternative representation.

The 7400 series members use only a single +5-V power supply. Each gate typi-

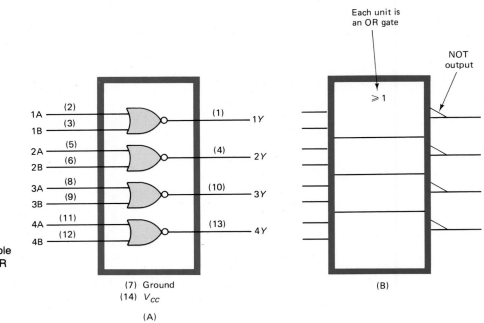

Each unit is
an OR gate

NOT
output

≥ 1

Fig. 17-3 Quadruple
2-input positive NOR
gates (7402 IC).
(A) Logic circuits.
(B) Alternative
representation.

1A (2)
1B (3)
(1) 1Y

2A (5)
2B (6)
(4) 2Y

3A (8)
3B (9)
(10) 3Y

4A (11)
4B (12)
(13) 4Y

(7) Ground
(14) V_{CC}

(A)

(B)

cally dissipates 10 mW and has a switching delay of 10 ns. The above ICs can be combined in many different ways to produce more complex logical functions. Two examples are discussed below.

Example 17-1 **Four-Input NOR Circuit**

Use the 7400 and 7402 ICs to design a 4-input NOR gate.

Solution Figure 17-4 shows one possible solution. Note that the implementation uses two of the NAND gates in a 7400 and three of the NOR gates in a 7402.

The logic function for Example 17-1 can also be implemented directly by using a 7425 IC, as shown in Fig. 17-5. This IC contains two 4-input positive NOR gates. In addition, it provides inputs G_1 and G_2 to allow an external strobe to control the output. ∎

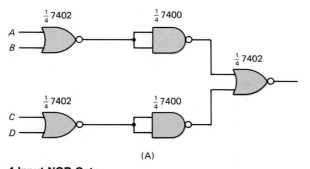

(A)

4-Input NOR Gate

A	B	C	D	AB	CD	E
0	0	0	0	0	0	1
1	0	0	0	1	0	0
0	1	0	0	1	0	0
. . .	. . .	. . .	. . .	. . .	. . .	. . .
1	1	0	0	1	0	0
. . .	. . .	. . .	. . .	. . .	. . .	. . .
1	1	1	1	1	1	0

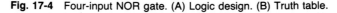

(B)

Fig. 17-4 Four-input NOR gate. (A) Logic design. (B) Truth table.

Example 17-2 **SR Flip-Flop**

Show how the 7402 IC can create an SR flip-flop, or latch. Show the pin connections that must be made.

Solution The SR flip-flop was introduced in Chapter 16, and a logic diagram is shown in Fig. 16-3. From a comparison between Figs. 16-3 and 17-3, the desired logic unit can be created by using one-half of the 7402 IC. The external pin connections are shown in Fig. 17-6. SR flip-flops are also available in other members of the 7400 family. The 7400 family includes a wide range of additional logic devices, as listed in Fig. 17-7. ∎

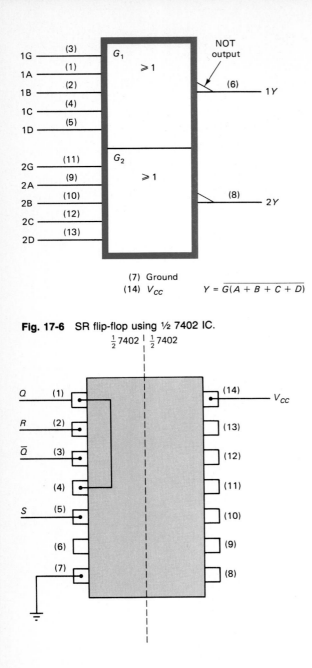

1G (3)

1A (1)

1B (2)

1C (4)

1D (5)

G_1 ≥1

NOT output

(6) 1Y

2G (11)

2A (9)

2B (10)

2C (12)

2D (13)

G_2 ≥1

(8) 2Y

(7) Ground
(14) V_{CC} $Y = \overline{G(A + B + C + D)}$

Fig. 17-5 Dual 4-input positive NOR gates with strobe (7425 IC).

Fig. 17-6 SR flip-flop using ½ 7402 IC.

$\frac{1}{2}$ 7402 | $\frac{1}{2}$ 7402

Q (1)

R (2)

$\overline{Q}$ (3)

(4)

S (5)

(6)

(7)

(14) V_{CC}

(13)

(12)

(11)

(10)

(9)

(8)

FIGURE 17-7
Types of ICs Available in the TI 7400 Series

Gates and inverters
Expanders, buffers, drivers, and transceivers
Clock generators
Flip-flops, latches, and multivibrators
Registers
Programmable logic arrays
Counters
Decoders and encoders
Data selectors, multiplexers, and shifters
Display decoders/drivers
Memory/microprocessor controllers
Voltage-controlled oscillators
Arithmetic circuits
Error detection circuits
Memories

17-3 THE FAIRCHILD 10K ECL FAMILY

The Fairchild 10K ECL family (along with similar families offered by other companies) uses the configurations described in Section 15-8. Because ECL circuits are intended to work at high switching speeds, the distributed inductance and capacitance along interconnections is important. The design strategies must allow for impedance matching problems at high frequencies, where interconnections are described in terms of transmission line properties. A basic introduction to high-frequency concepts is provided in microwave courses.

The basic OR/NOR gate and a possible AND/NAND gate design are discussed in Section 15-8. Figure 17-8 shows how both of these circuits can be represented. As discussed in Chapter 15 and shown in Fig. 17-8(C), combinations of OR gates can also provide the NAND function. It can be a substantial advantage to have both inverting and noninverting outputs present for each gate. A number of simplifications in logic design can result.

Two members of the Fairchild 10K series are shown in Fig. 17-9. The F10102 includes four dual-input NOR gates, with one of the gates also having an OR output.

(A)   (B)

Fig. 17-8 ECL gate representations.
(A) OR/NOR gate.
(B) AND/NAND gate.
(C) AND/NAND gate from OR/NOR gates.

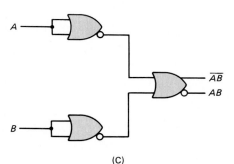

(C)

Fig. 17-9 Members of the Fairchild 10K family. (A) F10102. (B) F10104. (Courtesy of Fairchild Semiconductor Corporation)

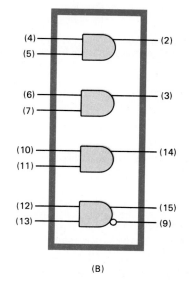

(1) V_{R1}
(2) V_{R2}
(8) V_{EE}

(A)

(B)

The F10104 provides four dual-input AND gates, with one of the gates also having a NAND output. The ECL gates shown here have a typical propagation delay of 2 ns and a power dissipation of 25 mW.

Example 17-3 **ECL NAND Gate**

Show how the F10102 can be connected to produce a 2-input NAND gate.

Solution The connections are shown in Fig. 17-10. Figure 17-11 lists the types of logic circuits that are available in the F10K series. ■

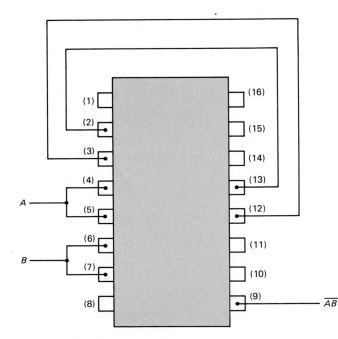

Fig. 17-10 NAND gate using the F10102.

FIGURE 17-11
Types of ICs Available in the Fairchild 10K Family

Shift registers and counters
Terminators
Gates
Line receivers
Drivers
TTL-to-ECL translators
Flip-flops, latches
Decoders, encoders, and multiplexers
Counters
Registers
Parity checkers/generators
Comparators
Carry lookahead generators
Arithmetic units

As discussed in Section 15-10, CMOS circuits dissipate only small amounts of power in the standby condition. A typical CMOS logic device dissipates only 100 to 400 nW as long as the device state remains unchanged.

If the device switches with an average frequency f, an additional power dissipation occurs. The level of dissipation can be estimated by considering that a gate circuit drives a capacitance C. If the voltage across C changes from V_{DD} to 0 V during each switching cycle, then the charge flow during each cycle is

$$\Delta Q = (I_{av})T = C \Delta V = CV_{DD} \qquad (17\text{-}1)$$

The average current is then

$$I_{av} = CV_{DD}/T \qquad (17\text{-}2)$$

and the power dissipation in the NMOS and PMOS transistors during the charging and discharging of C becomes

$$P = (CV_{DD}/T)V_{DD} = V_{DD}^2\, Cf \qquad (17\text{-}3)$$

Additional dissipation is experienced due to the voltage changes across internal device capacitances.

Figure 17-12 shows two of the ICs available as part of the RCA CD4000 series. The CD4000UB includes two 3-input NOR gates and an inverter, and the CD4011B includes four 2-input NAND gates.

The CD4000UB has a typical 30-ns propagation delay time with a capacitive load $C_L = 50$ pF and $V_{DD} = 10$ V. The dc standby current is a maximum of 1 μA at

Fig. 17-12 Members of the RCA CD4000 series. (A) CD4000UB. (Continued on p. 692).

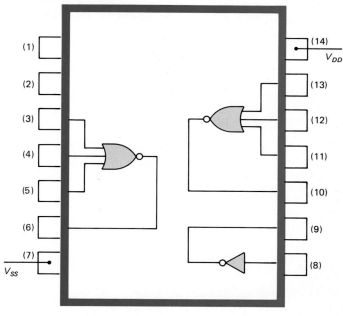

(A)

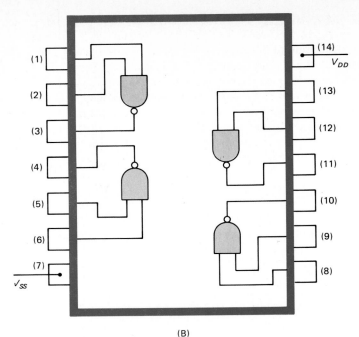

(B)

Fig. 17-12 (cont.) (B) CD4011B.

18 V. The recommended operating conditions are for a supply voltage between 3 and 18 V. Figure 17-13 shows the metallization pattern of the chip.

The CD4011B has a typical 60-ns propagation delay time with a capacitive load $C_L = 50$ pF and $V_{DD} = 10$ V. The maximum dc standby current is again 1 μA at 18 V.

Fig. 17-13 Metallization pattern for the CD4000UB IC.
Dimensions in parentheses are in mm, others in mils.
(Courtesy of RCA Corporation)

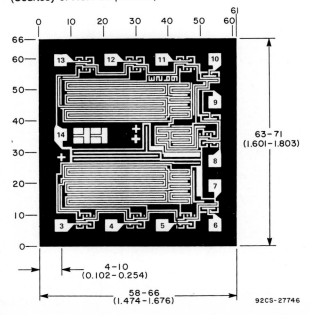

The RCA CD4000 family comes in a wide range of logic members, similar to those listed for the TTL and ECL families. And, as noted, equivalents to many members of the TTL 7400 family are available in CMOS. The RCA high-speed QMOS series is pin compatible with many existing 7400 and CD4000 devices. The system typically operates with 8-ns propagation delays with $C_L = 15$ pF. The series members are identified by the format SN74HC XX, where the digits XX identify the family member. For example, the logical diagram for the SN 74HC00 is identical with that of the 7400.

Example 17-4   **CMOS Power Dissipation**

Given $C = 20$ pF, $V_{DD} = 5.0$ V, and $I_{DD} = 20$ μA, find P_D for the SN74HC00 at an equivalent frequency $f = 10^5$ Hz. The dc standby current is 20 μA.

Solution   By substitution in Eq. (17-3),

$$P_D = (20 \text{ pF})(5.0 \text{ V})^2(10^5 \text{ Hz}) + (20 \text{ } \mu\text{A})(5 \text{ V})$$
$$= 0.15 \text{ mW}$$

For the four gates on the SN74HC00 chip,

$$P_D = 4 \times 0.15 \text{ W} = 0.60 \text{ mW/device}$$ ■

17-5   MICROPROCESSORS

When a microprocessor (μP) is designed, a set of operational or instruction codes is permanently stored in the device or in an accompanying ROM. These codes provide the fundamental logical operations that control the μP. Applications include moving a word from a particular memory location to a register in the ALU, adding two words, storing a word in memory, and many other logical operations.

A typical μP can have an instruction set consisting of several hundred codes. Each code is used as input to a decoder, and the output of the decoder produces the voltage levels (**1**s and **0**s) throughout the μP to cause the desired operation to take place. The decoders that implement the desired instruction set enable the μP to respond to external commands. A command to transfer a word from memory to the ALU might enable the READ control lines in the memory and enable the register input gates. The instruction command is followed by two address codes that enable the specific memory location and the ALU register to be used. The μP can then provide the desired operation. This aspect of μP design can be viewed as the interface between the logic and electronics levels of device operation.

It is not convenient to read these commands in binary because of the long word length. On the other hand, these fundamental machine instructions should be coded in a way that results in efficient μP response. The hexadecimal (base-16) representation is often used, since four binary bits form a convenient subunit and can represent 16 different digits. The 16 digits used for such a representation are 0 . . . 9 and A . . . F. The μP instruction sets are often defined in this hexadecimal system.

Higher-order program languages can be built to use combinations of the machine language codes. Assembly language combines several codes into a single instruction. Higher-order languages (such as BASIC) combine many more complex operations in a simple instruction. As the languages grow more powerful, the overhead time and space associated with their use grows, making further demands on performance.

Over the past two decades there has been an evolution in word length from 8 to 16 to 32 bits. For example, Motorola produced the 6800, 68000, and 68020 μPs in such a sequence. In general, each extension in word length is associated with improved technology and higher performance levels, including increased circuit density and switching speed and reduced power dissipation. In addition, longer word lengths result in more powerful systems in terms of the maximum number of memory locations that can be accessed as a single operation and the maximum word length that can be stored for single-step access. When an 8-bit μP is used to form 16-bit words, or a 16-bit μP is used to form 32-bit words, the speed of the computer system is slowed and performance degrades.

17-6 THE MC6800 MICROPROCESSOR

The Motorola MC6800 is an 8-bit (data bus) microprocessor. The μP can be used with other family members to create a digital system of the type shown in Fig. 16-1. The family units that are available include the μP itself, a 128×8 RAM (the MCM6810), a 1024×8 ROM (the MCM6830), the parallel and serial interface devices (the MC6820 and the MC6850, respectively). A complete system also includes a clock and start-up device. Each package operates on a single 5-V power supply and is TTL compatible.

The 16-bit address bus is used to address both memory and input/output devices. The user can converse with I/O using appropriate memory addresses. The 5-bit control bus carries clock, start-up, and read/write control signals.

The development of a μP-based digital system requires the ability to consider both hardware and software performance objectives and constraints. For many applications, tradeoffs exist between task implementation using hardware versus software design strategies.

The MC6800 μP is packaged in a 40-pin DIP configuration, as shown in Fig. 17-14. Pins 9 to 20 and 22 to 25 provide connections for the 16-bit address bus, and pins 26 to 33 provide connections to the 8-bit data bus. V_{DD} is connected to +5 V and V_{SS} to ground. The remaining connections are for the necessary control signals.

The μP requires a two-phase clock input, that is, one phase of the clock is high when the other is low. Pins 3 and 37 provide the two required clock connections. The RESET input is used to start the μP with appropriate initial conditions. The valid memory address (VMA) and read/write (R/W) lines are associated with the bus transfers of data. VMA goes high whenever a valid address is being placed on the address bus. The R/W line is high for input to the μP and low for data output from the μP. The interrupt request (IRQ) is an input connection that allows I/O devices to tell the μP that they need to be serviced. The remaining pins are associated with the direct transfer of data into memory, bypassing the μP, or are not used.

The MC6800 contains six registers that can hold binary words and provide the means for performing basic arithmetic operations. There are two accumulator registers used in math and logic operations, an index register, a program counter, a stack pointer (which is used to access a set of consecutive memory locations), and a condition code (status report) register.

The operation of the MC6800 can be examined as follows. A high signal is placed

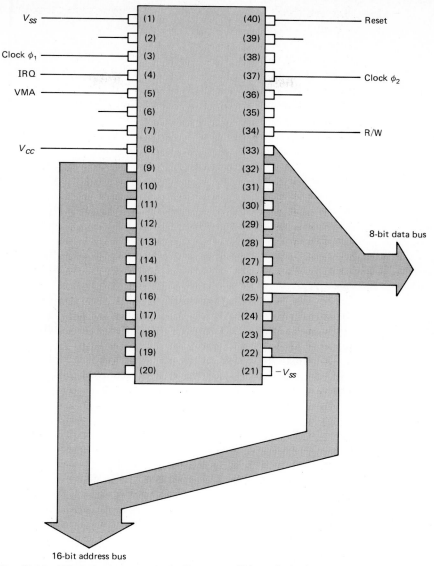

Fig. 17-14 MC6800 microprocessor. (Courtesy of Motorola, Inc.)

on the reset, causing the μP to initialize operations. The μP accesses an initial address in ROM, seeking data to start the desired operational sequence. If instructions are already placed in ROM, the μP begins to read and execute the instructions one by one.

Data is read from ROM and any appropriate subsystem. By shifting data back and forth among subsystems, the μP itself, and RAM, the μP performs the logical operations that are listed in the instruction sequence. The result of the analysis is then output to the desired subsystem.

The preparation of an instruction sequence for the μP requires writing a program. Selected μP operations are coded into ROM as part of the manufacturing process.

For the MC6800, 72 instruction words are available to the programmer. The accomplishment of any desired task must make use of the logical operations performed by this set.

The programmer prepares a list of instructions, then uses an input data subsystem (such as a keyboard) to read them into the μP. The instructions are individually addressed by the μP and placed in RAM for later availability. Data can be read through an input subsystem (for example, a keyboard or disk), addressed, and stored in memory by the μP. As the instruction list (program) is executed, the data can be accessed as necessary.

17-7 ANALOG-TO-DIGITAL CONVERTERS

In many signal processing applications, it is necessary to convert an analog signal to a digital equivalent. A device that performs this function is called an *analog-to-digital converter (ADC)*.

Figure 17-15 illustrates the basic procedure. An input analog signal is sampled at selected times. The value of v_{in} at each sample time becomes V_{sample} between samples. A *sample-and-hold* circuit is used for this purpose. The result is an approximation to the actual v_{in}, limited by the accuracy (number of bits) for each voltage sampled and by the interval between samples. Obtaining more significant figures for each sample requires a longer word length, and rapidly varying input signals require a faster sampling rate.

The dual-slope and successive-approximation ADCs are described below as representative of the strategies used to achieve the desired conversions. The performances of several types of IC ADCs are then compared.

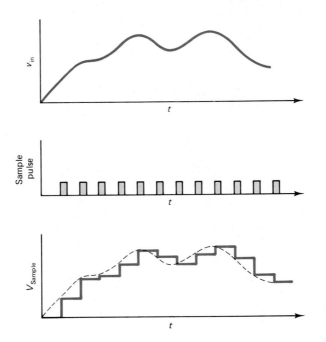

Fig. 17-15 Sampling of an analog input signal.

A dual-slope ADC is illustrated in Fig. 17-16. Once V_{sample} is available, switch S is closed to position 1. Operational amplifier 1 functions as an integrator, so that the voltage out increases to the value

$$V_1 = \frac{1}{RC} \int_{T_1}^{T_2} V_{sample}\, dt = \frac{1}{RC} V_{sample}(T_2 - T_1) \tag{17-4}$$

The time $T_2 - T_1$ is determined by a digital counter and recorded in a register.

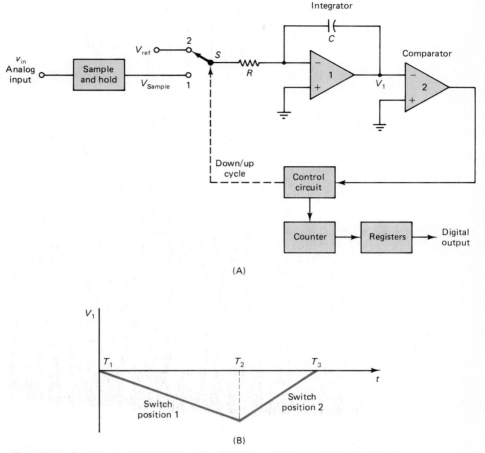

(A)

(B)

Fig. 17-16 Dual-slope ADC. (A) Functional diagram. (B) Output of dual integration.

At time T_2, the switch is changed to position 2, so that a reference voltage V_{ref} of the opposite polarity to v_{in} is applied to the op-amp input. The output voltage becomes

$$V_1 = \frac{1}{RC} V_{sample}(T_2 - T_1) - \frac{1}{RC} \int_{T_2}^{T_3} V_{ref}\, dt$$

$$= \frac{1}{RC} [V_{sample}(T_2 - T_1) - V_{ref}(T_3 - T_2)] \tag{17-5}$$

A digital counter is again used to determine $T_3 - T_2$. The comparator indicates when V_1 changes polarity and opens switch S. As a result,

$$V_{sample} = V_{ref}\frac{T_3 - T_2}{T_2 - T_1} \qquad (17\text{-}6)$$

within the accuracy limits introduced by loop time delays. As noted, the time intervals $T_2 - T_1$ and $T_3 - T_2$ are determined by a digital counter. Therefore, the output voltage can be represented in terms of these two binary words and the (predetermined) value of V_{ref}.

Dual-slope IC ADCs use an external integrating capacitor. As noted in the data sheet for the TL500C (see Fig. 17-17), this capacitor must be "within the recommended range (of values for C) and must have good voltage linearity and low dielectric absorption. A polypropylene-dielectric capacitor . . . is recommended. Stray coupling from the comparator output to any analog (device) pin must be minimized to avoid oscillations."

A successive-approximation ADC is shown in Fig. 17-18. The analog input signal v_{in} is again sampled periodically. A sequence of trial digital words is produced, as shown in Fig. 17-18(B). The first word has a **1** in the most significant bit, with all other bits set to zero. This word is converted to an analog reference signal equal to one-half of the maximum v_{in} allowed. If V_{sample} is greater than V_{ref}, the **1** is left in the trial word. If V_{sample} is less than V_{ref}, the **1** is removed, and the second trial word is used with a **1** in the next significant bit and the other bits equal to **0**.

The comparison continues, producing the series of trial and outcome words shown. The final result (for the case shown) is the digital word **1001**, which is the output of the conversion process. A 4-bit output can be used to count $2^4 = 16$ intervals. Therefore, the 4-bit output word represents V_{sample} to the nearest $(1/16)$ $(V_{sample})_{max}$. The binary word **1001** corresponds to the decimal value $8 + 1 = 9$, so that the digital output of the ADC is $(9/16)(V_{sample})_{max}$.

Many different types of ADCs are available in IC form. Figure 17-17 lists some of the properties of selected devices. The dual-slope approach tends to be slower because of the integration time required, but shows substantial noise protection for the same reason. Successive-approximation ADCs are faster because of the rapid convergence to the desired output. Very-high-speed ADCs can be developed by using many parallel comparators, each with an individual reference voltage.

Typical devices are available in 7- to 14-bit configurations. Conversion times vary widely, depending on the design technique and the number of bits used.

FIGURE 17-17
Performance Data for Selected IC ADCs

Manufacturer	Device Code Number	Type ADC	No. of Bits	Minimum Conversion Time
Texas Instruments	TL500C	Dual-slope	14	0.1 s
Texas Instruments	ADC0804C	Successive-approximation	8	100 μs
Signetics	NE5030	Variation on successive-approximation	10	2.5 μs
Signetics	PNA7507	Parallel comparators	7	0.1 μs

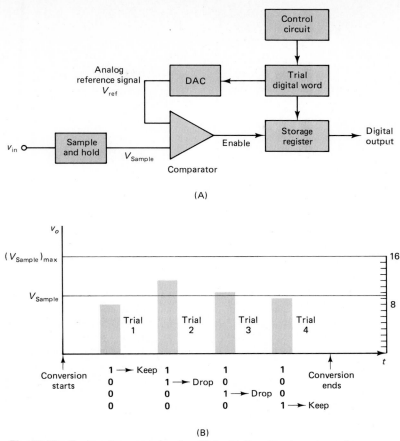

(A)

(B)

Fig. 17-18 Successive-approximation ADC. (A) Functional diagram. (B) Sequence of trial values for V_{ref}.

17-8 DIGITAL-TO-ANALOG CONVERTERS

In order to apply the output of a digital system to produce an analog signal, a *digital-to-analog converter (DAC)* is required. One simple approach uses the summing op-amp circuit introduced in Chapter 13. Applying Eq. (13-6),

$$V_o = -V_{ref}(R_f/R_1 + R_f/R_2 + \ldots)$$

for such a circuit. The DAC can be developed as shown in Fig. 17-19. The resistors (R, $R/2$, $R/4$, and $R/8$) are the input resistors. By direct substitution,

$$V_o = -V_{ref}\left(\frac{R}{R/8} + \frac{R}{R/4} + \frac{R}{R/2} + \frac{R}{R}\right) \tag{17-7}$$

$$= -V_{ref}(8 + 4 + 2 + 1)$$

$$= -V_{ref}(2^3 + 2^2 + 2^1 + 2^0) \tag{17-8}$$

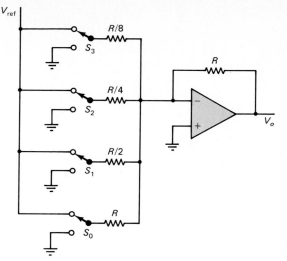

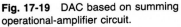

Fig. 17-19 DAC based on summing operational-amplifier circuit.

The resistor inputs are arranged so that V_o is formed by combinations of powers of 2. The sequence $2^3 + 2^2 + 2^1 + 2^0$ can be viewed as a binary number (**1111**). The switches S_0, S_1, S_2, and S_3 turn each input resistor *off* and *on*. Therefore, any one of the bits can be enabled or set equal to zero to produce an arbitrary output.

A binary word is thus converted to an analog equivalent by using the word to turn the switches (S_0, . . . , S_n) *off* and *on*. The summing op-amp forms an output that is proportional to the value represented by the word.

The output of a DAC is not continuous, but is divided into 2^n increments, where n is the number of bits used in the input. For a 4-bit word, the output increments will be $(1/16)V_{ref}$. Additional op-amp stages can be used to obtain the maximum voltage desired.

A more practical DAC circuit is shown in Fig. 17-20. This version uses only two resistor values, R and $2R$. The R-$2R$ DAC of Fig. 17-20 performs the same function as the resistor ladder shown in Fig. 17-19.

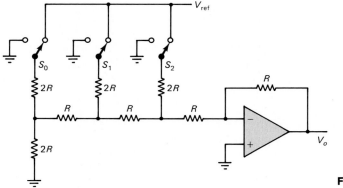

Fig. 17-20 R-$2R$ DAC.

If S_0 is closed to V_{ref} and S_1 and S_2 are connected to ground, the circuit of Fig. 17-21 is obtained. Figure 17-22 shows how a step-by-step application of Thevenin equivalent circuits can be used to calculate V_o.

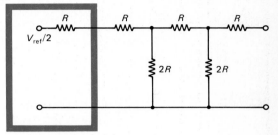

Fig. 17-21 *R-2R*
DAC with S_0 closed.

Fig. 17-22 Equivalent circuits for *R-2R* DAC.

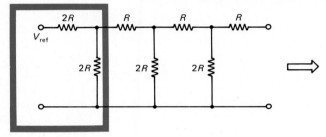

(A)

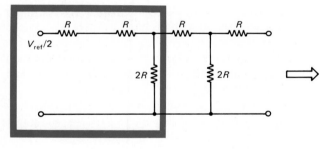

(B)

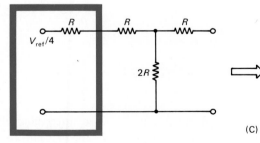

(C)

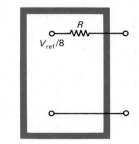

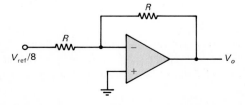

(D)

701

Looking into the circuit formed by V_{ref} and the first two resistors in the ladder (closest to V_{ref}), the Thevenin equivalents $V_{\text{TH}} = V_{\text{ref}}/2$, and $R_{\text{TH}} = R$ are found [Fig. 17-22(A)]. Adding the next two resistors to the circuit produces the equivalents $V_{\text{TH}} = V_{\text{ref}}/4$, and $R_{\text{TH}} = R$ [Fig. 17-22(B)]. Adding the last two resistors to the circuit, $V_{\text{TH}} = V_{\text{ref}}/8$, and $R_{\text{TH}} = R$ [Fig. 17-22(C)], giving the final result of Fig. 17-22(D).

Similarly, if S_1 is closed, $V = -V_{\text{ref}}/2$, and if S_2 is closed, $V_o = -V_{\text{ref}}/2^1$. If all three switches are closed,

$$V_o = -V_{\text{ref}}(1/2^3 + 1/2^2 + 1/2^1)$$
$$= -\frac{V_{\text{ref}}}{2^3}(2^0 + 2^1 + 2^2) \qquad (17\text{-}9)$$

by superposition. The switches can again be associated with binary words. The output is given by a sum of terms, as shown in Eq. (17-9). The IC DAC variations that are available generally make use of $R\text{-}2R$ resistor ladders. The number of significant figures available for the output depends on the number of ladder steps and the tolerances associated with the resistors used. For ladders with many bits (10 to 14 are typical), extreme care must be taken to provide resistors that are accurately matched.

17-9 MAP IMPLEMENTATION

The features of the MAP computer network are introduced in Chapter 16. As might be expected, the widespread implementation of such technology depends on the costs associated with the required interfaces.

Given the MAP layered approach to interface design, particular implementation requirements can be met by combining standard and custom computer hardware and software to match the individual station requirements, as shown in Fig. 17-23.

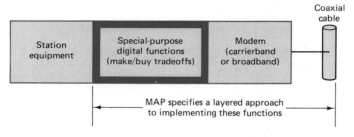

Fig. 17-23 The functional areas associated with a MAP network interface.

A necessary building block for any interface is the *modem*. Interchangeable carrierband and broadband printed circuit boards are available for this purpose. Figure 17-24 shows a broadband modem that can operate over three dual-frequency channels.

The layers between the modem and the internal station processor can lead to complex tradeoffs among alternative solutions. Printed circuit boards that provide complete solutions to all MAP levels can be purchased to connect between the modem and equipment with standard bus protocols. For example, the MVME372

Fig. 17-24 Broadband modem for use in MAP networks, with capability for three dual-frequency channels. The frequency of operation and power levels are software controllable. (Courtesy of Concord Communications, Inc.)

VME module by Motorola provides an interface between the modem and VME bus architecture.

As more IC chip sets become available, the designer can mix and match various levels of standard and custom hardware and software to allow the most cost-effective interface configuration for a given application. As can be seen, the introduction of protocols such as MAP can allow the rapid development of technology that can be used on a broad basis.

17-10   TRENDS IN IC DEVICES

The complexity of IC devices continues to change rapidly because of improvements in technology and opportunities for more sophisticated applications. Silicon-based circuits are dominant in the field today, but GaAs has the potential for major market growth as the technology matures. Many other possible materials are under evaluation in research laboratories to identify those with possible applications for IC fabrication.

Higher-level functional-design software aids and computer-controlled manufacturing processes are making it progressively easier to produce custom ICs. Analog and digital circuitry are being combined in building block units to produce ICs that can satisfy almost any set of technical performance specifications. Printed circuit boards and hybrid microcircuits continue to develop to complement the IC evolution and maintain a range of cost/performance tradeoffs for the designer. The trend is toward more programmable and μP-controlled analog, digital, and combined IC circuitry, with a matching emphasis on software development. As higher-level computer-aided software development techniques become available, the hardware and software design aspects become more in balance, enabling the designer to routinely explore simulations that make use of various mixes to obtain an optimum solution.

Questions
17-1. What four levels of IC manufacturing affect the digital systems designer?

17-2. What four scales of integration are commonly used to describe digital circuits?

17-3. What advantages are associated with ECL and CMOS circuits?

17-4. Why are CMOS versions of many bipolar and NMOS devices being produced?

17-5. What determines the power dissipation for CMOS digital circuits?

17-6. Why is the hexadecimal representation often used for μP instructions?

17-7. What are the advantages of 32-bit μPs over 8- and 16-bit μPs?

17-8. What are the relative advantages and disadvantages of the dual-slope and successive-approximation ADCs?

17-9. What is the major fabrication limitation to making very accurate DACs?

17-10. What tradeoffs are involved in deciding how to implement a MAP interface?

Answers
17-1. The digital systems designer is affected by decision making regarding: (1) materials, processes, and device fabrication; (2) manufacture of individual gate circuits; (3) fabrication of large numbers of gates onto a single IC; and (4) the fabrication of systems using multiple ICs, or chip sets.

17-2. Digital circuits are often described in terms of small-scale integration (SSI), midscale integration (MSI), large-scale integration (LSI), and very-large-scale integration (VLSI).

17-3. Among the circuit technologies discussed here, ECL digital circuits provide the minimum switching delays, whereas CMOS digital circuits provide the lowest power consumption.

17-4. In general, CMOS versions of bipolar and NMOS devices provide the same logic functions with reduced power consumption. Some sacrifice in speed can result.

17-5. For CMOS digital circuits, standby power dissipation is small (of the order of 10 to 20 μW for the CD4000UB). The principal power dissipation takes place during switching (of the order of V_{DD}^2Cf) as the NMOS and PMOS devices charge and discharge circuit capacitances.

17-6. The hexadecimal code is a convenient representation to work with because four binary bits can be used for each digit. For example, given the digits $0 \ldots 9$ and $A \ldots F$, the following associations can be made:

$$
\underbrace{1101}_{D} \qquad \underbrace{0010}_{2} \qquad \underbrace{0001}_{1}
$$

A long string of binary digits, which are in a machine-readable form, can be represented by much shorter sequences of digits in hexadecimal.

17-7. In general, 32-bit μPs are associated with improved technology and performance. In addition, longer words speed up the logical operations associated with the memory access and long-bit words (required for a large number of significant figures).

17-8. The dual-slope ADC provides noise protection because of the integrating function, but is slower because of the same reason. The successive-approximation ADC is faster, but provides less noise protection.

17-9. DACs are limited in accuracy by the ability to develop R-$2R$ resistive ladders that are highly matched. Hybrid microelectronics technology is often used to produce R-$2R$ ladders with tight tolerances.

17-10. In implementing a MAP interface, complex tradeoffs exist with respect to whether to buy component IC chips and design and fabricate the required interface hardware, or to buy printed circuit boards that provide a complete interface for linkage between the network and standard bus protocols. Additional tradeoffs must be made between the use of special-purpose hardware versus the use of μPs combined with special-purpose software.

EXERCISES AND SOLUTIONS

Exercises   *In exercises 17-1 to 17-5, show how combinations of the TTL 7400 and 7402 devices introduced in this chapter can be used to produce the desired logical operations.*

(17-2)   **17-1.** Produce the NOT function using a portion of the 7400.

(17-2)   **17-2.** Design a 4-input AND gate using the available devices.

(17-2)   **17-3.** Show how to produce the input logic and SR flip-flop for the D-type random-access memory cell of Fig. 16-4.

(17-2)   **17-4.** Show how to produce the gates associated with the encoder of Fig. 16-8.

(17-2)   **17-5.** Show how to obtain the gates associated with the decoder of Fig. 16-10.

(17-2)   **17-6.** The binary coded decimal (BCD) code uses four binary bits to represent the decimals 0 to 9. Since four bits can be used to count from zero to $2^4 - 1 = 15$, there are unused words whenever this code is applied. Figure 17-25 shows the data sheet for the SN7445 BCD-to-decimal decoder (a member of the TTL 7400 family). Develop a truth table for this device and show that it produces the desired logical operation.

(17-3)   **17-7.** Show how the F10102 can be connected to produce a 4-input NOR gate.

(17-4)   **17-8.** Give $C = 50$ pF and $f = 1.0$ MHz, find P_D for the SN74HC00 device.

(17-7)   **17-9.** A dual-slope ADC uses $V_{ref} = 15$ V. The counter output associated with $T_2 - T_1$ is **111011**, and the counter output associated with $T_3 - T_2$ is **110100**. Find the accuracy with which V_{sample} can be determined.

(17-7)   **17-10.** You are given a successive-approximation ADC with 6 bits, $(V_{sample})_{max} = 10.00$ V, and an input $V_{sample} = 8.00$ V. What is the sequence of trial values for V_{ref} and the final digital output?

(17-7)   **17-11.** Find the number of (decimal) digits of accuracy associated with the first ADC listed in Fig. 17-18. (The number of digits of accuracy is found by considering the number of significant figures that separates adjacent outputs.)

(17-8)   **17-12.** Show how an R-$2R$ DAC converts the binary word **1000100** to an analog output. Find the magnitude of the output, given $V_{ref} = 5.0$ V.

Solutions   **17-1.** By referring to Figs. 15-3 and 15-4, the NOT function can be produced from a NAND gate as shown in Fig. 17-26.

17-2. A 4-input AND gate can be produced as shown in Fig. 17-27.

17-3. By combining Figs. 16-3 and 16-4, the logic design of Fig. 17-28 can be found.

17-4. Each of the three 4-input OR gates in Fig. 16-8(A) can be produced as shown in Fig. 17-29.

17-5. The inverter/gate combination of Fig. 17-30 applies to output O_2. The other outputs take a similar form.

SN5445 . . . J or W package
SN7445 . . . J or N package
(Top view)

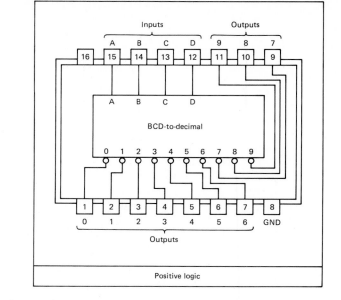

Featuring

· Full decoding of input logic
· 80-mA sink-current capability
· All outputs are off for invalid
  BCD input conditions

Description

These monolithic BCD-to-decimal decoders/drivers consist of eight inverters and ten 4-input NAND gates. The inverters are connected in pairs to make BCD input data available for decoding by the NAND gates. Full decoding of valid BCD input logic ensures that all outputs remain <u>off</u> for all invalid binary input conditions. These decoders feature TTL inputs and high-performance npn output transistors designed for use as indicator/relay drivers or as open-collector logic-circuit drivers. Each of the high-breakdown output transistors (30 V) will sink up to 80 mA of current. Each input is one normalized Series 54–74 load. Inputs and outputs are entirely compatible for use with TTL or DTL logic circuits, and the outputs are compatible for interfacing with most MOS integrated circuits. Power dissipation is typically 215 mW.

Functional block diagram

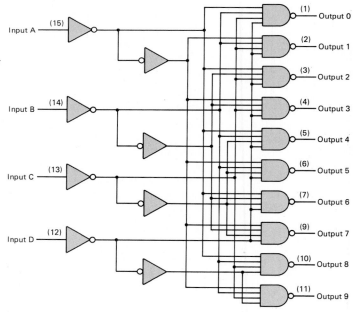

Fig. 17-25 The SN7445 BCD-to-decimal decoder. (Courtesy of Texas Instruments Incorporated)

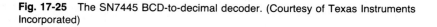

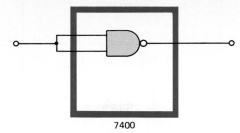

7400

Fig. 17-26 Producing the NOT function from a NAND gate (solution for Exercise 17-1).

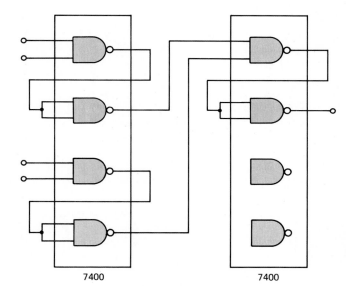

7400   7400

Fig. 17-27 Producing a 4-input AND gate from 2-input NAND gates (solution for Exercise 17-2).

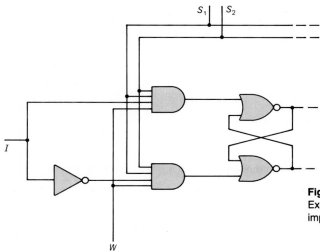

S_1 S_2

I

W

Fig. 17-28 Obtaining a logic function for Exercise 17-3. (A) Logic to be implemented. (Continued on p. 708)

(A)

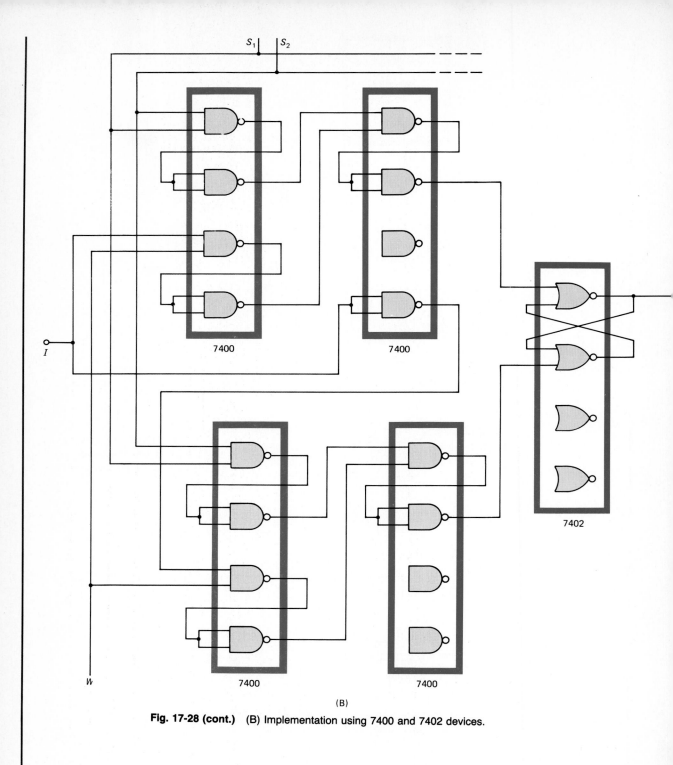

Fig. 17-28 (cont.) (B) Implementation using 7400 and 7402 devices.

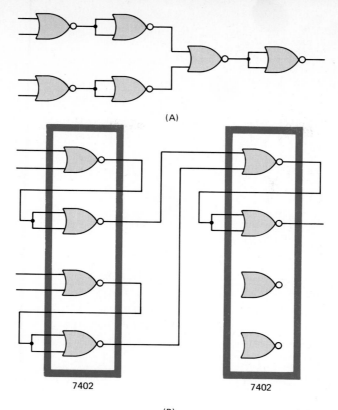

(A)

(B)

7402

7402

Fig. 17-29 Obtaining a logic function for Exercise 17-4. (A) Logic to be implemented. (B) Implementation using 7402 devices.

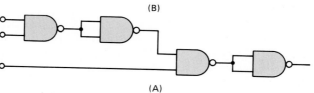

(A)

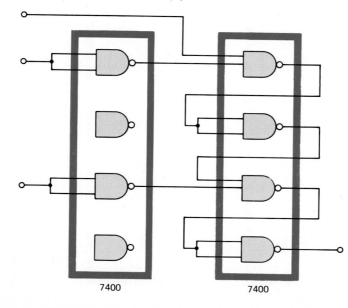

7400

7400

Fig. 17-30 Obtaining a logic function for Exercise 17-5. (A) Logic to be implemented. (B) Implementation using 7400 devices.

(B)

FIGURE 17-31
Solution to Exercise 17-6

No.	Inputs				Outputs									
	D	C	B	A	0	1	2	3	4	5	6	7	8	9
0	L	L	L	L	L	H	H	H	H	H	H	H	H	H
1	L	L	L	H	H	L	H	H	H	H	H	H	H	H
2	L	L	H	L	H	H	L	H	H	H	H	H	H	H
3	L	L	H	H	H	H	H	L	H	H	H	H	H	H
4	L	H	L	L	H	H	H	H	L	H	H	H	H	H
5	L	H	L	H	H	H	H	H	H	L	H	H	H	H
6	L	H	H	L	H	H	H	H	H	H	L	H	H	H
7	L	H	H	H	H	H	H	H	H	H	H	L	H	H
8	H	L	L	L	H	H	H	H	H	H	H	H	L	H
9	H	L	L	H	H	H	H	H	H	H	H	H	H	L
Invalid	H	L	H	L	H	H	H	H	H	H	H	H	H	H
	H	L	H	H	H	H	H	H	H	H	H	H	H	H
	H	H	L	L	H	H	H	H	H	H	H	H	H	H
	H	H	L	H	H	H	H	H	H	H	H	H	H	H
	H	H	H	L	H	H	H	H	H	H	H	H	H	H
	H	H	H	H	H	H	H	H	H	H	H	H	H	H

H = high level (*off*)
L = low level (*on*)

17-6. Figure 17-31 shows the resultant truth table, with a **1** represented by H (high) and a **0** represented by L (low). Since it is possible to count from 0 to 15 with four binary digits, and only the ability to count from 0 to 9 is required for BCD, the final six input contributions are invalid (not used).

17-7. From Fig. 17-4, the result of Fig. 17-32 is obtained.

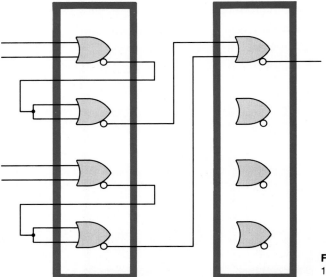

Fig. 17-32 Solution for Exercise 17-7.

17-8. The SN74HC00 uses a 5.0-V power supply. Therefore, from Eq. (17-3), the power dissipation per gate is

$$P_D = V_{DD}^2 \, Cf$$
$$= (5.0\,\text{V})^2 (50 \times 10^{-12}\,\text{F})(1.0 \times 10^6\,\text{Hz})$$
$$= 1.25\,\text{mW}$$

Given four gates per device,

$$(P_D)_{\text{TOT}} = 5.0\,\text{mW}$$

17-9. The binary number **111011** corresponds to the decimal value

$$32 + 16 + 8 + 2 + 1 = 59$$

and the binary number **110100** corresponds to the decimal value

$$32 + 16 + 4 = 52$$

From Eq. (17-6), the nominal value of V_{sample} is

$$V_{\text{sample}} = (15\,\text{V})(52/59) = 13.2\,\text{V}$$

Given 6-bit registers, the maximum accuracy is given by 1 part in 64. Therefore, V_{sample} can vary from the correct value by $\pm 1/64$:

$$\Delta V_{\text{sample}} = \pm 15(1/64) \cong \pm 0.2$$

17-10. Following the method shown in Fig. 17-17, the results of Fig. 17-33 are obtained. The final binary number is **110011**, with a decimal value of

$$32 + 16 + 2 + 1 = 51$$

so that

$$V_{\text{dec}} = (51/64)(10\,\text{V}) = 7.97\,\text{V}$$

is the decimal result of the conversion.

FIGURE 17-33
Solution for Exercise 17-10

Trial Value Step Number	Comparison	Binary Code	Result
1	>1/2	**100000**	Yes—keep binary digit
2	>3/4	**110000**	Yes—keep binary digit
3	>7/8	**111000**	No—drop binary digit
4	>13/16	**110100**	No—drop binary digit
5	>25/32	**110010**	Yes—keep binary digit
6	>49/64	**110011**	Yes—keep binary digit

17-11. A 14-bit binary number can be used to count from zero to $2^n - 1 = 16{,}383$. Therefore, the decimal accuracy associated with 14 binary bits is

$$1/16{,}383 = 0.00006 \text{ or } 6 \times 10^{-5}$$

17-12. From Eq. (17-9),

$$V_o = \frac{-V_{\text{ref}}}{2^7}(2^6 + 2^2)$$

$$= (-5.0\,\text{V})68/128 = -2.656\,\text{V}$$

PROBLEMS

(17-2) **17-1.** Show how part of a 7402 device can be used to produce the NOT function.

(17-2) **17-2.** Use one or more 7402 devices to produce a 4-input NOR circuit.

(17-2) **17-3.** Use one or more 7402 devices to produce a 4-input NAND circuit.

(17-2) **17-4.** Use the 7400 series devices discussed in this chapter to produce the circuit of Fig. 16-17.

(17-2) **17-5.** Use the 7400 series devices discussed in this chapter to produce the circuit of Fig. 16-19.

(17-2) **17-6.** Show how one or more 7400 devices can create an SR flip-flop.

(17-3) **17-7.** Show how one or more F10104 devices can produce a 4-input NOR gate.

(17-4) **17-8.** A CMOS circuit with $V_{DD} = 5.0$ V drives a 100-pF capacitive load. If the time between pulses is 1.0 μs, what is the average current flow associated with a change of state?

(17-4) **17-9.** A CMOS with $V_{DD} = 5.0$ V dissipates 5.0 mW at an operating frequency of 2.0 MHz. What is the effective capacitive load being driven?

(17-7) **17-10.** A dual-slope ADC produces the binary number **111000** in the $T_2 - T_1$ register and **110011** in the $T_3 - T_2$ register. If $V_{\text{ref}} = 10.00$ V, find V_{sample}.

(17-7) **17-11.** A dual-slope ADC produces $V_{\text{sample}} = 9.00$ V, given $V_{\text{ref}} = 12.00$ V. If **101010** is stored in the $T_2 - T_1$ register, what binary number is stored in the $T_3 - T_2$ register?

(17-7) **17-12.** A successive-approximation ADC has six bits, $(V_{\text{sample}})_{\text{max}} = 8.00$ V, and $V_{\text{sample}} = 2.50$ V. What is the sequence of trial values for V_{ref} and the final digital output?

(17-7) **17-13.** Find the number of decimal digits of accuracy associated with each ADC listed in Fig. 17-17.

(17-8) **17-14.** An R-$2R$ DAC converts the binary word **00111100** to an analog output. Given $V_{\text{ref}} = 10.00$ V, find the analog output.

EXPERIMENTAL APPLICATIONS

17-1. Obtain a TTL 7400 IC device and connect it for operation following the manufacturer's data sheet. Select one gate for study. Apply a **1** signal to one input and an appropriate square wave to the other. Observe and record the switching delay associated with the gate and compare with the rated value. Determine the power dissipation for the device (and therefore for each gate). Where does your circuit appear on the comparison of digital technologies shown in Fig. 15-23?

17-2. Obtain a Fairchild F10102 or F10104 IC device and determine the gate delay time and average power dissipation as discussed above. Place this gate circuit on the comparison of digital technologies shown in Fig. 15-23.

Design Network

PART 7
DESIGN AND FABRICATION TRADEOFFS

Previous chapters have emphasized the operation of semiconductor devices and strategies that can incorporate these devices into electronic circuits in order to achieve specified performance objectives. Given that circuit performance is ultimately dependent on device properties, which are in turn dependent on the fabrication methods used to create these devices, a systems approach to electronics requires that the designer gain an appreciation for the manufacturing methods used to create semiconductor devices and circuits, and how these methods link to device performance.

Chapter 18 begins by exploring the relationships between fabrication methods and semiconductor device parameters. Some of the basic tradeoffs involved in device performance are treated using the semiconductor models introduced earlier. The chapter then describes the techniques associated with several integration and packaging strategies that help determine system performance. By emphasizing performance constraints associated with semiconductor physics, device fabrication, integration and packaging, and circuit and system performance, the full scope of electronics can be considered.

Chapter 19 then discusses alternative settings in which the electronics design process can take place. Fragmented and systems-oriented strategies are contrasted, with emphasis in the latter case on computer-integrated design and manufacturing (CID/CIM). The objective of this chapter is to point out the importance of integrated approaches to electronics design.

CHAPTER 18
TECHNOLOGY
AND PERFORMANCE
IN MICROELECTRONICS

This chapter expands on earlier discussions regarding the properties of semiconductor materials and devices. An overview of common fabrication technologies is provided and then related to the device characteristics through modeling of performance parameters. The chapter thus illustrates how the foundation of semiconductor physics can be combined with fabrication concepts to define areas of interaction between manufacturing technology and device performance.

Technologies associated with printed circuit boards and hybrid microcircuits are then covered to indicate how subsystem- and system-level integration and packaging must be viewed as an essential determinant of system performance. Thermal and mechanical constraints are also described as they relate to the manufacturing of systems that meet all performance objectives.

18-1 FABRICATION OF INTEGRATED CIRCUITS

Integrated circuit (IC) technology begins with the manufacture of a single crystal of silicon, as discussed in Section 2-7. Purified silicon is melted in a furnace and then grown on a seed crystal to produce a long, cylindrical ingot that maintains the uniform crystal structure. The ingot, which can be several inches in diameter, is then sliced into thin wafers, which are polished and made ready for device fabrication. A typical wafer might be 0.2 mm thick. The IC circuits are built on the flat surface of a wafer, with hundreds or thousands of devices formed simultaneously on the same surface.

The creation of an array of device structures on the wafer surface requires the ability to change the properties of the surface in selected regions. This is accomplished by a *photolithography* process, as shown in Fig. 18-1.

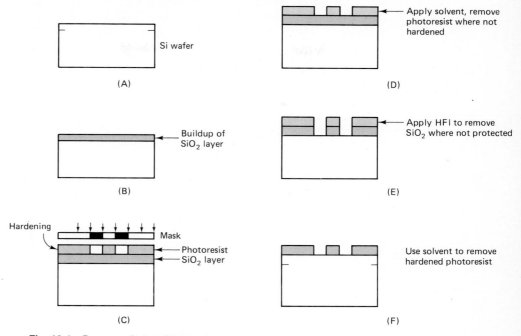

Fig. 18-1 Process of photolithography.

A layer of silicon oxide (SiO_2) is grown on the wafer by exposing the surface to an oxygen atmosphere with a moisture content. A thin film of photosensitive material is applied to the surface area as a liquid. This *photoresist* is a polymer, having the properties introduced in Section 2-7.

Each step in IC fabrication is accomplished by applying a process to selected surface areas on the wafer and preventing the process from being applied to all other locations. The photoresist must thus be exposed in selected areas and not in others. A photographic *mask,* consisting of transparent and opaque regions, is used for this purpose. An energy source, such as ultraviolet light, shines through the mask to illuminate the photoresist in the desired pattern, or *topography*. The energy activates the polymer, producing cross-linkages among the mers and effectively hardening the material where the energy is incident. The unexposed film can be removed with a solvent.

An acid is used to etch away the SiO_2 where there is no protective film. The result is that the SiO_2 layer remains only in the desired pattern that can be used to control the development of device doping and geometry.

To produce a BJT, the photolithography process is combined with epitaxial growth and diffusion to produce the desired collector, base, and emitter regions. When portions of the wafer surface are exposed to the correct environment, an epitaxial layer can be grown on the surface in such a way that the structure of a single crystal is maintained. If the surface is exposed to a gas at a defined pressure and temperature, impurity atoms diffuse into the surface, producing the desired doping levels.

Fabrication of an npn BJT can start with a p-type substrate, as shown in Fig. 18-2(A). A strongly doped n* region can be diffused into selected regions to produce the lower portion of the collector, followed by an epitaxial (n-type) growth to produce the upper portion of the collector. The base and emitter regions can then be produced by successive additional diffusions. A metal layer is applied to the surface to create the necessary contacts and interconnections.

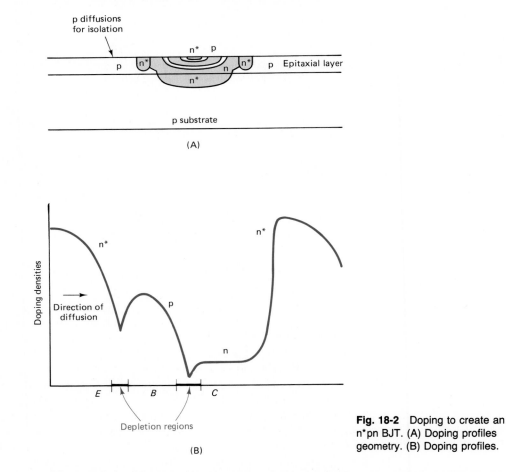

Fig. 18-2 Doping to create an n*pn BJT. (A) Doping profiles geometry. (B) Doping profiles.

The result is shown in Fig. 18-2(A) (and in Fig. 6-22). A typical profile through the center of the device is shown in Fig. 18-2(B). This profile is discussed in detail in the next section.

As may be gathered from the above description, IC fabrication is a complex technology involving numerous chemical processes. Fabrication is conducted at high temperatures with a wide range of pressures, and with great accuracy and control over microscopic dimensions. By-products include toxic gases and liquids. Successful IC operations can be maintained only with great care and at great cost.

The semiconductor manufacturing engineer must understand how to combine materials and processes to produce devices that have the most advantageous properties.

This requires an understanding of the relationships between device performance and the typical data shown in Fig. 18-2(B). The next section indicates how the links between device parameters and IC fabrication methods can be formed, making use of many of the concepts introduced in earlier chapters.

18-2 BJT DESIGN

The fabrication information provided in this chapter can be combined with the materials and semiconductor device discussions of Chapters 2, 4, 6, and 8 to explore some of the tradeoffs involved in the design of bipolar junction transistors (BJTs). By bringing together a number of different models associated with BJT operation, insight can be gained into the relationships between semiconductor properties, fabrication constraints, and circuit design considerations.

One of the most important performance objectives for a BJT is to obtain a large value of $\beta_{dc} = h_{FE}$ (typically in the range of 100 to 200 for active operation). The following development illustrates how basic semiconductor properties are linked to the design of a BJT that satisfies this design objective.

The following discussion makes use of the highly idealized "layer" model of the BJT introduced in Chapter 6. The emitter, base, and collector are represented by constant cross-section "slices" of doped semiconductor material with constant doping levels in each region. Obviously, based on the fabrication methods outlined above, this is a highly idealized case. However, given the simplicity of the model, it provides a useful starting point for gaining an understanding of the tradeoffs that arise in BJT design. More realistic geometries and variable doping levels can then be introduced for more accurate predictions of device characteristics.

Figure 18-3 shows the carrier flow in such an npn device (as introduced in Section 6-1). To obtain a high β_{dc}, all currents contributing to I_B must be minimized.

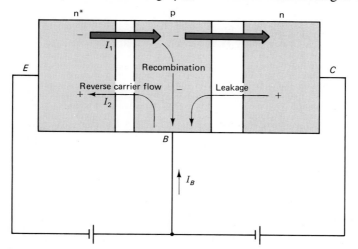

Fig. 18-3 Carrier flow in an n⁺pn BJT.

Figure 18-4 illustrates the excess carrier density levels introduced by injection from emitter to base and base to emitter for a forward biased junction. (The former injection level is much larger than the latter due to the heavy emitter doping.) If the

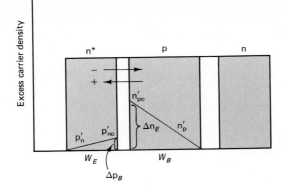

Fig. 18-4 Excess carrier density level in a BJT (with W_e and W_B small with respect to L_n and L_p.)

base width W_B is much smaller than the characteristic electron penetration distance L_n and if the emitter width W_E is much smaller than the characteristic hole penetration distance L_p, then the excess carrier levels will form the approximate linear functions as shown. If the injected carrier levels near the emitter-base boundary are much larger than the minority-carrier levels due to doping, then the total minority-carrier density on the base side of the emitter-base junction is approximately $n'_{po} \cong \Delta n_E$ and the total minority carrier density on the emitter side is approximately $p'_{no} \cong \Delta p_B$.

The first step in analysis considers device performance with an injection efficiency $\gamma = 1$ ($\Delta n_E \gg \Delta p_B$) and collector-base currents assumed negligible. (These factors can then be considered in turn.) For this case, β_{dc} is determined by the base transport factor B. (The parameters γ and B were introduced in Exercise 6-1 of Chapter 6.)

For the simple BJT model being followed,

$$\beta_{dc} \cong 2(L_n/W_B)^2 \qquad \gamma = 1 \tag{18-1}$$

as described previously (Exercise 6-7, Chapter 6) so that

$$\alpha_{dc} = \frac{2\left(\dfrac{L_n}{W_B}\right)^2}{1 + 2(L_n/W_B)^2} = \frac{1}{1 + \frac{1}{2}(W_B/L_n)^2} \qquad \gamma = 1 \tag{18-2}$$

This equation must now be modified for the general case of $\gamma < 1$.[1] This will be done by deriving an equation for γ, and then using this relationship to modify Eq. (18-2).

By definition, from Exercise 6-1 of Chapter 6, γ can be written

$$\gamma = \frac{I_1}{I_1 + I_2} = \frac{1}{1 + I_2/I_1} \tag{18-3}$$

where I_1 is due to carrier injection from emitter to base, and I_2 is due to carrier flow from base to emitter, as shown in Fig. 18-3. The excess carrier distributions that result are indicated in Fig. 18-4 (assuming that W_B and W_E are small with respect to L_n

---

[1] In this section, both α_{dc} and β_{dc} are treated as having positive values.

and L_p). The diffusion currents I_1 and I_2 [as given in Eq. (4-4)] can then be written in the form

$$I_1 \cong qAD_n \Delta n_E / W_B \cong qAD_n \frac{n'_{po}}{W_B}$$

$$I_2 \cong qAD_p \Delta p_B / W_E \cong qAD_p \frac{p'_{no}}{W_E}$$

(18-4)

where n_{po} and p_{no} are the total carrier densities determined at the junction boundary and the injected carrier densities are assumed large with respect to those present due to doping. Therefore,

$$\gamma = \frac{1}{1 + \dfrac{D_p}{D_n} \dfrac{p'_{no}}{W_E} \dfrac{W_B}{n'_{po}}} = \frac{1}{1 + \dfrac{D_p}{D_n} \dfrac{p'_{no}}{n'_{po}} \dfrac{W_B}{W_E}}$$

(18-5)

Given the relationship between minority and majority carriers [Eq. (4-3)],

$$p'_{no} n'_{no} = (n_i^2)_E \qquad \text{and} \qquad n'_{po} p'_{po} = (n_i^2)_B$$

(18-6)

and using the Einstein relation [Eq. (4-12)],

$$D_p / D_n = \mu_p / \mu_n$$

(18-7)

so that

$$\gamma \cong \frac{1}{1 + \dfrac{\mu_p (n_i^2)_E}{\mu_n (n_i^2)_B} \dfrac{p'_{po}}{n'_{no}} \dfrac{W_B}{W_E}} = \frac{1}{1 + \dfrac{\sigma_p}{\sigma_n} \dfrac{W_B}{W_E} \dfrac{(n_i^2)_E}{(n_i^2)_B}}$$

(18-8)

where, due to the straight-line carrier densities, σ_p and σ_n can be taken as the average conductivities in the two regions. As a result,

$$\sigma_n = (n_n)_{av} q\mu_n = N_E q\mu_E$$

$$\sigma_p = (p_p)_{av} q\mu_p = N_B q\mu_B$$

(18-9)

for the npn configuration being considered.

In previous chapters, the energy gap $\mathcal{E}_g$ is considered a constant, so that $(n_i^2)_E = (n_i^2)_B$. However, for high doping levels, the effective value of $\mathcal{E}_g$ decreases in a process called band-gap narrowing (mentioned in Exercise 6-1 of Chapter 6). From Eqs. (4-1) and (4-2),

$$(n_i^2)_E = (n_i^2)_B e^{\Delta \mathcal{E}_g / kT}$$

(18-10)

and

$$\gamma = \frac{1}{1 + \dfrac{\sigma_p}{\sigma_n} \dfrac{W_B}{W_E} e^{\Delta \mathcal{E}_g / kT}}$$

(18-11)

Given this value for γ, Eq. (18-2) can be generalized to

$$\alpha_{dc} = \frac{1}{1 + \dfrac{1}{2} \left(\dfrac{W_B}{L_n} \right)^2} \frac{1}{1 + \dfrac{\sigma_p}{\sigma_n} \dfrac{W_B}{W_E} e^{\Delta \mathcal{E}_g / kT}}$$

(18-12)

The desired value of β_{dc} can then be found from the relationship

$$\beta_{dc} = \frac{\alpha_{dc}}{1 - \alpha_{dc}} \qquad (18\text{-}13)$$

Equations (18-12) and (18-13) provide many insights into BJT design. First, it is clear that α_{dc} must be within less than 1 percent of unity to achieve $\beta_{dc} > 100$. Thus, the requirements

$$W_B \ll L_n \qquad (18\text{-}14)$$

$$\sigma_p W_B e^{\Delta \mathcal{E}_g / kT} \ll \sigma_n W_E \qquad (18\text{-}15)$$

exist. Equation (18-14) provides a constraint on the physical base width, whereas Eq. (18-15) requires that the emitter be doped much more strongly than the base. However, the emitter doping must not be sufficiently high to create an appreciable band gap narrowing $\Delta \mathcal{E}_g$, or the additional doping does not achieve the desired effect on γ. Figure 18-5 shows the relationship between $\Delta \mathcal{E}_g$ and the doping level. Given $\mathcal{E}_g \cong 1.11$ eV for Si, the effect of $\Delta \mathcal{E}_g$ becomes important for $N_D \gtrsim 10^{25}/m^3$.

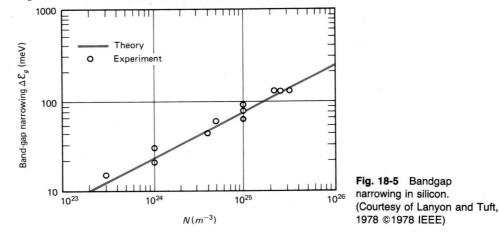

Fig. 18-5 Bandgap narrowing in silicon. (Courtesy of Lanyon and Tuft, 1978 ©1978 IEEE)

The value of σ_p is also affected at high current levels (discussed in Section 4-4 and Appendix 8) for which the density of injected carriers in the base is sufficient to increase σ_p. Therefore, it can be expected that β_{dc} drops at high current levels, an effect that is experimentally observed. The low-level carrier recombination in the depletion regions reduces the effective β_{dc} for low current levels. The combined current effects can be illustrated as shown in Fig. 18-6.

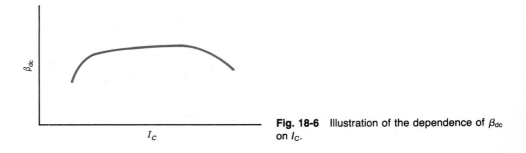

Fig. 18-6 Illustration of the dependence of β_{dc} on I_C.

Figure 18-7 illustrates some of the tradeoffs involved in BJT design. As noted above, N_E is limited by band-gap narrowing and $N_B \ll N_E$ is required to produce $\gamma \cong 1$. The base width W_B must also be much less than L_n to achieve a high value of B. Several other important factors are also illustrated in this figure and discussed below.

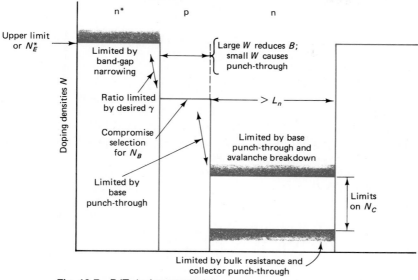

Fig. 18-7 BJT design constraints.

The behavior of the collector-base depletion region is an important consideration in device design. From Eq. (4-40), the thickness of the depletion region components in the base and collector are given by

$$t_B = \left[\frac{2\epsilon(V_t - V_{CB})}{qN_B(1 + N_B/N_C)} \right]^{1/2} \tag{18-16}$$

$$t_C = \left[\frac{2\epsilon(V_t - V_{CB})}{qN_C(1 + N_C/N_B)} \right]^{1/2} \tag{18-17}$$

where N_C is the doping density in the collector, and to achieve overall charge neutrality,

$$N_B t_B = N_C t_C \tag{18-18}$$

For a depletion layer, t decreases with forward bias and increases with reverse bias. The depletion layer associated with the collector-base junction can have a significant effect on device design due to the normal reverse-bias state used for linear operation.

If the depletion region extends through either the base or collector, punch-through occurs (as introduced in Exercise 6-8 in Chapter 6). Given the resultant n*n configuration, a large current flows in the BJT. To achieve normal operating conditions, punch-through must be prevented.

From Eqs. (18-16) and (18-17), the depletion region thickness at a junction extends furthest into the least-doped region. Since the base width W_B must be kept thin to produce B close to unity, the potential for base punch-through is high unless appropriate precautionary steps are taken. One possible strategy, as shown in Fig. 18-7, is to choose $N_C \ll N_B$, so that most of the collector-base depletion region extends into the collector.

Given the relationship $N_C \ll N_B$, Eqs. (18-16) and (18-17) become

$$t_B \cong \left[\frac{2\epsilon (V_t - V_{CB})}{qN_C} \left(\frac{N_C}{N_B} \right)^2 \right]^{1/2} \tag{18-19}$$

$$t_C \cong \left[\frac{2\epsilon (V_t - V_{CB})}{qN_C} \right]^{1/2} \tag{18-20}$$

As a result the total depletion region thickness becomes $t_T = t_B + t_C \cong t_C$, producing what is often called a *one-sided abrupt junction*. From Eq. (18-20), collector punch-through occurs for low collector doping values (N_C), providing a minimum level constraint for this parameter. On the other hand, if N_C approaches N_B, more of the depletion region moves into the base and base punch-through becomes a concern. Limits can thus be established for N_C, as shown in Fig. 18-7.

Minimum doping levels can also be specified to assure that the BJT does not have an excessive bulk resistance that would distort the characteristic curves of the device and produce excessive power dissipation.

Generalizing from Eq. (3-1), the bulk resistance is given by

$$R_b = \rho \frac{\ell}{A} = \frac{1}{qA} \left(\frac{W_E}{N_E \mu_E} + \frac{W_B}{N_B \mu_B} + \frac{W_C}{\mu_C N_C} \right) \tag{18-21}$$

For the configuration used here, N_C dominates the overall R_b and

$$R_b \cong \frac{W_C}{qAN_C \mu_C} \tag{18-22}$$

The selected maximum allowable value for R_b thus sets a lower limit on the doping levels that can be used, as shown in Fig. 18-7.

Based on linear operation, a reasonable constraint is to require

{Power dissipated in the bulk collector region}

$\ll$ {Power dissipated in the collector-base depletion region}

for all values of V_{CB} corresponding to linear operation. Therefore,

$$I_C^2 R_C \ll V_{CB} I_C$$

$$I_C R_C \ll V_{CB}$$

If $(V_{CB})_{min} = 2.0$ V and $I_C = 1.0$ mA,

$$R_C \ll 2.0 \text{ V}/1.0 \text{ mA} = 2.0 \text{ k}\Omega$$

Based on saturated switching operation, a reasonable constraint is to require

{Voltage drop across the bulk collector region}

$\ll$ {Saturated voltage drop across an ideal device}

$$I_C R_C \ll (V_{sat})_{ideal}$$

If $(V_{\text{sat}})_{\text{ideal}} = 0.3$ V and $I_C = 1.0$ mA,

$$R_C \ll 0.3 \text{ V}/1.0 \text{ mA} = 0.3 \text{ k}\Omega$$

The calculations above suggest order-of-magnitude limits to be placed on R_C.

To achieve the objectives of (1) a low value of N_C to prevent base punch-through and (2) a high value of N_C to minimize R_b, the solution of Fig. 18-7 is often used. Near the collector-base junction, the collector is doped at the lower level. Then the N_C level transitions to an n* level away from the junction. To prevent interaction between the n* collector region and the base, the collector-base junction and n* region must be separated by a distance greater than the diffusion length L_n.

Finally, *avalanche breakdown* is also a concern at the collector-base junction. A large electric field exists at the junction between the portions of the depletion region formed from n- and p-type semiconductors. This electric field can be sufficiently high to produce an avalanche effect in which electrons gain sufficient energy between collisions to form new carriers when collisions occur. Figure 18-8 shows the breakdown voltage for a one-sided abrupt junction appropriate to the model considered here.

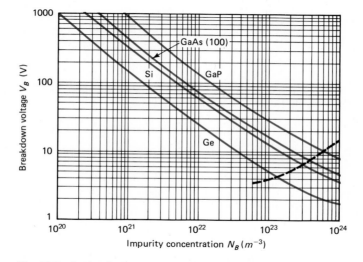

Fig. 18-8 Avalanche breakdown voltage for one-sided abrupt junctions. (Courtesy of Sze and Gibbons, 1966 ©1966 American Institute of Physics)

As the doping level increases, the depletion region thickness is reduced [Eqs. (18-16) and (18-17)]. Therefore, the resultant E field increases and breakdown occurs at a lower voltage. The curves of Fig. 18-8 thus illustrate an expected behavior. The collector doping must be maintained at a sufficiently low level to avoid breakdown.

The parameter constraints indicated in Fig. 18-7 (with associated data) can be combined with Eqs. (18-12) and (18-13) to explore some of the interactions between performance objectives and semiconductor properties.

Example 18-1 **BJT Design (Part 1)**

Design a Si BJT with $\beta_{\text{dc}} > 150$, subject to the constraints presented above in this section. Assume that $L_n = 10 \ \mu$m.

Solution To achieve $\beta_{dc} > 150$ requires

$$\alpha_{dc} > \frac{\beta_{dc}}{\beta_{dc} + 1} = \frac{150}{151} = 0.9934$$

The design starts with the satisfaction of Eq. (18-12). Select $B = 0.9980$. Then

$$\frac{1}{2}\left(\frac{W_B}{L_n}\right)^2 = \frac{1 - 0.9980}{0.9980} = 0.0020$$

and

$$W_B/L_n = 0.063$$

For $L_n = 10 \ \mu m$,

$$W_B = 0.63 \ \mu m$$

The injection efficiency γ must then be selected to have the value

$$0.9934/0.9980 = 0.9954$$

and

$$\frac{\sigma_p}{\sigma_n}\frac{W_B}{W_E} e^{\Delta \mathcal{E}_g/kT} = \frac{1 - 0.9954}{0.9954} = 0.0046$$

Try $W_E = 2.0 \ W_B = (2.0)(0.63 \ \mu m) \cong 1.3 \ \mu m$ and assume that doping levels are selected so that $e^{\Delta \mathcal{E}_g/kT}$ is negligible. Then

$$\sigma_p/\sigma_n = (0.0046)(2.0) = 0.0092$$

and

$$\frac{N_B \mu_B}{N_E \mu_E} = 0.0092$$

From Fig. 2-6,

$$\mu_E \cong 0.14 \ m^2/V \cdot s \qquad \text{and} \qquad \mu_B \cong 0.05 \ m^2/V \cdot s$$

so that

$$\frac{N_B}{N_E} < 0.0092(0.14/0.05) \cong 0.026$$

From Fig. 18-5, select $N_E = 1.0 \times 10^{24}/m^3$. Then

$$N_B = (0.026)(1.0 \times 10^{24}/m^3) = 2.6 \times 10^{22}/m^3$$

Choose the width of the collector region to be $W_C = L_n = 10 \ \mu m$ to prevent interaction with the n* collector region, and assume that the collector is to be doped much more lightly than the base. Then from Eq. (18-20), the voltage (V_{CB}) that produces collector punch-through is

$$V_{CB} \cong \frac{t_c^2 q N_C}{2\epsilon} = \frac{(10 \times 10^{-6} \ m)^2 (1.6 \times 10^{-19} \ C) N_C}{2 \times 12 \times 8.85 \times 10^{-12} \ F/m}$$

$$\cong (7.53 \times 10^{-20} m^3 \cdot V) N_C$$

To achieve $(V_{CB})_{max} = 50$ V, N_C must be

$$N_C > \frac{50V}{7.53 \times 10^{-20}m^3 \cdot V} \cong 6.64 \times 10^{20}/m^3$$

The ratio

$$\frac{N_C}{N_B} = \frac{6.64 \times 10^{20}/m^3}{2.6 \times 10^{22}/m^3} = 0.026$$

fulfills the requirement that $N_C \ll N_B$.

At $V_{CB} = 50$ V, the thickness of the depletion region in the base is

$$t_B = \frac{N_C t_C}{N_B} = (10 \ \mu m)(0.026) = 0.26 \ \mu m$$

which is

$$0.26 \ \mu m/0.63 \ \mu m \times 100\% \cong 41\%$$

of the total base thickness.

From Fig. 18-8, the choice $N_C = 6.64 \times 10^{20}/m^3$ does not cause excessive susceptibility to avalanche breakdown.

Choose $A = 20 \ \mu m \times 20 \ \mu m = 4.0 \times 10^{-10} \ m^2$. Then from Eq. (18-22),

$$R_b \cong \frac{10 \ \mu m}{(1.6 \times 10^{-19}C)(4.0 \times 10^{-10} \ m^2)(0.14 \ m^2/V \cdot s)(6.64 \times 10^{20}/m^3)}$$
$$\cong 1.7 \ k\Omega$$

The selected BJT parameters are thus

$\beta_{dc} = 150$	$W_B = 0.63 \ \mu m$
$\alpha_{dc} = 0.9934$	$W_E = 1.3 \ \mu m$
$L_n = 10 \ \mu m$	$W_C = 10 \ \mu m$
$N_E = 1.0 \times 10^{24}/m^3$	$V_{CB} < 50$ V
$N_B = 2.6 \times 10^{22}/m^3$	$A = 20 \ \mu m \times 20 \ \mu m$
$N_C = 6.64 \times 10^{20}/m^3$	$R_b = 1.72 \ k\Omega$

The above example indicates how difficult it is to design a BJT that meets all of the given criteria. The values selected barely meet the constraints in many cases. If a higher value of β_{dc} is required, if A is too large, or if V_{CB} must increase, efforts must be made to look more closely at manufacturing processes, detailed geometries and doping profiles, and modified design strategies. ∎

A number of other BJT design considerations can be added to the factors introduced above. The effective output resistance of the BJT can be calculated by finding the Early voltage V_E (refer to Exercise 6-8, Chapter 6). The configuration produced above can be further evaluated by calculating V_E and determining whether the result is acceptable.

Another important design issue is the current-carrying capability of the BJT. From Eqs. (4-27) and (4-28),

$$I = I_0(e^{V/V_0} - 1) \qquad (18\text{-}23\text{a})$$

$$I_0 \cong qA\left[\frac{D_p}{L_p N_E} + \frac{D_n}{L_n N_B}\right]n_i^2 \cong qA\frac{D_n n_i^2}{L_n N_B} \qquad (18\text{-}23\text{b})$$

For fixed doping levels, the current level is proportional to the cross-sectional area A of the device. Given the parameters introduced above, the area A can be adjusted to achieve the desired current flow. However, as A increases, the density of devices on an IC substrate decreases, providing another performance tradeoff.

The ac performance of the BJT is another essential consideration. The high-frequency half-power point f_2 of the device is determined by the effective capacitances associated with the emitter-base and base-collector junctions and the time required for the excess carrier-density distributions to become modified as the operating parameters change. Equation (4-41) allows calculation of the capacitances associated with the depletion regions themselves.

$$C = A\left(\frac{q\epsilon}{2}\frac{N_a N_d}{N_a + N_d}\frac{1}{V_t - V}\right)^{1/2} \qquad (18\text{-}24)$$

The values for C_{EB} are substantially increased due to the buildup of excess charge near the emitter-base junction.

By definition,

$$C = dQ/dV$$

and from Fig. 6-26 or 18-4,

$$Q = \frac{1}{2}\Delta p_E(W_B A)q$$

so that

$$C_{EB} = \frac{1}{2}(W_B A)q\frac{d\Delta p_E}{dV} \qquad (18\text{-}25)$$

From Eqs. (4-4) and (4-28),

$$I \cong qAD_n\frac{\Delta p_E}{W_B} = I_0(e^{V/V_0} - 1) \qquad (18\text{-}26)$$

and differentiating

$$\frac{dI}{dV} = \frac{qAD_n}{W_B}\frac{d\Delta p_E}{dV} \cong \frac{I}{V_0} \qquad (18\text{-}27)$$

so that

$$C_{EB} \cong \frac{W_B^2 q I_E}{2D_n kT} \qquad (18\text{-}28)$$

This value of C_{EB} is associated with small-signal operation around a bias current I_E, and is often the largest capacitance associated with the device.

Given the effective capacitances for the device, the approximate value of f_2 can be found by considering the parallel combination of C_{EB} and h_{ie} as forming the input section to the ac equivalent circuit for the BJT. From Eq. (10-38),

$$(f_2)_{\text{diff}} \cong \frac{1}{2\pi h_{ie}C_{EB}} \qquad (18\text{-}29)$$

with

$$h_{ie} \cong \frac{V_o}{I_B} = \frac{kT}{qI_B}$$

By substituting from Eq. (18-28),

$$(f_2)_{\text{diff}} \cong \frac{D_n}{\pi W_B^2 h_{FE}} \qquad (18\text{-}30)$$

The (large-signal) switching frequency response of the BJT can be found from

$$\tau_n = L_n^2/D_n \qquad (18\text{-}31)$$

which defines the recombination time τ_n in terms of L_n and D_n (as discussed in Exercise 4-6 of Chapter 4 and Exercise 6-7 of Chapter 6). The time τ_n describes how long it takes the BJT to turn *off* (for the excess carrier density in the base to fall to $1/e$ of its maximum value).

Let τ_n be the turn-off time constant in response to a square-wave input. Then, from Section 11-1,

$$(f_2)_{\text{recomb}} \cong \frac{1}{2\pi\tau_n} = \frac{D_n}{2\pi L_n^2} \qquad (18\text{-}32)$$

Thus, as efforts are made to increase L_n in order to achieve a higher value of B, the lifetime τ_n also increases, reducing the effective bandwidth f_2 of the device. The ac performance objectives for a BJT must be included in the evaluation of design alternatives, and interaction is often required to achieve balance among the many competing objectives.

This section brings together many of the device concepts introduced throughout the text and provides a brief introduction to some of the important tradeoffs that arise in producing a viable BJT. Because actual fabrication processes result in model properties that can differ substantially from those used here, more sophisticated methods of analysis are required to understand device properties in more detail.

Example 18-2   BJT Design (Part 2)

For the BJT discussed in Example 18-1, find (a) the Early voltage V_E; (b) the current-carrying capability of the device; (c) the base-collector junction capacitance at $V_{CB} = 10 \text{ V}$ reverse bias; (d) the emitter-base diffusion capacitance for $I_E = 1.0$ and 10 mA, and f_2 associated with the device in each case; and (e) the f_2 value associated with the large-signal switching operation.

Solution   (a) From Exercise 6-8 of Chapter 6,

$$V_E \cong \frac{1}{4} \frac{W_o^2 q N_B^2}{\epsilon N_C} \qquad N_B \gg N_C$$

and by substitution,

$$V_E \cong \frac{1}{4} \frac{(0.63 \times 10^{-6} \text{ m})^2 (1.6 \times 10^{-19} \text{ C})(2.6 \times 10^{22}/\text{m}^3)^2}{(12 \times 8.85 \times 10^{-12} \text{ F/m})(6.64 \times 10^{20}/\text{m}^3)}$$

$$\cong 1.5 \times 10^2 \text{ V}$$

This is a satisfactory result. Observe that the choice $N_B \gg N_C$ to reduce the size of the depletion region in the base increases the magnitude of V_E, a preferred outcome.

(b) The current-carrying capability can be found by first calculating the contact potential V_t for the emitter-base junction:

$$V_t = \frac{kT}{q} \ln \frac{N_E N_B}{n_i^2}$$

$$= (0.026 \text{ V}) \ln \frac{(1.0 \times 10^{24}/\text{m}^3)(2.6 \times 10^{22}/\text{m}^3)}{(1.6 \times 10^{16}\text{m}^3)^2}$$

$$\cong 0.84 \text{ V}$$

which represents an upper limit on V.

From Eq. (18-23b) and Fig. 2-6,

$$I_0 \cong \frac{(1.6 \times 10^{-19} \text{ C})(4 \times 10^{-10} \text{ m}^2)(35 \times 10^{-4} \text{ m}^2/\text{s})(1.6 \times 10^{16}/\text{m}^3)^2}{(10 \times 10^{-6} \text{ m})(2.6 \times 10^{22}/\text{m}^3)}$$

$$\cong 2.2 \times 10^{-16} \text{ A}$$

This is an unrealistically low value. Leakage currents and other processes typically result in an I_0 from 10^{-12} to 10^{-14} A.

For the moment, apply the value $I_0 = 2.2 \times 10^{-16}$ A and the (near) maximum $V_t = 0.8$ V. Then from Eq. (18-23a),

$$(I_E)_{\text{max}} = (2.2 \times 10^{-16} \text{ A})(e^{0.8 \text{ V}/0.026 \text{ V}} - 1)$$

$$\cong 5.1 \text{ mA}$$

For more realistic values $I_0 = 1.0 \times 10^{-13}$ A and $V = 0.7$ V,

$$(I_E)_{\text{max}} = (1.0 \times 10^{-13} \text{ A})(e^{0.7 \text{ V}/0.026 \text{ V}} - 1)$$

$$\cong 50 \text{ mA}$$

which is a more satisfactory result.

(c) From Eq. (18-24),

$$C_{CB} = A \left(\frac{q \epsilon N_C}{2 V_{CB}} \right)^{1/2}$$

$$= (4.0 \times 10^{-10} \text{ m}^2) \left[\frac{(1.6 \times 10^{-19} \text{ C}) \times (12 \times 8.85 \times 10^{-12} \text{ F/m})(6.64 \times 10^{20}/\text{m}^3)}{2(10 \text{ V})} \right]^{1/2}$$

$$\cong 0.01 \text{ pF}$$

Other distributed capacitances will typically be substantially larger than this value.

(d) From Eq. (18-28),

$$C_{EB} = \frac{(0.63 \times 10^{-6}\ \text{m})^2 (1.6 \times 10^{-19}\ \text{C})(1.0 \times 10^{-3}\ \text{A})}{2(35 \times 10^{-4}\ \text{m}^2/\text{s})(1.38 \times 10^{-23}\ \text{J}/^\circ\text{K})(3 \times 10^2\ ^\circ\text{K})}$$

$$\cong 2.0\ \text{pF}$$

From Eq. (18-30),

$$(f_2)_{\text{diff}} \cong \frac{35 \times 10^{-4}\ \text{m}^2/\text{s}}{\pi (0.63 \times 10^{-6}\ \text{m})^2 (150)} \cong 20\ \text{MHz}$$

If I_E increases to 10 mA,

$$C_{EB} \cong 20\ \text{pF}$$

and

$$(f_2)_{\text{diff}} \cong 2.0\ \text{MHz}$$

(e) From Eq. (18-32),

$$(f_2)_{\text{recomb}} = \frac{D_n}{2\pi L_n^2}$$

$$= \frac{(35 \times 10^{-4}\ \text{m}^2/\text{s})}{6.28(10 \times 10^{-6}\ \text{m})^2} \cong 5.6\ \text{MHz} \qquad \blacksquare$$

18-3 JFET DESIGN

A simple model for JFET design can be developed by drawing on the concepts of Chapter 8 and the BJT concepts introduced above. In Section 8-3, the following approximate characteristic equations describe the saturated transfer characteristic for an idealized and simplified JFET model [Eqs. (8-6) to (8-8)]:

$$I_D \cong I_{D0}(1 - V_{GS}/V_P)[1 - (V_{GS}/V_P)^{1/2}] \cong I_{D0}(1 - V_{GS}/V_P)^2 \qquad (18\text{-}33)$$

where

$$V_P = \left(\frac{W_o}{2K}\right)^2 = \frac{W_o^2 q N_{CH}}{8\epsilon} \qquad (18\text{-}34)$$

$$I_{D0} = \frac{V_P z W_o}{\rho L} = \frac{W_o^3 q N_{CH} z}{8\epsilon\rho L} = \frac{W_o^3 q^2 N_{CH}^2 z \mu}{8\epsilon L} \qquad (18\text{-}35)$$

introducing N_{CH} as the channel doping density. The resultant output characteristic for the JFET is presented in Fig. 8-9. This model provides a means for evaluating some of the tradeoffs associated with JFET design.

A useful strategy is to define the desired current and voltage levels to be associated with point P in Fig. 8-9. The output characteristic, transfer characteristic, and g_m are then determined. From Eqs. (8-34) and (8-35), these performance objectives require the selection of values for the dimensions z, L, W_o, and channel resistivity ρ.

The constraints associated with JFET design are illustrated in Fig. 18-9. Trial values for the channel doping level and geometry can be obtained by considering the factors shown.

The doping level and geometry for the gate structures can be found by requiring that the voltage drop along the gate from source to drain be negligible with respect

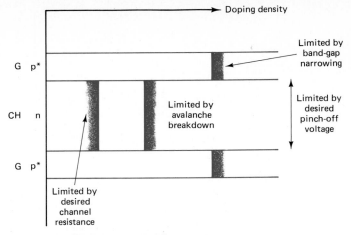

Fig. 18-9 JFET design constraints.

to the scale of values for V_{GS} and V_{DS}. The gate potential is then approximately constant, as required. The combination of doping levels N_{CH} and N_G can then be checked to ensure that gate-channel avalanche breakdown does not occur for the range of voltages to be applied.

The current-carrying capability of the JFET can be found from

$$I_D = \sigma EA = N_{CH}q\mu(V_P/L)\text{A} \qquad (18\text{-}36)$$

The ac device performance can be described in terms of the effective gate-channel capacitance. Once the geometries and doping levels are known, $C_{G\text{-}CH}$ can be calculated in terms of the properties of the gate-channel pn junction. Applying Eq. (4-41) used above for the study of the BJT depletion region,

$$C = A\left(\frac{q\epsilon}{2}\frac{N_{CH}}{V_t - V}\right)^{1/2} \qquad (18\text{-}37)$$

assuming the gate is present on only one side of the channel.

The JFET switching *off* delay is dominated by the time required for mobile carriers to return to the depletion region formed under bias conditions. The transit time for charges through the channel also serves as an upper limit to device frequency response.

Example 18-3   JFET Design

Select parameters for a Si JFET to achieve $V_P = 3.0$ V and $I_{D0} = 3.0$ mA for point P in Fig. 8-9.

Solution From Eqs. (8-34) and (8-35),

$$I_{D0} = 3.0 \text{ mA} = (3.0 \text{ V})\frac{z}{\rho L}W_o$$

$$V_P = 3.0 \text{ V} = W_o^2 N_{CH}\frac{1.6 \times 10^{-19} \text{ C}}{8(12)(8.85 \times 10^{-12} \text{ F/m})}$$

so that

$$\frac{z}{\rho L} W_o = 1.0 \times 10^{-3} \ \text{\mho}$$

$$W_o^2 N_{CH} = 1.6 \times 10^{10}/\text{m}$$

If $N_{CH} = 1.0 \times 10^{22}/\text{m}^3$, $W_o = 1.3 \ \mu\text{m}$. Given these values,

$$\rho = \frac{1}{N_{CH} q \mu_n} = \frac{1}{(1.0 \times 10^{22}/\text{m}^3)(1.6 \times 10^{-19} \ \text{C})(0.14 \ \text{m}^2/\text{V} \cdot \text{s})}$$
$$= 4.5 \times 10^{-3} \ \Omega \cdot \text{m}$$

and

$$\frac{z}{L} \cong \frac{(1.0 \times 10^{-3} \ \text{\mho}) (4.5 \times 10^{-3} \ \Omega \cdot \text{m})}{1.3 \times 10^{-6} \ \text{m}} \cong 3.5$$

If $L = 3.0 \ \mu\text{m}$, $z = 11 \ \mu\text{m}$.

If $N_G = 10 \ N_{CH} = 1.0 \times 10^{23}/\text{m}^3$, the avalanche breakdown voltage can be found from Fig. 18-8. For the value $N_{CH} = 1.0 \times 10^{22}/\text{m}^3$, the breakdown voltage is about 50 V.

The current-carrying capability is obtained from Eq. (8-36),

$$I = \frac{(0.22 \times 10^3)(3.0 \ \text{V})(11 \ \mu\text{m})(3.0 \ \mu\text{m})}{3.0 \ \mu\text{m}}$$

$$\cong 7.3 \ \text{mA}$$

From Eq. (8-37), the capacitance of the gate-channel junction at $V = 3.0$ V is about

$$C = (11 \ \mu\text{m})(3.0 \ \mu\text{m})\left[\frac{(1.6 \times 10^{-19} \ \text{C})(12)(8.85 \times 10^{-12} \ \text{F/m})(1.0 \times 10^{22}/\text{m}^3)}{(2)(3.7)}\right]^{1/2}$$

$$\cong 0.005 \ \text{pF} \qquad \blacksquare$$

18-4 D-MOSFET DESIGN

The depletion MOSFET (D-MOSFET) design can proceed after the discussion of Section 8-11. For a thin oxide layer t_{ox}, the dc design for the D-MOSFET is equivalent to the JFET discussion above. For a thick oxide layer, the same methods can be used, with appropriate adjustment in the relationships for I_{D0} and V_P, as given in Eqs. (8-38) and (8-39).

Given the channel thickness

$$t_{CH} = \left[(t_{ox})^2 + \frac{2\epsilon V_{GS}}{q N_{CH}}\right]^{1/2} - t_{ox} \qquad (18\text{-}38)$$

then the total charge Q in the channel is

$$Q = q N_{CH} A \left\{\left[(t_{ox})^2 + \left(\frac{2\epsilon V_{GS}}{q N_{CH}}\right)\right]^{1/2} - t_{ox}\right\} \qquad (18\text{-}39)$$

and the effective capacitance becomes

$$C = \frac{dQ}{dV_{GS}} = \frac{\epsilon A}{\left[(t_{ox})^2 + \left(\dfrac{2\epsilon V_{GS}}{qN_{CH}} \right) \right]^{1/2}} \qquad (18\text{-}40)$$

For small values of t_{ox},

$$C = A \left(\frac{\epsilon q N_{CH}}{2V_{GS}} \right)^{1/2} \qquad (18\text{-}41)$$

in agreement with the JFET pn junction calculation of Eq. (18-37). For large values of t_{ox},

$$C = \epsilon A / t_{ox} \qquad (18\text{-}42)$$

representing a parallel-plate capacitor formed between the conducting layer and the channel.

Example 18-4   D-MOSFET Capacitance

A Si D-MOSFET is designed with the same output characteristic as derived for the JFET of Example 18-3. If $t_{ox} = 1.0$ μm, what is the gate-channel capacitance of the device for $V_{GS} = 0$? Assume $\epsilon_r \cong 4.0$ for the SiO$_2$ layer.

Solution   From Eq. (18-40),

$$C = \frac{(4 \times 8.85 \times 10^{-12} \text{ F/m}) (11 \times 10^{-6} \text{ m}) (3.0 \times 10^{-6} \text{ m})}{1.0 \times 10^{-6} \text{ m}}$$

$$\cong 0.001 \text{ pF} \qquad \blacksquare$$

18-5   E-MOSFET DESIGN

The operation of the E-MOSFET is described in Section 8-13. From Eq. (8-49), trial values for I_{D0} and V_T can be used to find acceptable values for the term $\mu_n \epsilon z / L t_{ox}$. (The value of μ_n used in this term is, in general, different from the typical value for a bulk semiconductor due to the interaction with the oxide layer.) The effective gate-channel capacitance is that associated with a parallel-plate capacitor:

$$C = \epsilon z L / t_{ox} \qquad (18\text{-}43)$$

Example 18-5   E-MOSFET Capacitance

Find the capacitance for an E-MOSFET with length $z = 11$ μm, $L = 3.0$ μm, and $t_{ox} = 0.10$ μm.

Solution   From Eq. (18-43),

$$C = \frac{(4 \times 8.85 \times 10^{-12} \text{ F/m})(11 \times 10^{-6} \text{ m}) (3.0 \times 10^{-6} \text{ m})}{0.10 \times 10^{-6} \text{ m}}$$

$$\cong 0.01 \text{ pF} \qquad \blacksquare$$

The above discussion is concerned with semiconductor device properties and the tradeoffs that affect device design. As seen below, applications of these devices in realistic subsystems and systems introduce many additional restrictions on circuit performance.

Figure 18-10 compares the development of various integration technologies. These technologies are explained in further detail in the following sections.

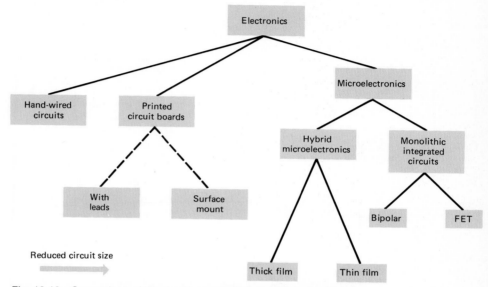

Fig. 18-10 Comparison of the development of electronics technologies.

Figure 18-11 shows the developmental path that has been followed by electronics packaging concepts. In the right-hand column of the figure are the building block components for electronics circuits. As shown, the most dense building blocks are the bare IC chips themselves, with complex circuits produced on a single crystal. Next lower in density are the bare diode and transistor chips, followed by the print and fire resistors of hybrid microelectronics (discussed below), miniaturized discrete passive components, and standard passive components.

In the center column of Fig. 18-11 are the individually packaged IC, transistor, and diode chips. As noted in the following discussion of printed circuit boards (Section 18-7), these packages can have leads or can be designed for surface-mount applications. The carrier package for a single chip can be viewed as a "minihybrid." The actual chip is attached to a substrate, wire bonded to external connectors, and encapsulated.

The left-hand column of Fig. 18-11 shows a range of packaging concepts. At the bottom (the least dense) are circuit breadboards using standard passive components and packaged chips. A higher density is achieved by the leaded printed circuit board, which uses chip carriers with leads and miniaturized discrete passive components. The surface-mount printed circuit board represents an additional increase in density, shifting to leadless chip carriers.

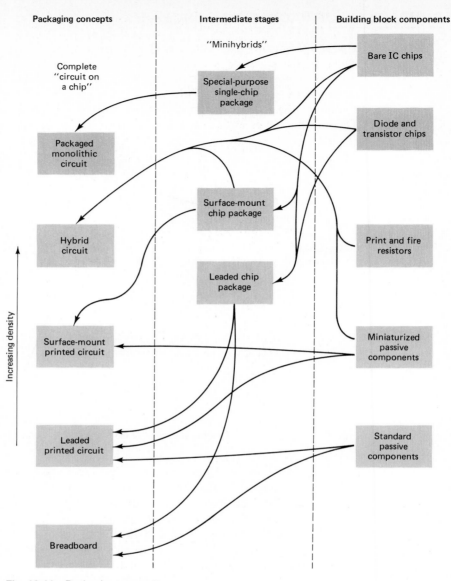

Fig. 18-11 Packaging concepts.

Hybrid microcircuits use bare IC and transistor chips, "print and fire resistors" fabricated directly on the hybrid substrate, and miniaturized discrete device components. In addition, some hybrids also use ICs and transistors in surface-mount packages.

Finally, the most dense circuits are fabricated when an entire circuit is developed on a single IC chip and a special-purpose single-chip package (a minihybrid) is used.

The electronics designer is faced with many different integration alternatives in deciding how to best meet performance objectives. A comprehensive systems approach to design requires an ability to appreciate these alternatives sufficiently to merge manufacturing tradeoffs with those associated with desired circuit function.

Thermal and mechanical properties are also important in contrasting packaging concepts. Thermal conduction and convection processes determine the maximum power-dissipation limits and cooling requirements for the circuit. Several aspects of thermal conduction and mechanical limits are considered below.

18-7 PRINTED CIRCUIT BOARDS

Printed circuit boards are widely used as a mounting surface for ICs (in DIP packages) and other building block elements.

A typical printed circuit board is shown in Fig. 1-13. Copper pathways are formed on a fiberglass board to act as circuit interconnections. Circuit components are then mounted on the board and soldered to the copper conductors to complete the circuit.

The process used to create this board is shown in the flowchart of Fig. 18-12. In step 1, the board is laminated with a thin layer of copper. In step 2, holes are drilled through the copper and board to provide for the future attachment of components.

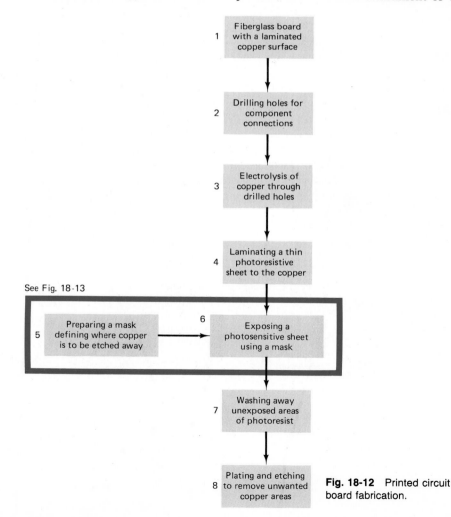

Fig. 18-12 Printed circuit board fabrication.

735

Electrolysis is then used to build up a continuous layer of copper from both sides through the holes (step 3). At this stage of development, no circuit pattern has yet been introduced. The copper sheets cover both sides of the substrate, except where holes have been drilled.

The circuit pattern can be created on the board through a photolithographic process (steps 4 to 7). The process starts with preparation of a large-scale layout of the pattern that is to appear on the board [step 5 of Figs. 18-12 and 18-13(A)]. The layout is made of two types of material, one that transmits light and another that is opaque. A photoreduction printer creates a reduced-size photographic print of the layout. Typically, the photoreduction is printed on a flexible sheet of transparent plastic (mylar). The original transparent areas on the large-scale drawing become black (opaque) on reduction.

The reduction is used as a mask to define where the copper is to be etched away from the printed circuit board substrate and where it is to remain to provide circuit interconnections. A thin *photoresist* is heat-bonded to the copper surface (step 4). The photosensitive layer is then exposed to blue or ultraviolet light through the mask to transfer the desired pattern [step 6, Figs. 18-12 and 18-13(B)]. The exposed photoresist layer becomes softened and is washed away, leaving a coating only where no exposure has taken place (step 7).

In step 8, etching removes the unwanted copper areas. The photoresist remaining on the substrate prevents chemical interaction with the copper on areas that are to remain plated. After etching, the original layout now appears in copper on the surface of the board. The copper pattern includes interconnecting lines that replace wires and larger terminating areas, called *pads,* where components are to be attached. The previously drilled holes are placed at the center of these pads.

Circuit components can be manufactured with leads for attachment to printed circuits. By using automatic equipment, components are inserted into the board with leads extending through the prepared holes. The board is passed over a flowing stream of hot solder to secure the electrical connections. The circuit is then complete, with the board as a mechanical base, copper conductors for interconnections, and individual components soldered to the conductors.

The reference type of printed circuit board discussed above is in wide use. However, there are also numerous variations available. For complex circuits, multilayer

Fig. 18-13 Printed circuit board photolithography. (A) Step 5, photoreduction. (B) Step 6, exposure of photosensitive sheet.

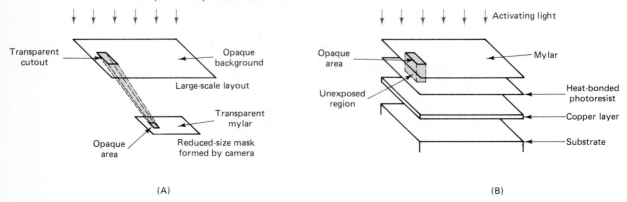

(A)

(B)

boards are often used, with up to six or eight printed circuit boards bonded to produce circuit pathways. Other types of boards and conducting materials are also applied for special purposes.

At present, substantial interest is being given to surface-mount printed circuit boards. Circuit components are no longer manufactured with leads, but instead make use of little flanges, as shown in Fig. 18-14. Holes are not drilled in the printed circuit board. Components are soldered directly to pads that have been prepared. This surface-mount printed circuit board can achieve about a 50 percent reduction in size below the leaded printed circuit board for a typical circuit. Figure 18-15 illustrates the high IC density that can result.

Fig. 18-14 Automated placement of surface mount IC packages on a printed circuit board. (Courtesy of Universal Instruments Corporation)

Fig. 18-15 Eight megabyte memory board fabricated using surface mount technology. (Courtesy of Texas Instruments Incorporated.)

Numerous tradeoffs are encountered in selecting a printed circuit board configuration as part of the design process. Multilayer boards and surface-mount alternatives must be evaluated by the designer in order to produce a system that best meets performance objectives. Circuit structure, cost, the availability of materials and components, the required capital investment, and expected volume of sales all enter into the configuration that is selected.

18-8 THICK-FILM HYBRID MICROELECTRONICS

Hybrid microcircuits represent a further evolution of technology toward smaller, more reliable circuits (Fig. 1-14). The hybrid circuit is built on a substrate that usually consists of a thin, flat layer of ceramic. Alumina (Al_2O_3) is a commonly used material for this purpose. To form a thick-film hybrid, multiple layers of material are built up on this substrate using a series of silk screen processes and electrical pastes, as shown in Fig. 18-16.

In the silk screen process, layers of material are deposited on a substrate through a very-fine-mesh stainless-steel screen. To create the desired printing pattern on a screen, a layer of photographic emulsion is bonded to the screen, then exposed to ultraviolet light through a mylar photoreduction of the circuit pattern (or topography) that is desired, as shown in Fig. 18-17. An emulsion is used that hardens when exposed to ultraviolet light and remains soft elsewhere.

The unexposed areas of the emulsion are then washed away to leave the desired pattern. This photolithographic process produces a screen master that can be used to print the desired material on a substrate. Each screening cycle uses a paste specifically designed to achieve a desired circuit function and a screen master that can apply the paste to selected regions of the substrate.

A thick-film paste is a complex liquid that can include many different constituents, including finely ground metal particles, glass powders, and an organic liquid. The powders are mixed and the liquid is added to provide the desired viscosity and screening properties. To produce a dielectric paste, only glass particles are added. To produce a conducting paste, a low-resistivity metal and glass are added. To produce a resistive paste, combinations of metals and metal oxides are used with the glass powder. The *rheology* of the paste (its performance under applied temperature and pressure) is obviously important. The ideal paste (1) flows through the screen under light pressure, (2) ceases to flow out of the target area when it reaches the surface of the substrate, and (3) allows the surface of the paste to level out and remove all shape distortions produced by the screen. The tension of the screen, the viscosity of the paste, and the speed of the application process are all important variables in producing the desired film properties. Modern hybrid production facilities use complex silk screen printers that are able to control these variables and create a reproducible product.

As shown in the flowchart of Fig. 18-16, development of a hybrid microcircuit begins with preparation of a screen master. The first printed layer usually forms desired conductor interconnections. For this case, conducting paste is applied through the silk screen master and the circuit is allowed to sit for 10 to 15 minutes until its surface levels. Then it is ready for further processing.

The conducting film on the substrate is dried in an oven at about 150°F. As the

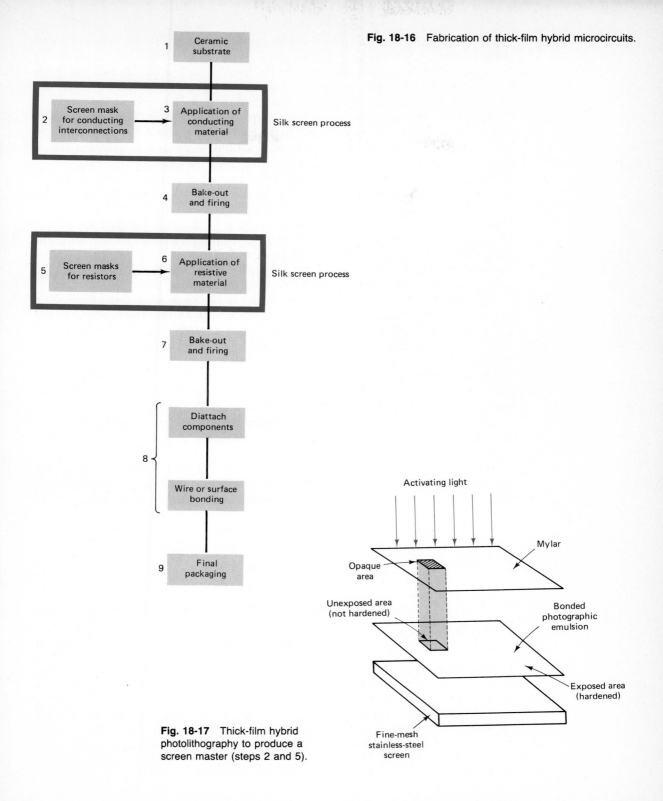

Fig. 18-16 Fabrication of thick-film hybrid microcircuits.

1 Ceramic substrate

2 Screen mask for conducting interconnections → 3 Application of conducting material   Silk screen process

4 Bake-out and firing

5 Screen masks for resistors → 6 Application of resistive material   Silk screen process

7 Bake-out and firing

8 Diattach components

Wire or surface bonding

9 Final packaging

Fig. 18-17 Thick-film hybrid photolithography to produce a screen master (steps 2 and 5).

Activating light

Mylar

Opaque area

Unexposed area (not hardened)

Bonded photographic emulsion

Exposed area (hardened)

Fine-mesh stainless-steel screen

paste is heated, most of the organic binder bakes out slowly, leaving behind the particles of glass and metal.

The circuit is now ready for firing in a furnace. Usually a belt furnace is used to move the circuit through a carefully controlled temperature profile. The result is a conducting material that is attached to the substrate by a glass-ceramic bond.

Once the conducting interconnections have been applied and fired, the resistor pastes can be applied using the same type of screening process. To provide insight into the nature of the circuit that is being produced, consider what happens after a resistive paste is baked out and ready for firing in the furnace.

As the temperature increases, any remaining organic filler is baked out of the film. Because the smaller metal particles were originally distributed around the larger glass particles, the metal particles are formed in chainlike arrays through the material. As the glass particles begin to *sinter*[2] (link), the particles are further pushed into a chain-like structure. When the temperature is sufficiently high, the glass *wets* the surfaces of the metal particles and forms thin layers of glass between the metal particles. *Liquid-phase sintering* takes place among the metal particles, introducing impurities into the thick glass layers between particles.

The film is now cooled slowly. The interior of the resistor consists of chainlike arrays of metal particles spaced through the material, with thin layers of glass (and impurities) between the particles. The glass forms a matrix structure that interlocks with the metal and mechanically attaches the metal to the ceramic substrate. Such materials are often called *cermets*.

The resistive properties of the final film are controlled by the nature of the metals used, the sintering characteristics, the shape of the interwoven metal structure, and the geometric shape of the film and substrate surfaces. Current research indicates that the dominant conduction process is a quantum-mechanical tunneling across the glass layers between particles, which is affected by the impurities that were introduced by diffusion.

The resistivity of the final product can be varied over a wide range by changing the fraction of metal particles in the composite mixture. And the complex structure and conduction processes result in a low temperature coefficient of resistance (TCR) that is desirable.

Thick-film resistors are usually printed in a rectangular shape. With a length L, width W, and thickness t, the resistance can be written as a function of the resistivity ρ of the material

$$R = \rho L/A = \rho L/Wt \qquad (18\text{-}44)$$

The thickness t is specified by the manufacturer of the resistive paste that is being used. Therefore,

$$R = (\rho/t)(L/W) \qquad (18\text{-}45)$$

The ratio L/W defines the number of *squares* of material present in the resistor. The ratio L/W can be adjusted to give a range of resistor values for a given ρ and t. A range of $1/4 < L/W < 4$ is typical.

Resistive pastes are marketed in terms of the ρ/t ratio they provide. This ratio is prescribed in units of ohms per square ($\Omega/\square$).

---

[1] The sintering and wetting processes are described in Section 18-14.

Example 18-6 Thick-Film Resistors

A resistive paste is available with 1.0 kΩ/□. A 2.5-kΩ thick-film resistor is to be produced with a width of 50 mils (1 mil = 0.001 in.). (a) How long should the resistor be? (b) If $t = 25$ μm according to specifications, what is the resistivity of the paste?

Solution (a) From Eq. (18-45),

$$L = \frac{2.5 \text{ k}\Omega}{1.0 \text{ k}\Omega}(50 \text{ mils}) \cong 1.3 \times 10^2 \text{ mils}$$

(b) By definition,

$$\rho/t = 1.0 \text{ k}\Omega$$

$$\rho = (1.0 \text{ k}\Omega)(25 \ \mu\text{m}) = 2.5 \times 10^{-2} \ \Omega \cdot \text{m}$$

This value of ρ lies in the semiconductor region (refer to Fig. 2-1). ∎

After all conducting, resistive, and dielectric layers are completed, circuit components are attached to the ceramic substrate. Typically, bare (unpackaged) transistors and integrated circuits can be glued with special epoxies directly to the substrate. Using *wire bonders,* very fine interconnections are made between these semiconductor chips and the surrounding circuit. (This type of interconnection is illustrated in Fig. 1-10.) Miniaturized resistors and capacitors (called "chip" components) can also be attached to the circuit using conducting epoxies or wire bonding. The final product is a small, highly reliable circuit that is feasible for manufacturing in small- or large-scale quantities.

The designer thus is provided with another area of selection, since tradeoffs exist among combinations of hybrid microcircuits and the alternative types of printed circuit boards. As technology develops, more variations on fabrication methods are becoming available, blurring distinctions and producing a more integrated continuum of choices.

18-9 THIN-FILM HYBRID MICROELECTRONICS

The use of thin-film technology is illustrated in Fig. 18-18. A substrate is placed in a vacuum container and layers of material are formed by *evaporation* or *sputtering*. Evaporation involves heating the upper electrode so that it emits atoms of the desired material. Sputtering involves forming ions in the container and applying an external voltage that causes the ions to hit the upper electrode and knock out atoms of the desired material.

Additive thin-film technology uses a metal mask to restrict areas where material lands on the substrate, according to a desired pattern. Substractive thin-film technology coats the entire substrate, applies a photosensitive material, exposes through the mask, and etches away portions of the material.

Thin-film technology minimizes the sideways migration of deposited material and can thus be used to produce fine circuit detail. However, where the thick-film technology can produce almost any desired resistor value, the possible resistance values using thin-film technology are more limited.

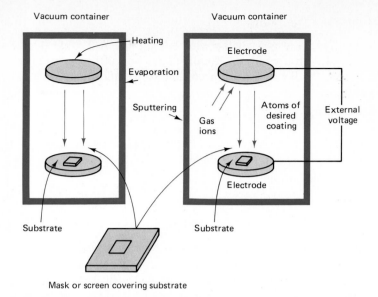

Vacuum container

Heating

Evaporation

Sputtering

Electrode

Atoms of
desired
coating

Gas
ions

External
voltage

Electrode

Substrate

Substrate

Mask or screen covering substrate

Fig. 18-18 Thin-film technology.

After the layered substrates are completed, discrete components are added with appropriate mechanical and electrical bonding and interconnection technologies. Finally, the entire unit is packaged for the anticipated environmental use. Thin- and thick-film hybrid microcircuits can be compared in terms of component characteristics, fabrication properties, systems performance, and cost.

18-10 DESIGN OF SUBSYSTEMS AND SYSTEMS USING PRINTED CIRCUIT BOARDS AND HYBRID MICROCIRCUIT TECHNOLOGY

Printed circuit boards and hybrid microcircuits represent alternative integration strategies for producing electronic subsystems and systems. Many fabrication alternatives exist for the designer to consider, and many different parameters must be evaluated in the tradeoffs that result.

Every fabrication alternative requires the selection of a substrate for use. The circuit components are then assembled on this substrate using leaded components or surface-mount technology. Printed circuit boards are usually made of polymers and are available in multilayer format (up to 20 to 30 layers) if required. Alternatively, hybrid microcircuits use alumina (Al_2O_3), ruby, or polymer substrates. The type of substrate affects the electrical, thermal, and mechanical properties of the resultant circuit.

The methods used to produce conducting runs (thick or thin film, in copper, silver, or gold) and types of resistors to be fabricated for hybrids (palladium- and ruthenium-based systems) are also important fabrication considerations.

Components to be mounted on the substrate are available in a wide variety of packages. Bare IC chips can be used as components for hybrid microcircuits, whereas single and multipackaged ICs in leaded or surface-mount configurations can be used for printed circuit and hybrid integration strategies. Discrete active and passive devices are available in a wide range of individual packages and combinations. Assembly uses highly automated equipment to reduce costs (Fig. 18-14) and circuit design reflects the capabilities and limitations of the equipment.

The types of bonding and interconnection technologies to be used affect both assembly and performance. Circuit design must often adapt to the bonding properties of the materials that are applied during fabrication.

In summary, the performance of an electronic system is directly influenced by the materials and processes that are used during fabrication, device characteristics, and methods of integration and packaging, as well as the circuit configurations selected.

18-11 THERMAL AND MECHANICAL DESIGN CONSIDERATIONS

In addition to the doping level and geometry constraints associated with the production of semiconductor devices, it is important to recognize that thermal and mechanical constraints can significantly shape final fabrication decisions. As circuits become more dense, adequate power-dissipation capability becomes a higher priority. The thermal conduction properties of the materials and packages used in device fabrication can be equally as important as are the electrical properties. Section 18-12 uses simplified models to develop an awareness of some of the thermal considerations that are encountered.

The mechanical integrity of the device and circuit is a fundamental requirement and one that is often difficult to obtain. Although an IC itself can place almost no restriction on mechanical performance objectives, the bonding and packaging often does. A complete systems approach to design requires an integration of the electrical, thermal, and mechanical properties of the device. Section 18-13 introduces some of the mechanical considerations that are often encountered in electronics.

18-12 THERMAL CONSIDERATIONS

The selection of a preferred technology for a given application often depends on thermal considerations. As circuits become more dense, heat dissipation becomes more of a problem. At the same time, the small dimensions involved can make it difficult to solve the heating problem in a way that maintains the advantages of the packaging technology.

As discussed in Chapters 15 and 16, there has been a strong move toward digital circuits that can operate with minimal power demands. The advantages of CMOS are discussed in this connection. The same need to reduce power dissipation also defines those components and circuits that are feasible in analog circuit applications.

A working knowledge of basic thermal relationships is thus important to the circuit designer. This section provides an overview of some of the considerations that are involved when thermal issues arise.

Of particular interest in electronics are design constraints associated with the thermal expansion, the thermal conductivity, the specific heat, and the melting point of the materials used. Thermal expansion is important when two materials are bonded and subjected to temperature variations. If they expand and contract at different rates, the bond between the materials can be broken or cracks and discontinuities can be introduced in the materials.

FIGURE 18-19
Typical Values for the Linear
Coefficient of Thermal Expansion α_ℓ

Material	$\alpha_\ell (\times 10^{-6 \circ} K^{-1})$
Fused silica	0.55
Alumina (Al_2O_3)	8.8
Copper wire	14

The *linear coefficient of thermal expansion* α_ℓ is defined as

$$\alpha_\ell = \frac{\Delta \ell}{\ell} \frac{1}{\Delta T} \tag{18-46}$$

where $\Delta \ell / \ell$ is the fractional change in the length of the material associated with a temperature change ΔT. Typical values for α_ℓ range from 10^{-6} to $10^{-5}/°K$ for metals and ceramics and 10^{-4} to $10^{-5}/°K$ for polymers. A difference factor of 10 to 100 can thus easily exist in the thermal expansion properties of two different materials. Typical values of α_ℓ are shown in Fig. 18-19.

Example 18-7 **Thermal Expansion**

A layer of fused silica is bonded to an alumina substrate. Over a 4.0 cm distance, what maximum difference in length is associated with the material if $\Delta T = 100°K$? See Fig. 18-20.

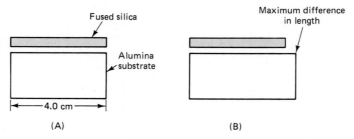

Fig. 18-20 Material arrangement for Example 18-7. (A) Before heating. (B) After heating.

Solution For fused silica,

$$\Delta \ell_1 = \alpha_\ell \ell \Delta T = (0.55 \times 10^{-6}/°K)(0.04 \text{ m})(100°K)$$
$$\cong 2.2 \ \mu m$$

For alumina,

$$\Delta \ell_2 = \alpha_l \ell \Delta T = (8.8 \times 10^{-6}/°K)(0.04 \text{ m})(100°K)$$
$$\cong 35 \ \mu m$$

The maximum difference in length is

$$\Delta \ell_{\text{diff}} = \Delta \ell_2 - \Delta \ell_1 \cong 33 \ \mu m$$

For the above example, cracks can develop in the fused silicon layer due to the stretching force exerted by the alumina substrate. ∎

Thermal conductivity is an important determinant of the ability of a circuit to conduct heat away from power-dissipating devices. The amount of heat that flows through a material depends on the temperature gradient and the material characteristics. The *thermal conductivity k* of a material is defined as

$$\frac{1}{A}\frac{\Delta Q}{\Delta t} = k\left(-\frac{\Delta T}{\Delta x}\right) \tag{18-47}$$

where $\Delta Q/\Delta t$ is the power flow through the material, A is the cross-sectional area, ΔT is the temperature difference across the material, and Δx is the material thickness. The minus sign indicates that the temperature decreases as you move away from the source of heat.

Figure 18-21 gives typical values for k. By comparing Figs. 2-1 and 18-21 observe that the electrical conductivity $\sigma = 1/\rho$ and the thermal conductivity k correlate with each other. High values for one are linked to high values for the other. Good electrical conductors are also good thermal conductors. This relationship makes it difficult to find substrates that are both good electrical insulators and good heat conductors.

FIGURE 18-21
Values for the Thermal Conductivity *k*

Material	k (W/cm·°K)	Material	k (W/cm·°K)
Silicon	1.4	Tin	0.65
Germanium	0.6	Lead	0.33
Silver	4.2	Alumina (Al_2O_3)	0.3
Copper	4.0	Polymers:	
Gold	3.0	cellular	0.2–0.4
Aluminum	2.4	noncellular	0.02–0.04
		Glass and mica	0.008
		Vacuum	0

Example 18-8   **Thermal Conductivity**

A power flow of 20 W exists through a material with $A = 10$ cm^2 and $\Delta x = 1.0$ mm. If the design objective is to limit ΔT to 20°C, what value of k is required?

Solution   From Eq. (18-47),

$$k > \frac{-\dfrac{1}{A}\dfrac{\Delta Q}{\Delta t}}{\dfrac{\Delta T}{\Delta x}} = \frac{\dfrac{1}{(0.10 \text{ m})^2}(20 \text{ W})}{\dfrac{20°\text{K}}{0.001 \text{ m}}}$$

$$k > 0.1 \text{ W/m·°K} = 0.001 \text{ W/cm·°K}$$

From Fig. 18-21, both polymers and alumina are realistic nonconducting candidates for the substrate material. ∎

The specific heat of a material is used to find out how the temperature of a material changes as heat is absorbed or released by a material. Where thermal conductivity is concerned with heat flow through materials and the resultant temperature gradient, specific heat is concerned with temperature increases for the whole material sample due to the absorption of energy by the sample.

The *specific heat c* of a material is defined so that

$$\Delta Q = mc\,\Delta T \tag{18-48}$$

where ΔQ is the heat absorbed by the material, m is the mass of the material, c is the specific heat, and ΔT is the resultant increase in temperature. Values of c are given in Fig. 18-22.

A final thermal property of concern is the *melting point* of materials. During fabrication of components and circuits, heat is often applied and the effects on materials must be anticipated. Figure 18-23 gives the melting points for some materials of interest.

FIGURE 18-22
Values for the Specific Heat c

Material	c(kJ/kg ·°K)
Brass	0.38
Aluminum	0.88
Copper	0.38
Glass	0.78
Water	4.2 (by definition)
Solid polymers	1–2

FIGURE 18-23
Melting Points

Material	Melting point (°C)
Polymers:	
Polyethylene	140
Nylon	200–300
Semiconductors:	
Silicon	1414
Germanium	937
Ceramics:	
Alumina (Al_2O_3)	2000
Glass	200–1600
Metals:	
Aluminum	660
Copper	1084
Tin	232
Gold	1064
Lead	328

Some alloys have interesting melting point properties. For example, consider the case of solder, which is widely used to join conductors together in circuits. The properties of solder can be described using a phase diagram, as shown in Fig. 18-24. In this diagram, temperature is plotted as a function of the ratio of lead (Pb) and tin (Sn) in the alloy. In region I, both metals are liquid. In regions IIa and IIb, partial liquid mixtures are obtained. In region III, the alloy is solid. Full melting is obtained at a minimum temperature corresponding to a specific ratio of materials. This is called a *eutectic mixture* and for solder exists at about 180°C.

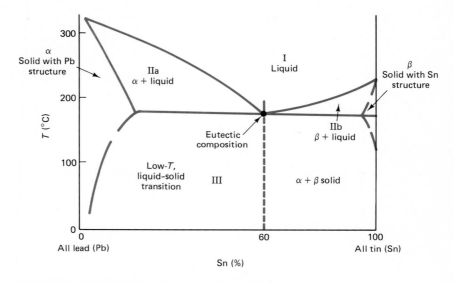

Fig. 18-24 Phase diagram for soft solder, a eutectic alloy.

By drawing phase diagrams for combinations of materials, mixtures can be obtained with lower melting points than those of the individual materials. The result is often of practical significance in the fabrication of circuits.

18-13 MECHANICAL PROPERTIES OF MATERIALS USED IN MICROELECTRONICS

The mechanical properties of the circuits that are designed are important to prevent mechanical failure. One of the most useful ways to study the mechanical properties of solid materials is illustrated in Fig. 18-25. The *stress* (force per unit area) that is applied to a material is shown on the vertical axis and the resulting *strain* (the fractional shift in length $\Delta\ell/\ell$) is shown on the horizontal axis. Curve A is for a material that linearly lengthens under an increasing applied force until it suddenly fractures (breaks). Since $\Delta\ell/\ell$ is proportional to F/A, this is called an elastic material. If the force is reduced at any point before fracture, the material recovers and returns to its initial state.

Many ceramics and other brittle materials have stress/strain curves of this shape due to the directional bonds that hold them.

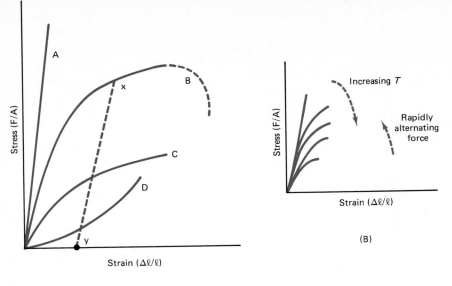

Fig. 18-25 Mechanical response of materials. (A) Types of materials. (B) Effect of temperature.

The relationship

$$\frac{\Delta\ell}{\ell}E = \frac{F}{A} \tag{18-49}$$

can be defined, where the proportionality constant E is Young's modulus. For curve A, E is a constant.

Curve B shows a material that responds elastically to an initial applied force, then begins to experience plastic deformation. If released from the applied force at point x, the material returns to point y, not to the origin. The material has been permanently stretched or bent. If the force continues past x, the material eventually fractures. Many metals show this behavior. The plastic shaping is due to the nondirective bonds.

Curve C shows a material that stretches or bends readily under an applied stress and permanently distorts under a small force. Many polymers fall into this category. Curve D describes a soft, rubbery material.

Typical values for Young's modulus E are

- Metals: 35 to 350×10^9 N/m^2
- Ceramics: Up to 1000×10^9 N/m^2
- Polymers: 3 to 4×10^9 N/m^2

Specific values for E are given in Figure 18-26.

If a material is heated, the shape of the curve shifts, as shown in Fig. 18-25(B). The material becomes less brittle and more ductile. If the force is alternated rapidly, a fatigue phenomenon is observed and the curve shifts upward. The material is less able to respond to the rapidly changing applied force.

FIGURE 18-26
Typical Values for Young's
Modulus E

Material	E ($\times 10^9$ N/m²)
Alumina (Al_2O_3)	300
Gold	80
Copper	100
Aluminum	70
Tungsten	400
Fused silicon	700

18-14 BONDING

The fabrication of electronic circuits often requires the bonding of different types of materials. In previous chapters, n-type and p-type semiconducting regions are joined by using a single crystal structure and varying the types of impurities introduced. The two semiconductor regions are held together by the covalent atomic bonds in the crystal.

A broader range of bonds is now to be considered, including those that enable the joining of metals, ceramics, and polymers in a variety of different arrangements. Material bonding can be formed through primary linkages (metal, ionic, or covalent bonds) or secondary linkages (Van der Waals bonds). Secondary bonds are important when chemical reactions do not take place when the two materials are placed in contact.

If two polycrystalline materials are placed in contact, they usually do not bond because irregularities in the two surfaces prevent close contact at the atomic level, as shown in Fig. 18-27. Bonding forces are not strong enough at a distance. Bonding often requires melting one of the materials. As the grains begin to lose their shape and a liquid forms, a flow fills in the surface gaps. Once the materials are in close contact, bonding takes place.

Bonding properties can be understood by considering the surface energies associ-

Fig. 18-27 Surface contact between two polycrystalline materials.

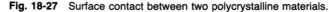
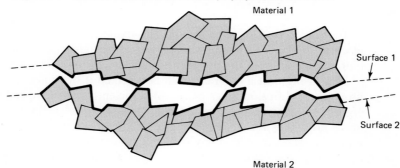

ated with a material. Consider the nature of solid–solid bonds. Work must be done to separate portions of the material, breaking the bonds that hold the two portions together. Therefore, the atoms along the separation boundary have a higher energy than their internal neighbors. The oxidation process, so important in electronics, can be understood from this perspective.

For metal surfaces in an oxygen atmosphere, electrons attached to surface atoms in the metal have higher energy states than those in the interior. When an oxygen atom approaches, it is predisposed to accept an extra electron. The oxygen atom tends to attach itself to the surface, lowering the total energy of the metal atoms and oxygen atom. This effect explains the formation of oxides on the surfaces of metals.

Such oxides can prevent the desired bonding of metals and can reduce electrical conduction across surfaces. To solder two metals together, a rosin is used with the eutectic mixture of tin and lead. Rosin is a turpentine-based additive that chemically removes the oxygen atoms from the metal surface to allow the formation of a metal–metal bond.

Bonding between solids and liquids can be discussed in terms of the contact angle θ that is formed between the solid surface and liquid, as shown in Fig. 18-28. To form a bond, the liquid must *wet* the solid surface. If the liquid completely wets the surface, $\theta \to 0°$. If it does not wet the surface at all, $\theta \to 180°$. The value of θ depends on the energy relationships between the surface atoms on the solid and the surface atoms on the outside of the liquid. If impurities are attached to the solid surface, the angle θ can be strongly affected. When solder is applied to a piece of wire without the use of rosin, the solder can refuse to wet the wire ($\theta \to 180°$) because of the oxide layers present.

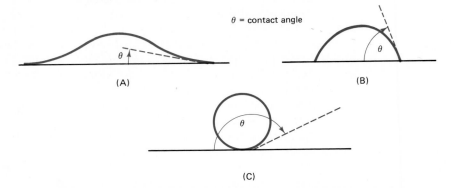

Fig. 18-28 The liquid–solid wetting process. (A) $\theta \longrightarrow 0$ for complete wetting. (B) Intermediate case. (C) $\theta \longrightarrow 180°$ for nonwetting.

If rosin is then used to remove the oxide layer, wetting and bonding can take place ($\theta \to 0°$). The bonding properties of various types of materials are an important consideration in the design and manufacture of electronics circuits.

The strength of a bond can also be affected if the junction temperature reaches a level where diffusion takes place. At sufficiently high temperatures, the thermal energy of the material is great enough to cause some of the atoms in the material to move around in the lattice structure. These energetic atoms can move *interstitially* (between other atoms) or by filling in where impurity vacancies are found in the crystal. As the result of diffusion, bond characteristics can change between solids and between solid–liquid interfaces. In microelectronics, bond strength is often measured by the use of a pull test that measures the force required to break the bond.

To form many materials that are important in electronics, one type of bonding is due to sintering. This is an important process in the fabrication of thick-film resistors. Consider what happens when solid particles of the same material are brought into contact. For large particles at low temperature, no interaction is observed. At sufficiently high temperatures, particles melt and flow together. However, for temperatures between the melting temperature T_m and about $T_{m/2}$, another type of bonding process is observed due to diffusion.

For two separate identical particles, the total volume of both particles is $V_1 + V_1 = 2V_1$, and the combined surface area of the two particles is $2(4\pi r_1^2)$. Compare this result with the situation that exists if the two particles are melted to form a single particle of volume $2V_1$. The surface area of this larger particle would be $4\pi r_2^2$. Therefore,

$$2V_1 = 2\left(\frac{4}{3}\pi r_1^3\right) = \frac{4}{3}\pi r_2^3 \tag{18-50}$$

so that

$$r_2 = r_1 2^{1/3} \tag{18-51}$$

and

$$\frac{\text{Surface area of two particles}}{\text{Surface area of combined particle}} = \frac{2(4\pi r_1^2)}{4\pi r_2^2}$$

$$= 2^{2/3} \cong 1.6 \tag{18-52}$$

Therefore, the total surface area is reduced if the two particles begin to combine. Since the surface energy states have extra energy, the combined form has fewer surface states and reduced energy. Thus, from thermodynamics, there is a tendency for the two materials to merge.

At low temperatures, diffusion processes are not strong enough to provide a mechanism for the growing together, or sintering, of particles. But at higher temperatures, diffusion grows rapidly. The maximum density of surface states exists where the two particles touch. Therefore, diffusion activity is a maximum at this point. Atoms migrate from the interior and along the surface of each particle to this point. A *neck* begins to grow, bonding the two particles. The degree of sintering depends on the duration of the elevated temperature.

Liquid-phase sintering can take place when two solid particles do not touch but are separated by a liquid that wets both surfaces. The diffusion of atoms can take place through the liquid interface, changing the liquid content and creating a bond.

18-15 TECHNOLOGIES, PROCESSES, AND MATERIALS IN ELECTRONICS DESIGN

This chapter describes some of the technologies and processes being used to fabricate electronic circuits and some of the tradeoffs that are important to device and circuit performance. Because of the close relationships that exist between circuit function, device parameters, and manufacturing technology, the designer must be able to understand the problems and solution strategies associated with each area.

Questions   18-1. What relationships must hold to produce $(1 - \gamma) \ll 1$?

18-2. What circumstance can produce different values of n_i within different portions of a BJT?

18-3. Why does β_{dc} drop at low and high current levels?

18-4. What are the advantages of doping the collector less than the base?

18-5. What is a one-sided abrupt junction?

18-6. Why is it necessary to set an upper limit on the bulk resistance of the BJT?

18-7. For a one-sided abrupt junction, when is avalanche breakdown a major problem?

18-8. Why is it desirable to have a large value for the Early voltage V_E?

18-9. Why are the device capacitances calculated in this chapter usually lower than those of actual devices?

18-10. For a JFET, how do the parameters I_{D0}, V_P, and R in the transition region change as the base width increases?

18-11. For a JFET, why is it necessary for the voltage drop within the gate to be small with respect to the scale of values for V_{GS} and V_{DS}?

18-12. How can the results of Section 8-3 for the JFET characteristic curve and Section 8-13 for the E-MOSFET characteristic curve be used to describe the saturation transfer characteristic for the D-MOSFET?

18-13. Why is it important to gain an understanding of printed circuit boards and hybrid microcircuits?

18-14. Why is it important to gain an understanding of thermal and mechanical considerations for electronic subsystems and systems?

Answers   18-1. To obtain $(1 - \gamma) \ll 1$, the device design must observe the constraint

$$\sigma_p W_B e^{\Delta \mathcal{E}_g / kT} \ll \sigma_n W_E$$

Constraints are thus placed on the ratios W_B/W_E and σ_p/σ_n and the band-gap narrowing $\Delta \mathcal{E}_g$.

18-2. High doping levels can produce $\mathcal{E}_g$ values that are dependent on doping. Therefore, regions of the BJT can have different effective band gaps, producing changes in n_i.

18-3. At low current levels, carrier recombination in the depletion regions reduces β_{dc}, whereas at high current levels the increase in excess carrier density in the base increases the base-emitter carrier flow and reduces β_{dc}.

18-4. By doping the collector less than the base, most of the collector-base depletion region extends into the collector, enabling a thinner base region to be used. Because the depletion region in the base is less dependent on the Early voltage, V_E for the BJT increases, which is another advantage.

18-5. A one-sided abrupt junction describes those junctions for which one side is doped much more strongly than the other, with the transition between sides small with respect to the diffusion lengths involved. For such cases, the properties of the junction are dominated by the doping on the least-doped side.

18-6. The bulk resistance of a BJT helps determine the slope of the characteristic curves, power dissipation in the device, and the minimum voltage level across the device for a given current. If the resistance is too high, the device characteristics deviate significantly from the ideal case discussed in previous chapters, power dissipation is excessive, and the saturated applications of the device are limited.

18-7. For a one-sided abrupt junction, avalanche breakdown is a major problem as the doping density increases in the less-doped side of the junction.

18-8. A large value of the Early voltage V_E results in a high output resistance $1/h_{oe}$. As noted in Chapter 12, high-gain op amps require high values for this parameter.

18-9. For actual devices, effective capacitances are partially due to the junction and carrier properties introduced in this chapter. Additional distributed capacitances between metallization layers and device regions and due to surface effects can dominate the values for C_{eff} in typical cases.

18-10. From Eq. (18-34) and (18-35), I_{D0} increases in proportion to W_o^3 and V_P increases in proportion to W_o^2. In the transition region, $R = V_P/I_{D0}$ decreases as $1/W_o$.

18-11. If the source-to-drain voltage drop in the gate is appreciable, then V_{DS} has less effect on the channel because the voltage across the depletion region is the difference between the gate and channel voltages. As a result, the JFET is less sensitive to changes in V_{DS} and has a lower gain.

18-12. The saturated transfer characteristic curve for the JFET can be described by Eq. (8-10):

$$I_D = I_{D0}(1 - V_{GS}/V_P)^2 \qquad 0 < V_{GS}/V_P < 1$$

and the saturation transfer characteristic for the E-MOSFET can be described by Eq. (8-48):

$$I_D = I_{D0}(V_{GS}/V_T - 1)^2 \qquad 1 < V_{GS}/V_T$$

For the D-MOSFET, additional charge is drawn into the channel for $V_{GS}/V_P \geq 1$, which produces the same conduction mechanism associated with the E-MOSFET. Therefore, the relationship predicted by the E-MOSFET transfer characteristic can also be used to describe the current flow in the D-MOSFET for $V_{GS}/V_P \geq 1$. If the transition $V_{GS} \rightarrow -V_{GS}$, $V_P \rightarrow -V_T$ is made for Eq. (8-48), the equation is transformed into Eq. (8-10). Therefore, Eq. (8-10) can be extended to values of $V_{GS}/V_P \geq 1$ to produce the D-MOSFET transfer characteristic curve.

18-13. Printed circuit boards and hybrids represent integration technologies that are used to create electronic subsystems and systems from components, and the specific implementation characteristics of these technologies interact with device and idealized circuit properties to affect overall performance.

18-14. Thermal and mechanical constraints are important factors for the design of practical electronic subsystems and systems. If these constraints are not considered during the circuit design process, the final system configuration will likely not represent the best solution to the problem. Performance objectives are not satisfied in the preferred way.

EXERCISES AND SOLUTIONS

Exercises
(18-2)

18-1. A p*np BJT has a base width of 1.5 μm. If $L_p = 8.0 \times 10^{-4}$ cm, what is the maximum value of h_{FE} for $(\gamma = 1)$?

(18-2) **18-2.** For the p*np of Exercise 18-1, $I_2 = 0.005\,I_1$. Find h_{FE} for the device.

(18-2) **18-3.** The emitter of an n*pn BJT is doped with $N_E = 10^{25}/\text{m}^3$. Estimate the ratio $(n_i^2)_E/(n_i^2)_B$ at $300°\text{K}$.

(18-2) **18-4.** An npn BJT is doped with $N_C = 5.0 \times 10^{20}/\text{m}^3$. When voltage V_{CB} is applied across the base-collector junction, the depletion region just punches through the 12-μm-wide collector region and extends 1.0 μm into the base. What is the base doping level?

(18-2) **18-5.** A collector region is doped so that $\rho_C = 6.0 \times 10^{-3}\ \Omega \cdot \text{m}$. If the collector is 12-μm wide and 100 $(\mu\text{m})^2$ in cross-sectional area, what is the bulk resistance of the collector?

18-6. A collector-base junction is doped with $N_C = 5.0 \times 10^{23}/m^3$ and $N_B = 2.0 \times 10^{21}/m^3$. Estimate the avalanche breakdown voltage for the simple planar abrupt junction used in this chapter.

18-7. Estimate the resistance of the $V_{GS}/V_P = 0.20$ characteristic curve for a Si JFET in the transition region (near the origin) if $N_{CH} = 5.5 \times 10^{22}/m^3$, the width-to-length ratio is $z/L = 4.0$, and $W_o = 1.0$ μm.

18-8. A Si n-channel JFET is to be doped so that $V_P = 3.0$ V. If $W_o = 0.8$ μm is selected, what doping level should be used in the channel?

18-9. An n-channel Si D-MOSFET is designed with the oxide thickness much less than the thickness of the depletion region in the channel. If $V_P = 2.5$ V, $z/L = 5.0$, and $W_o = 0.5$ μm, what doping level is required to produce $I_D = 6.0$ mA at $V_{GS} = 0$?

18-10. An n-channel Si E-MOSFET is designed with $V_T = 2.5$ V, $z/L = 10$, and $t_{ox} = 1.0$ μm. Estimate the current-carrying capability for the device.

18-11. A 1.0-kΩ thick-film resistor is fabricated with a length of 50 mils, a width of 25 mils, and a thickness of 25 μm. Find the $\Omega/\square$ and ρ of the resistive material.

18-12. A resistive paste of 10 k$\Omega/\square$ is to be used to make a 25-kΩ thick-film resistor. From power considerations, the total resistor area must be at least 5×10^3 mil^2. Find the dimensions of the smallest resistor size that can be used.

18-13. An alumina substrate initially at room temperature (20°C) is heated to 150°C. If the substrate length was 2.0 in. before heating, what is the difference in length in mils after heating? Use $\alpha_\ell = 8.8 \times 10^{-6}/°$K.

18-14. A resistor is in contact with an alumina substrate so that the contact area = 50 mm^2. The alumina is 2 mm thick and is mounted on a metal heat sink that is kept at 70°C. If 10 W is dissipated in the resistor, what is the temperature of the resistor?

18-15. A hybrid microcircuit is fabricated on a 1.0-in. × 2.0-in. alumina substrate that is 1 mm thick. If 50 W is distributed (approximately) uniformly over the top area of the substrate, what is the maximum (steady-state) temperature diffential between the top and bottom of the substrate?

18-16. A silver-filled conducting epoxy has a pull strength of 2000 lb/in.^{2} How much force is required to remove an IC from a substrate after it has been glued, if the IC measures 40 × 50 mils?

Solutions **18-1.** From Eq. (18-2),

$$\alpha_{dc} = \cfrac{1}{1 + \cfrac{1}{2}(1.5/8.0)^2} \cong 0.983$$

so that

$$h_{FE} = \frac{\alpha_{dc}}{1 - \alpha_{dc}} \cong \frac{0.983}{0.017} \cong 58$$

18-2. From Eq. (18-3),

$$\gamma = \frac{1}{1 + 0.005} \cong 0.995$$

so from Eq. (18-12) and Example 18-1,

$$\alpha = (0.995)(0.983) = 0.978$$

$$h_{FE} = \frac{0.978}{1 - 0.978} \cong 44$$

18-3. From Fig. 18-5,

$$\Delta\mathcal{E}_g \cong 0.08 \text{ eV}$$

Therefore, from Eq. (18-10),

$$(n_i^2)_E/(n_i^2)_B = e^{0.08\,\text{eV}/0.026\,\text{eV}} \cong 22$$

18-4. From Eq. (18-18),

$$N_B = N_C\, t_C/t_B = (5.0 \times 10^{20}/\text{m}^3)(12 \text{ } \mu\text{m}/1.0 \text{ } \mu\text{m}) = 6.0 \times 10^{21}/\text{m}^3$$

18-5. From Eq. (8-21),

$$R_b = \rho\frac{W_C}{A} = (6.0 \times 10^{-3} \text{ } \Omega \cdot \text{m}) \left(\frac{12 \times 10^{-6} \text{ m}}{100 \times 10^{-12} \text{ m}^2}\right)$$

$$\cong 0.72 \text{ k}\Omega$$

18-6. Since $N_C \gg N_B$, a one-sided abrupt junction results. From Fig. 18-8, the breakdown voltage is about 200 V for this model.

18-7. From Eqs. (18-34) and (18-35),

$$R_o \cong \frac{V_P}{I_{D0}} = \frac{1}{W_o(z/L)qN_{CH}\mu}$$

$$= \frac{1}{(1.0 \times 10^{-6}\text{m})(4.0)(1.6 \times 10^{-19}\text{C})(5.5 \times 10^{22}/\text{m}^3)(0.14 \text{ m}^2/\text{V} \cdot \text{s})}$$

$$\cong 0.20 \text{ k}\Omega$$

For the $V_{GS}/V_P = 0.20$ curve, from Eq. (18-33),

$$R = (0.20 \text{ k}\Omega)(1 - 0.2)[1 - (0.2)^{1/2}] \cong 0.09 \text{ k}\Omega$$

18-8. From Eq. (18-34),

$$N_{CH} = \frac{8V_p\epsilon}{W_o^2 q}$$

$$= \frac{(8)(3.0 \text{ V})(12 \times 8.85 \times 10^{-12} \text{ F/m})}{(0.8 \times 10^{-6} \text{ m})^2(1.6 \times 10^{-19} \text{ C})}$$

$$\cong 2.5 \times 10^{22}/\text{m}^3$$

18-9. From the discussion of Section 8-11, the characteristic curve for the D-MOSFET is given by Eq. (8-8), which was derived for the JFET. Therefore,

$$\rho = \frac{V_P z W_o}{I_D L}$$

$$= \frac{(2.5 \text{ V})(5.0)(0.5 \times 10^{-6} \text{ m})}{6.0 \times 10^{-3} \text{ A}}$$

$$\cong 1.04 \times 10^{-3} \text{ } \Omega \cdot \text{m}$$

and

$$N_{CH} = \frac{1}{\rho q\mu}$$

$$= \frac{1}{(1.04 \times 10^{-3} \text{ } \Omega \cdot \text{m})(1.6 \times 10^{-19} \text{ C})(0.14 \text{ m}^2/\text{V} \cdot \text{s})}$$

$$\cong 4.3 \times 10^{22}/\text{m}^3$$

18-10. From Eq. (8-49) in Section 8-13,

$$I_{D0} = \frac{V_T^2 \mu \epsilon z}{L t_{ox}}$$

$$= \frac{(2.5 \text{ V})^2 (0.14 \text{ m}^2/\text{V}\cdot\text{s})(12 \times 8.85 \times 10^{-12} \text{ F/m})(10)}{1.0 \times 10^{-6} \text{ m}}$$

$$\cong 0.93 \text{ mA}$$

18-11. From Eq. (18-45),

$$\rho = R\frac{Wt}{L}$$

$$= (1.0 \times 10^3 \text{ }\Omega)\frac{(25 \times 10^{-3} \text{ in.})(25 \times 10^{-6} \text{ m})}{50 \times 10^{-3} \text{ in.}}$$

$$\cong 12.5 \times 10^{-3} \text{ }\Omega\cdot\text{m}$$

and

$$\frac{\Omega}{\square} = \frac{\rho}{t} = \frac{12.5 \times 10^{-3} \text{ }\Omega\cdot\text{m}}{25 \times 10^{-6} \text{ m}} = 0.5 \text{ k}\Omega/\square$$

18-12. From Eq. (18-45),

$$R = (\rho/t)(L/W) = (10 \text{ k}\Omega/\square)(L/W) = 25 \text{ k}\Omega$$

so that

$$L/W = 2.5$$

To achieve the required power dissipation,

$$LW = 5 \times 10^3 \text{ mil}^2$$

Therefore,

$$(2.5W)(W) = 5 \times 10^3 \text{ mil}^2$$

$$W \cong 45 \text{ mil}$$

and

$$L = \frac{5 \times 10^3 \text{ mil}^2}{45 \text{ mil}} \cong 1.1 \times 10^2 \text{ mil}$$

18-13. From Eq. (18-46),

$$\Delta\ell = \alpha_\ell \ell \Delta T$$

$$= (8.8 \times 10^{-6}/°\text{K})(2.0 \text{ in.})(150°\text{C} - 20°\text{C})$$

$$\cong 2.3 \text{ mil}$$

18-14. From Eq. (18-47),

$$\Delta T = \frac{\Delta x}{kA}\left(-\frac{\Delta Q}{\Delta t}\right)$$

$$= \frac{2.0 \times 10^{-3} \text{ m}(10 \text{ W})}{(0.3 \text{ W/cm}\cdot°\text{K})(10^2 \text{ cm/m})(50 \times 10^{-6} \text{ m}^2)}$$

$$\cong 13°\text{C}$$

so that

$$T_{res} = 70°C + 13°C = 83°C$$

18-15. From Eq. (18-47),

$$\Delta T = \frac{\Delta x}{kA}\left(\frac{-\Delta Q}{\Delta t}\right)$$

$$= \frac{(1.0 \times 10^{-3} \text{ m})(50 \text{ W})}{(0.3 \text{ W/cm} \cdot °K)(10^2 \text{ cm/m})(1.0 \text{ in.} \times 2.0 \text{ in.})(2.54 \times 10^{-2} \text{ m/in.})^2}$$

$$\cong 1.3°K$$

18-16. From the information given,

$$(2000 \text{ lb/in.}^2)(40 \times 10^{-3} \text{ in.})(50 \times 10^{-3} \text{ in.}) = 4.0 \text{ lb}$$

PROBLEMS

(18-2) **18-1.** An n*pn BJT with $L_n = 1.0$ μm has a base width of 0.08 μm. What is the maximum value of h_{FE} for $\gamma = 1$?

(18-2) **18-2.** For an n*pn BJT, $I_2 = 0.003I_1$. Find γ for the device.

(18-2) **18-3.** Find β_{dc} for the device described by Problems 18-1 and 18-2 above with $\gamma < 1$.

(18-2) **18-4.** If $(n_i^2)_E/(n_i^2)_B = 50$, estimate the doping level for the emitter of the BJT.

(18-2) **18-5.** A BJT is doped with $N_B = 2.0 \times 10^{21}/\text{m}^3$ and $N_C = 0.8 \times 10^{21}/\text{m}^3$. (a) If the base-collector junction is reverse-biased by 2.0 V, find the widths of the depletion region in the base and collector. (b) Find the widths if a 10-V reverse bias is used.

(18-2) **18-6.** The base region in an n*pn BJT is doped so that $N_B = 5.5 \times 10^{21}/\text{m}^3$. If the base is 10 μm wide and has a base resistance of 0.10 kΩ, what is the area of the base?

(18-2) **18-7.** A collector-base junction is observed to produce avalanche breakdown with $V_{CB} = 100$ V. If $N_C \ll N_B$, estimate the doping level in the base.

(18-2) **18-8.** A BJT is designed with $W_o = 1.5$ μm and $N_C = 1.0 \times 10^{20}/\text{m}^3$. What value of N_B is required to produce $V_E = 150$ V?

(18-2) **18-9.** When a BJT conducts with $I_B = 100$ μA, the observed f_2 is 5.0 MHz. Estimate the (dominant) emitter-base diffusion capacitance.

(18-2) **18-10.** A Si p*np BJT has $L_p = 1.5 \times 10^{-4}$ cm. Estimate the recombination time τ_p for the excess charge distribution in the base.

(18-3) **18-11.** A Ge n-channel JFET is to be doped so that $V_P = 2.5$ V. If $W_o = 1.5$ μm is selected, what doping level should be used for the channel? (For Ge, $\epsilon_r \cong 16$).

(18-3) **18-12.** For a Ge n-channel JFET, $W_o = 1.5$ μm and $z/L = 10$. Estimate the maximum current that flows in the device if $N_{CH} = 1.0 \times 10^{21}/\text{m}^3$.

(18-4) **18-13.** A p-channel Si D-MOSFET is doped with $t_{ox} \ll (2\epsilon V_{GS}/qN_{CH})^{1/2}$. Given $z = 20$ μm, $L = 10$ μm, $W_o = 1.0$ μm, $V_P = 3.5$ V, and $N_{CH} = 1.5 \times 10^{22}/\text{m}^3$, what current flows in the device if $V_{GS} = 2.0$ V?

(18-5) **18-14.** A p-channel Si E-MOSFET is fabricated with $V_T = 3.0$ V, $z = 20$ μm, $L = 10$ μm, and $t_{ox} = 0.2$ μm. Estimate the current that flows in the device if $V_{GS} = 4.5$ V.

(18-8) **18-15.** A 1.0 kΩ thick-film resistor has a length of 25 mils and a width of 10 mils. For a 25 μm thickness, find the resistivity and $\Omega/\square$ rating of the material.

(18-8) **18-16.** A resistive paste of 1 k$\Omega/\square$ is used to make a 2.5-kΩ resistor that must have an area greater than 1.0×10^4 mil^2. Find the appropriate resistor dimensions to minimize the required substrate area.

(18-12) **18-17.** Rework Exercise 18-14 using a polymer substrate with $k = 0.03$ W/cm · °K and with a heat sink maintained at 40°C.

(18-12) **18-18.** Rework Exercise 18-15 using a polymer substrate with $k = 0.03$ W/cm · °K and 5.0 W dissipation.

(18-12) **18-19.** Assume that the heat flow out of the circuit of Exercise 18-15 takes place through a region that can be represented by an insulating layer of materials with thickness of 1.0 mm and $k = 0.01$ W/cm°K. Assuming the temperature of the substrate is approximately constant throughout, what is the maximum temperature of the circuit if a heat sink below the insulating layer is maintained at 20°C?

(18-12) **18-20.** In a given design situation, a shift is made from a glass barrier to an alumina barrier (the glass is removed and the alumina is added in its place). If all other aspects of the problem remain unchanged, what change in the temperature ΔT across the barrier is observed? (Expressed as a ratio.)

(18-12) **18-21.** A printed circuit board measures 4.0 in. × 6.0 in. and dissipates 10.0 W averaged across the upper surface. The bottom surface is placed on a heat sink at 20°C. If the board is 3 mm thick, what is the temperature of the top of the board? Use $k = 0.20$ W/cm · °K.

(18-13) **18-22.** A sample of alumina with cross-sectional area 0.10 cm^2 is placed under longitudinal stress so that a strain of 1.0×10^{-4} results. Assuming the sample does not fracture, what force is being applied?

(18-12) **18-23.** For Problem 18-22, what change in temperature ΔT is required to produce the same (18-13) fractional change in length?

CHAPTER 19
COMPUTER-INTEGRATED DESIGN AND MANUFACTURING IN ELECTRONICS

The design and manufacturing of electronic circuits can take place by drawing on many different strategies. These various strategies can be contrasted by considering the degree to which a systems approach is followed and the degree to which automation is used in the design and manufacturing processes.

A systems perspective of the electronics design process is introduced in Chapter 1. The flowchart of Fig. 1-1 illustrates how the setting of performance objectives is linked to modeling, breadboard development, prototype development, and final manufacturing. This chart emphasizes the importance of communications and feedback among the different steps of the process. The implementation of this process can be accomplished to varying degrees, and the strategy used strongly affects how the designer proceeds with the product development task.

Automation can be minimized or maximized in both design and manufacturing. Minimum-automation strategies depend on individual human effort, whereas maximum-automation strategies use sophisticated computer-supported techniques wherever possible.

In any given case, the preferred approaches to design and manufacturing depend on estimated capabilities and procedures, available resources and technology, and the competitive environment. However, during recent years, there has been a clear trend toward systems-oriented, highly automated strategies for the design and manufacturing processes. Driven by intense competition and supported by rapid developments in automation, the nature of the electronics environment is changing rapidly.

One trend is toward computer-integrated manufacturing (CIM), in which CAD and CAM are integrated and viewed from a systems perspective. In such an environment, the designer interacts with a complex computer network that provides design support (through expert systems) and considers the manufacturing process at each stage of design. Circuits must be designed for both function and production, and the computer system automatically forces the designer to consider the broader tradeoffs involved from such a perspective.

The separation of the design and production processes that has prevailed for many years, due to the way in which companies are organized, is being eliminated as organizations adapt to the opportunities associated with an integrated CID/CIM setting. The result is that the engineer is beginning to design circuits in a new context that requires a broader range of skills.

Although the era of a CIM-dominated industry is not yet here, it is important to understand present trends and to appreciate how they reflect on design efforts. The ways in which the materials in this book are applied depend on the dominant industrial strategies that are employed.

This chapter explores the development of design and manufacturing by first describing a fundamental, established approach that is often followed. Some of the failings of this approach are identified. With this background, the nature of a highly sophisticated CIM system is then pursued as an indication of the types of systems toward which the industry is evolving. Some of the limitations and problems associated with this sophisticated strategy are then discussed.

19-1 INTRODUCTORY MODEL OF THE PRODUCT-DEVELOPMENT PROCESS

Figure 19-1 shows a model of a technology-based company that is assumed to make only limited use of computer resources. The marketing unit has the responsibility for assessing the market setting in which the organization functions, suggesting new product areas that might be successful and evaluating product ideas that are brought forward. In many companies, this marketing function extends into the other functional areas.

The research and development (R&D) unit has the responsibility for maintaining a state-of-the-art awareness of technologies that affect corporate product lines. Tech-

Fig. 19-1 Model of a technology-based company.

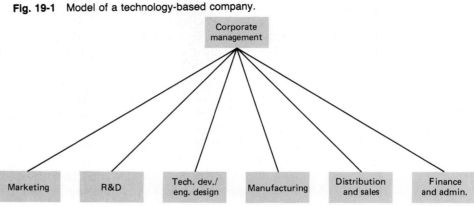

nical development/engineering design is responsible for producing new product concepts and detailed designs that enable the corporation to survive in a competitive environment.

Manufacturing must be able to produce competitive products at a reasonable price, and the distribution and sales unit must function to maintain the desired sales volume. Finance and administration is responsible for maintaining records, filing necessary reports with outside agencies, and tracking all uses of funds. Corporate management must guide all of these units into a cooperative activity that enhances performance objectives.

The idea for a new product can originate in several different ways. The marketing unit may discern a business need or opportunity based on customer relationships and pass along the information to the technical development group. On the other hand, the technical development group may identify an opportunity based on past areas of activity, professional associations, or first-hand customer contact. The idea for a new product may also originate in manufacturing, based on experience with fabrication efforts.

In any case, the product idea is usually developed into a full-fledged proposal before a reasonable evaluation of its value can be determined. Only then can corporate management decide if the idea is worth the investment of time and resources.

The product proposals that are developed are thus a key factor in determining the fate of such a technology-based corporation. If corporate management receives a steady flow of creative, well-prepared proposals, then the company is likely to prosper. Too few and inadequate proposals produce a high-risk situation.

A typical product proposal might consist of the following sections:

1. Product description
2. Market analysis
3. Manufacturing analysis
4. Schedule plan
5. Financial plan

The product description must be complete enough to define the product being considered. There should be sufficient drawings and technical analyses to describe the technical operation and overall functioning of the product. Technical performance objectives are a key part of the product description.

The market analysis addresses who will buy the product and why. What is the total available market for this type of product, what is the nature of the competition, and what reasonable share of the market can be captured? The prediction of market share depends on a composite strategy that considers the technical preference of the product, price, advertising, and distribution. The result of this analysis is a time-phased estimate of the volume of product sales over the predicted lifetime of the product.

The manufacturing analysis must address the ability to produce the desired product, with all desired performance objectives, at a cost that is compatible with the planned selling price. It is obvious that successful product development rests on close cooperation between the marketing, technical development, and manufacturing units of an organization. However, in many cases, this link is quite weak. Manufacturing may become involved in proposal preparation only through feedback regarding schedules and costs based on the product description that is given.

The schedule plan defines when necessary research and development activities are to start and end, the preparation necessary before manufacturing can begin to produce the product, and distribution and sales. All of these activities are brought together on a single time line to define the relationships among tasks. The financial plan is drawn from the schedule and provides a month-by-month prediction of the necessary flow of investment funds (company expenses) linked to the income to be earned.

The corporate organizational chart of Fig. 19-1 typically consists of groups of people formed into functional units, with communications between groups through written and verbal reports, committee meetings, and other exchanges. A product proposal developed in this setting reflects the structure and operational modes of the organization.

The preparation of the proposal itself begins the design process because decisions already have been made in order to describe the planned product. As noted above, there may have been minimal involvement by manufacturing in these decisions.

Detailed circuit and system design begins after the proposal is approved. For many cases, emphasis is on circuit configuration and the devices and components to be used, with little or no attention to the way in which the product is to be manufactured. Once the design is complete, manufacturing must decide how to fabricate the product. This sequential procedure prevents meaningful tradeoffs regarding the materials, processes, technologies, and manufacturing strategies to be used. As noted below, a "design-to-production" strategy is quite different.

19-2 PROPERTIES OF A CID/CIM-BASED CORPORATION

Computer-aided design (CAD) is discussed throughout this book. Computer Applications at the end of most chapters are used to encourage an awareness of the computer as a design tool. The SPICE simulation program is discussed in some detail as a commonly available software package. These few examples serve to illustrate some of the ways in which interaction with a computer can simplify and enhance the design of a new product. However, these examples do not indicate the full breadth of the shift in design activity that it taking place today.

The most helpful strategy for computer support requires an ongoing interactive exchange between the designer and the computer system. Some more recent versions of SPICE allow for the design engineer to draw a circuit schematic on a monitor, access various models to predict circuit performance, then go back and forth between circuit modification and resultant performance changes in order to satisfy the desired performance objectives. This mode of operation is an improvement, but is still primitive in many ways.

As noted in earlier chapters, IC technology is increasingly using functional compilers that can translate schematics to detailed IC fabrication without any further human intervention. This high-level software is making the design of custom ICs a feasible task. In the same way, high-level software packages are being developed to produce hybrid microcircuits and printed circuit boards. Building on the SPICE concepts and more sophisticated methods of interaction, the designer can test new concepts on the computer in an interactive mode.

As the capabilities of expert systems using artificial intelligence (AI) increase rapidly, so do the abilities of the computer work station to become an integral part of the design activity from start to finish. It is reasonable to look forward to expert systems that can suggest circuit configurations based on desired performance objectives as well as describe how a given circuit can be expected to perform.

The evolution of computer-integrated design (CID) is being matched by rapid changes in manufacturing. Manufacturing automation is a common feature of industry. Yet, the approach to automation has changed significantly over the years, from fixed-purpose machines with human operators to similar machines with computer-controlled operators. The current shift is to much more flexible computer-controlled machines (robots) and to linking these machines through computer networks to produce a high level of automation.

In its ultimate form, a CIM factory consists of collections of computer-driven equipment without any human operators. Such a factory can run 24 hours a day and is not subject to many of the constraints of more typical industry facilities.

There is another important aspect of a highly computerized factory. Since the machines in use are very flexible and can be assigned new tasks simply by changing the software instructions, the concept of the learning curve (introduced in Exercise 3-16 of Chapter 3) is no longer necessarily true. Small volumes of production become feasible if the total factory remains at a satisfactory level of production.

If a sophisticated CAD system is linked to a highly automated manufacturing system, in such a way that the two systems interactively optimize the operations of the company, the result is computer-integrated design and manufacturing.

Figure 19-2 indicates how a CID/CIM corporation might function. As illustrated, most company communications take place by means of several computer networks. The CIM factory consists of work cell controllers, each of which may be responsible for operating several robots or other types of computer-integrated machinery. The necessary exchanges among work cells take place via the manufacturing network communication channel. Messages are passed back and forth along this network channel using a protocol that defines the message content and length and assures that messages are placed on the channel and received from the channel through an appropriate cycle. The channel may consist of only two conductors in the form of a twisted pair of wires or a coaxial cable. Alternatively, the channel may include more conductors or may use optical fiber communications. The selection depends on the types of messages required to operate the factory. The manufacturing work cells and network shown in Fig. 19-2 are often said to constitute a local area network (LAN).

It is clearly important to adopt standard network protocols for the manufacture of machines that are intended for CIM operation, and major steps are being taken in this direction. The development of IC-based interface modules for work cells is proceeding rapidly.

The operational units of the company are linked through the operations network communication channel and a series of work stations. The essential functions of each unit are conducted by using the information flow on this channel. Arrangements are made to purchase access to external databases that are needed to support unit operations. Protocols for operations networks have been proposed by several groups and IC-based interface modules are available.

Circuit and system design activities are performed by the technical development group through interactions with this network. An interface is provided between the operations and manufacturing networks to allow an interactive exchange that improves the effectiveness and efficiency of the corporation.

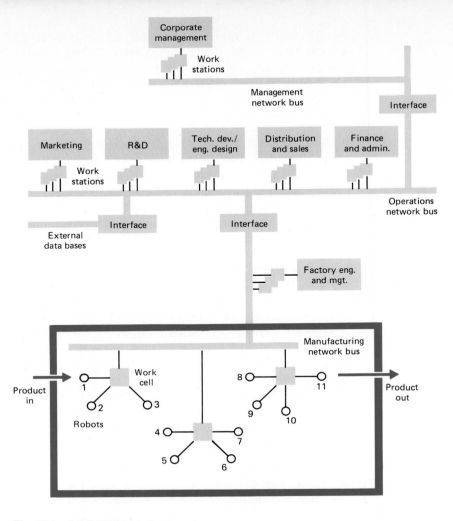

Fig. 19-2 A CID/CIM-based corporation.

A CID/CIM system of the type described in Fig. 19-2 changes the ways in which products are developed. The following section describes how a typical electronics design engineer might function in such a setting.

19-3 PRODUCT DEVELOPMENT IN A CID/CIM SETTING

There are many different ways in which CID/CIM systems will affect electronics design in a corporate setting. Obviously, the variations depend on the past organizational history, the specific business line, corporate resources, and management strategy. However, many of the essential features of such systems can be expected to take on similar form and function because of a common external market environment and similar technologies used to fabricate the CID/CIM networks, work stations, and available software.

The following example describes how one such organization might function, to allow a contrast to be made between present organizations and those anticipated for the future. The results are intended to be descriptive, not definitive. They do suggest some of the ways in which electronics design may evolve.

Example 19-1   **Electronics Design in a CID/CIM Organization**

The ABC Electronics Company is structured along the lines indicated in Fig. 19-2. How might the company sense and respond to changing market conditions for an existing product?

Discussion   The example starts with the situation shown in Fig. 19-3. The marketing unit of ABC has purchased the right to access an external database that provides sales predictions for all parts of the U.S. economy, with monthly updates. On October 1, the

Fig. 19-3   Marketing activities for Example 19-1.

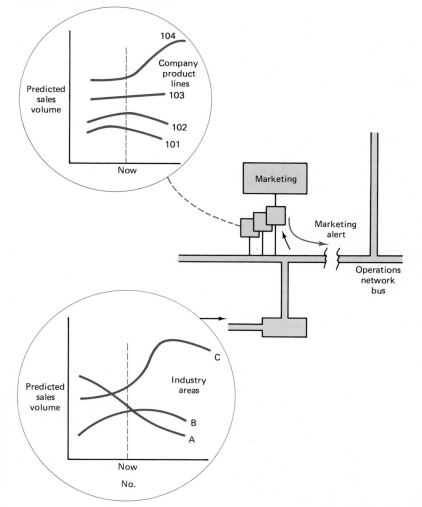

marketing staff reviews those projections that relate to ABC's product areas and notes that a significant shift has occurred. The external database prediction is that industry area C is going to grow rapidly in sales volume over the next 12 months, then decline.

The database input is evaluated by a standard software package available to the marketing group that interprets industry-wide projections in terms of the major product areas for the company. The result is a forecast predicting that ABC products 101 and 102 will begin to decline in sales, 103 will stay flat, and 104 has the potential for rapid growth.

Once this information is obtained, the computer network automatically creates a "market alert" that is electronically transferred to predefined work station addresses throughout ABC. One of the work station addresses for the alert is in the R&D area, as shown in Fig. 19-4. The alert automatically logs onto the work station task list with a self-established high priority. The work station sends a confirmation message back to marketing, noting that the message has been received and appropriately acted upon.

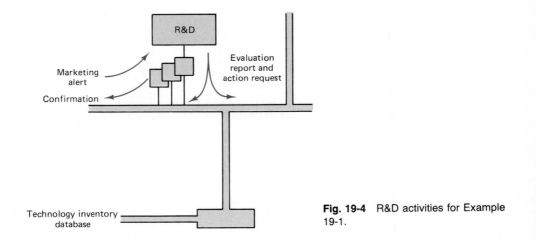

Fig. 19-4 R&D activities for Example 19-1.

The R&D staff reads the forecasts and interpretive analysis contained in the alert, then accesses an external database that maintains a current inventory of all major manufacturing technologies used in the United States. The staff searches the database for current or imminent changes in technology that can be expected to impact the market for each of the product lines (101 to 104) in the alert. An AI software package studies the relationships between sales volume forecasts, ABC product lines, and technology evolution. The external sales data used by marketing are reaccessed for further details.

The R&D unit then accesses an on-line inventory of past and current internal R&D projects for insight into possible areas of application. The total review and analysis process indicates that the potential of product line 104 will be achieved only with an upgrading of technical capability to meet anticipated competitive responses. Related R&D projects are suggested as possible resources. As a result of these efforts, an "evaluation report and action request" is entered into the R&D work station and placed on the network.

The technical development engineering design unit receives the action report. The group begins to develop an improved design for the 104 product line, as shown in Fig. 19-5.

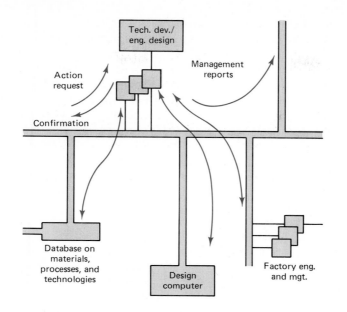

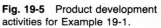
Fig. 19-5 Product development activities for Example 19-1.

The design process takes place interactively with the networks that link the company. Information regarding the manufacturing capabilities of the company, the available materials and processes, and costs associated with all aspects of production are integrated into the design software. One of the major performance objectives is to produce a circuit that is matched to corporation manufacturing capabilities. This design-to-production philosophy results in tradeoffs among circuit configurations, devices, and interconnection and packaging strategies. As the designer interacts with the computer system, decisions must be made with respect to all of these aspects of design.

The design specifications are automatically transferred through the network to factory engineering and management. The work station produces a list of materials and devices that are required for a breadboard of the improved 104 product (assuming that such a step is included in the design sequence). Based on the data sheet received, the stock room ships the needed items to the engineering design laboratories, purchasing nonstock items as required. The materials/parts list is also forwarded to factory engineering to begin developing the necessary hardware and software changes and supply lines for production of the modified 104 product.

Once the breadboard system is operational, the design computer automatically sets the test parameters for all needed automatic test equipment (ATE). The breadboard is tested and operation is confirmed. The results of all tests are integrated into the design computer for future reference and design support. The above activities are brought to the attention of all other operating units in ABC by continuing management reports provided over the network.

The prototype is produced in the CIM factory, using input from the design computer. The ATE system checks the performance of the prototype. Any differences between the breadboard and prototype are evaluated, and changes in the CIM fabrication instructions are made as necessary. The analysis forms part of the continuing database for the design computer.

While the technical design is being completed, along with breadboard and prototype evaluation, the distribution and sales unit of ABC is preparing to bring the new product to market. Based on the earlier market projections, a production schedule is made.

The factory receives approval for the technical product from engineering design and the production schedule from distribution and sales. During this developmental period, the finance and administration unit is preparing production budgets. Corporate management is tracking the evolution of the new product through automatic management information system (MIS) reports. At any step, suggestions and instructions can be placed into the system by this management group. The system is set to continue all activities unless intervention occurs.

At this final stage, management authorization must be provided to the factory to proceed with production. All units receive copies of this approval and the product development cycle is complete. ■

Since a shared database and communications network exists for a CID/CIM company, the product proposal takes place in a distributed, interactive way. The final proposal reflects a broader database, ongoing familiarity with the data, interactive support from AI software, a higher level of intracompany communications, and reduced delays. Product proposals are likely to consist of a collection of work station printouts bound together to provide a focus for decision making. The result is an improved use of design resources.

The above example is intended to illustrate how CID/CIM capabilities can change the way in which products are conceptualized, designed, and produced. For this particular case, performance objectives for the new product were set to meet the projections of market competition. The issue of setting performance objectives is sufficiently important to warrant further discussion in the following section.

19-4 SETTING PERFORMANCE OBJECTIVES

The purpose of every electronics design project is to achieve a specific set of objectives. As illustrated in Chapter 1 (Fig. 1-1), the design process begins with the setting of performance objectives and ends with the final desired product.

In previous chapters, technical circuit objectives are formulated by selecting a performance variable, then defining both a nominal value and tolerance range for the variable. The design that results depends on (1) the variables that are selected and (2) the nominal and tolerance values assigned to the variable.

In any realistic product setting, nontechnical objectives are an important aspect of system development. A few problems in earlier chapters explore the concepts of cost, reliability, ease of production, ease of repair, volume of sales as a determinant of unit costs, and other parameters. Decisions regarding technical objectives are usually closely linked to these nontechnical aspects and, in turn, to the larger setting in which design and production take place.

For the examples and problems in this text, performance objectives are largely assumed to be predetermined. The desired technical and nontechnical features of circuits are specified to produce a meaningful introduction to design procedure. Yet, it is clear that an introduction to electronics design is incomplete without a further consideration of the ways in which objectives are determined.

As may be expected, the nature of performance objectives and the procedures through which they are determined differ in a CID/CIM setting. To set performance objectives, the first task is to decide on the parameters that describe the performance. This is often a difficult task. The selected parameters must be carefully thought out to assure that the product satisfies its intended purpose. At the same time, care must be taken to minimize unnecessary constraints. Objectives must be appropriate to the available databases and methods of manufacturing applied.

The selection of the performance variables to be used is a critical stage in the design process. If inappropriate variables are chosen, the resulting product may not satisfy the intended application even if the product satisfies all specifications. Skill, experience, and resources are required to produce an optimum mix of technical and nontechnical parameters and objectives.

As noted in the discussion above, the data and computer interactive support available for the design process in a CID/CIM system allows a company-wide product strategy and removes many of the cost/volume constraints that can limit design in other settings. Electronics design in a CID/CIM system thus tends to change the types of technical and nontechnical performance parameters that are used and to change the ranges of performance values that are chosen. The result is to enhance the match between the design process and the product market.

19-5 DESIGN AND MANUFACTURING IN ELECTRONICS

This chapter is concerned with the ways in which electronic products are designed and manufactured. As corporations are restructured around computer networks, the nature of this process continues to change in many important ways.

A CIM system is usually implemented to meet competitive pressures. Such systems are able to respond rapidly to market changes, achieve cost reductions, improve quality, and obtain product flexibility within defined system boundaries.

Product decisions are linked to the marketing environment on a continuing basis. External database systems project evolutionary market patterns that influence corporation decisions. Due to rapid technology change and the limited ability of most companies to fund a broadly diversified R&D program, external databases are becoming steadily more important in product design. The technical development/engineering design, manufacturing, and distribution and sales functions are becoming more intertwined in order to improve product success. CIM factories are adaptive, reliable, and closely linked to the computer networks throughout the corporation. These capabilities allow design and manufacturing to become integrated efforts.

The CID/CIM capability affects the nature of performance specifications and product proposals. Final products depend on an interactive exchange among marketing, R&D, design engineering, manufacturing, distribution and sales, and finance. Constant monitoring results in a corporation that is always in the process of adapting.

On the other hand, it is important to note that a number of problem areas are encountered when a CIM system is considered. Capital investment costs can be high, placing a financial burden on the company. Many of the needed expert systems do not yet exist, and a limited knowledge base exists regarding the best ways to implement such a design and manufacturing facility. The product line is always limited by the defined boundaries of flexibility, and the training of engineers and other organizational members to use the system can be time consuming and difficult. Once these problems are anticipated, they can be dealt with as an expected part of the evolutionary process.

In the types of design and manufacturing settings that are developing today, the electronics designer must develop many different skills. A systems understanding of both technology and company operations must underlie effective interaction with the organization. The designer must develop a fundamental understanding of electronic circuits and systems. Simulation systems and AI design support are effective only if the designer has been able to develop a strong appreciation for the ways in which basic circuits and subsystems function and integrate into larger systems. There must be an ability to see the patterns that exist and to use these patterns to interpret the feedback from computer design support.

The designer must be comfortable with the types of tradeoffs that are experienced due to options in materials, processes, and technologies, and the ways to shift toward a preferred solution. The ability to set appropriate performance objectives and work iteratively within the organizational network is necessary to link technical skills to the design task.

This text emphasizes the fundamental patterns and problem-solving methods of electronics that can be used as a starting point for design. Initial experience is gained in applying these insights toward different types of problems. The value of computer support is illustrated, along with indications as to how simulation results can be applied. The importance of nontechnical factors in determining performance is explored. A systems approach to design provides a framework for best understanding and meeting the complex performance objectives that are encountered in realistic settings.

APPENDIX 1
AC CIRCUIT ANALYSIS

A sine-wave voltage that oscillates in time can be described by the equation

$$v = V_m \sin (2\pi f t + \phi) \qquad \text{(A1-1)}$$

where V_m is the amplitude of the wave, f is the frequency, and ϕ is the phase angle. Further,

$$\omega = 2\pi f \qquad \text{and} \qquad T = 1/f \qquad \text{(A1-2)}$$

can be defined as the angular frequency ω and period T of the signal.

A useful graphical representation of this wave is shown in Fig. A1-1(A). A vector of length V_m rotates in the counterclockwise direction, with an angle $(\omega t + \phi)$ mea-

Fig. A1-1 Vector representations of sine waves.

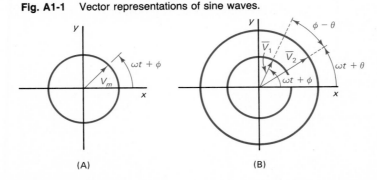

(A) (B)

sured with respect to the $+x$ axis. The projection of the vector along the y axis is given by Eq. (A1-1). For two ac waves of the same frequency, the representation of Fig. A1-1(B) can be used. The two vectors rotate in time with the same angular frequency ω, so that the relative angle between the two vectors $\phi - \theta$ remains constant.

The projections of these vectors on the x or y axis can represent voltages and currents of interest. As time passes, these projections trace sine-wave oscillations. Because the primary concern is usually with the relative angles between these vectors, the vectors and their relationship can be examined at any convenient time. The result is a vector representation, as shown in Fig. A1-2(A).

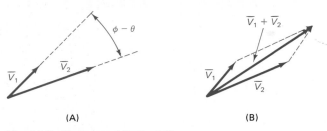

(A)   (B)

Fig. A1-2 Vectors and their addition.

Multiple voltages and currents for the same circuit can be combined on the same graph. AC voltages or ac currents can be combined by using the methods of vector addition, as shown in Fig. A1-2(B), and the angular relationships between all vectors can then be found graphically.

A more analytical approach to ac circuits results from the use of complex numbers. Each voltage and current vector is represented by a complex number of the form $\mathbf{A} = a + jb$, where the real part of $\mathbf{A}$ (equal to a) lies along the x axis, and the imaginary part of $\mathbf{A}$ (equal to b) lies along the y axis.

The symbol $j = \sqrt{-1}$ is an operator that can be used to represent angular relationships. If a vector is multiplied by j, the vector is rotated through an angle of $+90°$, as shown in Fig. A1-3(A). This can be understood by calculating

$$j\mathbf{A}_1 = j(a + jb) = ja - b = -b + ja$$

As shown in the figure, $j\mathbf{A}_1$ is perpendicular to $\mathbf{A}_1$. Multiplication by $j^2 = -1$ rotates the vector through an angle of $180°$, as shown in Fig. A1-3(B). Similar results can be obtained for other combinations.

Fig. A1-3 Rotations using the j operator. (A) Rotation by $+j$. (B) Rotation by j^2.

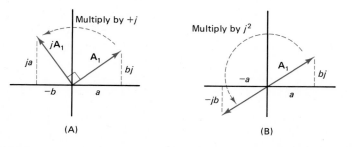

(A)   (B)

The addition of vectors using complex numbers is equivalent to graphical vector addition. Because it lends itself to straightforward analytic methods, the complex notation is useful for the analysis of ac circuit performance.

Conversion between complex numbers written in the rectangular form $\mathbf{A} = a + jb$ and in the polar form $\mathbf{A} = A_m e^{j\theta}$ can be accomplished by noting the following equivalencies:

$$a = A_m \cos \theta$$

$$b = A_m \sin \theta \qquad (A1\text{-}3)$$

and

$$|A_m| = \sqrt{a^2 + b^2} \qquad (A1\text{-}4)$$

$$\theta = \tan^{-1}(b/a) \qquad (A1\text{-}5)$$

By convention, the real part of $\mathbf{A}$ (the projection along the x axis) is used to represent currents and voltages.

In order to apply Eq. (A1-5), the graph of the tangent function must be used. As shown in Fig. A1-4, the tan θ curve passes through the origin and is asymptotic to $+\infty$ and $-\infty$ as θ approaches $\pi/2$ and $-\pi/2$. When taking the inverse tangent (or arc tangent), the range $-\pi/2 \leq \theta \leq \pi/2$ is defined for the principal values of θ.

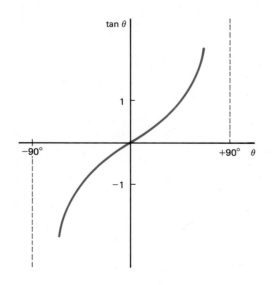

Fig. A1-4 Plot of the function tan θ.

By drawing on the above foundation, the phasors $(\mathbf{v}, \mathbf{i})$ can represent the nontime-dependent relationships between voltages and currents. The sinusoidal time dependence can then be expressed through the relationships

$$\mathbf{v}(t) = \mathbf{v}e^{j\omega t} \qquad \text{and} \qquad \mathbf{i}(t) = \mathbf{i}e^{j\omega t}$$

The average value of a complex time-varying current $\mathbf{i}(t)$ or voltage $\mathbf{v}(t)$ is given by

$$I_{\text{av}} = \frac{1}{T}\int_o^T [\text{Re } \mathbf{i}(t)]\, dt \qquad \text{and} \qquad V_{\text{av}} = \frac{1}{T}\int_o^T [\text{Re } \mathbf{v}(t)]\, dt \qquad (A1\text{-}6)$$

where T is the period of the wave. For any waveform that has an equal area above and below the axis Re $\mathbf{v}(t) = 0$, the average value of the wave is equal to zero, as shown in Fig. A1-5(A).

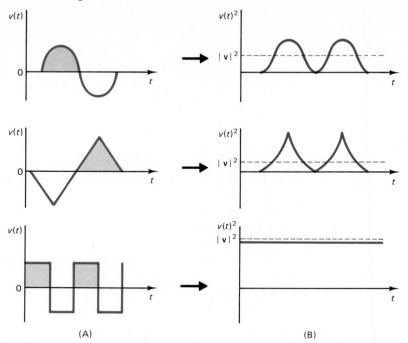

Fig. A1-5 Average and effective values. (A) $V_{av} = 0$ if areas above and below the $v = 0$ axis are equal. (B) $|\mathbf{v}|^2 \neq 0$ for all the waveforms shown in part (A).

It is often important to find the average power dissipated in a resistor R by an arbitrary voltage or current signal in order to calculate the level of heating to take place.

The effective values $\mathbf{i}$ and $\mathbf{v}$ are defined so that

$$P_{av} = |\mathbf{i}|^2 R = |\mathbf{v}|^2/R \tag{A1-7}$$

where

$$|\mathbf{i}| = \left\{ \frac{1}{T} \int_o^T [\text{Re } \mathbf{i}(t)]^2 \, dt \right\}^{1/2}$$

$$|\mathbf{v}| = \left\{ \frac{1}{T} \int_o^T [\text{Re } \mathbf{v}(t)]^2 \, dt \right\}^{1/2} \tag{A1-8}$$

For a sine wave,

$$|\mathbf{i}| = \left[\frac{1}{T} \int_o^T I_m^2 \cos^2 (\omega t) \, dt \right]^{1/2} = \frac{I_m}{\sqrt{2}} \tag{A1-9}$$

$$|\mathbf{v}| = \left[\frac{1}{T} \int_o^T V_m^2 \cos^2 (\omega t) \, dt \right]^{1/2} = \frac{V_m}{\sqrt{2}} \tag{A1-10}$$

The angles associated with $\mathbf{i}$ and $\mathbf{v}$ are those associated with $\mathbf{i}(t)$ and $\mathbf{v}(t)$.

The waveforms of Fig. A1-5(B) illustrate that $|\mathbf{v}|^2 \neq 0$ even when $V_{av} = 0$. When discussing ac circuits and using phasors, effective values of the voltages and current are used unless otherwise noted.

Consider what happens when an ac voltage $\mathbf{v}$ is applied to a resistor R. A current $\mathbf{i} = \mathbf{v}/R$ flows. Because R is a real, positive number, the resultant current $\mathbf{i}$ is in phase with the applied voltage $\mathbf{v}$. Figure A1-6(A) illustrates the relationship between these two vectors.

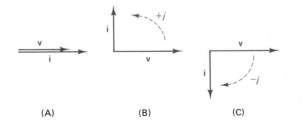

Fig. A1-6 Complex representation of voltages and currents showing phase relationships. (A) Resistor. (B) Capacitor. (C) Inductor.

If an ac voltage is applied to a capacitor,

$$\mathbf{i}(t) = C\frac{d\mathbf{v}(t)}{dt} = C\frac{d}{dt}\,\mathbf{v}e^{j\omega t} = j\omega C\,\mathbf{v}(t) \tag{A1-11}$$

so that

$$\mathbf{i} = j\omega C\,\mathbf{v} \tag{A1-12}$$

Since $\mathbf{v}$ is multiplied by $+j$ to find the current $\mathbf{i}$, $\mathbf{i}$ is 90° ahead of the applied voltage, as shown in Fig. A1-6(B). The magnitude of $\mathbf{i}$ is given by

$$|\mathbf{i}| = |\mathbf{v}|/X_C \tag{A1-13}$$

where $X_C = 1/\omega C$ is the capacitive reactance with the units of resistance.

If an ac voltage $\mathbf{v}$ is applied to an inductor,

$$\mathbf{v}(t) = L\frac{d\mathbf{i}(t)}{dt} \tag{A 1- 14}$$

$$\mathbf{i}(t) = \frac{1}{L}\int \mathbf{v}(t)\, dt = \frac{1}{L}\int \mathbf{v}e^{j\omega t}\, dt = \frac{\mathbf{v}(t)}{j\omega L} = \frac{-j\mathbf{v}(t)}{\omega L}$$

so that

$$\mathbf{i} = -j\mathbf{v}/\omega L \tag{A1-15}$$

Since $\mathbf{v}_L$ is multiplied by $-j$ to find the current $\mathbf{i}$, $\mathbf{i}$ is 90° behind the applied voltage, as shown in Fig. A1-6(C). The magnitude of $\mathbf{i}$ is given by

$$|\mathbf{i}| = |\mathbf{v}|/X_L \tag{A1-16}$$

where $X_L = \omega L$ is the inductive reactance also with the units of resistance.

For combinations of resistors, capacitors, and inductors, the equivalent impedance Z of the combination can be calculated in complex notation. The current produced in an arbitrary circuit with an impedance Z due to an applied voltage $\mathbf{v}$ is then

$$\mathbf{i} = \mathbf{v}/Z \tag{A1-17}$$

By using complex notation, circuit elements can be combined as if they were resistors.

The impedance of a series RLC circuit becomes

$$Z = R + jX_L - jX_C \qquad \text{(A1-18)}$$

and

$$\mathbf{i} = \mathbf{v}/Z$$

Similarly, the impedance of a parallel RLC circuit becomes

$$\frac{1}{Z} = \frac{1}{R} + \frac{1}{jX_L} + \frac{1}{-jX_C} \qquad \text{(A1-19)}$$

and again

$$\mathbf{i} = \mathbf{v}/Z$$

This method of circuit analysis can be applied to any combination or arrangement of circuit elements.

APPENDIX 2
RC FILTERS

Consider Fig. A2-1(A), which can be viewed as a voltage divider using complex components. By applying the concepts of Appendix 1,

$$\mathbf{i} = \frac{\mathbf{v}_{in}}{R - jX_C}$$

$$\mathbf{v}_{out} = \mathbf{i}R = \mathbf{v}_{in}\frac{R}{R - jX_C}$$

$$= \mathbf{v}_{in}\frac{\omega RC(\omega RC + j)}{1 + \omega^2 R^2 C^2} \qquad \text{(A2-1)}$$

so that

$$\left|\frac{\mathbf{v}_{out}}{\mathbf{v}_{in}}\right| = \frac{\omega RC\sqrt{(\omega RC)^2 + 1}}{1 + \omega^2 R^2 C^2} = \frac{\omega RC}{\sqrt{1 + \omega^2 R^2 C^2}} \qquad \text{(A2-2)}$$

As $\omega \to 0$, $|\mathbf{v}_{out}/\mathbf{v}_{in}| \to 0$, and as $\omega \to \infty$, $|\mathbf{v}_{out}/\mathbf{v}_{in}| \to 1$. Further,

$$\theta = \tan^{-1}(1/\omega RC) \qquad \text{(A2-3)}$$

As $\omega \to 0$, $\theta \to 90°$, and as $\omega \to \infty$, $\theta \to 0°$.

If the magnitude $|\mathbf{v}_{out}/\mathbf{v}_{in}|$ and the phase θ for this circuit are plotted as a function of ω, the results of Figs. A2-1(B) and (C) are obtained. This RC combination acts as a *high-pass filter*. High frequencies pass through the circuit unaffected, whereas low frequencies are reduced in magnitude and experience a phase shift of $+j$ (or $+90°$).

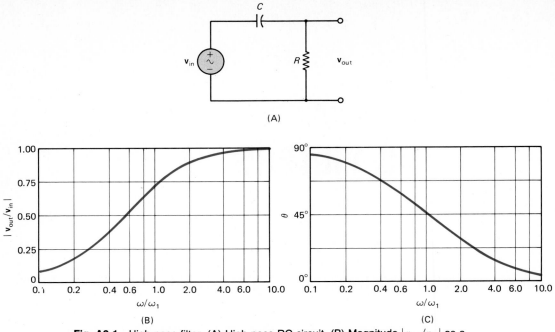

(A)

(B)   (C)

Fig. A2-1   High-pass filter. (A) High-pass RC circuit. (B) Magnitude $|v_{out}/v_{in}|$ as a function of frequency. (C) Circuit phase shift as a function of frequency.

These characteristics can be understood by remembering the properties of capacitors. At high frequencies, the capacitor acts as a short, so that $\mathbf{v}_{out} = \mathbf{v}_{in}$. At low frequencies, the capacitor begins to act as an open circuit and $\mathbf{v}_{out}$ drops.

In this text, f_1 and θ_1 describe the performance of high-pass RC filters. Frequency f_1 occurs in the transition region between high frequencies (where C acts as a short circuit) and low frequencies (where C acts as an open circuit). Therefore, f_1 is useful in identifying the boundary between these two regions.

At frequency f_1, the voltage delivered to R is $\dfrac{1}{\sqrt{2}}$ times its maximum value, so the power delivered to the resistor is $\frac{1}{2}$ times its maximum value. Frequency f_1 is called the half-power frequency. Frequencies below f_1 are attenuated and frequencies above f_1 pass through the filter.

Now consider the circuit of Fig. A2-2(A), reversing the placement of the resistor and capacitor in the complex divider. By inspection

$$\mathbf{i} = \frac{\mathbf{v}_{in}}{R - jX_C}$$

$$\mathbf{v}_{out} = \mathbf{i}(-jX_C) = \mathbf{v}_{in}\frac{-jX_C}{R - jX_C}$$

$$= \mathbf{v}_{in}\frac{1 - j\omega RC}{1 + \omega^2 R^2 C^2}$$

so that

$$\left|\frac{\mathbf{v}_{out}}{\mathbf{v}_{in}}\right| = \frac{1}{\sqrt{1 + \omega^2 R^2 C^2}} \tag{A2-4}$$

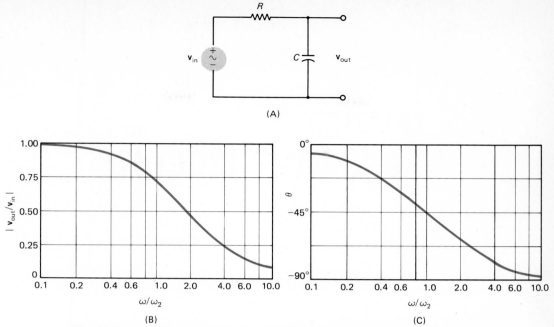

Fig. A2-2 Low-pass filter. (A) Low-pass RC circuit. (B) Magnitude $|v_{out}/v_{in}|$ as a function of frequency. (C) Circuit phase shift as a function of frequency.

As $\omega \to 0$, $|\mathbf{v}_{out}/\mathbf{v}_{in}| \to 1$. And as $\omega \to \infty$, $|\mathbf{v}_{out}/\mathbf{v}_{in}| \to 0$. If θ is the angle between the input and output voltages,

$$\theta = \tan^{-1}(-\omega RC) \tag{A2-5}$$

As $\omega \to 0$, $\theta \to 0$, and as $\omega \to \infty$, $\theta \to -\pi/2$.

If the magnitude $|\mathbf{v}_{out}/\mathbf{v}_{in}|$ and the phase θ are plotted, the results of Figs. A2-2(B) and (C) are obtained. This RC divider acts as a *low-pass filter*. Low frequencies pass through the circuit unaffected in magnitude or phase, whereas high frequencies are reduced in magnitude and experience a phase shift of $-j$ (or $-90°$). At high frequencies, the capacitor shorts out the output, and at low frequencies, the capacitor behaves as an open circuit and does not affect the output.

The frequency f_2 and phase angle θ_2 define the half-power frequency for the low-pass filter. Frequencies below f_2 pass through the filter and frequencies above f_2 are attenuated and shifted in phase.

APPENDIX 3
RESONANT CIRCUITS

Consider the series RLC circuit shown in Fig. A3-1(A). From Appendix 1,

$$Z = R + j(X_L - X_C)$$

so that

$$\mathbf{i} = \frac{\mathbf{v}}{Z} = \frac{\mathbf{v}}{R + j(X_L - X_C)}$$

$$= \mathbf{v}\frac{R - j(X_L - X_C)}{R^2 + (X_L - X_C)^2} \tag{A3-1}$$

and

$$|\mathbf{i}| = \frac{|\mathbf{v}|}{\sqrt{R^2 + (X_L - X_C)^2}} \tag{A3-2}$$

$$\theta = \tan^{-1}\left[-\left(\frac{X_L - X_C}{R}\right)\right] \tag{A3-3}$$

These results are plotted in Figs. A3-2(B) and (C). The minimum current flows through the circuit at the resonant frequency

$$\omega_o = 1/\sqrt{LC}$$

and the phase shift introduced by the circuit is zero at this frequency.

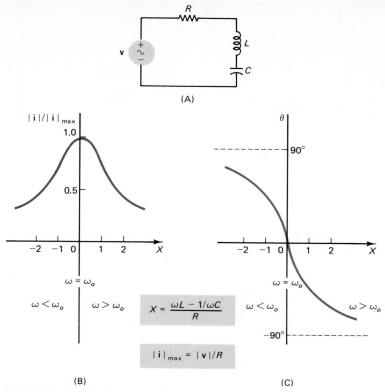

$$X = \frac{\omega L - 1/\omega C}{R}$$

$$|\mathbf{i}|_{max} = |\mathbf{v}|/R$$

Fig. A3-1 Series resonant circuit.

Consider the parallel RLC circuit shown in Fig. A3-2(A). By inspection,

$$1/Z = 1/R + 1/jX_L + 1/-jX_C$$

$$Z = \frac{R(jX_L)(-jX_C)}{R(jX_L - jX_C) + X_L X_C}$$

so that

$$\mathbf{i} = \frac{\mathbf{v}}{Z} = \mathbf{v}\frac{X_L X_C + jR(X_L - X_C)}{RX_L X_C} \tag{A3-4}$$

$$|\mathbf{i}| = |\mathbf{v}|\frac{[(X_L X_C)^2 + R^2(X_L - X_C)^2]^{1/2}}{RX_L X_C} \tag{A3-5}$$

$$\theta = \tan^{-1}\left[\frac{R(X_L - X_C)}{X_L X_C}\right] \tag{A3-6}$$

These results are plotted in Figs. A3-2(B) and (C). The minimum current flows through the circuit at the resonant frequency

$$\omega_o = 1/\sqrt{LC}$$

and the phase shift is again zero at this frequency.

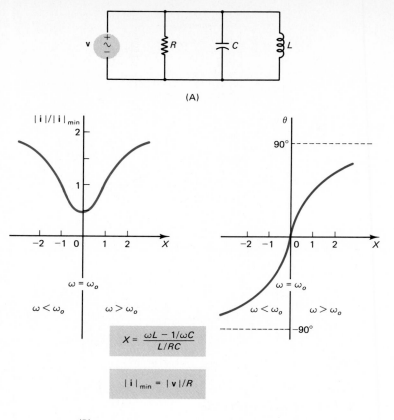

$$X = \frac{\omega L - 1/\omega C}{L/RC}$$

$$|\,i\,|_{min} = |\,v\,|/R$$

(B)   (C)

Fig. A3-2 Parallel resonant circuit.

APPENDIX 4
LOGARITHMIC SCALES

In electronics design, it is often helpful to use graphs that display one or both variables on a logarithmic scale. Figure A4-1 illustrates a *semilog plot,* with a linear scale on the vertical axis and a logarithmic scale on the horizontal axis.

Fig. A4-1 Semilog graph.

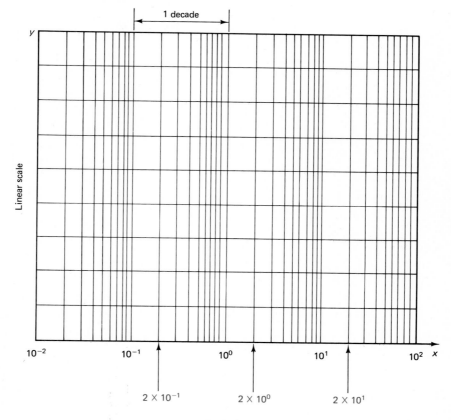

The logarithmic scale repeats its basic cycle every time the variable x increases or decreases by a factor of 10. This axis can be labeled using those powers of 10 that are appropriate for the problem of interest. For the graph shown, the scale runs from 10^{-2} to 10^{+2}. Each time the variable x increases by a factor of 10, the increase is a *decade*. Intermediate values between the even powers of 10 can be read using the scale marks. Values of $2 \times 10^{-1} = 0.2$, $2 \times 10^0 = 2$, and $2 \times 10^1 = 20$ are shown to illustrate the scale use.

In some applications, a *log-log plot* is applied, with logarithmic scales on both axes. For this case, the same method for developing a log scale is applied to both x and y variables.

APPENDIX 5
SIGNIFICANT FIGURES

The functional elements used to develop electronic circuits are based on complex materials and processes, and approximate models are necessary to help produce manageable problems in design. For example, it is often found useful to consider resistors, capacitors, and inductors as ideal models with constant (known) values, neglecting distributed effects, tolerances, temperature dependence, and other effects. Similarly, diodes can be treated according to several different models with increasing complexity, sometimes neglecting bulk resistance, variations in the ideality factor n, variations due to temperature, and other factors.

Electronics design thus requires many different types of approximate models. An awareness must be maintained of the limits of these models. One step that can be taken to reinforce constantly an awareness of this situation is to be concerned with the number of significant figures used in calculations.

Given an arbitrary number, the number of significant figures is defined by starting with the most left-hand nonzero digit. Count this digit and all digits to the right to determine the number of significant figures. When scientific notation is used (multiplication by a power of 10), the power of 10 does not affect the number of significant figures.

Example A5-1   **Finding the Number of Significant Figures**

How many significant figures are present in the following numbers: (a) 5.0050, (b) 10×10^5, and (c) 00.0006?

Solution   Following the rule given in the discussion, (a) five significant figures, (b) two significant figures, and (c) one significant figure.   ■

In dealing with significant figures, there is always the issue of balance. How exact should calculations be? The objective is to strike a balance between too much emphasis on significant figures on the one hand and the total disregard for significant figures on the other. Particularly, an effort must be made to avoid the use of long strings of digits in solutions where they are inappropriate and misleading.

A reasonable decision can be made based on the purpose of the text and the characteristics of electronics devices. In electronic design applications, the accuracy of predictive techniques is often limited to two to three significant figures by component tolerances and variations among device parameters, except where special design effort and methods are used to increase the accuracy of the prediction. Therefore, two to three significant figures are used throughout most calculations. Exceptions can be noted and discussed as they occur.

Example A5-2 **Dividing and Rounding Off**

Given the above strategy, suppose a given analysis requires the calculation of $R = V/I$, with $V = 10$ V and $I = 0.06$ mA.

Solution Using a calculator, the result

$$R = 166666.6667$$

is obtained. In order to use only two significant figures, as indicated, this solution must be written as

$$R = 1.7 \times 10^5 \ \Omega$$

Observe that by using exponential notation, the decimal can be placed where desired. ∎

Example A5-3 **Uncertainty Levels**

For another use, consider that for a specific circuit the voltage at a given node is $V = 3.51$ V. What does this indicate?

Solution The number has three significant figures. It lies between 3.50 and 3.52 V, so the uncertainty in the value is of the order of ± 0.01 V.

In contrast, if the number were

$$3.500 \text{ V}$$

it would have four significant figures and an uncertainty of the order of ± 0.001. ∎

APPENDIX 6
EXPERIMENTAL PROCEDURE

The experiments in this text provide a coordinated introduction to electronic devices and circuits. This appendix outlines an approach for conducting the experiments that has been proven effective. The emphasis is on developing a systematic strategy for design, test, and interpretation of a wide range of applications. Many variations on the material presented here can be used to meet specific needs.

A6-1 KEEPING A RESEARCH NOTEBOOK

It is essential for every research engineer and scientist to learn how to keep a complete and accurate research notebook in order to:

- Carefully document the objectives and designs for all research activities
- Provide a detailed research record and database
- Provide a logical flow of ideas and analysis to trace the path of learning experiences
- Summarize and interpret findings
- Note ideas for future research directions that might prove productive

The experiments in this text are intended to encourage the development of good practices with respect to research activities and to prepare students for realistic research and development settings.

In completing the experiments described here, it is recommended that each student make use of two research notebooks. Once an experiment is assigned, preparation for the experiment can be entered into one of the notebooks. All laboratory work, including data, analysis, and conclusions, is entered in the same notebook. The notebook is turned in for grading, and the second notebook is used for the preparation of the next lab. This cycle repeats throughout the course.

The integrity of the research notebook must be strictly maintained in order to assure its completeness and accuracy. Work should be in pen, not pencil, with no erasures. Pages should be numbered sequentially, and each page should be dated and initialled as it is completed. (In a research setting, significant results should be supported by the dated signatures of witnesses to support later patent-application efforts.)

The research notebook is the major document of research activities. Pages should never be torn out or otherwise removed. Work should be as neat as possible. Sections that seem to be in error can be carefully crossed out with comments to explain the action. Wide margins are needed for later comments and notes. In summary, the research notebook should provide a complete record of the scope of studies, experimental designs, data, interpretations, learning experiences, and insights.

A6-2 PREPARING FOR THE LABORATORY

A week before each lab, preparation can begin for an assigned experiment using the materials contained in this text.

Then, during work in the lab, all activities can be documented in the notebook, along with findings and conclusions. The notebook is turned in for grading at the end of the lab. Work can begin on the next assignment, using the second notebook.

A6-3 SUGGESTED FORMAT

The first page of the notebook should be for a table of contents. Each experiment should start on a new page.

The report procedure for each experiment is described here in terms of nine interrelated activities, which can become the section headings for the lab report. Sections 1 to 5 are to be completed prior to the lab and 6 to 9 are to be completed during the lab.

The following directions can be used for the preparation and completion of each section:

1. *Heading*. Give the title of the experiment, starting on a new page. List any partners who assist during the lab preparation.
2. *Purpose*. Include a statement of objectives for the lab and a brief inventory of critical issues and areas of concern in order to provide overall purpose, direction, and guidance for the work to follow. Use your own ideas and words.
3. *Theory/Approach*. Briefly note the conceptual approach to be used in addressing your objectives, key references, and general theoretical basis for your planned work. Include general equations and methods that will be used in the experiment.

4. *Preliminary Analysis*. The analysis is to be based on any data provided prior to the lab itself. If no data are provided before the lab, write "no initial data provided" in this section.

5. *Procedure Plan*. This section provides a step-by-step procedure for planned activities in the lab. Include schematics for circuits you plan to construct, show how general equations are to be used in specific applications, and describe the specific steps to be followed in data collection and analysis. Note questions to be examined and areas of concern. Be very specific, giving details regarding your choices and decisions for application of the general theory to the design process. Link your resultant design to the original objectives you set for yourself.

The following sections are then completed in the research notebook during the laboratory session:

6. *Preliminary Data and Analysis*. When appropriate, record data given to you at the beginning of the lab period. Perform preliminary calculations as necessary, following your procedure plan. If this section does not apply in a given lab, write "does not apply" in this section.

7. *Data Collection*. Collect the data called for in your design. Always record your raw data directly. All data must be carefully organized in tables, charts, and so on. No analysis should be included in this step.

8. *Analysis*. Follow the planned procedure to analyze your data. Compare theoretical and experimental results and develop graphs as appropriate. Label variables and sketches carefully.

In some design settings, it may be necessary to cycle through sections 7 and 8 more than once. You may collect data, perform analysis, collect data, perform analysis, etc. In this case, label the appropriate sections 7a, 8a, 7b, 8b, and so on.

9. *Conclusions*. Interpret your outcomes with respect to your original objectives. Discuss your specific findings and the significance of your results with respect to these objectives. What have you learned from the lab/design process?

All entries should be completed according to the assignment and organized in the specified format. If sections are out of order, provide clear instructions to explain how they are linked. The final structure must be clear to another reader.

All figures must include titles and all graph axes must be labeled. Interpret all plots with appropriate asymptotes and identifying comments. When plotting data, carefully indicate all data points. Label curves (ideal, theoretical, curve fit, predicted, and so on). Do not connect the points, but provide your best estimate of the function associated with your data. See Fig. A6-1. Use semilog paper when appropriate (see Appendix 4). Use multicolored pens where helpful in explaining and organizing material.

If any paper is added to your notebook (such as a graph or computer printouts), it must be cut to size and glued in or completely taped on all edges. No multilayer sheets or loosely attached sheets are acceptable. The use of significant figures should be carefully considered for each calculation (see Appendix 5).

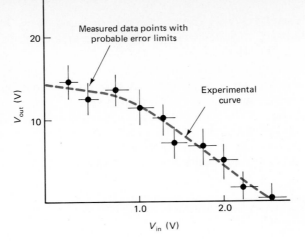

Measured data points with
probable error limits

Experimental
curve

V_{out} (V)

20

10

1.0   2.0

V_{in} (V)

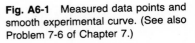

Fig. A6-1 Measured data points and
smooth experimental curve. (See also
Problem 7-6 of Chapter 7.)

A6-4   OVERVIEW

The experiments in the text and the procedure described here require that students
be able to:

- Understand and apply basic theories regarding device operation and circuit design.
- Understand and appreciate the nature of the tradeoffs and selection processes involved in circuit design
- Build and test the circuits that they have designed
- Evaluate and interpret circuit performance in a systematic way
- Redesign circuits as necessary based on experimental data
- Document all activities in a detailed research notebook
- Communicate these findings to others in an effective way.

The experiments build on one another and constitute an integrated approach to
the study of electronic devices and circuits.

APPENDIX 7
CONDUCTION
IN SEMICONDUCTORS

In an intrinsic semiconductor, the total number of electrons in the conduction band ($\hat{n}$) is given by

$$\hat{n} = \int_{\mathcal{E}_C}^{\infty} P(\mathcal{E}) g(\mathcal{E}) \, d\mathcal{E} \tag{A7-1}$$

where $g(\mathcal{E})$ is the number of possible states at energy $\mathcal{E}$ that can be filled by electrons, and $P(\mathcal{E})$ is the probability that an electron has energy $\mathcal{E}$. The integration starts at the edge of the conduction band and continues to all higher energies available to the electron.

In a similar way, the number of holes in the valence band is given by

$$\hat{p} = \int_{-\infty}^{\mathcal{E}_V} P(\mathcal{E}) g(\mathcal{E}) \, d\mathcal{E} \tag{A7-2}$$

For an intrinsic semiconductor, $\hat{n} = \hat{p} = \hat{n}_i$.

To find $g(\mathcal{E})$, the electrons and holes must be treated from a quantum-mechanical perspective, as having wavelike properties in the crystal. In a confined volume $\mathcal{V}$, only a finite number of different waves can exist. The maximum number of states is determined by how many half-wavelengths (associated with the particle) can fit into the given volume. For a cube with side L,

$$\frac{L}{\lambda/2}$$

waves are possible in each direction.

The number of states is found by considering all of the possible directions of propagation that can be developed using combinations of the above waves. This is equivalent to visualizing a sphere of radius $L/\lambda/2$ and finding the volume of the fraction of this sphere that is along the positive axes. The result is

$$\frac{1}{8}\left(\frac{4}{3}\pi\right)\left(\frac{L}{\lambda/2}\right)^3 = \frac{1}{6}\mathcal{V}\frac{k^3}{\pi^2} \tag{A7-3}$$

where $k = 2\pi/\lambda$ is the *wave number* associated with the wavelength λ.

To find the density of states $g(\mathcal{E})$, electrons are now treated as particles moving in the crystal with an effective mass m_n^*, momentum p, and with a kinetic energy given by $\mathcal{E} - \mathcal{E}_C$. Therefore,

$$\mathcal{E} - \mathcal{E}_c = \frac{p^2}{2m_n^*} = \frac{(\hbar k)^2}{2m_n^*} \tag{A7-4}$$

where $\hbar = h/2\pi$, h is Planck's constant, defined in Chapter 2, and $p = \hbar k$ is a basic relationship from quantum mechanics.

Substituting from Eq. (A7-4) into (A7-3), the number of states N_C is

$$N_C = \frac{1}{6}\frac{\mathcal{V}}{\pi^2}\left[\frac{2m_n^*(\mathcal{E} - \mathcal{E}_C)}{\hbar^2}\right]^{3/2} \times 2 \text{ (directions of propagation for each wave)}$$

$$\times 2 \text{ (spin states possible for each wave)} = \frac{2}{3}\frac{\mathcal{V}}{\pi^2}\left[\frac{2m_n^*(\mathcal{E} - \mathcal{E}_C)}{\hbar^2}\right]^{3/2} \tag{A7-5}$$

The density of states is thus

$$g_C(\mathcal{E}) = \frac{dN_C}{d\mathcal{E}} = \frac{2}{3}\frac{\mathcal{V}}{\pi^2}\frac{3}{2}\left[\frac{2m_n^*(\mathcal{E} - \mathcal{E}_C)}{\hbar^2}\right]^{1/2}\left(\frac{2m_n^*}{\hbar^2}\right)$$

$$= \frac{4\pi\mathcal{V}}{h^3}(2m_n^*)^{3/2}(\mathcal{E} - \mathcal{E}_C)^{1/2} \tag{A7-6}$$

Similarly, for the valence band,

$$\mathcal{E}_V - \mathcal{E} = \frac{(\hbar k)^2}{2m_p^*} \tag{A7-7}$$

and

$$g_V(\mathcal{E}) = \frac{4\pi\mathcal{V}}{h^3}(2m_p^*)^{3/2}(\mathcal{E}_V - \mathcal{E})^{1/2} \tag{A7-8}$$

From basic probability arguments and the Pauli exclusion principle, it can be shown that the probability $P(\mathcal{E})$ that an available state is occupied by an electron is given by

$$P_C(\mathcal{E}) = \frac{1}{\exp\left[(\mathcal{E} - \mathcal{E}_F)/kT\right] + 1} \tag{A7-9}$$

$$P_V(\mathcal{E}) = \frac{1}{\exp\left[(\mathcal{E}_F - \mathcal{E})/kT\right] + 1} \tag{A7-10}$$

where $\mathcal{E}_F$ is the Fermi energy. As can be noted, an energy state at $\mathcal{E} = \mathcal{E}_F$ has a 50 percent probability of being occupied.

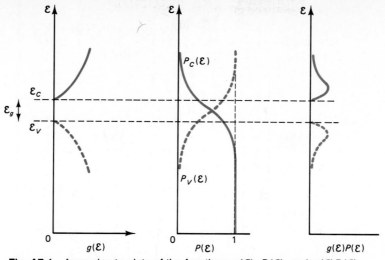

Fig. A7-1 Approximate plots of the functions $g(\mathcal{E})$, $P(\mathcal{E})$, and $g(\mathcal{E})P(\mathcal{E})$.

The functions $g(\mathcal{E})$ and $P(\mathcal{E})$ take the general shape shown in Fig. A7-1. The integrals of Eqs. (A7-1) and (A7-2) thus involve multiplying these two functions together and then finding the area under the product.

$$\hat{n} = C_n \int_{\mathcal{E}_C}^{\infty} \frac{(\mathcal{E} - \mathcal{E}_C)^{1/2} d\mathcal{E}}{\exp\left[(\mathcal{E} - \mathcal{E}_F)/kT\right] + 1} \quad \text{and} \quad C_n = \frac{4\pi \mathcal{V}}{h^3} (2m_n^*)^{3/2} \qquad \text{(A7-11)}$$

$$\hat{p} = C_p \int_{-\infty}^{\mathcal{E}_V} \frac{(\mathcal{E}_V - \mathcal{E})^{1/2} d\mathcal{E}}{\exp\left[(\mathcal{E}_F - \mathcal{E})/kT\right] + 1} \quad \text{and} \quad C_p = \frac{4\pi \mathcal{V}}{h^3} (2m_p^*)^{3/2} \qquad \text{(A7-12)}$$

From symmetry arguments, $\mathcal{E}_F$ is located between $\mathcal{E}_C$ and $\mathcal{E}_V$, and the $+1$ factor in the denominators can be neglected. With the change of variable,

$$u = \frac{\mathcal{E} - \mathcal{E}_C}{kT} \quad \text{and} \quad du = \frac{d\mathcal{E}}{kT}$$

Eq. (A7-11) can be rewritten

$$\hat{n} = C_n(kT)^{3/2} \exp\left[(\mathcal{E}_F - \mathcal{E}_C)/kT\right] \int_0^{\infty} u^{1/2} c^{-u} \, du \qquad \text{(A7-13)}$$

This integral is a standard one with the value $\dfrac{\sqrt{\pi}}{2}$. Therefore,

$$n = \frac{\hat{n}}{\mathcal{V}} = \frac{2}{h^3} (2\pi m_n^* kT)^{3/2} \exp\left[-(\mathcal{E}_C - \mathcal{E}_F)/kT\right] \qquad \text{(A7-14)}$$

and for holes,

$$p = \frac{\hat{p}}{\mathcal{V}} = \frac{2}{h^3} (2\pi m_p^* kT)^{3/2} \exp\left[-(\mathcal{E}_F - \mathcal{E}_V)/kT\right] \qquad \text{(A7-15)}$$

At equilibrium, $n = p = n_i$, so that

$$n_i = (np)^{1/2} = \frac{2}{h^3}(2\pi m_o kT)^{3/2}\left(\frac{m_n^*}{m_o}\frac{m_p^*}{m_o}\right)^{3/4}\exp\left[-(\mathcal{E}_C - \mathcal{E}_V)/2kT\right] \qquad \text{(A7-16)}$$

which is introduced as Eq. (2-2) in Chapter 2. By forming the ratio $n/p = 1$,

$$\mathcal{E}_F = \tfrac{1}{2}(\mathcal{E}_C + \mathcal{E}_V) - \tfrac{3}{4}kT\ln m_n^*/m_p^* \qquad \text{(A7-17)}$$

which is introduced as Eq. (2-3) in Chapter 2. [Note that Eq. (A7-17) confirms that $\mathcal{E}_F$ lies between $\mathcal{E}_C$ and $\mathcal{E}_V$.]

Let $\mathcal{E}_F = \mathcal{E}_i$ be the location of the Fermi level in an intrinsic semiconductor. Then Eq. (A7-14) can be written

$$n_i = n_o \exp\left[-(\mathcal{E}_C - \mathcal{E}_i)/kT\right] \qquad \text{(A7-18)}$$

With doping,

$$n = n_o \exp\left[-(\mathcal{E}_C - \mathcal{E}_F)/kT\right] \qquad \text{(A7-19)}$$

so that

$$\frac{n}{n_i} = \frac{\exp\left[-(\mathcal{E}_C - \mathcal{E}_F)/kT\right]}{\exp\left[-(\mathcal{E}_C - \mathcal{E}_i)/kT\right]} = \exp\left[(\mathcal{E}_F - \mathcal{E}_i)/kT\right] \qquad \text{(A7-20)}$$

which is introduced as Eq. (4-1) in Chapter 4.

Similarly,

$$\frac{p}{n_i} = \frac{\exp\left[-(\mathcal{E}_F - \mathcal{E}_V)/kT\right]}{\exp\left[-(\mathcal{E}_i - \mathcal{E}_V)/kT\right]} = \exp\left[(\mathcal{E}_i - \mathcal{E}_F)/kT\right] \qquad \text{(A7-21)}$$

which is introduced as Eq. (4-2) in Chapter 4.

APPENDIX 8
PN JUNCTION PROPERTIES WITH HIGH CURRENT FLOW

This appendix is modeled after an analysis by Streetman, 1980.

Assume a current flow exists across a pn junction, with the following definitions:

- Δn_p is the change in the number of electrons in the p-type semiconductor
- Δp_p is the change in the number of holes in the p-type semiconductor
- Δn_n is the change in the number of electrons in the n-type semiconductor
- Δp_n is the change in the number of holes in the n-type semiconductor

To maintain space-charge neutrality, require at all times that

$$\Delta p_p = \Delta n_p \qquad \text{and} \qquad \Delta p_n = \Delta n_n \tag{A8-1}$$

From Chapter 4,

$$\frac{n_n + \Delta n_n}{n_p + \Delta n_p} = \exp\left[q(V_t - V)/kT\right] \tag{A8-2}$$

$$\frac{p_p + \Delta p_p}{p_n + \Delta p_n} = \exp\left[q(V_t - V)/kT\right] \tag{A8-3}$$

The Δ values are treated as unknowns. With four equations and four unknowns, all but (Δp_n) can be eliminated by substitution:

$$\Delta p_n = \frac{p_p + \dfrac{n_n + \Delta p_n}{\exp\left[q(V_t - V)/kT\right]} - n_p}{\exp\left[q(V_t - V)/kT\right]} - p_n$$

$$\Delta p_n = [\exp(qV/kT) - 1]\{p_n + n_p \exp[-q(V_t - V)/kT]\} \tag{A8-4}$$

Because $n_p p_p = n_i^2$, Δp_n can be rewritten

$$\Delta p_n = \frac{[\exp{(qV/kT)} - 1] \, p_n}{1 - \exp{[-2q(V_t - V)/kT]}} \left[1 + \frac{n_i^2}{p_p^2} \exp{(qV/kT)} \right] \qquad (A8\text{-}5)$$

where use has again been made of the relationship

$$p_p = p_n \exp{(qV_t/kT)}$$

Finally,

$$I = I_o \frac{\exp{(qV/kT)} - 1}{1 - \exp{[-2q(V_t - V)/kT]}} \qquad (A8\text{-}6)$$

As V tends toward V_t, I approaches

$$\lim_{V \to V_t} I = \frac{\text{constant}}{V_t - V} \qquad (A8\text{-}7)$$

APPENDIX 9
FOURIER SERIES REPRESENTATION

$\mathbf{A}$ny periodic signal $s(t)$ can be resolved into a harmonic series of sinusoidal components. A general equation for this expansion can be written

$$s(t) = b_0 + \sum_{n=1}^{\infty} [a_n \sin (n\omega t) + b_n \cos (n\omega t)] \qquad \text{(A9-1)}$$

To find the a_n coefficients given $s(t)$, multiply both sides of Eq. (A9-1) by $\sin (m\omega t)$ and integrate from 0 to 2π:

$$\int_0^{2\pi} s(t) \sin (m\omega t) \, d(\omega t) = \int_0^{2\pi} b_0 \sin (m\omega t) \, d(\omega t)$$

$$+ \sum_{n=1}^{\infty} \int_0^{2\pi} a_n \sin (n\omega t) \sin (m\omega t) \, d(\omega t)$$

$$+ \sum_{n=1}^{\infty} \int_0^{2\pi} b_n \cos (n\omega t) \sin (m\omega t) \, d(\omega t) \qquad \text{(A9-2)}$$

The sine functions are orthogonal over 2π, so that

$$\int_0^{2\pi} \cos (n\omega t) \sin (m\omega t) \, d(\omega t) = 0 \qquad \text{for all } m \text{ and } n \qquad \text{(A9-3)}$$

$$\int_0^{2\pi} \cos (n\omega t) \cos (m\omega t) \, d(\omega t) = 0 \qquad \text{for } m \neq n \qquad \text{(A9-4)}$$

$$= \pi \qquad \text{for } m = n$$

$$\int_0^{2\pi} \sin(n\omega t) \sin(m\omega t)\, d(\omega t) = 0 \qquad \text{for } m \neq n \qquad \text{(A9-5)}$$

$$= \pi \qquad \text{for } m = n$$

Equation (A9-2) thus becomes

$$a_m = \frac{1}{\pi} \int_0^{2\pi} s(t) \sin(m\omega t) d(\omega t) \qquad \text{(A9-6)}$$

Similarly, multiplying through by cos $(m\omega t)$ and integrating from 0 to 2π produces

$$b_m = \frac{1}{\pi} \int_0^{2\pi} s(t) \cos(m\omega t)\, d(\omega t) \qquad \text{(A9-7)}$$

$$b_0 = \frac{1}{2\pi} \int_0^{2\pi} s(t)\, d(\omega t) \qquad \text{(A9-8)}$$

Example A9-1 Fourier Expansion for a Half-Wave Rectified Output

Assume that $s(t) = V_0 \sin\theta$, $\theta = \omega t$ for θ between 0 and π, and $s(t) = 0$ over π to 2π. Find the Fourier coefficients for $s(t)$.

Solution To obtain the solution, the following trigonometric relationships are used:

$$\sin A \cos B = \frac{1}{2}[\sin(A + B) - \sin(B - A)] \qquad \text{(A9-9)}$$

$$\sin A \sin B = \frac{1}{2}[\cos(B - A) - \cos(B + A)] \qquad \text{(A9-10)}$$

Setting $\theta = \omega t$, the a_1 coefficient can be found:

$$a_1 = \frac{1}{\pi} \int_0^{\pi} V_0 \sin\theta \sin\theta\, d\theta = \frac{V_0}{2\pi} \theta \Big|_0^{\pi} = \frac{V_0}{2} \qquad \text{(A9-11)}$$

and for a_n, $n \geq 2$,

$$a_n = \frac{1}{\pi} \int_0^{\pi} V_0 \sin\theta \sin(n\theta)\, d\theta = 0 \qquad \text{(A9-12)}$$

because the $\sin\theta$ and $\sin(n\theta)$ functions are orthogonal.
The b_n coefficients can be found as follows:

$$b_0 = \frac{1}{2\pi} \int_0^{\pi} V_0 \sin\theta\, d\theta = \frac{-V_0}{2\pi} \cos\theta \Big|_0^{\pi} = \frac{V_0}{\pi} \qquad \text{(A9-13)}$$

$$b_1 = \frac{1}{\pi} \int_0^{\pi} V_0 \sin\theta \cos\theta\, d\theta = \frac{-V_0}{4\pi} \cos(2\theta) \Big|_0^{\pi} = 0 \qquad \text{(A9-14)}$$

$$b_2 = \frac{1}{\pi} \int_0^{\pi} V_0 \sin\theta \cos(2\theta)\, d\theta = \frac{V_0}{2\pi} \left[-\frac{\cos(3\theta)}{3} + \cos\theta \right]_0^{\pi} = -\frac{2V_0}{3\pi} \qquad \text{(A9-15)}$$

$$b_3 = \frac{1}{\pi} \int_0^{\pi} V_0 \sin\theta \cos(3\theta)\, d\theta = \frac{V_0}{2\pi} \left[-\frac{\cos(4\theta)}{4} + \frac{\cos(2\theta)}{2} \right]_0^{\pi} = 0 \qquad \text{(A9-16)}$$

$$b_4 = \frac{1}{\pi} \int_0^{\pi} V_0 \sin\theta \cos(4\theta)\, d\theta = \frac{V_0}{2\pi}\left[-\frac{\cos(5\theta)}{5} + \frac{\cos(3\theta)}{3}\right]_0^{\pi} = \frac{-2V_0}{15\pi} \quad \text{(A9-17)}$$

and similarly for higher-order terms.
Based on the above terms,

$$s(t) = V_0\left[\frac{1}{2}\sin(\omega t) + \frac{1}{\pi} - \frac{2}{3\pi}\cos(2\omega t) - \frac{2}{15\pi}\cos(4\omega t) + \cdots +\right]$$

$$\text{(A9-18)}$$

This equation is introduced as Eq. (5-15) in Chapter 5. ∎

Example A9-2 Fourier Expansion for a Square Wave

Assume that $s(t)$ is given by the square wave of Fig. A9-1. Find the Fourier expansion.

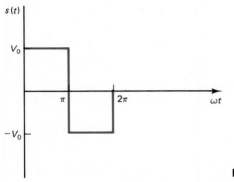

Fig. A9-1 Square wave for Example A9-2.

Solution

$$a_n = \frac{1}{\pi} \int_0^{\pi} V_0 \sin(m\omega t)\, d(\omega t) + \int_{\pi}^{2\pi} (-V_0)\sin(m\omega t)\, d(\omega t)$$

$$= \frac{-V_0}{m\pi}\cos(m\omega t)\Big|_0^{\pi} + \frac{V_0}{m\pi}\cos(m\omega t)\Big|_{\pi}^{2\pi}$$

$$= \frac{V_0}{m\pi}[2 - 2\cos(m\pi)] = \frac{2V_0}{m\pi}(-1)^m$$

$$b_0 = 0$$

$$b_m = \frac{1}{\pi} \int_0^{\pi} \left(-V_0\right)\cos(m\omega t)\, d(\omega t) + \int_{\pi}^{2\pi} V_0 \cos(m\omega t)\, d(\omega t) = 0$$

Applying the above results,

$$s(t) = \frac{4V_0}{\pi}\left[\sin(\omega t) + \frac{1}{3}\sin(3\omega t) + \frac{1}{5}\sin(5\omega t) + \cdots +\right]$$

This result is used as Eq. (11-1) in Chapter 11. ∎

APPENDIX 10
THE DECIBEL

When studying circuit performance, it is useful to compare power, voltage, and current levels using a logarithmic unit called the *decibel*. The relationship between two power levels P_1 and P_2 can be described in terms of the ratio P_2/P_1. This relationship can also be expressed in terms of the decibel (dB) through the equation

$$N \text{ (dB)} = N_{\text{dB}} = 10 \log P_2/P_1 \qquad \text{(A10-1)}$$

If P_2 is dissipated across resistance R_2 and P_1 is dissipated across R_1,

$$N \text{ (dB)} = 10 \log \frac{V_2^2/R_2}{V_1^2/R_1} = 20 \log V_2/V_1 - 10 \log R_2/R_1 \qquad \text{(A10-2)}$$

$$N \text{ (dB)} = 10 \log \frac{I_2^2 R_2}{I_1^2 R_1} = 20 \log I_2/I_1 + 10 \log R_2/R_1 \qquad \text{(A10-3)}$$

And if $R_1 = R_2$,

$$N \text{ (dB)} = 20 \log V_2/V_1$$
$$N \text{ (dB)} = 20 \log I_2/I_1 \qquad \text{(A10-4)}$$

The term *decibel* has meaning only when the ratio of two numbers is involved. The appropriate reference level depends on the nature of the problem.

In many applications, it is common (if inaccurate) usage to apply Eq. (A10-4) to circuits for which R_2 and R_1 are not the same and perhaps not even known. When this application is made, these equations are used as *definitions* to relate V_2/V_1 and I_2/I_1 to dB equivalents.

The power gain A_P, voltage gain A_V, and current gain A_i of a circuit can be defined as

$$A_P = \frac{\Delta p_2}{\Delta p_1} \qquad A_V = \frac{\Delta v_2}{\Delta v_1} \qquad A_I = \frac{\Delta i_2}{\Delta i_1} \qquad \text{(A10-5)}$$

so that the circuit gain relationships can also be expressed in dB:

$$A_P \text{ (dB)} = 10 \log A_P$$

$$A_V \text{ (dB)} = 20 \log A_V$$

$$A_I \text{ (dB)} = 20 \log A_I \qquad \text{(A10-6)}$$

APPENDIX 11
MULTISTAGE AC AMPLIFIERS

$\mathbf{T}$he frequency response of a multistage ac amplifier is determined by multiple low-pass and high-pass filters. The lower half-power point for an n-stage ac amplifier is determined by n high-pass filters. From Eq. (10-21), the value of f_1 is determined by

$$\left| \frac{1}{1 - jf_1/f_{1n}} \right|^n = \frac{1}{2^{1/2}} \qquad \text{(A11-1)}$$

so that

$$2^{1/2} = [1 + (f_1/f_{1n})^2]^{n/2}$$

$$f_{1n} = \frac{f_1}{(2^{1/n} - 1)^{1/2}} \qquad \text{(A11-2)}$$

The upper half-power point for an n-stage ac amplifier is determined by n low-pass filters. From Eq. (10-32), the value of f_2 is determined by

$$\left| \frac{1}{1 + jf_{2n}/f_2} \right|^n = \frac{1}{2^{1/2}} \qquad \text{(A11-3)}$$

By an analysis similar to that given above,

$$f_{2n} = f_2(2^{1/n} - 1)^{1/2} \qquad \text{(A11-4)}$$

At the lower half-power point, the phase shift associated with n amplifier stages is given by

$$\tan \theta = \frac{\text{Im} \left(\dfrac{1}{1 - jf_1/f_{1n}} \right)^n}{\text{Re} \left(\dfrac{1}{1 - jf_1/f_{1n}} \right)^n} = \frac{\text{Im} \left(1 + jf_1/f_{1n} \right)^n}{\text{Re} \left(1 + jf_1/f_{1n} \right)^n} \qquad \text{(A11-5)}$$

For $n = 1$,

$$\tan \theta = f_1/f_{11} = 1 \qquad \text{and} \qquad \theta = 45°$$

For $n = 2$,

$$f_1/f_{12} = (2^{1/2} - 1)^{1/2} = 0.64$$

$$\tan \theta = \frac{\text{Im} \, (1 + j0.64)^2}{\text{Re} \, (1 + j0.64)^2} = \frac{1.28}{0.59} \cong 2.17$$

$$\theta \cong 65°$$

Summarizing and extending the above analysis,

$n = 1$	$f_{11} = f_1, \; \theta = 45°$	$f_{21} = f_2, \; \theta = -45°$
$n = 2$	$f_{12} = f_1/0.64, \; \theta \cong 65°$	$f_{22} \cong 0.64 f_2, \; \theta \cong -65°$
$n = 3$	$f_{13} = f_1/0.51, \; \theta \cong 80°$	$f_{23} \cong 0.51 f_2, \; \theta \cong -80°$
$n = 4$	$f_{14} = f_1/0.44, \; \theta \cong 94°$	$f_{24} \cong 0.44 f_2, \; \theta \cong -94°$

APPENDIX 12
OSCILLATION FREQUENCY FOR THE PHASE-SHIFT OSCILLATOR

The objective of this appendix is to show that

$$f = \frac{1}{2\pi\sqrt{6}RC} \quad \text{and} \quad X_C = \sqrt{6}R$$

provides the oscillation frequency for the phase-shift oscillator. From Fig. 14-6,

$$X_C = \sqrt{6}R$$

$$Z_a = \frac{R(1 - j\sqrt{6})}{2 - j\sqrt{6}} \tag{A12-1}$$

$$Z_b = \frac{R(-3j\sqrt{6} - 5)}{-4j\sqrt{6} - 3}$$

The gain of the feedback network of Fig. 14-5 is thus

$$B = \frac{R}{R - j\sqrt{6}\,R} \frac{1}{1 - \dfrac{j\sqrt{6}\,R(2 - j\sqrt{6})}{R(1 - j\sqrt{6})}} \frac{1}{1 - \dfrac{j\sqrt{6}\,R(-4j\sqrt{6} - 3)}{R(-3j\sqrt{6} - 5)}} \tag{A12-2}$$

$$= \frac{1}{1 - j\sqrt{6}} \frac{1 - j\sqrt{6}}{(1 - j\sqrt{6}) - j\sqrt{6}(2 - j\sqrt{6})}$$

$$\times \frac{-3j\sqrt{6} - 5}{(-3j\sqrt{6} - 5) - j\sqrt{6}(-4j\sqrt{6} - 3)} \tag{A12-3}$$

$$= -1/29$$

The imaginary part of B vanishes and the real part is equal to $-1/29$.

APPENDIX 13
TABLE OF CONSTANTS

Constant	Symbol and Value
Planck's constant	$h = 6.62 \times 10^{-34}$ J·s $= 4.14 \times 10^{-15}$ eV·s
Boltzmann's constant	$k = 1.38 \times 10^{-23}$ J/°K $= 8.62 \times 10^{-5}$ eV/°K
Electron rest mass	$m_o = 9.1 \times 10^{-31}$ kg
Permittivity of free space	$\epsilon_o = 8.85 \times 10^{-12}$ F/m
Permeability of free space	$\mu_0 = 4\pi \times 10^{-7}$ H/m or W/Am
Electron charge	$q = 1.6 \times 10^{-19}$ C
Speed of light in free space	$c = 3 \times 10^{8}$ m/s
Room temperature	$T = 300$°K
$V_0 = kT/q$ at 300°K	$V_0 = 0.026$ V

REFERENCES

GENERAL

AHMED, H., and P. J. SPREADBURY. 1973. *Electronics for engineers*. Cambridge, England: Cambridge University Press.

BROPHY, J. J. *Basic electronics for scientists*. 1983. New York: McGraw-Hill.

CASASENT, D. 1973. *Electronic circuits*. New York: Quantum.

CHIRLIAN, P. M. 1981. *Analysis and design of integrated electronic circuits*. New York: Harper & Row.

COLCLASER, R. A., D. A. NEAMEN, and C. F. HAWKINS. 1984. *Electronic circuit analysis: Basic principles*. New York: John Wiley & Sons.

GHAUSI, M. S. 1985. *Electronic devices and circuits: Discrete and integrated*. New York: Holt, Rinehart & Winston.

GRAY, D. E., and C. L. SEARLE. 1969. *Electronic principles: Physics, models and circuits*. New York: John Wiley & Sons.

GRINICH, V. H., and H. G. JACKSON. 1975. *Introduction to integrated circuits*. New York: McGraw-Hill.

HAYT, W. H., JR., and G. W. NEUDECK. 1984. *Electronic circuit analysis and design* (2d ed.). Boston: Houghton Mifflin.

HENRY, R. W. 1978. *Electronic systems and instrumentation*. New York: John Wiley & Sons.

HOLT, C. A. 1978. *Electronic circuits: Digital and analog*. New York: John Wiley & Sons.

HOROWITZ, P., and W. HILL. 1981. *The art of electronics*. Cambridge, England: Cambridge University Press.

MATTSON, R. H. 1963. *Basic junction devices and circuits*. New York: John Wiley & Sons.

MILLMAN, J. 1979. *Microelectronics: Digital and analog circuits and systems*. New York: McGraw-Hill.

NAVON, D. H. 1986. *Semiconductor microdevices and materials*. New York: Holt, Rinehart & Winston.

NEUDECK, G. W. 1983a. *The pn junction diode*. Reading, MA: Addison-Wesley.

NEUDECK, G. W. 1983b. *The bipolar junction transistor*. Reading, MA: Addison-Wesley.

RIPS, E. M. 1986. *Discrete and integrated electronics*. Englewood Cliffs, NJ: Prentice-Hall.

SEDRA, A. S., and K. C. SMITH. 1982. *Microelectronic circuits*. New York: Holt, Rinehart & Winston.

STREETMAN, B. G. 1980. *Solid state electronic devices*. (2d ed.). Englewood Cliffs, NJ: Prentice-Hall.

SZE, S. M. 1981. *Physics of semiconductor devices*. New York: John Wiley & Sons.

SUPPLEMENTARY REFERENCES BY CHAPTER

Preface

ACCREDITATION BOARD FOR ENGINEERING AND TECHNOLOGY. 1985. *1985 Annual Report*. New York: Accreditation Board for Engineering and Technology.

TUINENGA, P. 1988. *SPICE: A Guide to Circuit Simulation and Analysis Using PSPICE*. Englewood Cliffs, NJ: Prentice-Hall.

Chapter 1

CASWELL, G. 1984. *Surface mount technology*. Silver Spring, MD: International Society for Hybrid Microelectronics.

HOWARD, R. T., and S. S. FURKAY, R. F. KILBURN, and G. MONTI, JR. (Eds.). 1985. *Thermal management concepts for microelectronics packaging*. Silver Spring, MD: International Society for Hybrid Microelectronics.

JONES, R. D. 1982. *Hybrid circuit design and manufacture*. New York: Marcel Dekker.

RUST, R. D., and D. A. DOANE. June 1985. Growing interdependence within the microelectronics industry: An overview perspective. *Solid State Technology*, pp. 97, 125–128.

SAVAGE, C. M. (Ed.). 1985. *A program guide for CIM implementation*. Dearborn, MI: Computer and Automated Systems Association of the Society for Manufacturing Engineers.

TEXAS INSTRUMENTS, INC. June 1986. Today's designers abandon bread boards for CAD/CAE approach. *FYI newsletter*, p. 1.

VAN GIGCH, J. P. 1978. *Applied general systems theory* (2d ed.). New York: Harper & Row.

Chapter 2

DALVEN, R. 1980. *Introduction to applied solid state physics*. New York: Plenum.

KITTEL, C. 1986. *Introduction to solid state physics* (6th ed.). New York: John Wiley & Sons.

MADAN, A. September 1986. Amorphous silicon: From promise to practice. *IEEE Spectrum* 23(9):38–43.

PIERRET, R. F. 1983. *Semiconductor fundamentals*. Reading, MA: Addison-Wesley.

RUNYAN, W. R. 1965. *Silicon semiconductor technology*. New York: McGraw-Hill.

RUOFF, A. L. 1979. *Introduction to materials science*. Huntington, NY: Krieger.

VAN DER ZIEL, A. 1968. *Solid state physical electronics* (2d ed.). Englewood Cliffs, NJ: Prentice-Hall.

THURMOND, C. D. August 1975. The standard thermodynamic function for the formation of electrons and holes in Ge, Si, GaAs, and GaP. *J. Electrochem. Soc.* 122(8):1133–1141.

VAN DER ZIEL, A. 1968. *Solid state physical electronics* (2d ed.). Englewood Cliffs, NJ: Prentice-Hall.

VAN VLACK, L. H. 1980. *Elements of materials science and engineering* (4th ed.). Reading, MA: Addison-Wesley.

WYATT, O. H., and D. DEW-HUGHES. 1974. *Metals, ceramics and polymers*. London: Cambridge University Press.

Chapter 3

HARPER, C. A. 1977. *Handbook of components for electronics*. New York: McGraw-Hill.

LEE, R. 1955. *Electronic transformers and circuits*. New York: John Wiley & Sons.

LYMAN, W. T. 1979. *Transformer and inductor design handbook*. New York: Marcel Dekker.

MCGRAW-HILL BOOK CO. 1977. *McGraw-Hill encyclopedia of science and technology*. New York: McGraw-Hill.

D. VAN NOSTRAND CO. 1968. *Van Nostrand's scientific encyclopedia*. Princeton: Van Nostrand.

WARRING, R. H. 1983. *Electronic components handbook for circuit designers*. Blue Ridge Summit, PA: TAB Books.

WINCH, R. P. 1963. *Electricity and magnetism* (2d ed.). Englewood Cliffs, NJ: Prentice-Hall.

Chapter 5

A. B. ASSOCIATES. n.d. *IGSPICE*. Tampa, FL: A. B. Associates.

HEWLETT-PACKARD CORP. n.d. *User's guide for HP-SPICE*. Cupertino, CA: Hewlett-Packard Corp.

MICROSIM CORP. n.d. *PSPICE*. Laguna Hills, CA: MicroSim Corp.

NAGEL, L. 1975. *SPICE: A computer program to simulate semiconductor circuits* (Report M520). Berkeley, CA: University of California, Department of Electrical Engineering and Computer Sciences.

TUINENGA, P. 1988. *SPICE: A Guide to Circuit Simulation and Analysis Using PSPICE*. Englewood Cliffs, NJ: Prentice-Hall.

VLADIMIRESCU, A., and S. LIU. 1980. *Simulations of MOS integrated circuits using SPICE2* (Report M80/7). Berkeley, CA: University of California, Department of Electrical Engineering and Computer Sciences.

VLADIMIRESCU, A., A. R. NEWTON, and D. O. PEDERSON. 1980. *SPICE version 2G user's guide*. Berkeley, CA: University of California, Department of Electrical Engineering and Computer Sciences.

Chapter 6

DEMAN, H. J. J. October 1971. The influence of heavy doping on the emitter efficiency of a bipolar transistor. *IEEE Trans. Elec. Dev.* V. ED-18(10):833–835.

EARLY, J. M. November 1952. Effects of space-charge layer widening in junction transistors. *Proc. IEEE,* pp. 1401–1406.

EBERS. J. J., and J. L. MOLL. December 1954. Large-signal behavior of junction transistors. *Proc. IEEE,* pp. 1761–1772.

GARTNER, W. W. May 1957. Temperature dependence of junction transistor parameters. *Proc. IRE,* pp. 662–680.

Chapter 8

EVANS, A. D. 1981. *Designing with field-effect transistors*. New York: McGraw-Hill.

MIDDLEBROOK, R. D. August 1963. A simple derivation of field-effect transistor characteristics. *Proc. IEEE,* pp. 1146–1147.

PIERRET, R. F. 1983. *Field-effect devices*. Reading, MA: Addison-Wesley.

Chapters 12 and 13

FITZGERALD, K. March 1987. Whatever happened to analog computers. *IEEE Spectrum,* p. 18.

GOODENOUGH, F. March 20, 1986. First GaAs op amp hits 150 MHz at unity gain. *Electronic Design,* p. 61.

GRAY, P. R., and R. G. MEYER. 1984. *Analysis and design of analog integrated circuits* (2d ed.). New York: John Wiley & Sons.

MEYER, R. G. (Ed.). 1978. *Integrated-circuit operational amplifiers*. New York: IEEE Press.

SIGNETICS CORP. 1985a. *Linear LSI data and applications manual*. Sunnyvale, CA: Signetics Corp.

Signetics Corp. 1985b. *Linear data and applications manual*. Sunnyvale, CA: Signetics Corp.

Texas Instruments, Inc. 1984. *Linear circuits data book*. Dallas: Texas Instruments, Inc.

Chapter 14

Carlson, A.B. 1986. *Communication systems: An introduction to signals and noise in electrical communication* (3d ed.). New York: McGraw-Hill.

Guio, P., Jr. March 1987. Crystal oscillators: Today's answer for tomorrow's communications equipment. *Hybrid Circuit Technology*, p. 13.

Stremler, F. G. 1982. *Introduction to communication systems* (2d ed.). Reading, MA: Addison-Wesley.

Young, P. H. 1985. *Electronic communication techniques*. Columbus, OH: Merrill.

Chapters 15 to 17

Bartee, T. C. 1985. *Digital computer fundamentals* (6th ed.). New York: McGraw-Hill.

Blakeslee, T. R. 1979. *Digital design with standard MSI and LSI*. New York: John Wiley & Sons.

Brywater, R. E. H. 1981. *Hardware/software design of digital systems*. Englewood Cliffs, NJ: Prentice-Hall.

Cahill, S. J. 1982. *Digital and microprocessor engineering*. New York: John Wiley & Sons.

Casasent, D. 1974. *Digital electronics*. New York: Quantum.

Coli, V. J. January 1987. Introduction to programmable array logic. *Byte*, p. 207.

Fairchild Semiconductor. 1977. *ECL data book*. Mountain View, CA: Fairchild Semiconductor Corp.

Fritz, J. S., C. F. Kaldenbach, and L. M. Progar. 1985. *Local area networks: Selection guidelines*. Englewood Cliffs, NJ: Prentice-Hall.

Josephson Computer Technology. March 1986. *Physics Today* 39(3):46–52.

Mead, C., and L. Conway. 1980. *Introduction to VLSI systems*. Reading, MA: Addison-Wesley.

Morris, R. L., and J. R. Miller (Eds.). 1971. *Designing with TTL integrated circuits*. New York: McGraw-Hill.

Motorola, Inc. 1975. *MC6800 microprocessor applications manual*. Phoenix: Motorola, Inc.

Myers, G. J. 1980. *Digital system design with LSI bit-slice logic*. New York: John Wiley & Sons.

RCA Corp. 1983. *RCA CMOS integrated circuits (data book)*. Somerville, NJ: RCA Corp.

Tanenbaum, A. S. 1981. *Computer networks*. Englewood Cliffs, NJ: Prentice-Hall.

Texas Instruments, Inc. 1984a. *The TTL data book for design engineers* (2d ed.). Dallas: Texas Instruments, Inc.

Texas Instruments, Inc. 1984b. *High-speed CMOS logic data book*. Dallas: Texas Instruments, Inc.

Chapter 18

Brodie, I., and J. J. Muray. 1982. *The physics of microfabrication*. New York: Plenum.

Colclaser, R. A. 1980. *Microelectronics processing and device design*. New York: John Wiley & Sons.

Hall, C. 1981. *Polymer materials: An introduction for technologists and scientists*. New York: John Wiley & Sons.

Lanyon, H. P. D., and R. A. Taft. 1978. Bandgap narrowing in heavily doped silicon. *IEEE Technical Digest*, Int. Electron. Device Mtg, p. 316.

Smallman, R. E. 1980. *Modern physical metallurgy* (3d ed.). London: Butterworths.

Sze, S. M. and G. Gibbons. 1966. Avalance breakdown voltages of abrupt and linearly graded pn junctions in Ge, Si, GaAs, and GaP. *APL* 8:111.

Van Vlack, L. H. 1964. *Physical ceramics for engineers*. Reading, MA: Addison-Wesley.

INDEX

Semiconductors
    definition of, 38, 50–51
    extrinsic, 43, 100
    intrinsic, 42, 100, 791–794
Sense amplifier, 644
Shift register, 643
Short-circuit output current, 500
Shot noise, 456
Signal-to-noise ratio, 456–457, 574–577, 663
Significant figures, 16, 68, 785–786
Silicon, 36, 43–45, 53, 67, 100–101, 143,
        314, 703, 714
Sintering, 740
Slew rate, 489, 500, 528
Smoothing filters. *See* Filters
Source, of FET, 281, 303
Source follower, 356–358
Specific heat, 156, 746
SPICE simulations
    ac amplifiers, 426–433
    BJT, 242
    dc-coupled amplifiers, 277–279, 326–328
    definition of, 99, 146
    description, 180–183
    full-wave rectifiers, 164–167, 195–198
    gate circuits, 633
    half-wave rectifiers, 164–167, 195–198
    push-pull amplifiers, 466–467
Spurious oscillations, 567–570
Square-wave response
    Fourier components, 435–437, 461
    SPICE simulation, 467–468
    testing method, 488–489
Standby power dissipation, 500
Subtracting amplifier, 520
Summing amplifier, 520
Supply current, 500. *See also* Power supplies
Supply voltage, 500. *See also* Power supplies
Supply-voltage sensitivity, 500
Systems approach, 3, 21, 23

T

TCC, 76
TCR, 66, 68, 84, 92
Temperature coefficient of capacitance. *See*
    TCC
Temperature coefficient of resistance. *See* TCR
Thermal conductivity, 745
Thermal drift, 154–156, 182
Thermal effects
    amplifier thermal noise, 453–457
    on BJT characteristic, 219, 242–243
    on carrier density, 45

on carrier scattering, 47, 49
on circuit design, 743–747
due to component density, 7–8, 25
impact on crystal, 42–43
diffusion, 103–105
on diode junction, 117–143
on diode Q-point, 154–155, 198–200
drift of Q-point, 156, 255–258
on JFET characteristic, 290
TCC, 76
TCR, 66, 68, 84, 92
temperature range, 500
Thermal noise, 453–457
Thermal runaway, 266
Thermal scattering, 47, 49
Thermistor, 63, 67, 553
Three-terminal circuits, 6, 205
Threshold voltage, for E-MOSFET, 307
Total noise voltage, 455
Totem pole, 606–621
Tradeoffs
    circuit, 5–10, 244, 263–265, 298–299,
        437–441, 499–501, 566–567, 618–620,
        662–665, 682–685, 733–734
    cost, 86, 92, 140, 184–185, 239, 250–251,
        274–275, 366–367, 584, 625, 669
Transconductance, 288, 317, 336, 364, 417
Transconductance amplifier, 438
Transfer characteristic, 284
Transfer function, 408
Transfer variable, 406
Transformers, 81–83, 158
Transition region, of JFET, 283
Transmission channel, 571
Transmission lines, 571
Transmitter, 571
Tristate drivers, 657
Truth table, 592
Tuned amplifier, 452, 454
Tunnel diode, 26, 29, 33, 93, 123–124, 132,
        138
Tunnel diode circuits, 178–179, 192, 564–566
Tunneling, 123–124

U

Unity follower, 520, 534
Unity-gain bandwidth, 486–488

V

Valence band, 41
Van der Waals bonds, 52, 749